# An Introduction to Modern Stellar Astrophysics

# An Introduction to Modern Stellar Astrophysics

Dale A. Ostlie
Bradley W. Carroll

Weber State University

ADDISON-WESLEY PUBLISHING COMPANY, INC.
Reading, Massachusetts · Menlo Park, California · New York · Don Mills, Ontario
Wokingham, England · Amsterdam · Bonn · Singapore · Tokyo
Madrid · San Juan · Sydney · Milan · Paris

This book was reproduced from PostScript files supplied by the authors. The authors typeset the book using the LaTeX document preparation system.

*Sponsoring Editor:* Julia Berrisford
*Associate Editor:* Jennifer Albanese
*Senior Production Coordinator:* Kathleen A. Manley
*Senior Marketing Manager:* Kate Derrick
*Cover Designer:* Diana C. Coe
*Cover Design Supervisor:* Eileen R. Hoff
*Technical Illustrator:* Scientific Illustrators
*Art Editors:* Susan London-Payne, Joseph Vetere
*Manufacturing Supervisor:* Hugh Crawford
*Senior Manufacturing Manager:* Roy E. Logan

The photograph on the cover is copyright © Anglo-Australian Observatory, photography by David Malin.

Carroll, Bradley W.
    An introduction to modern stellar astrophysics / Bradley W. Carroll, Dale A. Ostlie.
       p.    cm.
    Includes bibliographical references and index.
    ISBN 0-201-59880-9
    1. Stars.  2. Astrophysics.   I. Ostlie, Dale A.   II. Title.
QB801.C25 1996
523.8—dc20                                                                       95-45143
                                                                                                                     CIP

Copyright © 1996 by Addison-Wesley Publishing Company, Inc.

All rights reserved. No part of this publication may be reproduced, stored in a retrieval system, or transmitted, in any form or by any means, electronic, mechanical, photocopying, recording, or otherwise, without the prior written permission of the publisher. Printed in the United States of America.

1 2 3 4 5 6 7 8 9 10—DOC—9998979695

# *Preface*

There has never been a more exciting time to study modern astrophysics. Orbiting observatories such as the Hubble Space Telescope have extended our vision, allowing us to see vistas with remarkable clarity that previously were only imagined. On Earth, instruments using adaptive optics and other revolutionary techniques have exceeded previous expectations of what can be accomplished from the ground. Powerful computers have aided the advance from imagination to image by simulating exotic processes that are hidden from view, and events that occur so slowly or rapidly as to defy direct observation. Discoveries follow each other so rapidly that even the Internet can barely manage to disseminate the news quickly enough.

Our goal in writing this book was to open the field of modern astrophysics to the reader by using only the basic tools of physics. Nothing is more satisfying than appreciating the drama of the universe through an understanding of its underlying physical principles. The advantages of a mathematical approach to understanding the heavenly spectacle were obvious to Plato, as manifested in his *Epinomis*:

> Are you unaware that the true astronomer must be a person of great wisdom? Hence there will be a need for several sciences. The first and most important is that which treats of pure numbers. To those who pursue their studies in the proper way, all geometric constructions, all systems of numbers, all duly constituted melodic progressions, the single ordered scheme of all celestial revolutions should disclose themselves. And, believe me, no one will ever behold that spectacle without the studies we have described, and so be able to boast that they have won it by an easy route.

Now, 24 centuries later, the application of a little physics and mathematics still leads to deep insights.

This text was also born of the frustration encountered while teaching our junior-level astrophysics course. Most of the available astronomy texts seemed more descriptive than mathematical. Students who were learning about the Schrödinger equation, partition functions, and multipole expansions in other courses felt handicapped because their astrophysics text did not take advantage of their physics background. It seemed a double shame to us because a course in astrophysics offers students the unique opportunity of actually using the physics they have learned to appreciate many of astronomy's fascinating phenomena. Furthermore, as a discipline, astrophysics draws on essentially every aspect of physics. Thus astrophysics provides students the chance to review and extend their knowledge.

Anyone who has had an introductory calculus-based physics course (usually called something like "Physics for Scientists and Engineers") is ready to understand nearly all the major concepts of modern astrophysics. The amount of modern physics covered in such a course varies widely, so we have included a chapter on the theory of special relativity and one on quantum physics, which will provide the necessary background in these areas. Everything else in the text is self-contained and generously cross-referenced, so the reader will not loose sight of the chain of reasoning that leads to some of the most astounding ideas in all of science.[1] Although we have attempted to be fairly rigorous, we have tended to favor the sort of back-of-the-envelope calculation that uses a simple model of the system being studied. The payoff-to-effort ratio is so high, yielding 80% of the understanding for 20% of the effort, that these quick calculations should be a part of every astrophysicist's toolkit. In fact, while writing this book we were constantly surprised by the number of phenomena that could be described in this way. Above all, we have tried to be honest with the reader; we remained determined not to simplify the material beyond recognition. Stellar interiors, stellar atmospheres, and general relativity, all are described with a depth that is more satisfying than mere hand-waving description.

Computational astrophysics has become a third branch of astronomy (along with observation and traditional theory), and we have included numerous computer problems as well as complete programs that are integrated with the text material. Readers can calculate their own planetary orbits and make their own models of stars. These programs favor simplicity over sophistication for pedagogical reasons; readers can easily expand on the conceptually transparent codes that we have provided. Astrophysicists have traditionally led the way in large-scale computation and visualization, and we have tried to provide a gentle introduction to this blend of science and art.

Instructors can use this text to create courses tailored to their particular

---

[1] Footnotes are used when we don't want to interrupt the main flow of a paragraph.

needs by approaching its contents as an astrophysical smorgasbord. By judiciously selecting topics, we have used this text to teach a ten-week course in stellar astrophysics. (Of course, much was omitted from the first 16 chapters, but the text is designed to accommodate such surgery.) A few of the more advanced topics are labeled as "optional" sections and are preceded by brief summaries. Otherwise, instructors may follow their own as well as their students' interests and design a course to their liking.

Astronomy and astrophysics is a human endeavor and owes its advances to the insights, hard work, and sacrifices of countless men and women throughout its history. To inject some of the subject's humanity, we have sprinkled the names of its participants throughout the text. Undoubtedly, we have inadvertently omitted a few major figures while including lesser ones, but we are not and do not pretend to be historians. Several of the people named in the text have won the Nobel prize, but in general we have allowed their outstanding work to speak for itself. Although astronomers are properly pleased when one of their own is so honored, this has deliberately gone unmentioned because students often mistake the Nobel prize for either a competition or a sanctification of scientific results.

By now, everyone with an interest in astronomy should be familiar with the World Wide Web. Today, the lag time from a discovery to its dissemination is measured in days, if not hours. For example, the image of protostellar jets (Fig. 12.13) was inserted into the text on the same day it was released to the public. To learn what has happened since the book's contents were set in stone in October 1995, the reader may wish to visit some of our favorite Web sites, such as the homepages for the Space Telescope Science Institute, the National Space Science Data Center, NASA, and the Jet Propulsion Laboratory. (With the efficient search engines now available on the Web, we need not give specific addresses since they are subject to change.) We also invite the reader to our own Modern Astrophysics homepage here at Weber State University (`http://www.weber.edu/physics/modastro/modastro.htm`). There you will find a short (we hope) list of corrections for the text, answers to FAQs (frequently asked questions) we have received about the book, links to some sites with the latest astronomical news, and information about downloading the computer codes found in the text (both uncompiled and executable).

We started to write this book six years ago, in what now seems like the technological Stone Age. Since then, a revolution in publishing has been underway that requires a redefinition of the relationships among authors, editors, and production staff. We are grateful to everyone at Addison-Wesley for their support and willingness to experiment with the assembly of this book. The result was a mixture of freedom and responsibility that, at times, must have seemed as frightening to them as it seemed intimidating to us. The book was typeset by the authors using LaTeX, which is becoming a standard for scientific publication. We have included many figures from the scientific literature, and we thank everyone who granted permission to reproduce their work. Some figures were electronically scanned from journals and review articles or provided by the researchers in electronic format, while others were downloaded from various public-domain sources via Internet ftp and the World Wide Web.

Throughout the process our editors at Addison-Wesley have maintained a positive attitude that has sustained us throughout. From the very beginning Stuart Johnson has shared in our vision of what we wanted this book to be, and he encouraged us as it evolved. Although we must have sorely tried the patience of both Jennifer Albanese and Kathy Manley, their ongoing confidence in this project on both the editorial and production sides kept us moving forward. We must also thank Kim Woods, book rep extraordinaire, for starting the manuscript on its long journey toward publication.

We have certainly been fortunate in our professional associations throughout the years. Our gratitude and appreciation is expressed to Art Cox, John Cox, Carl Hansen, Hugh Van Horn, and Lee Anne Willson, whose profound influence on us has remained and, we hope, shines through the pages ahead. Our good fortune has now been extended to include the expert reviewers who cast a merciless eye on our chapters and gave us invaluable advice on how improve them. We owe a great debt for their careful reading to Robert Antonucci, Martin Burkhead, Peter Foukal, David Friend, Carl Hansen, Steven Kawaler, Judith Pipher, Lawrence Pinsky, and J. Allyn Smith. Several generations of students gave us a different perspective, and several chapters were rewritten in response to their suggestions. Furthermore, a number of friends and colleagues read preliminary versions of parts of the text and spurred us on in the early stages of this mammoth project, and we are grateful for their informal input. Unfortunately, no matter how fine the sieve, some mistakes are sure to slip through. The responsibility for the remaining errors is entirely ours, and we invite the reader to submit comments and corrections to us at our e-mail address: `modastro@weber.edu`.

Unfortunately, the burden of writing has not been confined to the authors but was unavoidably shared by our families. Six years of unskied snow, unhiked trails, and unfished rivers is a long time. With the completion of this text, Brad is looking forward to spending his vacations admiring the scenery instead of reading the pages of *The Annual Review of Astronomy and Astrophysics*, and Dale is anxious to spend more cherished time with his wonderful children, Michael and Megan. We wish to thank our parents, Dean and Dorothy Ostlie, and Wayne and Marjorie Carroll, for raising us to be intellectual explorers of this fascinating universe. Finally, it is to the two people who make our universe so wondrous that we dedicate this book: our wives, Candy Ostlie and Lynn Carroll. Without their love, patience, encouragement, and constant support, this project would never have been completed.

**Dale A. Ostlie**
**Bradley W. Carroll**
Department of Physics
Weber State University
Ogden, UT

# Contents

## I  The Tools of Astronomy — 1

**1  The Celestial Sphere** — 3
1.1  The Greek Tradition — 3
1.2  The Copernican Revolution — 6
1.3  Positions on the Celestial Sphere — 10
1.4  Physics and Astronomy — 21

**2  Celestial Mechanics** — 25
2.1  Elliptical Orbits — 25
2.2  Newtonian Mechanics — 31
2.3  Kepler's Laws Derived — 43
2.4  The Virial Theorem — 53

**3  The Continuous Spectrum of Light** — 63
3.1  Stellar Parallax — 63
3.2  The Magnitude Scale — 65
3.3  The Wave Nature of Light — 69
3.4  Blackbody Radiation — 75
3.5  The Quantization of Energy — 79
3.6  The Color Index — 82

**4  The Theory of Special Relativity** — 93
4.1  The Failure of the Galilean Transformations — 93
4.2  The Lorentz Transformations — 96
4.3  Time and Space in Special Relativity — 102
4.4  Relativistic Momentum and Energy — 113

**5  The Interaction of Light and Matter** — 125
5.1  Spectral Lines — 125

|     | 5.2 | Photons | 130 |
|---|---|---|---|
|     | 5.3 | The Bohr Model of the Atom | 134 |
|     | 5.4 | Quantum Mechanics and Wave–Particle Duality | 143 |

## 6 Telescopes — 159
- 6.1 Basic Optics . . . . . . . . . . . . . . . . . . . . . . . . . 159
- 6.2 Optical Telescopes . . . . . . . . . . . . . . . . . . . . . 173
- 6.3 Radio Telescopes . . . . . . . . . . . . . . . . . . . . . . 181
- 6.4 Infrared, Ultraviolet, and X-Ray Astronomy . . . . . . . 187

## II  The Nature of Stars — 199

## 7 Binary Stars and Stellar Parameters — 201
- 7.1 The Classification of Binary Stars . . . . . . . . . . . . 201
- 7.2 Mass Determination Using Visual Binaries . . . . . . . 205
- 7.3 Eclipsing, Spectroscopic Binaries . . . . . . . . . . . . . 208

## 8 The Classification of Stellar Spectra — 223
- 8.1 The Formation of Spectral Lines . . . . . . . . . . . . . 223
- 8.2 The Hertzsprung–Russell Diagram . . . . . . . . . . . . 241

## 9 Stellar Atmospheres — 255
- 9.1 The Description of the Radiation Field . . . . . . . . . 255
- 9.2 Stellar Opacity . . . . . . . . . . . . . . . . . . . . . . . 261
- 9.3 Radiative Transfer . . . . . . . . . . . . . . . . . . . . . 276
- 9.4 The Structure of Spectral Lines . . . . . . . . . . . . . . 293

## 10 The Interiors of Stars — 315
- 10.1 Hydrostatic Equilibrium . . . . . . . . . . . . . . . . . . 315
- 10.2 Pressure Equation of State . . . . . . . . . . . . . . . . 320
- 10.3 Stellar Energy Sources . . . . . . . . . . . . . . . . . . . 329
- 10.4 Energy Transport and Thermodynamics . . . . . . . . . 350
- 10.5 Stellar Model Building . . . . . . . . . . . . . . . . . . . 365
- 10.6 The Main Sequence . . . . . . . . . . . . . . . . . . . . 371

## 11 The Sun — 381
- 11.1 The Solar Interior . . . . . . . . . . . . . . . . . . . . . 381
- 11.2 The Solar Atmosphere . . . . . . . . . . . . . . . . . . . 394
- 11.3 The Solar Cycle . . . . . . . . . . . . . . . . . . . . . . 416

## Contents

### 12 The Process of Star Formation — 437
- 12.1 Interstellar Dust and Gas — 437
- 12.2 The Formation of Protostars — 447
- 12.3 Pre-Main-Sequence Evolution — 458

### 13 Post-Main-Sequence Stellar Evolution — 483
- 13.1 Evolution on the Main Sequence — 483
- 13.2 Late Stages of Stellar Evolution — 494
- 13.3 The Fate of Massive Stars — 510
- 13.4 Stellar Clusters — 529

### 14 Stellar Pulsation — 541
- 14.1 Observations of Pulsating Stars — 541
- 14.2 The Physics of Stellar Pulsation — 548
- 14.3 Modeling Stellar Pulsation — 557
- 14.4 Nonradial Stellar Pulsation — 561
- 14.5 Helioseismology — 567

### 15 The Degenerate Remnants of Stars — 577
- 15.1 The Discovery of Sirius B — 577
- 15.2 White Dwarfs — 579
- 15.3 The Physics of Degenerate Matter — 583
- 15.4 The Chandrasekhar Limit — 588
- 15.5 The Cooling of White Dwarfs — 592
- 15.6 Neutron Stars — 598
- 15.7 Pulsars — 608

### 16 Black Holes — 633
- 16.1 The General Theory of Relativity — 633
- 16.2 Intervals and Geodesics — 648
- 16.3 Black Holes — 661

### 17 Close Binary Star Systems — 683
- 17.1 Gravity in a Close Binary Star System — 683
- 17.2 Accretion Disks — 692
- 17.3 A Survey of Close Binary Systems — 704
- 17.4 White Dwarfs in Semidetached Binaries — 710
- 17.5 Neutron Stars and Black Holes in Binaries — 723

| | |
|---|---|
| A  Astronomical and Physical Constants | A-1 |
| B  Solar System Data | A-3 |
| C  The Constellations | A-5 |
| D  The Brightest Stars | A-11 |
| E  Stellar Data | A-13 |
| F  The Messier Catalog | A-19 |
| G  A Planetary Orbit Code | A-23 |
| H  STATSTAR, A Stellar Structure Code | A-27 |
| I  STATSTAR Stellar Models | A-51 |
| Index | I-1 |

# Part I

# THE TOOLS OF ASTRONOMY

# Chapter 1

# THE CELESTIAL SPHERE

## 1.1 The Greek Tradition

Human beings have long looked up at the sky and pondered its mysteries. Evidence of the long struggle to understand its secrets may be seen in remnants of cultures around the world: the great Stonehenge monument in England, the structures and the writings of the Maya and Aztecs, and the medicine wheels of the Native Americans. However, our modern scientific view of the universe traces its beginnings to the ancient Greek tradition of natural philosophy. Pythagoras (ca. 550 B.C.) first demonstrated the fundamental relationship between numbers and nature through his study of musical intervals and through his investigation of the geometry of the right angle. The Greeks continued their study of the universe for hundreds of years using the natural language employed by Pythagoras, namely mathematics. The modern discipline of astronomy depends heavily on a mathematical formulation of its physical theories, following the process begun by the ancient Greeks.

In an initial investigation of the night sky, perhaps its most obvious feature to a careful observer is the fact that it is constantly changing. Not only do the stars move steadily from east to west during the course of a night, but different stars are visible in the evening sky, depending upon the season. Of course, the Moon also changes, both in its position in the sky and its phase. More subtle, yet more complex, are the movements of the planets, or "wandering stars."

Plato (ca. 350 B.C.) suggested that, to understand the motions of the heavens, one must first begin with a set of workable assumptions, or hypotheses. It seemed without question that the stars of the night sky revolve about a fixed Earth and that the heavens ought to obey the purest possible form of motion. Plato therefore proposed that celestial bodies should move about Earth with

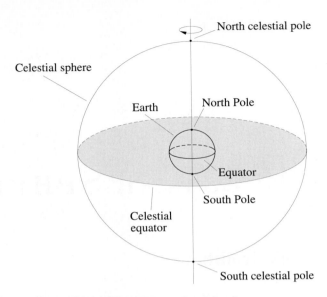

**Figure 1.1** The celestial sphere.

a uniform (or constant) speed and follow a circular motion with Earth at the center of that motion. This concept of a **geocentric universe** was a natural consequence of the apparently unchanging relationship of the stars to one another in fixed constellations. If the stars were simply attached to a **celestial sphere** that rotated about an axis passing through the North and South poles of Earth and intersecting the celestial sphere at the **north** and **south celestial poles** respectively (Fig. 1.1), all of the stars' known motions could be described.

The wandering stars posed a somewhat more difficult problem. A planet such as Mars moves slowly from west to east against the fixed background stars and then mysteriously reverses direction for a period of time before resuming its previous path (Fig. 1.2). Attempting to understand this backward, or **retrograde**, **motion** became the principal problem in astronomy for nearly 2000 years! Eudoxus of Cnidus, a student of Plato's and an exceptional mathematician, suggested that each of the wandering stars occupied its own sphere and that all the spheres were connected through axes oriented at different angles and rotating at various speeds. Although this theory of a complex system of spheres initially was marginally successful at explaining retrograde motion, predictions began to deviate significantly from the observations as more data were obtained.

Hipparchus (ca. 150 B.C.), perhaps the most notable of the Greek as-

## 1.1 The Greek Tradition

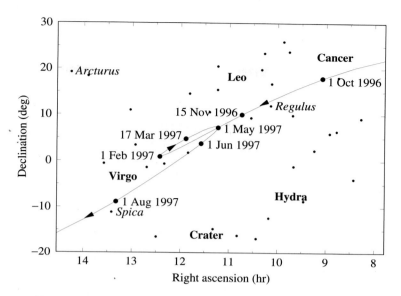

**Figure 1.2** The retrograde motion of Mars. The coordinates of right ascension and declination are discussed on page 14 and in Fig. 1.13.

tronomers, proposed a system of circles to explain retrograde motion. By placing a planet on a small, rotating **epicycle** that in turn moved on a larger **deferent**, he was able to reproduce the behavior of the wandering stars. Furthermore, this system was able to explain the increased brightness of the planets during their retrograde phases as resulting from changes in their distances from Earth. Hipparchus also created the first catalog of the stars, developed a magnitude system for describing the brightness of stars that is still in use today, and contributed to the development of trigonometry.

During the next two hundred years, the model of planetary motion put forth by Hipparchus also proved increasingly unsatisfactory in explaining many of the details of the observations. Claudius Ptolemy (ca. A.D. 100) introduced refinements to the epicycle/deferent system by adding **equants** (Fig. 1.3), resulting in a constant *angular* speed of the epicycle about the deferent ($d\theta/dt$ was assumed to be constant). He also moved Earth away from the deferent center and even allowed for a wobble of the deferent itself. Predictions of the Ptolemaic model did agree more closely with observations than any previously devised scheme, but the original philosophical tenets of Plato (uniform and circular motion) were significantly compromised.

Despite its shortcomings, the Ptolemaic model became almost universally accepted as the correct explanation of the motion of the wandering stars. When

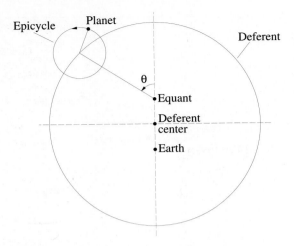

**Figure 1.3** The Ptolemaic model of planetary motion.

a disagreement between the model and observations would develop, the model was modified slightly by the addition of another circle. This process of "fixing" the existing theory led to an increasingly complex theoretical description of observable phenomena.

## 1.2 The Copernican Revolution

By the sixteenth century the inherent simplicity of the Ptolemaic model was gone. Polish-born astronomer Nicolaus Copernicus (1473–1543), hoping to return the science to a less cumbersome, more elegant view of the universe, suggested a **heliocentric** (Sun-centered) model of planetary motion (Fig. 1.4).[1] His bold proposal led immediately to a much less complicated description of the relationships between the planets and the stars. Fearing severe criticism from the Catholic Church, whose doctrine then declared that Earth was the center of the universe, Copernicus postponed publication of his ideas until late in life. *De Revolutionibus Orbium Coelestium* (*On the Revolution of the Celestial Sphere*) first appeared in the year of his death. Faced with a radical new view of the universe, along with Earth's location in it, even some supporters of Copernicus argued that the heliocentric model merely represented a mathematical improvement in calculating planetary positions but did not actually

---

[1] Actually, Aristarchus proposed a heliocentric model of the universe in 280 B.C. At the time his theory was presented, however, there was no compelling evidence to suggest that Earth itself was in motion.

## 1.2 The Copernican Revolution

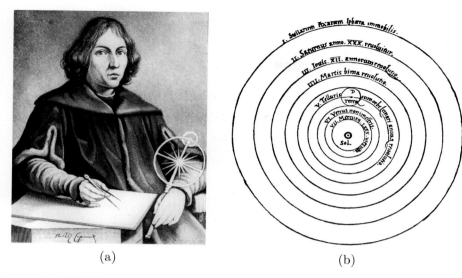

**Figure 1.4** (a) Nicolaus Copernicus (1473–1543). (b) The Copernican model of planetary motion: Planets travel in circles with the Sun at the center of motion. (Courtesy of Yerkes Observatory.)

represent the true geometry of the universe. In fact, a preface to that effect was added by Osiander, the priest who acted as the book's publisher.

One immediate consequence of the Copernican model was the ability to establish the order of all of the planets from the Sun, along with their relative distances and orbital periods. The fact that Mercury and Venus are never seen more than 28° and 47°, respectively, east or west of the Sun, clearly establishes that their orbits are located inside the orbit of Earth. These planets are referred to as **inferior planets**, and their maximum angular separations east or west of the Sun are known as **greatest eastern elongation** and **greatest western elongation**, respectively (see Fig. 1.5). Mars, Jupiter, and Saturn (the most distant planets known to Copernicus) can be seen as much as 180° from the Sun, an alignment known as **opposition**. This could only occur if these **superior planets** have orbits outside Earth's orbit. The Copernican model also predicts that only inferior planets can pass in front of the solar disk (**inferior conjunction**), as observed.

The great long-standing problem of astronomy—retrograde motion—was also easily explained through the Copernican model. Consider the case of a superior planet such as Mars. Assuming, as Copernicus did, that the farther a planet is from the Sun, the more slowly it moves in its orbit, Mars will then be overtaken by the faster-moving Earth. As a result, the apparent position of

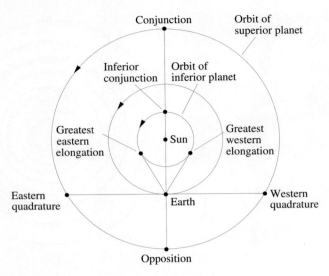

**Figure 1.5** Orbital configurations of the planets.

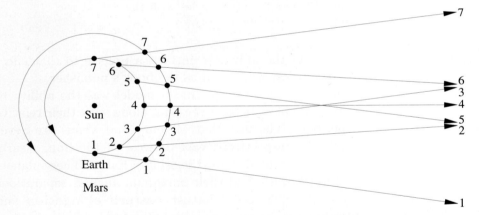

**Figure 1.6** Retrograde motion in the Copernican model.

Mars will shift against the relatively fixed background stars, with the planet seemingly moving backward near opposition, where it is closest to Earth and at its brightest (see Fig. 1.6). Since the orbits of all of the planets are not in the same plane, retrograde loops will occur. The same analysis works equally well for all other planets, superior and inferior.

The relative orbital motions of Earth and the other planets means that the time interval between successive oppositions or conjunctions can differ significantly from the amount of time necessary to make one complete orbit

## 1.2 The Copernican Revolution

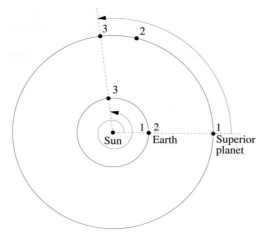

**Figure 1.7** The relationship between sidereal and synodic periods. The two periods do not agree due to the motion of Earth.

relative to the background stars (Fig. 1.7). The former time interval (between oppositions) is known as the **synodic period** ($S$), and the latter time interval (measured relative to the background stars) is referred to as the **sidereal period** ($P$). It is left as an exercise to show that the relationship between the two periods is given by

$$1/S = \begin{cases} 1/P - 1/P_\oplus & \text{(inferior)} \\ 1/P_\oplus - 1/P & \text{(superior),} \end{cases} \qquad (1.1)$$

where $P_\oplus$ is the sidereal period of Earth (365.26 days).

Although the Copernican model did represent a simpler, more elegant model of planetary motion, it was not successful in predicting positions any more accurately than the Ptolemaic model. This lack of improvement was due to Copernicus's inability to relinquish the 2000-year-old concept that planetary motion required circles, the human notion of perfection. As a consequence, Copernicus was forced (as were the Greeks) to introduce the concept of epicycles to "fix" his model.

Perhaps the quintessential example of a scientific revolution was the revolution begun by Copernicus. What we think of today as the obvious solution to the problem of planetary motion—a heliocentric universe—was perceived as a very strange and even rebellious notion during a time of major upheaval, when Columbus had recently sailed to the "new world" and Martin Luther had proposed radical revisions in Christianity. Thomas Kuhn has suggested that an established scientific theory is much more than just a framework for guiding the study of natural phenomena. The present **paradigm** (or prevailing scientific theory) is actually a way of *seeing* the universe around us. We ask questions, pose new research problems, and interpret the results of experiments and observations in the context of the paradigm. Viewing the universe in any other way requires a complete shift from the current paradigm. To suggest that Earth actually orbits the Sun instead of believing that the Sun inexorably rises and sets about a fixed Earth is to argue for a change in the very structure of the universe, a structure that was believed to be correct and beyond question for nearly 2000 years. Not until the complexity of the old Ptolemaic scheme became too unwieldy could the intellectual environment reach a point where the concept of a heliocentric universe was even possible.

## 1.3 Positions on the Celestial Sphere

The Copernican revolution has shown us that the notion of a geocentric universe is incorrect. Nevertheless, with the exception of a small number of planetary probes, our observations of the heavens are still based on a reference frame centered on Earth. The daily (or **diurnal**) rotation of Earth, coupled with its annual motion around the Sun and the slow wobble of its rotation axis, together with relative motions of the stars, planets, and other objects, result in the constantly changing positions of celestial objects. To catalog the locations of objects such as the Crab supernova remnant in Taurus or the great spiral galaxy of Andromeda, coordinates must be specified. Moreover, the coordinate system should not be sensitive to the short-term manifestations of Earth's motions; otherwise the specified coordinates would constantly change.

Viewing objects in the night sky requires only directions to them, not their distances. We can imagine that all objects are located on a celestial sphere, just as the ancient Greeks believed. It then becomes sufficient to specify only two coordinates. The most straightforward coordinate system one might devise is based on the observer's local horizon. The **altitude–azimuth** (or **horizon**) **coordinate system** is based on the measurement of the azimuth angle along the horizon and the altitude angle above the horizon (Fig. 1.8). The **altitude** $h$ is defined as that angle measured from the horizon to the object along a great circle[2] that passes through that object and the point on the celestial sphere directly above the observer, a point known as the **zenith**. Equivalently, the **zenith distance** $z$ is the angle measured from the zenith to

---

[2] A great circle is the curve resulting from the intersection of a sphere with a plane passing through the *center* of that sphere.

## 1.3 Positions on the Celestial Sphere

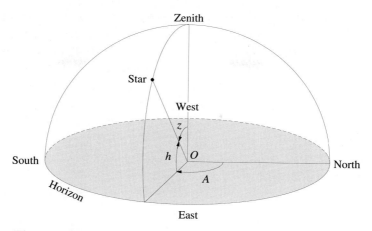

**Figure 1.8** The altitude–azimuth coordinate system. $h$, $z$, and $A$ are the altitude, zenith distance, and azimuth, respectively.

the object, so $z + h = 90°$. The **azimuth** $A$ is simply the angle measured along the horizon eastward from north to the great circle used for the measure of altitude. (The **meridian** is another frequently used great circle; it is defined as passing through the observer's zenith and intersecting the horizon due north and south.)

Although simple to define, the altitude–azimuth system is difficult to use in practice. Coordinates of celestial objects in this system are specific to the local latitude and longitude of the observer and are difficult to transform to other locations on Earth. Also, since Earth is rotating, stars appear to move constantly across the sky, meaning that the coordinates of each object are constantly changing, even for the local observer. Complicating the problem still further, the stars rise approximately 4 minutes earlier on each successive night, so that even when viewed from the same location at a specified time, the coordinates change from day to day.

To understand the problem of these day-to-day changes in altitude–azimuth coordinates, we must consider the orbital motion of Earth about the Sun (see Fig. 1.9). As Earth orbits the Sun, our view of the distant stars is constantly changing. Our line of sight to the Sun sweeps through the constellations during the seasons; consequently, we see the Sun apparently move through those constellations along a path referred to as the **ecliptic**. During the spring the Sun appears to travel across the constellation of Virgo, in the summer it moves through Taurus, during the autumn months it enters Aquarius, and in the winter the Sun is located near Sagittarius. As a consequence, those constellations become obscured in the glare of daylight and other constellations appear in

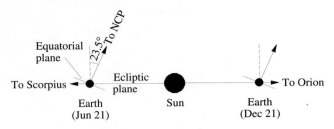

**Figure 1.9** The plane of Earth's orbit seen edge-on. The tilt of Earth's rotation axis relative to the ecliptic is also shown.

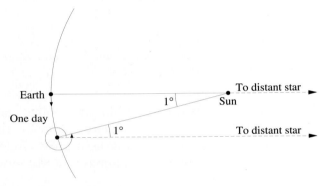

**Figure 1.10** Earth must rotate nearly 361° per solar day and only 360° per sidereal day.

our night sky. This seasonal change in the constellations is directly related to the fact that a given star rises approximately 4 minutes earlier each day. Since Earth completes one sidereal period in 365.26 days, it moves slightly less than 1° around its orbit in 24 hours. Thus Earth must actually rotate nearly 361° to bring the Sun to the meridian on two successive days (Fig. 1.10). Because of the much greater distances to the stars, they do not shift their positions significantly as Earth orbits the Sun. As a result, placing a star on the meridian on successive nights requires only a 360° rotation. It takes approximately 4 minutes for Earth to rotate the extra 1°. Therefore a given star rises 4 minutes earlier each night. **Solar time** is defined as an *average* interval of 24 hours between meridian crossings of the Sun, and **sidereal time** is based on consecutive meridian crossings of a star.

Seasonal climatic variations are also due to the orbital motion of Earth, coupled with the approximately 23.5° tilt of its rotation axis. As a result of the tilt, the ecliptic moves north and south of the **celestial equator** (Fig. 1.11), which is defined by passing a plane through Earth at its equator and extending

## 1.3 Positions on the Celestial Sphere

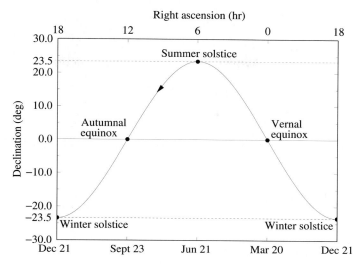

**Figure 1.11** The ecliptic is the annual path of the Sun across the celestial sphere and is sinusoidal about the celestial equator. See Fig. 1.13, page 15, for definitions of right ascension and declination.

that plane out to the celestial sphere. The sinusoidal shape of the ecliptic occurs because the Northern Hemisphere alternately points toward and then away from the Sun during Earth's annual orbit. Twice during the year the Sun crosses the celestial equator, once moving northward along the ecliptic and later moving to the south. In the first case, the point of intersection is called the **vernal equinox** and the southern crossing occurs at the **autumnal equinox**. Spring officially begins when the center of the Sun is precisely on the vernal equinox; similarly, fall begins when the center of the Sun crosses the autumnal equinox. The most northern excursion of the Sun along the ecliptic occurs at the **summer solstice**, representing the official start of summer, and the southernmost position of the Sun is defined as the **winter solstice**.

The seasonal variations in weather are due to the position of the Sun relative to the celestial equator. During the summer months in the Northern Hemisphere, the Sun's northern declination causes it to appear higher in the sky, producing longer days and more intense sunlight. During the winter months the declination of the Sun is below the celestial equator, its path above the horizon is shorter, and its rays are less intense (see Fig. 1.12). The more direct the Sun's rays, the more energy per unit area strikes Earth's surface and the higher the resulting surface temperature.

A coordinate system that results in nearly constant values for the positions of celestial objects, despite the complexities of diurnal and annual motions, is

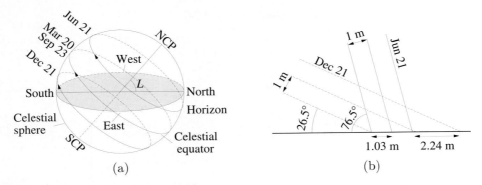

**Figure 1.12** (a) The diurnal path of the Sun across the celestial sphere when the Sun is located at the vernal equinox (March), the summer solstice (June), the autumnal equinox (September), and the winter solstice (December) for an observer at latitude $L$. NCP and SCP designate the north and south celestial poles, respectively. (b) The direction of the Sun's rays at noon at the summer solstice (solid lines) and at the winter solstice (dashed lines).

necessarily less straightforward than the altitude–azimuth system. The **equatorial coordinate system** (see Fig. 1.13) is based on the latitude–longitude system of Earth but does not participate in the planet's rotation. **Declination** $\delta$ is the equivalent of latitude and is measured in degrees north or south of the celestial equator. **Right ascension** $\alpha$ is analogous to longitude and is measured eastward along the celestial equator from the vernal equinox ($\Upsilon$) to its intersection with the object's **hour circle** (the great circle passing through the object being considered and through the north celestial pole). Right ascension is traditionally measured in hours, minutes, and seconds; 24 hours of right ascension is equivalent to 360°, or 1 hour = 15°. The rationale for this unit of measure is based on the 24 hours (sidereal time) necessary for an object to make two successive crossings of the observer's local meridian. The coordinates of right ascension and declination are also indicated in Figs. 1.2 and 1.11. Since the equatorial coordinate system is based on the celestial equator and the vernal equinox, changes in the latitude and longitude of the observer do not affect the values of right ascension and declination. Values of $\alpha$ and $\delta$ are similarly unaffected by the annual motion of Earth around the Sun.

The local sidereal time of the observer is defined as the amount of time elapsing since the vernal equinox last traversed the meridian. Local sidereal time is also equivalent to the **hour angle** $H$ of the vernal equinox, where hour angle is defined as the angle between a celestial object and the observer's

## 1.3 Positions on the Celestial Sphere

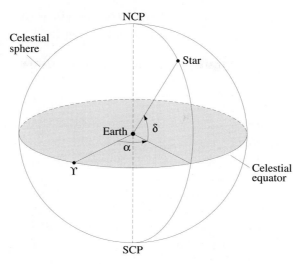

**Figure 1.13** The equatorial coordinate system. $\alpha$, $\delta$, $\Upsilon$ designate right ascension, declination, and the position of the vernal equinox, respectively.

meridian, measured in the direction of the object's motion around the celestial sphere.

Despite referencing the equatorial coordinate system to the celestial equator and its intersection with the ecliptic (the vernal equinox), **precession** causes the right ascension and declination of celestial objects to change, albeit very slowly. Precession is the slow wobble of Earth's rotation axis due to the planet's nonspherical shape and its gravitational interaction with the Sun and the Moon. It was Hipparchus who first observed the effects of precession. Although we will not discuss the physical cause of this phenomenon in detail, it is completely analogous to the well-known precession of a child's toy top. Earth's precession period is 25,770 years and causes the north celestial pole to make a slow circle through the heavens. Although Polaris (the North Star) is currently within 1° of the north celestial pole, in 13,000 years it will be nearly 47° away from that point. The same effect also causes a $50.26''$ yr$^{-1}$ westward motion of the vernal equinox along the ecliptic.[3] An additional precession effect due to Earth–planet interactions results in an eastward motion of the vernal equinox of $0.12''$ yr$^{-1}$.

Because precession alters the position of the vernal equinox along the ecliptic, it is necessary to refer to a specific epoch when listing the right ascension and declination of a celestial object. The current values of $\alpha$ and $\delta$ may then be

---

[3] 1 arcminute = $1'$ = 1/60 degree; 1 arcsecond = $1''$ = 1/60 arcminute.

calculated, based upon the amount of time elapsed since the reference epoch. The increase in the values of the coordinates is given by

$$\Delta\alpha = [m + n\sin\alpha\tan\delta]\,N \quad (1.2)$$

$$\Delta\delta = [n\cos\alpha]\,N, \quad (1.3)$$

where $N$ is the number of years between the desired time and the reference epoch (either positive or negative). If the reference epoch is 1950.0 (January 1, 1950), then $m = 3.07327^s\ \text{yr}^{-1}$ and $n = 20.0426''\ \text{yr}^{-1}$.

---

**Example 1.1** Altair, the brightest star in the summer constellation of Aquila, has the epoch 1950.0 coordinates, $\alpha = 19^h 48.3^m$ and $\delta = +08°44'$. Using Eqs. (1.2) and (1.3), we may precess the star's coordinates to July 30, 1989.

From the relations between time and the angular measure of right ascension,

$$1^h = 15°$$
$$1^m = 15'$$
$$1^s = 15''$$

and the amount of time elapsed since 1950.0, $N = 39.58$ years, we have that

$$\Delta\alpha = [3.07327 + 1.33617\sin(297.075°)\tan(8.73333°)] \times 39.58\ \text{sec}$$
$$= 114.406^s$$
$$\simeq 1.91^m$$

and

$$\Delta\delta = [20.0426\cos(297.075°)] \times 39.58\ \text{arcsec}$$
$$= 361.07''$$
$$\simeq 6.02'.$$

Thus Altair's precessed coordinates become $\alpha = 19^h 50.2^s$ and $\delta = +08°50'$.

---

Archaeoastronomy, the study of the astronomy of past cultures, relies heavily on the alignments of ancient structures toward celestial objects. Because of the long periods of time since construction, care must be given to the proper precession of celestial coordinates if any proposed alignments are to be meaningful. The Great Pyramid at Giza (Fig. 1.14), one of the "seven wonders of

## 1.3 Positions on the Celestial Sphere

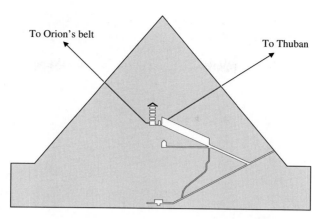

**Figure 1.14** The astronomical alignments of the Great Pyramid at Giza. (Adaptation of figure from Griffith Observatory.)

the world," is an example of such a structure. Believed to have been erected about 2600 B.C., the Great Pyramid has long been the subject of speculation. Although many of the proposals concerning this amazing monument are more than somewhat fanciful, there can be no doubt about its careful orientation with the four cardinal positions, north, south, east, and west. The greatest misalignment of any side from a true cardinal direction is no more than $5\frac{1}{2}'$. Equally astounding is the nearly perfect square formed by its base; no two sides differ in length by more than 7.8 inches.

Perhaps the most demanding alignments discovered so far are associated with the "air shafts" leading from the King's Chamber (the main chamber of the pyramid) to the outside. These air shafts seem too poorly designed to circulate fresh air into the tomb of Pharaoh, and it is now thought that they served another function. The Egyptians believed that when their pharaohs died, their souls would travel to the sky to join Osiris, the god of life, death, and rebirth. Osiris was associated with the constellation we now know as Orion. Allowing for over one-sixth of a precession period since the construction of the Great Pyramid, Virginia Trimble has shown that one of the air shafts pointed directly to Orion's belt. The other air shaft pointed toward Thuban, the star that was *then* closest to the north celestial pole, the point in the sky about which all else turns.

As a modern scientific culture, we trace our study of astronomy to the ancient Greeks, but it has become apparent that many cultures carefully studied the sky and its mysterious points of light. Archaeological structures worldwide apparently exhibit astronomical alignments. Although some of these

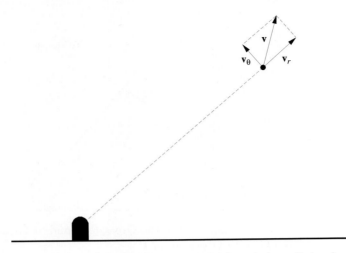

**Figure 1.15** The components of velocity. $\mathbf{v}_r$ is the star's radial velocity and $\mathbf{v}_\theta$ is the star's transverse velocity.

alignments may be coincidental, it is clear that many of them were by design.

Another effect contributing to the change in equatorial coordinates is due to the intrinsic velocity of the objects themselves.[4] As we have already discussed, the Sun, the Moon, and the planets exhibit relatively rapid and complex motions through the heavens. The stars also move with respect to one another. Even though their actual speeds may be very large, the apparent relative motions of stars are generally very difficult to measure due to their enormous distances.

Consider the velocity of a star relative to an observer (Fig. 1.15). The velocity vector may be decomposed into two mutually perpendicular components, one lying along the line of sight and the other perpendicular to it. The line-of-sight component is the star's **radial velocity**, $\mathbf{v}_r$, and will be discussed in Section 4.3; the second component is the star's **transverse** or **tangential velocity**, $\mathbf{v}_\theta$, along the celestial sphere. This transverse velocity appears as a *slow, angular change* in its equatorial coordinates, known as **proper motion** (usually expressed in seconds of arc per year). In a time interval $\Delta t$ the star will have moved in a direction perpendicular to the observer's line of sight a distance

$$\Delta d = v_\theta \Delta t.$$

If the distance from the observer to the star is $r$, then the angular change in

---

[4]Parallax, an important, periodic motion of the stars resulting from the motion of Earth about the Sun, will be discussed in detail in Section 3.1.

## 1.3 Positions on the Celestial Sphere

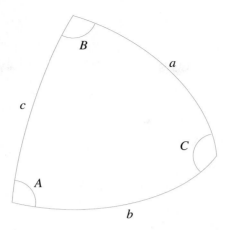

**Figure 1.16** A spherical triangle. Each leg is a segment of a great circle on the surface of a sphere, and all angles are less than 180°.

its position along the celestial sphere is given by

$$\Delta\theta = \frac{\Delta d}{r} = \frac{v_\theta}{r}\Delta t.$$

Thus the star's proper motion, $\mu$, is related to its transverse velocity by

$$\mu \equiv \frac{d\theta}{dt} = \frac{v_\theta}{r}. \tag{1.4}$$

---

*Optional:* The laws of spherical trigonometry must be employed in order to find the relationship between $\Delta\theta$ and changes in the equatorial coordinates, $\Delta\alpha$ and $\Delta\delta$. For the spherical triangle shown in Fig. 1.16, composed of three intersecting segments of great circles, the following relationships hold:

*Law of sines*

$$\frac{\sin a}{\sin A} = \frac{\sin b}{\sin B} = \frac{\sin c}{\sin C} \tag{1.5}$$

*Law of cosines for sides*

$$\cos a = \cos b \cos c + \sin b \sin c \cos A \tag{1.6}$$

*Law of cosines for angles*

$$\cos A = -\cos B \cos C + \sin B \sin C \cos a \tag{1.7}$$

where all sides are measured in arc length.

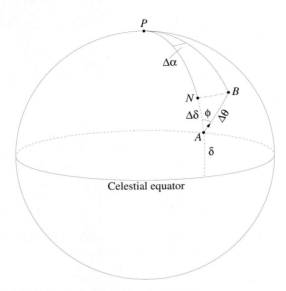

**Figure 1.17** The proper motion of a star across the celestial sphere. The star is assumed to be moving from $A$ to $B$ along the position angle $\phi$.

Figure 1.17 shows the motion of a star on the celestial sphere from point $A$ to point $B$ along the **position angle** $\phi$ ($\angle PAB$), measured from the north celestial pole. The angular distance traveled is $\Delta\theta$. Let point $P$ be located at the north celestial pole so that the arcs $AP$, $AB$, and $BP$ form segments of great circles. Now, construct a segment of a circle $NB$ such that $N$ is at the same declination as $B$ and $\angle BNP = 90°$. If the coordinates of the star at point $A$ are $(\alpha, \delta)$ and its new coordinates at point $B$ are $(\alpha + \Delta\alpha, \delta + \Delta\delta)$, then $\angle APB = \Delta\alpha$, $\overline{AP} = 90° - \delta$, and $\overline{NP} = \overline{BP} = 90° - (\delta + \Delta\delta)$. Using the law of sines,

$$\frac{\sin(\Delta\alpha)}{\sin(\Delta\theta)} = \frac{\sin\phi}{\sin[90° - (\delta + \Delta\delta)]},$$

or

$$\sin(\Delta\alpha)\cos(\delta + \Delta\delta) = \sin(\Delta\theta)\sin\phi.$$

Assuming that the changes in position are much less than one radian, we may use the small angle approximations $\sin\epsilon \sim \epsilon$ and $\cos\epsilon \sim 1$. Employing the appropriate trigonometric identity and neglecting all terms of second order or higher, the previous equation reduces to

$$\Delta\alpha = \Delta\theta \frac{\sin\phi}{\cos\delta}. \tag{1.8}$$

The law of cosines for sides may also be used to find an expression for the change in the declination:

$$\cos\left[90° - (\delta + \Delta\delta)\right] = \cos\left(90° - \delta\right)\cos\left(\Delta\theta\right) + \sin\left(90° - \delta\right)\sin\left(\Delta\theta\right)\cos\phi.$$

This immediately simplifies to

$$\Delta\delta = \Delta\theta\cos\phi. \tag{1.9}$$

(Note that this is the same result that would be obtained if we had used plane trigonometry. This should be expected, however, since we have assumed that the triangle being considered has an area much smaller than the total area of the sphere and should therefore appear essentially flat.) Combining Eqs. (1.8) and (1.9), we arrive at the expression for the angular distance traveled in terms of the changes in right ascension and declination:

$$(\Delta\theta)^2 = (\Delta\alpha\cos\delta)^2 + (\Delta\delta)^2. \tag{1.10}$$

## 1.4 Physics and Astronomy

The mathematical view of nature first proposed by Pythagoras and the Greeks led ultimately to the Copernican revolution. The inability of astronomers to accurately fit the observed positions of the "wandering stars" with mathematical models resulted in a dramatic change in our perception of Earth's location in the universe. However, an equally important step still remained in the development of science, the search for *physical causes* of observable phenomena. As we will see constantly throughout this book, the modern study of astronomy relies heavily on an understanding of the physical nature of the universe. The application of physics to astronomy, *astrophysics*, has proven very successful in explaining a wide range of observations, including strange and exotic objects and events, such as pulsating stars, supernovae, variable x-ray sources, black holes, and quasars.

As a part of our investigation of the science of astronomy, it will be necessary to study the details of celestial motions, the nature of light, the structure of the atom, and the shape of space itself. Rapid advances in astronomy over the past several decades have occurred because of advances in our understanding of fundamental physics and because of improvements in the tools we use to study the heavens: telescopes and computers.

Essentially every area of physics plays an important role in some aspect of astronomy. Particle physics and astrophysics merge in the study of the Big Bang; the basic question of the origin of the zoo of elementary particles, as well as the very nature of the fundamental forces, is intimately linked to how the universe was formed. Nuclear physics provides information about the types of reactions that are possible in the interiors of stars, and atomic physics describes how individual atoms interact with one another and with light, processes that are very important in all astrophysical phenomena. Even electronics plays an important role in the development of new detectors capable of giving a clearer view of the universe around us.

With the advent of modern technology and the space age, telescopes have been built to study the heavens with ever-increasing sensitivity. No longer limited to detecting visible light, telescopes are now capable of "seeing" x-rays, ultraviolet light, infrared radiation, and radio signals. Many of these telescopes require operation above Earth's atmosphere to carry out their missions. Other types of telescopes, very different in nature, detect elementary particles instead of light and are often placed below ground to study the heavens.

Computers have provided us with the power to carry out the enormous number of calculations necessary to build mathematical models from fundamental physical principles. The birth of high-speed computing machines has allowed astronomers to calculate the evolution of a star and compare those calculations with observations; it is also possible to study the rotation of a galaxy and its interaction with neighboring galaxies. Processes that require billions of years (significantly longer than any National Science Foundation grant) cannot possibly be observed directly but may be investigated using the modern supercomputer.

All of these tools and related disciplines are used to look at the heavens with a probing eye. The study of astronomy is a natural extension of human curiosity in its purest form. Just as a small child is always asking why this or that is the way it is, the goal of an astronomer is to attempt to understand the nature of the universe in all of its complexity, simply for the sake of understanding—the ultimate end of any intellectual adventure. In a very real sense the true beauty of the heavens lies not only in observing the stars on a dark night but in considering the delicate interplay between the physical processes that cause the stars to exist at all.

*The most incomprehensible thing about the universe is that it is comprehensible.* — Albert Einstein

## Suggested Readings

### General

Aveni, Anthony, *Skywatchers of Ancient Mexico*, The University of Texas Press, Austin, 1980.

Bronowski, J., *The Ascent of Man*, Little, Brown, Boston, 1973.

Casper, Barry M., and Noer, Richard J., *Revolutions in Physics*, W. W. Norton, New York, 1972.

Hadingham, Evan, *Early Man and the Cosmos*, Walker and Company, New York, 1984.

Krupp, E. C., *Echos of the Ancient Skies: The Astronomy of Lost Civilizations*, Harper and Row, New York, 1983.

Kuhn, Thomas S., *The Structure of Scientific Revolutions*, Second Edition, The University of Chicago Press, Chicago, 1970.

Sagan, Carl, *Cosmos*, Random House, New York, 1980.

### Technical

Acker, Agnes, and Jaschek, Carlos, *Astronomical Methods and Calculations*, John Wiley and Sons, Chichester, 1986.

Lang, Kenneth R., *Astrophysical Formulae*, Second Edition, Springer-Verlag, Berlin, 1980.

Smart, W. M., *Textbook on Spherical Astronomy*, Sixth Edition, Cambridge University Press, Cambridge, 1977.

## Problems

1.1 Derive the relationship between a planet's synodic period and its sidereal period (Eq. 1.1). Consider both inferior and superior planets.

1.2 Devise methods to determine the *relative* distances of each of the planets from the Sun given the information available to Copernicus (orbital configurations and synodic periods).

1.3 (a) The observed orbital synodic periods of Venus and Mars are 583.9 days and 779.9 days, respectively. Calculate their sidereal periods.

(b) Which planet in the solar system has the shortest synodic period? Why?

1.4 List the right ascension and declination of the Sun when it is located at the vernal equinox, the summer solstice, the autumnal equinox, and the winter solstice.

1.5 (a) Calculate the altitude of the Sun along the meridian on the first day of summer for an observer at a latitude of 42° north.

(b) What is the maximum altitude of the Sun on the first day of winter at the same latitude?

1.6 (a) Circumpolar stars are defined as stars that never set below the horizon of the local observer or stars that are never visible above the horizon. Calculate the range of declinations for these two groups of stars for an observer at the latitude $L$.

(b) At what latitude(s) on Earth will the Sun never set when it is at the summer solstice?

(c) Is there any latitude on Earth where the Sun will never set when it is at the vernal equinox? Where?

1.7 Proxima Centauri ($\alpha$ Centauri C) is the closest star to the Sun and is a part of a triple star system. It has the epoch 1950.0 coordinates $(\alpha, \delta) = (14^h 26.3^m, -62°28')$ while the center of the system is located at $(\alpha, \delta) = (14^h 36.2^m, -60°38')$.

(a) What is the angular separation of Proxima Centauri from the center of the triple star system?

(b) If the distance to Proxima Centauri is $4.0 \times 10^{18}$ cm, how far is the star from the center of the triple system?

1.8 (a) Using the information in Problem 1.7, precess the coordinates of Proxima Centauri to epoch 1990.0.

(b) The proper motion of Proxima Centauri is 3.84" yr$^{-1}$ with the position angle 282°. Calculate the change in $\alpha$ and $\delta$ due to proper motion between 1950.0 and 1990.0.

(c) Which effect makes the largest contribution to changes in the coordinates of Proxima Centauri: precession or proper motion?

# Chapter 2

# Celestial Mechanics

## 2.1 Elliptical Orbits

Although the inherent simplicity of the Copernican model was aesthetically pleasing, the idea of a heliocentric universe was not immediately accepted; it lacked the support of observations capable of unambiguously demonstrating that a geocentric model was wrong. After the death of Copernicus, Tycho Brahe (1546–1601), the foremost naked-eye observer, carefully followed the motions of the "wandering stars" and other celestial objects. He carried out his work at the observatory, Uraniborg, on the island of Hveen (a facility provided for him by King Frederick II of Denmark). To improve the accuracy of his observations, Tycho used large measuring instruments, such as the quadrant depicted in the mural in Fig. 2.1(a). Tycho's observations were so meticulous that he was able to measure the position of an object in the heavens to an accuracy of better than $4'$, approximately one-eighth the angular diameter of a full moon. Through the accuracy of his observations he demonstrated for the first time that comets must be very distant, well beyond the Moon, rather than being some form of atmospheric phenomenon. Tycho is also credited with observing the supernova of 1572, which clearly demonstrated that the heavens were not unchanging as Church doctrine held. (This observation prompted King Frederick to build Uraniborg.) Despite the great care with which he carried out his work, Tycho was not able to find any clear evidence of the motion of Earth through the heavens, and he therefore concluded that the Copernican model must be false (see Section 3.1).

At Tycho's invitation, Johannes Kepler (1571–1630), a German mathematician, joined him at Uraniborg (Fig. 2.1b). Unlike Tycho, Kepler was a heliocentrist, and it was his desire to find a geometrical model of the universe

(a)            (b)

**Figure 2.1** (a) Mural of Tycho Brahe (1546–1601). (b) Johannes Kepler (1571–1630). (Courtesy of Yerkes Observatory.)

that would be consistent with the best observations then available, namely Tycho's. After Tycho's death, Kepler inherited the mass of observations accumulated over the years and began a painstaking analysis of the data. His initial, almost mystic, idea was that the universe is arranged with five perfect solids, nested to support the six known naked-eye planets (including Earth) on crystalline spheres, with the entire system centered on the Sun. After this model proved unsuccessful, he attempted to devise an accurate set of circular planetary orbits about the Sun, focusing specifically on Mars. Through his very clever use of offset circles and equants,[1] Kepler was able to obtain excellent agreement with Tycho's data for all but two of the points available. In particular, the discrepant points were each off by approximately $8'$, or twice the accuracy of Tycho's data. Believing that Tycho would not have made observational errors of this magnitude, Kepler felt forced to dismiss the idea of purely circular motion.

Rejecting the last fundamental assumption of the Ptolemaic model, Kepler began to consider the possibility that planetary orbits were elliptical in shape rather than circular. Through this relatively minor mathematical (though monumental philosophical) change, he was finally able to bring all of Tycho's observations into agreement with a model for planetary motion. This paradigm shift also allowed Kepler to discover that the orbital speed of a planet is not

---

[1] Recall the geocentric use of circles and equants by Ptolemy; see Fig. 1.3.

## 2.1 Elliptical Orbits

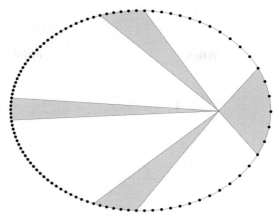

**Figure 2.2** Kepler's second law states that the area swept out by a line between a planet and the focus of an ellipse is always the same for a given time interval, regardless of the planet's position in its orbit. The dots are evenly spaced in time.

constant but varies in a precise way depending on its location in its orbit. In 1609 Kepler published the first two of his three laws of planetary motion in the book, *Astronomica Nova*, or *The New Astronomy*:

**Kepler's First Law** A planet orbits the Sun in an ellipse, with the Sun at one focus of the ellipse.

**Kepler's Second Law** A line connecting a planet to the Sun sweeps out equal areas in equal time intervals.

Kepler's first and second laws are illustrated in Fig. 2.2, where each dot on the ellipse represents the position of the planet during evenly spaced time intervals.

Kepler's third law (also known as the harmonic law) was published ten years later in the book *Harmonica Mundi* (*The Harmony of the World*). His final law relates the average orbital distance of a planet from the Sun to its sidereal period:

**Kepler's Third Law** $P^2 = a^3$,

where $P$ is the orbital period of the planet, measured in *years*, and $a$ is the average distance of the planet from the Sun, in *astronomical units*, or AU. An **astronomical unit** is, by definition, the average distance between Earth and the Sun, $1.496 \times 10^{13}$ cm. A graph of Kepler's third law is shown in Fig. 2.3 using data for each planet in our solar system as given in Appendix B.

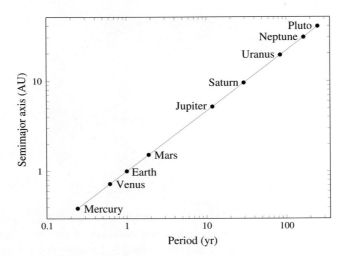

**Figure 2.3** Kepler's third law for planets orbiting the Sun.

In retrospect it is easy to understand why the assumption of uniform and circular motion first proposed nearly 2000 years earlier was not found to be wrong much sooner; in most cases, planetary motion differs little from purely circular motion. In fact, it was actually fortuitous that Kepler chose to focus on Mars, since the data for that planet were particularly good and Mars deviates from circular motion more than most of the others.

To appreciate the significance of Kepler's laws, we must first understand the nature of the **ellipse**. An ellipse (see Fig. 2.4) is defined by that set of points that satisfies the equation

$$r + r' = 2a, \qquad (2.1)$$

where $a$ is a constant, known as the **semimajor axis** (half the length of the long axis of the ellipse) and $r$ and $r'$ represent the distances to the ellipse from the two **focal points**, $F$ and $F'$, respectively. Notice that if $F$ and $F'$ were located at the same point, then $r' = r$ and the previous equation would reduce to $r = r' = a$, the equation for a circle. Thus a circle is simply a special case of an ellipse. The distance $b$ is known as the **semiminor axis**. The distance of either focal point from the center of the ellipse may be expressed as $ae$, where $e$ is defined to be the **eccentricity** of the ellipse ($0 \leq e < 1$). For a circle, $e = 0$.

A convenient relationship among $a$, $b$, and $e$ may be found. Consider one of the two points at either end of the semiminor axis of an ellipse, where $r = r'$. In this case, $r = a$ and, by the Pythagorean theorem, $r^2 = b^2 + a^2 e^2$. Substitution

## 2.1 Elliptical Orbits

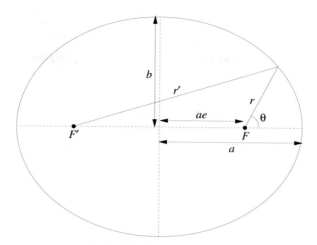

**Figure 2.4** The geometry of an elliptical orbit.

leads immediately to the expression

$$b^2 = a^2(1 - e^2). \tag{2.2}$$

According to Kepler's first law, a planet orbits the Sun in an ellipse, with the Sun located at one focus of the ellipse, the **principal focus** (the other focus is empty space). The second law states that the orbital speed of a planet depends on its *location* in that orbit. To describe in detail the orbital behavior of a planet, it is necessary to specify where that planet is (its position vector) as well as how fast, and in what direction, the planet is moving (its velocity vector).

It is often most convenient to express a planet's orbit in polar coordinates, indicating its distance $r$ from the principal focus in terms of an angle $\theta$ measured counterclockwise from the major axis of the ellipse (see Fig. 2.4). Using the Pythagorean theorem,

$$r'^2 = r^2 \sin^2 \theta + (2ae + r \cos \theta)^2,$$

which reduces to

$$r'^2 = r^2 + 4ae(ae + r \cos \theta).$$

Using the definition of an ellipse, $r + r' = 2a$, we find that

$$r = \frac{a(1 - e^2)}{1 + e \cos \theta} \qquad (0 \leq e < 1). \tag{2.3}$$

It is left as an exercise to show that the total area of an ellipse is given by

$$A = \pi ab. \tag{2.4}$$

**Example 2.1** Using Eq. (2.3), it is possible to determine the variation in distance of a planet from the principal focus throughout its orbit. The semimajor axis of Mars' orbit is 1.5237 AU (or $2.2794 \times 10^{13}$ cm) and the planet's orbital eccentricity is 0.0934. When $\theta = 0°$, the planet is closest to the Sun, a point known as **perihelion**, and is at a distance given by

$$r_p = \frac{a(1-e^2)}{1+e}$$
$$= a(1-e) \qquad (2.5)$$
$$= 1.3814 \text{ AU}.$$

Similarly, **aphelion** ($\theta = 180°$), the point where Mars is farthest from the Sun, is at a distance given by

$$r_a = \frac{a(1-e^2)}{1-e}$$
$$= a(1+e) \qquad (2.6)$$
$$= 1.6660 \text{ AU}.$$

The variation in Mars' orbital distance from the Sun amounts to approximately 19% between perihelion and aphelion.

---

An ellipse is actually one of a class of curves known as conic sections, found by passing a plane through a cone (see Fig. 2.5). Each type of conic section has its own characteristic range of eccentricities. As already mentioned, a circle is a conic section having $e = 0$, and an ellipse has $0 \leq e < 1$. A curve having $e = 1$ is known as a **parabola** and is described by the equation

$$r = \frac{2p}{1 + \cos\theta} \qquad (e = 1), \qquad (2.7)$$

where $p$ is the distance of closest approach to the parabola's *one* focus, at $\theta = 0$. Curves having eccentricities greater than unity, $e > 1$, are **hyperbolas** and have the form

$$r = \frac{a(e^2 - 1)}{1 + e\cos\theta} \qquad (e > 1). \qquad (2.8)$$

Each type of conic section is related to a specific form of celestial motion.

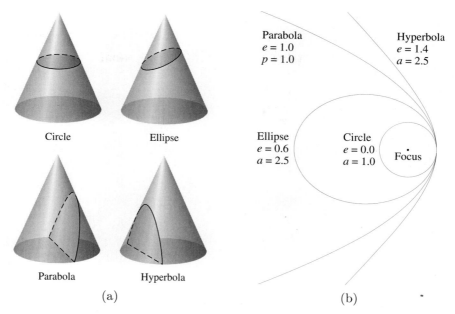

**Figure 2.5** (a) Conic sections. (b) Related orbital motions.

## 2.2 Newtonian Mechanics

At the time Kepler was developing his three laws of planetary motion, Galileo Galilei (1564–1642), perhaps the first of the true experimental physicists, was studying the motion of objects on Earth (Fig. 2.6a). It was Galileo who proposed the earliest formulation of the concept of inertia; he had also developed an understanding of acceleration. In particular, he realized that objects near the surface of Earth fell with the same acceleration, independent of their weight. Whether Galileo publicly proved this fact by dropping objects of differing weights from the Leaning Tower of Pisa is a matter of some debate.

Galileo is also the father of modern observational astronomy. Shortly after learning about the 1608 invention of the first crude spyglass, he thought through its design and constructed his own. Using his new telescope to carefully observe the heavens, Galileo quickly made a number of important observations in support of the heliocentric model of the universe. In particular, he discovered that the band of light known as the Milky Way, which runs from horizon to horizon, was not merely a cloud, as had previously been supposed, but actually contains an enormous number of individual stars not resolvable by the naked eye. Galileo also observed that the Moon contained craters and

(a)  (b)

**Figure 2.6** (a) Galileo Galilei (1564–1642). (b) Isaac Newton (1642–1727). (Courtesy of Yerkes Observatory.)

therefore was not a perfect sphere. Observations of the varying phases of Venus implied that the planet did not shine by its own power, but must be reflecting sunlight from varying positions in its orbit about the Sun. He also discovered that the Sun itself is blemished, possessing sunspots that varied in number and location. But perhaps the most damaging observation for the geocentric model, a model still strongly supported by the Church, was the discovery of four moons in orbit about Jupiter, indicating the existence of at least one other center of motion in the universe.

Many of Galileo's first observations were published in his book *Sidereus Nuncius* (*The Starry Messenger*) in 1610. By 1616 the Church forced him to withdraw his support of the Copernican model, although he was able to continue his study of astronomy for some years. In 1632 Galileo published another work, *The Dialogue on the Two Chief World Systems*, in which a three-character play was staged. In the play Salviati was the proponent of Galileo's views, Simplicio believed in the old Aristotelian view, and Sagredo acted as the neutral third party who was invariably swayed by Salviati's arguments. In a strong reaction, Galileo was called before the Roman Inquisition and his book was heavily censored. The book was then placed on the *Index* of banned books, a collection of titles that included works of Copernicus and Kepler. Galileo was put under house arrest for the remainder of his life, serving out his term at his home in Florence.

## 2.2 Newtonian Mechanics

In 1992, after a 13-year study by Vatican experts, Pope John Paul II officially announced that, because of a "tragic mutual incomprehension," the Roman Catholic Church had erred in its condemnation of Galileo some 360 years earlier. By reevaluating its position, the Church demonstrated that, at least on this issue, there is room for the philosophical views of both science and religion.

In the year of Galileo's death, on Christmas day, Isaac Newton was born (1642–1727), arguably the greatest of all scientific minds (Fig. 2.6b). At age 18, Newton enrolled at Cambridge University and subsequently obtained his bachelor's degree. In the two years following the completion of his formal studies, and while living at home in Woolsthorpe, in rural England, away from the immediate dangers of the Plague, Newton engaged in what was likely the most productive period of scientific work ever carried out by one individual. During that interval, he made significant discoveries and theoretical advances in understanding motion, astronomy, optics, and mathematics. Although his work was not published immediately, the *Philosophiae Naturalis Principia Mathematica* (*Mathematical Principles of Natural Philosophy*), now simply known as the *Principia*, finally appeared in 1687 and contained much of his work on mechanics, gravitation, and the calculus. The publication of the *Principia* came about largely as a result of the urging of Edmond Halley, who paid for its printing. Another book, *Optiks*, appeared separately in 1704 and contained Newton's ideas about the nature of light and some of his early experiments in optics. Although many of his ideas concerning the particle nature of light were later shown to be in error (see Section 3.3), much of Newton's other work is still used extensively today.

Newton's great intellect is evidenced in his solution of the so-called brachistochrone problem posed by Johann Bernoulli, the Swiss mathematician, as a challenge to his colleagues. The brachistochrone problem amounts to finding the curve along which a bead could slide over a frictionless wire in the least amount of time while only under the influence of gravity. The deadline for finding a solution was set at a year and a half. The problem was presented to Newton late one afternoon; by the next morning he had found the answer by inventing a new area of mathematics known as the calculus of variations. Although the solution was published anonymously at Newton's request, Bernoulli commented, "By the claw, the lion is revealed."

Concerning the successes of his own career, Newton wrote:

> I do not know what I may appear to the world; but to myself I seem to have been only like a boy, playing on the seashore, and diverting myself, in now and then finding a smoother pebble or a

prettier shell than ordinary, while the great ocean of truth lay all undiscovered before me.

Classical mechanics is described by Newton's three laws of motion, along with his universal *law of gravity*. Outside of the realms of atomic dimensions, velocities approaching the speed of light, or extreme gravitational forces, Newtonian physics has proven very successful in explaining the results of observations and experiments. Those regimes where Newtonian mechanics have been shown to be unsatisfactory will be discussed in later chapters.

Newton's first law of motion may be stated as:

> **Newton's First Law** The Law of Inertia. An object at rest will remain at rest and an object in motion will remain in motion in a straight line at a constant speed unless acted upon by an unbalanced force.

To establish whether an object is actually moving, a reference frame must be established. We will refer later to reference frames that have the special property that the first law is valid; all such frames are known as **inertial reference frames**. Noninertial reference frames are accelerated with respect to inertial frames.

The first law may be restated in terms of the momentum of an object, $\mathbf{p} = m\mathbf{v}$, where $m$ and $\mathbf{v}$ are mass and velocity, respectively.[2] Thus Newton's first law may be expressed as *"the momentum of an object remains constant unless it experiences an unbalanced force."*[3]

The second law is actually a definition of the concept of force:

> **Newton's Second Law** The *net* force (the sum of all forces) acting on an object is proportional to the object's mass and its resultant acceleration.

If an object is experiencing $n$ forces, then the net force is given by

$$\mathbf{F}_{\text{net}} = \sum_{i=1}^{n} \mathbf{F}_i = m\mathbf{a}. \tag{2.9}$$

However, since $\mathbf{a} = d\mathbf{v}/dt$, Newton's second law may also be expressed as

$$\mathbf{F}_{\text{net}} = m\frac{d\mathbf{v}}{dt} = \frac{d(m\mathbf{v})}{dt} = \frac{d\mathbf{p}}{dt}; \tag{2.10}$$

---

[2] Hereafter, all vectors will be indicated by boldface type. Vectors are quantities described by both a magnitude and a direction.

[3] The law of inertia is an extension of the original concept developed by Galileo.

## 2.2 Newtonian Mechanics

**Figure 2.7** Newton's third law.

the net force on an object is equal to the time rate of change of its momentum, **p**. $\mathbf{F}_{\text{net}} = d\mathbf{p}/dt$ represents the most general statement of the second law, allowing for a time variation in the mass of the object such as occurs with rocket propulsion.

The third law of motion is generally expressed as:

**Newton's Third Law** For every action there is an equal and opposite reaction.

In this law, *action* and *reaction* are to be interpreted as forces acting on different objects. Consider the force exerted *on* one object (object 1) *by* a second object (object 2), $\mathbf{F}_{12}$. Newton's third law states that the force on object 2 due to object 1, $\mathbf{F}_{21}$, must necessarily be of the same magnitude but in the opposite direction (see Fig. 2.7). Mathematically, the third law can be represented as

$$\mathbf{F}_{12} = -\mathbf{F}_{21}.$$

Using his three laws of motion along with Kepler's third law, Newton was able to find an expression describing the force that holds planets in their orbits. Consider the case of *circular* orbital motion of a mass $m$ about a much larger mass $M$ ($M \gg m$). Allowing for a system of units other than years and astronomical units, Kepler's third law may be written as

$$P^2 = kr^3,$$

where $r$ is the distance between the two objects and $k$ is a constant of proportionality. Thus the period of the orbit may be found from the constant velocity of $m$ by

$$P = \frac{2\pi r}{v}.$$

Substituting into the previous equation gives

$$\frac{4\pi^2 r^2}{v^2} = kr^3.$$

Rearranging terms and multiplying both sides by $m$ leads to the expression

$$m\frac{v^2}{r} = \frac{4\pi^2 m}{kr^2}.$$

The left-hand side of the equation may be recognized as the centripetal force for circular motion; thus

$$F = \frac{4\pi^2 m}{kr^2}$$

must be the gravitational force keeping $m$ in its orbit about $M$. However, Newton's third law states that the magnitude of the force exerted on $M$ by $m$ must equal the magnitude of the force exerted on $m$ by $M$. Therefore the form of the equation ought to be symmetric with respect to exchange of $m$ and $M$. Expressing this symmetry explicitly and grouping the remaining constants into a new constant $G$, we arrive at the form of the *law of universal gravitation* found by Newton,

$$F = G\frac{Mm}{r^2}, \tag{2.11}$$

where $G = 6.67259 \times 10^{-8}$ dyne cm$^2$ g$^{-2}$ (the *universal gravitational constant*).[4]

Newton's law of gravity applies to any two objects having mass. In particular, for an extended object (as opposed to a point mass), the force exerted by that object on another extended object may be found by integrating over each of their mass distributions.

---

**Example 2.2** The force exerted by a *spherically symmetric* object of mass $M$ on a point mass $m$ may be found by integrating over rings centered along a line connecting the point mass to the center of the extended object (see Fig. 2.8). In this way all points on a specific ring are located at the same distance from $m$. Furthermore, due to the symmetry of the ring, the gravitational force vector associated with it is oriented along the ring's central axis. Once a general description of the force due to one ring is determined, it is possible to add up the individual contributions from all such rings throughout the entire volume of the mass $M$. The result will be the force on $m$ due to $M$.

Let $r$ be the distance between the centers of the two masses, $M$ and $m$. $R_0$ is the radius of the large mass, and $s$ is the distance from the point mass to a point on the ring. Due to the symmetry of the problem, only the component of the gravitational force vector along the line connecting the centers of the two objects needs to be calculated; the perpendicular components will cancel.

---

[4]Dyne is the measure of force in the cgs system of units: 1 dyne = 1 g cm s$^{-2}$ = $10^{-5}$ newton.

## 2.2 Newtonian Mechanics

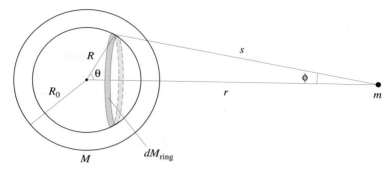

**Figure 2.8** The gravitational effect of a spherically symmetric mass distribution.

If $dM_{\text{ring}}$ is the mass of the *ring* being considered, the force exerted by that ring on $m$ is given by

$$dF_{\text{ring}} = G\frac{m\,dM_{\text{ring}}}{s^2}\cos\phi.$$

Assuming that the mass density, $\rho(R)$, of the extended object is a function of radius only and that the volume of the ring of thickness $dR$ is $dV_{\text{ring}}$, then

$$dM_{\text{ring}} = \rho(R)\,dV_{\text{ring}}$$
$$= \rho(R)2\pi R\sin\theta R\,d\theta\,dR$$
$$= 2\pi R^2\rho(R)\sin\theta\,dR\,d\theta.$$

The cosine is given by

$$\cos\phi = \frac{r - R\cos\theta}{s},$$

where $s$ may be found by the Pythagorean theorem,

$$s = \sqrt{(r - R\cos\theta)^2 + R^2\sin^2\theta}$$
$$= \sqrt{r^2 - 2rR\cos\theta + R^2}.$$

Substituting into the expression for $dF_{\text{ring}}$, summing over all *rings* located at a distance $R$ from the center of the mass $M$ (i.e., integrating over all $\theta$ from 0 to $\pi$ for constant $R$) and then summing over all resultant *shells* of radius $R$ from $R = 0$ to $R = R_0$ gives the total force of gravity acting on the small mass

$\dot{m}$ along the system's line of symmetry:

$$F = Gm \int_0^{R_0} \int_0^{\pi} \frac{(r - R\cos\theta)\rho(R) 2\pi R^2 \sin\theta}{s^3} \, d\theta \, dR$$

$$= 2\pi Gm \int_0^{R_0} \int_0^{\pi} \frac{rR^2 \rho(R) \sin\theta}{(r^2 + R^2 - 2rR\cos\theta)^{3/2}} \, d\theta \, dR$$

$$- 2\pi Gm \int_0^{R_0} \int_0^{\pi} \frac{R^3 \rho(R) \sin\theta \cos\theta}{(r^2 + R^2 - 2rR\cos\theta)^{3/2}} \, d\theta \, dR.$$

The integrations over $\theta$ may be carried out by making the change of variable, $u \equiv s^2 = r^2 + R^2 - 2rR\cos\theta$. Then $\cos\theta = (r^2 + R^2 - u)/2rR$ and $\sin\theta \, d\theta = du/2rR$. After the appropriate substitutions and integration over the new variable $u$, the equation for the force becomes

$$F = \frac{Gm}{r^2} \int_0^{R_0} 4\pi R^2 \rho(R) \, dR.$$

Notice that the integrand is just the mass of a *shell* of thickness $dR$, having a volume $dV_{\text{shell}}$, or

$$dM_{\text{shell}} = 4\pi R^2 \rho(R) \, dR$$
$$= \rho(R) \, dV_{\text{shell}}.$$

Therefore the integrand gives the force on $m$ due to a spherically symmetric mass shell of mass $dM_{\text{shell}}$ as

$$dF_{\text{shell}} = \frac{Gm \, dM_{\text{shell}}}{r^2}.$$

*The shell acts gravitationally as if its mass were located entirely at its center.* Finally, integrating over the mass shells, we have that the force exerted on $m$ by an extended, spherically symmetric mass distribution is directed along the line of symmetry between the two objects and is given by

$$F = G\frac{Mm}{r^2},$$

just the equation for the force of gravity between two point masses.

When an object is dropped near the surface of Earth, it accelerates toward the center of Earth at the rate $g = 980$ cm s$^{-2}$, the local acceleration of gravity. Using Newton's second law and his law of gravity, an expression for the acceleration of gravity may be found. If $m$ is the mass of the falling object, $M_\oplus$ and $R_\oplus$ are the mass and radius of Earth, respectively, and $h$ is the height of the object above Earth, then the force of gravity on $m$ due to Earth is given by

$$F = G\frac{M_\oplus m}{(R_\oplus + h)^2}.$$

Assuming that $m$ is near Earth's surface, then $h \ll R_\oplus$ and

$$F \simeq G\frac{M_\oplus m}{R_\oplus^2}.$$

However, $F = ma = mg$; thus

$$g = G\frac{M_\oplus}{R_\oplus^2}. \tag{2.12}$$

Substituting the values $M_\oplus = 5.974 \times 10^{27}$ g and $R_\oplus = 6.378 \times 10^8$ cm gives a value for $g$ in agreement with the measured value.

The famous story that an apple falling on Newton's head allowed him to immediately realize that gravity holds the Moon in its orbit is probably somewhat fanciful and inaccurate. However, he did demonstrate that, along with the acceleration of the falling apple, gravity was responsible for the motion of Earth's closest neighbor.

---

**Example 2.3** Assuming for simplicity that the Moon's orbit is exactly circular, the centripetal acceleration of the Moon may be calculated rapidly. Recall that the centripetal acceleration of an object moving in a perfect circle is given by

$$a_c = \frac{v^2}{r}.$$

In this case $r$ is the distance from the center of Earth to the center of the Moon, $r = 3.84403 \times 10^{10}$ cm, and $v$ is the Moon's orbital velocity, given by,

$$v = \frac{2\pi r}{P},$$

where $P = 27.3$ days $= 2.36 \times 10^6$ s is the sidereal orbital period of the Moon. Finding $v = 1.02 \times 10^5$ cm s$^{-1}$ gives a value for the centripetal acceleration of

$$a_c = 0.27 \text{ cm s}^{-2}.$$

The acceleration of the Moon caused by Earth's gravitational pull may also be calculated directly from

$$a_g = G\frac{M_\oplus}{r^2} = 0.27 \text{ cm s}^{-2},$$

in agreement with the value for the centripetal acceleration.

In astrophysics, as in any area of physics, it is often very helpful to have some understanding of the energetics of specific physical phenomena in order to determine whether these processes are important in certain systems. Some models may be ruled out immediately if they are incapable of producing the amount of energy observed. Energy arguments also often result in simpler solutions to particular problems. For example, in the evolution of a planetary atmosphere, the possibility of a particular component of the atmosphere escaping must be considered. Such a consideration is based on a calculation of the escape velocity of the gas particles.

The amount of energy (the work) necessary to raise an object of mass $m$ a height $h$ against a gravitational force is equal to the change in the *potential energy* of the system. Generally, the change in potential energy resulting from a change in position between two points is given by

$$U_f - U_i = \Delta U = -\int_{\mathbf{r}_i}^{\mathbf{r}_f} \mathbf{F} \cdot d\mathbf{r}, \tag{2.13}$$

where $\mathbf{F}$ is the force vector, $\mathbf{r}_i$ and $\mathbf{r}_f$ are the initial and final position vectors, respectively, and $d\mathbf{r}$ is the infinitesimal change in the position vector for some general coordinate system (see Fig. 2.9). If the gravitational force on $m$ is due to a mass $M$ located at the origin, then $\mathbf{F}$ is directed inward toward $M$, $d\mathbf{r}$ is directed outward, $\mathbf{F} \cdot d\mathbf{r} = -F dr$, and the change in potential energy becomes

$$\Delta U = \int_{r_i}^{r_f} G\frac{Mm}{r^2} dr.$$

Evaluating the integral, we have

$$U_f - U_i = -GMm \left(\frac{1}{r_f} - \frac{1}{r_i}\right).$$

Since only *relative changes* in potential energy are physically meaningful, a reference position where the potential energy is identically zero may be chosen. If, for a specific gravitational system, it is assumed that the potential energy

## 2.2 Newtonian Mechanics

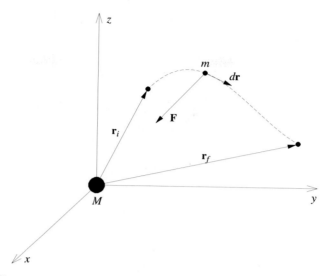

**Figure 2.9** Gravitational potential energy. The amount of work done depends on the direction of motion relative to the direction of the force vector.

goes to zero at infinity, letting $r_f$ approach infinity ($r_f \to \infty$) and dropping the subscripts for simplicity, gives

$$U = -G\frac{Mm}{r}. \tag{2.14}$$

Of course, the process may be reversed, the force may be found by *differentiating* the potential. For forces that depend only on $r$,

$$F = -\frac{\partial U}{\partial r}. \tag{2.15}$$

In a general three-dimensional description, $\mathbf{F} = -\nabla U$, where $\nabla U$ represents the *gradient* of $U$. In rectangular coordinates this becomes

$$\mathbf{F} = -\frac{\partial U}{\partial x}\hat{\mathbf{i}} - \frac{\partial U}{\partial y}\hat{\mathbf{j}} - \frac{\partial U}{\partial z}\hat{\mathbf{k}}.$$

Work must be performed on a massive object if its *speed*, $|\mathbf{v}|$, is to be changed. This can be seen by rewriting the work integral, first in terms of time, then speed:

$$W \equiv -\nabla U$$
$$= \int_{\mathbf{r}_i}^{\mathbf{r}_f} \mathbf{F} \cdot d\mathbf{r}$$
$$= \int_{t_i}^{t_f} \frac{d\mathbf{p}}{dt} \cdot (\mathbf{v}\, dt)$$
$$= \int_{t_i}^{t_f} m \frac{d\mathbf{v}}{dt} \cdot (\mathbf{v}\, dt)$$
$$= \int_{t_i}^{t_f} m \left( \mathbf{v} \cdot \frac{d\mathbf{v}}{dt} \right) dt$$
$$= \int_{t_i}^{t_f} m \frac{d\left(\frac{1}{2}v^2\right)}{dt} dt$$
$$= \int_{v_i}^{v_f} m\, d\left(\frac{1}{2}v^2\right)$$
$$= \frac{1}{2}mv_f^2 - \frac{1}{2}mv_i^2.$$

We may now identify the quantity

$$K = \frac{1}{2}mv^2 \tag{2.16}$$

as the *kinetic energy* of the object. Thus work done on the particle results in an equivalent change in the particle's kinetic energy. This statement is simply one example of the *conservation of energy*, a concept that is encountered frequently in all areas of physics.

Consider a particle of mass $m$, having an initial velocity $\mathbf{v}$ that is at a distance $r$ from the center of a larger mass $M$, such as Earth. How fast must the mass be moving upward to escape completely the pull of gravity? To calculate the *escape velocity*, energy conservation may be used directly. The total initial mechanical energy of the particle (both kinetic and potential) is given by

$$E = \frac{1}{2}mv^2 - G\frac{Mm}{r}.$$

Assume that, in the critical case, the final velocity of the mass will be zero at a position infinitely far from $M$, implying that both the kinetic and potential energy will become zero. Clearly, by conservation of energy, the *total* energy of the particle must be identically zero at all times. Therefore

$$\frac{1}{2}mv^2 = G\frac{Mm}{r},$$

which may be solved immediately for the initial speed of $m$ to give

$$v_{\text{esc}} = \sqrt{2GM/r}. \tag{2.17}$$

Notice that the mass of the escaping object does not enter into the final expression for the escape velocity. Near the surface of Earth, $v_{\text{esc}} = 11.2$ km s$^{-1}$.

## 2.3 Kepler's Laws Derived

Although Kepler did finally determine that the geometry of planetary motion was in the more general form of an ellipse rather than circular motion, he was unable to explain the nature of the force that kept the planets moving in their precise patterns. Not only was Newton successful in quantifying that force, he was also able to generalize Kepler's work, deriving the empirical laws of planetary motion from the gravitational force law. The derivation of Kepler's laws represented a crucial step in the development of modern astrophysics.

Although it is beyond the scope of this book, it can be shown that an elliptical orbit results from an attractive $r^{-2}$ central force law such as gravity, when the total energy of the system is less than zero (a bound system). It can also be shown that a parabolic path is obtained when the energy is identically zero and that a hyperbolic path results from an unbounded system with an energy that is greater than zero. Newton was able to demonstrate the elliptical behavior of planetary motion and found that Kepler's first law must be generalized somewhat: The **center of mass** of the system, rather than the exact center of the Sun, is actually located at the focus of the ellipse. For our solar system, such a mistake is understandable, since the largest of the planets, Jupiter, has only 1/1000 the mass of the Sun. This places the center of mass of the Sun–Jupiter system near the surface of the Sun. Having used the naked-eye data of Tycho, Kepler can be forgiven for not realizing his error.

Before proceeding onward to derive Kepler's second and third laws, it will be useful to examine more closely the dynamics of orbital motion. An interacting two-body problem, such as binary orbits, or the more general many-body problem (often called the $N$-body problem), is most easily done in the reference frame of the center of mass.

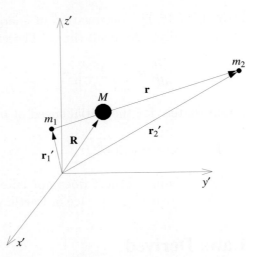

**Figure 2.10** A general Cartesian coordinate system indicating the positions of $m_1$, $m_2$, and the center of mass (located at $M$).

Figure 2.10 shows two objects of masses $m_1$ and $m_2$ at positions $\mathbf{r}_1'$ and $\mathbf{r}_2'$, respectively, with the displacement vector from $\mathbf{r}_1'$ to $\mathbf{r}_2'$ given by

$$\mathbf{r} = \mathbf{r}_2' - \mathbf{r}_1'.$$

Define a position vector $\mathbf{R}$ to be a *weighted* average of the position vectors of the individual masses,

$$\mathbf{R} \equiv \frac{m_1 \mathbf{r}_1' + m_2 \mathbf{r}_2'}{m_1 + m_2}. \tag{2.18}$$

Of course, this definition could be immediately generalized to the case of $n$ objects,

$$\mathbf{R} \equiv \frac{\sum_{i=1}^{n} m_i \mathbf{r}_i'}{\sum_{i=1}^{n} m_i}. \tag{2.19}$$

Rewriting the equation, we have

$$\sum_{i=1}^{n} m_i \mathbf{R} = \sum_{i=1}^{n} m_i \mathbf{r}_i'.$$

Then, if we define $M$ to be the total mass of the system, $M \equiv \sum_{i=1}^{n} m_i$, the previous equation becomes

$$M\mathbf{R} = \sum_{i=1}^{n} m_i \mathbf{r}_i'.$$

## 2.3 Kepler's Laws Derived

Assuming that the individual masses do not change, differentiating both sides with respect to time gives

$$M \frac{d\mathbf{R}}{dt} = \sum_{i=1}^{n} m_i \frac{d\mathbf{r}_i'}{dt}$$

or

$$M\mathbf{V} = \sum_{i=1}^{n} m_i \mathbf{v}_i'.$$

The right-hand side is the sum of the linear momenta of every particle in the system, so the total linear momentum of the system may be treated as though *all* of the mass were located at $\mathbf{R}$, moving with a velocity $\mathbf{V}$. Thus $\mathbf{R}$ is the position of the center of mass of the system and $\mathbf{V}$ is the center-of-mass velocity. Letting $\mathbf{P} \equiv M\mathbf{V}$ be the linear momentum of the center of mass and $\mathbf{p}_i' \equiv m_i \mathbf{v}_i'$ be the linear momentum of an individual particle $i$, and again differentiating both sides with respect to time yields,

$$\frac{d\mathbf{P}}{dt} = \sum_{i=1}^{n} \frac{d\mathbf{p}_i'}{dt}.$$

If we assume that *all* of the forces acting on individual particles in the system are due to other particles contained within the system, Newton's third law requires that the total force must be zero. This constraint exists because of the equal magnitudes of action–reaction pairs. Of course, the momentum of individual masses may change. Using center-of-mass quantities, the total (or net) force on the system is

$$\mathbf{F} = \frac{d\mathbf{P}}{dt} = M \frac{d^2\mathbf{R}}{dt^2} = 0.$$

Therefore the center of mass will not accelerate if no external forces exist. This implies that a reference frame associated with the center of mass must be an inertial reference frame and the $N$-body problem may be simplified by choosing a coordinate system for which the center of mass is at rest at $\mathbf{R} = 0$.

**Kepler's First Law (revisited)** Both objects in a binary orbit move about the center of mass in ellipses, with the center of mass occupying one focus of each ellipse.

This is just Kepler's first law, generalized to binary orbits. (See Fig. 7.3 for an example of the position of the center of mass in a binary star system.)

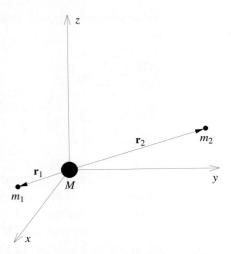

**Figure 2.11** The center-of-mass reference frame for a binary orbit, with the center of mass fixed at the origin of the coordinate system.

If we choose a center-of-mass reference frame for a binary system, depicted in Fig. 2.11 ($\mathbf{R} = 0$), Eq. (2.18) becomes

$$\frac{m_1 \mathbf{r}_1 + m_2 \mathbf{r}_2}{m_1 + m_2} = 0, \tag{2.20}$$

where the primes have been dropped, indicating center-of-mass coordinates. Both $\mathbf{r}_1$ and $\mathbf{r}_2$ may now be rewritten in terms of the displacement vector, $\mathbf{r}$. Substituting $\mathbf{r}_2 = \mathbf{r}_1 + \mathbf{r}$ gives

$$\mathbf{r}_1 = -\frac{m_2}{m_1 + m_2} \mathbf{r} \tag{2.21}$$

$$\mathbf{r}_2 = \frac{m_1}{m_1 + m_2} \mathbf{r}. \tag{2.22}$$

Next, define the *reduced mass* to be

$$\mu \equiv \frac{m_1 m_2}{m_1 + m_2}. \tag{2.23}$$

Then $\mathbf{r}_1$ and $\mathbf{r}_2$ become

$$\mathbf{r}_1 = -\frac{\mu}{m_1} \mathbf{r}$$

$$\mathbf{r}_2 = \frac{\mu}{m_2} \mathbf{r}.$$

## 2.3 Kepler's Laws Derived

The convenience of the center-of-mass reference frame becomes evident when the total energy and orbital angular momentum of the system are considered. Including the necessary kinetic energy and gravitational potential energy terms, the total energy may be expressed as

$$E = \frac{1}{2} m_1 |\mathbf{v}_1|^2 + \frac{1}{2} m_2 |\mathbf{v}_2|^2 - G \frac{m_1 m_2}{|\mathbf{r}_2 - \mathbf{r}_1|}.$$

Substituting the relations for $\mathbf{r}_1$ and $\mathbf{r}_2$, along with the expression for the total mass of the system and the definition for the reduced mass, gives

$$E = \frac{1}{2} \mu v^2 - G \frac{M \mu}{r}, \tag{2.24}$$

where $v = |\mathbf{v}|$ and $\mathbf{v} \equiv d\mathbf{r}/dt$. We have also used the notation $r = |\mathbf{r}_2 - \mathbf{r}_1|$. The total energy of the system is equal to the kinetic energy of the reduced mass, plus the potential energy of the reduced mass moving about a mass $M$, assumed to be located at the origin. The distance between $\mu$ and $M$ is equal to the separation between the objects of masses $m_1$ and $m_2$.

Similarly, the total orbital angular momentum,

$$\mathbf{L} = m_1 \mathbf{r}_1 \times \mathbf{v}_1 + m_2 \mathbf{r}_2 \times \mathbf{v}_2$$

becomes

$$\mathbf{L} = \mu \mathbf{r} \times \mathbf{v} = \mathbf{r} \times \mathbf{p}. \tag{2.25}$$

The total orbital angular momentum equals the angular momentum of the reduced mass only. *In general, the two-body problem may be treated as an equivalent one-body problem with the reduced mass moving about a fixed mass $M$ at a distance $r$* (see Fig. 2.12).

To obtain Kepler's second law, it is necessary to consider the effect of gravitation on the orbital angular momentum of a planet. Using center-of-mass coordinates and evaluating the time derivative of the orbital angular momentum of the reduced mass (Eq. 2.25) gives

$$\frac{d\mathbf{L}}{dt} = \frac{d\mathbf{r}}{dt} \times \mathbf{p} + \mathbf{r} \times \frac{d\mathbf{p}}{dt}$$
$$= \mathbf{v} \times \mathbf{p} + \mathbf{r} \times \mathbf{F},$$

the second expression arising from the definition of velocity and Newton's second law. Notice that, because $\mathbf{v}$ and $\mathbf{p}$ are in the same direction, their cross product is identically zero. Similarly, since $\mathbf{F}$ is a central force, directed inward

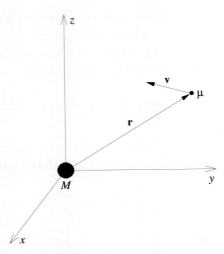

**Figure 2.12** A binary orbit may be reduced to the equivalent problem of calculating the motion of the reduced mass, $\mu$, about the total mass, $M$, located at the origin.

along $\mathbf{r}$, the cross product of $\mathbf{r}$ and $\mathbf{F}$ is also zero. The result is an important general statement concerning angular momentum:

$$\frac{d\mathbf{L}}{dt} = 0, \qquad (2.26)$$

*the angular momentum of a system is a constant for a central force law.*

Since Kepler's second law relates the area of a section of an ellipse to a time interval, consider the infinitesimal area element in polar coordinates, as shown in Fig. 2.13:

$$dA = dr\,(r\,d\theta) = r\,dr\,d\theta.$$

If we integrate from the principal focus of the ellipse to a specific distance, $r$, the area swept out by an infinitesimal change in $\theta$ becomes

$$dA = \frac{1}{2}r^2\,d\theta.$$

Therefore the time rate of change in area swept out by a line joining a point on the ellipse to the focus becomes

$$\frac{dA}{dt} = \frac{1}{2}r^2\frac{d\theta}{dt}. \qquad (2.27)$$

Now, the orbital velocity, $\mathbf{v}$, may be expressed in two components, one directed along $\mathbf{r}$ and the other perpendicular to $\mathbf{r}$. Letting $\hat{\mathbf{r}}$ and $\hat{\theta}$ be the unit vectors

## 2.3 Kepler's Laws Derived

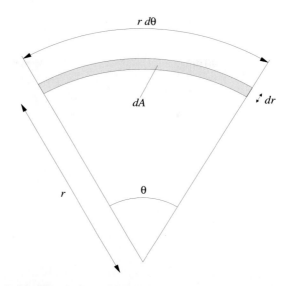

**Figure 2.13** The infinitesimal area element in polar coordinates.

along **r** and its normal, respectively, **v** may be written as (see Fig. 2.14)

$$\mathbf{v} = \mathbf{v}_r + \mathbf{v}_\theta$$
$$= \frac{dr}{dt}\hat{\mathbf{r}} + r\frac{d\theta}{dt}\hat{\theta}. \tag{2.28}$$

Substituting $v_\theta$ into Eq. (2.27) gives

$$\frac{dA}{dt} = \frac{1}{2}rv_\theta.$$

Since **r** and $\mathbf{v}_\theta$ are perpendicular,

$$rv_\theta = |\mathbf{r} \times \mathbf{v}| = \left|\frac{\mathbf{L}}{\mu}\right| = \frac{L}{\mu}.$$

Finally, the time derivative of the area becomes

**Kepler's Second Law (revisited)**

$$\frac{dA}{dt} = \frac{1}{2}\frac{L}{\mu}. \tag{2.29}$$

It has already been shown that, because the orbital angular momentum is a constant, *the time rate of change of the area swept out by a line connecting a*

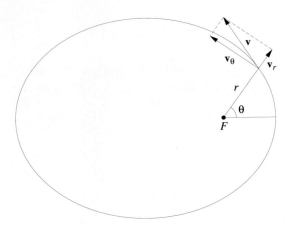

**Figure 2.14** The velocity vector for elliptical motion in polar coordinates.

*planet to the focus of an ellipse is a constant*, one-half of the orbital angular momentum per unit mass. This is just Kepler's second law.

Simple expressions for both the total orbital angular momentum and the total energy may be found by computing their values at perihelion and aphelion and by invoking the appropriate conservation laws. Since at both perihelion and aphelion, **r** and **v** are perpendicular, the magnitude of the angular momentum simply becomes

$$L = \mu r v.$$

Recalling from Example 2.1 that $r_p = a(1-e)$ at perihelion and $r_a = a(1+e)$ at aphelion, and using the conservation of angular momentum,

$$\mu r_p v_p = \mu r_a v_a,$$

immediately reduces to

$$\frac{v_p}{v_a} = \frac{1+e}{1-e}.$$

Similarly, equating the total energy at both perihelion and aphelion provides a second expression relating $v_p$ and $v_a$:

$$\frac{1}{2}\mu v_p^2 - G\frac{M\mu}{a(1-e)} = \frac{1}{2}\mu v_a^2 - G\frac{M\mu}{a(1+e)}.$$

Combining the two previous equations gives

$$v_p^2 = \frac{GM}{a}\left(\frac{1+e}{1-e}\right) \tag{2.30}$$

## 2.3 Kepler's Laws Derived

and
$$v_a^2 = \frac{GM}{a}\left(\frac{1-e}{1+e}\right). \tag{2.31}$$

Having found expressions for the orbital velocity at both perihelion and aphelion, the total orbital angular momentum may be easily calculated from
$$L = \mu r_p v_p,$$
which becomes
$$L = \mu\sqrt{GMa(1-e^2)}. \tag{2.32}$$

Note that $L$ is a maximum for purely circular motion ($e = 0$) and goes to zero as the eccentricity approaches unity, as expected.

The total orbital energy may be found as well:
$$E = \frac{1}{2}\mu v_p^2 - G\frac{M\mu}{r_p}.$$

Making the appropriate substitutions, and after some rearrangement,
$$E = -G\frac{M\mu}{2a} = -G\frac{m_1 m_2}{2a}. \tag{2.33}$$

The total energy of a binary orbit depends only on the semimajor axis $a$ and is exactly one-half the time-averaged potential energy of the system,
$$E = \frac{1}{2}\langle U \rangle,$$
where $\langle U \rangle$ denotes an average over one orbital period.[5] This is one example of the *virial theorem*, a general property of gravitationally bound systems. The virial theorem will be discussed in detail in Section 2.4.

A useful expression for the velocity of the reduced mass (or the relative velocity of $m_1$ and $m_2$) may be found directly by using the conservation of energy and equating the total orbital energy to the sum of the kinetic and potential energies:
$$-G\frac{M\mu}{2a} = \frac{1}{2}\mu v^2 - G\frac{M\mu}{r}.$$

Using the identity $M = m_1 + m_2$, this simplifies to give
$$v^2 = G(m_1 + m_2)\left(\frac{2}{r} - \frac{1}{a}\right). \tag{2.34}$$

---

[5] The proof that $\langle U \rangle = -GM\mu/a$ is left as an exercise.

This expression could also have been obtained directly by adding the vector components of orbital velocity. Calculating $\mathbf{v}_r$, $\mathbf{v}_\theta$, and $v^2$ will be left as exercises.

We are finally in a position to derive the last of Kepler's laws. Integrating the mathematical expression for Kepler's second law (Eq. 2.29) over one orbital period, $P$, gives the result

$$A = \frac{1}{2}\frac{L}{\mu}P.$$

Here the mass $m$ orbiting about a much larger *fixed* mass $M$ has been replaced by the more general reduced mass $\mu$ orbiting about the center of mass. Substituting the area of an ellipse, $A = \pi ab$, squaring the equation, and rearranging, produces the expression

$$P^2 = \frac{4\pi^2 a^2 b^2 \mu^2}{L^2}.$$

Finally, using Eq. (2.2) and the expression for the total orbital angular momentum (Eq. 2.32), the last equation simplifies to become

### Kepler's Third Law (revisited)

$$P^2 = \frac{4\pi^2}{G(m_1 + m_2)} a^3. \tag{2.35}$$

This is the general form of Kepler's third law. Not only did Newton demonstrate the relationship between the semimajor axis of an elliptical orbit and the orbital period, he also found a term not discovered empirically by Kepler, the square of the orbital period is inversely proportional to the total mass of the system. Once again Kepler can be forgiven for not noticing the effect. Tycho's data were for our solar system only and, because the Sun's mass $M_\odot$ is so much greater than the mass of any of the planets, $M_\odot + m_{\text{planet}} \simeq M_\odot$. Expressing $P$ in years and $a$ in astronomical units gives a value of unity for the collection of constants (including the Sun's mass).[6]

The importance to astronomy of Newton's form of Kepler's third law cannot be overstated. This law provides the most direct way of obtaining masses of celestial objects, a critical parameter in understanding a wide range of phenomena. Kepler's laws, as derived by Newton, apply equally well to planets orbiting the Sun, moons orbiting planets, stars in orbit about one another, and galaxy–galaxy orbits. Knowledge of the period of an orbit and the semimajor axis of the ellipse yields the total mass of the system. If relative distances to

---

[6]In 1621 Kepler was able to demonstrate that the four Galilean moons also obeyed his third law in the form $P^2 = ka^3$, where the constant $k$ differed from unity. He did not attribute the fact that $k \neq 1$ to mass, however.

the center of mass are also known, the individual masses may be determined using Eq. (2.20).

---

**Example 2.4** The orbital sidereal period of Io, one of the four Galilean moons of Jupiter, is 1.77 days $= 1.53 \times 10^5$ s and the semimajor axis of its orbit is $4.22 \times 10^{10}$ cm. Assuming that the mass of Io is insignificant compared to that of Jupiter, the mass of the planet may be estimated using Kepler's third law:

$$M_{\text{Jupiter}} = \frac{4\pi^2}{G} \frac{a^3}{P^2}$$

$$= 1.90 \times 10^{30} \text{ g}$$

$$= 318 \text{ M}_\oplus.$$

---

Appendix G contains a simple FORTRAN computer program, ORBIT, that makes use of many of the ideas discussed in this chapter. ORBIT will calculate, as a function of time, the location of a small mass that is orbiting about a much larger star (or it may be thought of as calculating the motion of the reduced mass about the total mass). Data generated by ORBIT were used to produce Fig. 2.2.

## 2.4 The Virial Theorem

In the last section we found that the total energy of the binary orbit was just one-half of the time-averaged gravitational potential energy (Eq. 2.33), or $E = \langle U \rangle / 2$. Since the total energy of the system is negative, the system is necessarily bound. For gravitationally bound systems in equilibrium, it can be shown that the total energy is always one-half of the time-averaged potential energy; this is known as the **virial theorem**.

To prove the virial theorem, begin by considering the quantity

$$Q \equiv \sum_i \mathbf{p}_i \cdot \mathbf{r}_i,$$

where $\mathbf{p}_i$ and $\mathbf{r}_i$ are the linear momentum and position vectors for particle $i$ in some inertial reference frame and the sum is taken to be over all particles in the system. The time derivative of $Q$ is

$$\frac{dQ}{dt} = \sum_i \left( \frac{d\mathbf{p}_i}{dt} \cdot \mathbf{r}_i + \mathbf{p}_i \cdot \frac{d\mathbf{r}_i}{dt} \right).$$

Now, the left-hand side of the expression is just

$$\frac{dQ}{dt} = \frac{d}{dt} \sum_i m_i \frac{d\mathbf{r}_i}{dt} \cdot \mathbf{r}_i$$

$$= \frac{d}{dt} \sum_i \frac{1}{2} \frac{d}{dt} \left( m_i r_i^2 \right)$$

$$= \frac{1}{2} \frac{d^2 I}{dt^2},$$

where

$$I = \sum_i m_i r_i^2$$

is the *moment of inertia* of the system of particles. So now

$$\frac{1}{2} \frac{d^2 I}{dt^2} - \sum_i \mathbf{p}_i \cdot \frac{d\mathbf{r}_i}{dt} = \sum_i \frac{d\mathbf{p}_i}{dt} \cdot \mathbf{r}_i. \tag{2.36}$$

The second term on the left-hand side is just

$$-\sum_i \mathbf{p}_i \cdot \frac{d\mathbf{r}_i}{dt} = -\sum_i m_i \mathbf{v}_i \cdot \mathbf{v}_i$$

$$= -2 \sum_i \frac{1}{2} m_i v_i^2$$

$$= -2K,$$

twice the negative of the total kinetic energy of the system. If we use Newton's second law, Eq. (2.36) becomes

$$\frac{1}{2} \frac{d^2 I}{dt^2} - 2K = \sum_i \mathbf{F}_i \cdot \mathbf{r}_i. \tag{2.37}$$

The right-hand side of this expression is known as the *virial of Clausius*, named after the physicist who first found this important energy relation.

If $\mathbf{F}_{ij}$ represents the force of interaction between two particles in the system (actually the force on $i$ due to $j$), then, considering all of the possible forces acting on $i$,

$$\sum_i \mathbf{F}_i \cdot \mathbf{r}_i = \sum_i \left( \sum_{\substack{j \\ j \neq i}} \mathbf{F}_{ij} \right) \cdot \mathbf{r}_i.$$

## 2.4 The Virial Theorem

If we use Newton's third law, $\mathbf{F}_{ij} = -\mathbf{F}_{ji}$, so that

$$\sum_i \mathbf{F}_i \cdot \mathbf{r}_i = \frac{1}{2} \sum_i \left[ \sum_{\substack{j \\ j \neq i}} (\mathbf{F}_{ij} - \mathbf{F}_{ji}) \right] \cdot \mathbf{r}_i. \tag{2.38}$$

After some manipulation, it can be shown that the virial of Clausius may be expressed as

$$\sum_i \mathbf{F}_i \cdot \mathbf{r}_i = \frac{1}{2} \sum_i \sum_{\substack{j \\ j \neq i}} \mathbf{F}_{ij} \cdot (\mathbf{r}_i - \mathbf{r}_j). \tag{2.39}$$

If it is assumed that the only contribution to the force is the result of the gravitational interaction between massive particles included in the system, then $\mathbf{F}_{ij}$ is

$$\mathbf{F}_{ij} = G \frac{m_i m_j}{r_{ij}^2} \hat{\mathbf{r}}_{ij},$$

where $r_{ij} = |\mathbf{r}_j - \mathbf{r}_i|$ is the separation between particles $i$ and $j$ and $\hat{\mathbf{r}}_{ij}$ is the unit vector directed from $i$ to $j$:

$$\hat{\mathbf{r}}_{ij} \equiv \frac{\mathbf{r}_j - \mathbf{r}_i}{r_{ij}}.$$

Substituting the gravitational force into Eq. (2.39) gives

$$\sum_i \mathbf{F}_i \cdot \mathbf{r}_i = -\frac{1}{2} \sum_i \sum_{\substack{j \\ j \neq i}} G \frac{m_i m_j}{r_{ij}^3} (\mathbf{r}_j - \mathbf{r}_i)^2$$

$$= -\frac{1}{2} \sum_i \sum_{\substack{j \\ j \neq i}} G \frac{m_i m_j}{r_{ij}}. \tag{2.40}$$

The quantity

$$-G \frac{m_i m_j}{r_{ij}}$$

is just the potential energy $U_{ij}$ between particles $i$ and $j$. Note, however, that

$$-G \frac{m_j m_i}{r_{ji}}$$

also represents the same potential energy term and is included in the double sum as well, so the right-hand side of Eq. (2.40) includes the potential interaction between each pair of particles twice. Considering the factor of 1/2, Eq. (2.40) simply becomes

$$\sum_i \mathbf{F}_i \cdot \mathbf{r}_i = -\frac{1}{2} \sum_i \sum_{\substack{j \\ j \neq i}} G \frac{m_i m_j}{r_{ij}} = \frac{1}{2} \sum_i \sum_{\substack{j \\ j \neq i}} U_{ij} = U, \tag{2.41}$$

the total potential energy of the system of particles. Finally, substituting into Eq. (2.37) and taking the average with respect to time gives

$$\frac{1}{2} \left\langle \frac{d^2 I}{dt^2} \right\rangle - 2 \langle K \rangle = \langle U \rangle. \tag{2.42}$$

The average of $d^2 I/dt^2$ over some time interval $\tau$ is just

$$\left\langle \frac{d^2 I}{dt^2} \right\rangle = \frac{1}{\tau} \int_0^\tau \frac{d^2 I}{dt^2} \, dt \tag{2.43}$$

$$= \frac{1}{\tau} \left( \left. \frac{dI}{dt} \right|_\tau - \left. \frac{dI}{dt} \right|_0 \right). \tag{2.44}$$

If the system is periodic, as in the case for orbital motion, then

$$\left. \frac{dI}{dt} \right|_\tau = \left. \frac{dI}{dt} \right|_0$$

and the average over one period will be zero. Even if the system being considered is not strictly periodic, the average will still approach zero when evaluated over a sufficiently long period of time (i.e., $\tau \to \infty$), assuming of course that $dI/dt$ is bounded. This would describe, for example, a system that has reached an equilibrium or steady-state configuration. In either case, we now have $\langle d^2 I/dt^2 \rangle = 0$, so

$$-2 \langle K \rangle = \langle U \rangle. \tag{2.45}$$

This result is one form of the virial theorem. The theorem may also be expressed in terms of the total energy of the system by using the relation $\langle E \rangle = \langle K \rangle + \langle U \rangle$. Thus

$$\langle E \rangle = \frac{1}{2} \langle U \rangle, \tag{2.46}$$

just what we found for the binary orbit problem.

The virial theorem applies to a wide variety of systems, from an ideal gas to a cluster of galaxies. For instance, consider the case of a static star. In equilibrium a star must obey the virial theorem, implying that its total energy is negative, one-half of the total potential energy. Assuming that the star formed as a result of the gravitational collapse of a large cloud (a nebula), the potential energy of the system must have changed from an initial value of nearly zero to its negative static value. This implies that the star must have lost energy in the process, meaning that gravitational energy must have been radiated into space during the collapse. Applications of the virial theorem will be described in more detail in later chapters.

## Suggested Readings

### GENERAL

Kuhn, Thomas S., *The Structure of Scientific Revolutions*, Second Edition, Enlarged, University of Chicago Press, Chicago, 1970.

Westfall, Richard S., *Never at Rest: A Biography of Isaac Newton*, Cambridge University Press, Cambridge, 1980.

### TECHNICAL

Arya, Atam P., *Introduction to Classical Mechanics*, Prentice Hall, Englewood Cliffs, NJ, 1990.

Clayton, Donald D., *Principles of Stellar Evolution and Nucleosynthesis*, McGraw-Hill, New York, 1968.

Fowles, Grant R., and Cassiday, George L., *Analytical Mechanics*, Fifth Edition, Harcourt Brace and Company, Fort Worth, 1993.

Marion, Jerry B., and Thornton, Stephen T., *Classical Dynamics of Particles and Systems*, Fourth Edition, Saunders College Publishing, Fort Worth, 1995.

## Problems

2.1 Assume that a rectangular coordinate system has its origin at the center of an elliptical planetary orbit and that the coordinate system's $x$ axis lies along the major axis of the ellipse. Show that the equation for the ellipse is given by

$$\frac{x^2}{a^2} + \frac{y^2}{b^2} = 1,$$

where $a$ and $b$ are the lengths of the semimajor axis and the semiminor axis, respectively.

2.2 Using the result of Problem 2.1, prove that the area of an ellipse is given by $A = \pi ab$.

2.3 (a) Beginning with Eq. (2.3) and Kepler's second law, derive general expressions for $\mathbf{v}_r$ and $\mathbf{v}_\theta$ for a mass $m_1$ in an elliptical orbit about a second mass $m_2$. Your final answers should be functions of $P$, $e$, $a$, and $\theta$ only.

(b) Using the expressions for $\mathbf{v}_r$ and $\mathbf{v}_\theta$ that you derived in part (a), verify Eq. (2.34) directly from $v^2 = v_r^2 + v_\theta^2$.

2.4 Derive Eq. (2.24) from the sum of the kinetic and potential energy terms for the masses, $m_1$ and $m_2$.

2.5 Derive Eq. (2.25) from the total angular momentum of the masses, $m_1$ and $m_2$.

2.6 By expanding Eq. (2.38) and rearranging, obtain Eq. (2.39).

2.7 (a) Assuming that the Sun interacts only with Jupiter, calculate the total orbital angular momentum of the Sun–Jupiter system. The semimajor axis of Jupiter's orbit is $a = 5.2$ AU, its orbital eccentricity is $e = 0.048$, and its orbital period is $P = 11.86$ yr.

(b) Estimate the contribution the Sun makes to the total orbital angular momentum of the Sun–Jupiter system. For simplicity, assume that the Sun's orbital eccentricity is $e = 0$, rather than $e = 0.048$. *Hint:* First find the distance of the center of the Sun from the center of mass.

(c) Making the approximation that the orbit of Jupiter is a perfect circle, estimate the contribution it makes to the total orbital angular momentum of the Sun–Jupiter system. Compare your answer with the difference between the two values found in parts (a) and (b).

(d) Recall that the moment of inertia of a solid sphere of mass $m$ and radius $r$ is given by $I = \frac{2}{5}mr^2$, and that when the sphere spins on an axis passing through its center, its rotational angular momentum may be written as

$$L = I\omega,$$

where $\omega$ is the angular frequency measured in rad s$^{-1}$. Assuming (incorrectly) that both the Sun and Jupiter rotate as solid spheres, calculate approximate values for the rotational angular momenta of the Sun and Jupiter. Take the rotation periods of the Sun and Jupiter to be 26 days and 10 hours, respectively. The radius of the Sun is $6.96 \times 10^{10}$ cm, and the radius of Jupiter is $6.9 \times 10^9$ cm.

(e) What part of the Sun–Jupiter system makes the largest contribution to the total angular momentum?

2.8 (a) Using data contained in Problem 2.7 and in the chapter, calculate the escape velocity at the surface of Jupiter.

(b) Calculate the escape velocity from the solar system, starting from Earth's orbit. Assume that the Sun constitutes all of the mass of the solar system.

2.9 (a) The Hubble Space Telescope is in a nearly circular orbit, approximately 380 miles above the surface of Earth. Estimate its orbital period.

(b) Communications and weather satellites are often placed in *geosynchronous* "parking" orbits above Earth. These are orbits where satellites can remain fixed above a specific point on the surface of Earth. At what altitude must these satellites be located?

(c) Is it possible for a satellite in a geosynchronous orbit to remain "parked" over any location on the surface of Earth? Why or why not?

2.10 In general, an *integral average* of some continuous function $f(t)$ over an interval $\tau$ is given by

$$\langle f(t) \rangle = \frac{1}{\tau} \int_0^\tau f(t)\, dt.$$

Beginning with an expression for the integral average, prove that

$$\langle U \rangle = -G\frac{M\mu}{a},$$

a binary system's gravitational potential energy, averaged over one period, equals the value of the instantaneous potential energy of the system when the two masses are separated by the distance $a$, the semimajor axis of the orbit of the reduced mass about the center of mass. *Hint:* You may find the following definite integral useful;

$$\int_0^{2\pi} \frac{d\theta}{1 + e\cos\theta} = \frac{2\pi}{\sqrt{1-e^2}}.$$

2.11 Cometary orbits usually have very large eccentricities, often approaching (or even exceeding) unity. Halley's comet has an orbital period of 76 yr and an orbital eccentricity of $e = 0.9673$.

(a) What is the semimajor axis of Comet Halley's orbit?

(b) Use the orbital data of Comet Halley to estimate the mass of the Sun.

(c) Calculate the distance of Comet Halley from the Sun at perihelion and aphelion.

(d) Determine the orbital speed of the comet when at perihelion, at aphelion, and on the semiminor axis of its orbit.

(e) How many times larger is the kinetic energy of Halley's comet at perihelion when compared to aphelion?

2.12 **Computer Problem** Using ORBIT, the FORTRAN computer code found in Appendix G, together with the data given in Problem 2.11, estimate the amount of time required for Halley's comet to move from perihelion to a distance of 1 AU away from the principal focus.

2.13 **Computer Problem** The computer code ORBIT (Appendix G) can be used to generate orbital positions, given the mass of the central star, the semimajor axis of the orbit, and the orbital eccentricity. Using ORBIT to generate the data, plot on a single sheet of graph paper the orbits for three hypothetical objects orbiting our Sun. Assume that the semimajor axis of each orbit is 1 AU and that the orbital eccentricities are:

(a) 0.0.

(b) 0.4.

(c) 0.9.

*Note:* Indicate the principal focus, located at $x = 0.0, y = 0.0$.

2.14 **Computer Problem**

(a) From the data given in Example 2.1, use ORBIT (Appendix G) to generate an orbit for Mars. Plot at least 25 points, evenly spaced in time, on a sheet of graph paper and clearly indicate the principal focus.

(b) Using a compass, draw a perfect circle on top of the elliptical orbit for Mars, choosing the radius of the circle and its center carefully in order to make the best possible approximation of the orbit. Be sure to mark the center of the circle you chose.

(c) What can you conclude about the merit of Kepler's first attempts to use offset circles and equants to model the orbit of Mars?

2.15 Given that a geocentric universe is (mathematically) only a matter of the choice of a reference frame, explain why the Ptolemaic model of the universe was able to survive scrutiny for such a long period of time.

# Chapter 3

# THE CONTINUOUS SPECTRUM OF LIGHT

## 3.1 Stellar Parallax

Measuring the intrinsic brightness of stars is inextricably linked with determining their distances. This chapter on the light emitted by stars therefore begins with the problem of finding the distance to astronomical objects, one of the most important and most difficult tasks faced by astronomers. Kepler's laws in their original form describe the *relative* sizes of the planets' orbits in terms of astronomical units; their actual dimensions were unknown to Kepler and his contemporaries. The true scale of the solar system was first revealed in 1761 when the distance to Venus was measured as it crossed the disk of the Sun in a rare transit during inferior conjunction. The method used was **trigonometric parallax**, the familiar surveyor's technique of triangulation. On Earth, the distance to the peak of a remote mountain can be determined by measuring that peak's angular position from two observation points separated by a known baseline distance. Simple trigonometry then supplies the distance to the peak; see Fig. 3.1. Similarly, the distances to the planets can be measured from two widely separated observation sites on Earth.

Finding the distance even to the nearest stars requires a longer baseline than Earth's diameter. As Earth orbits the Sun, two observations of the same star made 6 months apart employ a baseline equal to the diameter of Earth's orbit. These measurements reveal that a nearby star exhibits an annual back-and-forth change in its position against the stationary background of much more distant stars. (As mentioned in Section 1.3, a star may also change its position due to its own motion through space. However, this *proper motion*,

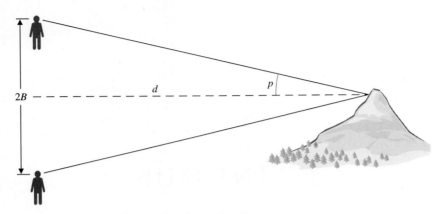

**Figure 3.1** Trigonometric parallax: $d = B/\tan p$.

seen from Earth, is not periodic, and so can be distinguished from the star's periodic displacement caused by Earth's orbital motion.) As shown in Fig. 3.2, a measurement of the **parallax angle** $p$ (one-half of the maximum change in angular position) allows the calculation of the distance $d$ to the star.

$$d = \frac{1 \text{ AU}}{\tan p} \simeq \frac{1}{p} \text{ AU},$$

where the small angle approximation $\tan p \simeq p$ has been employed for the parallax angle $p$ measured in *radians*. Using 1 radian = $57.3° = 2.063 \times 10^{5\prime\prime}$ to convert $p$ to $p''$ in units of *arcseconds* produces

$$d = \frac{2.063 \times 10^5}{p''} \text{ AU}.$$

Defining a new unit of distance, the **parsec** (**par**allax-**sec**ond, abbreviated pc), as 1 pc = $2.063 \times 10^5$ AU = $3.086 \times 10^{18}$ cm leads to

$$d = \frac{1}{p''} \text{ pc}. \tag{3.1}$$

By definition, when the parallax angle $p = 1''$, the distance to the star is 1 pc. Thus 1 parsec is the distance from which the radius of Earth's orbit, 1 AU, subtends an angle of $1''$. Another unit of distance often encountered is the **light-year** (abbreviated ly), the distance traveled by light through a vacuum in one year: 1 ly = $9.461 \times 10^{17}$ cm. One parsec is equivalent to 3.262 ly.

Even Proxima Centauri, the nearest star other than the Sun, has a parallax angle of less than $1''$. (Proxima Centauri is a member of the triple star system

## 3.2 The Magnitude Scale

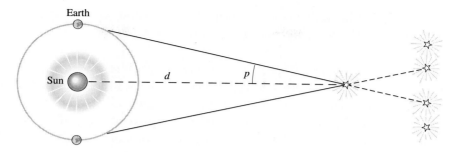

**Figure 3.2** Stellar parallax: $d = 1/p''$ pc.

$\alpha$ Centauri, and has a parallax angle of 0.77″. If Earth's orbit around the Sun were represented by a dime, then Proxima Centauri would be located 1.5 miles away!) In fact, this cyclic change in a star's position is so difficult to detect that it was not until 1838 that it was first measured, by Friedrich Wilhelm Bessel (1784–1846), a German mathematician and astronomer.[1] Using spacecraft high above Earth's distorting atmosphere, parallax angles approaching 0.001″ have been measured, corresponding to a distance of 1000 pc ≡ 1 kiloparsec (kpc). This distance is still quite small compared to the 8 kpc distance to the center of our Milky Way Galaxy, so stellar parallax is useful only for surveying the local neighborhood of the Sun.

**Example 3.1** In 1838, after 4 years of observing 61 Cygni, Bessel announced his measurement of a parallax angle of 0.316″ for that star. This corresponds to a distance of

$$d = \frac{1}{p''} \text{ pc} = \frac{1}{0.316} \text{ pc} = 3.16 \text{ pc} = 10.3 \text{ ly},$$

within 10% of the modern value 11.1 ly. 61 Cygni is one of the Sun's nearest neighbors.

## 3.2 The Magnitude Scale

Nearly all of the information astronomers have received about the universe beyond our solar system has come from the careful study of the light emitted by stars, galaxies, and interstellar clouds of gas and dust. Our modern

---

[1] Tycho Brahe had searched for stellar parallax 250 years earlier, but his instruments were too imprecise to find it. Tycho concluded that Earth does not move through space, and he was thus unable to accept Copernicus's model of a heliocentric solar system.

understanding of the universe has been made possible by the quantitative measurement of the intensity and polarization of light in every part of the electromagnetic spectrum.

The Greek astronomer Hipparchus was one of the first skywatchers to catalog the stars that he saw. In addition to compiling a list of the positions of some 850 stars, Hipparchus invented a numerical scale to describe how bright each star appeared in the sky. He assigned an **apparent magnitude** $m = 1$ to the brightest stars in the sky, and he gave the dimmest stars visible to the naked eye an apparent magnitude of $m = 6$. Note that a smaller apparent magnitude means a brighter-appearing star.

Since Hipparchus's time, astronomers have extended and refined his apparent magnitude scale. In the nineteenth century, it was thought that the human eye responded to the difference in the *logarithms* of the brightness of two luminous objects. This theory led to a scale in which a difference of one magnitude between two stars implies a constant *ratio* between their brightness. By the modern definition, a difference of 5 magnitudes corresponds exactly to a factor of 100 in brightness, so a difference of one magnitude corresponds exactly to a brightness ratio of $100^{1/5} \approx 2.512$. Thus a first magnitude star appears 2.512 times brighter than a second magnitude star, $2.512^2 = 6.310$ times brighter than a third magnitude star, and 100 times brighter than a sixth magnitude star.

Using sensitive instruments called **photometers**, astronomers can measure the apparent magnitude of an object with an accuracy of ±0.01 magnitude, and *differences* in magnitudes with an accuracy of ±0.002 magnitude. Hipparchus's scale has been extended in both directions, from $m = -26.81$ for the Sun to approximately $m = 29$ for the faintest object detectable. The total range of more than 55 magnitudes corresponds to over $100^{55/5} = (10^2)^{11} = 10^{22}$ for the ratio of the apparent brightness of the Sun to that of the faintest star or galaxy yet observed.

The "brightness" of a star is actually measured in terms of the **radiant flux** $F$ received from the star. The radiant flux is the total amount of light energy of all wavelengths that crosses a unit area oriented perpendicular to the direction of the light's travel in unit time; that is, it is the number of ergs of starlight energy arriving per second at one square centimeter of a detector aimed at the star.[2] Of course, the radiant flux received from an object depends on both its intrinsic **luminosity** (energy emitted per second) and its distance from the observer. The same star, if located farther from Earth, would appear less bright in the sky.

---

[2] 1 erg = $10^{-7}$ joule.

## 3.2 The Magnitude Scale

Imagine a star of luminosity $L$ surrounded by a huge spherical shell of radius $r$. Then, assuming that no light is absorbed during its journey out to the shell, the radiant flux, $F$, measured at distance $r$ is related to the star's luminosity by

$$F = \frac{L}{4\pi r^2}. \tag{3.2}$$

Since $L$ does not depend on $r$, the radiant flux is inversely proportional to the square of the distance from the star. This is the well-known **inverse square law** for light.[3]

---

**Example 3.2** The luminosity of the Sun is $L_\odot = 3.826 \times 10^{33}$ erg s$^{-1}$. At a distance of 1 AU $= 1.496 \times 10^{13}$ cm, Earth receives a radiant flux above its absorbing atmosphere of

$$F = \frac{L}{4\pi r^2} = 1.360 \times 10^6 \text{ erg s}^{-1} \text{ cm}^{-2}.$$

This value of the solar flux is known as the **solar constant**. At a distance of 10 pc $= 2.063 \times 10^6$ AU, an observer would measure the radiant flux to be only $(1/2.063 \times 10^6)^2$ as large. That is, the radiant flux from the Sun would be $3.196 \times 10^{-7}$ erg s$^{-1}$ cm$^{-2}$ at a distance of 10 pc.

---

Using the inverse square law, astronomers can assign an **absolute magnitude**, $M$, to each star.[4] This is defined to be the apparent magnitude a star would have *if* it were located at a distance of 10 pc. Recall that a difference of 5 magnitudes between the apparent magnitudes of two stars corresponds to the smaller-magnitude star being 100 times brighter than the larger-magnitude star. This allows us to specify their flux ratio as

$$\frac{F_2}{F_1} = 100^{(m_1 - m_2)/5}. \tag{3.3}$$

Taking the logarithm of both sides leads to the alternate form:

$$m_1 - m_2 = -2.5 \log_{10}\left(\frac{F_1}{F_2}\right). \tag{3.4}$$

The connection between a star's apparent and absolute magnitudes and its distance may be found by combining Eqs. (3.2) and (3.3):

$$100^{(m-M)/5} = \frac{F_{10}}{F} = \left(\frac{d}{10 \text{ pc}}\right)^2,$$

---

[3]If the star is moving with a speed near that of light, the inverse square law must be modified slightly.

[4]The magnitudes discussed hereafter are actually *bolometric* magnitudes, measured over all wavelengths of light; see page 82.

where $F_{10}$ is the flux that would be received if the star were at a distance of 10 pc, and $d$ is the star's distance, measured in *parsecs*. Solving for $d$ gives

$$d = 10^{(m-M+5)/5} \text{ pc}. \tag{3.5}$$

The quantity $m - M$ is therefore a measure of the distance to a star and is called the star's **distance modulus**:

$$m - M = 5\log_{10}(d) - 5 = 5\log_{10}\left(\frac{d}{10 \text{ pc}}\right). \tag{3.6}$$

---

**Example 3.3** The apparent magnitude of the Sun is $m_{\text{Sun}} = -26.81$, and its distance is 1 AU $= 4.848 \times 10^{-6}$ pc. Equation (3.6) shows that the absolute magnitude of the Sun is

$$M_{\text{Sun}} = m_{\text{Sun}} - 5\log_{10}(d) + 5 = 4.76,$$

as already given. The Sun's distance modulus is thus $m_{\text{Sun}} - M_{\text{Sun}} = -31.57$.[5]

For two stars at the same distance, Eq. (3.2) shows that the ratio of their radiant fluxes is equal to the ratio of their luminosities. Thus Eq. (3.3) for absolute magnitudes becomes

$$100^{(M_1 - M_2)/5} = \frac{L_2}{L_1}. \tag{3.7}$$

Letting one of these stars be the Sun reveals the direct relation between a star's absolute magnitude and its luminosity:

$$M = M_{\text{Sun}} - 2.5\log_{10}\left(\frac{L}{L_\odot}\right), \tag{3.8}$$

where the absolute magnitude and luminosity of the Sun are $M_{\text{Sun}} = 4.76$ and $L_\odot = 3.826 \times 10^{33}$ erg s$^{-1}$, respectively. It is left as an exercise for the reader to show that a star's apparent magnitude $m$ is related to the radiant flux $F$ received from the star by

$$m = M_{\text{Sun}} - 2.5\log_{10}\left(\frac{F}{F_{10,\odot}}\right), \tag{3.9}$$

---

[5] The magnitudes $m$ and $M$ for the Sun have a "Sun" subscript (instead of "$\odot$") to avoid confusion with $M_\odot$, the standard symbol for the Sun's mass.

where $F_{10,\odot}$ is the radiant flux received from the Sun at a distance of 10 pc (see Example 3.2).

The inverse square law for light, Eq. (3.2), relates the intrinsic properties of a star (luminosity $L$ and absolute magnitude $M$) to the quantities measured at a distance from that star (radiant flux $F$ and apparent magnitude $m$). At first glance, it may seem that astronomers must start with the measurable quantities $F$ and $m$ and then use the distance to the star (if known) to determine the star's intrinsic properties. However, if the star belongs to an important class of objects known as *pulsating variable stars*, its intrinsic luminosity $L$ and absolute magnitude $M$ can be determined *without* any knowledge of its distance. Equation (3.5) then gives the distance to the variable star. As will be discussed in Section 14.1, these stars act as beacons that illuminate the fundamental distance scale of the universe.

## 3.3 The Wave Nature of Light

Much of the history of physics is concerned with the evolution of our ideas about the nature of light. The speed of light was first measured with some accuracy in 1675, by the Danish astronomer Ole Roemer (1644–1710). Roemer observed the moons of Jupiter as they passed into the giant planet's shadow, and he was able to calculate when future eclipses of the moons should occur by using Kepler's laws. However, Roemer discovered that when Earth was moving closer to Jupiter, the eclipses occurred earlier than expected. Similarly, when Earth was moving away from Jupiter, the eclipses occurred behind schedule. Roemer realized that the discrepancy was caused by the differing amounts of time it took for light to travel the changing distance between the two planets, and he concluded that 22 minutes was required for light to cross the diameter of Earth's orbit.[6] The resulting value of $2.2 \times 10^{10}$ cm s$^{-1}$ was close to the modern value of the speed of light. In 1983 the speed of light *in vacuo* was recognized as a fundamental constant of nature whose value is, *by definition*, $c = 2.99792458 \times 10^{10}$ cm s$^{-1}$.

Even the fundamental nature of light has long been debated. Isaac Newton, for example, believed that light must consist of a rectilinear stream of particles, because only such a stream could account for the sharpness of shadows. Christian Huygens (1629–1695), a contemporary of Newton, advanced the idea that light must consist of waves. According to Huygens, light is described by the usual quantities appropriate for a wave. The distance between two successive wave crests is the **wavelength** $\lambda$, and the number of waves per second that

---

[6]We now know that it takes light about 16.5 minutes to travel 2 AU.

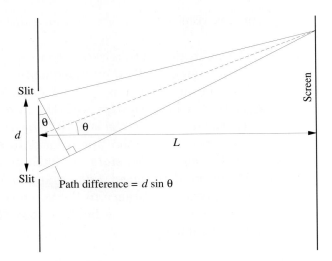

**Figure 3.3** Double-slit experiment.

pass a point in space is the **frequency** $\nu$ of the wave. Then the speed of the light wave is given by

$$c = \lambda \nu. \tag{3.10}$$

Both the particle and wave models could explain the familiar phenomena of the reflection and refraction of light. However, the particle model of light prevailed, primarily on the strength of Newton's reputation, until its wave nature was conclusively demonstrated by Thomas Young's (1773–1829) famous double-slit experiment.

In a double-slit experiment, monochromatic light of wavelength $\lambda$ from a single source passes through two narrow, parallel slits that are separated by a distance $d$. The light then falls upon a screen a distance $L$ beyond the two slits (see Fig. 3.3). The series of light and dark *interference fringes* that Young observed on the screen could be explained only by a wave model of light. As the light waves pass through the narrow slits,[7] they spread out (diffract) radially in a succession of crests and troughs. Light obeys a *superposition principle*, so when two waves meet, they add algebraically; see Fig. 3.4. At the screen, if a wave crest from one slit meets a wave crest from the other slit, a bright fringe or maximum is produced by the resulting **constructive interference**. But if a wave crest from one slit meets a wave trough from the other slit, they cancel each other, and a dark fringe or minimum results from this **destructive interference**.

---

[7]Actually, Young used pinholes in his original experiment.

## 3.3 The Wave Nature of Light

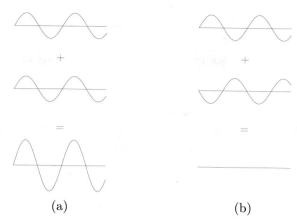

**Figure 3.4** Superposition principle for light waves. (a) Constructive interference. (b) Destructive interference.

The interference pattern observed thus depends on the difference in the lengths of the paths traveled by the light waves from the two slits to the screen. As shown in Fig. 3.3, if $L \gg d$, then to a good approximation this path difference is $d \sin \theta$. The light waves will arrive at the screen *in phase* if the path difference is equal to an integral number of wavelengths. On the other hand, the light waves will arrive 180° *out of phase* if the path difference is equal to an odd integral number of half-wavelengths. So for $L \gg d$, the angular positions of the bright and dark fringes for **double-slit interference** are given by

$$d \sin \theta = \begin{cases} n\lambda & (n = 0, 1, 2, \ldots \text{ for bright fringes}) \\ (n - \tfrac{1}{2})\lambda & (n = 1, 2, 3, \ldots \text{ for dark fringes}). \end{cases} \quad (3.11)$$

In either case, $n$ is called the **order** of the maximum or minimum. From the measured positions of the light and dark fringes on the screen, Young was able to determine the wavelength of the light. Measured in units of **angstroms**, abbreviated Å, Young obtained a wavelength of 4000 Å for violet light, and 7000 Å for red light.[8] The diffraction of light goes unnoticed under everyday conditions for these short wavelengths, thus explaining Newton's sharp shadows.

The nature of these waves of light remained elusive until the early 1860s, when the Scottish mathematical physicist James Clerk Maxwell (1831–1879) succeeded in condensing everything known about electric and magnetic fields into the four equations that today bear his name. Maxwell found that his

---

[8] 1 angstrom = $10^{-8}$ cm.

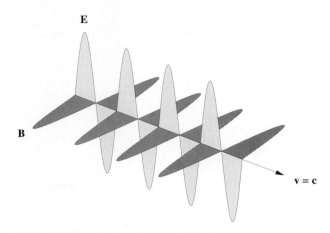

**Figure 3.5** Electromagnetic wave.

equations could be manipulated to produce wave equations for the electric and magnetic field vectors **E** and **B**. These wave equations predicted the existence of *electromagnetic waves* that travel through a vacuum with a speed $v = 1/\sqrt{\varepsilon_o \mu_o}$. Upon inserting the values of $\varepsilon_o$ and $\mu_o$, Maxwell was amazed to discover that electromagnetic waves travel at the speed of light.[9] Furthermore, these equations implied that electromagnetic waves are *transverse* waves, with the oscillatory electric and magnetic fields perpendicular to each other *and* to the direction of the wave's propagation (see Fig. 3.5); such waves could exhibit the polarization[10] known to occur for light. Maxwell wrote that "we can scarcely avoid the inference that light consists in the transverse modulations of the same medium which is the cause of electric and magnetic phenomena."

Maxwell did not live to see the experimental verification of his prediction of electromagnetic waves. Ten years after Maxwell's death, the German physicist Heinrich Hertz (1857–1894) succeeded in producing radio waves in his laboratory. Hertz determined that these electromagnetic waves do indeed travel at the speed of light, and he confirmed their reflection, refraction, and polarization properties. In 1889, Hertz wrote:

> What is light? Since the time of Young and Fresnel we know that it is wave motion. We know the velocity of the waves, we know their lengths, and we know that they are transverse; in short, our

---

[9]$\varepsilon_o$ and $\mu_o$ are fundamental constants in the SI system of units.
[10]The electromagnetic wave shown in Fig. 3.5 is *plane-polarized*, with its electric and magnetic fields oscillating in planes. Because **E** and **B** are always perpendicular, their respective planes of polarization are perpendicular as well.

## 3.3 The Wave Nature of Light

| Region | Wavelength |
|---|---|
| Gamma ray | $\lambda < 0.1$ Å |
| X-ray | $0.1$ Å $< \lambda < 100$ Å |
| Ultraviolet | $100$ Å $< \lambda < 4000$ Å |
| Visible | $4000$ Å $< \lambda < 7000$ Å |
| Infrared | $7000$ Å $< \lambda < 1$ mm |
| Microwave | $1$ mm $< \lambda < 10$ cm |
| Radio | $10$ cm $< \lambda$ |

**Table 3.1** The Electromagnetic Spectrum.

knowledge of the geometrical conditions of the motion is complete. A doubt about these things is no longer possible; a refutation of these views is inconceivable to the physicist. The wave theory of light is, from the point of view of human beings, certainty.

Today, astronomers utilize light from every part of the **electromagnetic spectrum**. The total spectrum of light consists of electromagnetic waves of all wavelengths, ranging from very short wavelength gamma rays to very long wavelength radio waves. Table 3.1 shows how the electromagnetic spectrum has been arbitrarily divided into various wavelength regions.

Like all waves, electromagnetic waves carry both energy and momentum in the direction of propagation. The amount of energy carried by a light wave is described by the **Poynting vector**, **S**. The Poynting vector[11] points in the direction of the electromagnetic wave's propagation and has a magnitude equal to the amount of energy per unit time that crosses a unit area oriented perpendicular to the direction of the propagation of the wave. Because the magnitudes of the fields **E** and **B** vary harmonically with time, the quantity of practical interest is the *time-averaged* value of the Poynting vector over one cycle of the electromagnetic wave. In a vacuum the magnitude of the time-averaged Poynting vector, $\langle S \rangle$, is

$$\langle S \rangle = \frac{c}{8\pi} E_\circ B_\circ \quad \text{(cgs units of erg s}^{-1}\text{ cm}^{-2}\text{)} \quad (3.12)$$

$$= \frac{1}{2\mu_\circ} E_\circ B_\circ \quad \text{(SI units of watt m}^{-2}\text{)},$$

where $E_\circ$ and $B_\circ$ are the maximum magnitudes (amplitudes) of the electric

---

[11]The Poynting vector is named after John Henry Poynting (1852–1914), the physicist who first described it. In cgs units $\mathbf{S} = c\mathbf{E} \times \mathbf{B}/4\pi$, and in SI units $\mathbf{S} = \mathbf{E} \times \mathbf{B}/\mu_\circ$.

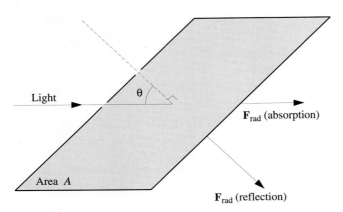

**Figure 3.6** Radiation pressure force.

and magnetic fields.[12] The time-averaged Poynting vector thus provides a description of the radiant flux in terms of the electric and magnetic fields of the light waves. However, it should be remembered that the radiant flux discussed in Section 3.2 involves the amount of energy received *at all wavelengths* from a star, whereas $E_\circ$ and $B_\circ$ describe an electromagnetic wave of a specified wavelength.

Because an electromagnetic wave carries momentum, it can exert a force on a surface hit by the light. The resulting **radiation pressure** depends on whether the light is reflected from or absorbed by the surface. If the light is completely absorbed, then the force due to radiation pressure is in the direction of the light's propagation and has magnitude

$$F_{\rm rad} = \frac{\langle S \rangle A}{c} \cos\theta \qquad \text{(absorption)}, \tag{3.13}$$

where $\theta$ is the angle of incidence of the light as measured from the direction perpendicular to the surface of area $A$ (see Fig. 3.6). Alternatively, if the light is completely reflected, then the radiation pressure force must act in a direction perpendicular to the surface; the reflected light cannot exert a force parallel to the surface. Then the magnitude of the force is

$$F_{\rm rad} = \frac{2\langle S \rangle A}{c} \cos^2\theta \qquad \text{(reflection)}. \tag{3.14}$$

Radiation pressure has a negligible effect on physical systems under everyday conditions. However, radiation pressure may play a dominant role in

---

[12]For an electromagnetic wave in a vacuum, $E_\circ$ and $B_\circ$ are related by $E_\circ = B_\circ$ (cgs units) or $E_\circ = cB_\circ$ (SI units).

## 3.4 Blackbody Radiation

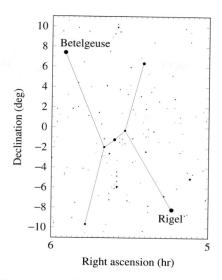

**Figure 3.7** The constellation of Orion.

determining some aspects of the behavior of extremely luminous objects such as early main-sequence stars, red supergiants, or accreting compact stars. It may also have a significant effect on the small particles of dust found throughout the interstellar medium.

## 3.4 Blackbody Radiation

Anyone who has looked at the constellation of Orion on a clear winter night has noticed the strikingly different colors of red Betelgeuse (in Orion's northeast shoulder) and blue-white Rigel (in the southwest leg); see Fig. 3.7. These colors betray the difference in the surface temperatures of the two stars. Betelgeuse has a surface temperature of about 3400 K, significantly cooler than the 10,100 K surface of Rigel.[13]

The connection between the color of light emitted by a hot object and its temperature was first noticed in 1792 by the English maker of fine porcelain, Thomas Wedgewood. All of his ovens became red-hot at the same temperature, independent of their size, shape, and construction. Subsequent investigations by many physicists revealed that any object with a temperature above absolute zero emits light of all wavelengths with varying degrees of efficiency; an *ideal emitter* is an object that absorbs *all* of the light energy incident upon it,

---

[13]Both of these stars are pulsating variables (Chapter 14), so the values quoted are *average* temperatures.

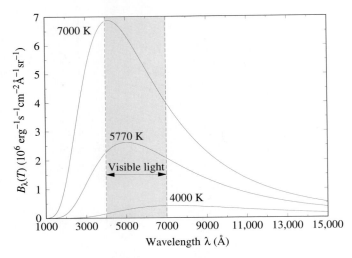

**Figure 3.8** Blackbody spectrum [Planck function $B_\lambda(T)$].

and reradiates this energy with the characteristic spectrum shown in Fig. 3.8. Because an ideal emitter reflects no light, it is known as a **blackbody**, and the radiation it emits is called **blackbody radiation**. Stars and planets are blackbodies, at least to a rough first approximation.

Figure 3.8 shows that a blackbody of temperature $T$ emits a **continuous spectrum** with some energy at all wavelengths and that this blackbody spectrum peaks at a wavelength $\lambda_{\max}$, which becomes shorter with increasing temperature. The relation between $\lambda_{\max}$ and $T$ is known as **Wien's displacement law**:[14]

$$\lambda_{\max} T = 0.290 \text{ cm K}. \tag{3.15}$$

---

**Example 3.4** Betelgeuse has a surface temperature of 3400 K. If we treat Betelgeuse as a blackbody, Wien's displacement law shows that its continuous spectrum peaks at a wavelength of

$$\lambda_{\max} = \frac{0.290 \text{ cm K}}{3400 \text{ K}} = 8.53 \times 10^{-5} \text{ cm} = 8530 \text{ Å},$$

which is in the infrared region of the electromagnetic spectrum. Rigel, with a surface temperature of 10,100 K, has a continuous spectrum that peaks at a

---

[14]In 1911, the German physicist Wilhelm Wien received the Nobel Prize for his theoretical contributions to understanding the blackbody spectrum.

## 3.4 Blackbody Radiation

wavelength of

$$\lambda_{\max} = \frac{0.290 \text{ cm K}}{10,100 \text{ K}} = 2.87 \times 10^{-5} \text{ cm} = 2870 \text{ Å},$$

in the ultraviolet region.

---

Figure 3.8 also shows that as the temperature of a blackbody increases, it emits more energy per second at *all* wavelengths. Experiments performed by the Austrian physicist Josef Stefan in 1879 showed that the luminosity, $L$, of a blackbody of area $A$ and temperature $T$ (in kelvin) is given by

$$L = A\sigma T^4. \tag{3.16}$$

Five years later another Austrian physicist, Ludwig Boltzmann, derived this equation, now called the **Stefan–Boltzmann equation**, using the laws of thermodynamics and Maxwell's formula for radiation pressure. The Stefan–Boltzmann constant, $\sigma$, has the value

$$\sigma = 5.670 \times 10^{-5} \text{ erg s}^{-1} \text{ cm}^{-2} \text{ K}^{-4}.$$

For a spherical star of radius $R$ and surface area $A = 4\pi R^2$, the Stefan–Boltzmann equation takes the form

$$L = 4\pi R^2 \sigma T_e^4. \tag{3.17}$$

Since stars are not perfect blackbodies, we use this equation to *define* the **effective temperature** $T_e$ of a star's surface. Combining this with the inverse square law, Eq. (3.2), shows that at the surface of the star ($r = R$), the *surface flux* is

$$F_{\text{surf}} = \sigma T_e^4. \tag{3.18}$$

---

**Example 3.5** The luminosity of the Sun is $L_\odot = 3.826 \times 10^{33}$ erg s$^{-1}$ and its radius is $R_\odot = 6.960 \times 10^{10}$ cm. The effective temperature of the Sun's surface is then

$$T_\odot = \left( \frac{L_\odot}{4\pi R_\odot^2 \sigma} \right)^{\frac{1}{4}} = 5770 \text{ K}.$$

The radiant flux at the solar surface is

$$F_{\text{surf}} = \sigma T_\odot^4 = 6.285 \times 10^{10} \text{ erg s}^{-1} \text{ cm}^{-2}.$$

According to Wien's displacement law, the Sun's continuous spectrum peaks at a wavelength of

$$\lambda_{\max} = \frac{0.290 \text{ cm K}}{5770 \text{ K}} = 5.03 \times 10^{-5} \text{ cm} = 5030 \text{ Å}.$$

This wavelength falls in the *green* region (4910 Å $< \lambda <$ 5750 Å) of the spectrum of visible light. However, the Sun emits a continuum of wavelengths both shorter and longer than $\lambda_{\max}$, and the human eye perceives the Sun's color as yellow. Because the Sun emits most of its energy at visible wavelengths (see Fig. 3.8), and because Earth's atmosphere is transparent at these wavelengths, the evolutionary process of natural selection has produced a human eye sensitive to this wavelength region of the electromagnetic spectrum.

Rounding off $\lambda_{\max}$ and $T_\odot$ to the more easily remembered values of 5000 Å and 5800 K, respectively, permits Wien's displacement law to be written in the convenient form

$$\lambda_{\max} T = (5000 \text{ Å})(5800 \text{ K}). \tag{3.19}$$

---

This section draws to a close at the end of the nineteenth century. The physicists and astronomers of the time believed that all of the principles that govern the physical world had finally been discovered. Their scientific world view, the *Newtonian paradigm*, was the culmination of the heroic, golden age of classical physics that had flourished for over three hundred years. The construction of this paradigm began with the brilliant observations of Galileo and the subtle insights of Newton. Its architecture was framed by Newton's laws, supported by the twin pillars of the conservation of energy and momentum and illuminated by Maxwell's electromagnetic waves. Its legacy was a deterministic description of a universe that ran like clockwork, with wheels turning inside of wheels, all of its gears perfectly meshed. Physics was in danger of becoming a victim of its own success. There were no challenges remaining. All of the great discoveries apparently had been made, and the only task remaining for the men and women of science at the turn of the century was the filling in of details.

However, as the twentieth century opened, it became increasingly apparent that a crisis was brewing. Physicists were frustrated by their inability to answer some of the simplest questions concerning light. What is the medium through which light waves travel the vast distances between the stars, and what is Earth's speed through this medium? What determines the continuous spectrum of blackbody radiation and the characteristic, discrete colors of tubes filled with hot glowing gases? Astronomers were tantalized by hints of a treasure of knowledge just beyond their grasp.

It took a physicist of the stature of Albert Einstein to topple the Newtonian paradigm and bring about two revolutions in physics. One transformed our ideas about space and time, and the other changed our basic concepts of matter and energy. The rigid clockwork universe of the golden era was found to be an illusion and was replaced by a random universe governed by the laws of probability and statistics. The following four lines aptly summarize the situation. The first two lines were written by the English poet Alexander Pope, a contemporary of Newton; the last two, by J. C. Squire, are of a more recent vintage.

> Nature and Nature's laws lay hid in night:
> God said, *Let Newton be!* and all was light.
>
> It did not last: the Devil howling "Ho!
> Let Einstein be!" restored the status quo.

## 3.5 The Quantization of Energy

By late 1900 the German physicist Max Planck (1858–1947) had discovered an empirical formula that fit the blackbody spectra shown in Fig. 3.8:

$$B_\lambda(T) = \frac{a/\lambda^5}{e^{b/\lambda T} - 1},$$

where $a$ and $b$ are constants. In spherical coordinates, the amount of energy per unit time of radiation having wavelength between $\lambda$ and $\lambda + d\lambda$ emitted by a blackbody of temperature $T$ and surface area $dA$ into a solid angle $d\Omega \equiv \sin\theta\, d\theta\, d\phi$ is given by

$$B_\lambda(T)\, d\lambda\, dA\, \cos\theta\, d\Omega = B_\lambda(T)\, d\lambda\, dA\, \cos\theta\, \sin\theta\, d\theta\, d\phi;$$

see Fig. 3.9.[15] The units of $B_\lambda$ are therefore erg s$^{-1}$ cm$^{-3}$ sr$^{-1}$. Unfortunately, these units can be misleading. The reader should note that "erg cm$^{-3}$" indicates an energy per unit area per unit wavelength interval, erg cm$^{-2}$ cm$^{-1}$, *not* an energy per unit volume. To help avoid confusion, the units of the wavelength interval $d\lambda$ are sometimes expressed in angstroms rather than centimeters, so the units of the Planck function become erg s$^{-1}$ cm$^{-2}$ Å$^{-1}$ sr$^{-1}$, as in Fig. 3.8.[16]

---

[15] Note that $dA\cos\theta$ is the area $dA$ projected onto a plane perpendicular to the direction in which the radiation is traveling. The concept of a solid angle will be fully described in Section 6.1.

[16] The value of the Planck function thus depends on the units of the wavelength interval. The conversion of $d\lambda$ from centimeters to angstroms means that the values of $B_\lambda$ obtained by evaluating Eq. (3.20) must be divided by $10^8$.

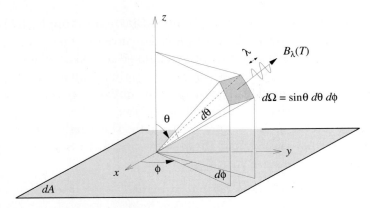

**Figure 3.9** Blackbody radiation from an element of surface area $dA$.

To evaluate the constants $a$ and $b$, Planck considered a cavity of temperature $T$ filled with blackbody radiation. This may be thought of as a hot oven filled with standing waves of electromagnetic radiation. If $L$ is the distance between the oven's walls, then the permitted wavelengths of the radiation are $\lambda = 2L, L, 2L/3, 2L/4, 2L/5, \ldots$, extending forever to increasingly shorter wavelengths.[17] Planck's dilemma was that, according to classical physics, each of these permitted wavelengths should receive an amount of energy equal to $kT$, where $k = 1.381 \times 10^{-16}$ erg K$^{-1}$ is Boltzmann's constant, familiar from the ideal gas law $PV = NkT$. The infinite number of infinitesimally short wavelengths implied that an unlimited amount of blackbody radiation energy was contained in the oven, a theoretical result so absurd it was dubbed the "ultraviolet catastrophe."

Planck attempted to circumvent this problem by using a clever mathematical trick. He assumed that a standing electromagnetic wave of wavelength $\lambda$ and frequency $\nu = c/\lambda$ could not acquire just any arbitrary amount of energy. Instead, the wave could have only specific allowed energy values that were integral multiples of a minimum wave energy.[18] This minimum energy, a **quantum** of energy, is given by $h\nu$ or $hc/\lambda$, where $h$ is a constant. Thus the energy of an electromagnetic wave is $nh\nu$ or $nhc/\lambda$, where $n$ (an integer) is the number of quanta in the wave. Given this assumption of quantized wave energy with a minimum energy proportional to the frequency of the wave, the entire oven could not contain enough energy to supply even one quantum of

---

[17]This is analogous to standing waves on a string of length $L$ that is held fixed at both ends. The permitted wavelengths are the same as those of the standing electromagnetic waves.

[18]Actually, Planck restricted the possible energies of hypothetical electromagnetic oscillators in the oven walls that emit the electromagnetic radiation.

## 3.5 The Quantization of Energy

energy for the short-wavelength, high-frequency waves. Thus the ultraviolet catastrophe would be avoided. Planck hoped that, at the end of his derivation, the constant $h$ could be set to zero; certainly, an artificial constant should not remain in his final result for $B_\lambda(T)$.

Planck's stratagem worked! His formula agreed wonderfully with experiment, but *only* if the constant $h$ remained in the equation, now known as the **Planck function**:

$$B_\lambda(T) = \frac{2hc^2/\lambda^5}{e^{hc/\lambda kT} - 1}. \tag{3.20}$$

The constant $h$, now called **Planck's constant**, has the value $h = 6.626 \times 10^{-27}$ erg s. At times it is more convenient to deal with frequency intervals $d\nu$ rather than with wavelength intervals $d\lambda$. In this case the Planck function has the form

$$B_\nu(T) = \frac{2h\nu^3/c^2}{e^{h\nu/kT} - 1}. \tag{3.21}$$

So in spherical coordinates

$$B_\nu \, d\nu \, dA \cos\theta \, d\Omega = B_\nu \, d\nu \, dA \cos\theta \sin\theta \, d\theta \, d\phi$$

is the amount of energy per unit time of blackbody radiation having frequency between $\nu$ and $\nu + d\nu$ emitted by a blackbody of temperature $T$ and surface area $dA$ into a solid angle $d\Omega = \sin\theta \, d\theta \, d\phi$.

The Planck function can be used to make the connection between the observed properties of a star (radiant flux, apparent magnitude) and its intrinsic properties (radius, temperature). Consider a model star consisting of a spherical blackbody of radius $R$ and temperature $T$. Assuming that each small patch of surface area $dA$ emits blackbody radiation *isotropically* (equally in all directions) over the outward hemisphere, the energy per second having wavelengths between $\lambda$ and $\lambda + d\lambda$ emitted by the star is

$$L_\lambda \, d\lambda = \int_{\phi=0}^{2\pi} \int_{\theta=0}^{\pi/2} \int_A B_\lambda \, d\lambda \, dA \cos\theta \sin\theta \, d\theta \, d\phi. \tag{3.22}$$

The angular integration yields a factor of $\pi$, and the integral over the area of the sphere produces a factor of $4\pi R^2$. The result is

$$L_\lambda \, d\lambda = 4\pi^2 R^2 B_\lambda \, d\lambda \tag{3.23}$$

$$= \frac{8\pi^2 R^2 hc^2/\lambda^5}{e^{hc/\lambda kT} - 1} \, d\lambda. \tag{3.24}$$

$L_\lambda$ is known as the **monochromatic luminosity**. Comparing the Stefan–Boltzmann equation (3.17) with the result of integrating Eq. (3.23) over all wavelengths shows that

$$\int_0^\infty B_\lambda(T)\, d\lambda = \frac{\sigma T^4}{\pi}. \tag{3.25}$$

In Problem 3.12, you will use Eq. (3.24) to express the Stefan–Boltzmann constant, $\sigma$, in terms of the fundamental constants $c$, $h$, and $k$. The monochromatic luminosity is related to the **monochromatic flux**, $F_\lambda$, by the inverse square law for light, Eq. (3.2):

$$F_\lambda\, d\lambda = \frac{L_\lambda}{4\pi r^2}\, d\lambda = \frac{2\pi h c^2/\lambda^5}{e^{hc/\lambda kT} - 1}\left(\frac{R}{r}\right)^2 d\lambda, \tag{3.26}$$

where $r$ is the distance to the model star. Thus $F_\lambda\, d\lambda$ is the number of ergs of starlight energy with wavelength between $\lambda$ and $\lambda + d\lambda$ that arrive per second at one square centimeter of a detector aimed at the model star, assuming that no light has been absorbed or scattered during its journey from the star to the detector. Of course, Earth's atmosphere absorbs some starlight, but measurements of fluxes and apparent magnitudes can be corrected to account for this absorption; see Section 9.2. The values of these quantities usually quoted for stars are in fact corrected values and would be the results of measurements above Earth's absorbing atmosphere.

## 3.6 The Color Index

The apparent and absolute magnitudes discussed in Section 3.2, measured over all wavelengths of light emitted by a star, are known as **bolometric magnitudes** and are denoted by $m_{\text{bol}}$ and $M_{\text{bol}}$, respectively.[19] In practice, however, most detectors measure the radiant flux of a star only within a certain wavelength region defined by the sensitivity of the detector. The *color* of a star may be precisely determined by using filters that transmit the star's light only within certain narrow wavelength bands. In the standard *UBV* system, a star's apparent magnitude is measured through three filters and is designated by three capital letters:

- $U$, the star's *ultraviolet* magnitude, is measured through a filter centered at 3650 Å with an effective bandwidth of 680 Å.

---

[19] A *bolometer* is an instrument that measures the increase in temperature caused by the radiant flux it receives at all wavelengths.

## 3.6 The Color Index

- $B$, the star's *blue* magnitude, is measured through a filter centered at 4400 Å with an effective bandwidth of 980 Å.

- $V$, the star's *visual* magnitude, is measured through a filter centered at 5500 Å with an effective bandwidth of 890 Å.

Using Eq. (3.6), a star's absolute color magnitudes $M_U$, $M_B$, and $M_V$ may be determined if its distance $d$ is known.[20] A star's $(U-B)$ **color index** is the difference between its ultraviolet and blue magnitudes, and a star's $(B-V)$ color index is the difference between its blue and visual magnitudes:

$$U - B = M_U - M_B$$

and

$$B - V = M_B - M_V.$$

Stellar magnitudes *decrease* with increasing brightness; consequently, a star with a smaller $(B-V)$ color index is *bluer* than a star with a larger value of $B-V$. Because a color index is the difference between two magnitudes, Eq. (3.6) shows that it is independent of the star's distance. The difference between a star's bolometric magnitude and its visual magnitude is its **bolometric correction** $BC$:

$$BC = m_{\text{bol}} - V = M_{\text{bol}} - M_V.$$

---

**Example 3.6** Sirius, the brightest appearing star in the sky, has $U$, $B$, and $V$ apparent magnitudes of $U = -1.50$, $B = -1.46$, and $V = -1.46$. Thus for Sirius,

$$U - B = -1.50 - (-1.46) = -0.04$$

and

$$B - V = -1.46 - (-1.46) = 0.00.$$

Sirius is brightest at ultraviolet wavelengths, as expected for a star with an effective temperature of $T_e = 9910$ K. For this surface temperature,

$$\lambda_{\text{max}} = \frac{(5000 \text{ Å})(5800 \text{ K})}{9910 \text{ K}} = 2930 \text{ Å},$$

which is in the ultraviolet portion of the electromagnetic spectrum. The bolometric correction for Sirius is $BC = -0.09$, so its apparent bolometric magnitude is

$$m_{\text{bol}} = V + BC = -1.46 + (-0.09) = -1.55.$$

---

[20]Note that although apparent magnitude is not denoted by a subscripted "$m$" in the $UBV$ system, the absolute magnitude is denoted by a subscripted "$M$."

The relation between apparent magnitude and radiant flux, Eq. (3.4), can be used to derive expressions for the ultraviolet, blue, and visual magnitudes measured (above Earth's atmosphere) for a star. A *sensitivity function* $\mathcal{S}(\lambda)$ is used to describe the fraction of the star's flux that is detected at wavelength $\lambda$. $\mathcal{S}$ depends on the reflectivity of the telescope mirrors, the bandwidth of the $U$, $B$, and $V$ filters, and the response of the photometer. Thus, for example, a star's ultraviolet magnitude $U$ is given by

$$U = -2.5 \log_{10} \left( \int_0^\infty F_\lambda \mathcal{S}_U \, d\lambda \right) + C_U, \tag{3.27}$$

where $C_U$ is a constant. Similar expressions are used for a star's apparent magnitude within other wavelength bands. The constants $C$ in the equations for $U$, $B$, and $V$ differ for each of these wavelength regions and are chosen so that the star Vega ($\alpha$ Lyrae) has a magnitude of *zero* as seen through each filter.[21] This is a completely arbitrary choice and does *not* imply that Vega would appear equally bright when viewed through the $U$, $B$, and $V$ filters. However, the resulting values for the visual magnitudes of stars are about the same as those recorded by Hipparchus two thousand years ago.[22]

A *different* method is used to determine the constant $C_{\text{bol}}$ in the expression for the bolometric magnitude, measured over all wavelengths of light emitted by a star. For a *perfect* bolometer, capable of detecting 100 percent of the light arriving from a star, we set $\mathcal{S}(\lambda) \equiv 1$:

$$m_{\text{bol}} = -2.5 \log_{10} \left( \int_0^\infty F_\lambda \, d\lambda \right) + C_{\text{bol}}. \tag{3.28}$$

The value for $C_{\text{bol}}$ originated in the wish of astronomers that the value of the bolometric correction

$$BC = m_{\text{bol}} - V$$

be negative for all stars (since a star's radiant flux over all wavelengths is greater than its flux in any specified wavelength band) while still being as close to zero as possible. After a value of $C_{\text{bol}}$ was agreed upon, it was discovered that some supergiant stars have *positive* bolometric corrections. Nevertheless, astronomers have chosen to continue using this unphysical method of measuring magnitudes.[23] It is left as an exercise for the reader to evaluate the constant $C_{\text{bol}}$ by using the value of $m_{\text{bol}}$ assigned to the Sun: $m_{\text{Sun}} = -26.81$.

---

[21] Actually, the average magnitude of several stars is used for this calibration.

[22] See Chapter 1 of Böhm-Vitense (1989b) for a further discussion of the vagaries of the magnitude system used by astronomers.

[23] Some authors, such as Böhm-Vitense (1989a, 1989b), prefer to define the bolometric correction as $BC = V - m_{\text{bol}}$, so their values of $BC$ will usually be positive.

## 3.6 The Color Index

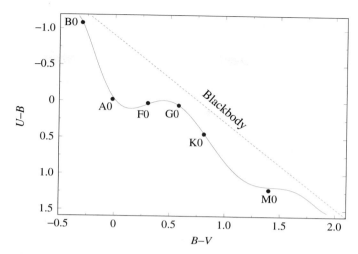

**Figure 3.10** Color–color diagram for main-sequence stars. The dashed line is for a blackbody.

The color indices $U - B$ and $B - V$ are immediately seen to be

$$U - B = -2.5 \log_{10}\left(\frac{\int F_\lambda \mathcal{S}_U \, d\lambda}{\int F_\lambda \mathcal{S}_B \, d\lambda}\right) + C_{U-B}, \tag{3.29}$$

where $C_{U-B} \equiv C_U - C_B$. A similar relation holds for $B - V$. From Eq. (3.26), note that although the apparent magnitudes depend on the radius $R$ of the model star and its distance $r$, the color indices do not, because the factor of $(R/r)^2$ cancels in Eq. (3.29). Thus the color index is a measure solely of the temperature of a model blackbody star.

Figure 3.10 is a **color–color diagram** showing the relation between the $(U - B)$ and $(B - V)$ color indices for main-sequence stars.[24] Astronomers face the difficult task of connecting a star's position on a color–color diagram with the physical properties of the star itself. If stars actually behaved as blackbodies, the color–color diagram would be the straight dashed line shown in Fig. 3.10. However, stars are not true blackbodies. As will be discussed in detail in Chapter 9, some light is absorbed as it travels through a star's atmosphere, and the amount of light absorbed depends on both the wavelength of the light and the temperature of the star. Other factors also play a role, causing the color indices of main sequence and supergiant stars of the same

---

[24] As will be discussed in Section 10.6, main-sequence stars are powered by the nuclear fusion of hydrogen nuclei in their centers. Approximately 80% to 90% of all stars are main-sequence stars. The letter labels in Fig. 3.10 are *spectral types*; see Section 8.1.

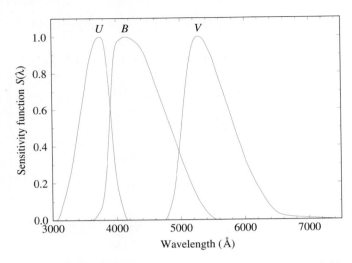

**Figure 3.11** Sensitivity functions $S(\lambda)$ for $U$, $B$, and $V$ filters.

temperature to be slightly different. The color–color diagram in Fig. 3.10 shows that the agreement between actual stars and model blackbody stars is best for very hot stars.

---

**Example 3.7** A star of spectral type O5 (to be defined in Section 8.1) has a surface temperature of 44,500 K and color indices $U - B = -1.19$ and $B - V = -0.33$. The large negative value of $U - B$ indicates that this star appears brightest at ultraviolet wavelengths, as can be confirmed using Wien's displacement law, Eq. (3.19). The spectrum of a 44,500 K blackbody peaks at

$$\lambda_{\max} = \frac{(5000 \text{ Å})(5800 \text{ K})}{44,500 \text{ K}} = 652 \text{ Å},$$

in the ultraviolet region of the electromagnetic spectrum. This wavelength is much shorter than the wavelengths transmitted by the $U$, $B$, and $V$ filters (see Fig. 3.11), so we will be dealing with the smoothly declining long wavelength "tail" of the Planck function $B_\lambda(T)$.

We can use the values of the color indices to estimate the constant $C_{U-B}$ in Eq. (3.29), and $C_{B-V}$ in a similar equation for the color index $B - V$. In this estimate, we will use a step function to represent the sensitivity function: $S(\lambda) = 1$ inside the filter's bandwidth, and $S(\lambda) = 0$ otherwise. The integrals in Eq. (3.29) may then be approximated by the value of the Planck function $B_\lambda$ at the center of the filter bandwidth, multiplied by that bandwidth. Thus,

## 3.6 The Color Index

for the wavelengths and bandwidths $\Delta\lambda$ listed on pages 82–83,

$$U - B = -2.5 \log_{10}\left(\frac{B_{3650}\,\Delta\lambda_U}{B_{4400}\,\Delta\lambda_B}\right) + C_{U-B}$$

$$-1.19 = -0.32 + C_{U-B}$$

$$C_{U-B} = -0.87,$$

and

$$B - V = -2.5 \log_{10}\left(\frac{B_{4400}\,\Delta\lambda_B}{B_{5500}\,\Delta\lambda_V}\right) + C_{B-V}$$

$$-0.33 = -0.99 + C_{B-V}$$

$$C_{B-V} = 0.66.$$

The latter value is in good agreement with the value $C_{B-V} = 0.67$ found in Allen (1972), considering the roughness of the approximations involved.

It is left as an exercise for the reader to use these values of $C_{U-B}$ and $C_{B-V}$ to estimate the color indices for a model blackbody Sun with a surface temperature of 5770 K. Although the resulting value of $B - V = +0.58$ is in fair agreement with the measured value of $B - V = +0.64$ for the Sun, the estimate of $U - B = -0.22$ is quite different than the measured value of $U - B = +0.16$. The reason for this large discrepancy at ultraviolet wavelengths will be discussed in Example 9.4.

## Suggested Readings

**GENERAL**

Ferris, Timothy, *Coming of Age in the Milky Way*, William Morrow, New York, 1988.

Griffin, Roger, "The Radial-Velocity Revolution," *Sky and Telescope*, September 1989.

Hearnshaw, John B., "Origins of the Stellar Magnitude Scale," *Sky and Telescope*, November 1992.

Herrmann, Dieter B., *The History of Astronomy from Hershel to Hertzsprung*, Cambridge University Press, Cambridge, 1984.

Segre, Emilio, *From Falling Bodies to Radio Waves*, W. H. Freeman and Company, New York, 1984.

**TECHNICAL**

Allen, C. W., *Astrophysical Quantities*, Third Edition, Athlone Press, London, 1972.

Arp, Halton, "$U - B$ and $B - V$ Colors of Black Bodies," *The Astrophysical Journal, 133*, 874, 1961.

Böhm-Vitense, Erika, *Introduction to Stellar Astrophysics, Volume 1: Basic Stellar Observations and Data*, Cambridge University Press, Cambridge, 1989a.

Böhm-Vitense, Erika, *Introduction to Stellar Astrophysics, Volume 2: Stellar Atmospheres*, Cambridge University Press, Cambridge, 1989b.

Harwit, Martin, *Astrophysical Concepts*, Second Edition, John Wiley and Sons, New York, 1989.

Lang, Kenneth R., *Astrophysical Formulae*, Second Edition, Springer-Verlag, Berlin, 1980.

Van Helden, Albert, *Measuring the Universe*, The University of Chicago Press, Chicago, 1985.

## Problems

3.1 In 1672, an international effort was made to measure the parallax angle of Mars at the time of opposition, when it was closest to Earth; see Fig. 1.6.

(a) Consider two observers who are separated by a baseline equal to Earth's diameter. If the difference in their measurements of Mars' angular position is 33.6″, what is the distance between Earth and Mars at the time of opposition? Express your answer both in units of cm and AU.

(b) If the distance to Mars is to be measured to within 10%, how closely must the clocks used by the two observers be synchronized? *Hint:* Ignore the rotation of Earth. The average orbital velocities of Earth and Mars are 29.79 km s$^{-1}$ and 24.13 km s$^{-1}$, respectively.

3.2 At what distance from a 100-watt light bulb is the radiant flux equal to the solar constant?

3.3 The parallax angle for Sirius is 0.377″.

(a) Find the distance to Sirius in units of (i) parsecs; (ii) light-years; (iii) AU; (iv) cm.

(b) Determine the distance modulus for Sirius.

3.4 Using the information in Example 3.6 and Problem 3.3, determine the absolute bolometric magnitude of Sirius and compare it with that of the Sun. What is the ratio of Sirius' luminosity to that of the Sun?

3.5 Derive the relation

$$m = M_{\text{Sun}} - 2.5 \log_{10}\left(\frac{F}{F_{10,\odot}}\right).$$

3.6 A $1.2 \times 10^4$ kg spacecraft is launched from Earth and is to be accelerated radially away from the Sun using a circular solar sail. The initial acceleration of the spacecraft is to be $1g$. Assuming a flat sail, determine the radius of the sail if it is

(a) black, so it absorbs the Sun's light.

(b) shiny, so it reflects the Sun's light.

*Hint:* The spacecraft, like Earth, is orbiting the Sun. Should you include the Sun's gravity in your calculation?

3.7 The average person has 1.4 m² of skin at a skin temperature of roughly 92°F (306 K). Consider the average person to be an ideal radiator standing in a room at a temperature of 68°F (293 K).

(a) Calculate the energy per second radiated by the average person in the form of blackbody radiation. Express your answer both in units of erg s$^{-1}$ and in watts.

(b) Determine the peak wavelength $\lambda_{max}$ of the blackbody radiation emitted by the average person. In what region of the electromagnetic spectrum is this wavelength found?

(c) A blackbody also absorbs energy from its environment, in this case from the 293-K room. The equation describing the absorption is the same as the equation describing the emission of blackbody radiation, Eq. (3.16). Calculate the energy per second absorbed by the average person, expressed both in units of erg s$^{-1}$ and in watts.

(d) Calculate the net energy per second lost by the average person due to blackbody radiation.

3.8 Consider a model of a star consisting of a spherical blackbody with a surface temperature of 28,000 K and a radius of $5.16 \times 10^{11}$ cm. Let this model star be located at a distance of 180 pc from Earth. Determine the following for the star:

(a) Luminosity.
(b) Absolute bolometric magnitude.
(c) Apparent bolometric magnitude.
(d) Distance modulus.
(e) Radiant flux at the star's surface.
(f) Radiant flux at Earth's surface (compare this with the solar constant).
(g) Peak wavelength $\lambda_{max}$.

This is a model of the star Dschubba, the center star in the head of the constellation Scorpius.

3.9 Before Planck discovered the correct description of the spectrum of blackbody radiation, a formulation that was valid only for *long wavelengths* was found by two English physicists, Lord Rayleigh and James Jeans.

(a) Derive the Rayleigh–Jeans law by considering the Planck function $B_\lambda$ in the limit of $\lambda \gg hc/kT$. (The first-order expansion $e^x \approx 1+x$ for $x \ll 1$ will be useful.) Notice that Planck's constant is not present in your answer. The Rayleigh–Jeans law is a *classical* result, so the "ultraviolet catastrophe" at short wavelengths, produced by the $\lambda^4$ in the denominator, cannot be avoided.

(b) On the same graph, plot the Planck function $B_\lambda$ and the Rayleigh–Jeans law for the Sun ($T_\odot = 5770$ K). At roughly what wavelength is the Rayleigh–Jeans value twice as large as the Planck function?

3.10 Derive Wien's displacement law, Eq. (3.15), by setting $dB_\lambda/d\lambda = 0$. *Hint:* You will encounter an equation that must be solved numerically, not algebraically.

3.11 (a) Use Eq. (3.21) to find an expression for the frequency $\nu_{\max}$ at which the Planck function $B_\nu$ attains its maximum value. (*Warning:* $\nu_{\max} \neq c/\lambda_{\max}$.)

(b) What is the value of $\nu_{\max}$ for the Sun?

(c) Find the wavelength of a light wave having frequency $\nu_{\max}$. In what region of the electromagnetic spectrum is this wavelength found?

3.12 (a) Integrate Eq. (3.24) over all wavelengths to obtain an expression for the total luminosity of a blackbody model star. *Hint:*

$$\int_0^\infty \frac{u^3\,du}{e^u - 1} = \frac{\pi^4}{15}.$$

(b) Compare your result with the Stefan–Boltzmann equation (3.17), and show that the Stefan–Boltzmann constant $\sigma$ is given by

$$\sigma = \frac{2\pi^5 k^4}{15 c^2 h^3}.$$

(c) Calculate the value of $\sigma$ from this expression, and compare with the value listed in Appendix A.

3.13 Use the data in Appendix E to answer the following questions.

(a) Calculate the absolute and apparent visual magnitudes, $M_V$ and $V$, for the Sun.

(b) Determine the magnitudes $M_B$, $B$, $M_U$, and $U$ for the Sun.

(c) Locate the Sun and Sirius on the color–color diagram in Fig. 3.10. Refer to Example 3.6 for the data on Sirius.

3.14 Use the filter bandwidths for the $UBV$ system on pages 82–83 and the effective temperature of 9500 K for Vega to determine through which filter Vega would appear brightest to a photometer [i.e., ignore the constant $C$ in Eq. (3.27)]. Assume that $S(\lambda) = 1$ inside the filter bandwidth and that $S(\lambda) = 0$ outside the filter bandwidth.

3.15 Evaluate the constant $C_{\text{bol}}$ in Eq. (3.28) by using $m_{\text{Sun}} = -26.81$.

3.16 Use the values of the constants $C_{U-B}$ and $C_{B-V}$ found in Example 3.7 to estimate the color indices $U-B$ and $B-V$ for the Sun.

3.17 Shaula ($\lambda$ Scorpii) is a bright ($V = 1.62$) blue-white subgiant star located at the tip of the scorpion's tail. Its surface temperature is about 22,000 K. Use the values of the constants $C_{U-B}$ and $C_{B-V}$ found in Example 3.7 to estimate the color indices $U-B$ and $B-V$ for Shaula. Compare your answers with the measured values of $U-B = -0.90$ and $B-V = -0.22$. (Shaula is a pulsating star, belonging to the class of Beta Cephei variables; see Section 14.2. As its magnitude varies between $V = 1.59$ and $V = 1.65$ with a period of 5 hours 8 minutes, its color indices also change slightly.)

# Chapter 4

# The Theory of Special Relativity

## 4.1 The Failure of the Galilean Transformations

A *wave* is a disturbance that travels through a medium. Water waves are disturbances traveling through water, and sound waves are disturbances traveling through air. James Clerk Maxwell predicted that light consists of "modulations of the same medium which is the cause of electric and magnetic phenomena," but what was the medium through which light waves traveled? At the time, physicists believed that light waves moved through a medium called the **luminiferous ether**. This idea of an all-pervading ether had its roots in the science of early Greece. In addition to the four earthly elements of Earth, Air, Fire, and Water, the Greeks believed that the heavens were composed of a fifth perfect element: the ether. Maxwell echoed their ancient belief when he wrote:

> There can be no doubt that the interplanetary and interstellar spaces are not empty, but are occupied by a material substance or body, which is certainly the largest, and probably the most uniform body of which we have any knowledge.

This modern reincarnation of the ether had been proposed for the sole purpose of transporting light waves; an object moving through the ether would experience no mechanical resistance, so Earth's velocity through the ether could not be directly measured.

In fact, *no* mechanical experiment is capable of determining the absolute velocity of an observer. It is impossible to tell whether you are at rest or in uniform motion (not accelerating). This general principle was recognized very

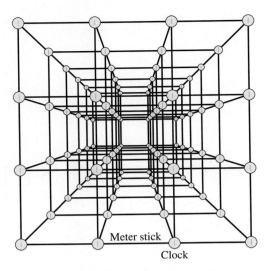

**Figure 4.1** Inertial reference frame.

early. Galileo described a laboratory completely enclosed below the deck of a smoothly sailing ship and argued that no experiment done in this uniformly moving laboratory could measure the ship's velocity. To see why, consider two *inertial reference frames*, $S$ and $S'$. As discussed in Section 2.2, an inertial reference frame may be thought of as a laboratory in which Newton's first law is valid: An object at rest will remain at rest and an object in motion will remain in motion in a straight line at constant speed unless acted upon by an unbalanced force. As shown in Fig. 4.1, the laboratory consists of (in principle) an infinite collection of meter sticks and synchronized clocks that can record the position and time of any event that occurs in the laboratory, *at the location of that event*; this removes the time delay involved in relaying information about an event to a distant recording device. With no loss of generality, the frame $S'$ can be taken as moving in the positive $x$-direction (relative to the frame $S$) with constant velocity $\mathbf{u}$, as shown in Fig. 4.2.[1] Furthermore, the clocks in the two frames can be started when the origins of the coordinate systems, $O$ and $O'$, coincide at time $t = t' = 0$.

Observers in the two frames $S$ and $S'$ measure the same moving object, recording its position $(x, y, z)$ and $(x', y', z')$ at time $t$ and $t'$, respectively. An appeal to common sense and intuition shows that these measurements are

---

[1]This does *not* imply that the frame $S$ is at rest, and that $S'$ is moving. $S'$ could be at rest while $S$ moves in the negative $x'$-direction, or both frames may be moving. The point of the following argument is that there is *no way to tell*; only the *relative velocity* $\mathbf{u}$ is meaningful.

## 4.1 The Failure of the Galilean Transformations

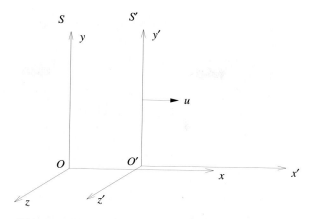

**Figure 4.2** Inertial reference frames $S$ and $S'$.

related by the **Galilean transformation equations**:

$$x' = x - ut \tag{4.1}$$

$$y' = y \tag{4.2}$$

$$z' = z \tag{4.3}$$

$$t' = t. \tag{4.4}$$

Taking time derivatives with respect to either $t$ or $t'$ (since they are always equal) shows how the components of the object's velocity $\mathbf{v}$ and $\mathbf{v}'$ measured in the two frames are related:

$$v_x' = v_x - u$$

$$v_y' = v_y$$

$$v_z' = v_z,$$

or, in vector form,

$$\mathbf{v}' = \mathbf{v} - \mathbf{u}. \tag{4.5}$$

Since $\mathbf{u}$ is constant, another time derivative shows that the *same* acceleration is obtained for the object as measured in both reference frames:

$$\mathbf{a}' = \mathbf{a}.$$

Thus $\mathbf{F} = m\mathbf{a} = m\mathbf{a}'$ for the object of mass $m$; Newton's laws are obeyed in both reference frames. Whether a laboratory is located in the hold of Galileo's ship or anywhere else in the universe, no mechanical experiment can be done to measure the laboratory's absolute velocity.

Maxwell's discovery that electromagnetic waves move through the ether with a speed of $c \simeq 3 \times 10^{10}$ cm s$^{-1}$ seemed to open the possibility of detecting Earth's *absolute* motion through the ether by measuring the speed of light from Earth's frame of reference[2] and comparing it with Maxwell's theoretical value of $c$. In 1887 two Americans, the physicist Albert A. Michelson (1852–1931) and his chemist colleague Edward W. Morley (1838–1923), performed a classic experiment that attempted this measurement of Earth's absolute velocity. Although Earth orbits the Sun at approximately 30 km s$^{-1}$, the results of the Michelson–Morley experiment were consistent with a velocity of Earth through the ether of *zero*! Furthermore, as Earth spins on its axis and orbits the Sun, a laboratory's speed through the ether should be constantly changing. The constantly shifting "ether wind" should easily be detected. However, all of the many physicists who have since repeated the Michelson–Morley experiment with increasing precision have reported the same null result. Everyone measures *exactly* the same value for the speed of light, regardless of the velocity of the laboratory on Earth or the velocity of the source of the light.

On the other hand, Eq. (4.5) implies that two observers moving with a relative velocity **u** should obtain *different* values for the speed of light. The contradiction between the commonsense expectation of Eq. (4.5) and the experimentally determined constancy of the speed of light means that this equation, and the equations from which it was derived (the Galilean transformation equations, 4.1–4.4), cannot be correct. Although the Galilean transformations adequately describe the familiar low-speed world of everyday life where $v/c \ll 1$, they are in sharp disagreement with the results of experiments involving velocities near the speed of light. A crisis in the Newtonian paradigm was developing.

## 4.2 The Lorentz Transformations

The young Albert Einstein (1875–1955; see Fig. 4.3) enjoyed discussing a puzzle with his friends: What would you see if you looked in a mirror while moving at the speed of light? Would you see your image in the mirror, or not? This was the beginning of Einstein's search for a simple, consistent picture of the universe, a quest that would culminate in his theories of relativity. After much

---

[2]Strictly speaking, a laboratory on Earth is not in an inertial frame of reference, because Earth both spins on its axis and accelerates as it orbits the Sun. However, these noninertial effects are unimportant for the Michelson–Morley experiment.

## 4.2 The Lorentz Transformations

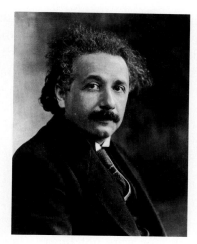

**Figure 4.3** Albert Einstein (1875–1955). (Courtesy of Yerkes Observatory.)

reflection, Einstein finally rejected the notion of an all-pervading ether. In 1905 he introduced his two postulates of special relativity[3] in a remarkable paper, "On the Electrodynamics of Moving Bodies."

> The phenomena of electrodynamics as well as of mechanics possesses no properties corresponding to the idea of absolute rest. They suggest rather that ... the same laws of electrodynamics and optics will be valid for all frames of reference for which the equations of mechanics hold good. We will raise this conjecture (the purport of which will hereafter be called the "Principle of Relativity") to the status of a postulate, and also introduce another postulate, which is only apparently irreconcilable to the former, namely, that light is always propagated in empty space with a definite speed $c$ which is independent of the state of motion of the emitting body.

In other words, **Einstein's postulates** are:

> **The Principle of Relativity** The laws of physics are the same in all inertial reference frames.
>
> **The Constancy of the Speed of Light** Light travels through a vacuum at a constant speed $c$ that is independent of the motion of the light source.

---

[3]The theory of *special* relativity deals only with inertial reference frames, while the *general* theory includes accelerating frames.

Einstein then went on to derive the equations that lie at the heart of his theory of special relativity, the **Lorentz transformations**.[4] For the two inertial reference frames shown in Fig. 4.2, the most general set of linear transformation equations between the space and time coordinates $(x, y, z, t)$ and $(x', y', z', t')$ of the *same event* measured from $S$ and $S'$ are

$$x' = a_{11}x + a_{12}y + a_{13}z + a_{14}t \tag{4.6}$$

$$y' = a_{21}x + a_{22}y + a_{23}z + a_{24}t \tag{4.7}$$

$$z' = a_{31}x + a_{32}y + a_{33}z + a_{34}t \tag{4.8}$$

$$t' = a_{41}x + a_{42}y + a_{43}z + a_{44}t. \tag{4.9}$$

If the transformation equations were not linear, then the length of a moving object or the time interval between two events would depend on the choice of origin for the frames $S$ and $S'$. This is unacceptable, since the laws of physics cannot depend on the numerical coordinates of an arbitrarily chosen coordinate system.

The coefficients $a_{ij}$ can be determined by using Einstein's two postulates and some simple symmetry arguments. Einstein's first postulate, the Principle of Relativity, implies that lengths *perpendicular* to **u**, the velocity of frame $S$ relative to $S'$, are unchanged. To see this, imagine that each frame has a meter stick oriented along the $y$- and $y'$-axes, with one end of each meter stick located at the origin of its respective frame; see Fig. 4.4. Paintbrushes are mounted perpendicular at both ends of each meter stick, and the frames are separated by a sheet of glass that extends to infinity in the $x$–$y$ plane. Each brush paints a line on the glass sheet as the two frames pass each other. Let's say that frame $S$ uses blue paint, and frame $S'$ uses red paint. If an observer in the frame $S$ measures the meter stick in frame $S'$ to be shorter than his own meter stick, he will see the red lines painted *inside* his blue lines on the glass. But by the Principle of Relativity, an observer in the frame $S'$ would measure the meter stick in frame $S$ as being shorter than her own meter stick, and would see the blue lines painted *inside* her red lines. Both color lines cannot lie inside the other; the only conclusion is that blue and red lines must overlap. The lengths of the meter sticks, perpendicular to **u**, are unchanged. Thus $y' = y$ and $z' = z$, so that $a_{22} = a_{33} = 1$, while $a_{21}$, $a_{23}$, $a_{24}$, $a_{31}$, $a_{32}$, and $a_{34}$ are all zero.

---

[4]These equations were first derived by Hendrik A. Lorentz (1853–1928) of the Netherlands, but were applied to a different situation involving a reference frame at absolute rest with respect to the ether.

## 4.2 The Lorentz Transformations

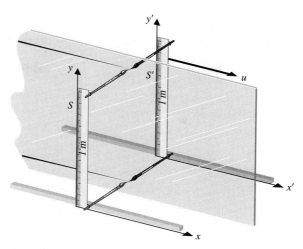

**Figure 4.4** Paintbrush demonstration that $y' = y$.

Another simplification comes from requiring that Eq. (4.9) give the same result if $y$ is replaced by $-y$ or $z$ is replaced by $-z$. This must be true because rotational symmetry about the axis parallel to the relative velocity **u** implies that a time measurement cannot depend on which side of the $x$-axis an event occurs. Thus $a_{42} = a_{43} = 0$.

Finally, consider the motion of the origin $O'$ of the frame $S'$. Since the frames' clocks are assumed to be synchronized at $t = t' = 0$ when the origins $O$ and $O'$ coincide, the $x$-coordinate of $O'$ is given by $x = ut$ in the frame $S$ and by $x' = 0$ in the frame $S'$. Thus Eq. (4.6) becomes

$$0 = a_{11}ut + a_{12}y + a_{13}z + a_{14}t,$$

which implies that $a_{12} = a_{13} = 0$ and $a_{11}u = -a_{14}$. Collecting the results found thus far reveals that Eqs. (4.6)–(4.9) have been reduced to

$$x' = a_{11}(x - ut) \tag{4.10}$$

$$y' = y \tag{4.11}$$

$$z' = z \tag{4.12}$$

$$t' = a_{41}x + a_{44}t. \tag{4.13}$$

At this point, these equations would be consistent with the commonsense Galilean transformation equations (4.1)–(4.4) if $a_{11} = a_{44} = 1$ and $a_{41} = 0$. However, only one of Einstein's postulates has been employed in the derivation thus far: the Principle of Relativity championed by Galileo himself.

Now the argument introduces the second of Einstein's postulates: Everyone measures exactly the same value for the speed of light. Suppose that when the origins $O$ and $O'$ coincide at time $t = t' = 0$, a flashbulb is set off at the common origins. At a later time $t$, an observer in frame $S$ will measure a spherical wavefront of light with radius $ct$, moving away from the origin $O$ with speed $c$ and satisfying

$$x^2 + y^2 + z^2 = (ct)^2. \tag{4.14}$$

Similarly, at a time $t'$, an observer in frame $S'$ will measure a spherical wavefront of light with radius $ct'$, moving away from the origin $O'$ with speed $c$ and satisfying

$$x'^2 + y'^2 + z'^2 = (ct')^2. \tag{4.15}$$

Inserting Eqs. (4.10)–(4.13) into Eq. (4.15) and comparing the result with Eq. (4.14) reveals that $a_{11} = a_{44} = 1/\sqrt{1 - u^2/c^2}$ and $a_{41} = -ua_{11}/c^2$. So the Lorentz transformation equations linking the space and time coordinates $(x, y, z, t)$ and $(x', y', z', t')$ of the *same event* measured from $S$ and $S'$ are

$$x' = \frac{x - ut}{\sqrt{1 - u^2/c^2}} \tag{4.16}$$

$$y' = y \tag{4.17}$$

$$z' = z \tag{4.18}$$

$$t' = \frac{t - ux/c^2}{\sqrt{1 - u^2/c^2}}. \tag{4.19}$$

Whenever the Lorentz transformations are used, the reader should be certain that the situation is consistent with the geometry of Fig. 4.2, where the inertial reference frame $S'$ is moving in the positive $x$-direction with velocity **u** relative to the frame $S$. The ubiquitous factor of

$$\gamma \equiv \frac{1}{\sqrt{1 - u^2/c^2}}, \tag{4.20}$$

called the **Lorentz factor**, may be used to estimate the importance of relativistic effects. Roughly speaking, relativity differs from Newtonian mechanics by 1% ($\gamma = 1.01$) when $u/c \simeq 1/7$, and by 10% when $u/c \simeq 5/12$; see Fig. 4.5. In the low-speed Newtonian world, the Lorentz transformations reduce to the Galilean transformation equations (4.1)–(4.4). A similar requirement holds for

## 4.2 The Lorentz Transformations

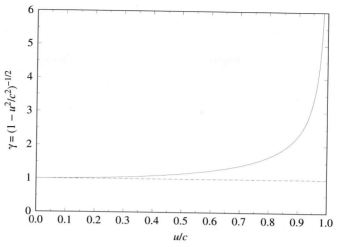

**Figure 4.5** Lorentz factor $\gamma$.

all relativistic formulas; they must agree with the Newtonian equations in the low-speed limit of $u/c \to 0$.

The inverse Lorentz transformations can be derived algebraically, or they can be obtained more easily by switching primed and unprimed quantities and by replacing $u$ with $-u$. (Be sure you understand the physical basis for these substitutions.) Either way, the inverse transformations are found to be

$$x = \frac{x' + ut'}{\sqrt{1 - u^2/c^2}} \quad (4.21)$$

$$y = y' \quad (4.22)$$

$$z = z' \quad (4.23)$$

$$t = \frac{t' + ux'/c^2}{\sqrt{1 - u^2/c^2}}. \quad (4.24)$$

The Lorentz transformation equations form the core of the theory of special relativity, and they have many surprising and unusual implications. The most obvious surprise is the intertwining roles of spatial and temporal coordinates in the transformations. In the words of Einstein's professor, Hermann Minkowski (1864–1909), "Henceforth space by itself, and time by itself, are doomed to fade away into mere shadows, and only a kind of union between the two will preserve an independent reality." The drama of the physical world unfolds on the stage of a four-dimensional **spacetime**, where events are identified by their spacetime coordinates $(x, y, z, t)$.

## 4.3  Time and Space in Special Relativity

Suppose an observer in frame $S$ measures two flashbulbs going off at the *same time* $t$ but at *different* $x$-coordinates $x_1$ and $x_2$. Then an observer in frame $S'$ measures the time interval $t_1' - t_2'$ between the flashbulbs going off to be (see Eq. 4.19)

$$t_1' - t_2' = \frac{(x_2 - x_1)\,u/c^2}{\sqrt{1 - u^2/c^2}}. \tag{4.25}$$

*According to the observer in frame $S'$, if $x_1 \neq x_2$, then the flashbulbs do not go off at the same time!* Events that occur simultaneously in one inertial reference frame do not occur simultaneously in all other inertial reference frames. There is no such thing as two events that occur at different locations happening *absolutely* at the same time. Equation (4.25) shows that if $x_1 < x_2$, then $t_1' - t_2' > 0$ for positive $u$; flashbulb 1 is measured to go off *after* flashbulb 2. An observer moving at the same speed in the opposite direction ($u$ changed to $-u$) will come to the opposite conclusion: Flashbulb 2 goes off *after* flashbulb 1. The situation is symmetric; an observer in frame $S'$ will conclude that the flashbulb he or she passes first goes off *after* the other flashbulb. It is tempting to ask, "Which observer is *really* correct?" However, this question is meaningless, and is equivalent to asking, "Which observer is *really* moving?" Neither question has an answer because "really" has no meaning in this situation. There is no absolute simultaneity just as there is no absolute motion. Each observer's measurement is correct, as made from his or her own frame of reference.

The implications of this **downfall of simultaneity** are far-reaching. The absence of a universal simultaneity means that clocks in relative motion will not stay synchronized. Newton's idea of an absolute universal time that "of itself and from its own nature flows equably without regard to anything external" has been overthrown. Different observers in relative motion will measure *different* time intervals between the *same* two events!

Imagine that a strobe light located at rest relative to the frame $S'$ produces a flash of light every $\Delta t'$ seconds; see Fig. 4.6. If one flash is emitted at time $t_1'$, then the next flash will be emitted at time $t_2' = t_1' + \Delta t'$, as measured by a clock in the frame $S'$. Using Eq. (4.24) with $x_1' = x_2'$, the time interval

## 4.3 Time and Space in Special Relativity

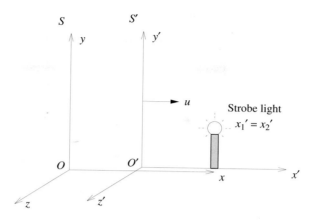

**Figure 4.6** A strobe light at rest ($x'$ = constant) in frame $S'$.

$\Delta t \equiv t_2 - t_1$ between the *same* two flashes measured by a clock in the frame $S$ is

$$t_2 - t_1 = \frac{(t_2' - t_1') + (x_2' - x_1')\,u/c^2}{\sqrt{1 - u^2/c^2}}$$

or

$$\Delta t = \frac{\Delta t'}{\sqrt{1 - u^2/c^2}}. \tag{4.26}$$

Because the clock in the frame $S'$ is *at rest* relative to the strobe light, $\Delta t'$ will be called $\Delta t_{\text{rest}}$. The frame $S'$ is called the clock's **rest frame**. Similarly, because the clock in the frame $S$ is *moving* relative to the strobe light, $\Delta t$ will be called $\Delta t_{\text{moving}}$. Thus Eq. (4.26) becomes

$$\Delta t_{\text{moving}} = \frac{\Delta t_{\text{rest}}}{\sqrt{1 - u^2/c^2}}. \tag{4.27}$$

This equation shows the effect of **time dilation** on a moving clock. It says that the time interval between two events is measured differently by different observers in relative motion. The *shortest time interval* is measured by a clock *at rest* relative to the two events. This clock measures the **proper time** between the two events. Any other clock moving relative to the two events will measure a longer time interval between them.

The effect of time dilation is often described by the phrase "moving clocks run slower" without explicitly identifying the two events involved. This easily leads to confusion, since the moving and rest subscripts in Eq. (4.27) mean "moving" or "at rest" *relative to the two events*. To gain insight into this

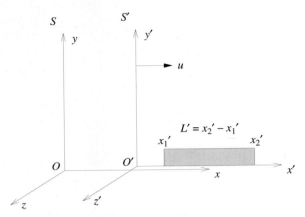

**Figure 4.7** A rod at rest in frame $S'$.

phrase, imagine that you are holding clock $C$ while it ticks once each second and at the same time are measuring the ticks of an identical clock $C'$ moving relative to you. The two events to be measured are consecutive ticks of clock $C'$. Since clock $C'$ is at rest relative to itself, it measures a time $\Delta t_{\text{rest}} = 1$ s between its own ticks. However, using your clock $C$, you measure a time

$$\Delta t_{\text{moving}} = \frac{\Delta t_{\text{rest}}}{\sqrt{1 - u^2/c^2}} = \frac{1 \text{ s}}{\sqrt{1 - u^2/c^2}} > 1 \text{ s}$$

between the ticks of clock $C'$. Because you measure clock $C'$ to be ticking slower than once per second, you conclude that clock $C'$, which is moving relative to you, is running more slowly than your clock $C$. Very accurate atomic clocks have been flown around the world on jet airliners and have confirmed that moving clocks do indeed run slower, in agreement with relativity.[5]

Both time dilation and the downfall of simultaneity contradict Newton's belief in absolute time. Instead, the time measured between two events differs for different observers in relative motion. Newton also believed that "absolute space, in its own nature, without relation to anything external, remains always similar and immovable." However, the Lorentz transformation equations require that different observers in relative motion will measure space differently as well.

Imagine that a rod lies along the $x'$-axis of the frame $S'$, at rest relative to that frame; $S'$ is the rod's rest frame (see Fig. 4.7). Let the left end of the rod have coordinate $x_1'$, and the right end of the rod have coordinate $x_2'$. Then

---
[5] See Hafele and Keating (1972a, 1972b) for the details of this test of time dilation.

## 4.3 Time and Space in Special Relativity

the length of the rod as measured in the frame $S'$ is $L' = x_2' - x_1'$. What is the length of the rod measured from $S$? Because the rod is moving relative to $S$, care must be taken to measure the $x$-coordinates $x_1$ and $x_2$ of the ends of the rod *at the same time*. Then Eq. (4.16), with $t_1 = t_2$, shows that the length $L = x_2 - x_1$ measured in $S$ may be found from

$$x_2' - x_1' = \frac{(x_2 - x_1) - u(t_2 - t_1)}{\sqrt{1 - u^2/c^2}}$$

or

$$L' = \frac{L}{\sqrt{1 - u^2/c^2}}. \tag{4.28}$$

Because the rod is *at rest* relative to $S'$, $L'$ will be called $L_{\text{rest}}$. Similarly, because the rod is *moving* relative to $S$, $L$ will be called $L_{\text{moving}}$. Thus Eq. (4.28) becomes

$$L_{\text{moving}} = L_{\text{rest}} \sqrt{1 - u^2/c^2}. \tag{4.29}$$

This equation shows the effect of **length contraction** on a moving rod. It says that length or distance is measured differently by two observers in relative motion. If a rod is moving relative to an observer, that observer will measure a shorter rod than will an observer at rest relative to it. The *longest length*, called the rod's **proper length**, is measured in the rod's rest frame. Only lengths or distances *parallel* to the direction of the relative motion are affected by length contraction; distances perpendicular to the direction of the relative motion are unchanged (c.f. Eqs. 4.17–4.18).

---

**Example 4.1**  Cosmic rays from space collide with the nuclei of atoms in Earth's upper atmosphere, producing elementary particles called *muons*. Muons are unstable and decay after an average lifetime $\tau = 2.20 \times 10^{-6}$ s, as measured in a laboratory where the muons are at rest. That is, the number of muons in a given sample should decrease with time according to $N(t) = N_0 e^{-t/\tau}$, where $N_0$ is the number of muons originally in the sample at time $t = 0$. At the top of Mt. Washington in New Hampshire, a detector counted 563 muons hr$^{-1}$ moving downward at a speed $u = 0.9952c$. At sea level, 1907 m below the first detector, another detector counted 408 muons hr$^{-1}$.[6]

The muons take $(1.907 \times 10^5 \text{ cm})/(0.9952c) = 6.39 \times 10^{-6}$ s to travel from the top of Mt. Washington to sea level. Thus it might be expected that the number of muons detected per hour at sea level would have been

$$N = N_0 e^{-t/\tau} = 563 e^{-(6.39 \times 10^{-6} \text{ s})/(2.20 \times 10^{-6} \text{ s})} = 31 \text{ muons hr}^{-1}.$$

---
[6]Details of this experiment can be found in Frisch and Smith (1963).

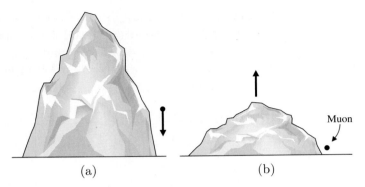

**Figure 4.8** Muons moving downward past Mt. Washington. (a) Mountain frame. (b) Muon frame.

This is much less than the 408 muons hr$^{-1}$ actually measured at sea level! How did the muons live long enough to reach the lower detector? The problem with the preceding calculation is that the lifetime of $2.20 \times 10^{-6}$ s is measured in the muon's rest frame as $\Delta t_{\text{rest}}$, but the experimenter's clocks on Mt. Washington and below are moving relative to the muons. They measure the muon's lifetime to be

$$\Delta t_{\text{moving}} = \frac{\Delta t_{\text{rest}}}{\sqrt{1 - u^2/c^2}} = \frac{2.20 \times 10^{-6} \text{ s}}{\sqrt{1 - (0.9952)^2}} = 2.25 \times 10^{-5} \text{ s},$$

more than *ten* times a muon's lifetime when measured in its own rest frame. The moving muons' clocks run slower, so more of them survive long enough to reach sea level. Repeating the preceding calculation using the muon lifetime as measured by the experimenters gives

$$N = N_0 \, e^{-t/\tau} = 563 \, e^{-(6.39 \times 10^{-6} \text{ s})/(2.25 \times 10^{-5} \text{ s})} = 424 \text{ muons hr}^{-1}.$$

When the effects of time dilation are included, the theoretical prediction is in excellent agreement with the experimental result.

From a muon's rest frame, its lifetime is only $2.20 \times 10^{-6}$ s. How would an observer riding along with the muons, as shown in Fig. 4.8, explain their ability to reach sea level? The observer would measure a severely length-contracted Mt. Washington (in the direction of the relative motion only). The distance traveled by the muons would not be $L_{\text{rest}} = 1907$ m, but rather

$$L_{\text{moving}} = L_{\text{rest}} \sqrt{1 - u^2/c^2} = 1907 \text{ m} \sqrt{1 - (0.9952)^2} = 186.6 \text{ m}.$$

Thus it would take $(1.866 \times 10^4 \text{ cm})/(0.9952c) = 6.25 \times 10^{-7}$ s for the muons to travel the length-contracted distance to the detector at sea level, as measured

## 4.3 Time and Space in Special Relativity

by an observer in the muons' rest frame. That observer would then calculate the number of muons reaching the lower detector to be

$$N = N_0 \, e^{-t/\tau} = 563 \, e^{-(6.25 \times 10^{-7} \text{ s})/(2.20 \times 10^{-6} \text{ s})} = 424 \text{ muons hr}^{-1},$$

in agreement with the previous result. This shows that an effect due to time dilation as measured in one frame may instead be attributed to length contraction as measured in another frame.

---

The effects of time dilation and length contraction are both symmetric between two observers in relative motion. Imagine two identical spaceships that move in opposite directions, passing each other at some relativistic speed. Observers onboard each spaceship will measure the other ship's length as being shorter than their own, and the other ship's clocks as running slower. *Both observers are right*, having made correct measurements from their respective frames of reference.

The reader should not think of these effects as being due to some sort of "optical illusion" caused by light taking different amounts of time to reach an observer from different parts of a moving object. The language used in the preceding discussions have involved the *measurement* of an event's spacetime coordinates $(x, y, z, t)$ using meter sticks and clocks located *at that event*, so there is no time delay. Of course, no actual laboratory has an infinite collection of meter sticks and clocks, and the time delays caused by finite light-travel times must be taken into consideration. This will be important in determining the relativistic Doppler shift formula, which follows.

In 1842 the Austrian physicist Christian Doppler showed that as a source of sound moves through a medium (such as air), the wavelength is compressed in the forward direction and expanded in the backward direction. This change in wavelength of any type of wave caused by the motion of the source or the observer is called a **Doppler shift**. Doppler deduced that the difference between the wavelength $\lambda_{\text{obs}}$ observed for a moving source of sound and the wavelength $\lambda_{\text{rest}}$ measured in the laboratory for a reference source at rest is related to the radial velocity $v_r$ (the component of the velocity directly toward or away from the observer; see Fig. 1.15) of the source through the medium by

$$\frac{\lambda_{\text{obs}} - \lambda_{\text{rest}}}{\lambda_{\text{rest}}} = \frac{\Delta \lambda}{\lambda_{\text{rest}}} = \frac{v_r}{v_s}, \qquad (4.30)$$

where $v_s$ is the speed of sound in the medium. However, this expression cannot be precisely correct for light. Experimental results such as those of Michelson and Morley led Einstein to abandon the ether concept, and they demonstrated

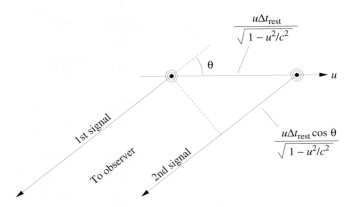

**Figure 4.9** Relativistic Doppler shift.

that no medium is involved in the propagation of light waves. The Doppler shift for light is a qualitatively different phenomenon than its counterpart for sound waves.

Consider a distant source of light that emits a light signal at time $t_{\text{rest},1}$ and another signal at time $t_{\text{rest},2} = t_{\text{rest},1} + \Delta t_{\text{rest}}$ as measured by a clock *at rest* relative to the source. If this light source is moving relative to an observer with velocity **u**, as shown in Fig. 4.9, then the time between receiving the light signals at the observer's location will depend on both the effect of time dilation and the different distances traveled by the signals from the source to the observer. (The light source is assumed to be sufficiently far away that the signals travel along parallel paths to the observer.) Using Eq. (4.27), we find that the time between the emission of the light signals as measured in the observer's frame is $\Delta t_{\text{rest}}/\sqrt{1 - u^2/c^2}$. In this time, the observer determines that the distance to the light source has changed by an amount $u\Delta t_{\text{rest}} \cos\theta / \sqrt{1 - u^2/c^2}$. Thus the time interval $\Delta t_{\text{obs}}$ between the arrival of the two light signals at the observer's location is

$$\Delta t_{\text{obs}} = \frac{\Delta t_{\text{rest}}}{\sqrt{1 - u^2/c^2}} \left[1 + (u/c)\cos\theta\right]. \tag{4.31}$$

If $\Delta t_{\text{rest}}$ is taken to be the time between the emission of the light wave crests, and $\Delta t_{\text{obs}}$ is the time between their arrival, then the frequencies of the light wave $\nu_{\text{rest}} = 1/\Delta t_{\text{rest}}$ and $\nu_{\text{obs}} = 1/\Delta t_{\text{obs}}$. The equation describing the **relativistic Doppler shift** is thus

$$\nu_{\text{obs}} = \frac{\nu_{\text{rest}}\sqrt{1 - u^2/c^2}}{1 + (u/c)\cos\theta} = \frac{\nu_{\text{rest}}\sqrt{1 - u^2/c^2}}{1 + v_r/c}, \tag{4.32}$$

where $v_r = u\cos\theta$ is the *radial velocity* of the light source. If the light source is moving directly away from the observer ($\theta = 0°$, $v_r = u$) or toward the observer ($\theta = 180°$, $v_r = -u$), then the relativistic Doppler shift reduces to

$$\nu_{\text{obs}} = \nu_{\text{rest}}\sqrt{\frac{1 - v_r/c}{1 + v_r/c}} \quad \text{(radial motion)}. \tag{4.33}$$

There is also a **transverse Doppler shift** for motion perpendicular to the observer's line of sight ($\theta = 90°$, $v_r = 0$). This transverse shift is entirely due to the effect of time dilation. Note that, unlike formulas describing the Doppler shift for sound, Eqs. (4.32) and (4.33) do not distinguish between the velocity of the source and the velocity of the observer. Only the relative velocity is important.

When astronomers observe a star or galaxy moving away from or toward Earth, the wavelength of the light they receive is shifted toward longer or shorter wavelengths, respectively. If the source of light is moving *away* from the observer ($v_r > 0$), then $\lambda_{\text{obs}} > \lambda_{\text{rest}}$. This shift to a longer wavelength is called a **redshift**. Similarly, if the source is moving *toward* the observer ($v_r < 0$), then there is a shift to a shorter wavelength, a **blueshift**.[7] Because most of the objects in the universe are moving away from us, redshifts are commonly measured by astronomers. A **redshift parameter** $z$ is used to describe the change in wavelength; it is defined as

$$z = \frac{\lambda_{\text{obs}} - \lambda_{\text{rest}}}{\lambda_{\text{rest}}} = \frac{\Delta\lambda}{\lambda_{\text{rest}}}. \tag{4.34}$$

The observed wavelength $\lambda_{\text{obs}}$ is obtained from Eq. (4.33) and $c = \lambda\nu$,

$$\lambda_{\text{obs}} = \lambda_{\text{rest}}\sqrt{\frac{1 + v_r/c}{1 - v_r/c}} \quad \text{(radial motion)}, \tag{4.35}$$

and the redshift parameter becomes

$$z = \sqrt{\frac{1 + v_r/c}{1 - v_r/c}} - 1 \quad \text{(radial motion)}. \tag{4.36}$$

In general, Eq. (4.34), together with $\lambda = \nu/c$, shows that

$$z + 1 = \frac{\Delta t_{\text{obs}}}{\Delta t_{\text{rest}}}. \tag{4.37}$$

---

[7]Doppler himself maintained that all stars would be white if they were at rest, and that the different colors of the stars are due to their Doppler shifts. However, the stars move much too slowly for their Doppler shifts to significantly change their colors.

This expression indicates that if the luminosity of an astrophysical source with redshift parameter $z > 0$ (receding) is observed to vary during a time $\Delta t_{\text{obs}}$, then the change in luminosity occurred over a *shorter* time $\Delta t_{\text{rest}} = \Delta t_{\text{obs}}/(z+1)$ in the rest frame of the source.

---

**Example 4.2** In its rest frame, the quasar PC 1247+3406 produces a hydrogen emission line of wavelength $\lambda_{\text{rest}} = 1216$ Å. On Earth, this emission line is observed to have a wavelength of $\lambda_{\text{obs}} = 7214$ Å. The redshift parameter for this quasar is thus

$$z = \frac{\lambda_{\text{obs}} - \lambda_{\text{rest}}}{\lambda_{\text{rest}}} = 4.93.$$

Using Eq. (4.36), we may calculate the speed of recession of the quasar:

$$z = \sqrt{\frac{1 + v_r/c}{1 - v_r/c}} - 1$$

$$\frac{v_r}{c} = \frac{(z+1)^2 - 1}{(z+1)^2 + 1} \tag{4.38}$$

$$= 0.945.$$

Quasar PC 1247+3406 is moving away from us at over 94% of the speed of light! See Schneider et al. (1991) for an account of the discovery of this quasar. As of this writing, this is the most distant object in the known universe; its redshift is $4.897 \pm 0.011$, an average calculated from several spectral lines.

---

Suppose the speed $u$ of the light source is small compared to that of light ($u/c \ll 1$). Using the expansion (to first order)

$$(1 + v_r/c)^{\pm 1/2} \simeq 1 \pm \frac{v_r}{2c}$$

together with Eqs. (4.34) and (4.35) for radial motion then shows for low speeds that

$$z = \frac{\Delta\lambda}{\lambda_{\text{rest}}} \simeq \frac{v_r}{c}, \tag{4.39}$$

where $v_r > 0$ for a receding source ($\Delta\lambda > 0$) and $v_r < 0$ for an approaching source ($\Delta\lambda < 0$). Although this equation is similar to Eq. (4.30), the reader should bear in mind that Eq. (4.39) is an *approximation*, valid only for low speeds. Misapplying this equation to Example 4.2 of the relativistic quasar PC 1247+3406 would lead to the erroneous conclusion that the quasar is moving away from us at 4.9 times the speed of light!

## 4.3 Time and Space in Special Relativity

Because space and time intervals are measured differently by different observers in relative motion, velocities must be transformed as well. The equations describing the relativistic transformation of velocities may be easily found from the Lorentz transformation equations (4.16–4.19) by writing them as differentials. Then dividing the $dx'$, $dy'$, and $dz'$ equations by the $dt'$ equation gives the **relativistic velocity transformations**:

$$v_x' = \frac{v_x - u}{1 - uv_x/c^2} \tag{4.40}$$

$$v_y' = \frac{v_y\sqrt{1 - u^2/c^2}}{1 - uv_x/c^2} \tag{4.41}$$

$$v_z' = \frac{v_z\sqrt{1 - u^2/c^2}}{1 - uv_x/c^2}. \tag{4.42}$$

As with the inverse Lorentz transformations, the inverse velocity transformations may be obtained by switching primed and unprimed quantities and by replacing $u$ with $-u$. It is left as an exercise to show that these equations do satisfy the second of Einstein's postulates: Light travels through a vacuum at a constant speed that is independent of the motion of the light source. From Eqs. (4.40)–(4.42), if **v** has a magnitude of $c$, so does **v**′ (see Problem 4.12).

---

**Example 4.3** As measured in the reference frame $S'$, a light source is at rest and radiates light equally in all directions. In particular, half of the light is emitted into the forward (positive $x'$) hemisphere. How is this situation measured from the frame $S$, which measures the light source traveling in the positive $x$-direction with a relativistic speed $u$?

Consider a light ray whose velocity components measured in $S'$ are $v_x' = 0$, $v_y' = c$, and $v_z' = 0$. This ray travels along the boundary between the forward and backward hemispheres of light as measured in $S'$. However, as measured in frame $S$, this light ray has the velocity components given by the inverse transformations of Eqs. (4.40)–(4.42):

$$v_x = \frac{v_x' + u}{1 + uv_x'/c^2} = u$$

$$v_y = \frac{v_y'\sqrt{1 - u^2/c^2}}{1 + uv_x'/c^2} = c\sqrt{1 - u^2/c^2}$$

$$v_z = \frac{v_z'\sqrt{1 - u^2/c^2}}{1 + uv_x'/c^2} = 0.$$

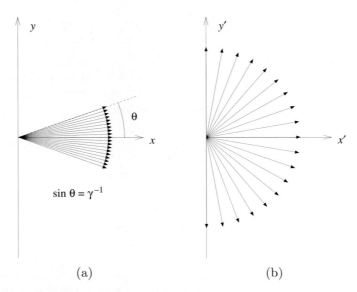

**Figure 4.10** Relativistic headlight effect. (a) Frame $S$. (b) Frame $S'$.

As measured in the frame $S$, the light ray is not traveling perpendicular to the $x$-axis; see Fig. 4.10.

In fact, for $u/c$ close to 1, the angle $\theta$ measured between the light ray and the $x$-axis may be found from $\sin\theta = v_y/v$, where

$$v = \sqrt{v_x^2 + v_y^2 + v_z^2} = c$$

is the speed of the light ray measured in the frame $S$. Thus

$$\sin\theta = \frac{v_y}{v} = \sqrt{1 - u^2/c^2} = \gamma^{-1}, \qquad (4.43)$$

where $\gamma$ is the Lorentz factor defined by Eq. (4.20). For relativistic speeds $u \approx c$, $\gamma$ is very large, so $\sin\theta$ (and hence $\theta$) become very small. All of the light emitted into the forward hemisphere, as measured in $S'$, is concentrated into a narrow cone in the direction of the light source's motion when measured in the frame $S$. Called the **headlight effect**, this result plays an important role in many areas of astrophysics. For example, as relativistic electrons spiral around magnetic field lines, they emit light in the form of **synchrotron radiation**. The radiation is concentrated in the direction of the electron's motion and is strongly plane-polarized. Synchrotron radiation is an important electromagnetic radiation process in the Sun, Jupiter's magnetosphere, pulsars, and active galaxies.

## 4.4 Relativistic Momentum and Energy

Up to this point, only relativistic kinematics has been considered. Einstein's theory of special relativity also requires new definitions for the concepts of momentum and energy. The ideas of conservation of linear momentum and energy are two of the cornerstones of physics. According to the Principle of Relativity, if momentum is conserved in one inertial frame of reference, then it must be conserved in all inertial frames. At the end of this section, it is shown that this requirement leads to a definition of the **relativistic momentum vector p**:

$$\mathbf{p} = \frac{m\mathbf{v}}{\sqrt{1 - v^2/c^2}} = \gamma m\mathbf{v}, \qquad (4.44)$$

where $\gamma$ is the Lorentz factor defined by Eq. (4.20). *Warning*: Some authors prefer to separate the "$m$" and the "$\mathbf{v}$" in this formula by defining a "relativistic mass," $m/\sqrt{1 - v^2/c^2}$. There is no compelling reason for this separation, and it can be misleading. In this book, the mass $m$ of a particle is measured to have the same value in *all* inertial reference frames; it is **invariant** under a Lorentz transformation, and so there is no reason to qualify the term as a "rest mass." Thus the mass of a moving particle *does not* increase with increasing speed, although its *momentum* approaches infinity as $v \to c$. Also note that the "$v$" in the denominator is the magnitude of the particle's velocity relative to the observer, *not* the relative velocity $u$ between two arbitrary frames of reference.

Using Eq. (4.44) and the relation between kinetic energy and work from Section 2.2, an expression for the relativistic kinetic energy can be derived. The starting point is Newton's second law, $\mathbf{F} = d\mathbf{p}/dt$, applied to a particle of mass $m$ that is initially at rest.[8] Consider a force of magnitude $F$ that acts on the particle in the $x$-direction. The particle's final kinetic energy $K$ equals the total work done by the force on the particle as it travels from its initial position $x_i$ to its final position $x_f$:

$$K = \int_{x_i}^{x_f} F\, dx = \int_{x_i}^{x_f} \frac{dp}{dt}\, dx$$

$$= \int_{p_i}^{p_f} \frac{dx}{dt}\, dp = \int_{p_i}^{p_f} v\, dp,$$

where $p_i$ and $p_f$ are the initial and final momenta of the particle, respectively.

---

[8]It is left as an exercise to show that $\mathbf{F} = m\mathbf{a}$ is *not correct*, since at relativistic speeds the force and the acceleration need *not* be in the same direction!

Integrating the last expression by parts and using the initial condition $p_i = 0$ gives

$$K = p_f v_f - \int_0^{v_f} p\, dv$$

$$= \frac{mv_f^2}{\sqrt{1 - v_f^2/c^2}} - \int_0^{v_f} \frac{mv}{\sqrt{1 - v^2/c^2}}\, dv$$

$$= \frac{mv_f^2}{\sqrt{1 - v_f^2/c^2}} + mc^2 \left( \sqrt{1 - v_f^2/c^2} - 1 \right).$$

If we drop the $f$ subscript, the expression for the **relativistic kinetic energy** becomes

$$K = mc^2 \left( \frac{1}{\sqrt{1 - v^2/c^2}} - 1 \right) = mc^2(\gamma - 1). \tag{4.45}$$

Although it is not apparent that this formula for the kinetic energy reduces to either of the familiar forms $K = \frac{1}{2}mv^2$ or $K = p^2/2m$ in the low-speed Newtonian limit, both forms must be true if Eq. (4.45) is to be correct. The proofs will be left as exercises.

The right-hand side of this expression for the kinetic energy consists of the difference between two energy terms. The first is identified as the **total relativistic energy** $E$,

$$E = \frac{mc^2}{\sqrt{1 - v^2/c^2}} = \gamma mc^2. \tag{4.46}$$

The second term is an energy that does not depend on the speed of the particle; the particle has this energy even when it is at rest. The term $mc^2$ is called the **rest energy** of the particle:

$$E_{\text{rest}} = mc^2. \tag{4.47}$$

The particle's kinetic energy is its total energy minus its rest energy. When the energy of a particle is given as (for example) 40 MeV, the implicit meaning is that the particle's *kinetic energy* is 40 MeV; the rest energy is not included. Finally, there is a very useful expression relating a particle's total energy $E$, the magnitude of its momentum $p$, and its rest energy $mc^2$. It states that

$$E^2 = p^2 c^2 + m^2 c^4. \tag{4.48}$$

## 4.4 Relativistic Momentum and Energy

As will be discussed in Section 5.2, this equation is valid even for particles that have no mass, such as photons.

For a *system* of $n$ particles, the total energy $E_{sys}$ of the system is the sum of the total energies $E_i$ of the individual particles: $E_{sys} = \sum_{i=1}^{n} E_i$. Similarly, the vector momentum $\mathbf{p}_{sys}$ of the system is the sum of the momenta $\mathbf{p}_i$ of the individual particles: $\mathbf{p}_{sys} = \sum_{i=1}^{n} \mathbf{p}_i$. If the momentum of the system of particles is conserved, then the total energy is also conserved, *even for inelastic collisions* in which the kinetic energy of the system, $K_{sys} = \sum_{i=1}^{n} K_i$, is reduced. The kinetic energy lost in the inelastic collisions goes into increasing the rest energy, and hence the mass, of the particles. This increase in rest energy allows the total energy of the system to be conserved. Mass and energy are two sides of the same coin; one can be transformed into the other.

---

**Example 4.4** In a one-dimensional completely inelastic collision, two identical particles of mass $m$ and speed $v$ approach each other, collide head-on, and merge to form a single particle of mass $M$. The initial energy of the system of particles is

$$E_{sys,i} = \frac{2mc^2}{\sqrt{1 - v^2/c^2}}.$$

Since the initial momenta of the particles are equal in magnitude and opposite in direction, the momentum of the system $\mathbf{p}_{sys} = 0$ before and after the collision. Thus after the collision, the particle is at rest and its final energy is

$$E_{sys,f} = Mc^2.$$

Equating the initial and final energies of the system shows that the mass $M$ of the conglomerate particle is

$$M = \frac{2m}{\sqrt{1 - v^2/c^2}}.$$

Thus the particle mass has increased by an amount

$$\Delta m = M - 2m = \frac{2m}{\sqrt{1 - v^2/c^2}} - 2m = 2m\left(\frac{1}{\sqrt{1 - v^2/c^2}} - 1\right).$$

The origin of this mass increase may be found by comparing the initial and final values of the kinetic energy. The initial kinetic energy of the system is

$$K_{sys,i} = 2mc^2\left(\frac{1}{\sqrt{1 - v^2/c^2}} - 1\right)$$

and the final kinetic energy $K_{\text{sys},f} = 0$. Dividing the kinetic energy lost in this inelastic collision by $c^2$ equals the particle mass increase, $\Delta m$.

---

*Optional*: To justify Eq. (4.44) for the relativistic momentum, we will consider a glancing elastic collision between two identical particles of mass $m$. This collision will be observed from three carefully chosen inertial reference frames, as shown in Fig. 4.11. When measured in an inertial reference frame $S''$, the two particles $A$ and $B$ have velocities and momenta that are equal in magnitude and opposite in direction, both before and after the collision. As a result, the total momentum must be zero both before and after the collision; momentum is conserved. This collision can also be measured from two other reference frames, $S$ and $S'$. From Fig. 4.11, if $S$ moves in the negative $x''$-direction with a velocity equal to the $x''$-component of particle $A$ in $S''$, then as measured from frame $S$, the velocity of particle $A$ only has a $y$-component. Similarly, if $S'$ moves in the positive $x''$-direction with a velocity equal to the $x''$-component of particle $B$ in $S''$, then as measured from frame $S'$, the velocity of particle $B$ only has a $y$-component. Actually, the figures for frames $S$ and $S'$ would be *identical* if the figures for one of these frames were rotated by 180° and the $A$ and $B$ labels were reversed. This means that the change in the $y$-component of particle $A$'s momentum as measured in frame $S$ is the same as the change in the $y'$-component of particle $B$'s momentum as measured in the frame $S'$, except for a change in sign (due to the 180° rotation): $\Delta p_{A,y} = -\Delta p'_{B,y}$. On the other hand, momentum must be conserved in frames $S$ and $S'$, just as it is in frame $S''$. This means that, measured in frame $S'$, the sum of the changes in the $y'$-components of particle $A$'s and $B$'s momenta must be zero: $\Delta p'_{A,y} + \Delta p'_{B,y} = 0$. Combining these results gives

$$\Delta p'_{A,y} = \Delta p_{A,y}. \tag{4.49}$$

So far, the argument has been independent of a specific formula for the relativistic momentum vector **p**. Let's assume that the relativistic momentum vector has the form $\mathbf{p} = fm\mathbf{v}$, where $f$ is a relativistic factor that depends on the *magnitude* of the particle's velocity, but not its direction.[9] As the particle's speed $v \to 0$, it is required that the factor $f \to 1$ to obtain agreement with the Newtonian result.

A second assumption allows the relativistic factor $f$ to be determined: the $y$- and $y'$-components of each particle's velocity are chosen to be arbitrarily

---

[9]There is no requirement that relativistic formulas appear similar to their low-speed Newtonian counterparts (c.f. Eq. 4.45). However, this simple argument produces the correct result.

## 4.4 Relativistic Momentum and Energy

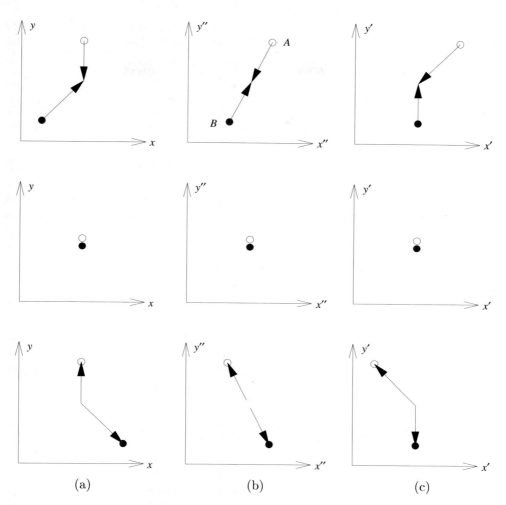

**Figure 4.11** An elastic collision measured in frames (a) $S$, (b) $S''$, and (c) $S'$. As observed from frame $S''$, the frame $S$ moves in the negative $x''$-direction, along with particle $A$, and the frame $S'$ moves in the positive $x''$-direction, along with particle $B$. For each reference frame, a vertical sequence of three figures shows the situation before (top), during, and after the collision.

small compared to the speed of light. Thus the $y$- and $y'$-components of particle $A$'s velocity in frames $S$ and $S'$ are extremely small, and the $x'$-component of particle $A$'s velocity in frame $S'$ is taken to be relativistic. Since $v_A' = \sqrt{v_{A,x}'^2 + v_{A,y}'^2} \approx c$ in frame $S'$, the relativistic factor $f'$ for particle $A$ in frame $S'$ is not equal to 1, whereas in frame $S$, $f$ is arbitrarily close to unity. If $v_{A,y}$ is the final $y$-component of particle $A$'s velocity, and similarly for $v_{A,y}'$, then Eq. (4.49) becomes

$$2f' m v_{A,y}' = 2m v_{A,y}. \tag{4.50}$$

The relative velocity $u$ of the frames $S$ and $S'$ is needed to relate $v_{A,y}'$ and $v_{A,y}$ using Eq. (4.41). Because $v_{A,x} = 0$ in frame $S$, Eq. (4.40) shows that $u = -v_{A,x}'$; that is, the relative velocity $u$ of the frame $S'$ relative to the frame $S$ is just the negative of the $x'$-component of particle $A$'s velocity in frame $S'$. Furthermore, because the $y'$-component of particle $A$'s velocity is arbitrarily small, we can set $v_{A,x}' = v_A'$, the magnitude of particle $A$'s velocity as measured in frame $S'$, and so use $u = -v_A'$. Inserting this into Eq. (4.41) with $v_{A,x} = 0$ gives

$$v_{A,y}' = v_{A,y}\sqrt{1 - v_A'^2/c^2}.$$

Finally, inserting this relation between $v_{A,y}'$ and $v_{A,y}$ into Eq. (4.50) and canceling terms reveals the relativistic factor $f$ to be

$$f = \frac{1}{\sqrt{1 - v_A'^2/c^2}},$$

as measured in frame $S'$. Dropping the prime superscript and the $A$ subscript (which merely identify the reference frame and particle involved) gives

$$f = \frac{1}{\sqrt{1 - v^2/c^2}}.$$

The formula for the **relativistic momentum vector** $\mathbf{p} = f m \mathbf{v}$ is thus

$$\mathbf{p} = \frac{m\mathbf{v}}{\sqrt{1 - v^2/c^2}} = \gamma m \mathbf{v}.$$

# Suggested Readings

### GENERAL

French, A. P. (ed.), *Einstein: A Centenary Volume*, Harvard University Press, Cambridge, MA, 1979.

Gardner, Martin, *The Relativity Explosion*, Vintage Books, New York, 1976.

### TECHNICAL

Bregman, Joel N. et al., "Multifrequency Observations of the Optically Violent Variable Quasar 3C 446," *The Astrophysical Journal*, *331*, 746, 1988.

Frisch, David H., and Smith, James H., "Measurement of the Relativistic Time Dilation Using $\mu$-Mesons," *American Journal of Physics*, *31*, 342, 1963.

Hafele, J. C., and Keating, Richard E., "Around-the-World Atomic Clocks: Predicted Relativistic Time Gains," *Science*, *177*, 166, 1972a.

Hafele, J. C., and Keating, Richard E., "Around-the-World Atomic Clocks: Observed Relativistic Time Gains," *Science*, *177*, 168, 1972b.

McCarthy, Patrick J. et al., "Serendipitous Discovery of a Redshift 4.4 QSO," *The Astrophysical Journal Letters*, *328*, L29, 1988.

Resnick, Robert, and Halliday, David, *Basic Concepts in Relativity and Early Quantum Theory*, Second Edition, John Wiley and Sons, New York, 1985.

Schneider, Donald P., Schmidt, Maarten, and Gunn, James E., "PC 1247+3406: An Optically Selected Quasar with a Redshift of 4.897," *The Astronomical Journal*, *102*, 837, 1991.

Taylor, Edwin F., and Wheeler, John A., *Spacetime Physics*, Second Edition, W. H. Freeman, San Francisco, 1992.

# Problems

**4.1** Use Eqs. (4.14) and (4.15) to derive the Lorentz transformation equations from Eqs. (4.10)–(4.13).

**4.2** Because there is no such thing as absolute simultaneity, two observers in relative motion may disagree on which of two events $A$ and $B$ occurred first. Suppose, however, that an observer in reference frame $S$ measures that event $A$ occurred first and *caused* event $B$. For example, event $A$ might be pushing a light switch, and event $B$ might be a light bulb turning on. Prove that an observer in another frame $S'$ cannot measure event $B$ (the effect) occurring before event $A$ (the cause). The temporal order of cause and effect are preserved by the Lorentz transformation equations. *Hint:* For event $A$ to cause event $B$, information must have traveled from $A$ to $B$, and the fastest that *anything* can travel is the speed of light.

**4.3** Consider the special *light clock* shown in Fig. 4.12. The light clock is at rest in the frame $S'$ and consists of two perfectly reflecting mirrors separated by a vertical distance $d$. As measured by an observer in frame $S'$, a light pulse bounces vertically back and forth between the two mirrors; the time interval between the pulse leaving and subsequently returning to the bottom mirror is $\Delta t'$. However, an observer in the frame $S$ sees a moving clock and determines that the time interval between the light pulse leaving and returning to the bottom mirror is $\Delta t$. Use the fact that both observers must measure that the light pulse moves with speed $c$, plus some simple geometry, to derive the time-dilation equation (4.27).

**4.4** A rod moving relative to an observer is measured to have its length $L_{\text{moving}}$ contracted to one-half of its length when measured at rest. Find the value of $u/c$ for the rod's rest frame relative to the observer's frame of reference.

**4.5** An observer $P$ stands on a train station platform as a high-speed train passes by at $u/c = 0.8$. The observer $P$, who measures the platform to be 60 m long, notices that the front and back ends of the train line up exactly with the ends of the platform at the same time.

(a) How long does it take the train to pass $P$ as he stands on the platform, as measured by his watch?

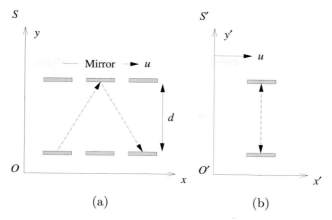

**Figure 4.12** (a) A light clock that is moving in frame $S$, and (b) at rest in frame $S'$.

    (b) According to a rider $T$ on the train, how long is the train?

    (c) According to a rider $T$ on the train, what is the length of the train station platform?

    (d) According to a rider $T$ on the train, how much time does it take for the train to pass observer $P$ standing on the train station platform?

    (e) According to a rider $T$ on the train, the ends of the train will *not* simultaneously line up with the ends of the platform. What time interval does $T$ measure between when the front end of the train lines up with the front end of the platform, and when the back end of the train lines up with the back end of the platform?

4.6   An astronaut in a starship travels to $\alpha$ Centauri, a distance of approximately 4 ly as measured from Earth, at a speed of $u/c = 0.8$.

    (a) How long does the trip to $\alpha$ Centauri take, as measured by a clock on Earth?

    (b) How long does the trip to $\alpha$ Centauri take, measured by the starship pilot?

    (c) What is the distance between Earth and $\alpha$ Centauri, measured by the starship pilot?

    (d) A radio signal is sent from Earth to the starship every 6 months, as measured by a clock on Earth. What is the time interval between the reception of the signals aboard the starship?

(e) A radio signal is sent from the starship to Earth every 6 months, as measured by a clock aboard the starship. What is the time interval between the reception of the signals on Earth?

(f) If the wavelength of the radio signal sent from Earth is $\lambda = 15$ cm, to what wavelength must the starship's receiver be tuned?

4.7 Upon reaching $\alpha$ Centauri, the starship in Problem 4.6 immediately reverses direction and travels back to Earth at a speed of $u/c = 0.8$. (Assume that the turnaround itself takes *zero* time.) Both Earth and the starship continue to emit radio signals at 6-month intervals, as measured by their respective clocks. Make a table for the entire trip showing at what times Earth receives the signals from the starship. Do the same for the times when the starship receives the signals from Earth. Thus an Earth observer and the starship pilot will agree that the pilot has aged 4 years less than the Earth observer during the round-trip voyage to $\alpha$ Centauri.

4.8 In its rest frame, quasar Q2203+29 produces a hydrogen emission line of wavelength 1216 Å. Astronomers on Earth measure a wavelength of 6568 Å for this line. Determine the redshift parameter and the speed of recession for this quasar. (For more information about this quasar, see McCarthy et al. 1988.)

4.9 Quasar 3C 446 is violently variable; its luminosity at optical wavelengths has been observed to change by a factor of 40 in as little as 10 days. Using the redshift parameter $z = 1.404$ measured for 3C 446, determine the time for the luminosity variation as measured in the quasar's rest frame. (For more details, see Bregman et al. 1988.)

4.10 Use the Lorentz transformation equations (4.16)–(4.19) to derive the velocity transformation equations (4.40)–(4.42).

4.11 The *spacetime interval*, $\Delta s$, between two events with coordinates

$$(x_1, y_1, z_1, t_1) \quad \text{and} \quad (x_2, y_2, z_2, t_2)$$

is defined by

$$(\Delta s)^2 = (c\Delta t)^2 - (\Delta x)^2 - (\Delta y)^2 - (\Delta z)^2.$$

(a) Use the Lorentz transformation equations (4.16)–(4.19) to show that $\Delta s$ has the same value in all reference frames. The spacetime interval is said to be *invariant* under a Lorentz transformation.

(b) If $(\Delta s)^2 > 0$, then the interval is *timelike*. Show that in this case,

$$\Delta \tau \equiv \frac{\Delta s}{c}$$

is the proper time between the two events. Assuming that $t_1 < t_2$, could the first event have possibly caused the second event?

(c) If $(\Delta s)^2 = 0$, then the interval is *lightlike* or *null*. Show that only light could have traveled between the two events. Could the first event have possibly caused the second event?

(d) If $(\Delta s)^2 < 0$, then the interval is *spacelike*. What is the physical significance of $\sqrt{-(\Delta s)^2}$? Could the first event have possibly caused the second event?

The concept of a spacetime interval will play a key role in the discussion of general relativity in Chapter 16.

4.12 General expressions for the components of a light ray's velocity as measured in reference frame $S$ are

$$v_x = c \sin\theta \cos\phi$$
$$v_y = c \sin\theta \sin\phi$$
$$v_z = c \cos\theta,$$

where $\theta$ and $\phi$ are the angular coordinates in a spherical coordinate system.

(a) Show that
$$v = \sqrt{v_x^2 + v_y^2 + v_z^2} = c.$$

(b) Use the velocity transformation equations to show that, as measured in reference frame $S'$,

$$v' = \sqrt{v_x'^2 + v_y'^2 + v_z'^2} = c,$$

and so confirm that the speed of light has the constant value $c$ in all frames of reference.

4.13 Starship $A$ moves away from Earth with a speed of $v_A/c = 0.8$. Starship $B$ moves away from Earth in the opposite direction with a speed of $v_B/c = 0.6$. What is the speed of starship $A$ as measured by starship $B$? What is the speed of starship $B$ as measured by starship $A$?

**4.14** Use Newton's second law $\mathbf{F} = d\mathbf{p}/dt$ and the formula for relativistic momentum, Eq. (4.44), to show that the acceleration vector $\mathbf{a} = d\mathbf{v}/dt$ produced by a force $\mathbf{F}$ acting on a particle of mass $m$ is

$$\mathbf{a} = \frac{\mathbf{F}}{\gamma m} - \frac{\mathbf{v}}{\gamma m c^2}(\mathbf{F} \cdot \mathbf{v}),$$

where $\mathbf{F} \cdot \mathbf{v}$ is the vector dot product between the force $\mathbf{F}$ and the particle velocity $\mathbf{v}$. Thus the acceleration depends on the particle's velocity and is not in general in the same direction as the force.

**4.15** Suppose a constant force of magnitude $F$ acts on a particle of mass $m$ initially at rest.

(a) Integrate the formula for the acceleration found in Problem 4.14 to show that the speed of the particle after time $t$ is given by

$$\frac{v}{c} = \frac{(F/m)t}{\sqrt{(F/m)^2 t^2 + c^2}}.$$

(b) Rearrange this equation to express the time $t$ as a function of $v/c$. If the particle's initial acceleration at time $t = 0$ is $a = g = 980$ cm s$^{-2}$, how much time is required for the particle to reach a speed of $v/c = 0.9$? $v/c = 0.99$? $v/c = 0.999$? $v/c = 0.9999$? $v/c = 1$?

**4.16** Find the value of $v/c$ when a particle's kinetic energy equals its rest energy.

**4.17** Prove that in the low-speed Newtonian limit of $v/c \ll 1$, Eq. (4.45) does reduce to the familiar form $K = \frac{1}{2}mv^2$.

**4.18** Show that the relativistic kinetic energy of a particle can be written as

$$K = \frac{p^2}{(1+\gamma)m},$$

where $p$ is the magnitude of the particle's relativistic momentum. This demonstrates that in the low-speed Newtonian limit of $v/c \ll 1$, $K = p^2/2m$ (as expected).

**4.19** Derive Eq. (4.48).

*Chapter 5*

# THE INTERACTION OF LIGHT AND MATTER

## 5.1 Spectral Lines

In 1835 a French philosopher, Auguste Comte, considered the limits of human knowledge. In his book *Positive Philosophy*, Comte wrote of the stars, "We see how we may determine their forms, their distances, their bulk, their motions, but we can never know anything of their chemical or mineralogical structure." Thirty-three years earlier, however, William Wollaston, like Newton before him, passed sunlight through a prism to produce a rainbowlike spectrum. He discovered that a number of dark **spectral lines** were superimposed on the continuous spectrum where the Sun's light had been absorbed at certain discrete wavelengths. By 1814, the German optician Joseph Fraunhofer had cataloged 475 of these dark lines (today called *Fraunhofer lines*) in the solar spectrum. While measuring the wavelengths of these lines, Fraunhofer made the first observation capable of proving Comte wrong. Fraunhofer determined that the wavelength of one prominent dark line in the Sun's spectrum corresponds to the wavelength of the yellow light emitted when salt is sprinkled in a flame. The new science of *spectroscopy* was born with the identification of this sodium line.

The foundations of spectroscopy were established by a German chemist, Robert Bunsen, and a Prussian theoretical physicist, Gustav Kirchhoff. Bunsen's burner produced a colorless flame that was ideal for studying the spectra of heated substances. He and Kirchhoff then designed a *spectroscope* that passed the light of a flame spectrum through a prism to be analyzed. The wavelengths of light absorbed and emitted by an element were found to be the *same*; Kirchhoff determined that 70 dark lines in the solar spectrum correspond to 70 bright lines emitted by iron vapor. In 1860 Kirchhoff and Bunsen published their classic work, *Chemical Analysis by Spectral Observations*, in which they developed the idea that every element produces its own pattern of spectral lines and thus may be identified by its unique spectral line "fingerprint." Kirchhoff summarized the production of spectral lines in three laws, which are now known as **Kirchhoff's laws**:

- A hot, dense gas or hot solid object produces a continuous spectrum with no dark spectral lines.[1]

- A hot, diffuse gas produces bright spectral lines (**emission lines**).

- A cool, diffuse gas in front of a source of a continuous spectrum produces dark spectral lines (**absorption lines**) in the continuous spectrum.

An immediate application of these results was the identification of elements found in the Sun and other stars. A new element previously unknown on Earth, *helium*,[2] was discovered spectroscopically on the Sun in 1868; it was not found on Earth until 1895. Figure 5.1 shows the visible portion of the solar spectrum and Table 5.1 lists some of the elements responsible for producing the dark absorption lines.

Another rich line of investigation was pursued by measuring the Doppler shifts of spectral lines. For individual stars, $v_r \ll c$, and so the *low-speed approximation* of Eq. (4.39),

$$\frac{\lambda_{\text{obs}} - \lambda_{\text{rest}}}{\lambda_{\text{rest}}} = \frac{\Delta\lambda}{\lambda_{\text{rest}}} = \frac{v_r}{c}, \tag{5.1}$$

can be utilized to determine their radial velocities. By 1887 the radial velocities of Sirius, Procyon, Rigel, and Arcturus had been measured with an accuracy of a few km s$^{-1}$.

---

**Example 5.1** The rest wavelength $\lambda_{\text{rest}}$ for an important spectral line of hydrogen (known as H$_\alpha$) is 6562.80 Å. However, the wavelength of the H$_\alpha$ absorption line in the spectrum of the star Vega in the constellation Lyra is

---

[1] In the first of Kirchhoff's laws, "hot" actually means any temperature above 0 K. However, according to Wien's displacement law (3.19), a temperature of several thousand degrees K is required for $\lambda_{\max}$ to fall in the visible portion of the electromagnetic spectrum. As will be discussed in Chapter 9, it is the opacity or *optical depth* of the gas that is responsible for the continuous blackbody spectrum.

[2] The name "helium" comes from *Helios*, a Greek sun god.

## 5.1 Spectral Lines

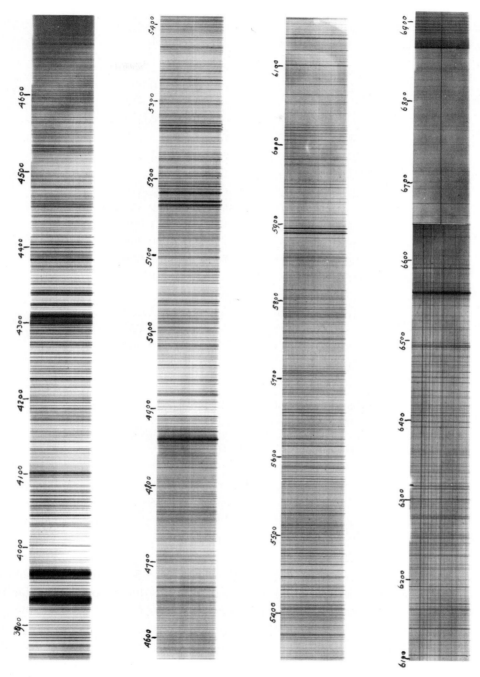

**Figure 5.1** Solar spectrum with Fraunhofer lines. (Courtesy of The Observatories of the Carnegie Institution of Washington.)

| Wavelength (Å) | Name | Atom | Equivalent Width (Å) |
|---|---|---|---|
| 3859.922 |  | Fe I | 1.554 |
| 3886.294 |  | Fe I | 0.920 |
| 3905.532 |  | Si I | 0.816 |
| 3933.682 | K | Ca II | 20.253 |
| 3968.492 | H | Ca II | 15.467 |
| 4045.825 |  | Fe I | 1.174 |
| 4101.748 | h, $H_\delta$ | H I | 3.133 |
| 4226.740 | g | Ca I | 1.476 |
| 4340.475 | G', $H_\gamma$ | H I | 2.855 |
| 4383.557 | d | Fe I | 1.008 |
| 4404.761 |  | Fe I | 0.898 |
| 4861.342 | F, $H_\beta$ | H I | 3.680 |
| 5167.327 | $b_4$ | Mg I | 0.935 |
| 5172.698 | $b_2$ | Mg I | 1.259 |
| 5183.619 | $b_1$ | Mg I | 1.584 |
| 5889.973 | $D_2$ | Na I | 0.752 |
| 5895.940 | $D_1$ | Na I | 0.564 |
| 6562.808 | C, $H_\alpha$ | H I | 4.020 |

**Table 5.1** Wavelengths of the Strong Fraunhofer Lines. The atomic notation is explained in Section 8.1, and the equivalent width of a spectral line is defined in Section 9.4. (Data from Lang, *Astrophysical Formulae*, Second Edition, Springer-Verlag, New York, 1980.)

measured to be 6562.50 Å. Equation (5.1) shows that the radial velocity of Vega is

$$v_r = \frac{c(\lambda_{\text{obs}} - \lambda_{\text{rest}})}{\lambda_{\text{rest}}} = -1.4 \times 10^6 \text{ cm s}^{-1} = -14 \text{ km s}^{-1};$$

the minus sign means that Vega is approaching the Sun. Recall from Section 1.3, however, that stars also have a *proper motion*, $\mu$, perpendicular to the line of sight. Vega's angular position in the sky changes by $\mu = 0.345''$ yr$^{-1}$. At a distance of $r = 8.0$ pc, this proper motion is related to the star's transverse velocity, $v_\theta$, by Eq. (1.4). Expressing $r$ in cm and $\mu$ in rad s$^{-1}$ results in

$$v_\theta = r\mu = 1.3 \times 10^6 \text{ cm s}^{-1}.$$

This transverse velocity of 13 km s$^{-1}$ is comparable to Vega's radial velocity.

## 5.1 Spectral Lines

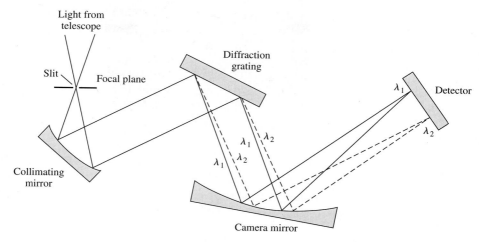

**Figure 5.2** Spectrograph.

Vega's speed through space relative to the Sun is thus

$$v = \sqrt{v_r^2 + v_\theta^2} = 19 \text{ km s}^{-1}.$$

The average speed of stars in the solar neighborhood is about 25 km s$^{-1}$. In reality, the measurement of a star's radial velocity is complicated by the 29.8 km s$^{-1}$ motion of Earth around the Sun, which causes the observed wavelength $\lambda_{\text{obs}}$ of a spectral line to vary sinusoidally over the course of a year. This effect of Earth's speed may be easily compensated for by subtracting the component of Earth's orbital velocity along the line of sight from the star's measured radial velocity.

---

Modern methods can measure radial velocities with an accuracy of nearly ±10 m s$^{-1}$! Today astronomers use *spectrographs* to measure the spectra of stars and galaxies; see Fig. 5.2.[3] After passing through a narrow slit, the starlight is collimated by a mirror and directed onto a *diffraction grating*. A diffraction grating is a piece of glass onto which narrow, closely spaced lines have been evenly ruled (typically several thousand lines per millimeter); the grating may be made to transmit the light (a *transmission grating*) or reflect the light (a *reflection grating*). In either case, the grating acts like a long series of neighboring double slits. Different wavelengths of light have their maxima

---

[3] As will be discussed in Chapter 7, measuring the radial velocities of stars in binary star systems allows the *masses* of the stars to be determined.

occurring at different angles $\theta$ given by Eq. (3.11):
$$d \sin \theta = n\lambda \quad (n = 0, 1, 2, \ldots),$$
where $d$ is the distance between adjacent lines of the grating, $n$ is the order of the spectrum, and $\theta$ is measured from the line normal (or perpendicular) to the grating. ($n = 0$ corresponds to $\theta = 0$ for all wavelengths, so the light is not dispersed into a spectrum in this case.) The spectrum is then focused onto a photographic plate or electronic detector for recording.

The ability of a spectrograph to resolve two closely spaced wavelengths separated by an amount $\Delta\lambda$ depends on the order of the spectrum, $n$, and the total number of lines of the grating that are illuminated, $N$. The smallest difference in wavelength that the grating can resolve is
$$\Delta\lambda = \frac{\lambda}{nN}, \tag{5.2}$$
where $\lambda$ is either of the closely spaced wavelengths being measured. The ratio $\lambda/\Delta\lambda$ is the *resolving power* of the grating.[4]

Astronomers recognized the great potential for uncovering the secrets of the stars in the empirical rules that had been obtained for the spectrum of light: Wien's law, the Stefan–Boltzmann equation, Kirchhoff's laws, and the new science of spectroscopy. By 1880 Gustav Wiedemann found that a detailed investigation of the Fraunhofer lines could reveal the temperature, pressure, and density of the layer of the Sun's atmosphere that produces the lines. The splitting of spectral lines by a magnetic field was discovered by Pieter Zeeman of the Netherlands in 1897, raising the possibility of measuring stellar magnetic fields. But a serious problem blocked further progress: However impressive, these results lacked the solid theoretical foundation required for the interpretation of stellar spectra. For example, the absorption lines produced by hydrogen are much stronger for Vega than for the Sun. Does this mean that Vega's composition contains significantly more hydrogen than the Sun's? The answer is no, but how can this information be gleaned from the dark absorption lines of a stellar spectrum recorded on a photographic plate? The answer required a new understanding of the nature of light itself.

## 5.2 Photons

Despite Heinrich Hertz's absolute certainty in the wave nature of light, the solution to the riddle of the continuous spectrum of blackbody radiation led

---

[4]In some cases, the resolving power of a spectrograph may be determined by other factors, for example, the slit width.

to a complementary description, and ultimately to new conceptions of matter and energy. Planck's constant $h$ (see Section 3.5) is the basis of the modern description of matter and energy known as **quantum mechanics**. Today $h$ is recognized as a fundamental constant of nature, like the speed of light $c$ and the universal gravitational constant $G$. Although Planck himself was uncomfortable with the implications of his discovery of energy quantization, quantum theory was to develop into a spectacularly successful description of the physical world. The next step forward was taken by Einstein, who convincingly demonstrated the reality of Planck's quantum bundles of energy.

When light shines on a metal surface, electrons are ejected from the surface, a result called the **photoelectric effect**. The electrons are emitted with a range of energies, but those originating closest to the surface have the maximum kinetic energy, $K_{max}$. A surprising feature of the photoelectric effect is that the value of $K_{max}$ does *not* depend on the brightness of the light shining on the metal. Increasing the intensity of a monochromatic light source will eject *more* electrons but will not increase their maximum kinetic energy. Instead, $K_{max}$ varies with the *frequency* of the light illuminating the metal surface. In fact, each metal has a characteristic *cutoff frequency* $\nu_c$ (and *cutoff wavelength* $\lambda_c = c/\nu_c$); electrons will be emitted only if the frequency $\nu$ of the light satisfies $\nu > \nu_c$ (or the wavelength satisfies $\lambda < \lambda_c$). This puzzling frequency dependence is nowhere to be found in Maxwell's classic description of electromagnetic waves. Equation (3.12) for the Poynting vector admits no role for the frequency in describing the energy carried by a light wave.

Einstein's bold solution was to take seriously Planck's assumption of the quantized energy of electromagnetic waves. According to Einstein's explanation of the photoelectric effect, the light striking the metal surface consists of a stream of massless particles called **photons**.[5] The energy of a single photon of frequency $\nu$ and wavelength $\lambda$ is just the energy of Planck's quantum of energy:

$$E_{\text{photon}} = h\nu = \frac{hc}{\lambda}. \tag{5.3}$$

---

**Example 5.2** The energy of a single photon of visible light is small by everyday standards. For red light of wavelength $\lambda = 7000$ Å, the energy of a single photon is

$$E_{\text{photon}} = \frac{hc}{\lambda} = \frac{12400 \text{ eV Å}}{7000 \text{ Å}} = 1.77 \text{ eV}.$$

---

[5] Only a massless particle can move with the speed of light, since a massive particle would have infinite energy; see Eq. (4.45). The term photon was first used in 1926 by the physicist G. N. Lewis.

Here, the product $hc$ has been expressed in the convenient units of (electron volts) × (angstroms); recall that $1 \text{ eV} = 1.602 \times 10^{-12}$ erg. For a single photon of blue light ($\lambda = 4000$ Å),

$$E_{\text{photon}} = \frac{hc}{\lambda} = \frac{12400 \text{ eV Å}}{4000 \text{ Å}} = 3.10 \text{ eV}.$$

How many visible photons ($\lambda = 5000$ Å) are emitted each second by a 100-watt light bulb (assuming that it is monochromatic)? The energy of each photon is

$$E_{\text{photon}} = \frac{hc}{\lambda} = \frac{12400 \text{ eV Å}}{5000 \text{ Å}} = 2.48 \text{ eV} = 3.97 \times 10^{-12} \text{ erg}.$$

Using 100 watts = $10^9$ erg s$^{-1}$, this means that the light bulb emits $2.52 \times 10^{20}$ photons per second. As this huge number illustrates, with so many photons nature does not appear "grainy." We see the world as a continuum of light, illuminated by a flood of photons.

---

Einstein reasoned that when a photon strikes the metal surface in the photoelectric effect, its energy may be absorbed by a single electron. The electron uses the photon's energy to overcome the binding energy of the metal and so escape from the surface. If the *minimum* binding energy of electrons in a metal (called the **work function** of the metal, usually a few eV) is $\phi$, then the maximum kinetic energy of the ejected electrons is

$$K_{\text{max}} = E_{\text{photon}} - \phi = h\nu - \phi = \frac{hc}{\lambda} - \phi. \tag{5.4}$$

Setting $K_{\text{max}} = 0$, the cutoff frequency and wavelength for a metal are seen to be $\nu_c = \phi/h$ and $\lambda_c = hc/\phi$, respectively.

The photoelectric effect established the reality of Planck's quanta. Albert Einstein was awarded the 1921 Nobel Prize, not for his theories of special and general relativity, but "for his services to theoretical physics, and especially for his discovery of the law of the photoelectric effect."[6] Today astronomers take advantage of the quantum nature of light in various instruments and detectors, such as the CCDs (charge-coupled devices) that will be described in Section 6.2.

In 1922, the American physicist Arthur Holly Compton provided the most convincing evidence that light does in fact manifest its particlelike nature when

---

[6] Partly in recognition of his determination of an accurate value of Planck's constant $h$ using Eq. (5.4), the American physicist Robert A. Millikan also received a Nobel Prize (1923) for the photoelectric effect.

## 5.2 Photons

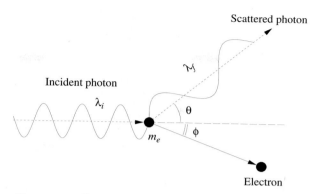

**Figure 5.3** The Compton effect: The scattering of a photon by a free electron.

interacting with matter. Compton measured the change in the wavelength of x-ray photons as they were scattered by free electrons. Because photons are massless particles that move at the speed of light, the relativistic energy equation, Eq. (4.48) (with mass $m = 0$ for photons), shows that the energy of a photon is related to its momentum $p$ by

$$E_{\text{photon}} = h\nu = \frac{hc}{\lambda} = pc. \tag{5.5}$$

Compton considered the "collision" between a photon and a free electron, initially at rest. As shown in Fig. 5.3, the electron recoils in the direction $\phi$. The photon is scattered by an angle $\theta$. Because the photon has lost energy to the electron, the wavelength of the photon has increased.

In this collision, both (relativistic) momentum and energy are conserved. It is left as an exercise to show that the final wavelength of the photon, $\lambda_f$, is greater than its initial wavelength, $\lambda_i$, by an amount

$$\Delta\lambda = \lambda_f - \lambda_i = \frac{h}{m_e c}(1 - \cos\theta), \tag{5.6}$$

where $m_e$ is the mass of the electron. Today, this change in wavelength is known as the **Compton effect**. The term $h/m_e c$ in Eq. (5.6), called the *Compton wavelength*, $\lambda_C$, is the characteristic change in the wavelength of the scattered photon and has the value $\lambda_C = 0.0243$ Å, 30 times smaller than the wavelength of the x-ray photons used by Compton. Compton's experimental verification of this formula provided convincing evidence that photons are indeed massless particles that nonetheless carry momentum, as described by Eq. (5.5). This is the physical basis for the force exerted by radiation upon matter, already discussed in Section 3.3 in terms of radiation pressure.

## 5.3 The Bohr Model of the Atom

The pioneering work of Planck, Einstein, and others at the beginning of the twentieth century revealed the **wave–particle duality** of light. Light exhibits its wave properties as it *propagates* through space, as demonstrated by its double-slit interference pattern. On the other hand, light manifests its particle nature when it *interacts* with matter, as in the photoelectric and Compton effects. Planck's formula describing the energy distribution of blackbody radiation explained many of the features of the continuous spectrum of light emitted by stars. But what physical process was responsible for the dark absorption lines scattered throughout the continuous spectrum of a star, or the bright emission lines produced by a hot diffuse gas in the laboratory?

In the very last years of the nineteenth century, working at Cambridge University's Cavendish Laboratory, Joseph John Thomson (1856–1940) had discovered the **electron**. Because bulk matter is electrically neutral, atoms were deduced to consist of negatively charged electrons and an equal positive charge of uncertain distribution. Ernest Rutherford (1871–1937) of New Zealand, working at England's University of Manchester, discovered in 1911 that an atom's positive charge was concentrated in a tiny, massive nucleus. Rutherford directed high-speed alpha particles (now known to be helium nuclei) onto thin metal foils. He was amazed to observe that a few of the alpha particles were bounced backward by the foils, instead of plowing through them with only a slight deviation. Rutherford later wrote: "It was quite the most incredible event that has ever happened to me in my life. It was almost as incredible as if you fired a 15-inch shell at a piece of tissue paper and it came back and hit you." Such an event could only occur as the result of a single collision of the alpha particle with a minute, massive, positively charged nucleus. Rutherford calculated that the radius of the nucleus was 10,000 times smaller than the radius of the atom itself, showing that ordinary matter is mostly empty space! He established that an electrically neutral atom consists of $Z$ electrons (where $Z$ is an integer), with $Z$ positive elementary charges confined to the nucleus. Rutherford coined the term **proton** to refer to the nucleus of the hydrogen atom ($Z = 1$), 1836 times more massive than the electron. But how were these charges arranged?

The experimental data were abundant. The wavelengths of 14 spectral lines of hydrogen had been precisely determined. Those in the visible region of the electromagnetic spectrum are 6563 Å (red, known as $H_\alpha$); 4861 Å (turquoise, $H_\beta$); 4340 Å (blue, $H_\gamma$); and 4102 Å (violet, $H_\delta$). In 1885, a Swiss school teacher, Johann Balmer, had found, by trial and error, a formula to reproduce

## 5.3 The Bohr Model of the Atom

the wavelengths of these spectral lines of hydrogen, today called the **Balmer series** or **Balmer lines**:

$$\frac{1}{\lambda} = R_H \left( \frac{1}{4} - \frac{1}{n^2} \right), \tag{5.7}$$

where $n = 3, 4, 5, \ldots$, and $R_H = 109677.585 \pm 0.008$ cm$^{-1}$ is the experimentally determined Rydberg constant for hydrogen.[7] Balmer's formula was very accurate, to within a fraction of a percent. Inserting $n = 3$ gives the wavelength of the H$_\alpha$ Balmer line, $n = 4$ gives H$_\beta$, and so on. Furthermore, Balmer realized that since $2^2 = 4$, his formula could be generalized to

$$\frac{1}{\lambda} = R_H \left( \frac{1}{m^2} - \frac{1}{n^2} \right), \tag{5.8}$$

with $m < n$ (both integers). Many nonvisible spectral lines of hydrogen were found later, just as Balmer had predicted. Today, the lines corresponding to $m = 1$ are called *Lyman lines*. The Lyman series of lines is found in the ultraviolet region of the electromagnetic spectrum. Similarly, inserting $m = 3$ into Eq. (5.8) produces the wavelengths of the *Paschen* series of lines, which lie entirely in the infrared portion of the spectrum.

Yet all of this was sheer numerology, with no foundation in the physics of the day. Physicists were frustrated by their inability to construct a model of even this simplest of atoms. A planetary model of the hydrogen atom, consisting of a central proton and one electron held together by their mutual electrical attraction, should have been most amenable to analysis. However, a single electron and proton moving around their common center of mass suffers from a basic instability. According to Maxwell's equations of electricity and magnetism, an accelerating electric charge emits electromagnetic radiation. The orbiting electron should thus lose energy by emitting light with a continuously increasing frequency (the orbital frequency) as it spirals down into the nucleus. This theoretical prediction of a continuous spectrum disagreed with the discrete emission lines actually observed. Even worse was the calculated time scale: The electron should plunge into the nucleus in only $10^{-12}$ seconds. Obviously, matter is stable over much longer periods of time!

Theoretical physicists hoped that the answer might be found among the exciting new ideas of photons and quantized energy. A Danish physicist, Niels Bohr (1885–1962; see Fig. 5.4) came to the rescue in 1913 with a daring proposal. The dimensions of Planck's constant, ergs × seconds, are equivalent to g × cm s$^{-1}$ × cm, the units of angular momentum. Perhaps the angular momentum of the orbiting electron was quantized. This quantization

---
[7]$R_H$ is named in honor of Johannes Rydberg, a Swedish spectroscopist.

**Figure 5.4** Niels Bohr (1885–1962). (Courtesy of The Niels Bohr Archive, Copenhagen.)

had been previously introduced into atomic models by the British astronomer J. W. Nicholson. Although Bohr knew that Nicholson's models were flawed, he recognized the possible significance of the quantization of angular momentum. Just as an electromagnetic wave of frequency $\nu$ could have the energy of only an integral number of quanta, $E = nh\nu$, suppose the value of the angular momentum of the hydrogen atom could assume only integral multiples of Planck's constant divided by $2\pi$: $L = nh/2\pi = n\hbar$.[8] Bohr hypothesized that, in orbits with precisely these allowed values of the angular momentum, the electron would be stable and would not radiate *in spite of its centripetal acceleration*. What would be the result of such a bold departure from classical physics?

To analyze the mechanical motion of the atomic electron–proton system, we start with the mathematical description of their electrical attraction: **Coulomb's law**. However, a few words concerning Coulomb's law are required before proceeding with the derivation of the Bohr atom: For two charges $q_1$ and $q_2$ separated by a distance $r$, the electric force on charge 2 due to charge 1 has the familiar form

$$\mathbf{F} = k_C \frac{q_1 q_2}{r^2} \hat{\mathbf{r}},$$

where $\hat{\mathbf{r}}$ is a unit vector directed from charge 1 toward charge 2. In the SI system of units used by most introductory physics texts, electric charge is

---

[8]The quantity $\hbar \equiv h/2\pi = 1.0545727 \times 10^{-27}$ erg s, and is pronounced "h-bar."

## 5.3 The Bohr Model of the Atom

measured in coulombs (C) and the Coulomb constant $k_C$ has the value

$$k_C = \frac{1}{4\pi\varepsilon_o} = 8.988 \times 10^9 \text{ N m}^2 \text{ C}^{-2},$$

where $\varepsilon_o$ is the *permittivity of free space* in SI units. However, in this text we have followed the tradition of generations of astronomers who use the cgs (*grams-centimeters-seconds*) system of units in their writing and research. In the cgs system, charge is measured in units of esu (*electrostatic units*) with

$$1 \text{ C} \text{ equivalent to } 2.998 \times 10^9 \text{ esu}.$$

Thus in cgs units the fundamental charge is $e = 4.803 \times 10^{-10}$ esu, and in SI units it is $e = 1.602 \times 10^{-19}$ C. In the cgs system of units, the Coulomb constant $k_C \equiv 1$ and is unitless. Thus in cgs units Coulomb's law is just

$$\mathbf{F} = \frac{q_1 q_2}{r^2} \hat{\mathbf{r}}. \tag{5.9}$$

If the charges $q_1$ and $q_2$ are measured in esu and the separation $r$ is measured in centimeters, then the force has units of dynes.

Now we can return to Bohr's model of the hydrogen atom. Consider an electron (mass $m_e$ and charge $-e$) and a proton (mass $m_p$ and charge $+e$) in circular orbits around their common center of mass, under the influence of their mutual electrical attraction. Recall from Section 2.3 that this two-body problem may be treated as an equivalent one-body problem by using the reduced mass

$$\mu = \frac{m_e m_p}{m_e + m_p} = \frac{(m_e)(1836 m_e)}{m_e + 1836 m_e} = 0.9994556 \, m_e$$

and the total mass

$$M = m_e + m_p = m_e + 1836 \, m_e = 1837 \, m_e$$

of the system. Since $M \simeq m_p$ and $\mu \simeq m_e$, the hydrogen atom may be thought of as being composed of a proton of mass $M$ that is at rest and an electron of mass $\mu$ that follows a circular orbit of radius $r$ around the proton; see Fig. 5.5. The electrical attraction between the electron and the proton produces the electron's centripetal acceleration $v^2/r$, as described by Newton's second law:

$$\mathbf{F} = \mu \mathbf{a}$$

$$\frac{q_1 q_2}{r^2} \hat{\mathbf{r}} = -\mu \frac{v^2}{r} \hat{\mathbf{r}}$$

$$-\frac{e^2}{r^2} \hat{\mathbf{r}} = -\mu \frac{v^2}{r} \hat{\mathbf{r}}.$$

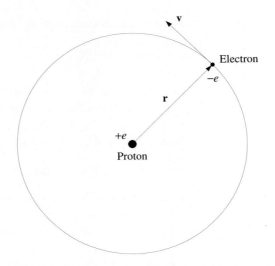

**Figure 5.5** The Bohr model of the hydrogen atom.

Canceling the minus sign and the unit vector $\hat{\mathbf{r}}$, this expression can be solved for the kinetic energy $\frac{1}{2}\mu v^2$:

$$K = \frac{1}{2}\mu v^2 = \frac{1}{2}\frac{e^2}{r}. \tag{5.10}$$

Now the electrical potential energy $U$ of the Bohr atom is (in cgs units)[9]

$$U = -\frac{e^2}{r} = -2K.$$

Thus the total energy $E = K + U$ of the atom is

$$E = K + U = K - 2K = -K = -\frac{1}{2}\frac{e^2}{r}. \tag{5.11}$$

Note that the relation between the kinetic, potential, and total energies is in accord with the virial theorem for an inverse-square force, as discussed for gravity in Section 2.4 $E = \frac{1}{2}U = -K$. Because the kinetic energy must be positive, the total energy $E$ is negative. This merely indicates that the electron and the proton are *bound*. To **ionize** the atom, removing the proton and electron to an infinite separation, an amount of energy of magnitude $|E|$ (or more) must be added to the atom.

---

[9]This is found from a derivation analogous to the one leading to the gravitational result, Eq. (2.14). The zero of potential energy is taken to be zero at $r = \infty$.

## 5.3 The Bohr Model of the Atom

Thus far the derivation has been completely classical in nature. At this point, however, we can use Bohr's quantization of angular momentum,

$$L = \mu v r = n\hbar, \tag{5.12}$$

to rewrite the kinetic energy, Eq. (5.10).

$$\frac{1}{2}\frac{e^2}{r} = \frac{1}{2}\mu v^2$$

$$= \frac{1}{2}\frac{(\mu v r)^2}{\mu r^2}$$

$$= \frac{1}{2}\frac{(n\hbar)^2}{\mu r^2}.$$

Solving this equation for the radius $r$ shows that the only values allowed by Bohr's quantization condition are

$$r_n = \frac{\hbar^2}{\mu e^2} n^2 = a_\circ n^2, \tag{5.13}$$

where $a_\circ = 5.29 \times 10^{-9}$ cm $= 0.529$ Å is known as the **Bohr radius**. Thus the electron can orbit at a distance of $a_\circ$, $4a_\circ$, $9a_\circ$, ... from the proton, but no other separations are allowed. According to Bohr's hypothesis, when the electron is in one of these orbits, the atom is stable and emits no radiation.

Inserting this expression for $r$ into Eq. (5.11) reveals that the allowed energies of the Bohr atom are

$$E_n = -\frac{\mu e^4}{2\hbar^2}\frac{1}{n^2} = -13.6 \text{ eV} \frac{1}{n^2}. \tag{5.14}$$

The integer $n$, known as the **principal quantum number**, completely determines the characteristics of each orbit of the Bohr atom. Thus, when the electron is in the lowest orbit (the *ground state*), with $n = 1$ and $r_1 = a_\circ$, its energy is $E_1 = -13.6$ eV. With the electron in the ground state, it would take at least 13.6 eV to ionize the atom. When the electron is in the *first excited state*, with $n = 2$ and $r_2 = 4a_\circ$, its energy is greater than it is in the ground state: $E_2 = -13.6/4$ eV $= -3.40$ eV.

If the electron does not radiate in any of its allowed orbits, then what is the origin of the spectral lines observed for hydrogen? Bohr proposed that a photon is emitted or absorbed when an electron makes a transition from one orbit to another. Consider an electron as it "falls" from a higher orbit, $n_{\text{high}}$, to a lower orbit, $n_{\text{low}}$, without stopping at any intermediate orbit. (This is *not* a fall in the classical sense; the electron is *never* observed between the two orbits.) The electron loses energy $\Delta E = E_{\text{high}} - E_{\text{low}}$, and this energy is carried away from the atom by a single photon. Equation (5.14) leads to an expression for the wavelength of the emitted photon,

$$E_{\text{photon}} = E_{\text{high}} - E_{\text{low}}$$

$$\frac{hc}{\lambda} = \left(-\frac{\mu e^4}{2\hbar^2}\frac{1}{n_{\text{high}}^2}\right) - \left(-\frac{\mu e^4}{2\hbar^2}\frac{1}{n_{\text{low}}^2}\right),$$

which gives

$$\frac{1}{\lambda} = \frac{\mu e^4}{4\pi \hbar^3 c}\left(\frac{1}{n_{\text{low}}^2} - \frac{1}{n_{\text{high}}^2}\right). \tag{5.15}$$

Comparing this with Eqs. (5.7) and (5.8) reveals that Eq. (5.15) is just the generalized Balmer formula for the spectral lines of hydrogen, with $n_{\text{low}} = 2$ for the Balmer series. Inserting values into the combination of constants in front of the parentheses shows that this term is exactly the Rydberg constant for hydrogen:

$$R_H = \frac{\mu e^4}{4\pi \hbar^3 c}$$

$$= \frac{0.9994556 m_e e^4}{4\pi \hbar^3 c}$$

$$= \frac{(0.9994556)(9.109390 \times 10^{-28}\text{ g})(4.803206 \times 10^{-10}\text{ esu})^4}{4\pi(1.0545727 \times 10^{-27}\text{ erg s}^{-1})^3(2.99792458 \times 10^{10}\text{ cm s}^{-1})}$$

$$= 109677.5 \text{ cm}^{-1}.$$

This value is in perfect agreement with the experimental value quoted following Eq. (5.7) for the hydrogen lines determined by Johann Balmer and illustrates the great success of Bohr's model of the hydrogen atom.[10]

---

**Example 5.3** What is the wavelength of the photon emitted when an electron makes a transition from the $n = 3$ to the $n = 2$ orbit of the Bohr hydrogen

---

[10]The slightly different Rydberg constant, $R_\infty$, found in many texts assumes an infinitely heavy nucleus. The reduced mass, $\mu$, in the expression for $R_H$ is replaced by the electron mass, $m_e$, in $R_\infty$.

## 5.3 The Bohr Model of the Atom

atom? The energy lost by the electron is carried away by the photon, so

$$E_{\text{photon}} = E_{\text{high}} - E_{\text{low}}$$

$$\frac{hc}{\lambda} = -13.6 \text{ eV} \, \frac{1}{n_{\text{high}}^2} - \left(-13.6 \text{ eV} \, \frac{1}{n_{\text{low}}^2}\right)$$

$$\frac{12400 \text{ eV Å}}{\lambda} = -13.6 \text{ eV} \left(\frac{1}{3^2} - \frac{1}{2^2}\right).$$

Solving for the wavelength gives $\lambda = 6565$ Å, within 0.03% of the measured value of the $H_\alpha$ spectral line.

---

The reverse process may also occur. If a photon has an energy equal to the *difference* in energy between two orbits (with the electron in the lower orbit), the photon may be absorbed by the atom. The electron uses the photon's energy to make an upward transition from the lower orbit to the higher orbit. The relation between the photon's wavelength and the quantum numbers of the two orbits is again given by Eq. (5.15).

After the quantum revolution, the physical processes responsible for Kirchhoff's laws (discussed in Section 5.1) finally became clear.

- A hot, dense gas or hot solid object produces a continuous spectrum with no dark spectral lines. This is the continuous spectrum of blackbody radiation, described by the Planck functions $B_\lambda(T)$ and $B_\nu(T)$, emitted at any temperature above absolute zero. The wavelength $\lambda_{\text{max}}$ at which the Planck function $B_\lambda(T)$ obtains its maximum value is given by Wien's displacement law, Eq. (3.15).

- A hot, diffuse gas produces bright emission lines. Emission lines are produced when an electron makes a downward transition from a higher to a lower orbit. The energy lost by the electron is carried away by a single photon. For example, the hydrogen Balmer emission lines are produced by electrons "falling" from higher orbits down to the $n = 2$ orbit; see Fig. 5.6.

- A cool, diffuse gas in front of a source of continuous spectrum produces dark absorption lines in the continuous spectrum. Absorption lines are produced when an electron makes a transition from a lower to a higher orbit. If an incident photon in the continuous spectrum has exactly the right amount of energy, equal to the difference in energy between a higher orbit and the electron's initial orbit, the photon is absorbed by

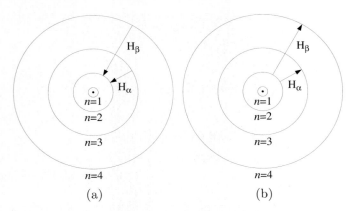

**Figure 5.6** Balmer lines produced by the Bohr hydrogen atom. (a) Emission lines. (b) Absorption lines.

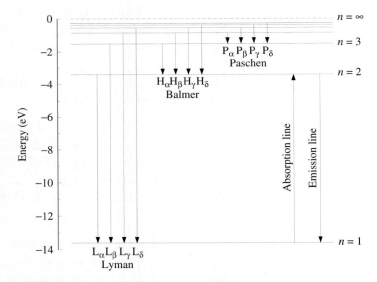

**Figure 5.7** Energy level diagram for the Balmer lines (downward arrows for emission lines, upward arrow for absorption lines).

the atom and the electron makes an upward transition to that higher orbit. For example, the hydrogen Balmer absorption lines are produced by atoms absorbing photons that cause electrons to make transitions from the ($n = 2$) orbit to higher orbits; see Figs. 5.6 and 5.7.

Despite the spectacular successes of Bohr's model of the hydrogen atom, it is not quite correct. Although angular momentum is quantized, it does not

**Figure 5.8** Louis de Broglie (1892–1987). (Courtesy of AIP Niels Bohr Library.)

have the values assigned by Bohr.[11] Bohr painted a *semiclassical* picture of the hydrogen atom, a miniature solar system with an electron circling the proton in a classical circular orbit. In fact, the electron orbits are not circular. They are not even orbits at all, in the classical sense of an electron at a precise location moving with a precise velocity. Instead, on an atomic level, nature is "fuzzy," with an attendant uncertainty that cannot be avoided. It was fortunate that Bohr's model, with all of its faults, led to the correct values for the energies of the orbits and to a correct interpretation of the formation of spectral lines. This intuitive, easily imagined model of the atom is what most astronomers have in mind when they visualize atomic processes.

## 5.4 Quantum Mechanics and Wave–Particle Duality

The last act of the quantum revolution began with the musings of a French prince, Louis de Broglie (1892–1987; see Fig. 5.8). Wondering about the recently discovered wave–particle duality for light, he posed a profound question: If light (classically thought to be a wave) could exhibit the characteristics of particles, might not particles sometimes manifest the properties of waves?

In his 1927 Ph.D. thesis, de Broglie extended the wave–particle duality to all of nature. Photons carry both energy $E$ and momentum $p$, and these

---

[11]As will be seen in the next section, instead of $L = n\hbar$, the actual values of the orbital angular momentum are $L = \sqrt{\ell(\ell+1)}\,\hbar$, where $\ell$, an integer, is a new quantum number.

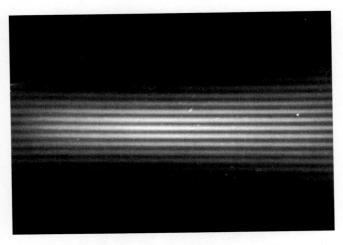

**Figure 5.9** Interference pattern from an electron double-slit experiment. (Figure from Jönsson, *Zeitschrift für Physik*, *161*, 454, 1961.)

quantities are related to the frequency $\nu$ and wavelength $\lambda$ of the light wave by Eq. (5.5):

$$\nu = \frac{E}{h} \tag{5.16}$$

$$\lambda = \frac{h}{p}. \tag{5.17}$$

De Broglie proposed that these equations be used to define a frequency and a wavelength for *all* particles. The **de Broglie wavelength and frequency** describe not only massless photons but massive electrons, protons, neutrons, atoms, molecules, people, planets, stars, and galaxies as well. This seemingly outrageous proposal of matter waves has been confirmed in countless experiments. Figure 5.9 shows the interference pattern produced by *electrons* in a double-slit experiment. Just as Thomas Young's double-slit experiment established the wave properties of light, the electron double-slit experiment can be explained only by the wavelike behavior of electrons, with *each* electron propagating through *both* slits.[12] The wave–particle duality applies to everything in the physical world; everything exhibits its wave properties in its *propagation*, and manifests its particle nature in its *interactions*.

---

[12] See Chapter 6 of Feynman (1965) for a fascinating description of the details and profound implications of the electron double-slit experiment.

## 5.4 Quantum Mechanics and Wave–Particle Duality

**Example 5.4** Let's compare the wavelengths of a free electron moving at $3 \times 10^8$ cm s$^{-1}$ and a 70 kg man jogging at 300 cm s$^{-1}$. For the electron,

$$\lambda = \frac{h}{p} = \frac{h}{m_e v} = 2.42 \times 10^{-8} \text{ cm},$$

about the size of an atom. This wavelength is much smaller than that of visible light. Electron microscopes utilize electrons with wavelengths one million times shorter than visible wavelengths to obtain a much higher resolution than is possible with optical microscopes.

The wavelength of the jogging man is

$$\lambda = \frac{h}{p} = \frac{h}{m_{\text{man}} v} = 3.16 \times 10^{-34} \text{ cm},$$

which is completely negligible on the scale of the everyday world, or even on an atomic scale. Thus the jogging gentleman need not worry about diffracting when returning home through his doorway!

---

Just what are the waves that are involved in the wave–particle duality of nature? In a double-slit experiment, each photon or electron must pass through *both* slits, since the interference pattern is produced by the constructive and destructive interference of the two waves. Thus the wave cannot convey information about where the photon or electron is, but only about where it *may* be. The wave is one of *probability*, and its amplitude is denoted by the Greek letter $\Psi$ (psi). The square of the wave amplitude, $|\Psi|^2$, at a certain location describes the probability of finding the photon or electron at that location. In the double-slit experiment, photons or electrons are never found where the waves from slits 1 and 2 have destructively interfered, that is, where $|\Psi_1 + \Psi_2|^2 = 0$.

The wave attributes of matter lead to some quite unexpected conclusions of paramount importance for the science of astronomy. For example, consider Fig. 5.10. The probability wave, $\Psi$, at the top is a sine wave, with a precise wavelength $\lambda$. Thus the momentum $p = h/\lambda$ of the particle described by this wave is known exactly. However, because $|\Psi|^2$ consists of a number of equally high peaks extending out to $x = \pm\infty$, the particle's location is perfectly uncertain. The particle's position can be narrowed down if several sine waves with different wavelengths are added together, so they destructively interfere with one another nearly everywhere. As shown at the bottom of Fig. 5.10, the resulting combination of waves, $\Psi$, is approximately zero everywhere except at one location. Now the particle's position may be determined with a greater

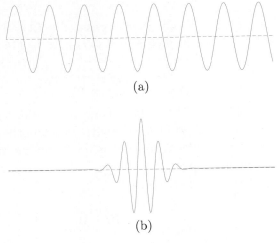

**Figure 5.10** Two examples of a probability wave, $\Psi$: (a) a single sine wave and (b) a pulse of several sine waves.

certainty because $|\Psi|^2$ is large only for a narrow range of values of $x$. However, the value of the particle's momentum has become more uncertain because $\Psi$ is now a combination of waves of various wavelengths. This is nature's intrinsic trade-off: The uncertainty in a particle's position, $\Delta x$, and the uncertainty in its momentum, $\Delta p$, are inversely related. As one decreases, the other must increase. This fundamental inability of a particle to *simultaneously* have a well-defined position and momentum is a direct result of the wave–particle duality of nature. A German physicist, Werner Heisenberg (1901–1976), placed this inherent "fuzziness" of the physical world in a firm theoretical framework. He stated that the uncertainty in a particle's position multiplied by the uncertainty in its momentum must be *at least* as large as $\hbar/2$:

$$\Delta x \, \Delta p \geq \frac{1}{2} \hbar.$$

Today this is known as **Heisenberg's uncertainty principle**. The equality is rarely realized in nature, and the form usually employed for making estimates is

$$\Delta x \, \Delta p \approx \hbar. \tag{5.18}$$

A similar statement relates the uncertainty of an energy measurement, $\Delta E$, and the time interval, $\Delta t$, over which the energy measurement is taken:

$$\Delta E \, \Delta t \approx \hbar. \tag{5.19}$$

## 5.4 Quantum Mechanics and Wave–Particle Duality

As the time available for an energy measurement increases, the inherent uncertainty in the result decreases. In Section 9.4, we will apply this version of the uncertainty principle to the sharpness of spectral lines.

---

**Example 5.5** Imagine an electron confined within a region of space the size of a hydrogen atom. We can estimate the minimum speed and kinetic energy of the electron using Heisenberg's uncertainty principle. Because we know only that the particle is within an atom-size region of space, we can take $\Delta x \approx a_o = 5.29 \times 10^{-9}$ cm. Thus the uncertainty in the electron's momentum is roughly

$$\Delta p \approx \frac{\hbar}{\Delta x} = 1.98 \times 10^{-19} \text{ g cm s}^{-1}.$$

Thus, if the magnitude of the momentum of the electron were repeatedly measured, the resulting values would vary within a range $\pm \Delta p$ around some average (or *expected*) value. Since this expected value, as well as the individual measurements, must be $\geq 0$, the expected value must be at least as large as $\Delta p$. Thus we can equate the minimum expected value of the momentum with its uncertainty: $p_{\min} \approx \Delta p$. Using $p_{\min} = m_e v_{\min}$, the minimum speed of the electron is estimated to be

$$v_{\min} = \frac{p_{\min}}{m_e} \approx \frac{\Delta p}{m_e} \approx 2.18 \times 10^8 \text{ cm s}^{-1}.$$

The minimum kinetic energy of the (nonrelativistic) electron is approximately

$$K_{\min} = \frac{1}{2} m_e v_{\min}^2 \approx 2.16 \times 10^{-11} \text{ erg} = 13.5 \text{ eV}.$$

This is in good agreement with the kinetic energy of the electron in the ground state of the hydrogen atom. An electron confined to such a small region *must* move rapidly with at least this speed and this energy. In Chapter 15, we will see that this subtle quantum effect is responsible for supporting white dwarf and neutron stars against the tremendous inward pull of gravity.

---

When a ray of light attempts to travel from a glass prism into air, it may undergo *total internal reflection* if it strikes the surface at an angle greater than the critical angle $\theta_c$, where the critical angle is related to the indices of refraction of the glass and air by

$$\tan \theta_c = \frac{n_{\text{air}}}{n_{\text{glass}}}.$$

This familiar result is nonetheless surprising because, even though the ray of light is totally reflected, the index of refraction of the outside air appears in this

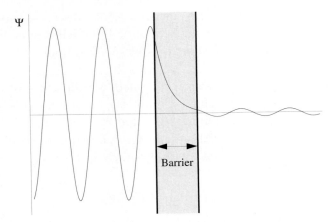

**Figure 5.11** Quantum mechanical tunneling or barrier penetration of a particle traveling to the right.

formula. In fact, the electromagnetic wave does enter the air, but it ceases to be oscillatory and instead dies away exponentially. In general, when a classical wave such as a water or light wave enters a medium through which it cannot propagate, it becomes *evanescent* and its amplitude decays exponentially with distance.

This total internal reflection can in fact be frustrated by placing another prism next to the first prism, so their surfaces nearly (but not quite) touch. Then the evanescent wave in the air may enter the second prism before its amplitude has completely died away. The electromagnetic wave once again becomes oscillatory upon entering the glass, and so the ray of light has traveled from one prism to another without passing through the air gap between the prisms. In the language of particles, photons have **tunneled** from one prism to another without traveling in the space between them.

The wave–particle duality of nature implies that particles can also tunnel through a region of space (a barrier) in which they cannot exist classically, as shown in Fig. 5.11. The barrier must not be too wide (not more than a few particle wavelengths) if tunneling is to take place; otherwise, the amplitude of the evanescent wave will have declined to nearly zero. This is consistent with Heisenberg's uncertainty principle, which implies that a particle's location cannot be determined with an uncertainty that is less than its wavelength. Thus, if the barrier is only a few wavelengths wide, the particle may suddenly appear on the other side of the barrier. **Barrier penetration** is extremely important in radioactive decay where alpha particles tunnel out of an atom's nucleus, in modern electronics where it is the basis for the "tunnel diode," and

## 5.4 Quantum Mechanics and Wave–Particle Duality

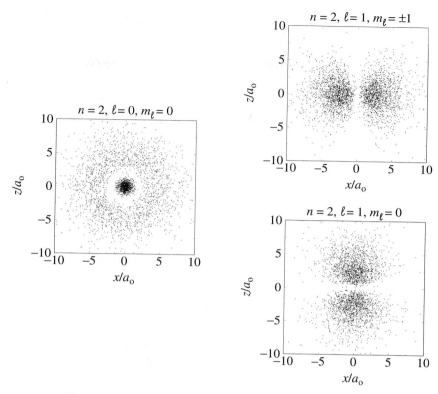

**Figure 5.12** Electron orbitals of the hydrogen atom.

inside stars where the rates of nuclear fusion reactions depend upon tunneling.

What are the implications for Bohr's model of the hydrogen atom? Heisenberg's uncertainty principle does not allow classical orbits, with their simultaneously precise values of the electron's position and momentum. Instead, the electron *orbitals* must be imagined as fuzzy clouds of probability, with the clouds being more "dense" in regions where the electron is more likely to be found (see Fig. 5.12). In 1925, a complete break from classical physics was imminent, one that would fully incorporate de Broglie's matter waves.

As described in Section 3.3, Maxwell's equations of electricity and magnetism can be manipulated to produce a wave equation for the electromagnetic waves that describe the propagation of photons. Similarly, a wave equation discovered in 1926 by Erwin Schrödinger (1877–1961), an Austrian physicist, led to a true **quantum mechanics**, the quantum analog of the classical mechanics that originated with Galileo and Newton. The Schrödinger equation can be solved for the probability waves that describe the allowed values of a particle's energy, momentum, and so on, as well as the particle's propagation through space. In particular, the Schrödinger equation can be solved analytically for the hydrogen atom, giving exactly the same set of allowed energies as those obtained by Bohr (c.f. Eq. 5.11). However, in addition to the principal quantum number $n$, Schrödinger found that two additional quantum numbers, $\ell$ and $m_\ell$, were required for a complete description of the electron orbitals. These additional numbers describe the angular momentum vector, **L**, of the atom. Instead of the quantization used by Bohr, $L = n\hbar$, the solution to the Schrödinger equation shows that the permitted values of the magnitude of the angular momentum $L$ are actually

$$L = \sqrt{\ell(\ell+1)}\,\hbar,$$

where $\ell = 0, 1, 2, \ldots, n-1$, and $n$ is the principal quantum number that determines the energy. The $z$-component of the angular momentum vector, $L_z$, can assume only the values $L_z = m_\ell \hbar$, with $m_\ell$ equal to any of the $(2\ell+1)$ integers between $-\ell$ and $+\ell$ inclusive. Thus the angular momentum vector can point in $(2\ell+1)$ different directions. For our purposes, the important point is that the values of the energy of an *isolated* hydrogen atom do not depend on $\ell$ and $m_\ell$. In the absence of a preferred direction in space, the direction of the angular momentum has no effect on the atom's energy. Different orbitals, labeled by different values of $\ell$ and $m_\ell$ (see Fig. 5.12), are said to be **degenerate** if they have the same value of the principal quantum number $n$ and so have the same energy. Electrons making a transition from a given orbital to one of several degenerate orbitals will produce the *same* spectral line, because they experience the same change in energy.

However, the atom's surroundings may single out one spatial direction as being different from another. For example, an electron in an atom will feel the effect of an external magnetic field. The magnitude of this effect will depend on the $(2\ell+1)$ possible orientations of the electron's motion, as given by $m_\ell$, and the magnetic field strength, $B$.[13] As the electron moves through the magnetic field, the normally degenerate orbitals acquire slightly different energies. Electrons making a transition between these formerly degenerate orbitals will thus produce spectral lines with slightly different frequencies. The splitting of spectral lines in a weak magnetic field is called the **Zeeman effect** and is shown in Fig. 5.13. The three frequencies of the split lines in the simplest

---

[13]In the cgs system, $B$ is measured in units of gauss (G). This is related to the SI unit, the tesla (T), by 1 tesla $= 10^4$ gauss.

## 5.4 Quantum Mechanics and Wave–Particle Duality

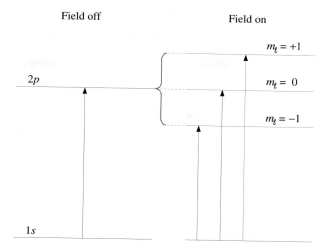

**Figure 5.13** Splitting of absorption lines by the Zeeman effect.

case (called the *normal Zeeman effect*) are

$$\nu = \nu_\circ \quad \text{and} \quad \nu_\circ \pm \frac{eB}{4\pi\mu c}, \tag{5.20}$$

where $\nu_\circ$ is the frequency of the spectral line in the absence of a magnetic field and $\mu$ is the reduced mass. Although the energy levels are split into $(2\ell + 1)$ components, electron transitions involving these levels produce just three spectral lines with different polarizations.[14] Viewed from different directions, all three lines may not be visible. For example, when looking parallel to the magnetic field (as when looking down on a sunspot), the unshifted line of frequency $\nu_\circ$ is absent.

Thus the Zeeman effect gives astronomers a probe of the magnetic fields observed around sunspots and on other stars. Even if the splitting of the spectral line is too small to be directly detected, the different polarizations across the closely spaced components can still be measured and the magnetic field strength deduced.

---

**Example 5.6** Interstellar clouds may contain very weak magnetic fields, as small as $B \approx 2 \times 10^{-6}$ G. Nevertheless, astronomers have been able to measure this magnetic field. Using radio telescopes, they detect the variation in

---

[14]Nature imposes a *selection rule*, requiring that $\Delta m_\ell = 0$ or $\pm 1$ and forbidding transitions between orbitals if both have $m_\ell = 0$. This restricts the possible electron transitions and permits the formation of only three spectral lines.

polarization that occurs across the blended Zeeman components of the absorption lines that are produced by these interstellar clouds of hydrogen gas. The change in frequency, $\Delta\nu$, produced by a magnetic field of this magnitude can be calculated from Eq. (5.20), using the mass of the electron, $m_e$, for the reduced mass $\mu$:

$$\Delta\nu = \frac{eB}{4\pi m_e c} = 2.8 \text{ Hz},$$

a minute change. The total difference in frequency from one side of this blended line to the other is twice this amount, or 5.6 Hz. For comparison, the frequency of the radio wave emitted by hydrogen with $\lambda = 21$ cm is $\nu = c/\lambda = 1.4 \times 10^9$ Hz, 250 million times larger!

---

Attempts to understand more complicated patterns of magnetic field splitting (the *anomalous Zeeman effect*), usually involving an even number of unequally spaced spectral lines, led physicists in 1925 to discover a *fourth* quantum number. In addition to its orbital motion, the electron possesses a **spin**. This is *not* a classical toplike rotation, but purely a quantum effect that endows the electron with a *spin angular momentum* **S**. **S** is a vector of constant magnitude $S = \sqrt{\frac{1}{2}(\frac{1}{2}+1)}\,\hbar = \sqrt{3}/2\,\hbar$, with a $z$-component $S_z = m_s \hbar$. The only values of the fourth quantum number, $m_s$, are $\pm\frac{1}{2}$.

With each orbital, or *quantum state*, labeled by four quantum numbers, physicists wondered how many electrons in a multielectron atom could occupy the same quantum state. The answer was supplied in 1925 by an Austrian theoretical physicist, Wolfgang Pauli (1900–1958): No two electrons can occupy the same quantum state. The **Pauli exclusion principle**, that *no two electrons can share the same set of four quantum numbers*, explained the electronic structure of atoms, thereby providing an explanation of the properties of the periodic table. Despite this success, Pauli was unhappy about the lack of a firm theoretical understanding of electron spin. Spin was stitched onto quantum theory in an ad hoc manner, and the seams showed. Pauli lamented this patchwork theory and asked, "How can one avoid despondency if one thinks of the anomalous Zeeman effect?"

The final synthesis arrived in 1928 from an unexpected source. A brilliant English theoretical physicist, Paul Adrien Maurice Dirac (1902–1984), was working at Cambridge to combine Schrödinger's wave equation with Einstein's theory of special relativity. When he finally succeeded in writing a relativistic wave equation for the electron, he was delighted to see that the mathematical solution automatically included the spin of the electron. It also explained and extended the Pauli exclusion principle by dividing the world of particles

## 5.4 Quantum Mechanics and Wave–Particle Duality

into two fundamental groups: fermions and bosons. **Fermions**[15] are particles such as electrons, protons, and neutrons[16] that have a spin of $\frac{1}{2}\hbar$ (or an odd integer times $\frac{1}{2}\hbar$, such as $\frac{3}{2}\hbar$, $\frac{5}{2}\hbar$, etc.). Fermions obey the Pauli exclusion principle, so no two fermions can have the same set of quantum numbers. The exclusion principle for fermions, along with Heisenberg's uncertainty relation, explains the structure of white dwarfs and neutron stars, as will be discussed in Chapter 15. **Bosons**[17] are particles such as photons that have an integral spin of $0$, $\hbar$, $2\hbar$, $3\hbar$, .... Bosons do not obey the Pauli exclusion principle, so any number of bosons can occupy the same quantum state.

As a final bonus, the Dirac equation predicted the existence of antiparticles. A particle and its antiparticle are identical except for their opposite electric charges and magnetic moments. Pairs of particles and antiparticles may be created from the energy of gamma-ray photons (according to $E = mc^2$). On the other hand, particle–antiparticle pairs may annihilate each other, with their mass converted back into the energy of two gamma-ray photons. As will be discussed in Section 16.3, pair creation and annihilation plays a major role in the evaporation of black holes.

The revolution in physics started by Max Planck culminated in the quantum atom and gave astronomers their most powerful tool: a theory that would allow them to analyze the spectral lines observed for stars, galaxies, and nebulae.[18] Different atoms, and combinations of atoms in molecules, have orbitals of distinctly different energies; thus they can be identified by their spectral line "fingerprints." The specific spectral lines produced by an atom or molecule depend on which orbitals are occupied by electrons. This, in turn, depends on its surroundings: the temperature, density, and pressure of its environment. These and other factors, such as the strength of a surrounding magnetic field, may be determined by a careful examination of spectral lines. Much of Chapters 8 and 9 will be devoted to the practical application of the quantum atom to stellar atmospheres.

---

[15] The fermion is named after the Italian physicist Enrico Fermi.

[16] The neutron was not discovered until 1932 by James Chadwick, the same year that the positron (antimatter electron) was discovered by Carl Anderson.

[17] The boson is named in honor of the Indian physicist S. N. Bose.

[18] Nearly all of the physicists mentioned in this chapter won the Nobel Prize for physics or chemistry in recognition of their work.

# Suggested Readings

## General

Feynman, Richard, *The Character of Physical Law*, The M.I.T. Press, Cambridge, MA, 1965.

French, A. P., and Kennedy, P. J. (eds.), *Niels Bohr: A Centenary Volume*, Harvard University Press, Cambridge, MA, 1985.

Hey, Tony, and Walters, Patrick, *The Quantum Universe*, Cambridge University Press, Cambridge, 1987.

Pagels, Heinz R., *The Cosmic Code*, Simon and Schuster, New York, 1982.

Segre, Emilio, *From X-Rays to Quarks*, W. H. Freeman and Company, San Francisco, 1980.

## Technical

Harwit, Martin, *Astrophysical Concepts*, Second Edition, John Wiley and Sons, New York, 1990.

Resnick, Robert, and Halliday, David, *Basic Concepts in Relativity and Early Quantum Theory*, Second Edition, John Wiley and Sons, New York, 1985.

# Problems

5.1 Barnard's star, named after the American astronomer Edward E. Barnard, is an orange star in the constellation Ophiuchus. It has the largest known proper motion ($\mu = 10.31''$ yr$^{-1}$) and the second-largest parallax angle ($p = 0.552''$). In the spectrum of Barnard's star, the H$_\alpha$ absorption line is observed to have a wavelength of 6560.44 Å.

   (a) Determine the radial velocity of Barnard's star.
   (b) Determine the transverse velocity of Barnard's star.
   (c) Calculate the speed of Barnard's star through space.

5.2 When salt is sprinkled on a flame, yellow light consisting of two closely spaced wavelengths, 5889.97 Å and 5895.94 Å, is produced. They are called the *sodium D lines* and were observed by Fraunhofer in the Sun's spectrum.

# Problems

(a) If this light falls on a diffraction grating with 300 lines per millimeter, what is the angle between the second-order spectra of these two wavelengths?

(b) How many lines of this grating must be illuminated for the sodium D lines to just be resolved?

**5.3** Prove that $hc = 12400$ eV Å.

**5.4** The photoelectric effect can be an important heating mechanism for the grains of dust found in interstellar clouds (see Section 12.1). The ejection of an electron leaves the grain with a positive charge, which affects the rates at which other electrons and ions collide with and stick to the grain to produce the heating. This process is particularly effective for ultraviolet photons ($\lambda \approx 1000$ Å) striking the smaller dust grains. If the average energy of the ejected electron is about 5 eV, estimate the work function of a typical dust grain.

**5.5** Use Eq. (5.5) for the momentum of a photon, plus the conservation of relativistic momentum and energy [Eqs. (4.44) and (4.48), respectively], to derive Eq. (5.6) for the change in wavelength of the scattered photon in the Compton effect.

**5.6** Consider the case of a "collision" between a photon and a free proton, initially at rest. What is the characteristic change in the wavelength of the scattered photon in units of angstroms? How does this compare with the Compton wavelength, $\lambda_C$?

**5.7** Verify that the units of Planck's constant are the units of angular momentum.

**5.8** A one-electron atom is an atom with $Z$ protons in the nucleus, and with all but one of its electrons lost to ionization.

(a) Starting with Coulomb's law, determine expressions for the orbital radii and energies for a Bohr model of the one-electron atom with $Z$ protons.

(b) Find the radius of the ground-state orbit, the ground-state energy, and the ionization energy of singly ionized helium (He II).

(c) Repeat part (b) for doubly ionized lithium (Li III).

**5.9** To demonstrate the relative strength of the electrical and gravitational forces of attraction between the electron and the proton in the Bohr

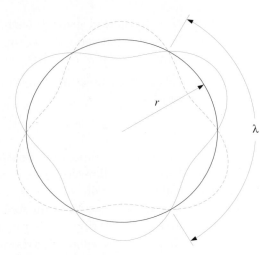

**Figure 5.14** Three de Broglie wavelengths spanning an electron's orbit in the Bohr atom.

atom, suppose the hydrogen atom were held together *solely* by the force of gravity. Determine the radius of the ground-state orbit (in units of Å and AU) and the energy of the ground state (in eV).

5.10 Calculate the energies and wavelengths of all possible photons that are emitted when the electron cascades from the $n = 3$ to the $n = 1$ orbit of the hydrogen atom.

5.11 Find the shortest wavelength photon emitted by a downward electron transition in the Lyman, Balmer, and Paschen series. These wavelengths are known as the *series limits*. In which regions of the electromagnetic spectrum are these wavelengths found?

5.12 An electron in a television set reaches a speed of about $5 \times 10^9$ cm s$^{-1}$ before it hits the screen. What is the wavelength of this electron?

5.13 Consider the de Broglie wave of the electron in the Bohr atom. The circumference of the electron's orbit must be an integral number of wavelengths, $n\lambda$; see Fig. 5.14. Otherwise, the electron wave will find itself *out of phase* and suffer destructive interference. Show that this requirement leads to Bohr's condition for the quantization of angular momentum, Eq. (5.12).

5.14 A white dwarf is a very dense star, with its ions and electrons packed extremely close together. Each electron may be considered to be located within a region of size $\Delta x \approx 1.5 \times 10^{-10}$ cm. Use Heisenberg's uncertainty principle, Eq. (5.18), to estimate the minimum speed of the electron. Do you think that the effects of relativity will be important for these stars?

5.15 An electron spends roughly $10^{-8}$ s in the first excited state of the hydrogen atom before making a spontaneous downward transition to the ground state.

(a) Use Heisenberg's uncertainty principle (Eq. 5.19) to determine the uncertainty $\Delta E$ in the energy of the first excited state.

(b) Calculate the uncertainty $\Delta \lambda$ in the wavelength of the photon involved in a transition (either upward or downward) between the ground and first excited states of the hydrogen atom. Why can you assume that $\Delta E = 0$ for the ground state?

This increase in the width of a spectral line is called *natural broadening*.

5.16 Each quantum state of the hydrogen atom is labeled by a set of four quantum numbers: $\{n, \ell, m_\ell, m_s\}$.

(a) List the sets of quantum numbers for the hydrogen atom having $n = 1$, $n = 2$, and $n = 3$.

(b) Show that the degeneracy of energy level $n$ is $2n^2$.

5.17 The members of a class of stars known as *Ap stars* are distinguished by their strong magnetic fields (usually a few thousand gauss).[19] The star HD215441 has an unusually strong magnetic field of 34,000 G. Find the frequencies and wavelengths of the three components of the $H_\alpha$ spectral line produced by the normal Zeeman effect for this magnetic field.

5.18 **Computer Problem** One of the most important ideas of the physics of waves is that *any* complex waveform can be expressed as the sum of the harmonics of simple cosine and sine waves. That is, any wave function $f(x)$ can be written as

$$f(x) = a_0 + a_1 \cos x + a_2 \cos 2x + a_3 \cos 3x + a_4 \cos 4x + \cdots$$
$$+ b_1 \sin x + b_2 \sin 2x + b_3 \sin 3x + b_4 \sin 4x + \cdots.$$

---

[19] The letter A is the star's spectral type (to be discussed in Section 8.1), and the letter p stands for "peculiar."

The coefficients $a_n$ and $b_n$ tell how much of each harmonic goes into the recipe for $f(x)$. This series of cosine and sine terms is called the *Fourier series* for $f(x)$. In general, both cosine and sine terms are needed, but in this problem you will use only the sine terms; all of the $a_n \equiv 0$.

On page 145, the process of constructing a wave pulse by adding a series of sine waves was described. The Fourier sine series that you will use to construct your wave employs only the *odd* harmonics, and is given by

$$\Psi = \frac{2}{N+1}(\sin x - \sin 3x + \sin 5x - \sin 7x + \cdots \pm \sin Nx)$$

$$= \frac{2}{N+1}\sum_{\substack{n=1 \\ n \text{ odd}}}^{N}(-1)^{(n-1)/2}\sin nx,$$

where $N$ is an odd integer. The leading factor of $2/(N+1)$ does not change the shape of $\Psi$, but scales the wave for convenience so its maximum value is equal to one for any choice of $N$.

(a) Graph $\Psi$ for $N = 5$, using values of $x$ (in radians) between 0 and $\pi$. What is the width, $\Delta x$, of the wave pulse?

(b) Repeat part (a) for $N = 11$.

(c) Repeat part (a) for $N = 21$.

(d) Repeat part (a) for $N = 41$.

(e) If $\Psi$ represents the probability wave of a particle, for which value of $N$ is the position of the particle known with the least uncertainty? For which value of $N$ is the momentum of the particle known with the least uncertainty?

# Chapter 6

# Telescopes

## 6.1 Basic Optics

From the beginning, astronomy has been an observational science. In comparison with what was previously possible with the naked eye, Galileo's use of the new optical device known as the telescope greatly enhanced our ability to observe the universe (see Section 2.2). Today we are still increasing our ability to "see" faint objects and to resolve them in greater detail. As a result, modern observational astronomy continues to supply scientists with more clues to the physical nature of our universe.

Although observational astronomy now covers the entire range of the electromagnetic spectrum, along with many areas of particle physics, the most familiar part of the field remains in the optical regime of the human eye (approximately 4000 Å to 7000 Å). Consequently, telescopes and detectors designed to investigate optical-wavelength radiation will be discussed in some detail.

Galileo's telescope was a **refracting** telescope that made use of lenses through which light would pass, ultimately forming an image. Later, Newton designed and built a **reflecting** telescope that made use of mirrors as the principal optical component. Both refractors and reflectors remain in use today.

To understand the effects of an optical system on the light coming from an astronomical object, we will focus first on refracting telescopes. The path of a light ray through a lens can be understood using **Snell's law** of refraction. Recall that as a light ray travels from one transparent medium to another, its path is bent at the interface. The amount that the ray is bent depends on the ratio of the wavelength-dependent indices of refraction $n_\lambda \equiv c/v_\lambda$ of each

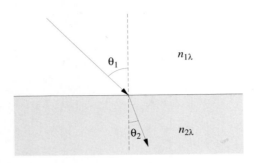

**Figure 6.1** Snell's law of refraction.

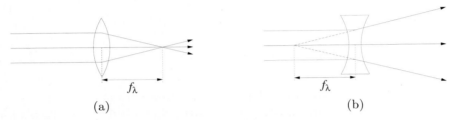

**Figure 6.2** (a) A converging lens, $f_\lambda > 0$. (b) A diverging lens, $f_\lambda < 0$.

material, where $v_\lambda$ represents the speed of light within the specific medium.[1] If $\theta_1$ is the angle of incidence, measured with respect to the normal to the interface between the two media, and $\theta_2$ is the angle of refraction, also measured relative to the normal to the interface (see Fig. 6.1), then Snell's law is given by

$$n_{1\lambda} \sin \theta_1 = n_{2\lambda} \sin \theta_2. \tag{6.1}$$

If the surfaces of the lens are shaped properly, a beam of light rays of a given wavelength, originally traveling parallel to the axis of symmetry of the lens, called the **optical axis** of the system, can be brought to a focus at a point along that axis by a *converging* lens (Fig. 6.2a). Alternatively, the light can be made to diverge by a *diverging* lens and the light rays will appear to originate from a single point along the axis (Fig. 6.2b). The unique point in either case is referred to as the **focal point** of the lens, and the distance to that point from the center of the lens is known as the **focal length**, $f$. For a converging lens the focal length is taken to be positive, and for a diverging lens the focal length is negative.

For an extended object, the image will also necessarily be extended. If a photographic plate or some other detector is to record this image, the detector

---

[1] It is only in a vacuum that $v_\lambda \equiv c$, independent of wavelength. The speed of light is wavelength-dependent in other environments.

## 6.1 Basic Optics

**Figure 6.3** The sign convention for the radii of curvature of a lens in the lensmaker's formula. (a) $R_1 > 0$, $R_2 > 0$. (b) $R_1 < 0$, $R_2 < 0$.

must be placed in the focal plane of the telescope. The **focal plane**, defined as that plane passing through the focal point, is perpendicular to the optical axis of the system. Since, for all practical purposes, any astronomical object can reasonably be assumed to be located infinitely far from the telescope,[2] all of the rays coming from that object are essentially parallel to one another, although not necessarily parallel to the optical axis. If the rays are not parallel to the optical axis, distortion of the image can result; this is just one of many forms of *aberration* discussed later.

The focal length of a given thin lens can be calculated directly from its index of refraction and geometry. If we assume that both surfaces of the lens are spheroidal, then it can be shown that the focal length $f_\lambda$ is given by the **lensmaker's formula**,

$$\frac{1}{f_\lambda} = (n_\lambda - 1)\left(\frac{1}{R_1} + \frac{1}{R_2}\right), \tag{6.2}$$

where $n_\lambda$ is the index of refraction of the lens and $R_1$ and $R_2$ are the radii of curvature of each surface, taken to be positive if the specific surface is convex and negative if it is concave (see Fig. 6.3).[3]

For mirrors $f$ is wavelength-independent, since reflection depends only on the fact that the angle of incidence always equals the angle of reflection ($\theta_1 = \theta_2$; see Fig. 6.4). Furthermore, in the case of a spheroidal mirror (Fig. 6.5), the focal length becomes $f = R/2$, where $R$ is the radius of curvature of the mirror, either positive (converging) or negative (diverging), a fact that can be demonstrated by simple geometry. Converging mirrors are generally used as the main mirrors in reflecting telescopes, although either diverging or flat mirrors may be used in other parts of the optical system.

---

[2]Technically, this implies that the distance to the astronomical object is much greater than the focal length of the telescope.

[3]It is worth noting that many authors choose to define the sign convention for the radii of curvature in terms of the direction of the incident light. This choice means that Eq. (6.2) must be expressed in terms of the *difference* in the reciprocals of the radii of curvature.

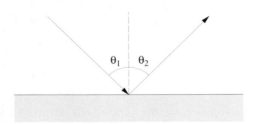

**Figure 6.4** The law of reflection, $\theta_1 = \theta_2$.

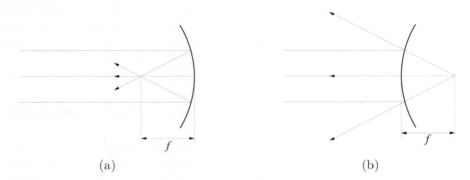

**Figure 6.5** (a) A converging mirror, $f > 0$. (b) A diverging mirror, $f < 0$.

The image separation of two point sources on the focal plane is related to the focal length of the lens being used. Figure 6.6 shows the rays of two point sources, the direction of one source being along the optical axis of a converging lens and the other located at an angle $\theta$ with respect to the optical axis. At the position of the focal plane, the rays from the on-axis source will converge at the focal point while the rays from the other will *approximately* meet at a distance $y$ from the focal point. Now from simple geometry, $y$ is given by

$$y = f \tan \theta$$

(the wavelength dependence of $f$ is implicitly assumed). If it is assumed that the field of view of the telescope is small, then $\theta$ must also be small. Using the small angle approximation, $\tan \theta \simeq \theta$, for $\theta$ expressed in radians, then

$$y = f\theta. \tag{6.3}$$

This immediately leads to the differential relation known as the **plate scale**, $d\theta/dy$,

$$\frac{d\theta}{dy} = \frac{1}{f}, \tag{6.4}$$

## 6.1 Basic Optics

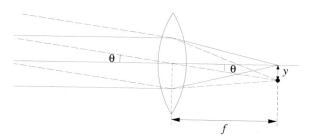

**Figure 6.6** The plate scale, determined by the focal length of the optical system.

which connects the angular separation of the objects with the linear separation of their images at the focal plane. As the focal length of the lens is increased, the linear separation of the image of two point sources separated by an angle $\theta$ also increases.

Unfortunately, the ability to "see" two objects in space that have a small angular separation $\theta$ is not simply a matter of choosing a sufficiently long focal length to produce the necessary plate scale. A fundamental limit exists in our ability to *resolve* those objects. This limitation arises due to diffraction produced by the advancing wavefronts of light coming from those objects. This phenomenon is closely related to the well-known single-slit diffraction pattern, as well as the Young double-slit interference pattern discussed in Section 3.3.

Consider for simplicity a single slit of width $D$. Assuming that the wavefronts are coherent, any ray passing through the opening (or **aperture**) and arriving at a specific point in the focal plane can be thought of as being associated with another ray passing through the aperture exactly one-half of a slit width away and arriving at the same point (see Fig. 6.7). If the two rays are one-half wavelength ($\lambda/2$) out of phase, then destructive interference will occur. This leads to the relation

$$\frac{D}{2}\sin\theta = \frac{1}{2}\lambda,$$

or

$$\sin\theta = \frac{\lambda}{D}.$$

We can next consider dividing the aperture into four equal segments, and pairing up a ray from the edge of the opening with one passing through a point one-quarter of a slit width away. For destructive interference to occur in this case it is necessary that

$$\frac{D}{4}\sin\theta = \frac{1}{2}\lambda,$$

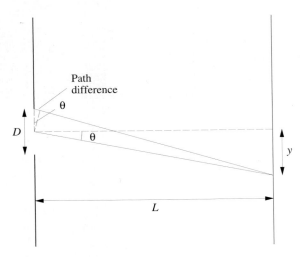

**Figure 6.7** For a minimum to occur, the path difference between paired rays must be a half-wavelength.

which gives
$$\sin\theta = 2\frac{\lambda}{D}.$$

This analysis may be continued by considering the aperture as being divided into six segments, then eight segments, then ten segments, and so on. We see, therefore, that the condition for minima to occur as a result of destructive interference from light passing through a single slit is given in general by

$$\sin\theta = m\frac{\lambda}{D}, \qquad (6.5)$$

where $m = 1, 2, 3, \ldots$ for dark fringes (as in Eq. 3.11). The intensity pattern produced by the light passing through a single slit is shown in Fig. 6.8.

The analysis for light passing through a circular aperture such as a telescope is similar, although somewhat more sophisticated. Due to the symmetry of the problem, the diffraction pattern appears as concentric rings (see Fig. 6.9). To evaluate this two-dimensional problem, it is necessary to perform a double integral over the aperture, considering the path differences of all possible pairs of rays passing through the aperture. The solution was first obtained in 1835 by Sir George Airy (1801–1892), Astronomer Royal of England; the central bright spot of the diffraction pattern is known as the **Airy disk**. Equation (6.5) remains appropriate for describing the locations of *both* the maxima and the minima, but $m$ is no longer an integer. Table 6.1 lists the values of $m$ along with the relative intensities of the maxima for the first three orders.

## 6.1 Basic Optics

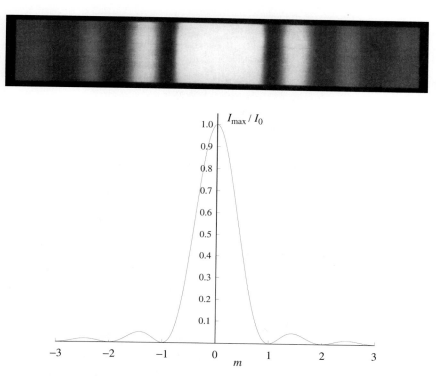

**Figure 6.8** The diffraction pattern produced by a single slit. (Photograph from Cagnet, Francon, and Thrierr, *Atlas of Optical Phenomena*, Springer-Verlag, Berlin, 1962.)

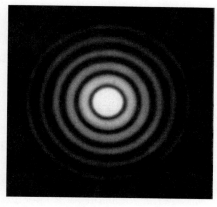

**Figure 6.9** The circular aperture diffraction pattern of a point source. (Photograph from Cagnet, Francon, and Thrierr, *Atlas of Optical Phenomena*, Springer-Verlag, Berlin, 1962.)

| Ring | $m$ | $I_{\max}/I_0$ |
|---|---|---|
| Central maximum | 0.000 | 1.00000 |
| First minimum | 1.220 | |
| Second maximum | 1.635 | 0.01750 |
| Second minimum | 2.233 | |
| Third maximum | 2.679 | 0.00416 |
| Third minimum | 3.238 | |

**Table 6.1** The Locations and Intensity Maxima of the Diffraction Rings Produced by a Circular Aperture.

As can be seen in Fig. 6.10, when the diffraction patterns of two sources are sufficiently close together (e.g., there is a very small angular separation, $\theta_{\min}$), the diffraction rings are no longer clearly distinguished and it becomes impossible to resolve the two sources. The two images are said to be unresolved when the central maximum of one pattern falls inside the location of the first minimum of the other. This *arbitrary* resolution condition is referred to as the **Rayleigh criterion**.[4] Assuming that $\theta_{\min}$ is quite small, and invoking the small-angle approximation, $\sin\theta_{\min} \simeq \theta_{\min}$, where $\theta_{\min}$ is expressed in radians, the Rayleigh criterion is given by

$$\theta_{\min} = 1.22\frac{\lambda}{D} \tag{6.6}$$

for a circular aperture. Therefore the resolution of a telescope improves with increasing size and when shorter wavelengths are observed, just as expected for diffraction phenomena.

Unfortunately, despite the implications of Eq. (6.6), the resolution of ground-based optical telescopes does not improve without limit as the size of the primary lens or mirror is increased. This property is due to the turbulent nature of Earth's atmosphere. Local changes in atmospheric temperature and density over small distances create regions where the light is refracted in nearly random directions, causing the image of a point source to become blurred. Since virtually all stars effectively appear as point sources, even when viewed through the largest telescopes, atmospheric turbulence produces the well-known "twinkling" of stellar images. The quality of the image of a stellar point source at a given observing location at a specific time is referred to as **seeing**. Some of the best seeing conditions found anywhere in the world are

---

[4]By undertaking a careful analysis of the diffraction patterns of the sources, it is possible to resolve objects that are somewhat more closely spaced than allowed by the Rayleigh criterion.

## 6.1 Basic Optics

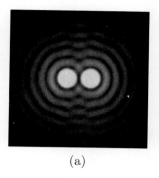
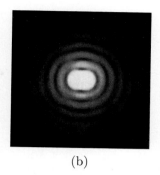

(a)  (b)

**Figure 6.10** The superimposed diffraction patterns from two point sources. (a) The sources are easily resolved. (b) The two sources are barely resolvable. (Photographs from Cagnet, Francon, and Thrierr, *Atlas of Optical Phenomena*, Springer-Verlag, Berlin, 1962.)

at **Mauna Kea Observatory** in Hawaii, located nearly 14,000 feet above sea level, where the resolution is between 0.5″ and 0.6″ approximately 50% of the time, improving to 0.25″ on the best nights. Other locations known for their excellent seeing are **Kitt Peak National Observatory** near Tucson, Arizona (Fig. 6.11), and **Cerro Tololo Inter-American Observatory** in Chile. As a result, these sites have become locations where large collections of optical telescopes have been built.

It is interesting to note that since the angular size of most planets is actually larger than the scale of atmospheric turbulence, distortions tend to be averaged out over the size of the image and the "twinkling" effect is removed.

---

**Example 6.1** After many years of delays, the **Hubble Space Telescope** (HST) was finally placed in a 380-mile-high orbit by the Space Shuttle *Discovery* in April 1990 (see Fig. 6.12). At this altitude HST is above the obscuring atmosphere of Earth, yet still accessible for needed repairs, instrument maintenance, or for a boost in its constantly decaying orbit.[5] HST is the most ambitious, and at a cost of approximately $2 billion, the most expensive scientific project ever completed.

HST has a 94-inch (2.4 m) primary mirror. When we observe at the ultraviolet wavelength of the hydrogen Lyman alpha ($L_\alpha$) line, 1216 Å, the Rayleigh

---

[5]Decaying orbits are caused by the drag produced by Earth's extended, residual atmosphere. The extent of the atmosphere is determined in part by the heating associated with the solar cycle; see Section 11.3.

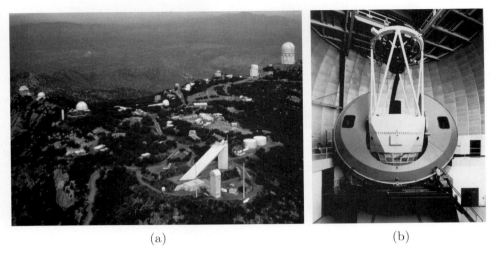

**Figure 6.11** (a) Kitt Peak National Observatory near Tucson, Arizona, is home to a large collection of optical telescopes. The 4.0-m Mayall reflector is in the upper right-hand corner of the photograph. (b) The 4.0-m Mayall reflector. (Courtesy of the National Optical Astronomy Observatories.)

**Figure 6.12** (a) The 1990 launch of the Hubble Space Telescope aboard the Space Shuttle *Discovery*. (b) HST and the Space Shuttle *Endeavour* during the December 1993 repair mission. (Courtesy of NASA.)

## 6.1 Basic Optics

criterion implies a resolution limit of

$$\theta = 1.22 \left( \frac{1216 \text{ Å}}{2.4 \text{ m}} \right) = 6.18 \times 10^{-8} \text{ rad} = 0.0127''.$$

This is roughly the equivalent of the angle subtended by a quarter from 400 km away! It was projected that HST would not quite be "diffraction-limited" in the ultraviolet region due to *extremely* small imperfections in the surfaces of the mirrors. Since resolution is proportional to wavelength and mirror defects become less significant as the wavelength increases, HST should have been nearly diffraction-limited at the red end of the visible spectrum. Unfortunately, due to an error in the grinding of the primary mirror, an optimal shape was not obtained. Consequently, those initial expectations were not reached until corrective optics packages were installed during a repair mission in December 1993.

---

Both lens and mirror systems suffer from inherent image distortions, known as **aberrations**. Often these aberrations are common to both types of systems, but **chromatic aberration** is unique to refracting telescopes. The problem stems from the fact that the focal length of a lens is wavelength-dependent. Equation (6.1) demonstrates that since the index of refraction varies with wavelength, the angle of refraction at the interface between two different media must also depend on wavelength. This translates to a wavelength-dependent focal length (Eq. 6.2) and, as a result, a focal point for blue light that differs from that for red light. The problem of chromatic aberration can be diminished somewhat by the addition of correcting lenses. The demonstration of this procedure is left as an exercise.

Several aberrations result from the shape of the reflecting or refracting surface(s). Although it is easier, and therefore cheaper, to grind lenses and mirrors into spheroids, not all areas of these surfaces will focus a parallel set of light rays to a single point. This effect, known as **spherical aberration**, can be overcome by producing carefully designed optical surfaces (paraboloids).

The cause of HST's initial imaging problems is a classic case of spherical aberration. A mistake that was made while grinding the primary left the center of the mirror too shallow by approximately two microns. The result of this minute error was that light reflected from near the edge of the mirror came to a focus almost 4 cm behind light reflected from the central portion. When the best possible compromise focal plane was used, the image of a point source (such as a distant star) had a definable central core and an extended, diffuse halo. Although the central core was quite small ($0.1''$ radius), it contained only 15% of the energy. The halo included more than half of the total energy and had a diameter of about 1.5″ (typical of ground-based telescopes). The remainder of the energy (approximately 30%) was spread out over an even larger area. Some of HST's original spherical aberration was compensated for by the use of computer programs designed to analyze the images produced by the flawed optical system and mathematically create a corrected version. During the repair mission in 1993, the COSTAR (Corrective Optics Space Telescope Axial Replacement) optics package was installed to compensate for the primary mirror's spherical aberration problem before the light reached many of the telescope's instruments. A second instrument package, Wide Field/Planetary Camera 2, has a self-contained corrective optics package (see Section 6.2). Today the spherical aberration problem of HST is only a bad memory of what can go wrong.

Even when paraboloids are used, mirrors are not necessarily free from aberrations. **Coma** produces elongated images of point sources that lie off the optical axis, because the focal lengths of paraboloids are a function of $\theta$, the angle between the direction of an incoming light ray and the optical axis. **Astigmatism** is a defect that derives from having different parts of a lens or mirror converge an image at slightly different locations on the focal plane. When a lens or mirror is designed to correct for astigmatism, **curvature of field** can then be a problem. Curvature of field is due to the focusing of images on a curve rather than on a plane. Yet another potential difficulty occurs when the plate scale (Eq. 6.4) depends on the distance from the optical axis; this effect is referred to as **distortion of field**.

It might be assumed that the **brightness** of an extended (resolved) image would increase with the area of the telescope lens, since more photons are collected as the aperture size increases; however, this assumption is not necessarily correct. To understand the brightness of an image, we begin by considering the **intensity** of the radiation. Some of the energy radiated from an infinitesimal portion of the surface of the source of area $d\sigma$ (shown in Fig. 6.13a) will enter a cone of differential **solid angle** $d\Omega \equiv dA_\perp/r^2$, where $dA_\perp$ is an infinitesimal amount of surface area that is located a distance $r$ from $d\sigma$ and oriented perpendicular to the position vector $\mathbf{r}$ (Fig. 6.13b).[6] The intensity is given by the amount of energy per unit time interval $dt$, and per unit wavelength interval $d\lambda$, radiated from $d\sigma$ into a differential solid angle $d\Omega$; the units of intensity are ergs s$^{-1}$ cm$^{-2}$ Å$^{-1}$ sr$^{-1}$.

---

[6]The unit of solid angle is the steradian, sr. It is left as an exercise to show that $\Omega_{\rm tot} = \oint d\Omega = 4\pi$ sr, the total solid angle about a point $P$, resulting from an integration over a closed surface containing that point, is $4\pi$ sr.

## 6.1 Basic Optics

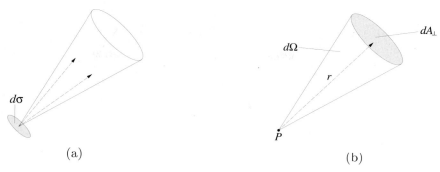

**Figure 6.13** (a) The geometry of intensity. (b) The definition of solid angle.

Consider an object located at a distance $r$ far from a telescope of focal length $f$. Assuming that the object is effectively infinitely far away (i.e., $r \gg f$), the image intensity $I_i$ may be determined from geometry. If an infinitesimal amount of surface area, $dA_0$, of the object has a surface intensity given by $I_0$, then the amount of energy per second per unit wavelength interval radiated into the solid angle defined by the telescope's aperture, $d\Omega_{T,0}$, is given by (see Fig. 6.14a)

$$I_0 \, d\Omega_{T,0} \, dA_0 = I_0 \left(\frac{A_T}{r^2}\right) dA_0,$$

where $A_T$ is the area of the telescope's aperture. Since an image will form from the photons emitted by the object, all of the photons coming from $dA_0$ must strike an area $dA_i$ on the focal plane. Therefore

$$I_0 \, d\Omega_{T,0} \, dA_0 = I_i \, d\Omega_{T,i} \, dA_i,$$

where $d\Omega_{T,i}$ is the solid angle defined by the telescope's aperture as seen from the image, or

$$I_0 \left(\frac{A_T}{r^2}\right) dA_0 = I_i \left(\frac{A_T}{f^2}\right) dA_i.$$

Solving for the image intensity gives

$$I_i = I_0 \left(\frac{dA_0/r^2}{dA_i/f^2}\right).$$

However, as can be seen in Fig. 6.14(b), the solid angle containing the object $d\Omega_{o,T}$ as seen from the center of the telescope's aperture must equal the solid angle of the image $d\Omega_{i,T}$, also seen from the telescope center, or $d\Omega_{o,T} = d\Omega_{i,T}$.

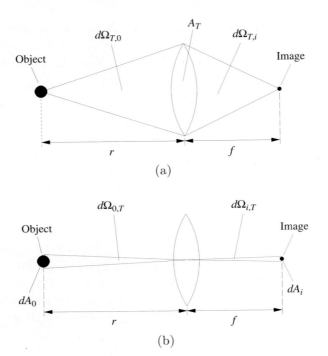

**Figure 6.14** The effect of telescopes on image intensity ($r \gg f$). (a) The solid angles subtended by the telescope, as measured from the object and the image. (b) The solid angle subtended by the object and the image, as measured from the center of the telescope.

This implies that

$$\frac{dA_0}{r^2} = \frac{dA_i}{f^2}.$$

Substituting into the expression for the image intensity gives the result that

$$I_i = I_0,$$

the image intensity is identical to the object intensity, independent of the area of the aperture. This result is completely analogous to the simple observation that a wall does not appear to get brighter when the observer walks toward it.

The concept that describes the effect of the light-gathering power of telescope is the **illumination** $J$, the amount of light energy per second focused onto a unit area of the resolved image. Since the amount of light collected from the source is proportional to the area of the aperture, the illumination $J \propto \pi(D/2)^2 = \pi D^2/4$, where $D$ is the diameter of the aperture. We have also shown that the linear size of the image is proportional to the focal length of the lens (Eq. 6.3); therefore the image area must be proportional to $f^2$ and correspondingly the illumination must be inversely proportional to $f^2$. Combining these results, the illumination must be proportional to the square of the ratio of the aperture diameter to the focal length. The inverse of this ratio is often referred to as the **focal ratio**,

$$F \equiv \frac{f}{D}. \tag{6.7}$$

Thus the illumination is related to the focal ratio by

$$J \propto \frac{1}{F^2}. \tag{6.8}$$

Since the number of photons per second striking a unit area of photographic plate or some other detector is described by the illumination, the illumination must indicate the amount of time required to collect the photons needed to form a sufficiently bright image for analysis.

---

**Example 6.2** The twin multimirror telescopes of the **Keck Observatory** at Mauna Kea have 10-m diameter primary mirrors with focal lengths of 17.5 m. The focal ratios of these mirrors are

$$F = \frac{f}{D} = 1.75.$$

It is standard to express focal ratios in the form $f/F$, where $f/$ signifies that the focal ratio is being referenced. Using this notation, the Keck telescopes have 10 m, $f/1.75$ primary mirrors.

---

We now see that the size of the aperture of a telescope is critical for two reasons: A larger aperture both improves resolution and increases the illumination. On the other hand, a longer focal length increases the linear size of the image but decreases the illumination. For a *fixed* focal ratio, increasing the diameter of the telescope results in greater spatial resolution, but the illumination remains constant. The proper design of a telescope must consider the principal applications that are intended for the instrument.

## 6.2 Optical Telescopes

The major optical component of a refracting telescope is the primary or *objective* lens of focal length $f_{\text{obj}}$. The purpose of the objective lens is to collect as much light as possible and with the greatest possible resolution, bringing the

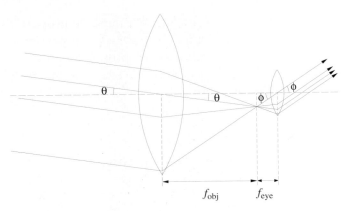

**Figure 6.15** A refracting telescope is composed of an objective lens and an eyepiece.

light to a focus at the focal plane. A photographic plate or other detector may be placed at the focal plane to record the image, or the image may be viewed with an eyepiece, which serves as a magnifying glass. The eyepiece would be placed at a distance from the focal plane equal to its focal length, $f_{\text{eye}}$, causing the light rays to be refocused at infinity. Figure 6.15 shows the path of rays coming from a point source lying off the optical axis at an angle $\theta$. The rays ultimately emerge from the eyepiece at an angle $\phi$ from the optical axis. The **angular magnification** produced by this arrangement of lenses can be shown to be (Problem 6.5)

$$m = \frac{f_{\text{obj}}}{f_{\text{eye}}}. \tag{6.9}$$

Clearly, eyepieces of different focal lengths can produce different angular magnifications. Viewing a large image requires a long objective focal length, in combination with a short focal length for the eyepiece.

Recall, however, that the illumination decreases with the square of the objective's focal length (Eq. 6.8). To compensate for the diminished illumination, a larger diameter objective is needed. Unfortunately, significant practical limitations exist for the size of the objective lens of a refracting telescope. Because light must pass through the objective lens, it is possible only to support the lens from its edges. As a result, when the size and weight of the lens are increased, deformation in its shape will occur due to gravity. The specific form of the deformation will depend upon the position of the objective, which changes as the orientation of the telescope changes.

Another problem related to size is the difficulty in constructing a lens that is sufficiently free of defects. Since light must pass through the lens, its entire

## 6.2 Optical Telescopes

**Figure 6.16** The 40-in telescope at Yerkes Observatory was built in 1897 and is the largest refractor in the world. (Courtesy of Yerkes Observatory.)

*volume* must be nearly optically perfect. Furthermore, *both* surfaces of the lens must be ground with great precision. Specifically, any defects in the material from which the lens is made or any deviations from the desired shape of the surface must be kept to less than some small fraction of the wavelength, typically $\lambda/20$. When observing at 5000 Å, this implies that any defects must be smaller than approximately 250 Å. (Recall that the size of an atom is on the order of 1 Å.)

Yet another difficulty with a large objective lens occurs because of its slow thermal response. When the dome is opened, the temperature of the telescope must adjust to its new surroundings. This produces thermally driven air currents around the telescope, significantly affecting seeing. The shape of the telescope will also change due to thermal expansion, making it advantageous to minimize the "thermal mass" of the telescope as much as possible.

A mechanical problem also arises with long focal-length refractors. Due to the long lever arm involved, placing a massive detector on the end of the telescope will create a large amount of torque that requires compensation.

We have already discussed the unique problem of chromatic aberration in lenses, a complication not shared by mirrors. Considering all of the challenges inherent in the design and construction of refracting telescopes, the vast majority of all large modern telescopes are reflectors. The largest refracting telescope in use today is at the **Yerkes Observatory** in Williams Bay, Wisconsin (Fig. 6.16). It was built in 1897 and has a 1.02-m (40 in) objective with a focal length of 19.36 m.

With the exception of chromatic aberration, most of the basic optical principles already discussed apply equally well to reflectors and refractors. A reflecting telescope is designed by replacing the objective lens with a mirror, significantly reducing or completely eliminating many of the problems already discussed. Because the light does not pass through a mirror, only the *one* reflecting surface needs to be ground with precision. Also, the weight of the mirror can be minimized by creating a honeycomb structure behind the reflecting surface, removing a large amount of unnecessary mass. In fact, because the mirror is supported from behind rather than along its edges, it is possible to design an active system of pressure pads that can help to eliminate distortions in the mirror's shape produced by the changes in the gravitational force on the mirror as the telescope moves.

Reflecting telescopes are not completely free of drawbacks, however. Since the objective mirror reflects light back along the direction from which it came, the focal point of the mirror, known as the **prime focus**, is in the path of the incoming light (see Fig. 6.17a). An observer or a detector can be placed at this position, but then some of the incident light is cut off (see Fig. 6.18). If the detector is too large, a substantial amount of light will be lost.

Isaac Newton first found a solution to the problem by placing a small, flat mirror in the reflected light's path, changing the location of the focal point; this arrangement is depicted in Fig. 6.17(b). Of course, the presence of this secondary mirror does block some of the incoming light from the primary, but if the ratio of the areas of the primary and secondary is sufficiently large, the effect of the lost light can be minimized. Such a **Newtonian** telescope suffers from the drawback that the eyepiece (or detector) must be placed at a significant distance from the center of mass of the system. If a massive detector were used, it would exert a significant torque on the telescope.

Since the region of the primary mirror located behind the secondary is effectively useless anyway, it is possible to bore a hole in the primary and use the secondary to reflect the light back through the hole. This **Cassegrain** design (Fig. 6.17c) allows the placement of heavy instrument packages near the center of mass of the telescope and permits an observer to stay near the bottom of the telescope, rather than near the top, as is the case for Newtonians. In this type of design the secondary mirror is usually convex, effectively increasing the focal length of the system.

If the instrument package is too massive, it is often more effective to bring the light directly to a special laboratory in which the detector is located. A **coudé** telescope (Fig. 6.17d) uses a series of mirrors to reflect the light down the telescope's mount to a *coudé room* located below the telescope. Because of the extended optical path, it is possible to create a very long focal length with

## 6.2 Optical Telescopes

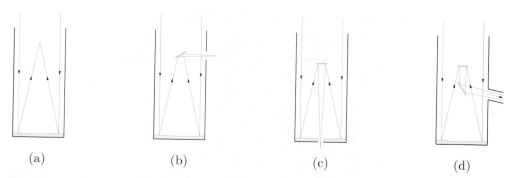

**Figure 6.17** Schematic drawings of various telescope optical systems: (a) Prime focus, (b) Newtonian, (c) Cassegrain, (d) coudé.

**Figure 6.18** Edwin Hubble working at the prime focus of the Hale reflecting telescope on Mount Palomar. (Courtesy of Palomar/Caltech.)

a coudé telescope. This can be particularly useful in high resolution work or in high dispersion spectral studies (see Section 5.1).

A unique instrument is the **Schmidt** telescope, specifically designed to provide a wide-angle field of view with low distortion. Schmidt telescopes are generally used as cameras, with the photographic plate located at the prime focus. To minimize coma, a spheroidal primary mirror is used, combined with a "correcting" lens to help remove spherical aberration. Whereas a large

Cassegrain telescope may have a field of view of a few arc minutes across, a Schmidt camera has a field of view of several degrees. These instruments provide important survey studies of large regions of the sky. For example, 1477 Schmidt plates have been scanned to produce the Guide Star Catalogue of 18,819,291 objects as faint as 15th magnitude. Of these, 15,169,873 are stars. This catalog is being used to supply the reference (or *guide*) stars needed to orient the Hubble Space Telescope.

Producing high resolution, deep-sky images of faint objects requires that the telescope be pointed at a fixed region of the sky for an extended period of time. For instance, if a photograph is being taken, enough photons must be collected to ensure that the desired object will be seen. Such *time integration* requires careful guiding of the telescope while compensating for the rotation of Earth.

The most common type of telescope mount is the **equatorial mount**. It incorporates a polar axis that is aligned to the north celestial pole, and the telescope simply rotates about that axis to compensate for the changing altitude and azimuth of the object of interest. With the equatorial mount, it is a simple matter to adjust the position of the telescope in both right ascension and declination. Unfortunately, the mount for a massive telescope can be extremely expensive and difficult to build. An alternative, more easily constructed mount for large telescopes, the **altitude–azimuth mount**, permits motion both parallel and perpendicular to the horizon. In this case, however, the tracking of a celestial object requires the continuous calculation of its altitude and azimuth from knowledge of the object's right ascension and declination, along with the local sidereal time and latitude of the telescope. A second difficulty with altitude–azimuth mounts is the effect of the continuous rotation of image fields. Without proper adjustment this can create complications when guiding the telescope during an extended exposure or when a spectrum is obtained by passing the light through a long slit. Fortunately, computers now allow compensation for all these effects.

The telescope at the **Special Astrophysical Observatory**, built in 1976 and located in the Caucus Mountains near Zelenchukskaya in the former Soviet Union, uses an unsegmented 6.0-m-diameter primary mirror. When the telescope was built, it was believed that the practical limit to telescope size had been reached. Despite efforts to decrease the mass of the mirror, problems with its shape persisted, as did the problem of convective turbulence affecting seeing. Today, however, telescopes are designed and built that make use of segmented primary mirrors. The mirror's components are constantly realigned using the interference patterns produced by beams of laser light that detect minute changes in position, a process known as **laser interferometry**. By

## 6.2 Optical Telescopes

counting interference fringes, the distance between mirror segments can be estimated to a small fraction of a wavelength. For instance, this technique is used to align the 36 hexagonal components that comprise each of the twin 10-m primary mirrors of the Keck Observatory. The two telescopes also work in tandem as a large optical interferometer (analogous to the technique of radio interferometry, discussed in Section 6.3).

In March 1988 the **New Technology Telescope** (NTT) at the **European Southern Observatory** (ESO) at La Silla in the Chilean Andes saw its first light. It has a 3.58-m primary mirror with a focal ratio of $f/2.2$. At a cost of \$13 million, making it one-third as expensive as another 3.6-m telescope at La Silla (which was built in 1976), the NTT has three times the resolution, requires one-third the integration time, and has one-third the mass of its older companion. The combined optics of the primary and secondary mirrors places approximately 80% of the light from a point source within the $0.125''$ diameter Airy disk. Combined with the superb seeing at the site, the NTT can achieve a resolution of $0.36''$. Furthermore, the NTT has also been able to image objects as faint as 29th magnitude, a phenomenal capability accomplished through the use of **adaptive optics**. The instrument employs 75 adjustable pressure pads on the back of the primary to modify automatically the shape of the mirror when it is in different positions. This is accomplished by analyzing the displacement of images of guide stars, with the entire process being carried out without the involvement of the observer.[7] The observatory also uses a special, wind-tunnel tested, octagonally shaped building that rotates with the telescope, together with an active climate control system, to minimize local seeing effects.

To overcome the inherent imaging problems imposed by Earth's atmosphere, observational astronomy is also carried out in space. The **Hubble Space Telescope** (Fig. 6.12) has a 2.4-m, $f/24$, primary that is the smoothest mirror ever constructed, with no surface imperfection larger than 1/50 of the 6328 Å test wavelength. Long duration exposures of 18 hours or more allow the telescope to "see" objects at least as faint as 29th magnitude. The optical system used by HST is of the Ritchey–Chrétien type, a modified Cassegrain system that operates from 1200 Å to 1 $\mu$m (ultraviolet to infrared, respectively).

As of 1995, the following complement of instrument and auxiliary optical packages are on board HST, located in the focal plane:

---

[7]Laser light, back scattered from air molecules, has also been used at some observatories to create *artificial* guide stars in the atmosphere that then provide the information required for the adaptive optics.

- *Faint Object Camera* (FOC). Using its internal optical system, this instrument can operate at either $f/48$ or $f/96$. The camera possesses a variety of filters, prisms, and polarizers that allow it to provide a range of functions, including very high resolution work. The FOC was built by the European Space Agency.

- *Faint Object Spectrograph.* This instrument is designed to produce moderate resolution spectra of very faint sources.

- *Fine Guidance Sensors.* These sensors track pairs of guide stars selected from the catalog of over 15 million guide stars produced by surveying Schmidt plates as discussed earlier. The sensors also make precise measurements of the positions of astronomical objects.[8]

- *Goddard High Resolution Spectrograph.* Built to operate in the ultraviolet, this instrument provides spectral resolution down to 0.01 Å for bright sources.

- *Wide-Field and Planetary Camera 2.* The wide-field and planetary camera (WF/PC 2), a second-generation instrument, replaced the original WF/PC. The only instrument located in the middle of the focal plane, it is designed to be the most versatile detector on board. With more than 2 1/2 million picture elements, or *pixels*, it has the capability of operating in two modes. In the "wide-field" mode it has a focal ratio of $f/12.9$ with a plate scale of $0.0966''$ pixel$^{-1}$, and in the "planetary" mode its focal ratio and plate scale are $f/28.3$ and $0.0455''$ pixel$^{-1}$, respectively. Along with generating many of the spectacular images returned by HST, WF/PC 2 can be used with a variety of filters to isolate specific wavelengths, investigate the amount of polarization present in the light, and measure spectra.

- *Corrective Optics Space Telescope Axial Replacement.* COSTAR is the optics package designed to correct image problems introduced by HST's flawed primary mirror (see the discussion on page 169). The 290-kg package contains five pairs of mirrors ranging from 12 to 25 mm in diameter that are used by the faint object camera, the faint object spectrograph, and the Goddard high resolution spectrograph. COSTAR replaced the high-speed photometer that was a part of the original complement of HST instrument packages.

---

[8]The study of astronomical positions is known as **astrometry**.

Although the human eye and photographic plates have traditionally been the tools of astronomers to record images and spectra, other, more efficient devices have recently been developed. In particular, the semiconductor detector known as the charge-coupled device (CCD) has revolutionized the way in which photons are counted. Whereas the human eye has a very low *quantum efficiency* of approximately 1% (one photon in one hundred is detected) and photographic plates do only slightly better, CCDs are able to detect nearly 100% of the incident photons. Moreover, CCDs are able to detect a very wide range of wavelengths. From soft (low-energy) x-rays to the infrared, they have a linear response: Ten times as many photons produce a signal ten times stronger. CCDs also have a wide dynamic range, and so can differentiate between very bright and very dim objects that are viewed simultaneously.

A CCD works by collecting electrons that are excited into higher energy states (conduction bands) when the detector is struck by a photon (a process similar to the photoelectric effect). The number of electrons collected in each pixel is then proportional to the brightness of the image at that location. The 2 1/2 million pixels of HST's WF/PC 2 are the individual elements of four $800 \times 800$ pixel CCD cameras, with each pixel capable of holding up to 70,000 electrons.

Given the rapid improvement in both ground-based and orbital telescopes, along with the development of CCDs, it is clear that the future of optical astronomy is indeed a bright one.

## 6.3 Radio Telescopes

In 1931 Karl Jansky (1905–1950) was conducting experiments for Bell Laboratories related to the production of radio-wavelength static from thunderstorms. During the course of his investigations Jansky discovered that some of the static in his receiver was of "extraterrestrial origin." By 1935 he had correctly concluded that much of the signal he was measuring originated in the plane of the Milky Way, with the strongest emission coming from the constellation Sagittarius, which lies in the direction of the center of our Galaxy. Jansky's pioneering work represented the birth of a whole new field of observational study, *radio astronomy*.

Today radio astronomy plays an important role in our investigation of the electromagnetic spectrum. Radio waves are produced by a variety of mechanisms related to a range of physical processes, such as the interactions of charged particles with magnetic fields. This new window on the universe

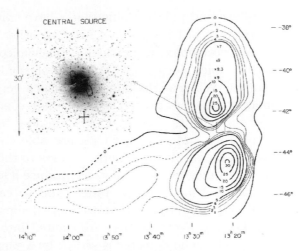

**Figure 6.19** A radio map of Centaurus A, together with an optical image of the same region. The contours show lines of constant radio power. (Figure from Matthews, Morgan, and Schmidt, *Ap. J.*, *140*, 35, 1964.)

provides astronomers and physicists with valuable clues to the inner workings of some of nature's most spectacular phenomena.

Since radio waves interact with matter differently than visible light does, the devices used to detect and measure it are necessarily very different from optical telescopes. The parabolic dish of a typical radio telescope reflects the radio energy of the source to an antenna. The signal is then amplified and processed to produce a *radio map* of the sky at a particular wavelength, like the one shown in Fig. 6.19.

The strength of a radio source is measured in terms of the **spectral flux density**, $S(\nu)$, the amount of energy per second, per unit frequency interval striking a unit area of the telescope. To determine the total amount of energy per second (the power) collected by the receiver, the spectral flux must be integrated over the telescope's collecting area and over the frequency interval for which the detector is sensitive, referred to as the *bandwidth*. If $f_\nu$ is a function describing the efficiency of the detector at the frequency $\nu$, then the amount of energy detected per second becomes[9]

$$P = \int_A \int_\nu S(\nu) f_\nu \, d\nu \, dA. \tag{6.10}$$

If the detector is 100% efficient over a frequency interval $\Delta\nu$, and $S(\nu)$ can be

---

[9] A similar expression applies to optical telescopes since filters and detectors (including the human eye) are frequency- or wavelength-dependent; see Section 3.6.

## 6.3 Radio Telescopes

considered to be constant over that interval, then the integral simplifies to give

$$P = SA\Delta\nu,$$

where $A$ is the effective area of the aperture.

A typical radio source has a spectral flux density $S(\nu)$ on the order of one Jansky (Jy), 1 Jy = $10^{-26}$ W m$^{-2}$ Hz$^{-1}$ = $10^{-23}$ erg s$^{-1}$ cm$^{-2}$ Hz$^{-1}$. Spectral flux density measurements of several mJy are not uncommon. With such weak sources, a large aperture is needed to collect enough photons to be measurable.

---

**Example 6.3** The second strongest radio source in the sky, after the Sun, is the galaxy Cygnus A. At 400 MHz (a wavelength of 75 cm), its spectral flux density is 4500 Jy. Assuming that a 25-m-diameter radio telescope is 100% efficient and is used to collect the radio energy of this source over a frequency bandwidth of 5 MHz, the total power detected by the receiver would be

$$P = S(\nu)\pi \left(\frac{D}{2}\right)^2 \Delta\nu = 1.1 \times 10^{-6} \text{ erg s}^{-1}.$$

---

One problem that radio telescopes share with optical telescopes is the need for greater resolution. Rayleigh's criterion (Eq. 6.6) applies to radio telescopes just as it does in the visible regime, except that radio wavelengths are much longer than those involved in optical work. Therefore, to obtain a level of resolution comparable to what is reached in the visible, much larger diameters are needed.

---

**Example 6.4** To obtain a resolution of 1″ at a wavelength of 21 cm using a single aperture, the dish diameter must be

$$D = 1.22 \frac{\lambda}{\theta} = 1.22 \left(\frac{0.21 \text{ m}}{4.85 \times 10^{-6} \text{ rad}}\right) = 52.8 \text{ km}.$$

For comparison, the largest single-dish radio telescope in the world is the fixed 300 m (1000 foot) diameter dish at **Arecibo Observatory**, Puerto Rico (see Fig. 6.20).

---

One advantage of working at such long wavelengths is that small deviations from an ideal parabolic shape are not nearly as crucial. Since the relevant criterion is to be within some small fraction of a wavelength (say $\lambda/20$) of what is considered to be a perfect shape, variations of 1 cm are tolerable when observing at 21 cm.

**Figure 6.20** The 300-m radio telescope at Arecibo Observatory, Puerto Rico. (Courtesy of the NAIC–Arecibo Observatory, which is operated by Cornell University for the National Science Foundation.)

Although it is clearly prohibitive to build individual dishes of sufficient size to produce the resolution at radio wavelengths that is anything like what is obtainable from the ground in the visible regime, astronomers have nevertheless been able to resolve radio images to better than $0.0015''$. This remarkable resolution is accomplished using a process not unlike the interference technique used in the Young double-slit experiment.

Figure 6.21 shows two radio telescopes separated by a baseline of distance $d$. Since the distance from telescope B to the source is greater than the distance from telescope A to the source by an amount $L$, a specific wavefront will arrive at B after it has reached A. The two signals will be *in phase* and their superposition will result in a maximum if $L$ is equal to an integral number of wavelengths ($L = n\lambda$, where $n = 0, 1, 2, \ldots$ for constructive interference). Similarly, if $L$ is an odd integral number of half-wavelengths, then the signals will be exactly *out of phase* and a superposition of signals will result in a minimum in the signal strength ($L = n - \frac{1}{2}$, where $n = 1, 2, \ldots$ for destructive interference). Since the pointing angle $\theta$ is related to $d$ and $L$ by

$$\sin\theta = \frac{L}{d}, \tag{6.11}$$

it is then possible to determine accurately the position of the source using the interference pattern that is produced by combining the signals of the two antennas. Equation (6.11) is completely analogous to Eq. (3.11) describing the Young double-slit experiment.

## 6.3 Radio Telescopes

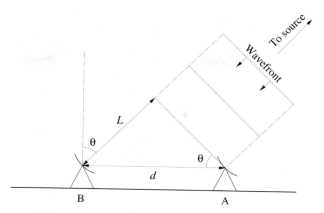

**Figure 6.21** The technique of radio interferometry.

Clearly the ability to resolve an image improves with a longer baseline $d$. Very long baseline interferometry (**VLBI**) is possible over the size of a continent or even between continents. In such cases the data can be recorded on site and delivered to a central location for processing at a later time. It is only necessary that the observations be simultaneous and that the exact time of data acquisition be recorded.

Recall that increasing the number of lines in a diffraction grating improves the ability of a spectrograph to resolve two closely spaced wavelengths (Eq. 5.2). Similarly, the resolving power of a radio interferometer also increases with the addition of more telescopes to the array. Although a single antenna is most sensitive in the direction in which it is pointing, the antenna can have a significant amount of sensitivity at angles far from the direction desired. Figure 6.22 shows a typical **antenna pattern** for a single radio telescope. It is a polar coordinate plot describing the direction of the antenna pattern along with the relative sensitivity in each direction; the longer the *lobe* the more sensitive the telescope is in that direction. Two characteristics are immediately noticeable: First, the main lobe is not infinitesimally thin (the directionality of the beam is not perfect), and second, side lobes exist that can result in the accidental detection of unwanted sources that are indistinguishable from the desired source.

The narrowness of the main lobe is described by specifying its angular width at half its length, referred to as the *half-power beam width* (HPBW). This width can be decreased, and the effect of the side lobes can be significantly reduced, by the addition of other telescopes to produce the desired diffraction pattern. This property is analogous to the increase in sharpness of a grating diffraction pattern as the number of grating lines is increased.

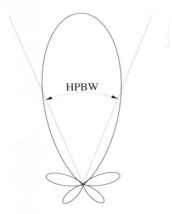

**Figure 6.22** A typical antenna pattern for a single radio telescope. The width of the main lobe is describable by the half-power beam width (HPBW).

**Figure 6.23** The Very Large Array (VLA) near Socorro, New Mexico. (Courtesy of the National Radio Astronomy Observatory, ©NRAO/AUI.)

The **Very Large Array** (VLA) located near Socorro, New Mexico, consists of 27 radio telescopes in a movable Y configuration with a maximum configuration diameter of 27 km. Each individual dish has a diameter of 25 m and uses receivers sensitive at a variety of frequencies (see Fig. 6.23). The signal from each of the separate telescopes is combined with all of the others and analyzed by computer to produce a high-resolution map of the sky. Of course, along with the resolution gain, the 27 telescopes combine to produce

## 6.4 Infrared, Ultraviolet, and X-Ray Astronomy

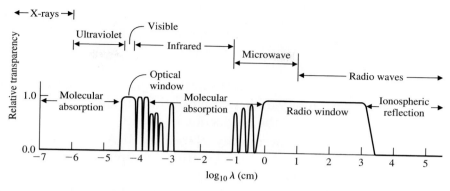

**Figure 6.24** The transparency of Earth's atmosphere as a function of wavelength.

an effective collecting area that is 27 times greater than that of an individual telescope.

To produce even higher resolution maps, the National Radio Astronomy Observatory operates the **Very Long Baseline Array**, composed of a series of telescopes scattered throughout the continental United States, Hawaii, and Puerto Rico.

## 6.4 Infrared, Ultraviolet, and X-Ray Astronomy

Given the enormous amount of data supplied by optical and radio observations, it is natural to consider studies in other wavelength regions. Unfortunately, such observations are either difficult or impossible to perform from the ground due to Earth's atmosphere being opaque to most wavelength regions outside of the visible and radio bands. Figure 6.24 shows the transparency of the atmosphere as a function of wavelength. Long wavelength ultraviolet radiation and some regions in the infrared are able to traverse the atmosphere with limited success but other wavelength regimes are completely blocked. For this reason, special measures must be taken to gather information at many photon energies.

The primary contributor to infrared absorption is water vapor. As a result, if an observatory can be placed above most of the atmospheric water vapor, some observations can be made from the ground. To this end, NASA operates a 3-m infrared telescope on Mauna Kea, where the humidity is quite low. Even at 14,000 feet, however, the problem is not completely solved. To get above more of the atmosphere, balloon and aircraft observations have also been used.

Besides atmospheric absorption, the situation in the infrared is complicated still further because steps must be taken to cool the detector, if not the entire telescope. Using Wien's displacement law (Eq. 3.15) the peak wavelength of a blackbody of temperature 300 K is found to be nearly 10 $\mu$m. Thus the telescope and its detectors can produce radiation in just the wavelength region the observer might be interested in. Of course, the atmosphere can radiate in the infrared as well, including the production of molecular IR emission lines.

In 1983 the **Infrared Astronomy Satellite** (IRAS) was placed in a 560-mile-high orbit, well above Earth's obscuring atmosphere. The 0.6-m imaging telescope was cooled to liquid helium temperatures, and its detectors were designed to observe at a variety of wavelengths, from 12 $\mu$m to 100 $\mu$m. Before its coolant was exhausted, IRAS proved to be very successful. Among its many accomplishments was the detection of dust in orbit around young stars, possibly indicating the formation of planetary systems. IRAS was also responsible for many important observations concerning the nature of galaxies.

Based upon the success of IRAS, the European Space Agency plans to launch the **Infrared Space Observatory** (ISO) in 1995. The observatory will be cooled, just as IRAS was, but to obtain nearly 1000 times the resolution of IRAS, ISO will be able to point toward a target for a much longer period of time, allowing it to collect a greater number of photons.[10]

ISO will make use of the revolutionary new detectors now becoming available for infrared work. Since infrared photons are not generally energetic enough to activate a photographic emulsion or eject electrons from most standard metals, detector technology differs from that used in the optical. Traditionally, *bolometers* have been the detectors of choice for infrared observations. These devices, whose properties (e.g., electrical resistance) change as a result of being heated by infrared radiation, are usually composed of only one element (pixel). To form an image of a source, the detector must scan the source one section at a time, a very time-consuming and inaccurate process. Today, semiconductor technology has begun to supply arrays of detectors that are much like a silicon CCD, differing only in their use of a variety of hybrid (or "doped") materials, each sensitive in a specific wavelength range.

Designed to investigate the electromagnetic spectrum at the longer wavelengths of the microwave regime, the **Cosmic Background Explorer** (COBE; Fig. 6.25a) was launched in 1989 and finally switched off in 1993. COBE made a number of important observations, including very precise mea-

---

[10]NASA has been developing plans to build its own successor to IRAS, called the **Space Infrared Telescope Facility** (SIRTF). However, at the time this book was written, SIRTF had not yet received funding from Congress.

## 6.4 Infrared, Ultraviolet, and X-Ray Astronomy

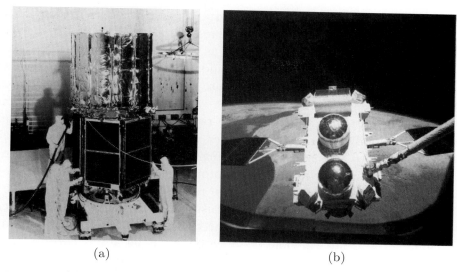

**Figure 6.25** (a) The Cosmic Background Explorer Satellite during assembly. (Courtesy of the COBE Science Working Group and NASA's Goddard Space Flight Center.) (b) The Compton Gamma Ray Observatory being deployed by the Space Shuttle *Atlantis* in 1991. (Courtesy of NASA.)

surements of the 2.7 K blackbody spectrum believed to be the remnant fireball of the Big Bang.

As in other wavelength regions, a number of challenges exist when observing in the ultraviolet portion of the electromagnetic spectrum. In this case, because of the short wavelengths involved (as compared to optical observations), great care must be taken to provide a very precise reflecting surface. As has already been mentioned, even the HST primary mirror has imperfections that prohibit shorter UV wavelengths from being observed at the theoretical resolution limit.

A second UV observing problem stems from the fact that glass is opaque to these short wavelength photons (as it is for much of the infrared). Consequently, glass lenses cannot be used in the optical system of a telescope designed to observe in the ultraviolet. Lenses made of crystal provide an appropriate substitute, however.

A real workhorse of ultraviolet astronomy has been the **International Ultraviolet Explorer**. Launched in 1978, the IUE proved to be a remarkably productive and durable instrument. Today, HST, with its sensitivity down to 1200 Å, provides another important window on the ultraviolet universe. At even shorter wavelengths, the **Extreme Ultraviolet Explorer**, launched in 1992, has made observations between 60 Å and 740 Å. The data from these telescopes have given astronomers important information concerning a vast array of astrophysical processes, including mass loss from hot stars, cataclysmic variable stars, and compact objects such as white dwarfs and pulsars.

At even shorter wavelengths, x-ray and gamma-ray astronomy yields information about very energetic phenomena, such as nuclear reaction processes and the environment around black holes. As a result of the very high photon energies involved, x-ray and gamma-ray observations require techniques that differ markedly from those at longer wavelengths. For instance, traditional glass mirrors are useless for forming images in this regime because of the great penetrating power of these photons. However, it is still possible to image sources by using grazing-incidence reflections (incident angle close to 90°). X-ray spectra can also be obtained using techniques such as Bragg scattering, an interference phenomenon produced by photon reflections from atoms in a regular crystal lattice. The distance between the atoms corresponds to the separation between slits in an optical diffraction grating.

In 1970 **UHURU** (also known as the Small Astronomy Satellite–1, SAS 1) made the first comprehensive survey of the x-ray sky. In the late 1970s the three **High Energy Astrophysical Observatories**, including the **Einstein Observatory**, discovered thousands of x-ray and gamma-ray sources. Launched in 1990, the x-ray observatory ROSAT (the **Roentgen Satellite**), a German–American–British satellite consisting of two detectors and an imaging telescope operating in the range of 5.1 Å to 124 Å, has investigated the hot coronas of stars, supernova remnants, and quasars. Japan's **Advanced Satellite for Cosmology and Astrophysics**, which began its mission in 1993, has also made valuable x-ray observations of the heavens. Planned for launch in 1998, NASA's **Advanced X-ray Astrophysics Facility** (AXAF) will carry a camera capable of producing x-ray images with a resolution of $1''$, comparable to ground-based optical observatories.

The **Compton Gamma Ray Observatory**[11] (CGRO; Fig. 6.25b) is designed to observe the heavens at wavelengths shorter than those measured by the x-ray telescopes. Placed into orbit by the Space Shuttle *Atlantis* in 1991, CGRO is composed of four experiments with overlapping energy ranges:

- *Burst and Transient Source Experiment* (BATSE). Designed to detect mysterious short-lived gamma-ray bursts in the energy range 20 to 600 keV, BATSE is able to alert the other instruments on board the spacecraft so that they can observe the event as well.

---

[11]The Compton Gamma Ray Observatory is the second of NASA's family of four Great Observatories to be placed in orbit, the first being the Hubble Space Telescope. The others are SIRTF (see footnote 10 in this chapter) and AXAF.

## 6.4 Infrared, Ultraviolet, and X-Ray Astronomy

- *Oriented Scintillation Spectrometer Experiment.* This instrument is sensitive to photons with energies between 100 keV and 10 MeV, in the range of those produced by radioactive elements.

- *Imaging Compton Telescope.* Photons with energies between 1 and 30 MeV are measured by this telescope. The detector is also capable of isolating the sources of the photons.

- *Energetic Gamma Ray Experiment Telescope* (EGRET). EGRET is sensitive to photons in the 20 MeV to 30 GeV range. These photons originate in very high energy processes such as stellar explosions and matter–antimatter annihilations.

Our ability to probe the heavens at wavelengths spanning the electromagnetic spectrum has provided an enormous amount of information not previously available from ground-based observations made in the visible wavelength region only. The change in the appearance of the sky when different wavelength regions are explored is illustrated in the radio, infrared, visible, ultraviolet, and gamma-ray images shown in Fig. 6.26. Notice that the plane of our Milky Way galaxy is clearly evident in each of the wavelength bands but that other features are not necessarily present in each image.

With the past successes of ground-based and orbital observatories, astronomers have been able to make great leaps in our understanding of the universe. Given the current advances in detectors and observational techniques, instruments of the future promise to provide significantly improved views of known objects in the heavens. However, perhaps the most exciting implications of these observational advances are to be found in as yet undiscovered and unanticipated phenomena.

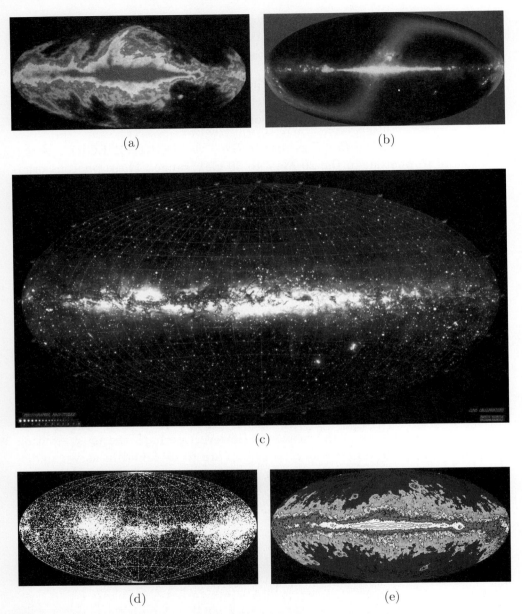

**Figure 6.26** Observations of the entire sky as seen in (a) radio (courtesy of the Max–Planck Institut für Radioastronomie), (b) infrared (courtesy of the COBE Science Working Group and NASA's Goddard Space Flight Center), (c) visible (courtesy of Lund Observatory), (d) ultraviolet (copyright Royal Observatory, Edinburgh), and (e) gamma-ray wavelengths (courtesy of NASA).

# Suggested Readings

## General

Fugate, Robert Q., and Wild, Walter J., "Untwinkling the Stars—Part I," *Sky and Telescope*, May 1994.

Janesick, James, "Sky on a Chip: The Fabulous CCD," *Sky and Telescope*, September 1987.

Martin, Buddy, Hill, John M., and Angel, Roger, "The New Ground-Based Optical Telescopes," *Physics Today*, March 1991.

Martinez, Patrick (ed.), *The Observer's Guide to Astronomy: Volume 1*, Cambridge University Press, Cambridge, 1994.

O'Dell, C. Robert, "Building the Hubble Space Telescope," *Sky and Telescope*, July 1989.

Roy, A. E., and Clarke, D., *Astronomy: Principles and Practice*, Third Edition, Adam Hilger, Bristol, 1988.

Sherrod, P. Clay, *A Complete Manual of Amateur Astronomy: Tools and Techniques for Astronomical Observations*, Prentice-Hall, Englewood Cliffs, NJ, 1981.

Stephens, Sally, "'We Nailed It!' A First Look at the New and Improved Hubble Space Telescope," *Mercury*, January/February 1994.

Wild, Walter J., and Fugate, Robert Q., "Untwinkling the Stars—Part II," *Sky and Telescope*, June 1994.

## Technical

Beckers, Jacques M., "Adaptive Optics for Astronomy: Principles, Performance, and Applications," *Annual Review of Astronomy and Astrophysics*, *31*, 1993.

Culhane, J. Leonard, and Sanford, Peter W., *X-ray Astronomy*, Faber and Faber, London, 1981.

Jenkins, Francis A., and White, Harvey E., *Fundamentals of Optics*, Fourth Edition, McGraw-Hill, New York, 1976.

Kraus, John D., *Radio Astronomy*, Second Edition, Cygnus-Quasar Books, Powell, Ohio, 1986.

## Problems

**6.1** For some point $P$ in space, show that, for any arbitrary closed surface surrounding $P$, the integral over a solid angle about $P$ gives,

$$\Omega_{\text{tot}} = \oint d\Omega = 4\pi.$$

**6.2** The light rays coming from an object do not, in general, travel parallel to the optical axis of a lens or mirror system. Consider an arrow to be the object, located a distance $p$ from the center of a simple converging lens of focal length $f$, such that $p > f$. Assume that the arrow is perpendicular to the optical axis of the system with the tail of the arrow located on the axis. To locate the image, draw two light rays coming from the tip of the arrow:

(i) One ray should follow a path *parallel* to the optical axis until it strikes the lens. It then bends toward the focal point of the side of the lens opposite the object.

(ii) A second ray should pass directly through the center of the lens undeflected. This assumes that the lens is sufficiently thin.

The intersection of the two rays is the location of the tip of the image arrow. All other rays coming from the tip of the object that pass through the lens will also pass through the image tip. The tail of the image is located on the optical axis, a distance $q$ from the center of the lens. The image should also be oriented perpendicular to the optical axis.

(a) Using similar triangles, prove the relation

$$\frac{1}{p} + \frac{1}{q} = \frac{1}{f}.$$

(b) Show that if the distance of the object is much larger than the focal length of the lens ($p \gg f$), then the image is effectively located on the focal plane. This is essentially always the situation for astronomical observations.

The analysis of a diverging lens or a mirror (either converging or diverging) is similar and leads to the same relation between object distance, image distance, and focal length.

# Problems

**6.3** Show that if two lenses of focal lengths $f_1$ and $f_2$ can be considered to have zero physical separation, then the effective focal length of the combination of lenses is

$$\frac{1}{f_{\text{eff}}} = \frac{1}{f_1} + \frac{1}{f_2}.$$

*Note:* assuming that the actual physical separation of the lenses is $x$, this approximation is strictly valid only when $f_1 \gg x$ and $f_2 \gg x$.

**6.4** (a) Using the result of Problem 6.3, show that a compound lens system can be constructed from two lenses of different indices of refraction, $n_{1\lambda}$ and $n_{2\lambda}$, having the property that the resultant focal lengths of the compound lens at two specific wavelengths $\lambda_1$ and $\lambda_2$, respectively, can be made equal, or

$$f_{\text{eff},\lambda_1} = f_{\text{eff},\lambda_2}.$$

(b) Argue qualitatively that this condition does not guarantee that the focal length will be constant for all wavelengths.

**6.5** Prove that the angular magnification of a telescope having an objective focal length of $f_{\text{obj}}$ and an eyepiece focal length of $f_{\text{eye}}$ is given by Eq. (6.9) when the objective and the eyepiece are separated by the sum of their focal lengths, $f_{\text{obj}} + f_{\text{eye}}$.

**6.6** The diffraction pattern for a single slit (Figs. 6.7 and 6.8) is given by

$$I(\theta) = I_0 \left[ \frac{\sin(\beta/2)}{\beta/2} \right]^2,$$

where $\beta \equiv 2\pi D \sin\theta/\lambda$.

(a) Using l'Hôpital's rule, prove that the intensity at $\theta = 0$ is given by $I(0) = I_0$.

(b) If the slit has an aperture of 1.0 $\mu$m, what angle $\theta$ corresponds to the first minimum if the wavelength of the light is 5000 Å? Express your answer in degrees.

**6.7 Computer Problem** Suppose that two identical slits are situated next to each other in such a way that the axes of the slits are parallel and oriented vertically. Assume also that the two slits are the same distance from a flat screen. Different light sources of identical intensity are placed behind each slit so that the two sources are incoherent, meaning that double-slit interference effects can be neglected.

(a) If the two slits are separated by a distance such that the central maximum of the diffraction pattern corresponding to the first slit is located at the second minimum of the second slit's diffraction pattern, plot the resulting superposition of intensities (i.e., the total intensity at each location). Include at least two minima to the left of the central maximum of the leftmost slit and at least two minima to the right of the central maximum of the rightmost slit. *Hint:* Refer to the equation given in Problem 6.6 and plot your results as a function of $\beta$.

(b) Repeat your calculations for the case when the two slits are separated by a distance such that the central maximum of one slit falls at the location of the first minimum of the second (the Rayleigh criterion for single slits).

(c) What can you conclude about the ability to resolve two individual sources (the slits) as the sources are brought progressively closer together?

6.8 (a) Using the Rayleigh criterion, estimate the angular resolution limit of the human eye at 5500 Å. Assume that the diameter of the pupil is 5 mm.

(b) Compare your answer in part (a) to the angular diameters of the Moon and Jupiter. You may find the data in Appendix B helpful.

(c) What can you conclude about the ability to resolve the Moon's disk and Jupiter's disk with the unaided eye?

6.9 (a) Using the Rayleigh criterion, estimate the theoretical diffraction limit for the angular resolution of a typical 8-inch amateur telescope at 5500 Å. Express your answer in arc seconds.

(b) Using the information in Appendix B, estimate the minimum size of a crater on the Moon that can be resolved by an 8-inch telescope.

(c) Is this resolution limit likely to be achieved? Why or why not?

6.10 (a) Using the information provided in the text, calculate the focal length of the primary mirror of the New Technology Telescope.

(b) What is the value of the plate scale of the NTT?

(c) $\epsilon$ Bootes is a double star system whose components are separated by 2.9″. Calculate the linear separation of the images on the primary mirror focal plane of the NTT.

# Problems

6.11 Based on the specifications for HST's WF/PC 2, estimate the angular size of the field of view of one CCD in the planetary mode.

6.12 Suppose that a radio telescope receiver has a bandwidth of 50 MHz centered at 1.430 GHz (1 GHz = 1000 MHz). Rather than being a perfect detector over the entire bandwidth, assume that receiver's frequency dependence is triangular, meaning that the sensitivity of the detector is 0% at the edges of the band and 100% at its center. This filter function can be expressed as

$$f_\nu = \begin{cases} \dfrac{\nu}{\nu_m - \nu_\ell} - \dfrac{\nu_\ell}{\nu_m - \nu_\ell} & \text{if } \nu_\ell \leq \nu \leq \nu_m \\ -\dfrac{\nu}{\nu_u - \nu_m} + \dfrac{\nu_u}{\nu_u - \nu_m} & \text{if } \nu_m \leq \nu \leq \nu_u \\ 0 & \text{elsewhere.} \end{cases}$$

(a) Find the values of $\nu_\ell$, $\nu_m$, and $\nu_u$.

(b) Assume that the radio dish is a 100% efficient reflector over the receiver's bandwidth and has a diameter of 100 m. Assume also that the source NGC 2558 (a spiral galaxy with an apparent visual magnitude of 13.8) has a constant spectral flux density of $S = 2.5$ mJy over the detector bandwidth. Calculate the total power *measured* at the receiver.

(c) If $d = 100$ Mpc, estimate the power emitted at the source in this frequency range. Assume that the source emits the signal isotropically.

6.13 What would the diameter of a single radio dish need to be to have the equivalent collecting area of the 27 telescopes of the VLA?

6.14 How much must the pointing angle of a two-element radio interferometer be changed in order to move from one interference maximum to the next? Assume that the two telescopes are separated by the diameter of Earth and that the observation is being made at a wavelength of 21 cm. Express your answer in arc seconds.

6.15 From the energy limits given in the text for the EGRET detector on board the Compton Gamma Ray Observatory, estimate the range of wavelengths that the instrument can measure.

# Part II
# The Nature of Stars

# Chapter 7

# Binary Stars and Stellar Parameters

## 7.1 The Classification of Binary Stars

A detailed understanding of the structure and evolution of stars (the goal of Part II) requires knowledge about their physical characteristics. We have seen that knowledge of blackbody radiation curves, spectra, and parallax enables us to determine a star's effective temperature, luminosity, radius, composition, and other parameters. However, the only direct way to determine the mass of a star is by studying its gravitational interaction with other objects.

In Chapter 2 Kepler's laws were used to calculate the masses of members of our solar system. However, the universality of the gravitational force allows Kepler's laws to be generalized to include the orbits of stars about one another or even the orbital interactions of galaxies, as long as proper care is taken to refer all orbits to the center of mass of the system.

Fortunately, nature has provided ample opportunity for astronomers to observe binary star systems. At least half of all "stars" in the sky are actually multiple systems, two or more stars in orbit about a common center of mass. Analysis of the orbital parameters of these systems provides vital information about a variety of stellar characteristics, including mass.

The methods used to analyze the orbital data vary somewhat depending on the geometry of the system, its distance from the observer, and the relative masses and luminosities of each component. Consequently, binary star systems are classified according to their specific observational characteristics.

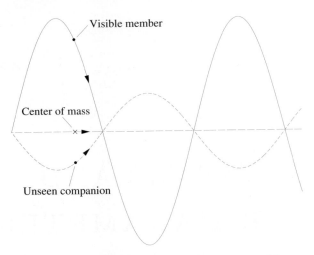

**Figure 7.1** An astrometric binary, which contains one visible member. The unseen component is implied by the oscillatory motion of the observed element of the system.

- **Optical double.** These systems are not true binaries but simply lie along the same line of sight (i.e., they have similar right ascensions and declinations). As a consequence of their large physical separations, they are not gravitationally bound and are not useful in determining stellar masses.

- **Visual binary.** Both stars in the binary can be resolved independently and, assuming that the orbital period is not prohibitively long, it is possible to monitor the motion of each member of the system. These systems provide important information about the angular separation of the stars from their mutual center of mass. If the distance to the binary is also known, the linear separations of the stars can then be calculated.

- **Astrometric binary.** If one member of a binary is significantly brighter than the other, it may not be possible to observe both members directly. In such a case the existence of the unseen member may be deduced by observing the oscillatory motion of the visible component. Since Newton's first law requires that a constant velocity be maintained by a mass unless a force is acting upon it, such an oscillatory behavior requires that another mass be present (see Fig. 7.1).

- **Eclipsing binary.** For binaries that have orbital planes oriented approximately along the line of sight of the observer, one star may pe-

## 7.1 The Classification of Binary Stars

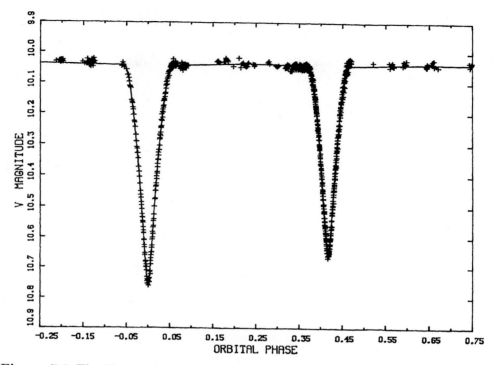

**Figure 7.2** The $V$ magnitude light curve of YY Sagittarii, an eclipsing binary star. The data from many of orbital periods have been plotted on this light curve as a function of phase, where the phase is defined to be 0.0 at the primary minimum. This system has an orbital period $P = 2.6284734$ d, an eccentricity $e = 0.1573$, and orbital inclination $i = 88.89°$ (see Section 7.2). (Figure from Lacy, C. H. S., *Astron. J.*, *105*, 637, 1993.)

riodically pass in front of the other, blocking the light of the eclipsed component (see Fig. 7.2). Such a system is recognizable by regular variations in the amount of light received at the telescope. Observations of these *light curves* not only reveal the presence of two stars but can also provide information about relative effective temperatures and radii of each component from the amount of light decrease and the length of the eclipse. Details of such an analysis will be discussed in Section 7.3.

- **Spectrum binary**. A spectrum binary is a system with two superimposed, independent, discernible spectra. The Doppler effect (Eq. 4.35) causes the spectral lines of a star to be shifted from their rest frame wavelengths if that star has a nonzero radial velocity. Since the stars in a binary system are constantly in motion about their mutual center of

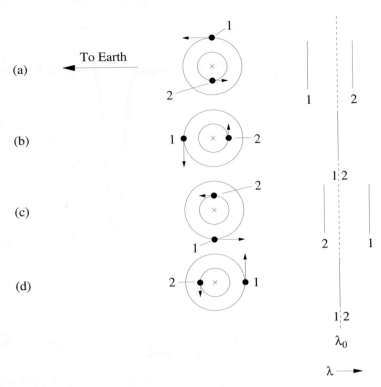

**Figure 7.3** The periodic shift in spectral features of a double-line spectroscopic binary. The relative wavelengths of the spectra of stars 1 and 2 are shown at four different phases during the orbit: (a) star 1 is moving toward the observer while star 2 is moving away, (b) both stars have velocities perpendicular to the line of sight, (c) star 1 is receding from the observer while star 2 is approaching, and (d) again both stars have velocities perpendicular to the line of sight.

mass, there must necessarily be periodic shifts in the wavelength of every spectral line of each star (unless the orbital plane is exactly perpendicular to the line of sight, of course). It is also apparent that when the lines of one star are blueshifted, the lines of the other must be redshifted, relative to the wavelengths that would be produced if the stars were moving with the *constant* velocity of the center of mass. However, it may be that the orbital period is so long that the time dependence of the spectral wavelengths may not be apparent. In any case, if one star is not overwhelmingly more luminous than its companion and if it is not possible to resolve each star separately, it may still be possible to recognize the object as a binary system by observing the superimposed and oppositely Doppler-shifted spectra.

Even if the Doppler shifts are not significant (if the orbital plane is perpendicular to the line of sight, for instance), it may still be possible to detect two sets of superimposed spectra if they originate from stars of differing spectral classes (see Section 8.1).

- **Spectroscopic binary.** If the period of a binary system is not prohibitively long and if the orbital motion has a component along the line of sight, a *periodic* shift in spectral lines will be observable. Assuming that the luminosities of each component are comparable, both spectra will be observable. However, if one star is much more luminous than the other, then the spectrum of the less luminous companion will be overwhelmed and only a single set of periodic spectral lines will be seen. In either situation, the existence of a binary star system is revealed. Figure 7.3 shows the relationship between spectra and orbital phase for a spectroscopic binary star system.

These specific classifications are not mutually exclusive. For instance, an unresolved system could be both an eclipsing and a spectroscopic binary. It is also true that some systems can be significantly more useful than others in providing information about stellar characteristics. Three types of systems can provide us with mass determinations: visual binaries combined with parallax information; visual binaries for which radial velocities are available over a complete orbit; and eclipsing, double-line, spectroscopic binaries.

## 7.2 Mass Determination Using Visual Binaries

When the angular separation between components of a binary system is greater than the resolution limit imposed by local seeing conditions and the fundamental diffraction limitation of the Rayleigh criterion, it becomes possible to analyze the orbital characteristics of the individual stars. From the orbital data, the orientation of the orbits and the mutual center of mass can be determined, providing knowledge of the ratio of the stars' masses. If the distance to the system is also known, from trigonometric parallax for instance, the linear separation of the stars can be determined, leading to the individual masses of the stars in the system.

To see how a visual binary can yield mass information, consider two stars in orbit about their mutual center of mass. Assuming that the orbital plane is perpendicular to the observer's line of sight, we see from the discussion of Section 2.3 that the ratio of masses may be found from the ratio of the angular separations of the stars from the center of mass. Using Eq. (2.20) and considering only the lengths of the vectors $\mathbf{r}_1$ and $\mathbf{r}_2$, we find that

$$\frac{m_1}{m_2} = \frac{r_2}{r_1} = \frac{a_2}{a_1}, \tag{7.1}$$

where $a_1$ and $a_2$ are the semimajor axes of the ellipses. If the distance from the observer to the binary star system is $d$, then the angles subtended by the semimajor axes are

$$\alpha_1 = \frac{a_1}{d} \quad \text{and} \quad \alpha_2 = \frac{a_2}{d},$$

where $\alpha_1$ and $\alpha_2$ are measured in radians. Substituting, the mass ratio simply becomes

$$\frac{m_1}{m_2} = \frac{\alpha_2}{\alpha_1}. \tag{7.2}$$

Even if the distance to the star system is not known, the mass ratio may still be determined. Note that since only the ratio of the subtended angles is needed, $\alpha_1$ and $\alpha_2$ may be expressed in arc seconds, the unit typically used for angular measure in astronomy.

The general form of Kepler's third law (Eq. 2.35),

$$P^2 = \frac{4\pi^2}{G(m_1 + m_2)} a^3,$$

gives the sum of the masses of the stars, provided that the semimajor axis of the orbit of the reduced mass is known. Since $a = a_1 + a_2$ (the proof of this is left as an exercise), the semimajor axis can be determined directly only if the distance to the system has been determined. Assuming that $d$ is known, $m_1 + m_2$ may be combined with $m_1/m_2$ to give each mass separately.

This process is complicated somewhat by the proper motion of the center of mass[1] (see Fig. 7.1) and by the fact that most orbits are not conveniently oriented with their planes perpendicular to the line of sight of the observer. Removing the proper motion of the center of mass from the observations is a relatively simple process since the center of mass must move at a constant velocity. Fortunately, estimating the orientation of the orbits is also possible and can be taken into consideration.

Let $i$ be the **angle of inclination** between the plane of an orbit and the plane of the sky, as shown in Fig. 7.4; note that the orbits of both stars are necessarily in the same plane. As a special case, assume that the orbital plane

---

[1]The annual wobble of stellar positions due to trigonometric parallax must also be considered, when significant.

## 7.2 Mass Determination Using Visual Binaries

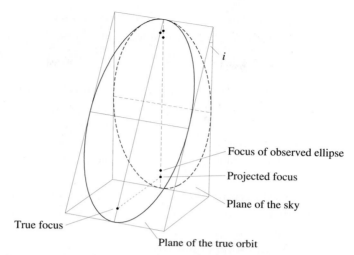

**Figure 7.4** An elliptical orbit projected onto the plane of the sky produces an observable elliptical orbit. The foci of the original ellipse do not project onto the foci of the observed ellipse, however.

and the plane of the sky (defined as being perpendicular to the line of sight) intersect along a line parallel to the minor axis, forming a **line of nodes**. The observer will not measure the actual angles subtended by the semimajor axes $\alpha_1$ and $\alpha_2$ but their projections onto the plane of the sky, $\tilde{\alpha}_1 = \alpha_1 \cos i$ and $\tilde{\alpha}_2 = \alpha_2 \cos i$. This geometrical effect plays no role in calculating the mass ratios since the $\cos i$ term will simply cancel in Eq. (7.2):

$$\frac{m_1}{m_2} = \frac{\alpha_2}{\alpha_1} = \frac{\alpha_2 \cos i}{\alpha_1 \cos i} = \frac{\tilde{\alpha}_2}{\tilde{\alpha}_1}.$$

However, this projection effect can make a significant difference when using Kepler's third law. Since $\alpha = a/d$ ($\alpha$ in radians), Kepler's third law may be solved for the sum of the masses to give

$$m_1 + m_2 = \frac{4\pi^2}{G} \frac{(\alpha d)^3}{P^2} = \frac{4\pi^2}{G} \left(\frac{d}{\cos i}\right)^3 \frac{\tilde{\alpha}^3}{P^2}, \tag{7.3}$$

where $\tilde{\alpha} = \tilde{\alpha}_1 + \tilde{\alpha}_2$.

To evaluate the sum of the masses properly, the angle of inclination must be deduced by carefully noting the apparent position of the center of mass of the system. As indicated in Fig. 7.4 the projection of an ellipse tilted at an angle $i$ with respect to the plane of the sky will result in an observed ellipse with a different eccentricity. However, the center of mass will not be located at one of the foci of the projection, a result that is inconsistent with Kepler's first law. By considering the projections of various ellipses onto the plane of the sky and comparing them with the observational data, the geometry of the true ellipse may be determined.

Of course, the problem of projection has been simplified here. Not only can the angle of inclination $i$ be nonzero, but the ellipse may be tilted about its major axis and rotated about the line of sight to produce any possible orientation. However, the general principles already mentioned still apply, making it possible to deduce the true shapes of the stars' elliptical orbits, as well as their masses.

It is also possible to determine the individual masses of members of visual binaries, even if the distance is not known. In this situation, detailed radial velocity data are needed. The projection of velocity vectors onto the line of sight, combined with information about the stars' positions and the orientation of their orbits, provides a means for determining the semimajor axes of the ellipses, as required by Kepler's third law.

## 7.3 Eclipsing, Spectroscopic Binaries

A wealth of information is available from a binary system even if it is not possible to resolve each of its stars individually. This is particularly true for a double-line, eclipsing, spectroscopic binary star system. In such a system, not only is it possible to determine the individual masses of the stars, but astronomers may be able to deduce other parameters as well, such as the stars' radii and the ratio of their fluxes (and hence the ratio of their effective temperatures).[2]

Consider a spectroscopic binary star system for which the spectra of both stars may be seen (a double-line, spectroscopic binary). Since the individual members of the system cannot be resolved, the techniques used to determine the orientation and eccentricity of the orbits of visual binaries are not applicable. Also, the inclination angle $i$ clearly plays a role in the solution obtained for the stars' masses because it directly influences the measured radial velocities. If $v_1$ is the velocity of the star of mass $m_1$ and $v_2$ is the velocity of the star of mass $m_2$ at some instant, then referring to Fig. 7.4, the observed radial velocities cannot exceed $v_{1r}^{\max} = v_1 \sin i$ and $v_{2r}^{\max} = v_2 \sin i$, respectively. In fact, the actual measured radial velocities depend upon the positions of the stars at that instant; if the directions of motion of the stars happen to

---

[2]Of course, eclipsing systems are not restricted to spectroscopic binaries but may apply to other types of binaries as well (i.e., visual binaries).

## 7.3 Eclipsing, Spectroscopic Binaries

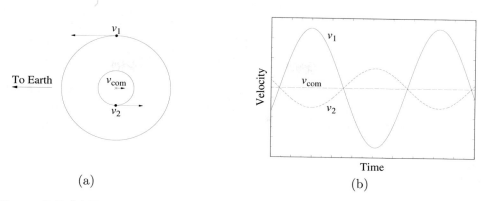

**Figure 7.5** (a) Two circular orbits whose plane lies along the line of sight of the observer. (b) The observed radial velocity curves: $v_1$, $v_2$, and $v_{\rm com}$ are the velocities of star 1, star 2, and the center of mass, respectively.

be perpendicular to the line of sight, then the observed radial velocities will be zero.

For a star system having circular orbits, the speed of each star will be constant. If the plane of their orbits lies in the line of sight of the observer ($i = 90°$), then the measured radial velocities will produce sinusoidal *velocity curves*, as in Fig. 7.5. Changing the orbital inclination does not alter the shape of the velocity curves; it merely changes their amplitudes by the factor $\sin i$. To estimate $i$ and the actual orbital velocities, therefore, other information about the system is necessary.

When the eccentricity, $e$, of the orbits is not zero, the observed velocity curves become skewed, as shown in Fig. 7.6. The exact shapes of the curves also depend strongly on the orientation of the orbits with respect to the observer, even for a given inclination angle.

In reality, many spectroscopic binaries possess nearly circular orbits, simplifying the analysis of the system somewhat. This occurs because close binaries tend to circularize their orbits due to tidal interactions between the components over time scales, which are short compared to the lifetimes of the stars involved.

If we assume that the orbital eccentricity is very small ($e \ll 1$), then the speeds of the stars are essentially constant and given by $v_1 = 2\pi a_1/P$ and $v_2 = 2\pi a_2/P$ for stars of mass $m_1$ and $m_2$, respectively, where $a_1$ and $a_2$ are the radii (semimajor axes) and $P$ is the period of the orbits. Solving for $a_1$ and $a_2$ and substituting into Eq. (7.1), we find that the ratio of the masses of

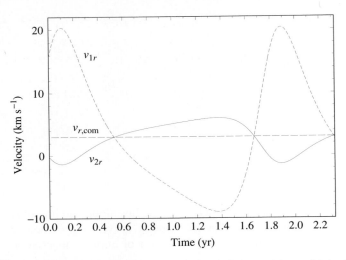

**Figure 7.6** The eccentricity and the orientation of the orbits, which determine the shape of the radial velocity curves. In this case $M_1 = 0.5$ M$_\odot$, $M_2 = 2.0$ M$_\odot$, $a = 2$ AU, $e = 0.3$, $i = 30°$, with an axis rotation= 45°.

the two stars becomes

$$\frac{m_1}{m_2} = \frac{v_2}{v_1}. \tag{7.4}$$

Recalling that $v_{1r} = v_1 \sin i$ and $v_{2r} = v_2 \sin i$, Eq. (7.4) can be written in terms of the observed radial velocities rather than actual orbital velocities:

$$\frac{m_1}{m_2} = \frac{v_{2r}/\sin i}{v_{1r}/\sin i} = \frac{v_{2r}}{v_{1r}}. \tag{7.5}$$

As is the situation with visual binaries, the ratio of the stellar masses can be determined without knowing the angle of inclination.

However, as is also the case with visual binaries, the sum of the masses does require knowledge of the angle of inclination. Replacing $a$ with

$$a = a_1 + a_2 = \frac{P}{2\pi}(v_1 + v_2)$$

in Kepler's third law (Eq. 2.35) and solving for the sum of the masses, we have

$$m_1 + m_2 = \frac{P}{2\pi G}(v_1 + v_2)^3. \tag{7.6}$$

Writing the actual radial velocities in terms of the observed values, we can express the sum of the masses as

$$m_1 + m_2 = \frac{P}{2\pi G}\frac{(v_{1r} + v_{2r})^3}{\sin^3 i}. \tag{7.7}$$

## 7.3 Eclipsing, Spectroscopic Binaries

It is clear from Eq. (7.7) that the sum of the masses can be obtained only if both $v_{1r}$ and $v_{2r}$ are measurable. Unfortunately, this is not always the case. If one star is much brighter than its companion, the spectrum of the dimmer member will be overwhelmed. Such a system is referred to as a *single-line spectroscopic binary*. If the spectrum of star 1 is observable while the spectrum of star 2 is not, Eq. (7.5) allows $v_{2r}$ to be replaced by the ratio of the stellar masses, giving a quantity that is dependent on both of the system masses and the angle of inclination. If we substitute, Eq. (7.7) becomes

$$m_1 + m_2 = \frac{P}{2\pi G} \frac{v_{1r}^3}{\sin^3 i} \left(1 + \frac{m_1}{m_2}\right)^3.$$

Rearranging terms gives

$$\frac{m_2^3}{(m_1 + m_2)^2} \sin^3 i = \frac{P}{2\pi G} v_{1r}^3. \tag{7.8}$$

The right-hand side of this expression, known as the **mass function**, depends only on the readily observable quantities: period and radial velocity. Since the spectrum of only one star is available, Eq. (7.5) cannot provide any information about mass ratios. As a result, the mass function is useful only for statistical studies or if an estimate of the mass of at least one component of the system already exists by some indirect means. For instance, if either $m_1$ or $\sin i$ is unknown, the mass function sets a lower limit for $m_2$, since the left-hand side is always less than $m_2$.

Even if both radial velocities are measurable, it is not possible to get exact values for $m_1$ and $m_2$ without knowing $i$. However, since stars can be grouped according to their effective temperatures and luminosities (see Section 8.2), and assuming that there is a relationship between these quantities and mass, then a statistical mass estimate for each class may be found by choosing an appropriately averaged value for $\sin^3 i$. An integral average of $\sin^3 i$ ($<\sin^3 i>$) evaluated between 0° and 90° has a value 0.42.[3] However, since no Doppler shift will be produced if $i = 0°$, it is more likely that a spectroscopic binary star system will be discovered if $i$ differs significantly from 0°. This *selection effect* suggests that a larger value of $<\sin^3 i> \sim 2/3$ would be more appropriate.

Evaluating masses of binaries has shown the existence of a well-defined **mass–luminosity relation** for the large majority of stars in the sky (see Fig. 7.7). One of the goals of the next several chapters is to understand the origin of this relation in terms of fundamental physical principles.

---

[3] The proof is left as an exercise.

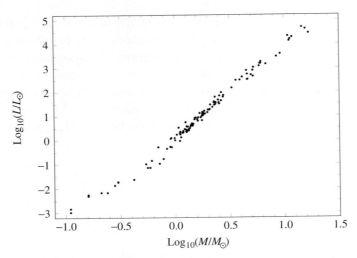

**Figure 7.7** The mass–luminosity relation. (Data from Popper, *Annu. Rev. Astron. Astrophys.*, **18**, 115, 1980.)

A good estimate of $i$ is possible in the special situation that a spectroscopic binary star system is observed to be an eclipsing system as well. Unless the distance of separation between the components of the binary is not much larger than the radii of the stars involved, an eclipsing system implies that $i$ must be close to $90°$, as suggested in Fig. 7.8. Even if it were assumed that $i = 90°$ while the actual value was closer to $75°$, an error of only 10% would result in the calculation of $\sin^3 i$ and in the determination of $m_1 + m_2$.

From the light curves produced by eclipsing binaries, it is possible to improve the estimate of $i$ still further. Figure 7.9 indicates that if the smaller star is completely eclipsed by the larger one, a nearly constant minimum will occur in the measured brightness of the system during the period of occultation. Similarly, even though the larger star will not be fully hidden from view when the smaller companion passes in front of it, a constant amount of area will still be obscured for a time and again a nearly constant, though diminished amount of light will be observed. When one star is not completely eclipsed by its companion (Fig. 7.10), the minima are no longer constant, implying that $i$ must be less than $90°$.

Using measurements of the duration of eclipses, it is also possible to find the radii of each member of an eclipsing, spectroscopic binary. Referring to Fig. 7.9, if we assume that $i \simeq 90°$, the amount of time between *first contact* ($t_a$) and minimum light ($t_b$), combined with the velocities of the stars, leads directly to the calculation of the radius of the smaller component. For example,

## 7.3 Eclipsing, Spectroscopic Binaries

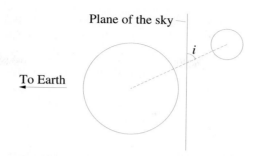

**Figure 7.8** The geometry of an eclipsing, spectroscopic binary, which implies that the angle of inclination $i$ must be close to $90°$.

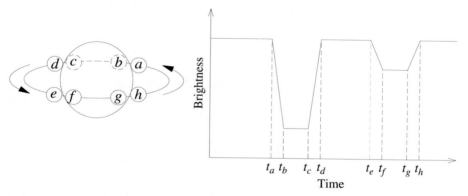

**Figure 7.9** The light curve of an eclipsing binary for which $i = 90°$. The times indicated on the light curve correspond to the positions of the smaller star relative to its larger companion. It is assumed in this example that the smaller star is hotter than the larger one.

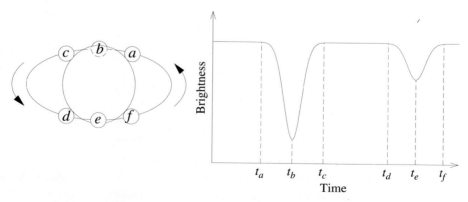

**Figure 7.10** The light curve of a partially eclipsing binary. It is assumed in this example that the smaller star is hotter than its companion.

if the semimajor axis of the smaller star's orbit is sufficiently large compared to either stellar radius and if the orbit is nearly circular, it can be approximately assumed that the smaller object is moving essentially perpendicular to the line of sight of the observer during the duration of the eclipse. In this case the radius of the smaller star is simply

$$r_s = \frac{v}{2}(t_b - t_a),  \qquad (7.9)$$

where $v = v_s + v_\ell$ is the *relative* velocity of the two stars ($v_s$ and $v_\ell$ are the velocities of the small and large stars, respectively). Similarly, if we consider the amount of time between $t_b$ and $t_c$, the size of the larger member can also be determined. It can be quickly shown that the radius of the larger star is just

$$r_\ell = \frac{v}{2}(t_c - t_a) = r_s + \frac{v}{2}(t_c - t_b).  \qquad (7.10)$$

---

**Example 7.1** An analysis of the spectrum of an eclipsing, double-line, spectroscopic binary having a period of $P = 8.6$ yr shows that the maximum Doppler shift of the hydrogen Balmer $H_\alpha$ (6562.8 Å) line is $\Delta\lambda_s = 0.72$ Å for the smaller member and only $\Delta\lambda_\ell = 0.068$ Å for its companion. From the sinusoidal shapes of the velocity curves, it is also apparent that the orbits are nearly circular. Using Eqs. (4.39) and (7.5), the mass ratio of the two stars must be

$$\frac{m_\ell}{m_s} = \frac{v_{rs}}{v_{r\ell}} = \frac{\Delta\lambda_s}{\Delta\lambda_\ell} = 10.6.$$

Assuming that the orbital inclination is $i = 90°$, the Doppler shift of the smaller star implies that the maximum measured radial velocity is

$$v_{rs} = \frac{\Delta\lambda_s}{\lambda}c = 3.3 \times 10^6 \text{ cm s}^{-1}$$

and the radius of its orbit must be

$$a_s = \frac{v_{rs}P}{2\pi} = 1.42 \times 10^{14} \text{ cm} = 9.5 \text{ AU}.$$

Similarly, the orbital velocity and radius of the other component are $v_{r\ell} = 3.1 \times 10^5$ cm s$^{-1}$ and $a_\ell = 0.90$ AU, respectively. Therefore the semimajor axis of the reduced mass becomes $a = a_s + a_\ell = 10.4$ AU.

The sum of the masses can now be determined from Kepler's third law. If Eq. (2.35) is written in terms of solar masses, astronomical units, and years, we have

$$m_s + m_\ell = a^3/P^2 = 15.2 \text{ M}_\odot.$$

## 7.3 Eclipsing, Spectroscopic Binaries

Solving for the masses independently, $m_s = 1.3$ M$_\odot$ and $m_\ell = 13.9$ M$_\odot$.

Furthermore, from the light curve for this system, it is found that $t_b - t_a = 11.7$ hours and $t_c - t_b = 164$ days. Using Eq. (7.9), the radius of the smaller star is

$$r_s = \frac{(v_{rs} + v_{r\ell})}{2}(t_b - t_a) = 7.6 \times 10^{10} \text{ cm} = 1.1 \text{ R}_\odot,$$

where one solar radius is 1 R$_\odot = 6.96 \times 10^{10}$ cm. Equation (7.10) now gives the radius of the larger star, which is found to be $r_\ell = 369$ R$_\odot$.

In this particular system the masses and radii of the stars are found to differ significantly.

---

The ratio of the effective temperatures of the two stars can also be obtained from the light curve of an eclipsing binary. This is accomplished by considering the objects as blackbody radiators and comparing the amount of light received during an eclipse with the amount received when both members are fully visible.

Referring again to Fig. 7.9, it can be seen that the dip in the light curve is deeper when the smaller, hotter star is passing behind its companion. To understand this effect, recall that the radiative surface flux is given by Eq. (3.18),

$$F_r = F_{\text{surf}} = \sigma T_e^4.$$

Regardless of whether the smaller star passes behind or in front of the larger one, the same total cross-sectional area is eclipsed. Now, assuming that the observed flux is constant across the disks,[4] when both stars are fully visible, the amount of light detected from the binary is given by

$$B_0 = k\left(\pi r_\ell^2 F_{r\ell} + \pi r_s^2 F_{rs}\right),$$

where $k$ is a constant that depends on the distance to the system and the nature of the detector. The deeper, or *primary*, minimum occurs when the hotter star passes behind the cooler one. Assuming that the smaller star is hotter and therefore has the larger surface flux, the amount of light detected during the primary minimum may be expressed as

$$B_p = k\pi r_\ell^2 F_{r\ell}$$

while the brightness of the *secondary* minimum is

$$B_s = k\left(\pi r_\ell^2 - \pi r_s^2\right)F_{r\ell} + k\pi r_s^2 F_{rs}.$$

---

[4]Stars often appear darker near the edges of their disks (a phenomenon referred to as *limb darkening*). This effect will be discussed in Section 9.3.

Since it is generally not possible to determine $k$ exactly, ratios are employed. Consider the ratio of the depth of the primary to the depth of the secondary. Using the expressions for $B_0$, $B_p$, and $B_s$, we find immediately that

$$\frac{B_0 - B_p}{B_0 - B_s} = \frac{F_{rs}}{F_{r\ell}} \tag{7.11}$$

or, from Eq. (3.18),

$$\frac{B_0 - B_p}{B_0 - B_s} = \left(\frac{T_s}{T_\ell}\right)^4. \tag{7.12}$$

---

**Example 7.2** Further examination of the light curve of the binary system discussed in Example 7.1 provides information on the relative temperatures of the two stars. Photometric observations show that at maximum light the bolometric magnitude is $m_{\text{bol},0} = 6.3$, that at the primary minimum $m_{\text{bol},p} = 9.6$, and that at the secondary minimum $m_{\text{bol},s} = 6.6$. From Eq. (3.3), the ratio of brightnesses between the primary minimum and maximum light is

$$\frac{B_p}{B_0} = 100^{(m_{\text{bol},0} - m_{\text{bol},p})/5} = 0.048.$$

Similarly, the ratio of brightnesses between the secondary minimum and maximum light is

$$\frac{B_s}{B_0} = 100^{(m_{\text{bol},0} - m_{\text{bol},s})/5} = 0.76.$$

Now, by rewriting Eq. (7.11), the ratio of the radiative fluxes is

$$\frac{F_{rs}}{F_{r\ell}} = \frac{1 - B_p/B_0}{1 - B_s/B_0} = 3.97.$$

Finally, from Eq. (3.18),

$$\frac{T_s}{T_\ell} = \left(\frac{F_{rs}}{F_{r\ell}}\right)^{1/4} = 1.41.$$

---

The modern approach to analyzing the data from binary star systems involves computing detailed models that can yield important information about a variety of physical parameters. Not only can masses, radii, and effective temperatures be determined, but for many systems other details can be described as well. For instance, gravitational forces, combined with the effects of rotation and orbital motion, alter the stars' shapes; they are no longer simply spherical objects but may become elongated (these effects will be discussed

## 7.3 Eclipsing, Spectroscopic Binaries

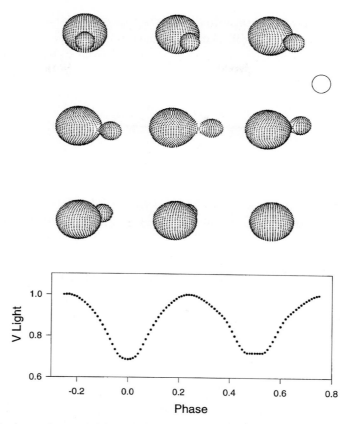

**Figure 7.11** A synthetic light curve of RR Centauri, an eclipsing binary star system for which the two components are in close contact. The open circle represents the size of the Sun. The orbital and physical characteristics of the RR Cen system are $P = 0.6057$ d, $e = 0.0$, $M_1 = 1.8\ M_\odot$, $M_2 = 0.37\ M_\odot$. The spectral classification of the primary is F0V (see Section 8.1 for a discussion of stellar spectral classifications). (Figure from R. E. Wilson, *Publ. Aston. Soc. Pac.*, *106*, 921, 1994; ©Astronomical Society of the Pacific.)

in more detail in Section 17.1). The models may also incorporate information about the nonuniform distribution of flux across the observed disks of the stars, variations in surface temperatures, and so on. Once the shapes of the gravitational equipotential surfaces and other parameters are determined, theoretical (*synthetic*) light curves can be computed for various wavelength bands ($U$, $B$, $V$, etc.), which are then compared the observational data. Adjustments in the model parameters are made until the light curves agree with the observations. One such model for the binary system, RR Centauri, is shown in Fig. 7.11. In this system the two stars are actually in contact with each other, producing interesting and subtle effects in the light curve.

The study of binary star systems provides valuable information about the characteristics of stars. These results are then employed in developing a theory of stellar structure and evolution.

## Suggested Readings

GENERAL

Burnham, Robert Jr., *Burnham's Celestial Handbook: An Observer's Guide to the Universe Beyond the Solar System*, Revised and Enlarged Edition, Dover Publications, New York, 1978.

Jones, Kenneth Glyn (ed.), *Webb Society Deep-Sky Observer's Handbook, Volume 1: Double Stars*, Enslow Publishers, Hillside, NJ, 1979.

Menzel, Donald H., and Pasachoff, Jay M., *Field Guide to the Stars and Planets*, Second Edition, Houghton Mifflin, Boston, 1983.

TECHNICAL

Batten, Alan H., Fletcher, J. Murray, and MacCarthy, D. G., "Eighth Catalogue of the Orbital Elements of Spectroscopic Binary Systems," *Publications of the Dominion Astrophysical Observatory*, *17*, 1989.

Böhm-Vitense, Erika, *Introduction to Stellar Astrophysics: Basic Stellar Observations and Data*, Volume 1, Cambridge University Press, Cambridge, 1989.

Eggen, O. J., "Masses of Visual Binary Stars," *Annual Review of Astronomy and Astrophysics*, *5*, 105, 1967.

Kitchin, C. R., *Astrophysical Techniques*, Second Edition, Adam Hilger, Philadelphia, 1991.

Popper, Daniel M., "Determination of Masses of Eclipsing Binary Stars," *Annual Review of Astronomy and Astrophysics*, *5*, 85, 1967.

Popper, Daniel M., "Stellar Masses," *Annual Review of Astronomy and Astrophysics*, *18*, 115, 1980.

Wilson, R. E., "Binary-Star Light-Curve Models," *Publications of the Astronomical Society of the Pacific*, *106*, 921, 1994.

# Problems

**7.1** Consider two stars in orbit about a mutual center of mass. If $a_1$ is the semimajor axis of the orbit of star of mass $m_1$ and $a_2$ is the semimajor axis of the orbit of star of mass $m_2$, prove that the semimajor axis of the orbit of the *reduced mass* is given by $a = a_1 + a_2$. *Hint:* Review the discussion of Section 2.3 and recall that $\mathbf{r} = \mathbf{r}_2 - \mathbf{r}_1$.

**7.2** Prove that $<\sin^3 i> = 0.42$ between 0 rad and $\frac{\pi}{2}$ rad (0° and 90°, respectively). Neglect the Doppler shift selection effect. *Hint:* Refer to the discussion of integral averages found in Problem 2.10.

**7.3** Assume that two stars are in circular orbits about a mutual center of mass and are separated by a distance $a$. Assume also that the angle of inclination is $i$ and their stellar radii are $r_1$ and $r_2$.

(a) Find an expression for the smallest angle of inclination that will just barely produce an eclipse. *Hint:* Refer to Fig. 7.8.

(b) If $a = 2$ AU, $r_1 = 10$ R$_\odot$, and $r_2 = 1$ R$_\odot$, what minimum value of $i$ will result in an eclipse?

**7.4** Sirius is a visual binary with a period of 49.94 yr. Its measured trigonometric parallax is 0.377″ and, assuming that the plane of the orbit is in the plane of the sky, the true angular extent of the semimajor axis of the reduced mass is 7.62″. The ratio of the distances of Sirius A and Sirius B from the center of mass is $a_A/a_B = 0.466$.

(a) Find the mass of each member of the system.

(b) The absolute bolometric magnitude of Sirius A is 1.33, and Sirius B has an absolute bolometric magnitude of 8.57. Determine their luminosities. Express your answers in terms of the luminosity of the Sun.

(c) The effective temperature of Sirius B is estimated to be approximately 27,000 K. Estimate its radius, and compare your answer to the radii of the Sun and Earth.

**7.5** ζ Phe is a 1.67-day spectroscopic binary with nearly circular orbits. The maximum measured Doppler shifts of the brighter and fainter components of the system are 121.4 km s$^{-1}$ and 247 km s$^{-1}$, respectively.

(a) Determine the quantity $m \sin^3 i$ for each star.

(b) Using a statistically chosen value for $\sin^3 i$ that takes into consideration the Doppler-shift selection effect, estimate the individual masses of the components of $\zeta$ Phe.

7.6 From the light and velocity curves of an eclipsing, spectroscopic binary star system, it is determined that the orbital period is 6.31 yr, and the maximum radial velocities of stars A and B are 5.4 km s$^{-1}$ and 22.4 km s$^{-1}$, respectively. Furthermore, the time period between first contact and minimum light $(t_b - t_a)$ is 0.58 d, the length of the primary minimum $(t_c - t_b)$ is 0.64 d, and the apparent bolometric magnitudes of maximum, primary minimum, and secondary minimum are 5.40 magnitudes, 9.20 magnitudes, and 5.44 magnitudes, respectively. From this information, and assuming circular orbits, find the

(a) Ratio of stellar masses.

(b) Sum of the masses (assume $i \simeq 90°$).

(c) Individual masses.

(d) Individual radii (assume that the orbits are circular).

(e) Ratio of the effective temperatures of the two stars.

7.7 The $V$-band light curve of YY Sgr is shown in Fig. 7.2.

(a) Assuming for simplicity that $e = 0.0$ and $i = 90°$, estimate the ratio of the radii of the two stars in the system.

(b) Neglecting bolometric corrections, estimate the ratio of the temperatures of the two stars in the system.

7.8 Referring to the synthetic light curve and model of RR Centauri shown in Fig. 7.11,

(a) Indicate the approximate points on the light curve (as a function of phase) that correspond to the orientations depicted.

(b) Explain qualitatively the shape of the light curve.

7.9 **Computer Problem**

(a) Using the information in Chapter 2 [including Eqs. (2.3) and (2.34)], write a short computer program to generate orbital radial velocity data similar to Fig. 7.6 for any choice of eccentricity. Assume that $M_1 = 0.5$ M$_\odot$, $M_2 = 2.0$ M$_\odot$, $a = 2.0$ AU, and $i = 30°$.

Plot your results for $e = 0, 0.2, 0.4$, and $0.5$. (You may assume that the center-of-mass velocity is zero and the orientation of the major axis is perpendicular to the line of sight.)

(b) Verify your results for $e = 0$ by using the equations developed in Section 7.3.

(c) Explain how you might determine the eccentricity of an orbital system.

# Chapter 8

# THE CLASSIFICATION OF STELLAR SPECTRA

## 8.1 The Formation of Spectral Lines

With the invention of photometry and spectroscopy, the new science of *astrophysics* progressed rapidly. As early as 1817 Joseph Fraunhofer had determined that different stars have different spectra. Stellar spectra were classified according to several schemes, the earliest of which recognized just three types of spectra. As instruments improved, increasingly subtle distinctions became possible. A spectral taxonomy developed at Harvard by Edward C. Pickering (1846–1919) and his assistant Williamina P. Fleming (1857–1911) in the 1890s labeled spectra with capital letters according to the strength of their hydrogen absorption lines, beginning with the letter A for the broadest lines. In 1901 Annie Jump Cannon[1] (1863–1941; see Fig. 8.1), another of Pickering's assistants, rearranged and consolidated the sequence of spectra by placing O and B before A and adding decimal subdivisions (e.g., A0–A9). With these changes, the Harvard classification scheme of "O B A F G K M" became a *temperature* sequence, running from the hottest blue O stars to the coolest red M stars. Generations of astronomy students have remembered this string of **spectral types** by memorizing the phrase, "Oh Be A Fine Girl/Guy, Kiss Me." Stars nearer the beginning of this sequence are referred to as *early-type* stars, and those closer to the end are called *late-type* stars. These labels also distinguish the stars within the spectral subdivisions, so astronomers may speak of a K0

---

[1]The Annie J. Cannon Award is bestowed annually by the American Association of University Women and the American Astronomical Society for distinguished contributions to astronomy by a woman.

**Figure 8.1** Annie Jump Cannon (1863–1941). (Courtesy of Harvard College Observatory.)

star as an "early K star," or refer to a B9 star as a "late B star," Cannon classified some 200,000 spectra between 1911 and 1914; the results were collected into the *Henry Draper Catalogue*.[2] Today, many stars are referred to by their HD numbers; the first star in the catalog, HD1, is Betelgeuse.

The physical basis of the Harvard spectral classification scheme remained obscure, however. Vega (spectral type A0) displays very strong hydrogen absorption lines, much stronger than the faint lines observed for the Sun (spectral type G2). On the other hand, the Sun's calcium absorption lines are much more intense than those of Vega. Is this a result of a variation in the *composition* of the two stars? Or are the different surface temperatures of Vega ($T_e = 9500$ K) and the Sun ($T_e = 5770$ K) responsible for the relative strengths of the absorption lines?

The theoretical understanding of the quantum atom achieved early in the twentieth century gave astronomers the key to the secrets of stellar spectra. As discussed in Section 5.3, absorption lines are created when an atom absorbs a photon with exactly the energy required for an electron to make an upward transition from a lower to a higher orbit. Emission lines are formed in the inverse process, when an electron makes a downward transition from a higher to a lower orbit and a single photon carries away the energy lost by the electron. The wavelength of the photon thus depends on the energies of the atomic

---

[2] In 1872 Henry Draper took the first photograph of a stellar spectrum. The catalog bearing his name was financed from his estate.

orbitals involved in these transitions. For example, the Balmer absorption and emission lines of hydrogen are caused by electrons making transitions between the $(n = 2)$ and higher orbitals.

The distinctions between the spectra of stars with different temperatures are due to electrons occupying different atomic orbitals in the atmospheres of these stars. The details of spectral line formation can be quite complicated because electrons can be found in any of an atom's orbitals. Furthermore, the atom can be in any of various stages of ionization and has a unique set of orbitals at each stage. An atom's stage of ionization is denoted by a Roman numeral following the symbol for the atom. For example, H I and He I are neutral (not ionized) hydrogen and helium; He II is singly ionized helium, and Si III and Si IV refer to a silicon atom that has lost two and three electrons, respectively.

In the Harvard system devised by Cannon, the Balmer lines reach their maximum intensity in the spectra of stars of type A0, which have an effective temperature (recall Eq. 3.17) of $T_e = 9520$ K. The visible spectral lines of neutral helium (He I) are strongest for B2 stars ($T_e = 22{,}000$ K), and the visible spectral lines of singly ionized calcium (Ca II) are most intense for K0 stars ($T_e = 5250$ K).[3] Table 8.1 lists some of the defining criteria for various spectral types, and Figs. 8.2 and 8.3 display some sample spectra.

To uncover the physical foundation of this classification system, two basic questions must be answered: In what orbitals are electrons most likely to be found? What are the relative numbers of atoms in various stages of ionization?

The answers to both questions are found in an area of physics known as *statistical mechanics*. This branch of physics studies the statistical properties of a system composed of many members. For example, a gas can contain a huge number of particles with a large range of speeds and energies. Although in practice it would be impossible to calculate the detailed behavior of any single particle, the gas as a whole does have certain well-defined properties, such as its temperature, pressure, and density. For such a gas in thermal equilibrium, the fraction of particles having a given speed remains stable and is described by the **Maxwell–Boltzmann distribution function**.[4] The number of gas particles per unit volume having a speed between $v$ and $v + dv$ is

$$n_v \, dv = n \left( \frac{m}{2\pi k T} \right)^{3/2} e^{-mv^2/2kT} 4\pi v^2 \, dv, \qquad (8.1)$$

---

[3] The two prominent spectral lines of Ca II are usually referred to as the H ($\lambda = 3968$ Å) and K ($\lambda = 3933$ Å) lines of calcium. The nomenclature for the H line is due to Fraunhofer; the K line was named by E. Mascart in the 1860s.

[4] This name honors James Clerk Maxwell and Ludwig Boltzmann (1844–1906), the latter considered the founder of statistical mechanics.

| Spectral Type | Characteristics |
|:---:|:---|
| O | Hottest blue-white stars with few lines. Strong He II absorption (sometimes emission) lines. He I absorption lines becoming stronger. |
| B | Hot blue-white stars. He I absorption lines strongest at B2. H I (Balmer) absorption lines becoming stronger. |
| A | White stars. Balmer absorption lines strongest at A0, becoming weaker later. Ca II absorption lines becoming stronger. |
| F | Yellow-white stars. Ca II lines continue to strengthen as Balmer lines continue to weaken. Neutral metal absorption lines (Fe I, Cr I). |
| G | Yellow stars. Solar-type spectra. Ca II lines continue becoming stronger. Fe I, other neutral metal lines becoming stronger. |
| K | Cool orange stars. Ca II H and K lines strongest at K0, becoming weaker later. Spectra dominated by metal absorption lines. |
| M | Coolest red stars. Spectra dominated by molecular absorption bands, especially titanium oxide (TiO). Neutral metal absorption lines remain strong. |

**Table 8.1** Harvard Spectral Classification.

## 8.1 The Formation of Spectral Lines

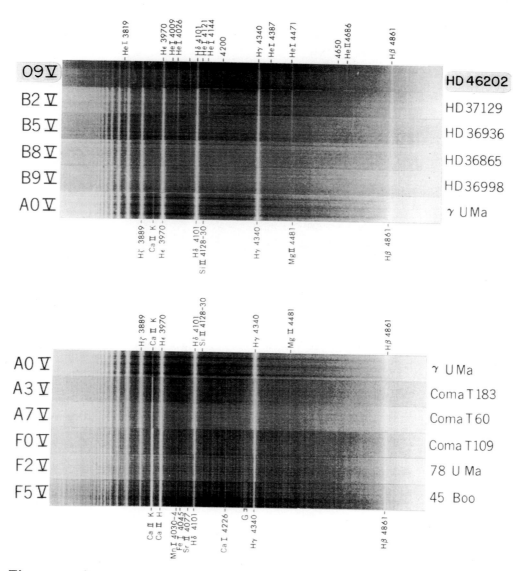

**Figure 8.2** Stellar spectra for main-sequence classes B0–F5. (Figure from Abt et al., *An Atlas of Low-Dispersion Grating Stellar Spectra*, Kitt Peak National Observatory, Tucson, AZ, 1968.)

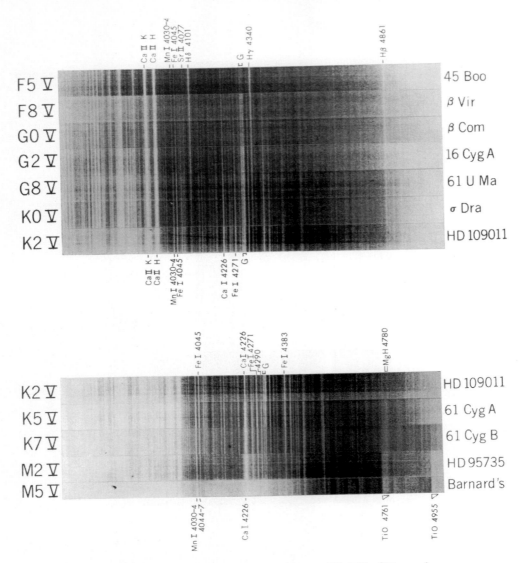

**Figure 8.3** Stellar spectra for main-sequence classes F5–M5. (Figure from Abt et al., *An Atlas of Low-Dispersion Grating Stellar Spectra*, Kitt Peak National Observatory, Tucson, AZ, 1968.)

## 8.1 The Formation of Spectral Lines

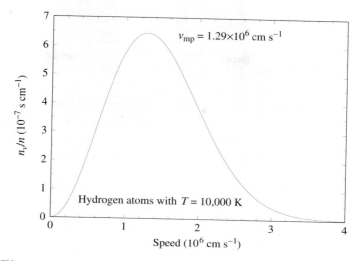

**Figure 8.4** Maxwell–Boltzmann distribution function, $n_v/n$.

where $n$ is the total number density (number of particles per unit volume), $m$ is a particle's mass, $k$ is Boltzmann's constant, and $T$ is the temperature of the gas *in kelvin*. Figure 8.4 shows the Maxwell–Boltzmann distribution of molecular speeds in terms of the *fraction* of molecules having a speed between $v$ and $v + dv$. The area under the curve between two speeds is equal to the fraction of gas particles in that range of speeds. The exponent of the distribution function is the ratio of a gas particle's kinetic energy, $\frac{1}{2}mv^2$, to the characteristic thermal energy, $kT$. It is difficult for a significant number of particles to have an energy much greater or less than the thermal energy; the distribution peaks when these energies are equal, at a *most probable* speed of

$$v_{\mathrm{mp}} = \sqrt{\frac{2kT}{m}}. \tag{8.2}$$

The high-speed exponential "tail" of the distribution function results in a somewhat higher average (*root-mean-square*) speed[5] of

$$v_{\mathrm{rms}} = \sqrt{\frac{3kT}{m}}. \tag{8.3}$$

In Section 10.3, we will see that this high-speed tail, which declines as

$$e^{-mv^2/2kT},$$

---

[5]The root-mean-square speed is the square root of the average (mean) value of $v^2$: $v_{\mathrm{rms}} = \sqrt{\overline{v^2}}$.

is of vital importance in the thermonuclear reactions that occur in stellar interiors.

The atoms of a gas gain and lose energy as they collide. As a result, the distribution in the speeds of the impacting atoms, given by Eq. (8.1), produces a definite distribution of the electrons among the atomic orbitals. This distribution of electrons is governed by a fundamental result of statistical mechanics: Orbitals of higher energy are less likely to be occupied by electrons.

Let $s_a$ stand for the specific set of quantum numbers that identifies a state of energy $E_a$ for a system of particles. Similarly, let $s_b$ identify the set of quantum numbers that identifies a state of energy $E_b$. (For example, $E_a = -13.6$ eV for the lowest orbit of the hydrogen atom, with $s_a = \{n = 1,\ \ell = 0,\ m_\ell = 0,\ m_s = +1/2\}$ identifying a specific state with that energy.) Then the ratio of the probability $P(s_b)$ that the system is in state $s_b$ to the probability $P(s_a)$ that the system is in state $s_a$ is given by

$$\frac{P(s_b)}{P(s_a)} = \frac{e^{-E_b/kT}}{e^{-E_a/kT}} = e^{-(E_b - E_a)/kT},$$

where $T$ is the common temperature of the two systems. The term $e^{-E/kT}$ is called the *Boltzmann factor*.[6]

The energy levels of the system may be degenerate, with more than one quantum state having the same energy. That is, if states $s_a$ and $s_b$ are degenerate, then $E_a = E_b$ but $s_a \neq s_b$. When taking averages, each of the degenerate states must be counted separately. To account properly for the number of states that have a given energy, define $g_a$ to be the number of states with energy $E_a$. Similarly, define $g_b$ to be the number of states with energy $E_b$. These are called the *statistical weights* of the energy levels.

---

**Example 8.1** The ground state of the hydrogen atom is twofold degenerate. In fact, although "ground state" is the standard terminology, the plural "ground state*s*" would be more precise because these are *two* quantum states that have the same energy of $-13.6$ eV.[7] In the same manner, the "first excited state" actually consists of *eight* degenerate quantum states with the same energy of $-3.40$ eV. Table 8.2 shows the sets of quantum numbers $\{n,\ \ell,\ m_\ell,\ m_s\}$ that identify each state; it also shows each state's energy. We will assign

---

[6]The energies encountered in this context are usually given in units of electron volts (eV), so it is useful to remember that, at a room temperature of 300 K, the product $kT$ is approximately 1/40 eV.

[7]This degeneracy is not perfect. As explained in Section 12.1, the two ground states of hydrogen actually have slightly different energies, enabling the hydrogen atom to emit 21-cm radio waves.

## 8.1 The Formation of Spectral Lines

| Ground States $s_1$ | | | | Energy $E_1$ (eV) |
|---|---|---|---|---|
| $n$ | $\ell$ | $m_\ell$ | $m_s$ | |
| 1 | 0 | 0 | $+1/2$ | $-13.6$ |
| 1 | 0 | 0 | $-1/2$ | $-13.6$ |
| First Excited States $s_2$ | | | | Energy $E_2$ (eV) |
| $n$ | $\ell$ | $m_\ell$ | $m_s$ | |
| 2 | 0 | 0 | $+1/2$ | $-3.40$ |
| 2 | 0 | 0 | $-1/2$ | $-3.40$ |
| 2 | 1 | 1 | $+1/2$ | $-3.40$ |
| 2 | 1 | 1 | $-1/2$ | $-3.40$ |
| 2 | 1 | 0 | $+1/2$ | $-3.40$ |
| 2 | 1 | 0 | $-1/2$ | $-3.40$ |
| 2 | 1 | $-1$ | $+1/2$ | $-3.40$ |
| 2 | 1 | $-1$ | $-1/2$ | $-3.40$ |

**Table 8.2** Quantum Numbers and Energies for the Hydrogen Atom.

$a = 1$ for the two states $s_1$ with an energy of $-13.6$ eV, and $b = 2$ for the eight states $s_2$ with an energy of $-3.40$ eV, corresponding to the value of the principal quantum number $n$ for these energies. Notice that there are $g_1 = 2$ ground states with the energy $E_1 = -13.6$ eV, and $g_2 = 8$ first excited states with the energy $E_2 = -3.40$ eV. This result agrees with that of Problem 5.16 that the degeneracy of energy level $n$ is $2n^2$.

The ratio of the probability $P(E_b)$ that the system will be found in *any* of the $g_b$ degenerate states with energy $E_b$ to the probability $P(E_a)$ that the system is in *any* of the $g_a$ degenerate states with energy $E_a$ is given by

$$\frac{P(E_b)}{P(E_a)} = \frac{g_b\, e^{-E_b/kT}}{g_a\, e^{-E_a/kT}} = \frac{g_b}{g_a} e^{-(E_b - E_a)/kT}.$$

Stellar atmospheres contain a vast number of atoms, so the ratio of probabilities is essentially equal to the ratio of the number of atoms. Thus, for the atoms of a given element in a specified state of ionization, the ratio of the number of atoms $N_b$ with energy $E_b$ to the number of atoms $N_a$ with energy $E_a$ in *different states of excitation* is given by the **Boltzmann equation**,

$$\frac{N_b}{N_a} = \frac{g_b\, e^{-E_b/kT}}{g_a\, e^{-E_a/kT}} = \frac{g_b}{g_a} e^{-(E_b - E_a)/kT}. \tag{8.4}$$

**Example 8.2** For a gas of neutral hydrogen atoms, at what temperature will equal numbers of atoms have electrons in the ground state ($n = 1$) and in the first excited state ($n = 2$)?[8] As in Example 8.1, we use $a = 1$ for the ground state energy and $b = 2$ for the first excited state energy, corresponding to the value of the principal quantum number $n$. Also recall from that example that the degeneracy of the $n$th energy level of the hydrogen atom is $g_n = 2n^2$. Setting $N_2 = N_1$ on the left-hand side of Eq. (8.4), and using Eq. (5.14) for the energy levels leads to

$$1 = \frac{2(2)^2}{2(1)^2} e^{-[(-13.6 \text{ eV}/2^2) - (-13.6 \text{ eV}/1^2)]/kT}$$

$$1 = \frac{8}{2} e^{-10.2 \text{ eV}/kT}$$

$$\frac{10.2 \text{ eV}}{kT} = \ln(4).$$

Expressing the Boltzmann constant in more convenient units,

$$k = 8.6174 \times 10^{-5} \text{ eV K}^{-1},$$

leads to

$$T = \frac{10.2 \text{ eV}}{k \ln(4)} = \frac{10.2 \text{ eV}}{(8.62 \times 10^{-5} \text{ eV K}^{-1}) \ln(4)} = 8.54 \times 10^4 \text{ K}.$$

High temperatures are required for a significant number of hydrogen atoms to have electrons in the first excited state. Figure 8.5 shows the relative occupancy of the ground and first excited states, $N_2/(N_1 + N_2)$, as a function of temperature.[9] This result is somewhat puzzling, however. Recall that the Balmer absorption lines are produced by electrons in hydrogen atoms making an upward transition from the $n = 2$ orbital. If temperatures on the order of 85,000 K are needed to provide electrons in the first excited state, then why do the Balmer lines reach their maximum intensity at a much lower temperature of 9520 K? At temperatures higher than 9520 K, an even greater proportion of

---

[8] We have reverted back to the standard practice of referring to the two degenerate states of lowest energy as the "ground state," and to the eight degenerate states of next-lowest energy as the "first excited state."

[9] For the remainder of this section, we will use $a = 1$ for the ground state energy, and $b = 2$ for the energy of the first excited state.

## 8.1 The Formation of Spectral Lines

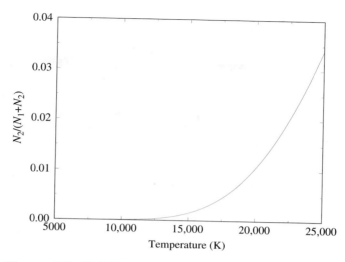

**Figure 8.5** $N_2/(N_1 + N_2)$ from the Boltzmann equation.

the electrons will be in the first excited state rather than in the ground state. If this is the case, then what is responsible for the diminishing strength of the Balmer lines at higher temperatures?

The Boltzmann factor also enters into the expression for the relative number of atoms *in different stages of ionization*. Let $\chi_i$ be the ionization energy needed to remove an electron from an atom in the ground state, taking it from ionization stage $i$ to stage $(i+1)$. For example, the ionization energy of hydrogen, the energy needed to convert it from H I to H II, is $\chi_I = 13.6$ eV. However, it may be that the initial and final atoms are not in the ground state. An average must be taken over the orbital energies to allow for the possible partitioning of the atom's electrons among its orbitals. This procedure involves calculating the *partition functions*, $Z$, for the initial and final atoms. The partition function is simply the weighted sum of the number of ways the atom can arrange its electrons with the same energy, with more energetic (and therefore less likely) configurations receiving less weight from the Boltzmann factor when taking the sum. If $E_j$ is the energy of the $j$th energy level and $g_j$ is the degeneracy of that level, then the partition function $Z$ is defined as

$$Z = g_1 + \sum_{j=2}^{\infty} g_j \, e^{-(E_j - E_1)/kT}. \tag{8.5}$$

If we use the partition functions $Z_i$ and $Z_{i+1}$ for the atom in its initial and final stages of ionization, the ratio of the number of atoms in stage $(i+1)$ to the number of atoms in stage $i$ is

$$\frac{N_{i+1}}{N_i} = \frac{2Z_{i+1}}{n_e Z_i} \left(\frac{2\pi m_e kT}{h^2}\right)^{3/2} e^{-\chi_i/kT}. \tag{8.6}$$

This equation is known as the **Saha equation**, after the Indian astrophysicist Meghnad Saha (1894–1956) who first derived it in 1920. Because a free electron is produced in the ionization process, it is not surprising to find the number density of free electrons (number of free electrons per unit volume), $n_e$, on the right-hand side of the Saha equation. Note that as the number density of free electrons increases, the number of atoms in the higher stage of ionization decreases, since there are more electrons with which the ion may recombine. The factor of 2 in front of the partition function $Z_{i+1}$ reflects the two possible spins of the free electron, with $m_s = \pm 1/2$. The term in parentheses is also related to the free electron, with $m_e$ being the electron mass.[10] Sometimes the pressure of the free electrons, $P_e$, is used in place of the electron number density; the two are related by the ideal gas law written in the form

$$P_e = n_e kT.$$

Then the Saha equation becomes

$$\frac{N_{i+1}}{N_i} = \frac{2kT Z_{i+1}}{P_e Z_i} \left(\frac{2\pi m_e kT}{h^2}\right)^{3/2} e^{-\chi_i/kT}. \tag{8.7}$$

The electron pressure in stellar atmospheres is typically somewhere between 1 dyne cm$^{-2}$ for cooler stars and 1000 dyne cm$^{-2}$ for hotter stars. In Section 9.4, we will describe how the electron pressure is determined for stellar atmospheres.

---

**Example 8.3** Consider the degree of ionization in a stellar atmosphere that is assumed to be composed of pure hydrogen. Assume for simplicity that the electron pressure is a constant $P_e = 200$ dyne cm$^{-2}$.

The Saha equation (8.7) will be used to calculate the fraction of atoms that are ionized, $N_{II}/(N_I + N_{II}) = N_{II}/N_{\text{total}}$, as the temperature $T$ varies between 5000 K and 25,000 K. However, the partition functions $Z_I$ and $Z_{II}$ must be determined first. A hydrogen ion is just a proton and so has no degeneracy; thus $Z_{II} = 1$. The energy of the first excited state of hydrogen is

---

[10]The term in parentheses is the number density of electrons for which the quantum energy (such as discussed in Example 5.5) is roughly equal to the characteristic thermal energy $kT$. For the classical conditions encountered in stellar atmospheres, this term is much greater than $n_e$.

## 8.1 The Formation of Spectral Lines

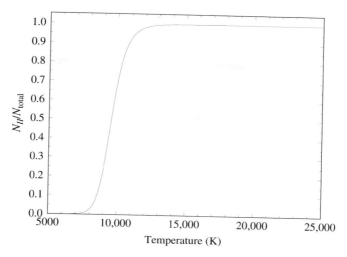

**Figure 8.6** $N_{II}/N_{\text{total}}$ for hydrogen from the Saha equation.

$E_2 - E_1 = 10.2$ eV above the ground state energy. Because 10.2 eV $\gg kT$ for the temperature regime under consideration, the Boltzmann factor $e^{-(E_2-E_1)/kT} \ll 1$. Nearly all of the H I atoms are therefore in the ground state (recall the previous example), so Eq. (8.5) for the partition function simplifies to $Z_I \simeq g_1 = 2(1)^2 = 2$.

Inserting these values into the Saha equation with $\chi_I = 13.6$ eV gives the ratio of ionized to neutral hydrogen, $N_{II}/N_I$. This ratio is then used to find the fraction of ionized hydrogen, $N_{II}/N_{\text{total}}$, by writing

$$\frac{N_{II}}{N_{\text{total}}} = \frac{N_{II}}{N_I + N_{II}} = \frac{N_{II}/N_I}{1 + N_{II}/N_I};$$

the results are displayed in Fig. 8.6. This figure shows that, when $T = 5000$ K, essentially none of the hydrogen atoms are ionized. At about 8300 K, 5% of the atoms have become ionized. Half of the hydrogen is ionized at a temperature of 9600 K, and when $T$ has risen to 11,300 K, all but 5% of the hydrogen is in the form of H II. Thus the ionization of hydrogen takes place within a temperature interval of approximately 3000 K. This range of temperatures is quite limited compared to the temperatures of tens of millions of degrees routinely encountered inside stars. The narrow region inside a star where hydrogen is partially ionized is called a hydrogen **partial ionization zone** and has a characteristic temperature of 10,000 K for a wide range of stellar parameters.

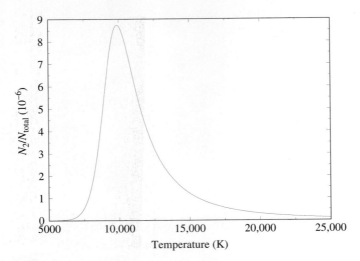

**Figure 8.7** $N_2/N_{\text{total}}$ for hydrogen from the Boltzmann and Saha equations.

Now we can see why the Balmer lines are observed to attain their maximum intensity at a temperature of 9520 K, instead of at the much higher characteristic temperatures (on the order of 85,000 K) required to excite electrons to the ($n = 2$) energy level of hydrogen. The strength of the Balmer lines depends on $N_2/N_{\text{total}}$, the fraction of *all* hydrogen atoms that are in the first excited state. This is found by combining the results of the Boltzmann and Saha equations. Because in fact practically all of the neutral hydrogen atoms are in either the ground or first excited states, we can set $N_1 + N_2 \simeq N_I$ and write

$$\frac{N_2}{N_{\text{total}}} = \left(\frac{N_2}{N_1 + N_2}\right)\left(\frac{N_I}{N_{\text{total}}}\right) = \left(\frac{N_2/N_1}{1 + N_2/N_1}\right)\left(\frac{1}{1 + N_{II}/N_I}\right).$$

Figure 8.7 shows that, in this example, the hydrogen gas would produce the most intense Balmer lines at a temperature of 9900 K, in good agreement with the observations. *The diminishing strength of the Balmer lines at higher temperatures is due to the rapid ionization of hydrogen above 10,000 K.* Figure 8.8 summarizes this situation.

---

Of course, stellar atmospheres are not composed of pure hydrogen, and the results obtained in Example 8.3 depended on an appropriate value for the electron pressure. In stellar atmospheres, there is typically one helium atom for every ten hydrogen atoms. The presence of ionized helium provides more electrons with which the hydrogen ions can recombine. Thus, when helium is

## 8.1 The Formation of Spectral Lines

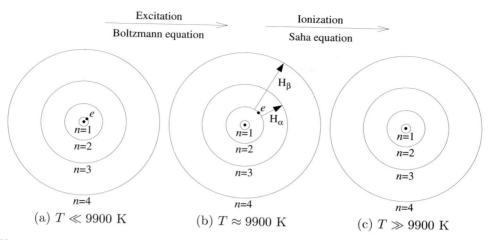

**Figure 8.8** The electron's position in the hydrogen atom at different temperatures. In (a), the electron is in the ground state. Balmer absorption lines are produced only when the electron is initially in the first excited state, as shown in (b). In (c), the atom has been ionized.

added, it takes a *higher* temperature to achieve the same degree of hydrogen ionization. It should also be emphasized that the Saha equation can be applied only to a gas in *thermal equilibrium*, so that the Maxwell–Boltzmann velocity distribution is obeyed. Furthermore, the density of the gas must not be too great (less than roughly $10^{-3}$ g cm$^{-3}$ for stellar material), or the presence of neighboring ions will distort an atom's orbitals and lower its ionization energy.

**Example 8.4** The "surface" of the Sun[11] has about 500,000 hydrogen atoms for each calcium atom. Estimate the relative strengths of the absorption lines due to hydrogen (the Balmer lines) and those due to calcium (the Ca II H and K lines).

We must compare the number of neutral hydrogen atoms with electrons in the first excited state (which produce the Balmer lines) to the number of singly ionized calcium atoms with electrons in the ground state (which produce the Ca II H and K lines). As in Example 8.3, we will use the Saha equation to determine the degree of ionization, and the Boltzmann equation to reveal the distribution of electrons between the ground and first excited states. The electron pressure in the solar photosphere[12] is about 15 dyne cm$^{-2}$.

---

[11]The Sun's "surface," where $T = T_e = 5770$ K, is a layer of the solar atmosphere called the *photosphere*; see Section 11.2.

[12]See Gingerich et al. (1971) for the *Harvard–Smithsonian Reference Atmosphere*.

Let's consider hydrogen first. If we substitute the partition functions found in Example 8.3 into the Saha equation, the ratio of ionized to neutral hydrogen is

$$\left[\frac{N_{II}}{N_I}\right]_{\text{H}} = \frac{2kT Z_{II}}{P_e Z_I} \left(\frac{2\pi m_e kT}{h^2}\right)^{3/2} e^{-\chi_I/kT}$$

$$= \frac{2(1.38 \times 10^{-16} \text{ erg K}^{-1})(5770 \text{ K})(1)}{(15 \text{ dyne cm}^{-2})(2)}$$

$$\times \left(\frac{2\pi(9.11 \times 10^{-28} \text{ g})(1.38 \times 10^{-16} \text{ erg K}^{-1})(5770 \text{ K})}{(6.63 \times 10^{-27} \text{ erg s})^2}\right)^{3/2}$$

$$\times e^{-13.6 \text{ eV}/[(8.62 \times 10^{-5} \text{ eV K}^{-1})(5770 \text{ K})]}$$

$$= 7.47 \times 10^{-5} = \frac{1}{13,400}.$$

Thus there is only one H II ion for every 13,000 neutral hydrogen atoms at the Sun's surface. Almost none of the hydrogen is ionized.

The Boltzmann equation (8.4) reveals how many of these neutral hydrogen atoms are in the first excited state:

$$\frac{N_2}{N_1} = \frac{g_2 \, e^{-E_2/kT}}{g_1 \, e^{-E_1/kT}} = \frac{g_2}{g_1} e^{-(E_2-E_1)/kT}.$$

Using $g_n = 2n^2$ for hydrogen, we have

$$\left[\frac{N_2}{N_1}\right]_{\text{H I}} = \frac{2(2)^2}{2(1)^2} e^{-[(-13.6 \text{ eV}/2^2)-(-13.6 \text{ eV}/1^2)]/[(8.62 \times 10^{-5} \text{ eV K}^{-1})(5770 \text{ K})]}$$

$$= 4.96 \times 10^{-9} = \frac{1}{202,000,000}.$$

The result is that only one of every 200 million hydrogen atoms is in the first excited state and capable of producing Balmer absorption lines:

$$\frac{N_2}{N_{\text{total}}} = \left(\frac{N_2}{N_1 + N_2}\right)\left(\frac{N_I}{N_{\text{total}}}\right) = 4.96 \times 10^{-9}.$$

We now turn to the calcium atoms. The ionization energy $\chi_I$ of Ca I is 6.11 eV, about half of the 13.6 eV ionization energy of hydrogen. We will soon see, however, that this small difference has a great effect on the ionization state of the atoms. Note that the Saha equation is very sensitive to the ionization

## 8.1 The Formation of Spectral Lines

energy because $\chi/kT$ appears as an *exponent* and $kT \approx 0.5$ eV $\ll \chi$. Thus a difference of several electron volts in the ionization energy produces a change of many powers of $e$ in the Saha equation.

Evaluating the partition functions $Z_I$ and $Z_{II}$ for calcium is a bit more complicated than for hydrogen, and the results have been tabulated elsewhere:[13] $Z_I = 1.32$ and $Z_{II} = 2.30$. Thus the ratio of ionized to un-ionized calcium is

$$\left[\frac{N_{II}}{N_I}\right]_{Ca} = \frac{2kTZ_{II}}{P_e Z_I}\left(\frac{2\pi m_e kT}{h^2}\right)^{3/2} e^{-\chi_I/kT} = 903.$$

Practically all of the calcium atoms are in the form of Ca II; only one atom out of 900 remains neutral. Now we can use the Boltzmann equation to estimate how many of these calcium ions are in the ground state, capable of forming the Ca II H and K absorption lines. The next calculation will consider the K ($\lambda = 3933$ Å) line; the results for the H ($\lambda = 3968$ Å) line are similar. The first excited state of Ca II is $E_2 - E_1 = 3.12$ eV above the ground state. The degeneracies for these states are $g_1 = 2$ and $g_2 = 4$. Then the ratio of the number of Ca II ions in the first excited state to those in the ground state is

$$\left[\frac{N_2}{N_1}\right]_{Ca\ II} = \frac{g_2}{g_1} e^{-(E_2-E_1)/kT} = 3.77 \times 10^{-3} = \frac{1}{265}.$$

Out of every 266 Ca II ions, all but one are in the ground state and are capable of producing the Ca II K line. Thus nearly *all* of the calcium atoms in the Sun's photosphere are singly ionized and in the ground state,[14] so that almost all of the calcium atoms are available for forming the H and K lines of calcium:

$$\left[\frac{N_1}{N_{total}}\right]_{Ca\ II} \simeq \left[\frac{N_1}{N_1+N_2}\right]_{Ca\ II}\left[\frac{N_{II}}{N_{total}}\right]_{Ca}$$

$$= \left(\frac{1}{1+[N_2/N_1]_{Ca\ II}}\right)\left(\frac{[N_{II}/N_I]_{Ca}}{1+[N_{II}/N_I]_{Ca}}\right)$$

$$= \left(\frac{1}{1+3.77\times 10^{-3}}\right)\left(\frac{903}{1+903}\right)$$

$$= 0.995.$$

Now it becomes clear why the Ca II H and K lines are so much stronger in the Sun's spectrum than are the Balmer lines. There are 500,000 hydrogen

---

[13]The values of the partition functions used here are from Aller (1963).

[14]It is left as an exercise to show that only a very small fraction of calcium atoms are doubly ionized (Ca III).

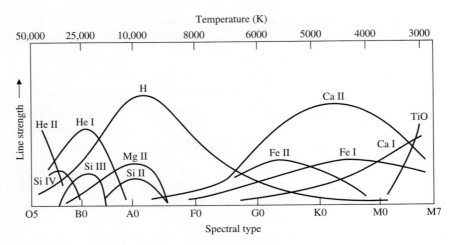

**Figure 8.9** The dependence of spectral line strengths on temperature.

atoms for every calcium atom in the solar photosphere, but only an extremely small fraction, $4.96 \times 10^{-9}$, of these hydrogen atoms are un-ionized and in the first excited state, capable of producing a Balmer line. Multiplying these two factors,

$$(500,000) \times (4.96 \times 10^{-9}) \approx 0.0025 = \frac{1}{400},$$

reveals that there are 400 times more Ca II ions with electrons in the ground state (to produce the Ca II H and K lines) than there are neutral hydrogen atoms with electrons in the first excited state (to produce the Balmer lines). The strength of the H and K lines is *not* due to a greater abundance of calcium in the Sun. The strength of these Ca II lines reflects the sensitive temperature dependence of the atomic states of excitation and ionization.

Figure 8.9 shows how the strength of various spectral lines varies with spectral type and temperature. As the temperature changes, a smooth variation from one spectral type to the next occurs, indicating that there are only minor differences in the composition of stars, as inferred from their spectra. The first person to determine the composition of the stars and discover the dominant role of hydrogen in the universe was Cecilia Payne (1900–1979). Her 1925 Ph.D. thesis, in which she calculated the relative abundances of 18 elements in stellar atmospheres, is widely recognized as the most brilliant ever done in astronomy. In Section 9.4, we will see just how the relative abundances of atoms and molecules in stellar atmospheres are measured.

## 8.2 The Hertzsprung–Russell Diagram

Early in the twentieth century, as astronomers accumulated data for an increasingly large sample of stars, they became aware of the wide range of stellar luminosities and absolute magnitudes. The O stars at one end of the Harvard sequence tended to be both brighter and hotter than the M stars at the other end. In addition, the empirical mass–luminosity relation (see Fig. 7.7), deduced from the study of binary stars, showed that O stars are more massive than M stars. These regularities led to a theory of stellar evolution[15] that described how stars might cool off as they age. This theory (no longer accepted) held that stars begin their lives as young, hot, bright blue O stars. It was suggested that, as they age, stars become less massive as they exhaust more and more of their "fuel" and that they then gradually become cooler and fainter until they fade away as old, dim red M stars. Although incorrect, a vestige of this idea remains in the terms *early* and *late* spectral types.

If this idea of stellar cooling were correct, then there should be a relation between a star's absolute magnitude and its spectral type. A Danish engineer and amateur astronomer, Ejnar Hertzsprung (1873–1967), analyzed stars whose absolute magnitudes and spectral types had been accurately determined. In 1905 he published a paper confirming the expected correlation between these quantities. However, he was puzzled by his discovery that stars of type G or later had a range of magnitudes, despite having the same spectral classification. Hertzsprung termed the brighter stars **giants**. This nomenclature was natural, since the Stefan–Boltzmann law (Eq. 3.17) shows that

$$R = \frac{1}{T_e^2}\sqrt{\frac{L}{4\pi\sigma}}. \tag{8.8}$$

If two stars have the same temperature (as inferred for stars having the same spectral type), then the more luminous star must be larger.

Hertzsprung presented his results in tabular form only. Meanwhile, at Princeton University, Henry Norris Russell (1877–1957) independently came to the same conclusions as Hertzsprung. Russell used the same term, *giant*, to describe the luminous stars of late spectral type, and the term **dwarf** stars for their dim counterparts. In 1913 Russell published the diagram shown in Fig. 8.10. It records a star's observed properties: absolute magnitude on the vertical axis (with brightness increasing upward), with spectral type running

---

[15]Stellar evolution describes the change in the structure and composition of an individual star as it ages. This usage of the term *evolution* differs from that in biology, where it describes the changes that occur over generations, rather than during the lifetime of a single individual.

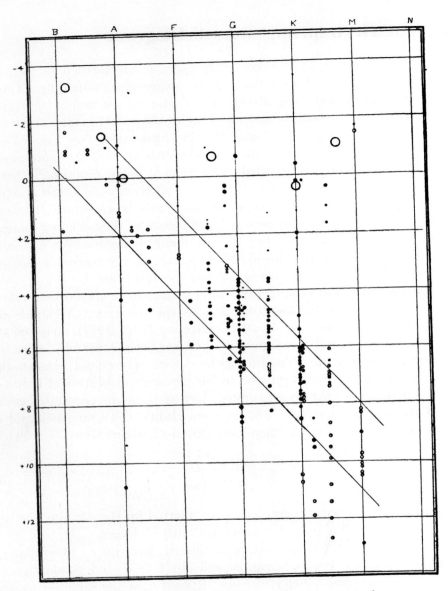

**Figure 8.10** Henry Norris Russell's first diagram, with spectral types listed along the top and absolute magnitudes on the left-hand side. (Figure from Russell, *Nature*, *93*, 252, 1914.)

## 8.2 The Hertzsprung–Russell Diagram

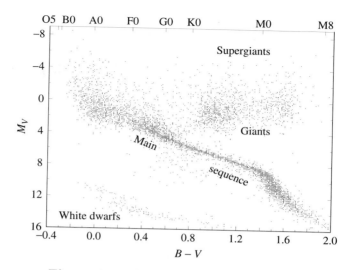

**Figure 8.11** An observer's H–R diagram.

horizontally (so temperature increases to the *left*). This first "Russell diagram" shows most of the features of its modern successor, the **Hertzsprung–Russell (H–R) diagram**.[16] More than 200 stars were plotted, most within a band reaching from the upper left-hand corner, home of the hot, bright O stars, to the lower right-hand corner, where the cool, dim M stars reside. This band is called the **main sequence** of the H–R diagram. Between 80% and 90% of all stars are main-sequence stars. In the upper right-hand corner are the giant stars. A single **white dwarf**, 40 Eridani B, sits at the lower left.[17] The vertical bands of stars in Russell's diagram are a result of the discrete classification of spectral types. A more recent version of an observational H–R diagram is shown in Fig. 8.11, with the absolute visual magnitude of each star plotted versus its color index and spectral type.

Figure 8.12 shows another version of the H–R diagram. Based on the average properties of main-sequence stars as listed in Appendix E, this diagram has a theorist's orientation: The luminosity and effective temperature are plotted for each star, rather than the observationally determined quantities of absolute

---

[16] The names of Hertzsprung and Russell were forever joined by another Danish astronomer, Bengt Strömgren (b. 1908), who suggested that the diagram be named after its two inventors. Strömgren's suggestion that star clusters be studied led to a clarification of the ideas of stellar evolution.

[17] Russell merely considered this star to be an extremely underluminous binary companion of the star 40 Eridani A; the extraordinary nature of white dwarfs was yet to be discovered. Note that the term *dwarf* refers to the stars on the main sequence and should not be confused with the *white dwarf* designation for stars lying well below the main sequence.

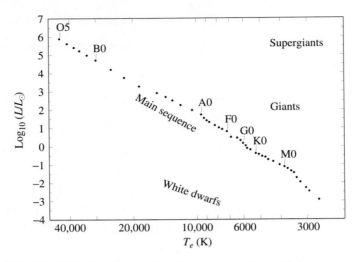

**Figure 8.12** The theorist's Hertzsprung–Russell diagram.

magnitude and color index or spectral type. The uneven nature of the main sequence is an artifact of the slight differences among the references used to compile the tables in this appendix. The Sun (G2) is found on the main sequence, as is Vega (A0). Both axes are scaled logarithmically to accommodate the huge span of stellar luminosities, ranging from about $5 \times 10^{-4}$ $L_\odot$ to nearly $10^6$ $L_\odot$.[18] Actually, the main sequence is not a line, but has a finite width, as shown in Figs. 8.10 and 8.11, due to the changes in a star's temperature and luminosity that occur while it is on the main sequence and to slight differences in the compositions of stars. The giant stars occupy the region above the lower main sequence, with the **supergiants**, such as Betelgeuse, in the extreme upper right-hand corner. The white dwarfs (which, despite their name, are often not white at all) lie well below the main sequence.

The radius of a star can be easily determined from its position on the H–R diagram. The Stefan–Boltzmann law in the form of Eq. (8.8) shows that if two stars have the same surface temperature, but one star is 100 times more luminous than the other, then the radius of the more luminous star is $\sqrt{100} = 10$ times larger. On a logarithmically plotted H–R diagram, the locations of stars having the same radii fall along diagonal lines that run roughly parallel to the main sequence, as shown in Fig. 8.13. The main-sequence stars show

---

[18]Extremely late and early spectral types are not included in Fig. 8.12. The dimmest main-sequence stars are difficult to find, and the brightest have very short lifetimes, making their detection unlikely. As a result, only a handful of stars belonging to these classifications are known, too few to establish their average properties.

## 8.2 The Hertzsprung–Russell Diagram

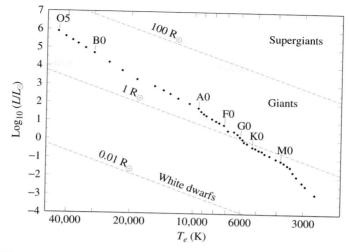

**Figure 8.13** Lines of constant radius on the H–R diagram.

some variation in their sizes, ranging from roughly 20 $R_\odot$ at the extreme upper end of the main sequence down to 0.1 $R_\odot$ at the lower end. The giant stars fall between roughly 10 $R_\odot$ and 100 $R_\odot$. For example, Aldebaran ($\alpha$ Tauri), the gleaming "eye" of the constellation Taurus (the Bull), is an orange giant star that is 45 times larger than the Sun.

The supergiant stars are even larger. Betelgeuse, a pulsating variable star, contracts and expands roughly between 700 and 1000 times the radius of the Sun with a period of 2070 days. If Betelgeuse were located at the Sun's position, its surface would at times extend past the orbit of Jupiter. The star $\mu$ Cephei in the constellation of Cepheus (a king of Ethiopia) is even larger and would swallow Saturn.[19]

The existence of such a simple relation between the luminosity and temperature of main-sequence stars is a valuable clue that the position of a star on the main sequence is governed by a single factor. This factor is the star's *mass*; in Chapter 10 we will see how mass determines the location of a star on the main sequence. The masses of stars along the main sequence are listed in Appendix E. The most massive O stars listed are observed to have masses of 60 $M_\odot$,[20] and the lower end of the main sequence is bounded by

---

[19] Like Betelgeuse, $\mu$ Cephei is a pulsating variable and has a period of 730 days. One of the reddest stars visible in the night sky, $\mu$ Cephei is known as the *Garnet Star*.

[20] Theoretical calculations indicate that main-sequence stars as massive as 90 $M_\odot$ may exist, and recent observations have been made of a few stars with masses estimated near 100 $M_\odot$; see Section 13.3.

M stars having at least 0.08 $M_\odot$. Combining the radii and masses known for main-sequence stars, we can calculate the average density of the stars. The result, perhaps surprising, is that main-sequence stars have roughly the same density as water. Moving up the main sequence, we find that the larger, more massive, early-type stars have a *lower* average density.

---

**Example 8.5** The Sun, a G2 star, has a mass of $1.99 \times 10^{33}$ g and a radius of $6.96 \times 10^{10}$ cm. Its average density is thus

$$\overline{\rho}_\odot = \frac{M_\odot}{\frac{4}{3}\pi R_\odot^3} = 1.41 \text{ g cm}^{-3}.$$

Sirius, the brightest-appearing star in the sky, is classified as an A1 star with a mass of 2.3 $M_\odot$ and a radius of 1.6 $R_\odot$. The average density of Sirius is

$$\overline{\rho} = \frac{2.3 \, M_\odot}{\frac{4}{3}\pi (1.6 \, R_\odot)^3} = 0.56 \left( \frac{M_\odot}{\frac{4}{3}\pi R_\odot^3} \right) = 0.56 \, \overline{\rho}_\odot = 0.79 \text{ g cm}^{-3},$$

about 80 percent of the density of water. However, this is enormously dense compared to a giant or supergiant star. The mass of Betelgeuse is estimated to lie between 10 and 15 $M_\odot$; we will adopt 10 $M_\odot$ here. If we take the maximum radius of this pulsating star to be about 1000 $R_\odot$, then the average density of Betelgeuse (at maximum size) is roughly

$$\overline{\rho} = \frac{10 \, M_\odot}{\frac{4}{3}\pi (1000 \, R_\odot)^3} = 10^{-8} \left( \frac{M_\odot}{\frac{4}{3}\pi R_\odot^3} \right) = 10^{-8} \, \overline{\rho}_\odot!$$

Thus Betelgeuse is a tenuous, ghostly object—a hundred thousand times less dense than the air we breathe. It is difficult even to define what is meant by the "surface" of such a wraithlike star.

---

Hertzsprung wondered whether there might be some difference in the spectra of giant and main-sequence stars of the same spectral type (or same effective temperature). Astronomers began a careful comparison of their spectra, discovering that there are indeed subtle differences in the relative strengths of spectral lines; see Fig. 8.14. This culminated in the 1943 publication of the *Atlas of Stellar Spectra* by William W. Morgan and Phillip C. Keenan of Yerkes Observatory. Their atlas consists of 55 prints of spectra that clearly display the effect of temperature and luminosity on stellar spectra and includes the criteria for the classification of each spectrum. The *MKK Atlas* established the

## 8.2 The Hertzsprung–Russell Diagram

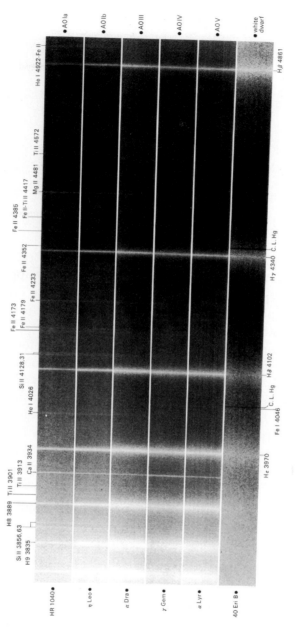

**Figure 8.14** A comparison of the strengths of the hydrogen Balmer lines in types A0 Ia, A0 Ib, A0 III, A0 IV, A0 V, and a white dwarf, showing the narrower lines found in supergiants. (Figure from Yamashita, Nariai, and Norimoto, *An Atlas of Representative Stellar Spectra*, University of Tokyo Press, Tokyo, 1978.)

| Class | Type of Star |
|---|---|
| Ia-O | Extreme, luminous supergiants |
| Ia | Luminous supergiants |
| Ib | Less luminous supergiants |
| II | Bright giants |
| III | Normal giants |
| IV | Subgiants |
| V | Main-sequence (dwarf) stars |
| VI, sd | Subdwarfs |
| D | White dwarfs |

**Table 8.3** Morgan–Keenan Luminosity Classes.

*two-dimensional* Morgan–Keenan (M–K) system of spectral classification.[21] A **luminosity class**, designated by a Roman numeral, is appended to a star's Harvard spectral type. The numeral "I" (subdivided into classes Ia and Ib) is reserved for the supergiant stars, and "V" denotes a main-sequence star. The ratio of the strengths of two closely spaced lines is often employed to place a star in the appropriate luminosity class. In general, for stars of the same spectral type, narrower lines are usually produced by more luminous stars.[22] The Sun is a G2 V star, and Betelgeuse is classified as M2 Ia.[23] The series of Roman numerals extends below the main sequence; the subdwarfs (class VI or "sd") reside slightly below the main sequence because they are deficient in **metals** (elements heavier than helium).[24] The M–K system does not extend to the white dwarfs, which are classified by the letter D. Table 8.3 lists the luminosity classes, and Fig. 8.15 shows the corresponding divisions on the H–R diagram and the location of a selection of specific stars.

The two-dimensional M–K classification scheme enables astronomers to locate a star's position on the Hertzsprung–Russell diagram based *entirely* on the appearance of its spectrum. Once the star's absolute magnitude, $M$, has been read from the vertical axis of the H–R diagram, the *distance* to the star

---

[21] Edith Kellman of Yerkes printed the 55 spectra and was a co-author of the atlas; hence the additional "K" in *MKK Atlas*.

[22] In Section 9.4, we will find that because the atmospheres of more luminous stars are less dense, there are fewer collisions between atoms to distort the energy of their orbitals and so broaden the spectral lines.

[23] Betelgeuse, a pulsating variable star, is sometimes given the intermediate classification M2 Iab.

[24] Astronomers simply refer to all elements heavier than helium as metals.

## 8.2 The Hertzsprung–Russell Diagram

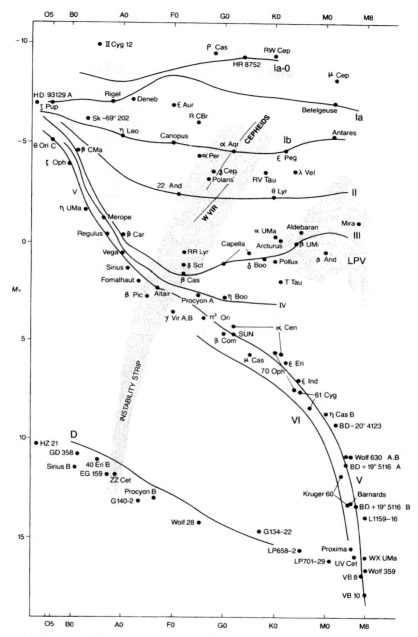

**Figure 8.15** Luminosity classes on the H–R diagram. (Figure from Kaler, *Stars and Stellar Spectra*, © Cambridge University Press 1989. Reprinted with the permission of Cambridge University Press.)

can be calculated from its apparent magnitude, $m$, via Eq. (3.5),

$$d = 10^{(m-M+5)/5},$$

where $d$ is in units of parsecs. This method of distance determination, called **spectroscopic parallax**, is responsible for most of the distances measured for stars,[25] but its accuracy is limited because there is not a perfect correlation between stellar absolute magnitudes and luminosity classes. The intrinsic scatter of roughly $\pm 1$ magnitude for a specific luminosity class renders $d$ uncertain by a factor of about $10^{1/5} = 1.6$.

## Suggested Readings

GENERAL

Aller, Lawrence H., *Atoms, Stars, and Nebulae*, Revised Edition, Harvard University Press, Cambridge, MA, 1971.

Dobson, Andrea K., and Bracher, Katherine, "A Historical Introduction to Women in Astronomy," *Mercury*, January/February 1992.

Hearnshaw, J. B., *The Analysis of Starlight*, Cambridge University Press, Cambridge, 1986.

Herrmann, Dieter B., *The History of Astronomy from Hershel to Hertzsprung*, Cambridge University Press, Cambridge, 1984.

Kaler, James B., *Stars and Their Spectra*, Cambridge University Press, Cambridge, 1989.

TECHNICAL

Aller, Lawrence H., *The Atmospheres of the Sun and Stars*, Ronald Press, New York, 1963.

Böhm-Vitense, Erika, *Stellar Astrophysics, Volume 2: Stellar Atmospheres*, Cambridge University Press, Cambridge, 1989.

Gingerich, O., Noyes, R. W., Kalkafen, W., and Cuny, Y., "The Harvard–Smithsonian Reference Atmosphere," *Sol. Phys.*, 18, 347, 1971.

Novotny, Eva, *Introduction to Stellar Atmospheres and Interiors*, Oxford University Press, New York, 1973.

---

[25]Since the technique of parallax is not involved, the term *spectroscopic parallax* is a misnomer, although the name does at least imply a distance determination.

# Problems

8.1 Show that, at room temperature, the thermal energy $kT \approx 1/40$ eV. At what temperature is $kT$ equal to 1 eV? to 13.6 eV?

8.2 Verify that Boltzmann's constant can be expressed as $k = 8.6174 \times 10^{-5}$ eV K$^{-1}$.

8.3 Use Fig. 8.4, the graph of the Maxwell–Boltzmann distribution for hydrogen gas at 10,000 K, to estimate the fraction of hydrogen atoms with a speed within $10^5$ cm s$^{-1}$ of the most probable speed, $v_{mp}$.

8.4 Show that the most probable speed of the Maxwell–Boltzmann distribution of molecular speeds (Eq. 8.1) is given by Eq. (8.2).

8.5 For a gas of neutral hydrogen atoms, at what temperature is the number of atoms in the first excited state only 1% of the number of atoms in the ground state? At what temperature is the number of atoms in the first excited state only 10% of the number of atoms in the ground state?

8.6 Consider a gas of neutral hydrogen atoms, as in Example 8.2.

(a) At what temperature will equal numbers of atoms have electrons in the ground state and in the second excited state ($n = 3$)?

(b) At a temperature of 85,400 K, when an equal number ($N$) of atoms are in the ground state and in the first excited state, how many atoms are in the second excited state ($n = 3$)? Express your answer in terms of $N$.

(c) As the temperature $T \to \infty$, how will the electrons in the hydrogen atoms be distributed, according to the Boltzmann equation? That is, what will be the relative numbers of electrons in the $n = 1, 2, 3, \ldots$ orbitals? Will this in fact be the distribution that actually occurs? Why or why not?

8.7 In Example 8.3, the statement was made that "nearly all of the H I atoms are in the ground state, so Eq. (8.5) for the partition function simplifies to $Z_I \simeq g_1 = 2(1)^2 = 2$." Verify that this statement is correct for a temperature of 10,000 K by evaluating the first three terms in Eq. (8.5) for the partition function.

8.8 Equation (8.5) for the partition function actually diverges as $n \to \infty$. Why can we ignore these large-$n$ terms?

8.9 Consider a box of electrically neutral hydrogen gas that is maintained at a constant volume $V$. In this simple situation, the number of electrons must equal the number of H II ions: $n_e V = N_{II}$. Also, the total number of hydrogen atoms (both neutral and ionized), $N_t$, is related to the density of the gas by $N_t = \rho V/(m_p + m_e) \simeq \rho V/m_p$, where $m_p$ is the mass of the proton. (The tiny mass of the electron may be safely ignored in this expression for $N_t$.) Let the density of the gas be $10^{-9}$ g cm$^{-3}$, typical of the photosphere of an A0 star.

(a) Make these substitutions into Eq. (8.6) to derive a quadratic equation for the fraction of ionized atoms,

$$\left(\frac{N_{II}}{N_t}\right)^2 + \left(\frac{N_{II}}{N_t}\right)\left(\frac{m_p}{\rho}\right)\left(\frac{2\pi m_e kT}{h^2}\right)^{3/2} e^{-\chi_I/kT}$$
$$- \left(\frac{m_p}{\rho}\right)\left(\frac{2\pi m_e kT}{h^2}\right)^{3/2} e^{-\chi_I/kT} = 0.$$

(b) Solve the quadratic equation in part (a) for the fraction of ionized hydrogen, $N_{II}/N_t$, for a range of temperatures between 5000 K and 25,000 K. Make a graph of your results, and compare it with Fig. 8.6.

8.10 In this problem, you will follow a procedure similar to that of Example 8.3 for the case of a stellar atmosphere composed of pure helium to find the temperature at the middle of the He I partial ionization zone, where half of the He I atoms have been ionized. (Such an atmosphere would be found on a white dwarf of spectral type DB; see Section 15.1.) The ionization energies of neutral helium and singly ionized hydrogen are $\chi_I = 24.6$ eV and $\chi_{II} = 54.4$ eV, respectively. The partition functions are $Z_I = 1$, $Z_{II} = 2$, and $Z_{III} = 1$ (as expected for a completely ionized atom). Use $P_e = 200$ dyne cm$^{-2}$ for the electron pressure.

(a) Use Eq. (8.7) to find $N_{II}/N_I$ and $N_{III}/N_{II}$ for temperatures of 5000 K, 15,000 K, and 25,000 K. How do they compare?

(b) Show that $N_{II}/N_{\text{total}} = N_{II}/(N_I + N_{II} + N_{III})$ can be expressed in terms of the ratios $N_{II}/N_I$ and $N_{III}/N_{II}$.

(c) Make a graph of $N_{II}/N_{\text{total}}$ similar to Fig. 8.6 for a range of temperatures from 5000 K to 25,000 K. What is the temperature at the middle of the He I partial ionization zone? Because the temperatures of the middle of the hydrogen and He I partial ionization zones are so similar, they are sometimes considered to be a single partial ionization zone with a characteristic temperature of $1$–$1.5 \times 10^4$ K.

8.11 Follow the procedure of Problem 8.10 to find the temperature at the middle of the He II partial ionization zone, where half of the He II atoms have been ionized. This ionization zone is found at a greater depth in the star, and so the electron pressure is larger—use a value of $P_e = 10^4$ dyne cm$^{-2}$. Let your temperatures range from 10,000 K to 60,000 K. This particular ionization zone plays a crucial role in pulsating stars, as will be discussed in Section 14.2.

8.12 Use the Saha equation to determine the fraction of hydrogen atoms that are ionized, $N_{II}/N_{\text{total}}$, at the center of the Sun. Here the temperature is 15.8 million K and the number density of electrons is about $n_e = 6.4 \times 10^{25}$ cm$^{-3}$. (Use $Z_I = 2$.) Does your result agree with the fact that practically *all* of the Sun's hydrogen is ionized at the Sun's center? What is the reason for any discrepancy?

8.13 Use the information in Example 8.4 to calculate the ratio of doubly to singly ionized calcium atoms (Ca III/Ca II) in the Sun's photosphere. The ionization energy of Ca II is $\chi_{II} = 11.9$ eV. Use $Z_{III} = 1$ for the partition function of Ca III. Is your result consistent with the statement in Example 8.4 that, in the solar photosphere, "nearly all of the calcium atoms are available for forming the H and K lines of calcium?"

8.14 Consider a giant star and a main-sequence star of the *same spectral type*. Appendix E shows that the giant star, which has a lower atmospheric density, has a slightly lower temperature than the main-sequence star. Use the Saha equation to explain why this is so. Note that this means that there is not a perfect correspondence between temperature and spectral type!

8.15 Figure 8.13 shows that a white dwarf star typically has a radius that is only 1% of the Sun's. Determine the average density of a 1-$M_\odot$ white dwarf.

8.16 The blue-white star Formalhaut ("the fish's mouth" in Arabic) is in the southern constellation of Pisces Austrinus. Formalhaut has an apparent visual magnitude of $V = 1.19$. Use the H–R diagram in Fig. 8.15 to determine the distance to this star.

# Chapter 9

# STELLAR ATMOSPHERES

## 9.1 The Description of the Radiation Field

The light that astronomers receive from a star comes from the star's atmosphere, the transparent layers of gas overlying the opaque interior. A flood of photons pours from these layers, releasing the energy produced by the thermonuclear reactions in the star's center. The temperature of the atmospheric layers from which these photons escape determines the features of the star's spectrum. To interpret the observed spectral lines properly, we must describe how light travels through the gas that makes up a star.

Figure 9.1 shows rays of light with a wavelength between $\lambda$ and $\lambda + d\lambda$ passing through a surface of area $dA$ at an angle $\theta$ into a cone of solid angle $d\Omega$.[1] The angle $\theta$ is measured from the direction perpendicular to the surface, so $dA \cos\theta$ is the area $dA$ projected onto a plane perpendicular to the direction in which the radiation is traveling. If $E_\lambda \, d\lambda$ is the amount of energy that these rays carry into the cone in a time interval $dt$, then the *average intensity* of the rays is defined as

$$\overline{I}_\lambda = \frac{E_\lambda \, d\lambda}{d\lambda \, dt \, dA \, \cos\theta \, d\Omega}.$$

The energy $E_\lambda \, d\lambda$ in the numerator becomes vanishingly small as the quantities in the denominator go to zero, but their ratio approaches a limiting value of $I_\lambda$, called the **specific intensity**, usually referred to simply as the *intensity*. Thus, in spherical coordinates,

$$E_\lambda \, d\lambda = I_\lambda \, d\lambda \, dt \, dA \, \cos\theta \, d\Omega = I_\lambda \, d\lambda \, dt \, dA \, \cos\theta \, \sin\theta \, d\theta \, d\phi$$

---

[1] The surface is a mathematical location in space and is not necessarily a real physical surface. The concept of a solid angle and its units of steradians (sr) was discussed in Section 6.1.

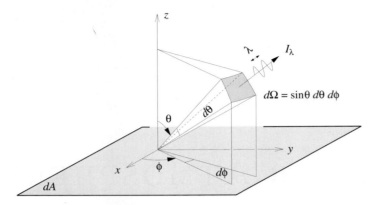

**Figure 9.1** Intensity $I_\lambda$.

is the amount of electromagnetic radiation energy having a wavelength between $\lambda$ and $\lambda+d\lambda$ that passes in time $dt$ through the area $dA$ into a solid angle $d\Omega = \sin\theta\, d\theta\, d\phi$. The specific intensity therefore has units of erg s$^{-1}$ cm$^{-3}$ sr$^{-1}$.[2] The Planck function $B_\lambda$, Eq. (3.20), is an example of the specific intensity for the special case of blackbody radiation. In general, however, the energy of the light need not vary with wavelength in the same way as it does for blackbody radiation. Later we will see under what circumstances we may set $I_\lambda = B_\lambda$.

Imagine a light ray of intensity $I_\lambda$ as it propagates through a vacuum. Because $I_\lambda$ is defined in the limit $d\Omega \to 0$, the energy of the ray *does not spread out (or diverge)*. The intensity is therefore *constant* along any ray traveling through empty space.

In general, the specific intensity $I_\lambda$ varies with direction. The **mean intensity** of the radiation is found by integrating the specific intensity over all directions and dividing the result by $4\pi$ sr, the solid angle enclosed by a sphere, to obtain an average value of $I_\lambda$. In spherical coordinates, this average value is[3]

$$\langle I_\lambda \rangle \equiv \frac{1}{4\pi} \int I_\lambda\, d\Omega = \frac{1}{4\pi} \int_{\phi=0}^{2\pi} \int_{\theta=0}^{\pi} I_\lambda \sin\theta\, d\theta\, d\phi. \tag{9.1}$$

For an isotropic radiation field (one with the same intensity in all directions), $\langle I_\lambda \rangle = I_\lambda$. Blackbody radiation is isotropic and has $\langle I_\lambda \rangle = B_\lambda$.

To determine how much energy is contained within the radiation field, we can use a "trap" consisting of a small cylinder of length $dL$, open at both ends, with perfectly reflecting walls inside; see Fig. 9.2. Light entering the trap at

---

[2] Recall from Section 3.5 that erg cm$^{-3}$ indicates an energy per unit area per unit wavelength interval, erg cm$^{-2}$ cm$^{-1}$, *not* an energy per unit volume.

[3] Many texts refer to the average intensity as $J_\lambda$ instead of $\langle I_\lambda \rangle$.

## 9.1 The Description of the Radiation Field

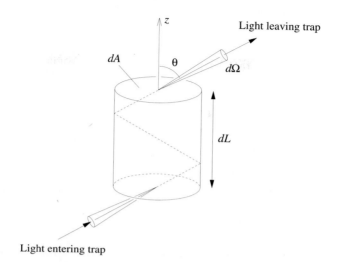

**Figure 9.2** Cylindrical "trap" for measuring energy density $u_\lambda$.

one end travels and (possibly) bounces back and forth until it exits the other end of the trap. The energy inside the trap is the same as what would be present at that location if the trap were removed. The radiation that enters the trap at an angle $\theta$ travels through the trap in a time $dt = dL/(c\cos\theta)$. Thus the amount of energy inside the trap with a wavelength between $\lambda$ and $\lambda + d\lambda$ that is due to the radiation that enters at angle $\theta$ is

$$E_\lambda\, d\lambda = I_\lambda\, d\lambda\, dt\, dA\, \cos\theta\, d\Omega = I_\lambda\, d\lambda\, dA\, d\Omega\, \frac{dL}{c}.$$

The quantity $dA\, dL$ is just the volume of the trap, so the specific **energy density** (energy per unit volume having a wavelength between $\lambda$ and $\lambda + d\lambda$) is found by dividing $E_\lambda\, d\lambda$ by $dL\, dA$, integrating over all solid angles, and using Eq. (9.1):

$$u_\lambda\, d\lambda = \frac{1}{c}\int I_\lambda\, d\lambda\, d\Omega = \frac{1}{c}\int_{\phi=0}^{2\pi}\int_{\theta=0}^{\pi} I_\lambda\, d\lambda\, \sin\theta\, d\theta\, d\phi = \frac{4\pi}{c}\langle I_\lambda\rangle\, d\lambda. \qquad (9.2)$$

For an isotropic radiation field, $u_\lambda\, d\lambda = (4\pi/c) I_\lambda\, d\lambda$, and for blackbody radiation,

$$u_\lambda\, d\lambda = \frac{4\pi}{c} B_\lambda\, d\lambda = \frac{8\pi hc/\lambda^5}{e^{hc/\lambda kT} - 1}\, d\lambda. \qquad (9.3)$$

At times it may be more useful to express the energy density in terms of the frequency, $\nu$, of the light:

$$u_\nu\, d\nu = \frac{4\pi}{c} B_\nu\, d\nu = \frac{8\pi h\nu^3/c^3}{e^{h\nu/kT} - 1}\, d\nu. \qquad (9.4)$$

Thus $u_\nu\, d\nu$ is the energy per unit volume with a frequency between $\nu$ and $\nu + d\nu$.

The total energy density, $u$, is found by integrating over all wavelengths:

$$u = \int_0^\infty u_\lambda\, d\lambda.$$

For blackbody radiation ($I_\lambda = B_\lambda$), Eq. (3.25) shows that

$$u = \frac{4\pi}{c}\int_0^\infty B_\lambda(T)\, d\lambda = \frac{4\sigma T^4}{c} = aT^4, \qquad (9.5)$$

where $a = 4\sigma/c$ is known as the *radiation constant* and has the value

$$a = 7.566 \times 10^{-15}\ \mathrm{erg\ cm^{-3}\ K^{-4}}.$$

Another quantity of interest is $F_\lambda$, the **radiative flux**. $F_\lambda\, d\lambda$ is the *net* energy having a wavelength between $\lambda$ and $\lambda + d\lambda$ that passes each second through a unit area in the direction of the $z$-axis:

$$F_\lambda\, d\lambda = \int I_\lambda\, d\lambda\, \cos\theta\, d\Omega = \int_{\phi=0}^{2\pi}\int_{\theta=0}^{\pi} I_\lambda\, d\lambda\, \cos\theta\, \sin\theta\, d\theta\, d\phi. \qquad (9.6)$$

The factor of $\cos\theta$ determines the $z$-component of a light ray and allows the cancellation of oppositely directed rays. For an isotropic radiation field there is no net transport of energy, and so $F_\lambda = 0$.

Both the radiative flux and the specific intensity describe the light received from a celestial source, and the reader may wonder which of these quantities is actually measured by a telescope's photometer, pointed at the source of light. The answer depends on whether the source is resolved by the telescope. Figure 9.3(a) shows a source of light, uniform over its entire surface,[4] that is resolved by the telescope; the angle $\theta$ subtended by the source as a whole is much larger than $\theta_{\min}$, the smallest angle resolvable according to Rayleigh's criterion. In this case, what is being measured is the *specific intensity*, the amount of energy passing through the aperture area into the solid angle $\Omega_{\min}$ defined by $\theta_{\min}$. For example, at a wavelength of 5010 Å, the measured value of the specific intensity at the center of the Sun's disk is

$$I_{5010} = 4.03 \times 10^{14}\ \mathrm{erg\ s^{-1}\ cm^{-3}\ sr^{-1}}.$$

Now imagine that the source is moved twice as far away. According to the inverse square law for light, Eq. (3.2), there will be only $(1/2)^2 = 1/4$ as much

---

[4]The assumption of a uniform light source precludes dimming effects such as limb darkening, which will be discussed later.

## 9.1 The Description of the Radiation Field

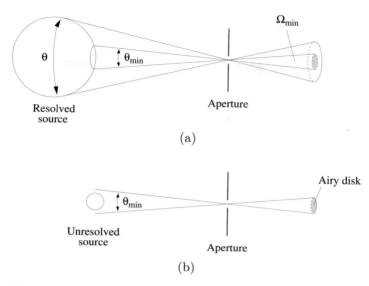

**Figure 9.3** The measurement of (a) the specific intensity for a resolved source and (b) the radiative flux for an unresolved source.

energy received from each square centimeter of the source. If the source is still resolved, however, then the amount of source area that contributes energy to the solid angle $\Omega_{\min}$ has increased by a factor of 4, resulting in the *same* amount of energy reaching each square centimeter of the detector. The specific intensity of light rays from the source is thus measured to be constant.[5]

However, it is the *radiative flux* that is measured for an unresolved source. As the source recedes farther and farther, it will eventually subtend an angle $\theta$ *smaller* than $\theta_{\min}$, and it can no longer be resolved by the telescope. When $\theta < \theta_{\min}$, the energy received from the *entire* source will disperse throughout the diffraction pattern (the Airy disk and rings; recall Section 6.1) determined by the telescope's aperture. Because the light arriving at the detector leaves the surface of the source at all angles [see Fig. 9.3(b)], the detector is effectively integrating the specific intensity over all directions. This is just the definition of the radiative flux, Eq. (9.6). As the distance $r$ to the source increases further, the amount of energy falling within the Airy disk (and consequently the value of the radiative flux) decreases as $1/r^2$, as expected.

A photon of energy $E$ carries a momentum of $p = E/c$ and thus can exert a **radiation pressure**. This radiation pressure can be derived in the same

---
[5]This argument has been encountered previously in the statement in Section 6.1 that the image and object intensities of a resolved object are the same.

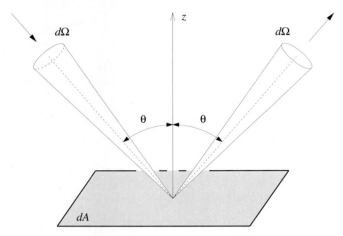

**Figure 9.4** Radiation pressure produced by incident photons from the solid angle $d\Omega$.

way that gas pressure is found for molecules bouncing off a wall. Figure 9.4 shows photons reflected at an angle $\theta$ from a perfectly reflecting surface of area $dA$ into a solid angle $d\Omega$. Because the angle of incidence equals the angle of reflection, the solid angles shown for the incident and reflected photons are the same size and inclined by the same angle $\theta$ on opposing sides of the $z$-axis. The change in the $z$-component of the momentum of photons with wavelengths between $\lambda$ and $\lambda + d\lambda$ that are reflected from the area $dA$ in a time interval $dt$ is

$$dp_\lambda \, d\lambda = [(p_\lambda)_{\text{final}} - (p_\lambda)_{\text{initial}}] \, d\lambda = \left[\frac{E_\lambda \cos\theta}{c} - \left(-\frac{E_\lambda \cos\theta}{c}\right)\right] d\lambda$$

$$= \frac{2\, E_\lambda \cos\theta}{c} \, d\lambda = \frac{2}{c} I_\lambda \, d\lambda \, dt \, dA \, \cos^2\theta \, d\Omega.$$

Dividing $dp_\lambda$ by $dt$ and $dA$ gives $(dp_\lambda/dt)/dA$. But from Newton's second and third laws, $-dp_\lambda/dt$ is the force exerted by the photons on the area $dA$.[6] Thus the radiation pressure is the force per unit area, $(dp_\lambda/dt)/dA$, produced by the photons within the solid angle $d\Omega$. Integrating over the hemisphere of all incident directions results in $P_{\text{rad},\lambda} \, d\lambda$, the radiation pressure exerted by those photons having a wavelength between $\lambda$ and $\lambda + d\lambda$:

$$P_{\text{rad},\lambda} \, d\lambda = \frac{2}{c} \int_{\text{hemisphere}} I_\lambda \, d\lambda \, \cos^2\theta \, d\Omega = \frac{2}{c} \int_{\phi=0}^{2\pi} \int_{\theta=0}^{\pi/2} I_\lambda \, d\lambda \, \cos^2\theta \, \sin\theta \, d\theta \, d\phi.$$

---

[6] We will ignore the minus sign, which merely says that the force is in the $-z$-direction.

## 9.2 Stellar Opacity

Just as the pressure of a gas exists throughout the volume of the gas and not just at the container walls, the radiation pressure of a "photon gas" exists everywhere in the radiation field. Imagine removing the reflecting surface $dA$ in Fig. 9.4 and replacing it with a mathematical surface. The incident photons will now keep on going through $dA$; instead of reflected photons, photons will be streaming through $dA$ from the other side. Thus, for an *isotropic radiation field*, there will be no change in the expression for the radiation pressure if the leading factor of 2 (which originated in the change in momentum upon reflection of the photons) is removed and the angular integration is extended over all solid angles:

$$P_{\text{rad},\lambda}\, d\lambda = \frac{1}{c} \int_{\text{sphere}} I_\lambda\, d\lambda \cos^2\theta\, d\Omega \tag{9.7}$$

$$= \frac{1}{c} \int_{\phi=0}^{2\pi} \int_{\theta=0}^{\pi} I_\lambda\, d\lambda \cos^2\theta\, \sin\theta\, d\theta\, d\phi$$

$$= \frac{4\pi}{3c} I_\lambda\, d\lambda. \tag{9.8}$$

However, it may be that the radiation field is *not* isotropic. In that case, Eq. (9.7) for the radiation pressure is still valid but the pressure depends on the orientation of the mathematical surface $dA$.

The total radiation pressure produced by photons of all wavelengths is found by integrating Eq. (9.8):

$$P_{\text{rad}} = \int_0^\infty P_{\text{rad},\lambda}\, d\lambda.$$

For blackbody radiation, it is left as a problem to show that

$$P_{\text{rad}} = \frac{4\pi}{3c} \int_0^\infty B_\lambda(T)\, d\lambda = \frac{4\sigma T^4}{3c} = \frac{1}{3} aT^4 = \frac{1}{3} u. \tag{9.9}$$

Thus the **blackbody radiation pressure** is one-third of the energy density. (For comparison, the pressure of an ideal monatomic gas is two-thirds of its energy density.)

## 9.2 Stellar Opacity

The classification of stellar spectra is an ongoing process. Even the most basic task, such as finding the surface temperature of a particular star, is complicated by the fact that stars are not actually blackbodies. The Stefan–Boltzmann relation, in the form of Eq. (3.17), defines a star's effective temperature, but some

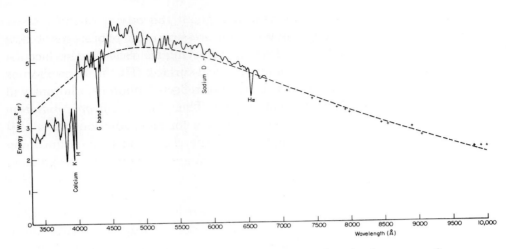

**Figure 9.5** The spectrum of the Sun. The dashed line is the curve of an ideal blackbody having the Sun's effective temperature. (Figure from Aller, *Atoms, Stars, and Nebulae*, Revised Edition, Harvard University Press, Cambridge, MA, 1971.)

effort is required to obtain a more accurate value of the surface temperature.[7] Figure 9.5 shows that the Sun's spectrum deviates substantially from the shape of the blackbody Planck function, $B_\lambda$, because solar absorption lines remove light from the Sun's continuous spectrum. The decrease in intensity produced by the dense series of metallic absorption lines in the solar spectrum is especially effective; this effect is called **line blanketing**.

In fact, there are many measures of a star's temperature. In addition to the effective temperature obtained from the Stefan–Boltzmann law, stellar temperature scales include the following:

- The **excitation temperature**, defined by the Boltzmann equation (8.4).

- The **ionization temperature**, defined by the Saha equation (8.6).

- The **kinetic temperature**, contained in the Maxwell–Boltzmann distribution, Eq. (8.1).

- The **color temperature**, obtained by fitting the shape of a star's continuous spectrum to the Planck function, Eq. (3.20).

These temperatures are the *same* for the simple case of a gas confined within a box. The confined gas particles and blackbody radiation will come into

---
[7] See Böhm-Vitense (1981) for more details concerning the determination of temperatures.

## 9.2 Stellar Opacity

equilibrium, individually and with each other, and can be described by a single well-defined temperature. In such a steady-state condition, no net flow of energy through the box or between the matter and the radiation occurs. Every process (e.g., the absorption of a photon) occurs at the same rate as its inverse process (e.g., the emission of a photon). This condition is called **thermodynamic equilibrium**.

However, a star cannot be in perfect thermodynamic equilibrium. A net outward flow of energy occurs through the star, and the temperature (however it is defined) varies with location. Gas particles and photons at one position in the star may have arrived there from other regions, either hotter or cooler. The distribution in particle speeds and photon energies thus reflects a range of temperatures. As the gas particles collide with one another and interact with the radiation field by absorbing and emitting photons, the description of the processes of excitation and ionization becomes quite complex. However, the idealized case of a single temperature can still be employed if the distance over which the temperature changes significantly is large compared with the distances traveled by the particles and photons between collisions (their *mean free paths*). In this case, referred to as **local thermodynamic equilibrium** (LTE), the particles and photons cannot escape the local environment and so are effectively confined to a limited volume (a "box") of nearly constant temperature.

---

**Example 9.1** The surface layer of the Sun's atmosphere, where the photons can escape into space, is called the photosphere (see Section 11.2). According to the "Harvard–Smithsonian Reference Atmosphere," the temperature in one region of the photosphere varies from 5650 K to 5890 K over a distance of 27.7 km. The characteristic distance over which the temperature varies, called the *temperature scale height*, $H_T$, is given by

$$H_T = \frac{T}{|dT/dr|} = \frac{5770 \text{ K}}{(5890 \text{ K} - 5650 \text{ K})/(2.77 \times 10^6 \text{ cm})} = 6.66 \times 10^7 \text{ cm},$$

where the average temperature has been used for the value of $T$.

How does the temperature scale height of 666 km compare with the average distance traveled by an atom before hitting another atom? The density of the photosphere at that level is about $\rho = 2.5 \times 10^{-7}$ g cm$^{-3}$, consisting primarily of neutral hydrogen atoms in the ground state. Assuming a pure hydrogen gas for convenience, the number of hydrogen atoms per cubic centimeter is

$$n = \frac{\rho}{m_H} = 1.50 \times 10^{17} \text{ cm}^{-3},$$

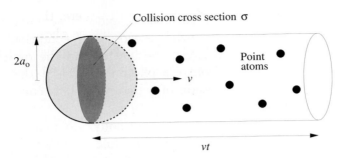

**Figure 9.6** Mean free path, $\ell$, of a hydrogen atom.

where $m_H$ is the mass of a hydrogen atom. Two of these atoms will collide if their centers pass within two Bohr radii, $2a_\circ$, of each other.[8] As shown in Fig. 9.6, we may consider the equivalent problem of a single atom of radius $2a_\circ$ moving with speed $v$ through a collection of stationary points that represent the centers of the other atoms. In an amount of time $t$, this atom has moved a distance $vt$ and has swept out a cylindrical volume $V = \pi(2a_\circ)^2 vt = \sigma vt$, where $\sigma \equiv \pi(2a_\circ)^2$ is the collision **cross section** of the atom.[9] Within this volume $V$ are $nV = n\sigma vt$ point atoms with which the moving atom has collided. Thus the average distance traveled between collisions is

$$\ell = \frac{vt}{n\sigma vt} = \frac{1}{n\sigma}. \tag{9.10}$$

The distance $\ell$ is the **mean free path** between collisions.[10] For a hydrogen atom,

$$\sigma = \pi(2a_\circ)^2 = 3.52 \times 10^{-16} \text{ cm}^2.$$

Thus the mean free path in this situation is

$$\ell = \frac{1}{n\sigma} = 1.89 \times 10^{-2} \text{ cm}.$$

The mean free path is several billion times smaller than the temperature scale height. As a result, the atoms in the gas see a nearly constant kinetic temperature between collisions. They are effectively confined within a limited volume of space in the photosphere. This cannot be true for the photons as well, since the Sun's photosphere is the visible layer of the solar surface that we observe

---

[8] This treats the atoms as solid spheres, a rough classical approximation to the quantum atom.
[9] The concept of *cross section* will be discussed in more detail in Section 10.3.
[10] A more careful calculation, using a Maxwellian velocity distribution for all of the atoms, results in a mean free path that is smaller by a factor of $\sqrt{2}$.

## 9.2 Stellar Opacity

from Earth. Thus, by the very definition of photosphere, the photons must be able to escape freely into space. To say more about the photon mean free path and the concept of LTE, and to better understand the solar spectrum shown in Fig. 9.5, we must examine the interaction of particles and photons in some detail.

---

We now turn to a consideration of a beam of parallel light rays traveling through a gas. Any process that removes photons from a beam of light will be called **absorption**. Absorption therefore includes the *scattering* of photons (such as Compton scattering, discussed in Section 5.2) as well as the true absorption of photons by atomic electrons making upward transitions. The change in the intensity, $dI_\lambda$, of a ray of wavelength $\lambda$ as it travels through a gas is proportional to its intensity, $I_\lambda$, the distance traveled, $ds$, and the density of the gas, $\rho$. That is,

$$dI_\lambda = -\kappa_\lambda \rho I_\lambda \, ds. \tag{9.11}$$

The distance $s$ is measured along the path traveled by the beam and increases in the direction that the beam travels; the minus sign in Eq. (9.11) shows that the intensity *decreases* with distance due to the absorption of photons. The quantity $\kappa_\lambda$ is called the **absorption coefficient**, or **opacity**.[11] The opacity is the cross section for absorbing photons of wavelength $\lambda$ per gram of stellar material and has units of cm$^2$ g$^{-1}$. In general, the opacity of a gas is a function of its composition, density, and temperature.

---

**Example 9.2** Consider a beam of light traveling through a gas in the $s$-direction with initial intensity $I_{\lambda,o}$ at $s = 0$. The final intensity $I_{\lambda,f}$ after the light has traveled a distance $s$ may be found by integrating Eq. (9.11):

$$\int_{I_{\lambda,o}}^{I_{\lambda,f}} \frac{dI_\lambda}{I_\lambda} = -\int_0^s \kappa_\lambda \rho \, ds.$$

This leads to

$$I_\lambda = I_{\lambda,o} e^{-\int_0^s \kappa_\lambda \rho \, ds}, \tag{9.12}$$

where the $f$ subscript has been dropped. For the specific case of a uniform gas of constant opacity and density,

$$I_\lambda = I_{\lambda,o} e^{-\kappa_\lambda \rho s}.$$

---

[11]The $\lambda$ subscript indicates that the opacity is wavelength-dependent; $\kappa_\lambda$ is sometimes referred to as a *monochromatic opacity*.

For pure absorption, there is no way of replenishing the photons lost from the beam. The intensity declines exponentially, falling by a factor of $e^{-1}$ over a characteristic distance of $\ell = 1/\kappa_\lambda \rho$. In the solar photosphere where the density is $\rho = 2.5 \times 10^{-7}$ g cm$^{-3}$, the opacity (at a wavelength of 5000 Å) is $\kappa_{5000} = 0.264$ cm$^2$ g$^{-1}$. Thus the characteristic distance traveled by a photon at this level of the photosphere before being removed from the beam is

$$\ell = \frac{1}{\kappa_{5000}\rho} = 1.52 \times 10^7 \text{ cm}.$$

Recalling Example 9.1, this distance is comparable to the temperature scale height $H_T = 6.66 \times 10^7$ cm. Thus the photospheric photons do *not* see a constant temperature, and so local thermodynamic equilibrium (LTE) is not strictly valid in the photosphere. The temperature of the regions from which the photons have traveled will be somewhat different from the local kinetic temperature of the gas. Nevertheless, LTE is a commonly invoked assumption in stellar atmospheres, but it must be used with caution.

For scattered photons, the characteristic distance $\ell$ is in fact the mean free path of the photons. From Eq. (9.10),

$$\ell = \frac{1}{\kappa_\lambda \rho} = \frac{1}{n\sigma_\lambda}.$$

Both $\kappa_\lambda \rho$ and $n\sigma_\lambda$ can be thought of as the gas target area encountered by a photon for every centimeter of distance it travels. Note that the mean free path is different for photons of different wavelengths.

---

When observing the light from a star, we are *looking back* along the path traveled by the photons. It is convenient to define an **optical depth**, $\tau_\lambda$, back along a light ray by

$$d\tau_\lambda = -\kappa_\lambda \rho \, ds, \qquad (9.13)$$

where $s$ is the distance measured along the photon's path in its direction of motion. The difference in optical depth between a light ray's initial position ($s = 0$) and its final position after traveling a distance $s$ is

$$\Delta \tau_\lambda = \tau_{\lambda,f} - \tau_{\lambda,\circ} = -\int_0^s \kappa_\lambda \rho \, ds. \qquad (9.14)$$

Note that $\Delta \tau_\lambda < 0$. As the light approaches an observer, it is traveling through smaller and smaller values of the optical depth. The outermost layers of a star may be taken to be at $\tau_\lambda = 0$ for all wavelengths, after which the light travels unimpeded to observers on Earth. With this definition of $\tau_\lambda = 0$,

## 9.2 Stellar Opacity

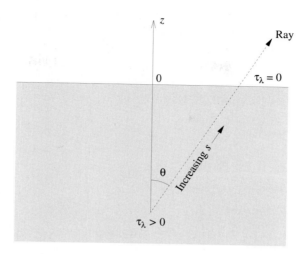

**Figure 9.7** Optical depth $\tau_\lambda$ measured back along a ray's path.

Eq. (9.14) gives the initial optical depth, $\tau_{\lambda,\circ}$, of a ray of light that traveled a distance $s$ to reach the top of the atmosphere:

$$0 - \tau_{\lambda,\circ} = -\int_0^s \kappa_\lambda \rho \, ds$$

$$\tau_\lambda = \int_0^s \kappa_\lambda \rho \, ds. \tag{9.15}$$

The "∘" subscript has been dropped with the understanding that $\tau_\lambda$ is the optical depth of the ray's initial position, a distance $s$ ($s > 0$) from the top of the atmosphere; see Fig. 9.7.

Combining Eq. (9.15) with Eq. (9.12) of Example 9.2 for the case of pure absorption, the decline in the intensity of a ray that travels through a gas from an optical depth $\tau_\lambda$ to reach the observer is given by

$$I_\lambda = I_{\lambda,\circ} e^{-\tau_\lambda}. \tag{9.16}$$

Thus, if the optical depth of the ray's starting point is $\tau_\lambda = 1$, the intensity of the ray will decline by a factor of $e^{-1}$ before escaping from the star. *The optical depth may be thought of as the number of mean free paths from the original position to the surface, as measured along the ray's path.* Of course, for pure absorption the intensity of the ray declines exponentially regardless of its direction of travel through the gas. But we can observe only those rays traveling toward us, and this is reflected in our choice of $\tau_\lambda = 0$ at the top of the atmosphere. Other choices of where $\tau_\lambda = 0$ may be more useful in some situations.

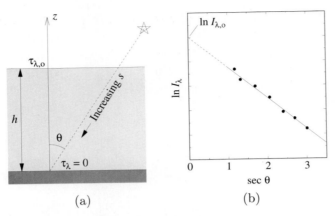

**Figure 9.8** (a) A light ray entering Earth's atmosphere at an angle $\theta$. (b) $\ln I_\lambda$ vs. $\sec \theta$.

If $\tau_\lambda \gg 1$ for a light ray passing through a volume of gas, the gas is said to be *optically thick*; if $\tau_\lambda \ll 1$, the gas is *optically thin*. Because the optical depth varies with wavelength, a gas may be optically thick at one wavelength, and optically thin at another. For example, Earth's atmosphere is optically thin at visible wavelengths (we can see the stars), but optically thick at x-ray wavelengths.

---

**Example 9.3** In Section 5.2, it was stated that measurements of a star's radiative flux and apparent magnitude are routinely corrected for the absorption of light by Earth's atmosphere. Figure 9.8 shows a ray of intensity $I_{\lambda,o}$ entering Earth's atmosphere at an angle $\theta$ and traveling to a telescope on the ground. The intensity of the light detected at the telescope is $I_\lambda$; the problem is to determine the value of $I_{\lambda,o}$. If we take $\tau_\lambda = 0$ at the telescope and $h$ to be the height of the atmosphere, then the optical depth of the light ray's path through the atmosphere may be found from Eq. (9.15). Using $ds = -dz/\cos\theta = -\sec\theta\, dz$,

$$\tau_\lambda = \int_0^s \kappa_\lambda \rho\, ds = -\int_h^0 \kappa_\lambda \rho \frac{dz}{\cos\theta} = \sec\theta \int_0^h \kappa_\lambda \rho\, dz = \tau_{\lambda,o} \sec\theta,$$

where $\tau_{\lambda,o}$ is the optical depth for a vertically traveling photon ($\theta = 0$).

The intensity of the light received at the telescope is therefore given by Eq. (9.16),

$$I_\lambda = I_{\lambda,o} e^{-\tau_{\lambda,o} \sec\theta}. \tag{9.17}$$

There are two unknowns in this equation, $I_{\lambda,o}$ and $\tau_{\lambda,o}$; neither can be determined by a single observation. However, as time passes and as Earth rotates on its axis, the angle $\theta$ will change, and a semilog graph of several measurements of the received intensity $I_\lambda$ as a function of $\sec\theta$ can be made. As shown in Fig. 9.8(b), the *slope* of the best-fitting straight line is $-\tau_{\lambda,0}$. Extrapolating the best-fitting line to $\sec\theta = 0$ provides the value of $I_{\lambda,0}$ at the point where the line intercepts the $I_\lambda$-axis.[12] In this way, measurements of the specific intensity or radiative flux can be corrected for absorption by Earth's atmosphere.

The opacity of the stellar material is determined by the details of how photons interact with particles (atoms, ions, and free electrons). If the photon passes within $\sigma_\lambda$ of the particle ($\sigma_\lambda$ is the particle's cross-sectional area, or effective target area), the photon may be either absorbed or scattered. In an absorption process, the photon ceases to exist, and in a scattering process it continues on in a different direction. Both absorption and scattering remove photons from a beam of light, and so contribute to the opacity, $\kappa_\lambda$, of the stellar material. If the opacity varies slowly with wavelength, it determines the star's continuous spectrum (or *continuum*). The dark absorption lines superimposed on the continuum are the result of a rapid variation in the opacity with wavelength.

In general, there are four primary sources of opacity available for removing stellar photons from a beam. Each involves a change in the quantum state of an electron, and the terms *bound* and *free* are used to describe whether the electron is bound to an atom or ion in its initial and final states.

1. **Bound–bound transitions** occur when an electron in an atom or ion makes a transition from one orbital to another. An electron can make an upward transition from a lower to a higher orbit when a photon of the appropriate energy is absorbed. Thus $\kappa_{\lambda,bb}$, the bound–bound opacity, is small except at those discrete wavelengths capable of producing an upward atomic transition. It is $\kappa_{\lambda,bb}$ that is responsible for forming the absorption lines in stellar spectra.

    The reverse emission process occurs when the electron makes a downward transition from a higher to a lower orbit. If the electron returns directly to its initial orbit (where it was before absorbing a photon), then a single photon is emitted in a random direction. The net result of this absorption–emission sequence is essentially a scattered photon. Otherwise, if the electron makes a transition to an orbit other than its initial one, the original photon is not recovered and the process is one of true absorption. An important by-product of this absorption process

---

[12]Because $\sec\theta \geq 1$, the best-fitting straight line must be extrapolated to the mathematically unavailable value of 0.

is degrading of the average energy of the photons in the radiation field. For example, if one photon is absorbed but two photons are emitted as the electron cascades down to its initial orbit, then the average photon energy has been reduced by half. There is no simple equation for bound–bound transitions that describes all of the contributions to the opacity by individual spectral lines.

2. **Bound–free absorption**, also known as photoionization, occurs when an incident photon has enough energy to ionize an atom. The resulting free electron can have any energy, so any photon with a wavelength $\lambda \leq hc/\chi_n$, where $\chi_n$ is the ionization energy of the $n$th orbital, can remove an electron from an atom. Thus $\kappa_{\lambda,bf}$, the bound–free opacity, is one source of the continuum opacity. The cross section for the photoionization of a hydrogen atom in quantum state $n$ by a photon of wavelength $\lambda$ is

$$\sigma_{bf} = 1.31 \times 10^{-15} \frac{1}{n^5} \left(\frac{\lambda}{5000 \text{ Å}}\right)^3 \text{ cm}^2,$$

which is comparable to the collision cross section for hydrogen found in Example 9.1. The inverse process of free–bound emission occurs when a free electron recombines with an ion, emitting one or more photons in random directions. As with bound–bound emission, this also contributes to reducing the average energy of the photons in the radiation field.

3. **Free–free absorption** is a scattering process, shown in Fig. 9.9, that takes place when a free electron in the vicinity of an ion absorbs a photon, causing the speed of the electron to increase.[13] This can occur for a continuous range of wavelengths, and so the free–free opacity, $\kappa_{\lambda,ff}$, is another contributor to the continuum opacity. It may also happen that the electron loses energy as it passes near an ion by emitting a photon, causing the electron to slow down as a result. This process of free–free emission is also known as *bremsstrahlung*, which means "braking radiation" in German.

4. **Electron scattering** is as advertised: A photon is scattered (*not absorbed*) by a free electron through the process of *Thomson scattering*. The electron is tiny and makes a poor target for an incident photon. The cross section for Thomson scattering has the same value for photons

---

[13] The nearby ion is required to conserve both energy and momentum. It is left as an exercise to show that an isolated free electron cannot absorb a photon.

## 9.2 Stellar Opacity

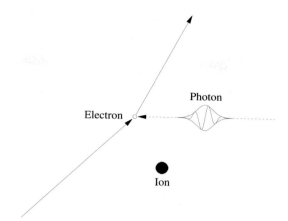

**Figure 9.9** Free–free absorption of a photon.

of all wavelengths:

$$\sigma_T = \frac{8\pi}{3}\left(\frac{e^2}{m_e c^2}\right)^2 = 6.65 \times 10^{-25} \text{ cm}^2. \tag{9.18}$$

This is typically two billion times smaller than the hydrogen cross section for photoionization, $\sigma_{bf}$. The small size of the Thomson cross section means that electron scattering is most effective as a source of opacity at high temperatures. In the atmospheres of the hottest stars (and in the interiors of all stars), where most of the gas is completely ionized, other sources of opacity that involve bound electrons are eliminated. In this high-temperature regime, the opacity due to electron scattering, $\kappa_{es}$, dominates the continuum opacity.

A photon may also be scattered by an electron that is loosely bound to an atomic nucleus. This result is called *Compton scattering* if the photon's wavelength is much smaller than the atom, and *Rayleigh scattering* if the photon's wavelength is much larger. In Compton scattering, the change in the wavelength and energy of the scattered photon is very small (recall the discussion of the Compton wavelength on page 133), so Compton scattering is usually lumped together with Thomson scattering. The cross section for Rayleigh scattering from a loosely bound electron is smaller than the Thomson cross section; it is proportional to $1/\lambda^4$ and so decreases with increasing photon wavelength. Rayleigh scattering can be neglected in most atmospheres, but it is important in the extended envelopes of supergiant stars. The scattering of photons from small particles is also responsible for the reddening of starlight as it passes through interstellar dust; see Section 12.1.

---

**Example 9.4** The energy of an electron in the $(n = 2)$ orbit of a hydrogen atom is given by Eq. (5.14):

$$E_2 = -13.6 \text{ eV} \frac{1}{2^2} = -3.40 \text{ eV}.$$

A photon must have an energy of at least $\chi_2 = 3.40$ eV to eject this electron from the atom. Thus any photon with a wavelength

$$\lambda \leq \frac{hc}{\chi_2} = \frac{12400 \text{ eV Å}}{3.40 \text{ eV}} = 3647 \text{ Å}$$

is capable of ionizing a hydrogen atom in the first excited state $(n = 2)$. The opacity of the stellar material suddenly increases at wavelengths $\lambda \leq 3647$ Å, and the radiative flux measured for the star accordingly decreases. The abrupt drop in the continuous spectrum of a star at this wavelength, called the **Balmer jump**, is clearly seen for the Sun's spectrum in Fig. 9.5. The size of the Balmer jump in hot stars depends on the fraction of hydrogen atoms that are in the first excited state. This fraction is determined by the temperature via the Boltzmann equation (Eq. 8.4), so a measurement of the size of the Balmer jump can be used to determine the temperature of the atmosphere. For cooler or very hot stars with other significant sources of opacity, the analysis is more complicated, but the size of the Balmer jump can still be used as a probe of atmospheric temperatures.

The wavelength 3647 Å is right in the middle of the bandwidth of the ultraviolet $(U)$ filter in the $UBV$ system, described on page 82. As a result, the Balmer jump will tend to decrease the amount of light received in the bandwidth of the $U$ filter and so *increase* both the ultraviolet magnitude $U$ and the color index $(U - B)$ observed for a star. This effect will be strongest when $N_2/N_{\text{total}}$, the fraction of all hydrogen atoms that are in the first excited state, is a maximum. From Example 8.3, this occurs at a temperature of 9600 K, about the temperature of an A0 star on the main sequence. A careful examination of the color–color diagram in Fig. 3.10 reveals that this is indeed the spectral type at which the value of $U - B$ differs most from its blackbody value.[14]

---

[14] The effect of line blanketing also affects the measured color indices, making the star appear more red than a model blackbody star of the same effective temperature, and thus increasing the values of both $U - B$ and $B - V$.

## 9.2 Stellar Opacity

The primary source of the continuum opacity in most stellar atmospheres is the photoionization of H⁻ ions. As will be discussed in Section 11.2, this fragile assemblage of two electrons orbiting a single proton has an ionization energy of only $\chi = 0.754$ eV. This means that any photon with wavelength

$$\lambda \leq \frac{hc}{\chi} = \frac{12400 \text{ eV Å}}{0.754 \text{ eV}} = 16400 \text{ Å}$$

can remove an electron from the ion. Consequently, H⁻ ions are an important source of continuum opacity at mid-infrared and all shorter wavelengths. However, the H⁻ ions become increasingly ionized at higher temperatures. For stars of spectral types B and A, the photoionization of hydrogen atoms and free–free absorption are the main sources of the continuum opacity. At the higher temperatures encountered for O stars, the ionization of atomic hydrogen means that electron scattering becomes more and more important, with the photoionization of helium also contributing to the opacity.

Molecules can survive in cooler stellar atmospheres and contribute to the bound–bound and bound–free opacities; the large number of discrete molecular absorption lines is an efficient impediment to the flow of photons. Molecules can also be broken apart into their constituent atoms by the absorption of photons in the process of *photodissociation*, which plays an important role in planetary atmospheres.

The total opacity is the sum of the opacities due to all of the preceding sources:

$$\kappa_\lambda = \kappa_{\lambda,bb} + \kappa_{\lambda,bf} + \kappa_{\lambda,ff} + \kappa_{es}.$$

The total opacity depends not only on the wavelength of the light being absorbed but also on the composition, density, and temperature of the stellar material.[15]

It is often useful to employ an opacity that has been averaged over all wavelengths to produce a function of only the composition, density, and temperature. This average opacity, $\bar{\kappa}$, is known as the **Rosseland mean opacity**, often simply referred to as the **Rosseland mean**.[16] Unfortunately, there is no simple equation capable of describing all of the contributions to the opacity by individual spectral lines in bound–bound transitions, and so the Rosseland mean cannot be given for this case. However, approximation formulae have

---

[15] The additional dependencies of the opacity on the electron number density, states of excitation and ionization of the atoms and ions, and other factors can be calculated from the composition, density, and temperature.

[16] This wavelength-averaged opacity was introduced in 1924 by S. Rosseland.

been developed that are adequate descriptions of both the average bound–free and free–free opacities:

$$\overline{\kappa}_{bf} = 4.34 \times 10^{25} \frac{g_{bf}}{t} Z(1+X) \frac{\rho}{T^{3.5}} \quad \text{cm}^2 \text{ g}^{-1} \tag{9.19}$$

$$\overline{\kappa}_{ff} = 3.68 \times 10^{22} g_{ff}(1-Z)(1+X) \frac{\rho}{T^{3.5}} \quad \text{cm}^2 \text{ g}^{-1}, \tag{9.20}$$

where $\rho$ is the density, $T$ is the temperature, and $X$ and $Z$ are the fractional abundances (by mass) of hydrogen and metals, respectively.[17] The *Gaunt factors*, $g_{bf}$ and $g_{ff}$, are quantum-mechanical correction terms calculated by J. A. Gaunt. These Gaunt factors are both $\approx 1$ for the visible and ultraviolet wavelengths of interest in stellar atmospheres. The additional correction factor, $t$, in the equation for the bound–free opacity is called the *guillotine factor* and describes the cutoff of an atom's contribution to the opacity after it has been ionized. Typical values of $t$ lie between 1 and 100.

Both of these formulae have the functional form $\overline{\kappa} = \kappa_\circ \rho / T^{3.5}$, where $\kappa_\circ$ is a constant. The first forms of these expressions were derived by H. A. Kramers in 1923 using classical physics and the Rosseland mean. Any opacity with this density and temperature dependence is called a *Kramers opacity law*.

Because the cross section for electron scattering is independent of wavelength, the Rosseland mean for this case has the particularly simple form

$$\overline{\kappa}_{es} = 0.2(1+X) \quad \text{cm}^2 \text{ g}^{-1}. \tag{9.21}$$

The total Rosseland mean opacity, $\overline{\kappa}$, is the average of the sum of the individual contributors to the opacity:

$$\overline{\kappa} = \overline{\kappa_{bb} + \kappa_{bf} + \kappa_{ff} + \kappa_{es}}. \tag{9.22}$$

Figure 9.10 shows the results of a calculation of the Rosseland mean opacity carried out by Forrest Rogers and Carlos Iglesias for a composition with $X = 0.70$ and $Z = 0.02$.[18] The values of $\overline{\kappa}$ are plotted as a function of the temperature for several densities. First, notice that the opacity increases with increasing density. Next, starting at the left-hand side of the figure, follow a single plot as it rises steeply with increasing temperature. This reflects the increase in the number of free electrons produced by the ionization of hydrogen and helium. (Recall from Example 8.3 that the hydrogen partial ionization

---

[17]$\rho$ is given in g cm$^{-3}$, and $T$ in kelvin; see Section 10.2 for a discussion of $X$ and $Z$.
[18]A specific mixture of elements known as the Anders–Grevesse abundances were used to calculate the opacities shown.

## 9.2 Stellar Opacity

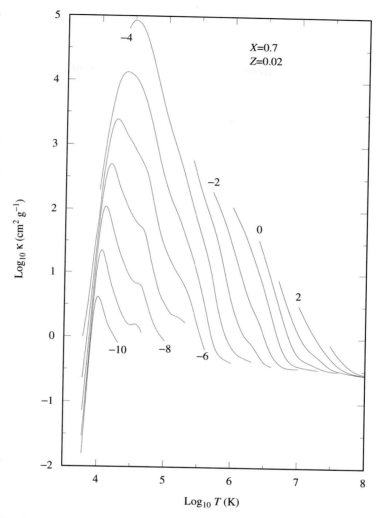

**Figure 9.10** Rosseland mean opacity. The curves are labeled by the value of the density ($\log_{10} \rho$ in g cm$^{-3}$). (Data from Rogers and Iglesias, *Ap. J. Suppl.*, **79**, 507, 1992.)

zone has a characteristic temperature of 10,000 K, and neutral helium is ionized at about the same temperature.) The decline of the plot after the peak in the opacity roughly follows a Kramers law, $\overline{\kappa} \propto T^{-3.5}$, and is due to the bound–free and free–free absorption of photons. The He II ion loses its remaining electron at a characteristic temperature of 40,000 K for a wide range of stellar parameters; the slight increase in the number of free electrons produces the "bump" seen near that temperature. Finally, the plot reaches a flat floor at the right-hand side of the figure. Electron scattering dominates at the highest temperatures, when nearly all of the stellar material is ionized and there are few bound electrons available for bound–bound and bound–free processes. The form of Eq. (9.21) for electron scattering, with no density or temperature dependence, requires that all of the curves in Fig. 9.10 converge to the same constant value in the high temperature limit.

## 9.3 Radiative Transfer

In an equilibrium, steady-state star, there can be no change in the total energy contained within any layer of the stellar atmosphere or interior.[19] In other words, the mechanisms involved in absorbing and emitting energy must be precisely in balance throughout the star. In this section, the competition between the absorption and emission processes will be described, first in qualitative terms and later in more quantitative detail.

Any process that adds photons to a beam of light will be called **emission**. Thus emission includes the scattering of photons into the beam, as well as the true emission of photons by electrons making downward atomic transitions. Each of the four primary sources of opacity listed in Section 9.2 has an inverse emission process: bound–bound and free–bound emission, free–free emission (bremsstrahlung), and electron scattering. The processes of absorption and emission hinder the flow of photons through the star by *redirecting* the paths of the photons and *redistributing* their energy. Thus in a star there is not a direct flow of photons streaming toward the surface, carrying energy outward at the speed of light. Instead, the individual photons travel only temporarily with the beam as they are repeatedly scattered in random directions following their encounters with gas particles.

As the photons diffuse upward through the stellar material, they follow a haphazard path called a **random walk**. Figure 9.11 shows a photon that undergoes a net displacement $\mathbf{d}$ as the result of making a large number $N$ of randomly directed steps, each of length $\ell$ (the mean free path):

$$\mathbf{d} = \boldsymbol{\ell}_1 + \boldsymbol{\ell}_2 + \boldsymbol{\ell}_3 + \cdots + \boldsymbol{\ell}_N.$$

---

[19]This is not the case for a star that is *not* in equilibrium. In the case of pulsating stars (to be discussed in Chapter 14), a periodic absorption or "damming up" of the outward flow of energy occurs that drives the oscillations.

## 9.3 Radiative Transfer

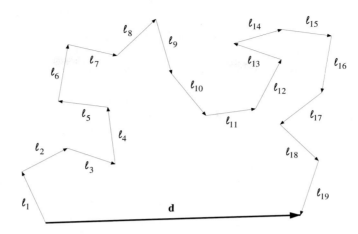

**Figure 9.11** Displacement **d** of a random-walking photon.

Taking the vector dot product of **d** with itself gives

$$\begin{aligned}\mathbf{d}\cdot\mathbf{d} =\ & \boldsymbol{\ell}_1\cdot\boldsymbol{\ell}_1 + \boldsymbol{\ell}_1\cdot\boldsymbol{\ell}_2 + \cdots + \boldsymbol{\ell}_1\cdot\boldsymbol{\ell}_N \\ & +\boldsymbol{\ell}_2\cdot\boldsymbol{\ell}_1 + \boldsymbol{\ell}_2\cdot\boldsymbol{\ell}_2 + \cdots + \boldsymbol{\ell}_2\cdot\boldsymbol{\ell}_N + \\ & \cdots + \boldsymbol{\ell}_N\cdot\boldsymbol{\ell}_1 + \boldsymbol{\ell}_N\cdot\boldsymbol{\ell}_2 + \cdots + \boldsymbol{\ell}_N\cdot\boldsymbol{\ell}_N,\end{aligned}$$

or

$$\begin{aligned}d^2 =\ & N\ell^2 + \ell^2[\cos\theta_{12} + \cos\theta_{13} + \cdots + \cos\theta_{1N} \\ & + \cos\theta_{21} + \cos\theta_{23} + \cdots + \cos\theta_{2N} + \\ & \cdots + \cos\theta_{N1} + \cos\theta_{N2} + \cdots + \cos\theta_{N(N-1)}],\end{aligned}$$

where $\theta_{ij}$ is the angle between the vectors $\boldsymbol{\ell}_i$ and $\boldsymbol{\ell}_j$. For a large number of randomly directed steps, the sum of all the cosine terms approaches zero. As a result, for a random walk, the displacement $d$ is related to the size of each step, $\ell$, by

$$d = \ell\sqrt{N}. \tag{9.23}$$

Thus the transport of energy through a star by radiation may be extremely inefficient. As a photon follows its tortuous path to the surface of a star, it takes 100 steps to travel a distance of $10\ell$; 10,000 steps to travel $100\ell$; and one million steps to travel $1000\ell$.[20] Because the optical depth at a point is

---

[20]As will be discussed in Section 10.4, the process of transporting energy by radiation is sometimes so inefficient that another method, *convection*, must take over.

roughly the number of photon mean free paths from that point to the surface (as measured along a light ray's straight path), Eq. (9.23) implies that the distance to the surface is $d = \tau_\lambda \ell = \ell\sqrt{N}$. The average number of steps a photon leaving the surface took to travel the distance $d$ is then

$$N = \tau_\lambda^2, \qquad (9.24)$$

for $\tau_\lambda \gg 1$. As might be expected, when $\tau_\lambda \approx 1$, a photon may escape from the surface of the star. A more careful analysis (performed later in this section) shows that the average level in the atmosphere from which photons of wavelength $\lambda$ escape is at an optical depth of $\tau_\lambda = 2/3$. *Looking into a star at any angle, we always look back to an optical depth of about $\tau_\lambda = 2/3$, as measured straight back along the line of sight.* In fact, a star's visible surface or **photosphere** is defined as the layer from which its visible light originates, that is, where $\tau_\lambda \approx 2/3$ for wavelengths in the star's continuum.

The realization that an observer looking vertically down on the surface of a star sees photons from $\tau_\lambda \approx 2/3$ offers an important insight into the formation of spectral lines. Recalling the definition of optical depth, Eq. (9.15),

$$\tau_\lambda = \int_0^s \kappa_\lambda \rho \, ds,$$

we see that, if the opacity $\kappa_\lambda$ increases at some wavelength, then the actual distance back along the ray to $\tau_\lambda = 2/3$ decreases for that wavelength. One cannot see as far into murky material, so an observer will not see as deeply into the star at wavelengths where the opacity is greater than average (i.e., greater than the continuum opacity). Thus, if the temperature of the stellar atmosphere decreases outward, then these higher regions of the atmosphere will be cooler. As a result, the intensity of the radiation at $\tau_\lambda \approx 2/3$ will decline the most for those wavelengths at which the opacity is greatest, resulting in absorption lines in the continuous spectrum. Thus the temperature *must* decrease outward for the formation of absorption lines. This is the analog for stellar atmospheres of Kirchhoff's law that a cool, diffuse gas in front of a source of a continuous spectrum produces dark spectral lines in the continuous spectrum.

Another implication of looking down to an optical depth of two-thirds is shown in Fig. 9.12. The line of sight of an observer on Earth viewing the Sun is vertically down at the center of the Sun's disk but makes an increasingly large angle $\theta$ with the vertical near the edge, or *limb*, of the Sun. Looking near the limb, the observer will not see as deeply into the solar atmosphere and will therefore see a lower temperature at an optical depth of two-thirds

## 9.3 Radiative Transfer

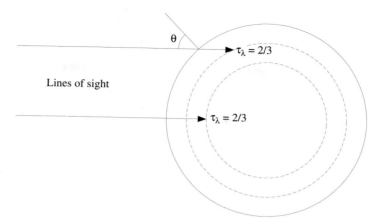

**Figure 9.12** Limb darkening.

(compared to looking at the center of the disk). As a result, the limb of the Sun appears darker than its center. This **limb darkening** is clearly seen in Fig. 11.13 (page 395) for the Sun and has also been observed in the light curves of some eclipsing binaries. More detailed information on limb darkening may be found later in this section.

Considering the meandering nature of a photon's journey to the surface, it is surprising that the energy created in the nuclear furnace of a star's center ever manages to escape into space. At great depth in the interior of the star, the photon's mean free path is only a fraction of a centimeter. After a few scattering encounters, the photon is traveling in a completely random direction, tens of billions of centimeters from the surface. This situation is analogous to the motions of air molecules in a closed room. An individual molecule moves about with a speed of nearly $5 \times 10^4$ cm s$^{-1}$, and it collides with other air molecules several billion times per second. As a result, the molecules are moving in random directions. Because there is no overall migration of the molecules in a closed room, a person standing in the room feels no wind. However, opening a window may cause a breeze if a pressure difference is established between one side of the room and the other. The air in the room responds to this pressure gradient, producing a net flux of molecules toward the area of lower pressure.

In a star the same mechanism causes a "breeze" of photons to move toward the surface of the star. Because the temperature in a star decreases outward, the radiation pressure is smaller at greater distances from the center [c.f., Eq. (9.9) for the blackbody radiation pressure]. This gradient in the radiation pressure produces the slight net movement of photons toward the surface that carries the radiative flux. As will be found later in this section, this process is described by

$$\frac{dP_{\rm rad}}{dr} = -\frac{\overline{\kappa}\rho}{c} F_{\rm rad}. \tag{9.25}$$

Thus the transfer of energy by radiation is a subtle process involving the slow upward diffusion of randomly walking photons, drifting toward the surface in response to minute differences in the radiation pressure. The description of a "beam" or a "ray" of light is only a convenient fiction, used to define the direction of motion momentarily shared by the photons that are continually absorbed and scattered into and out of the beam. Nevertheless, we will continue to use the language of photons traveling in a beam or ray of light, realizing that a specific photon is in the beam for only an instant.

---

*Optional:* The remainder of this section will be devoted to a more thorough examination of the flow of radiation through a stellar atmosphere. In this supplementary material, the basic equation of radiative transfer will be developed and then solved using several standard assumptions. The variation of temperature with optical depth in a simple model atmosphere will be derived before being applied to obtain a quantitative description of limb darkening.

In the following discussions of beams and light rays, the primary consideration is that of the net flow of energy in a given direction, not the specific path taken by individual photons. First, we will examine the emission process that increases the intensity of a ray of wavelength $\lambda$ as it travels through a gas. The increase in intensity $dI_\lambda$ is proportional to $ds$, the distance traveled in the direction of the ray, and to $\rho$, the density of the gas. For pure emission (no absorption of the radiation),

$$dI_\lambda = j_\lambda \rho\, ds, \tag{9.26}$$

where $j_\lambda$ is the **emission coefficient** of the gas. The emission coefficient, which has units of cm s$^{-3}$ sr$^{-1}$, varies with the wavelength of the light.

As a beam of light moves through the gas in a star, its specific intensity, $I_\lambda$, changes as photons traveling with the beam are removed by absorption and are replaced by photons emitted from the surrounding stellar material. Combining Eq. (9.11) for the decrease in intensity due to the absorption of radiation with Eq. (9.26) for the increase produced by emission gives the general result

$$dI_\lambda = -\kappa_\lambda \rho I_\lambda\, ds + j_\lambda \rho\, ds. \tag{9.27}$$

## 9.3 Radiative Transfer

The ratio of the rates at which the competing processes of emission and absorption occur determines how rapidly the intensity of the beam changes. This is similar to describing the flow of traffic on an interstate highway. Imagine following a group of cars as they leave Los Angeles, traveling north on I-15. Initially, nearly all of the cars on the road have California license plates. Driving north, the number of cars on the road declines as more individuals exit than enter the highway. Eventually approaching Las Vegas, the number of cars on the road increases again, but now the surrounding cars bear Nevada license plates. Continuing onward, the traffic fluctuates as the license plates eventually change to those of Utah, Idaho, and Montana. Most of the cars have the plates of the state they are in, with a few cars from neighboring states and even fewer from more distant locales. At any point along the way, the number of cars on the road reflects the local population density. Of course, this is to be expected; the surrounding area is the source of the traffic entering the highway, and the rate at which the traffic changes is determined by the ratio of the number of entering to exiting automobiles. This ratio determines how rapidly the cars on the road from elsewhere are replaced by the cars belonging to the local population. Thus the traffic constantly changes, always tending to resemble the number and types of automobiles driven by the people living nearby.

In a stellar atmosphere or interior, the same considerations describe the competition between the rates at which photons are plucked out of a beam of light by absorption, and introduced into the beam by emission processes. The ratio of the rates of emission and absorption determines how rapidly the intensity of the beam of light changes and describes the tendency of the population of photons in the beam to resemble the local source of photons in the surrounding stellar material. To introduce the ratio of emission to absorption, we divide Eq. (9.27) by $-\kappa_\lambda \rho \, ds$,

$$-\frac{1}{\kappa_\lambda \rho} \frac{dI_\lambda}{ds} = I_\lambda - \frac{j_\lambda}{\kappa_\lambda}.$$

The ratio of the emission coefficient to the absorption coefficient is called the **source function**, $S_\lambda \equiv j_\lambda / \kappa_\lambda$. It describes how photons originally traveling with the beam are removed and replaced by photons from the surrounding gas.[21] The source function, $S_\lambda$, has the same units as the intensity,

---

[21] As a ratio involving the inverse processes of absorption and emission, the source function is less sensitive to the detailed properties of the stellar material than are $j_\lambda$ and $\kappa_\lambda$ individually.

erg s$^{-1}$ cm$^{-3}$ sr$^{-1}$. Therefore, in terms of the source function,

$$-\frac{1}{\kappa_\lambda \rho}\frac{dI_\lambda}{ds} = I_\lambda - S_\lambda. \tag{9.28}$$

This is one form of the **equation of radiative transfer** (usually referred to as the **transfer equation**).[22] According to the transfer equation, if the intensity of the light does not vary (so that the left-hand side of the equation is zero), then the intensity is equal to the source function, $I_\lambda = S_\lambda$. If the intensity of the light is *greater* than the source function (the right-hand side of the transfer equation is greater than 0), then $dI_\lambda/ds$ is less than 0, and the intensity *decreases* with distance. On the other hand, if the intensity is *less* than the source function, the intensity *increases* with distance. This is merely a mathematical restatement of the tendency of the photons found in the beam to resemble the local source of photons in the surrounding gas. Thus *the intensity of the light tends to become equal to the local value of the source function*, although the source function itself may vary too rapidly with distance for an equality to be attained.

The source function for the special case of blackbody radiation can be found by considering a box of optically thick gas maintained at a constant temperature $T$. The confined particles and blackbody radiation are in thermodynamic equilibrium, with no net flow of energy through the box or between the gas particles and the radiation. With the particles and photons in equilibrium, individually and with each other, every process of absorption is balanced by an inverse process of emission. The intensity of the radiation is described by the Planck function, $I_\lambda = B_\lambda$. Furthermore, because the intensity is constant throughout the box, $dI_\lambda/ds = 0$, and so $I_\lambda = S_\lambda$. *For the case of thermodynamic equilibrium, the source function is equal to the Planck function,* $S_\lambda = B_\lambda$.

As mentioned in Section 9.2, a star cannot be in perfect thermodynamic equilibrium; there is a net flow of energy from the center to the surface. Deep in the atmosphere ($\tau_\lambda \gg 1$ as measured along a vertical ray), however, a random-walking photon will take at least $\tau_\lambda^2$ steps to reach the surface (recall Eq. 9.24) and so will suffer many scattering events before escaping from the star. At a depth at which the photon mean free path is small compared to the temperature scale height, the photons are effectively confined to a limited volume, a region of nearly constant temperature. The conditions for **local thermodynamic equilibrium** (LTE) are satisfied, and so, as already seen,

---

[22] It is assumed that the atmosphere is in a steady state, not changing with time. Otherwise, a time-derivative term would have to be included in the transfer equation.

## 9.3 Radiative Transfer

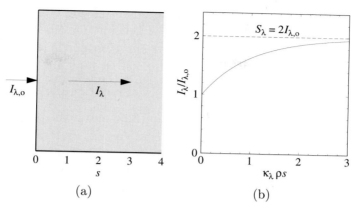

**Figure 9.13** Transformation of the intensity of a light ray traveling through a volume of gas. (a) A light ray entering a volume of gas. (b) Intensity of the light ray. The horizontal axis has units of $\kappa_\lambda \rho s$, the number of optical depths traveled into the gas.

the source function is equal to the Planck function, $S_\lambda = B_\lambda$. Making the assumption of LTE in a problem means setting $S_\lambda = B_\lambda$. However, even in LTE, the intensity of the radiation, $I_\lambda$, will not necessarily be equal to $B_\lambda$ unless $\tau_\lambda \gg 1$. In summary, saying that $I_\lambda = B_\lambda$ is a statement that the radiation field is described by the Planck function, while $S_\lambda = B_\lambda$ describes the physical source of the radiation, $j_\lambda/\kappa_\lambda$, as one that produces blackbody radiation.

**Example 9.5** To see how the intensity of a light ray tends to become equal to the local value of the source function, imagine a beam of light of initial intensity $I_{\lambda,o}$ at $s = 0$ entering a volume of gas of constant density, $\rho$, which has a *constant* opacity, $\kappa_\lambda$, and a *constant* source function, $S_\lambda$. Then it is left as an exercise to show that the transfer equation (Eq. 9.28) may be easily solved for the intensity of the light as a function of the distance $s$ traveled into the gas:

$$I_\lambda(s) = I_{\lambda,o}\, e^{-\kappa_\lambda \rho s} + S_\lambda (1 - e^{-\kappa_\lambda \rho s}). \tag{9.29}$$

As shown in Fig. 9.13 for the case of $S_\lambda = 2I_{\lambda,o}$, this solution describes the transformation of the intensity of the light ray from its initial value of $I_{\lambda,o}$ to $S_\lambda$, the value of the source function. The characteristic distance for this change to occur is $s = 1/\kappa_\lambda \rho$, which is one photon mean free path (recall Example 9.2), or one optical depth into the gas.

Although the transfer equation is the basic tool that describes the passage of light through a star's atmosphere, a reader seeing it for the first time may be prone to despair. In this troublesome equation the intensity of the light must depend on the direction of travel to account for the net outward flow of energy. And although absorption and emission coefficients are the same for light traveling in all directions (so the source function is independent of direction), the absorption and emission coefficients depend on the temperature and density in a rather complicated way.

However, if astronomers are to learn anything about the physical conditions in stellar atmospheres (temperature, density, etc.), they must know where (at what depth) a spectral line is formed. A vast amount of effort has therefore been devoted to solving and understanding the implications of the transfer equation, and several powerful techniques have been developed that simplify the analysis considerably.

We will begin by rewriting Eq. (9.28) in terms of the optical depth $\tau_\lambda$, defined by Eq. (9.13), resulting in

$$\frac{dI_\lambda}{d\tau_\lambda} = I_\lambda - S_\lambda. \tag{9.30}$$

Unfortunately, because the optical depth is measured along the path of the light ray, neither the optical depth nor the distance $s$ in Eq. (9.28) corresponds to a unique geometric depth in the atmosphere. Consequently the optical depth must be replaced by a meaningful measure of position.

To find a suitable replacement, we introduce the first of several standard approximations. The atmospheres of stars near the main sequence are physically thin compared with the size of the star, analogous to the skin of an onion. The atmosphere's radius of curvature is thus much larger than its thickness, and we may consider the atmosphere as a *plane-parallel slab*. As shown in Fig. 9.14, the $z$-axis is assumed to be in the vertical direction, with $z = 0$ at the top of this **plane-parallel atmosphere**.

Next, a **vertical optical depth**, $\tau_{\lambda,v}(z)$, is defined as

$$\tau_{\lambda,v}(z) = \int_z^0 \kappa_\lambda \rho \, dz. \tag{9.31}$$

A comparison with Eq. (9.15) reveals that this is just the initial optical depth of a ray traveling vertically upward from an initial position ($z < 0$) to the surface ($z = 0$) where $\tau_{\lambda,v} = 0$.[23] However, a ray that travels upward at an angle $\theta$ from the same initial position $z$ has farther to go through the same layers of the atmosphere to reach the surface. Therefore the optical depth

---

[23] Recall that as the light approaches the surface (and the observer on Earth), it is traveling through smaller and smaller values of the optical depth.

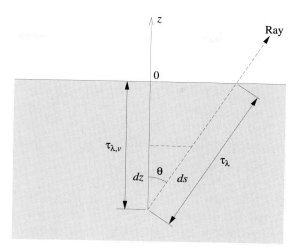

**Figure 9.14** Plane-parallel stellar atmosphere.

measured along this ray's path to the surface, $\tau_\lambda$, is *greater* than the vertical optical depth, $\tau_{\lambda,v}(z)$. Since $dz = ds \cos\theta$, the two optical depths are related by

$$\tau_\lambda = \frac{\tau_{\lambda,v}}{\cos\theta} = \tau_{\lambda,v} \sec\theta. \tag{9.32}$$

The vertical optical depth is a true vertical coordinate, analogous to $z$, that increases in the $-z$-direction. Its value does not depend on the direction of travel of a light ray, and so it can be used as a meaningful position coordinate in the transfer equation. Replacing $\tau_\lambda$ by $\tau_{\lambda,v}$ in Eq. (9.30) results in

$$\cos\theta \frac{dI_\lambda}{d\tau_{\lambda,v}} = I_\lambda - S_\lambda. \tag{9.33}$$

This form of the transfer equation is usually employed when dealing with the approximation of a plane-parallel atmosphere.

Of course, the value of the vertical optical depth at a level $z$ is wavelength-dependent because of the wavelength-dependent opacity in Eq. (9.31). To permit the identification of an atmospheric level with a unique value of $\tau_v$, the opacity is assumed to be *independent of wavelength* (usually taking it to be equal to the Rosseland mean opacity, $\overline{\kappa}$). A stellar atmosphere in which the opacity is assumed to be independent of wavelength is called a *gray atmosphere*, reflecting its indifference to the spectrum of wavelengths. The assumption of a gray atmosphere is a fairly good approximation for the majority of stars for which the photoionization of $H^-$ ions is the primary source of opacity, because this opacity does not vary rapidly with wavelength. If we write $\overline{\kappa}$ instead of

$\kappa_\lambda$ in Eq. (9.31), the vertical optical depth no longer depends on wavelength; we can therefore write $\tau_v$ instead of $\tau_{\lambda,v}$ in the transfer equation (Eq. 9.33). The remaining wavelength dependencies may be removed by integrating the transfer equation over all wavelengths, using

$$I = \int_0^\infty I_\lambda \, d\lambda \quad \text{and} \quad S = \int_0^\infty S_\lambda \, d\lambda.$$

With the preceding changes, the transfer equation appropriate for a plane-parallel gray atmosphere is

$$\cos\theta \frac{dI}{d\tau_v} = I - S. \tag{9.34}$$

This equation leads to two particularly useful relations between the various quantities describing the radiation field. First, integrating over all solid angles, we get

$$\frac{d}{d\tau_v} \int I \cos\theta \, d\Omega = \int I \, d\Omega - S \int d\Omega.$$

Using $\int d\Omega = 4\pi$ and the definitions of the radiative flux $F_{\text{rad}}$ (Eq. 9.6) and the mean intensity $\langle I \rangle$ (Eq. 9.1), both integrated over all wavelengths, we find

$$\frac{dF_{\text{rad}}}{d\tau_v} = 4\pi(\langle I \rangle - S). \tag{9.35}$$

The second relation is found by first multiplying the transfer equation (9.34) by $\cos\theta$ and again integrating over all solid angles:

$$\frac{d}{d\tau_v} \int I \cos^2\theta \, d\Omega = \int I \cos\theta \, d\Omega - S \int \cos\theta \, d\Omega.$$

The term on the left is the radiation pressure multiplied by the speed of light (recall Eq. 9.7). The first term on the right-hand side is the radiative flux. In spherical coordinates, the second term on the right-hand side is

$$\int \cos\theta \, d\Omega = \int_{\phi=0}^{2\pi} \int_{\theta=0}^{\pi} \cos\theta \sin\theta \, d\theta \, d\phi = 0.$$

Thus

$$\frac{dP_{\text{rad}}}{d\tau_v} = \frac{1}{c} F_{\text{rad}}. \tag{9.36}$$

In Problem 9.16, you will find that, in a spherical coordinate system with its origin at the center of the star, this equation is

$$\frac{dP_{\text{rad}}}{dr} = -\frac{\overline{\kappa}\rho}{c} F_{\text{rad}},$$

## 9.3 Radiative Transfer

which is just Eq. (9.25). As mentioned previously, this result can be interpreted as saying that the net radiative flux is driven by differences in the radiation pressure, with a "photon wind" blowing from high to low $P_{\text{rad}}$. Equation (9.25) will be employed in Chapter 10 to determine the temperature structure in the interior of a star.

In an equilibrium stellar atmosphere, every process of absorption is balanced by an inverse process of emission; no net energy is subtracted from or added to the radiation field. In a plane-parallel atmosphere, this means that the radiative flux must have the same value at every level of the atmosphere, including its surface. From Eq. (3.18),

$$F_{\text{rad}} = \text{constant} = F_{\text{surf}} = \sigma T_e^4. \tag{9.37}$$

Because the flux is a constant, $dF_{\text{rad}}/d\tau_v = 0$, and so Eq. (9.35) implies the mean intensity must be equal to the source function,

$$\langle I \rangle = S. \tag{9.38}$$

Equation (9.36) may now be integrated to find the radiation pressure as a function of the vertical optical depth:

$$P_{\text{rad}} = \frac{1}{c} F_{\text{rad}} \tau_v + C, \tag{9.39}$$

where $C$ is the constant of integration.

If we knew how the radiation pressure varied with temperature for the general case (and not just for blackbody radiation), we could use this equation to determine the temperature structure of our plane-parallel gray atmosphere. We would have to *assume* a description of the angular distribution of the intensity. In an approximation due to the brilliant English physicist Sir Arthur Stanley Eddington (1882–1944), the intensity of the radiation field is assigned one value, $I_{\text{out}}$, in the $+z$-direction (outward), and another value, $I_{\text{in}}$, in the $-z$-direction (inward); see Fig. 9.15. Both $I_{\text{out}}$ and $I_{\text{in}}$ vary with depth in the atmosphere, and in particular, $I_{\text{in}} = 0$ at the top of the atmosphere, where $\tau_v = 0$. It is left as an exercise to show that, with this **Eddington approximation**,[24] the mean intensity, radiative flux, and radiation pressure

---

[24] Actually, there are several, more mathematical, ways of implementing the Eddington approximation, but they are all equivalent.

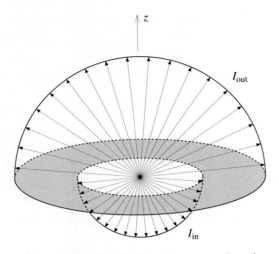

**Figure 9.15** The Eddington approximation.

are given by

$$\langle I \rangle = \frac{1}{2} \left( I_{\text{out}} + I_{\text{in}} \right) \tag{9.40}$$

$$F_{\text{rad}} = \pi \left( I_{\text{out}} - I_{\text{in}} \right) \tag{9.41}$$

$$P_{\text{rad}} = \frac{2\pi}{3c} \left( I_{\text{out}} + I_{\text{in}} \right) \tag{9.42}$$

$$= \frac{4\pi}{3c} \langle I \rangle. \tag{9.43}$$

(Note that, because the flux is a constant, Eq. (9.41) shows that there is a constant difference between $I_{\text{out}}$ and $I_{\text{in}}$ at any level of the atmosphere.)

Inserting the last relation for the radiation pressure into Eq. (9.39), we find that

$$\frac{4\pi}{3c} \langle I \rangle = \frac{1}{c} F_{\text{rad}} \tau_v + C. \tag{9.44}$$

The constant $C$ can be determined by evaluating Eqs. (9.40) and (9.41) at the top of the atmosphere, where $\tau_v = 0$ and $I_{\text{in}} = 0$. The result is that $\langle I(\tau_v = 0) \rangle = F_{\text{rad}}/2\pi$. Inserting this into Eq. (9.44) with $\tau_v = 0$ shows that

$$C = \frac{2}{3c} F_{\text{rad}}.$$

With this value of $C$, Eq. (9.44) becomes

$$\frac{4\pi}{3} \langle I \rangle = F_{\text{rad}} \left( \tau_v + \frac{2}{3} \right). \tag{9.45}$$

## 9.3 Radiative Transfer

Of course, we already know that the radiative flux is a constant, given by Eq. (9.37). Using this results in an expression for the mean intensity as a function of the vertical optical depth,

$$\langle I \rangle = \frac{3\sigma}{4\pi} T_e^4 \left( \tau_v + \frac{2}{3} \right). \tag{9.46}$$

We must now derive the final approximation to determine the temperature structure of our model atmosphere. If the atmosphere is assumed to be in local thermodynamic equilibrium, another expression for the mean intensity can be found and combined with Eq. (9.46). By the definition of LTE, the source function is equal to the Planck function, $S_\lambda = B_\lambda$. Integrating $B_\lambda$ over all wavelengths (see Eq. 3.25) shows that, for LTE,

$$S = B = \frac{\sigma T^4}{\pi},$$

and so, from Eq. (9.38),

$$\langle I \rangle = \frac{\sigma T^4}{\pi}. \tag{9.47}$$

Equating expressions (9.46) and (9.47) finally results in the variation of the temperature with vertical optical depth in a plane-parallel gray atmosphere in LTE, using the Eddington approximation:

$$T^4 = \frac{3}{4} T_e^4 \left( \tau_v + \frac{2}{3} \right). \tag{9.48}$$

This relation is well worth the effort of its derivation, because it reveals some important aspects of real stellar atmospheres. First, notice that $T = T_e$ at $\tau_v = 2/3$, *not* at $\tau_v = 0$. Thus the surface of a star, which by definition has temperature $T_e$ (recall the Stefan–Boltzmann equation, 3.17), is *not* at the top of the atmosphere, where $\tau_v = 0$, but deeper down, where $\tau_v = 2/3$. This result may be thought of as the average point of origin of the observed photons. Although this result came at the end of a string of assumptions, it can be generalized to the statement that *looking into a star at any angle, we always look back to an optical depth of $\tau_\lambda \approx 2/3$, as measured back along the ray*. The importance of this for the formation and interpretation of spectral lines was discussed earlier on page 278.

We now move on to take a closer look at limb darkening (recall Fig. 9.12). A comparison of theory and observations of limb darkening can provide valuable information about how the source function varies with depth in a star's atmosphere. To see how this is done, we first solve the general form of the transfer equation (Eq. 9.30),

$$\frac{dI_\lambda}{d\tau_\lambda} = I_\lambda - S_\lambda,$$

at least formally, rather than by making assumptions. (The inevitable assumptions will be required soon enough.) Multiplying both sides by $e^{-\tau_\lambda}$, we have

$$\frac{dI_\lambda}{d\tau_\lambda} e^{-\tau_\lambda} - I_\lambda e^{-\tau_\lambda} = -S_\lambda e^{-\tau_\lambda}$$

$$\frac{d}{d\tau_\lambda}(e^{-\tau_\lambda} I_\lambda) = -S_\lambda e^{-\tau_\lambda}$$

$$d(e^{-\tau_\lambda} I_\lambda) = -S_\lambda e^{-\tau_\lambda} d\tau_\lambda.$$

If we integrate from the initial position of the ray, at optical depth $\tau_{\lambda,\circ}$ where $I_\lambda = I_{\lambda,\circ}$, to the top of the atmosphere, at optical depth $\tau_\lambda = 0$ where $I_\lambda = I_\lambda(0)$, the result for the emergent intensity at the top of the atmosphere, $I_\lambda(0)$, is

$$I_\lambda(0) = I_{\lambda,\circ} e^{-\tau_{\lambda,\circ}} - \int_{\tau_{\lambda,\circ}}^{0} S_\lambda e^{-\tau_\lambda} d\tau_\lambda. \qquad (9.49)$$

This equation has a very straightforward interpretation. The emergent intensity on the left is the sum of two positive contributions. The first term on the right is the initial intensity of the ray, reduced by the effects of absorption along the path to the surface. The second term, also positive,[25] represents the emission at every point along the path, attenuated by the absorption between the point of emission and the surface.

We now return to the geometry of a plane-parallel atmosphere and the vertical optical depth, $\tau_v$. However, we do *not* assume a gray atmosphere, LTE, or make the Eddington approximation. As shown in Fig. 9.16, the problem of limb darkening amounts to determining the emergent intensity $I_\lambda(0)$ as a function of the angle $\theta$. Equation (9.49), the formal solution to the transfer equation, is easily converted to this situation by using Eq. (9.32) to replace $\tau_\lambda$ with $\tau_{\lambda,v} \sec\theta$ (the vertical optical depth) to get

$$I(0) = I_\circ e^{-\tau_{v,\circ} \sec\theta} - \int_{\tau_{v,\circ} \sec\theta}^{0} S \sec\theta \, e^{-\tau_v \sec\theta} d\tau_v.$$

---

[25] Remember that the optical depth, measured along the ray's path, decreases in the direction of travel, so $d\tau_\lambda$ is negative.

## 9.3 Radiative Transfer

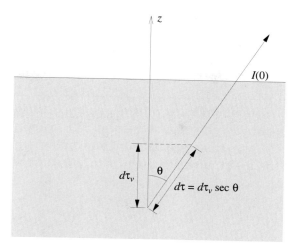

**Figure 9.16** Finding $I(0)$ as a function of $\theta$ for limb darkening in plane-parallel geometry.

Although both $I$ and $\tau_v$ depend on wavelength, the $\lambda$ subscript has been dropped to simplify the notation; the approximation of a gray atmosphere has *not* been made. To include the contributions to the emergent intensity from all layers of the atmosphere, the value of the initial position of the rays is taken to be at $\tau_{v,0} = \infty$. Then the first term on the right-hand side vanishes, leaving

$$I(0) = \int_0^\infty S \sec\theta \, e^{-\tau_v \sec\theta} \, d\tau_v. \tag{9.50}$$

If we knew how the source function depends on the vertical optical depth, this equation could be integrated to find the emergent intensity as a function of the direction of travel, $\theta$, of the ray. Although the form of the source function is not known, a good guess will be enough to estimate $I(0)$. Suppose that the source function has the form

$$S = a + b\tau_v, \tag{9.51}$$

where $a$ and $b$ are wavelength-dependent numbers to be determined. Inserting this into Eq. (9.50) and integrating (the details are left as an exercise) shows that the emergent intensity for this source function is

$$I_\lambda(0) = a_\lambda + b_\lambda \cos\theta, \tag{9.52}$$

where the $\lambda$ subscripts have been restored to the appropriate quantities to emphasize their wavelength dependence. By making careful measurements of the variation in the specific intensity across the disk of the Sun, the values of $a_\lambda$ and $b_\lambda$ for the solar source function can be determined for a range of wavelengths. For a wavelength of 5010 Å, Böhm-Vitense (1989) supplies values of $a_{5010} = 1.04 \times 10^{14}$ erg s$^{-1}$ cm$^{-3}$ sr$^{-1}$ and $b_{5010} = 3.52 \times 10^{14}$ erg s$^{-1}$ cm$^{-3}$ sr$^{-1}$.

---

**Example 9.6** Solar limb darkening provides an opportunity to test the accuracy of our "plane-parallel gray atmosphere in LTE using the Eddington approximation." In the preceding discussion of an equilibrium gray atmosphere, it was found that the mean intensity is equal to the source function,

$$\langle I \rangle = S$$

(Eq. 9.38). Then with the additional assumptions of the Eddington approximation and LTE, Eqs. (9.47) and (9.48) can be used to determine the mean intensity and thus the source function:

$$S = \langle I \rangle = \frac{\sigma T^4}{\pi} = \frac{3\sigma}{4\pi} T_e^4 \left( \tau_v + \frac{2}{3} \right).$$

The source function has the form of Eq. (9.51), $S = a + b\tau_v$, as used earlier for limb darkening (*after integrating over all wavelengths*). The values of the coefficients are

$$a = \frac{\sigma}{2\pi} T_e^4 \quad \text{and} \quad b = \frac{3\sigma}{4\pi} T_e^4.$$

The emergent intensity then will have the form of Eq. (9.52), $I(0) = a + b\cos\theta$ (again after integrating over all wavelengths). The ratio of the emergent intensity at angle $\theta$, $I(\theta)$, to that at the center of the star, $I(\theta = 0)$, is thus

$$\frac{I(\theta)}{I(\theta = 0)} = \frac{a + b\cos\theta}{a + b} = \frac{2}{5} + \frac{3}{5} \cos\theta. \tag{9.53}$$

We can compare the results of this calculation with observations of solar limb darkening in integrated light (made by summing over all wavelengths). Figure 9.17 shows both the observed values of $I(\theta)/I(\theta = 0)$ and the values from Eq. (9.53). The agreement is remarkably good, despite our numerous approximations. However, be forewarned that the agreement is much worse for observations made at a given wavelength (see Böhm-Vitense, 1989) due to wavelength-dependent opacity effects such as line blanketing.

---

## 9.4 The Structure of Spectral Lines

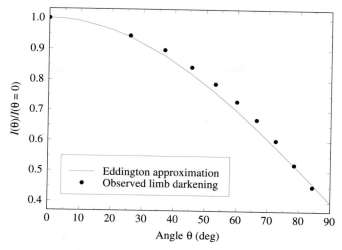

**Figure 9.17** Observed and theoretical solar limb darkening for light integrated over all wavelengths.

## 9.4 The Structure of Spectral Lines

We now have a formidable theoretical arsenal to bring to bear on the analysis of spectral lines. The shape of an individual spectral line contains a wealth of information about the environment in which it was formed. Figure 9.18 shows a graph of the radiant flux, $F_\lambda$, as a function of wavelength for a typical absorption line. In the figure, $F_\lambda$ is expressed as a fraction of $F_c$, the value of the flux from the continuous spectrum outside the spectral line. Near the central wavelength, $\lambda_o$, is the *core* of the line, and the sides sweeping upward to the continuum are the line's *wings*. Individual lines may be narrow or broad, shallow or deep. The quantity $(F_c - F_\lambda)/F_c$ is referred to as the *depth* of the line. The strength of a spectral line is measured in terms of its **equivalent width**. The equivalent width $W$ of a spectral line is defined as the width in angstroms of a box (shaded in Fig. 9.18) reaching up to the continuum that has the same area as the spectral line. That is,

$$W = \int \frac{F_c - F_\lambda}{F_c} \, d\lambda, \tag{9.54}$$

where the integral is taken from one side of the line to the other. The equivalent width of a line in the visible spectrum, shaded in Fig. 9.18, is usually on the order of 0.1 Å. Another measure of the width of a spectral line is the distance in angstroms from one side of the line to the other, where its depth

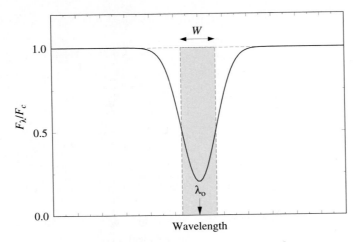

**Figure 9.18** The shape of a typical spectral line.

$(F_c - F_\lambda)/F_c = 1/2$; this is called the *full width at half-maximum* and will be denoted by $(\Delta\lambda)_{1/2}$.

The spectral line shown in Fig. 9.18 is termed *optically thin* because there is no wavelength at which the radiant flux has been completely blocked. The opacity $\kappa_\lambda$ of the stellar material is greatest at the wavelength $\lambda_o$ at the line's center and decreases moving into the wings. From the discussion on page 278, this means that the center of the line is formed at higher (and cooler) regions of the stellar atmosphere. Moving into the wings from $\lambda_o$, the line formation occurs at progressively deeper (and hotter) layers of the atmosphere. In Section 11.2 this idea will be applied to the absorption lines produced in the solar photosphere.

Three main processes responsible are for the broadening of spectral lines. Each of these mechanisms produces its own distinctive line shape or *line profile*.

1. **Natural broadening.** Spectral lines cannot be infinitely sharp, even for motionless, isolated atoms. According to Heisenberg's uncertainty principle, Eq. (5.19), as the time available for an energy measurement decreases, the inherent uncertainty of the result increases. Because an electron in an excited state occupies its orbital for only a brief instant, $\Delta t$, the orbital's energy, $E$, cannot have a precise value. Thus the uncertainty in the energy, $\Delta E$, of the orbital is

$$\Delta E \approx \frac{\hbar}{\Delta t}.$$

(The electron's lifetime in the ground state may be taken as infinite,

## 9.4 The Structure of Spectral Lines

so in that case $\Delta E = 0$.) Electrons can make transitions from and to anywhere within these "fuzzy" energy levels, producing an uncertainty in the wavelength of the photon absorbed or emitted in a transition. Using Eq. (5.3) for the energy of a photon, $E_{\text{photon}} = hc/\lambda$, we find that the uncertainty in the photon's wavelength has a magnitude of roughly

$$\Delta \lambda \approx \frac{\lambda^2}{2\pi c} \left( \frac{1}{\Delta t_i} + \frac{1}{\Delta t_f} \right), \qquad (9.55)$$

where $\Delta t_i$ is the lifetime of the electron in its initial state and $\Delta t_f$ is the lifetime in the final state. (The proof is left as a problem.)

---

**Example 9.7** The lifetime of an electron in the first and second excited states of hydrogen is about $\Delta t = 10^{-8}$ s. The natural broadening of the $H_\alpha$ line of hydrogen, $\lambda = 6563$ Å, is then

$$\Delta \lambda \approx 4.57 \times 10^{-12} \text{ cm} = 4.57 \times 10^{-4} \text{ Å}.$$

---

A more involved calculation shows that the full width at half-maximum of the line profile for natural broadening is

$$(\Delta \lambda)_{1/2} = \frac{\lambda^2}{\pi c} \frac{1}{\Delta t_o}, \qquad (9.56)$$

where $\Delta t_o$ is the average waiting time for a specific transition to occur. This results in a typical value of

$$(\Delta \lambda)_{1/2} \simeq 2.4 \times 10^{-4} \text{ Å},$$

in good agreement with the preceding estimate.

2. **Doppler broadening.** In thermal equilibrium, the atoms in a gas, each of mass $m$, are moving randomly about with a distribution of speeds that is described by the Maxwell–Boltzmann distribution function (Eq. 8.1), with the most probable speed given by Eq. (8.2), $v_{\text{mp}} = \sqrt{2kT/m}$. The wavelengths of the light absorbed or emitted by the atoms in the gas are Doppler-shifted according to (nonrelativistic) Eq. (4.30), $\Delta \lambda / \lambda = \pm |v_r|/c$. Thus the width of a spectral line due to Doppler broadening should be approximately

$$\Delta \lambda = \frac{2\lambda}{c} \sqrt{\frac{2kT}{m}}.$$

**Example 9.8** For hydrogen atoms in the Sun's photosphere ($T = 5770$ K), the Doppler broadening of the $H_\alpha$ line should be about

$$\Delta\lambda \approx 0.427 \text{ Å},$$

about 1000 times greater than for natural broadening.

---

An in-depth analysis, taking into account the different directions of the atoms' motions (with respect to one another and to the line of sight of the observer), shows that the full width at half-maximum of the line profile for Doppler broadening is

$$(\Delta\lambda)_{1/2} = \frac{2\lambda}{c}\sqrt{\frac{2kT \ln 2}{m}}. \tag{9.57}$$

Although the line profile for Doppler broadening is much wider at half-maximum than for natural broadening, the line depth for Doppler broadening decreases *exponentially* as the wavelength moves away from the central wavelength $\lambda_\circ$. This rapid decline is due to the high-speed exponential "tail" of the Maxwell–Boltzmann velocity distribution, and is a much faster falloff in strength than for natural broadening.

Doppler shifts caused by the large-scale turbulent motion of large masses of gas (as opposed to the random motion of the individual atoms) can also be accommodated by Eq. (9.57) if the distribution of turbulent velocities follows the Maxwell–Boltzmann distribution. In that case,

$$(\Delta\lambda)_{1/2} = \frac{2\lambda}{c}\sqrt{\left(\frac{2kT}{m} + v_{\text{turb}}^2\right)\ln 2}, \tag{9.58}$$

where $v_{\text{turb}}$ is the most probable turbulent speed. The effect of turbulence on line profiles is important in the atmospheres of giant and supergiant stars. In fact, the existence of turbulence in the atmospheres of these stars was deduced from the inordinately large effect of Doppler broadening on their spectra.

Other sources of Doppler broadening involve orderly, coherent mass motions, such as stellar rotation and pulsation. These phenomena can have a substantial effect on the shape and width of the line profiles but cannot be combined with the results of Doppler broadening produced by random thermal motions obeying the Maxwell–Boltzmann distribution.

3. **Pressure (and collisional) broadening.** The orbitals of an atom can be perturbed in a collision with a neutral atom or by a close encounter involving the electric field of an ion. The results of individual collisions are called *collisional broadening*, and the statistical effects of the electric fields of large numbers of closely passing ions can be termed *pressure broadening*.[26] In either case, the outcome depends on the average time between collisions or encounters with other atoms and ions.

Calculating the precise width and shape of a pressure-broadened line is quite complicated. Atoms and ions of the same or different elements, as well as free electrons, are involved in these collisions and close encounters. The general shape of the line, however, is like that found for natural broadening, Eq. (9.56), and the line profile shared by natural and pressure broadening is sometimes referred to as a *damping profile*.[27] The values of the full width at half-maximum for natural and pressure broadening usually prove to be comparable, although the pressure profile can at times be more than an order of magnitude wider.

An estimate of pressure broadening due to collisions with atoms of a single element can be obtained by taking the value of $\Delta t_o$ in Eq. (9.56) to be the average time between collisions. This time is approximately equal to the mean free path between collisions divided by the average speed of the atoms. Using Eq. (9.10) for the mean free path and Eq. (8.2) for the speed, we find that

$$\Delta t_o \approx \frac{\ell}{v} = \frac{1}{n\sigma\sqrt{2kT/m}},$$

where $m$ is the mass of an atom, $\sigma$ is its collision cross section, and $n$ is the number density of the atoms. Thus the width of the spectral line due to pressure broadening is on the order of

$$\Delta\lambda = \frac{\lambda^2}{c}\frac{1}{\pi\Delta t_o} \approx \frac{\lambda^2}{c}\frac{n\sigma}{\pi}\sqrt{\frac{2kT}{m}}. \tag{9.59}$$

Note that the width of the line is proportional to the number density $n$ of the atoms.

The physical reason for the Morgan–Keenan luminosity classes is now clear. The narrower lines observed for the more luminous giant and supergiant stars are due to the lower number densities in their extended

---

[26] In the following discussion, both of these effects are referred to as *pressure broadening*.

[27] The damping profile is so named because the shape is characteristic of the spectrum of radiation emitted by an electric charge undergoing simple harmonic motion; the oscillations are damped by the loss of energy.

atmospheres. Pressure broadening (with the width of the line profile proportional to $n$) broadens the lines formed in the denser atmospheres of main-sequence stars, where collisions occur more frequently.

---

**Example 9.9** Again, consider the hydrogen atoms in the Sun's photosphere, where the temperature is 5770 K and the number density of hydrogen atoms is about $1.5 \times 10^{17}$ cm$^{-3}$. Then the pressure broadening of the H$_\alpha$ line should be roughly

$$\Delta\lambda \approx 2.36 \times 10^{-4} \text{ Å},$$

which is comparable to the result for natural broadening found earlier. However, if the number density of the atoms in the atmosphere is larger, the line width will be larger as well, at times more than an order of magnitude larger.

---

The total line profile, called a **Voigt profile**, is due to the contributions of both the Doppler and damping profiles. The wider line profile for Doppler broadening dominates near the central wavelength $\lambda_\circ$. Farther from $\lambda_\circ$, however, the exponential decrease in the line depth for Doppler broadening means that there is a transition to a damping profile in the wings at a distance (in Å) of about 1.8 times the Doppler value of $(\Delta\lambda)_{1/2}$ from the center of the line. Thus line profiles tend to have *Doppler cores* and *damping wings*. Figure 9.19 schematically shows the Doppler and damping line profiles.

---

**Example 9.10** As a review of the ideas of spectral line formation discussed here and in Chapter 8, consider the subdwarfs of luminosity class VI or "sd," which reside about one magnitude below the main sequence (see Fig. 8.15). The spectra of these subdwarfs show that they are deficient in the atoms of metals. Because ionized metals are an important source of electrons in stellar atmospheres, the electron number density is reduced. As mentioned in Section 8.1, fewer electrons with which ions may recombine means that a higher degree of ionization for all atoms can be achieved at the same temperature. Specifically, this reduces the number of H$^-$ ions in the atmosphere by ionizing them, thereby diluting this dominant source of continuum opacity. As a consequence of a lower opacity, we can see longer distances into these stars before reaching an optical depth of $\tau_\lambda = 2/3$. The forest of metallic lines (which are already weakened by the low metal abundance of the subdwarfs) appears even weaker against the brighter continuum. Thus, as a result of an underabundance of metals, the spectrum of a subdwarf appears to be that of a star of

## 9.4 The Structure of Spectral Lines

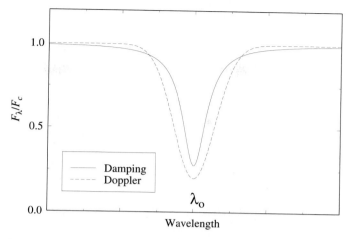

**Figure 9.19** Schematic damping and Doppler line profiles, scaled so they have the same equivalent width.

earlier spectral type with less prominent metal lines (see Table 8.1, page 226). It is more accurate to say that these stars are displaced to the *left* of the main sequence, toward higher temperatures, rather than one magnitude below the main sequence.

---

The simplest model used for calculating a line profile assumes that the star's photosphere acts as a source of blackbody radiation, and that the atoms above the photosphere remove photons from this continuous spectrum to form absorption lines.[28] Values for the temperature, density, and composition must be adopted for the region above the photosphere where the line is formed. The temperature and density determine the importance of Doppler and pressure broadening and are also used in the Boltzmann and Saha equations.

The calculation of a spectral line depends not only on the abundance of the element forming the line but also on the quantum-mechanical details of how atoms absorb photons. Let $N$ be the number of atoms of a certain element lying *above a unit area* of the photosphere. $N$ is a **column density** and has units of cm$^{-2}$. (In other words, suppose a hollow tube with a cross-section of 1 cm$^2$ was stretched from the observer to the photosphere; then the tube would contain $N$ atoms of the specified type.) To find the number of absorbing atoms per unit area, $N_a$, that have electrons in the proper orbital for absorbing a photon at the wavelength of the spectral line, the temperature and density

---

[28] Although this *Schuster–Schwarzschild model* is inconsistent with the idea that photons of wavelength $\lambda$ originate at an optical depth of $\tau_\lambda = 2/3$, it is still a useful approximation.

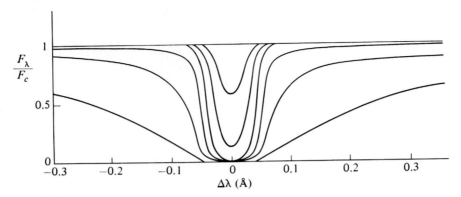

**Figure 9.20** Voigt profiles of the K line of Ca II. The shallowest line is produced by $N_a = 3.4 \times 10^{11}$ ions cm$^{-2}$, and the ions are ten times more abundant for each successively broader line. (Adapted from Novotny, *Introduction to Stellar Atmospheres and Interiors*, Oxford University Press, New York, 1973.)

are used in the Boltzmann and Saha equations to calculate the atomic states of excitation and ionization. Our goal is to determine the value of $N_a$ by comparing the calculated and observed line profiles.

This task is complicated by the fact that not all transitions between atomic orbitals are equally likely. For example, an electron initially in the $(n = 2)$ orbit of hydrogen is about five times more likely to absorb an H$_\alpha$ photon and make a transition to the $(n = 3)$ orbit than it is to absorb an H$_\beta$ photon and jump to the $(n = 4)$ orbit. The relative probabilities of an electron making a transition from the same initial orbital are given by the *f-values* or *oscillator strengths* for that orbital. For hydrogen, $f = 0.637$ for the H$_\alpha$ transition and $f = 0.119$ for H$_\beta$. The oscillator strengths may be calculated theoretically or measured in the laboratory, and they are defined so that the $f$-values for transitions from the same initial orbital add up to the number of electrons in the atom or ion. Thus the oscillator strength is the effective number of electrons per atom participating in a transition, and so multiplying the number of absorbing atoms per unit area by the $f$-value gives the number of atoms lying above each cm$^2$ of the photosphere that are actively involved in producing a given spectral line, $fN_a$. Figure 9.20 shows the Voigt profiles of the K line of Ca II ($\lambda = 3933$ Å) for various values of the number of absorbing calcium ions.

The **curve of growth** is an important tool that astronomers use to determine the value of $N_a$ and thus the abundances of elements in stellar atmospheres. As seen in Fig. 9.20, the equivalent width, $W$, of the line varies with $N_a$. A curve of growth, shown in Fig. 9.21, is a logarithmic graph of

## 9.4 The Structure of Spectral Lines

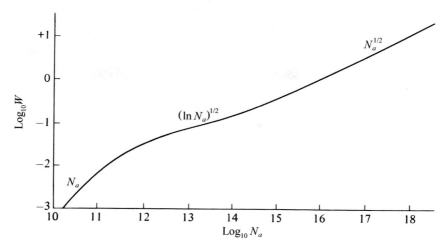

**Figure 9.21** The curve of growth for the K line of Ca II. As $N_a$ increases, the functional dependence of the equivalent width ($W$) changes. At various positions along the curve of growth, $W$ is proportional to the functional forms indicated. (Data from Aller, *The Atmospheres of the Sun and Stars*, Ronald Press, New York, 1963.)

the equivalent width, $W$, as a function of the number of absorbing atoms, $N_a$. To begin with, imagine that a specific element is not present in a stellar atmosphere. As some of that element is introduced, a weak absorption line appears that is initially optically thin. If the number of the absorbing atoms is doubled, twice as much light is removed and the equivalent width of the line is twice as great. So $W \propto N_a$, and the curve of growth is initially linear with $N_a$. As the number of absorbing atoms continues to increase, the center of the line becomes optically thick as the maximum amount of flux at the line's center is absorbed.[29] With the addition of still more atoms, the line bottoms out and becomes saturated. The wings of the line, which are still optically thin, continue to deepen. This occurs with little change in the line's equivalent width and produces a plateau on the curve of growth where $W \propto \sqrt{\ln N_a}$. Increasing the number of absorbing atoms still further increases the width of the pressure broadening profile (recall Eq. 9.59), enabling it to contribute to the wings of the line. The equivalent width grows more rapidly, although not as steeply as at first, with approximately $W \propto \sqrt{N_a}$ for the total line profile.

---

[29]The zero flux at the center of the line shown in Fig. 9.20 is a peculiarity of the Schuster–Schwarzschild model. Actually, there is always *some* flux received at the central wavelength, $\lambda_\circ$, even for very strong, optically thick lines. As a rule, the flux at any wavelength cannot fall below $F_\lambda = \pi S_\lambda(\tau_\lambda = 2/3)$, the value of the source function at an optical depth of 2/3; see Problem 9.20.

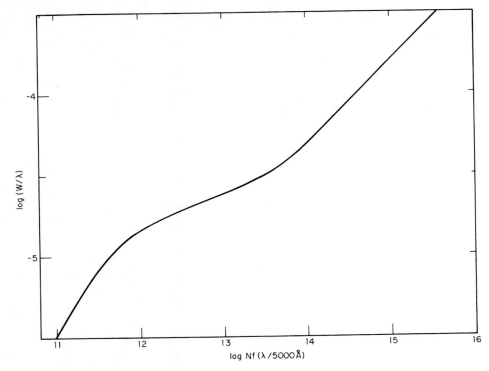

**Figure 9.22** A general curve of growth for the Sun. (Figure from Aller, *Atoms, Stars, and Nebulae*, Revised Edition, Harvard University Press, Cambridge, MA, 1971.)

Using the curve of growth and a measured equivalent width, we can obtain the number of absorbing atoms. The Boltzmann and Saha equations are then used to convert this value into the total number of atoms of that element lying above the photosphere.

To reduce the errors involved in using a single spectral line, it is advantageous to locate the positions of the equivalent widths of several lines on a single curve of growth. This can be accomplished by plotting $\log_{10}(W/\lambda)$ on the vertical axis and $\log_{10}[fN_a(\lambda/5000\ \text{Å})]$ on the horizontal axis. This scaling results in a general curve of growth that can be used for several lines formed by transitions from the same initial orbital.[30] Figure 9.22 shows a general curve of growth for the Sun. The use of such a curve of growth is best illustrated by an example.

---

[30]This is just one of several possible ways of scaling the curve of growth. The assumptions used to obtain such a scaling are not valid for all broad lines (such as hydrogen), and may lead to inaccurate results.

## 9.4 The Structure of Spectral Lines

| $\lambda$ (Å) | $W$ (Å) | $f$ | $\log_{10}(W/\lambda)$ | $\log_{10}[f(\lambda/5000\ \text{Å})]$ |
|---|---|---|---|---|
| 3302.38 | 0.088 | 0.0214 | $-4.58$ | $-1.85$ |
| 5889.97 | 0.730 | 0.645 | $-3.90$ | $-0.12$ |

**Table 9.1** Data for Solar Sodium Lines. (From Aller, *Atoms, Stars, and Nebulae*, Revised Edition, Harvard University Press, Cambridge, MA, 1971.)

---

**Example 9.11** We will use Fig. 9.22 to find the number of sodium atoms above each square centimeter of the Sun's photosphere from measurements of two sodium absorption lines. Values of $T = 5800$ K and $P_e = 10$ dyne cm$^{-2}$ for the temperature and electron pressure were used to construct this curve of growth and will be adopted in the calculations that follow.

Both of these lines are produced when an electron makes an upward transition from the ground state orbital of the neutral Na I atom and so these lines have the same value of $N_a$, the number of absorbing sodium atoms per unit area of the photosphere. This number can be found using the values of $\log_{10}(W/\lambda)$ (see Table 9.1) with the general curve of growth, Fig. 9.22, to obtain a value of $\log_{10}[fN_a(\lambda/5000\ \text{Å})]$ for each line. The results are

$$\log_{10}\left(\frac{fN_a\lambda}{5000\ \text{Å}}\right) = 13.20 \quad \text{for the 3302.38 Å line}$$

$$= 14.83 \quad \text{for the 5889.97 Å line}.$$

To obtain the value of the number of absorbing atoms per unit area, $N_a$, we use the measured values of $\log_{10}[f(\lambda/5000\ \text{Å})]$ together with

$$\log_{10} N_a = \log_{10}\left(\frac{fN_a\lambda}{5000\ \text{Å}}\right) - \log_{10}\left(\frac{f\lambda}{5000\ \text{Å}}\right),$$

to find

$$\log_{10} N_a = 13.20 - (-1.85) = 15.05 \quad \text{for the 3302.38 Å line}$$

and

$$\log_{10} N_a = 14.83 - (-0.12) = 14.95 \quad \text{for the 5889.97 Å line}.$$

The average value of $\log_{10} N_a$ is 15.0; thus there are about $10^{15}$ Na I atoms in the ground state per cm$^2$ of the photosphere.

To find the total number of sodium atoms, the Boltzmann and Saha equations must be used, Eqs. (8.4) and (8.7), respectively. The difference in energy between the final and initial states [$E_b - E_a$ in Eq. (8.4)] is just the energy of the emitted photon. Using Eq. (5.3), the exponential term in the Boltzmann equation is

$$e^{-(E_b - E_a)/kT} = e^{-hc/\lambda kT}$$

$$= 5.45 \times 10^{-4} \quad \text{for the 3302.38 Å line}$$

$$= 1.48 \times 10^{-2} \quad \text{for the 5889.97 Å line,}$$

and so nearly all of the neutral Na I atoms are in the ground state.

All that remains is to determine the total number of sodium atoms per unit area in all stages of ionization. If there are $N_I = 10^{15}$ neutral sodium atoms per cm$^2$, then the number of singly ionized atoms, $N_{II}$, comes from the Saha equation:

$$\frac{N_{II}}{N_I} = \frac{2kT Z_{II}}{P_e Z_I} \left( \frac{2\pi m_e kT}{h^2} \right)^{3/2} e^{-\chi_I / kT}.$$

Using $Z_I = 2.4$ and $Z_{II} = 1.0$ for the partition functions and $\chi_I = 5.14$ eV for the ionization energy of neutral sodium leads to $N_{II}/N_I = 2.43 \times 10^3$. There are about 2430 singly ionized sodium atoms for every neutral sodium atom in the Sun's photosphere,[31] so the total number of sodium atoms per unit area above the photosphere is about

$$N = 2430 N_I = 2.43 \times 10^{18} \text{ cm}^{-2}.$$

The mass of a sodium atom is $3.82 \times 10^{-23}$ kg, so the mass of sodium atoms above each square centimeter of the photosphere is roughly $9.3 \times 10^{-5}$ g cm$^{-2}$. (A more detailed analysis leads to a slightly lower value of $5.4 \times 10^{-5}$ g cm$^{-2}$.) For comparison, the mass of hydrogen atoms per unit area is about 1.1 g cm$^{-2}$.

---

Thus the number of absorbing atoms can be determined by comparing the equivalent widths measured for different absorption lines produced by atoms or ions initially in the same state (and so having the same column density in the stellar atmosphere) with a theoretical curve of growth. A curve-of-growth analysis can also be applied to lines originating from atoms or ions in different initial states; then applying the Boltzmann equation to the relative numbers of atoms and ions in these different states of excitation allows the excitation

---

[31]The ionization energy for Na II is 47.3 eV. This is sufficiently large to guarantee that $N_{III} \ll N_{II}$, so higher states of ionization can be neglected.

## 9.4 The Structure of Spectral Lines

| Element | Atomic Number | Log Relative Abundance | Column Density (g cm$^{-2}$) |
|---|---|---|---|
| Hydrogen | 1 | 12.00 | 1.1 |
| Helium | 2 | 10.99 | $4.3 \times 10^{-1}$ |
| Oxygen | 8 | 8.93 | $1.5 \times 10^{-2}$ |
| Carbon | 6 | 8.60 | $5.3 \times 10^{-3}$ |
| Neon | 10 | 8.09 | $2.7 \times 10^{-3}$ |
| Nitrogen | 7 | 8.00 | $1.5 \times 10^{-3}$ |
| Iron | 26 | 7.67 | $2.9 \times 10^{-3}$ |
| Magnesium | 12 | 7.58 | $1.0 \times 10^{-3}$ |
| Silicon | 14 | 7.55 | $1.1 \times 10^{-3}$ |
| Sulfur | 16 | 7.21 | $5.7 \times 10^{-4}$ |

**Table 9.2** The Most Abundant Elements in the Solar Photosphere. The relative abundance of an element is given by $\log_{10}(N_{\rm el}/N_{\rm H}) + 12$, and the column density is based on a value of 1.1 g cm$^{-2}$ for hydrogen. (Data from Grevesse and Anders, *Solar Atmosphere and Interior*, A. N. Cox, W. C. Livingston, and M. S. Matthews (eds.), The University of Arizona Press, Tucson, AZ, 1991.)

temperature to be calculated. Similarly, it is possible to use the Saha equation to find either the electron pressure or the ionization temperature (if the other is known) in the atmosphere from the relative numbers of atoms at various stages of ionization.

The ultimate refinement in the analysis of stellar atmospheres is the construction of a *model atmosphere* on a computer. Each atmospheric layer is involved in the formation of line profiles and contributes to the spectrum observed for the star. All of the ingredients of the preceding discussion, plus the equations of hydrostatic equilibrium, thermodynamics, statistical and quantum mechanics, and the transport of energy by radiation and convection, are combined with extensive libraries of opacities to calculate how the temperature, pressure, and density vary with depth below the surface.[32] Only when the variables of the model have been "fine-tuned" to obtain good agreement with the observations can astronomers finally claim to have decoded the information carried in the light from a star.

This basic procedure has led astronomers to an understanding of the abundances of the elements in the Sun (see Table 9.2) and other stars. Hydrogen and helium are by far the most common elements, followed by oxygen, carbon,

---
[32] Details of the construction of a model star will be deferred to Chapter 10.

and nitrogen; for every $10^{12}$ atoms of hydrogen, there are $10^{11}$ atoms of helium and about $10^9$ atoms of oxygen. These figures are in very good agreement with abundances obtained from meteorites, giving astronomers confidence in their results.[33] This knowledge of the basic ingredients of the universe provides invaluable observational tests and constraints for some of the most fundamental theories in astronomy: the nucleosynthesis of light elements as a result of stellar evolution, the production of heavier elements by supernovae, and the Big Bang that produced the primordial hydrogen and helium that started it all.

## Suggested Readings

GENERAL

Aller, Lawrence H., *Atoms, Stars, and Nebulae*, Revised Edition, Harvard University Press, Cambridge, MA, 1971.

Hearnshaw, J. B., *The Analysis of Starlight*, Cambridge University Press, Cambridge, 1986.

Kaler, James B., *Stars and Their Spectra*, Cambridge University Press, Cambridge, 1989.

TECHNICAL

Aller, Lawrence H., *The Atmospheres of the Sun and Stars*, Ronald Press, New York, 1963.

Böhm-Vitense, Erika, "The Effective Temperature Scale," *Annual Review of Astronomy and Astrophysics*, *19*, 295, 1981.

Böhm-Vitense, Erika, *Stellar Astrophysics, Volume 2: Stellar Atmospheres*, Cambridge University Press, Cambridge, 1989.

Novotny, Eva, *Introduction to Stellar Atmospheres and Interiors*, Oxford University Press, New York, 1973.

Rybicki, George B., and Lightman, Alan P., *Radiative Processes in Astrophysics*, John Wiley and Sons, New York, 1979.

---

[33] A notable exception is lithium, whose solar relative abundance of $10^{1.16}$ is significantly less than the value of $10^{3.31}$ obtained from meteorites. The efficient depletion of the Sun's lithium, sparing only one of every 140 lithium atoms, is not yet understood.

# Problems

9.1 Evaluate the energy of the blackbody photons inside your eye. Compare this with the visible energy inside your eye while looking at a 100-W ($10^9$ erg s$^{-1}$) light bulb that is 100 cm away. (You can assume that the light bulb is 100% efficient, although in reality it converts only a few percent of its 100 watts into visible photons. Take your eye to be a hollow sphere of radius 1.5 cm at a temperature of 37°C. The area of the eye's pupil is about 0.1 cm$^2$.) Why is it dark when you close your eyes?

9.2 (a) Find an expression for $n_\lambda \, d\lambda$, the number density of blackbody photons (the number of blackbody photons per cm$^3$) with a wavelength between $\lambda$ and $\lambda + d\lambda$.

(b) Find the total number of photons inside a kitchen oven set at 400°F, assuming a volume of 1 m$^3$.

9.3 (a) Use the results of Problem 9.2 to find the total number density, $n$, of blackbody photons of all wavelengths. Also show that the average energy per photon, $u/n$, is

$$\frac{u}{n} = \frac{\pi^4 kT}{15(2.404)} = 2.70 kT. \tag{9.60}$$

(b) Find the average energy per blackbody photon at the center of the Sun, where $T = 1.58 \times 10^7$ K, and in the solar photosphere, where $T = 5770$ K. Express your answers in units of electron volts (eV).

9.4 Derive Eq. (9.9) for the blackbody radiation pressure,

$$P_{\text{rad}} = \frac{4\pi}{3c} \int_0^\infty B_\lambda(T) \, d\lambda = \frac{4\sigma T^4}{3c} = \frac{1}{3} aT^4 = \frac{1}{3} u.$$

9.5 Consider a spherical blackbody of radius $R$ and temperature $T$. Integrate Eq. (9.6) for the radiative flux with $I_\lambda = B_\lambda$ over all outward directions to derive the Stefan–Boltzmann equation in the form of Eq. (3.17). (You will also have to integrate over all wavelengths and over the surface area of the sphere.)

9.6 Using the root-mean-square speed, $v_{\rm rms}$, estimate the mean free path of the nitrogen molecules in your classroom at room temperature (300 K). What is the average time between collisions? Take the radius of a nitrogen molecule to be 1 Å, and the density of air to be $1.2 \times 10^{-3}$ g cm$^{-3}$. A nitrogen molecule contains 28 nucleons (protons and neutrons).

9.7 Calculate how far you could see through Earth's atmosphere if it had the opacity of the solar photosphere. Use the value for the Sun's opacity from Example 9.2 and $1.2 \times 10^{-3}$ g cm$^{-3}$ for the density of Earth's atmosphere.

9.8 In Example 9.3, suppose that only two measurements of the specific intensity, $I_1$ and $I_2$, are available, made at angles $\theta_1$ and $\theta_2$. Determine expressions for the intensity $I_{\lambda,0}$ of the light above Earth's atmosphere and for the vertical optical depth of the atmosphere, $\tau_{\lambda,0}$, in terms of these two measurements.

9.9 Use the laws of conservation of relativistic energy and momentum to prove that an isolated electron cannot absorb a photon.

9.10 By measuring the slope of the curves in Fig. 9.10, verify that the decline of the curves after the peak in the opacity follows a Kramers law, $\overline{\kappa} \propto T^{-n}$, where $n \approx 3.5$.

9.11 According to a "standard model" of the Sun, the central density is 162 g cm$^{-3}$ and the Rosseland mean opacity at the center is 1.16 cm$^2$ g$^{-1}$.

(a) Calculate the mean free path of a photon at the center of the Sun.

(b) If this mean free path remained constant for the photon's journey to the surface, calculate the average time it would take for the photon to escape from the Sun.

9.12 If the temperature of a star's atmosphere is *increasing* outward, what type of spectral lines would you expect to find in the star's spectrum at those wavelengths where the opacity is greatest?

9.13 Consider a large hollow spherical shell of hot gas surrounding a star. Under what circumstances would you see the shell as a glowing *ring* around the star? What can you say about the optical thickness of the shell?

# Problems

**Problems 9.14 through 9.24 involve the optional material at the end of Section 9.3.**

9.14 Verify that the emission coefficient, $j_\lambda$, has units of cm s$^{-3}$ sr$^{-1}$.

9.15 Derive Eq. (9.29) in Example 9.5, which shows how the intensity of a light ray is converted from its initial intensity $I_\lambda$ to the value $S_\lambda$ of the source function.

9.16 The transfer equation, Eq. (9.28), is written in terms of the distance, $s$, measured along the path of a light ray. In different coordinate systems, the transfer equation will look slightly different, and care must be taken to include all of the necessary terms.

(a) Show that in a spherical coordinate system, with the center of the star at the origin, the transfer equation has the form

$$-\frac{\cos\theta}{\kappa_\lambda \rho}\frac{\partial I_\lambda}{\partial r} + \frac{\sin\theta}{\kappa_\lambda \rho r}\frac{\partial I_\lambda}{\partial \theta} = I_\lambda - S_\lambda.$$

You may assume that there is no $\phi$-dependence. Note that you cannot simply replace $s$ with $r$!

(b) Use this form of the transfer equation to derive Eq. (9.25).

9.17 Using the Eddington approximation for a plane-parallel atmosphere, show that in the Eddington approximation, the mean intensity, radiative flux, and radiation pressure are given by Eqs. (9.40)–(9.43).

9.18 Using the Eddington approximation for a plane-parallel atmosphere, determine the values of $I_{\text{in}}$ and $I_{\text{out}}$ as functions of the vertical optical depth. At what depth is the radiation isotropic to within 1%?

9.19 Using the results for the plane-parallel gray atmosphere in LTE, determine the ratio of the surface temperature of a star to its temperature at the top of the atmosphere. If $T_e = 5770$ K, what is the temperature at the top of the atmosphere?

9.20 Show that, for a plane-parallel gray atmosphere in LTE, the (constant) value of the radiative flux is equal to $\pi$ times the source function evaluated at an optical depth of 2/3:

$$F_{\text{rad}} = \pi S(\tau_v = 2/3).$$

This function, called the **Eddington–Barbier relation**, says that the radiative flux received from the surface of the star is determined by the value of the source function at $\tau_v = 2/3$.

9.21 Consider a horizontal plane-parallel slab of gas of thickness $L$ that is maintained at a constant temperature $T$. Assume that the gas has optical depth $\tau_{\lambda,o}$, with $\tau_\lambda = 0$ at the top surface of the slab. Assume further that no radiation enters the gas from outside. Use the general solution of the transfer equation (Eq. 9.49) to show that, when looking at the slab from above, you see blackbody radiation if $\tau_{\lambda,o} \gg 1$ and emission lines (where $j_\lambda$ is large) for $\tau_{\lambda,o} \ll 1$. You may assume that the source function, $S_\lambda$, does not vary with position inside the gas.

9.22 Consider a horizontal plane-parallel slab of gas of thickness $L$ that is maintained at a constant temperature $T$. Assume that the gas has optical depth $\tau_{\lambda,o}$, with $\tau_\lambda = 0$ at the top surface of the slab. Assume further that incident radiation of intensity $I_{\lambda,o}$ enters the bottom of the slab from outside. Use the general solution of the transfer equation (Eq. 9.49) to show that, when looking at the slab from above, you see blackbody radiation if $\tau_{\lambda,o} \gg 1$. If $\tau_{\lambda,o} \ll 1$, show that you see absorption lines superimposed on the spectrum of the incident radiation if $I_{\lambda,o} > S_\lambda$ and emission lines superimposed on the spectrum of the incident radiation if $I_{\lambda,o} < S_\lambda$. (These latter two cases correspond to the spectral lines formed in the Sun's photosphere and chromosphere, respectively; see Section 11.2.) You may assume that the source function, $S_\lambda$, does not vary with position inside the gas.

9.23 Verify that if the source function is $S_\lambda = a_\lambda + b_\lambda \tau_{\lambda,v}$, then the emergent intensity is given by Eq. (9.52),

$$I_\lambda(0) = a_\lambda + b_\lambda \cos\theta.$$

9.24 **Computer Problem** In this problem, you will use the values of the density and opacity at various points near the surface of the star to calculate the optical depth of these points. The data in Table 9.3 were obtained from the stellar model building program STATSTAR, described in Section 10.5 and found in Appendix I. The first point listed is at the surface of the stellar model.

(a) Find the optical depth at each point by numerically integrating Eq. (9.13). Use a simple trapezoidal rule such that

$$d\tau_\lambda = -\kappa\rho\, ds$$

becomes

$$\tau_{i+1} - \tau_i = -\left(\frac{\kappa_i \rho_i + \kappa_{i+1}\rho_{i+1}}{2}\right)(r_{i+1} - r_i).$$

| $i$ | $r$ (cm) | $T$ (K) | $\rho$ (g cm$^{-3}$) | $\kappa$ (cm$^2$ g$^{-1}$) |
|---|---|---|---|---|
| 1 | 7.10604E+10 | 0.00000E+00 | 0.00000E+00 | 0.00000E+00 |
| 2 | 7.09894E+10 | 3.28531E+03 | 1.74510E−11 | 2.30027E+02 |
| 3 | 7.09183E+10 | 6.57721E+03 | 2.51661E−10 | 1.71702E+02 |
| 4 | 7.08473E+10 | 9.87571E+03 | 1.23854E−09 | 1.48384E+02 |
| 5 | 7.07762E+10 | 1.31808E+04 | 3.72025E−09 | 1.30444E+02 |
| 6 | 7.07051E+10 | 1.64926E+04 | 8.61374E−09 | 1.16701E+02 |
| 7 | 7.06341E+10 | 1.98110E+04 | 1.70083E−08 | 1.06028E+02 |
| 8 | 7.05630E+10 | 2.31361E+04 | 3.01529E−08 | 9.75243E+01 |
| 9 | 7.04920E+10 | 2.64680E+04 | 4.94472E−08 | 9.05811E+01 |
| 10 | 7.04209E+10 | 2.98065E+04 | 7.64352E−08 | 8.47909E+01 |
| 11 | 7.03498E+10 | 3.31518E+04 | 1.12801E−07 | 7.98764E+01 |
| 12 | 7.02788E+10 | 3.62287E+04 | 1.55337E−07 | 7.57112E+01 |
| 13 | 7.02077E+10 | 3.93132E+04 | 2.08540E−07 | 7.20713E+01 |
| 14 | 7.01367E+10 | 4.24050E+04 | 2.73965E−07 | 6.88580E+01 |
| 15 | 7.00656E+10 | 4.55039E+04 | 3.53262E−07 | 6.59958E+01 |
| 16 | 6.99945E+10 | 4.86099E+04 | 4.48170E−07 | 6.34264E+01 |
| 17 | 6.99235E+10 | 5.17229E+04 | 5.60522E−07 | 6.11040E+01 |
| 18 | 6.98524E+10 | 5.48430E+04 | 6.92238E−07 | 5.89922E+01 |
| 19 | 6.97814E+10 | 5.79700E+04 | 8.45332E−07 | 5.70616E+01 |
| 20 | 6.97103E+10 | 6.11041E+04 | 1.02191E−06 | 5.52881E+01 |
| 21 | 6.96392E+10 | 6.42451E+04 | 1.22416E−06 | 5.36521E+01 |
| 22 | 6.95682E+10 | 6.73931E+04 | 1.45436E−06 | 5.21369E+01 |
| 23 | 6.94971E+10 | 7.05481E+04 | 1.71491E−06 | 5.07286E+01 |
| 24 | 6.94261E+10 | 7.37100E+04 | 2.00825E−06 | 4.94156E+01 |
| 25 | 6.93550E+10 | 7.68790E+04 | 2.33696E−06 | 4.81877E+01 |
| 26 | 6.92839E+10 | 8.00549E+04 | 2.70369E−06 | 4.70363E+01 |
| 27 | 6.92129E+10 | 8.32377E+04 | 3.11117E−06 | 4.59540E+01 |
| 28 | 6.91418E+10 | 8.64276E+04 | 3.56224E−06 | 4.49343E+01 |
| 29 | 6.90708E+10 | 8.96245E+04 | 4.05984E−06 | 4.39714E+01 |
| 30 | 6.89997E+10 | 9.28283E+04 | 4.60699E−06 | 4.30603E+01 |
| 31 | 6.89286E+10 | 9.60392E+04 | 5.20682E−06 | 4.21968E+01 |
| 32 | 6.88576E+10 | 9.92571E+04 | 5.86253E−06 | 4.13768E+01 |
| 33 | 6.87865E+10 | 1.02482E+05 | 6.57744E−06 | 4.05969E+01 |

**Table 9.3** A 1 M$_\odot$ STATSTAR Model for Problem 9.24. $T_e = 5500$ K.

| λ (Å)   | W (Å) | f      |
|---------|-------|--------|
| 3302.98 | 0.067 | 0.0049 |
| 5895.94 | 0.560 | 0.325  |

**Table 9.4** Data for Solar Sodium Lines for Problem 9.27. (Data from Aller, *Atoms, Stars, and Nebulae*, Revised Edition, Harvard University Press, Cambridge, MA, 1971.)

    Note that because $s$ is measured along the path traveled by the photons, $ds = dr$.

(b) Make a graph of the temperature (vertical axis) vs. the optical depth (horizontal axis).

(c) For each value of the optical depth, use Eq. (9.48) to calculate the temperature for a plane-parallel gray atmosphere in LTE. Plot these values of $T$ on the same graph.

(d) The STATSTAR program utilizes a simplifying assumption that the surface temperature is zero (see Appendix H). Comment on the validity of the surface value of $T$ that you found.

9.25 Suppose that the shape of a spectral line is fit with one-half of an ellipse, such that the semimajor axis $a$ is equal to the maximum depth of the line, and the minor axis $2b$ is equal to the maximum width of the line (where it joins the continuum). What is the equivalent width of this line? *Hint:* You may find Eq. (2.4) useful.

9.26 Derive Eq. (9.55) for the uncertainty in the wavelength of a spectral line due to Heisenberg's uncertainty principle.

9.27 The two solar absorption lines given in Table 9.4 are produced when an electron makes an upward transition from the ground state orbital of the neutral Na I atom.

(a) Using the general curve of growth for the Sun, Fig. 9.22, repeat the procedure of Example 9.11 to find $N_a$, the number of absorbing sodium atoms per unit area of the photosphere.

(b) Combine your results with those of Example 9.11 to find an average value of $N_a$. Use this value to plot the positions of the four sodium absorption lines on Fig. 9.22, and confirm that they do all lie on the curve of growth.

| λ (Å) | W (Å) | f |
|---|---|---|
| 10938 ($P_\gamma$) | 2.2 | 0.0554 |
| 10049 ($P_\delta$) | 1.6 | 0.0269 |

**Table 9.5** Data for Solar Hydrogen Lines for Problem 9.28. (Data from Aller, *Atoms, Stars, and Nebulae*, Revised Edition, Harvard University Press, Cambridge, MA, 1971.)

9.28 Pressure broadening (due to the presence of the electric fields of nearby ions) is unusually effective for the spectral lines of hydrogen. Using the general curve of growth for the Sun with these broad hydrogen absorption lines will result in an overestimate of the amount of hydrogen present. The following calculation nevertheless demonstrates just how abundant hydrogen is in the Sun.

The two solar absorption given in Table 9.5 belong to the Paschen series, produced when an electron makes an upward transition from the ($n = 3$) orbit of the hydrogen atom.

(a) Using the general curve of growth for the Sun, Fig. 9.22, repeat the procedure of Example 9.11 to find $N_a$, the number of absorbing hydrogen atoms per unit area of the photosphere (those with electrons initially in the $n = 3$ orbit).

(b) Use the Boltzmann and Saha equations to calculate the total number of hydrogen atoms above each square centimeter of the Sun's photosphere.

(c) Calculate the column density of hydrogen atoms, and compare your result with the value found in Table 9.2.

# Chapter 10

# The Interiors of Stars

## 10.1 Hydrostatic Equilibrium

In the last two chapters many of the observational details of stellar spectra were discussed along with the basic physical principles behind the production of the observed lines. Analysis of that light, collected by ground-based and orbital telescopes, allows astronomers to determine a variety of quantities related to the outer layers of stars, such as effective temperature, luminosity, and composition. However, with the exceptions of the detection of neutrinos from the Sun (which will be discussed later in this chapter and in Chapter 11) and from Supernova 1987A (Section 13.3), no direct way exists to observe the central regions of stars.

To deduce the detailed internal structure of stars requires the generation of computer models that are consistent with all known physical laws and that ultimately agree with observable surface features. Such calculations require very large and sophisticated software programs (often referred to as *codes*) and powerful computers. Although much of the theoretical foundation of stellar structure was understood in the first half of this century, not until the 1960s were sufficiently fast computing machines available to carry out all of the necessary calculations. Arguably the greatest success of theoretical astrophysics to date has been the computer modeling of stellar structure and evolution. Despite all of the successes of such calculations, however, numerous questions remain unanswered. The solution to many of these problems requires a more detailed theoretical understanding of the physical processes in operation in the interiors of stars, combined with significant computational power.

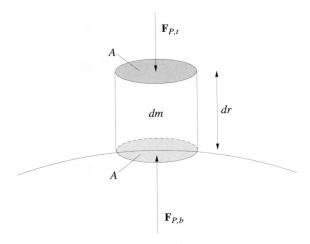

**Figure 10.1** In a static star the gravitational force on a mass element is exactly canceled by the outward force due to a pressure gradient in the star. A cylinder of mass $dm$ is located at a distance $r$ from the center of the star. The height of the cylinder is $dr$, and the areas of the top and bottom are both $A$. The density of the gas is assumed to be $\rho$ at that position.

The theoretical study of stellar structure, coupled with observational data, clearly shows that stars are dynamic objects, usually changing at an imperceptibly slow rate by human standards, although at other times changing in very rapid and dramatic ways, such as during a supernova explosion. That such changes must occur can be seen by simply considering the observed energy output of a star. In the Sun, $3.826 \times 10^{33}$ ergs of energy are emitted every second. This rate of energy output would be sufficient to melt a 0°C block of ice measuring 1 AU × 1 mile × 1 mile in only 0.3 s, assuming that the absorption of the energy were 100% efficient. Because stars do not have infinite supplies of energy, they must eventually use up their reserves and die.

Stellar evolution is the result of a constant fight against the relentless pull of gravity. Because the gravitational force is always attractive, an opposing force must exist if a star is to avoid collapse. This force is provided by pressure. To calculate how the pressure must vary with depth, consider a cylinder of mass $dm$ whose base is located a distance $r$ from the center of a spherical star (see Fig. 10.1). The areas of the top and bottom of the cylinder are each $A$ and the cylinder's height is $dr$. Furthermore, assume that the only forces acting on the cylinder are gravity and the pressure force, which is always normal to the surface and may vary with distance from the center of the star. Using

## 10.1 Hydrostatic Equilibrium

Newton's second law $\mathbf{F} = m\mathbf{a}$, we have the net force on the cylinder:

$$dm \frac{d^2 r}{dt^2} = F_g + F_{P,t} + F_{P,b},$$

where $F_g < 0$ is the gravitational force directed inward and $F_{P,t}$ and $F_{P,b}$ are the pressure forces on the top and bottom of the cylinder, respectively. Note that since the pressure forces on the side of the cylinder will cancel, they have been explicitly excluded from the expression. Because the pressure force is always normal to the surface, the force exerted on the top of the cylinder must necessarily be directed toward the center of the star ($F_{P,t} < 0$) while the force on the bottom is directed outward ($F_{P,b} > 0$). Writing $F_{P,t}$ in terms of $F_{P,b}$ and a correction term $dF_P$ that accounts for the change in force due to a change in $r$ results in

$$F_{P,t} = -(F_{P,b} + dF_P).$$

Substitution into the previous expression gives

$$dm \frac{d^2 r}{dt^2} = F_g - dF_P. \tag{10.1}$$

As was pointed out in Example 2.2, the gravitational force on a small mass $dm$ located at a distance $r$ from the center of a spherically symmetric mass is

$$F_g = -G \frac{M_r \, dm}{r^2}, \tag{10.2}$$

where $M_r$ is the mass inside the sphere of radius $r$, often referred to as the *interior mass*. The contribution to the gravitational force by spherically symmetric mass shells located outside $r$ is zero (the proof of this is left to Problem 10.2).

Pressure is defined as the amount of force per unit area exerted on a surface, or

$$P \equiv \frac{F}{A}. \tag{10.3}$$

Allowing for a difference in pressures $dP$ between the top of the cylinder and the bottom due to the different amount of force exerted on each surface, the differential force may be expressed as

$$dF_P = A \, dP. \tag{10.4}$$

Substituting Eqs. (10.2) and (10.4) into Eq. (10.1) gives

$$dm \frac{d^2 r}{dt^2} = -G \frac{M_r \, dm}{r^2} - A \, dP. \tag{10.5}$$

Assuming that the density of the cylinder of gas is $\rho$, its mass is just

$$dm = \rho A\, dr,$$

where $A\, dr$ is the cylinder's volume. Using this expression in Eq. (10.5) yields

$$\rho A\, dr \frac{d^2 r}{dt^2} = -G \frac{M_r \rho A\, dr}{r^2} - A\, dP.$$

Finally, dividing through by the volume of the cylinder, we have

$$\rho \frac{d^2 r}{dt^2} = -G \frac{M_r \rho}{r^2} - \frac{dP}{dr}. \tag{10.6}$$

This is the equation for the radial motion of the cylinder, assuming spherical symmetry.

If we assume further that the star is static, then the acceleration must be zero. In this case Eq. (10.6) reduces to

$$\frac{dP}{dr} = -G \frac{M_r \rho}{r^2} \tag{10.7}$$

$$= -\rho g,$$

where $g \equiv GM_r/r^2$ is the local acceleration of gravity at radius $r$. Equation (10.7), the condition of **hydrostatic equilibrium**, represents one of the fundamental equations of stellar structure for spherically symmetric objects under the assumption that accelerations are negligible. Equation (10.7) clearly indicates that in order for a star to be static, a *pressure gradient $dP/dr$* must exist to counteract the force of gravity. It is not the pressure that supports a star, but the change in pressure with radius. Furthermore, the pressure must *decrease* with increasing radius; the pressure is necessarily larger in the interior than it is near the surface.

---

**Example 10.1** To obtain a very crude estimate of the pressure at the center of the Sun, assume that $M_r = 1$ M$_\odot$, $r = 1$ R$_\odot$, and $\rho = \overline{\rho}_\odot = 1.41$ g cm$^{-3}$ is the *average* solar density (see Example 8.5). Assume also that the surface pressure is exactly zero. Then, converting the differential equation to a difference equation, the left hand side of Eq. (10.7) becomes

$$\frac{dP}{dr} \sim \frac{P_s - P_c}{R_s - 0} \sim -\frac{P_c}{R_\odot},$$

## 10.1 Hydrostatic Equilibrium

where $P_c$ is the central pressure, and $P_s$ and $R_s$ are the surface pressure and radius, respectively. Substituting into the equation of hydrostatic equilibrium and solving for the central pressure,

$$P_c \sim G\frac{M_\odot \bar{\rho}_\odot}{R_\odot} \sim 2.7 \times 10^{15} \text{ dynes cm}^{-2}.$$

To obtain a more accurate value, the hydrostatic equilibrium equation needs to be *integrated* from the surface to the center, taking into consideration the change in the interior mass $M_r$ at each point together with the variation of density with radius $\rho_r \equiv \rho(r)$, giving

$$\int_{P_s}^{P_c} dP = P_c = -\int_{R_s}^{R_c} \frac{GM_r \rho}{r^2} dr.$$

Actually carrying out the integration requires, however, functional forms of $M_r$ and $\rho$. Unfortunately, such explicit expressions are not available, implying that further relationships between such quantities must be developed.

From a more rigorous calculation, a standard solar model gives a central pressure of nearly $2.5 \times 10^{17}$ dynes cm$^{-2}$. This value is much larger than the one obtained from our crude estimate because of the increased density near the center of the Sun. As a reference, one atmosphere of pressure is 1 atm = $1.013 \times 10^6$ dynes cm$^{-2}$; therefore the more realistic model predicts a central pressure of $2.5 \times 10^{11}$ atm!

---

A second relationship between mass, radius, and density also exists. Again, for a spherically symmetric star, consider a *shell* of mass $dM_r$ and thickness $dr$, located a distance $r$ from the center, as in Fig. 10.2. Assuming that the shell is sufficiently thin (i.e., $dr \ll r$), the volume of the shell is approximately $dV = 4\pi r^2 dr$. If the local density of the gas is $\rho$, the shell's mass is given by

$$dM_r = \rho(4\pi r^2 dr).$$

Rewriting, we arrive at the **mass conservation equation**,

$$\frac{dM_r}{dr} = 4\pi r^2 \rho, \qquad (10.8)$$

which states how the interior mass of a star must change with distance from the center. Equation (10.8) is the second of the fundamental equations of stellar structure.

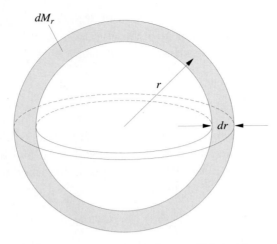

**Figure 10.2** A spherically symmetric shell of mass $dM_r$ having a thickness $dr$ and located a distance $r$ from the center of the star. The local density of the shell is $\rho$.

## 10.2 Pressure Equation of State

Up to this point no information has been provided about the origin of the pressure term required by Eq. (10.7). To describe this macroscopic manifestation of particle interactions, it is necessary to derive a pressure **equation of state** of the material. Such an equation of state relates the dependence of pressure on other fundamental parameters of the material. One well-known example of a pressure equation of state is the **ideal gas law**, often expressed as

$$PV = NkT,$$

where $V$ is the volume of the gas, $N$ is the number of particles, $T$ is the temperature, and $k$ is Boltzmann's constant.

Although this expression was first determined experimentally, it is informative to derive it from fundamental physical principles. The approach used here will also provide a general method for considering environments where the assumptions of the ideal gas law do not strictly apply, a situation frequently encountered in astrophysical problems.

Consider a cylinder of gas of length $\Delta x$ and cross-sectional area $A$, as in Fig. 10.3. The gas contained in the cylinder is assumed to be composed of point particles, each of mass $m$, that interact through perfectly elastic collisions only—in other words, as an ideal gas. To determine the pressure exerted on one of the ends of the container, examine the result of an impact on the right

## 10.2 Pressure Equation of State

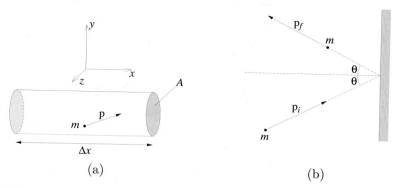

**Figure 10.3** (a) A cylinder of gas of length $\Delta x$ and cross-sectional area $A$. Assume that the gas contained in the cylinder is an ideal gas. (b) The collision of an individual point mass with one of the ends of the cylinder. For a perfectly elastic collision the angle of reflection must equal the angle of incidence.

wall by an individual particle. Since, for a perfectly elastic collision, the angle of reflection from the wall must be equal to the angle of incidence, the change in momentum of the particle is necessarily entirely in the $x$-direction, normal to the surface. From Newton's second law ($\mathbf{f} = m\mathbf{a} = d\mathbf{p}/dt$) and third law, the *impulse* $\mathbf{f}\,\Delta t$ delivered to the wall is just the negative of the change in momentum of the particle, or

$$\mathbf{f}\,\Delta t = -\Delta \mathbf{p} = 2p_x \hat{\mathbf{i}},$$

where $p_x$ is the component of the particle's initial momentum in the $x$-direction. Now the average force exerted by the particle over a period of time can be determined by evaluating the time interval between collisions with the right wall. Since the particle must traverse the length of the container twice before returning for a second reflection, the time interval between collisions with the same wall is given by

$$\Delta t = 2\frac{\Delta x}{v_x},$$

so that the average force exerted on the wall by a single particle over that time period is given by

$$f = \frac{2p_x}{\Delta t} = \frac{p_x v_x}{\Delta x},$$

where it is assumed that the direction of the force vector is normal to the surface.

Now, because $p_x \propto v_x$, the numerator is proportional to $v_x^2$. To evaluate this, recall that the magnitude of the velocity vector is given by $v^2 = v_x^2 + v_y^2 + v_z^2$. For a sufficiently large collection of particles in random motion, the likelihood of motion in each of the three directions is the same, or $\overline{v_x^2} = \overline{v_y^2} = \overline{v_z^2} = v^2/3$. Substituting $\frac{1}{3}pv$ for $p_x v_x$, the average force per particle having momentum $p$ is

$$f_p = \frac{1}{3}\frac{pv}{\Delta x}.$$

It is usually the case that the particles have a range of momenta. If the number of particles with momenta between $p$ and $p + dp$ is given by the expression $N_p\, dp$, then the total number of particles in the container is

$$N = \int_0^\infty N_p\, dp.$$

The contribution to the total force, $F_p$, by *all* particles in that momentum range is given by

$$dF_p = f_p N_p\, dp = \frac{1}{3}\frac{N_p}{\Delta x}pv\, dp.$$

Integrating over all possible values of the momentum, the total force exerted by particle collisions is

$$F = \frac{1}{3}\int_0^\infty \frac{N_p}{\Delta x}pv\, dp.$$

Dividing both sides of the expression by the surface area of the wall $A$ gives the pressure on the surface as $P = F/A$. Noting that $\Delta V = A\,\Delta x$ is just the volume of the cylinder and defining $n_p\, dp$ to be the number of particles *per unit volume* having momenta between $p$ and $p + dp$, or

$$n_p\, dp \equiv \frac{N_p}{\Delta V}\, dp,$$

the pressure exerted on the wall is found to be

$$P = \frac{1}{3}\int_0^\infty n_p pv\, dp. \qquad (10.9)$$

This expression is sometimes called the **pressure integral** and allows the pressure to be computed, given some distribution function, $n_p\, dp$.

Equation (10.9) is valid for both massive and massless particles (such as photons) traveling at any speed. For the special case of massive nonrelativistic particles, we may use $p = mv$ to write the pressure integral as

$$P = \frac{1}{3}\int_0^\infty mn_v v^2\, dv, \qquad (10.10)$$

## 10.2 Pressure Equation of State

where $n_v\, dv = n_p\, dp$ is the number of particles per unit volume having speeds between $v$ and $v + dv$.

The function $n_v\, dv$ is dependent on the physical nature of the system being described. In the case of an ideal gas, $n_v\, dv$ is the Maxwell–Boltzmann velocity distribution described in Chapter 8 (Eq. 8.1),

$$n_v\, dv = n \left(\frac{m}{2\pi kT}\right)^{3/2} e^{-mv^2/2kT} 4\pi v^2\, dv,$$

where $n = \int_0^\infty n_v\, dv$ is the particle number density. Substituting into the pressure integral finally gives

$$P_g = nkT. \tag{10.11}$$

Since $n \equiv N/V$, Eq. (10.11) is just the familiar ideal gas law.

In astrophysical applications it is often convenient to express the ideal gas law in an alternative form. Since $n$ is the particle number density, it is clear that it must be related to the mass density of the gas. Allowing for a variety of particles of different masses, it is then possible to express $n$ as

$$n = \frac{\rho}{\overline{m}}, \tag{10.12}$$

where $\overline{m}$ is the average mass of a gas particle. Substituting, the ideal gas law becomes

$$P_g = \frac{\rho kT}{\overline{m}}.$$

We now define a new quantity, the **mean molecular weight**, as

$$\mu \equiv \frac{\overline{m}}{m_H}, \tag{10.13}$$

where $m_H = 1.673525 \times 10^{-24}$ g is the mass of a hydrogen atom. The mean molecular weight is just the average mass of a free particle in the gas, in units of the mass of the hydrogen atom. The ideal gas law can now be written in terms of the mean molecular weight as

$$P_g = \frac{\rho kT}{\mu m_H}. \tag{10.14}$$

The mean molecular weight is dependent on the composition of the gas as well as the state of ionization of each species. The level of ionization enters because free electrons must be included in the average mass per particle $\overline{m}$. This implies that a detailed analysis of the Saha equation (Eq. 8.6) is necessary to calculate the relative numbers of ionization states. When the gas is either completely neutral or completely ionized, however, the calculation simplifies significantly.

Consider first the determination of $\overline{m}$. For a completely neutral gas

$$\overline{m}_n = \frac{\sum_j N_j m_j}{\sum_j N_j}, \qquad (10.15)$$

where $m_j$ and $N_j$ are the mass and the total number of atoms of type $j$ that are present in the gas and the sums are assumed to be carried out over all types of atoms. Dividing by the mass of the hydrogen atom,

$$\mu_n = \frac{\sum_j N_j A_j}{\sum_j N_j}, \qquad (10.16)$$

where $A_j \equiv m_j/m_H$. Similarly, for a completely ionized gas

$$\mu_i \simeq \frac{\sum_j N_j A_j}{\sum_j N_j(1 + z_j)}, \qquad (10.17)$$

where $z_j$ is the number of free electrons that result from completely ionizing an atom of type $j$.

Generally it is more useful to express the mean molecular weight in terms of mass ratios, known as **mass fractions**, rather than numbers of particles. Consequently, we will define the mass fractions of hydrogen ($X$), helium ($Y$), and metals ($Z$) to be[1]

$$X \equiv \frac{\text{total mass of hydrogen}}{\text{total mass of gas}}$$

$$Y \equiv \frac{\text{total mass of helium}}{\text{total mass of gas}}$$

$$Z \equiv \frac{\text{total mass of metals}}{\text{total mass of gas}}.$$

---

[1] As was already mentioned on page 248, since the primary components of most stellar gases are hydrogen and helium, all other constituents are frequently lumped together and referred to as *metals*. In certain applications, however, it is necessary to specify the composition in greater detail. In these cases each species is represented by its own mass fraction. For instance, in the next section we will find it useful to identify the combined mass fractions of carbon, nitrogen, and oxygen.

## 10.2 Pressure Equation of State

Clearly, $X + Y + Z = 1$. (Do not confuse $Z$ with $z_j$, which was used earlier to denote the number of electrons freed from a totally ionized atom of type $j$.)

By inverting the expression for $\overline{m}$, it is possible to write alternate equations for $\mu$ in terms of the mass fractions. Recalling that $\overline{m} = \mu m_H$, Eq. (10.15) for a neutral gas gives

$$\frac{1}{\mu_n m_H} = \frac{\sum_j N_j}{\sum_j N_j m_j}$$

$$= \frac{\text{total number of particles}}{\text{total mass of gas}}$$

$$= \sum_j \frac{\text{number of particles from } j}{\text{mass of particles from } j} \times \frac{\text{mass of particles from } j}{\text{total mass of gas}}$$

$$= \sum_j \frac{N_j}{N_j A_j m_H} X_j$$

$$= \sum_j \frac{1}{A_j m_H} X_j,$$

where $X_j$ is the mass fraction of atoms of type $j$. Solving for $1/\mu$, we have

$$\frac{1}{\mu_n} = \sum_j \frac{1}{A_j} X_j. \qquad (10.18)$$

Thus, for a neutral gas,

$$\frac{1}{\mu_n} \simeq X + \frac{1}{4}Y + \left\langle \frac{1}{A} \right\rangle_n Z. \qquad (10.19)$$

$\langle 1/A \rangle_n$ is a weighted average of all elements in the gas heavier than helium. For solar abundances, $\langle 1/A \rangle_n \sim 1/15.5$.

The mean molecular weight of a completely ionized gas may be determined in a similar way. It is necessary only to include the *total* number of particles contained in the sample, both nuclei and electrons. For instance, each hydrogen atom contributes one free electron, together with its nucleus, to the total number of particles. Similarly, one helium atom contributes two free electrons plus its nucleus to the total number of particles. Therefore, for a completely ionized gas, Eq. (10.18) becomes

$$\frac{1}{\mu_i} = \sum_j \frac{1 + z_j}{A_j} X_j. \qquad (10.20)$$

Including hydrogen and helium explicitly, we have

$$\frac{1}{\mu_i} \simeq 2X + \frac{3}{4}Y + \left\langle \frac{1+z}{A} \right\rangle_i Z. \tag{10.21}$$

For elements much heavier than helium, $1 + z_j \simeq z_j$, where $z_j \gg 1$ represents the number of protons (or electrons) in an atom of type $j$. It also holds that $A_j \simeq 2z_j$, the relation being based on the fact that sufficiently massive atoms have approximately the same number of protons and neutrons in their nuclei, and that protons and neutrons have very similar masses (see page 331). Thus

$$\left\langle \frac{1+z}{A} \right\rangle_i \simeq \frac{1}{2}.$$

If we assume that $X = 0.70$, $Y = 0.28$, and $Z = 0.02$, a composition typical of younger stars, then with these expressions for the mean molecular weight, $\mu_n = 1.30$ and $\mu_i = 0.62$.

It is also possible to combine the ideal gas law [in the form of Eq. (10.11)] with the pressure integral (Eq. 10.10) to find the average kinetic energy per particle. Equating, we see that

$$nkT = \frac{1}{3} \int_0^\infty mn_v v^2 \, dv.$$

This expression can be rewritten to give

$$\frac{1}{n} \int_0^\infty n_v v^2 \, dv = \frac{3kT}{m}.$$

However, the left-hand side of this expression is the integral average of $v^2$ weighted by the Maxwell–Boltzmann distribution function. Therefore

$$\overline{v^2} = \frac{3kT}{m}$$

or

$$\frac{1}{2} m \overline{v^2} = \frac{3}{2} kT. \tag{10.22}$$

It is worth noting that the factor of 3 arose from averaging particle velocities over the three coordinate directions (or *degrees of freedom*), which was performed on page 322. Thus the average kinetic energy of a particle is $\frac{1}{2}kT$ per degree of freedom.

As has already been mentioned, stellar environments exist where the assumptions of the ideal gas law do not hold. For instance, in the pressure integral it was assumed that the upper limit of integration for velocity was infinity. Of course, this cannot be the case since, from Einstein's special theory of relativity, the maximum possible value of velocity is $c$, the speed of light. Furthermore, the effects of quantum mechanics were not included in the derivation of the ideal gas law either. When the Heisenberg uncertainty principle and the Pauli exclusion principle are considered, a distribution function different from the Maxwell–Boltzmann distribution results. The **Fermi–Dirac** distribution function considers these important principles and leads to a very different pressure equation of state when applied to extremely dense matter such as that found in white dwarf stars and neutron stars. These exotic objects will be discussed in detail in Chapter 15. As mentioned in Section 5.4, particles such as electrons, protons, and neutrons that obey Fermi–Dirac statistics are called *fermions*.

Another statistical distribution function is obtained if it is assumed that the presence of some particles in a particular state enhances the likelihood of others being in the same state, an effect somewhat opposite to that of the Pauli exclusion principle. **Bose–Einstein** statistics has a variety of applications, including understanding the behavior of photons. Particles that obey Bose–Einstein statistics are known as *bosons*.

Just as special relativity and quantum mechanics must give classical results in the appropriate limits, Fermi–Dirac and Bose–Einstein statistics also approach the classical regime at sufficiently low densities and velocities. In these limits both distribution functions become indistinguishable from the classical Maxwell–Boltzmann distribution function.

Because photons inherently possess an amount of momentum $p_\gamma = h\nu/c$ (Eq. 5.5), they are capable of delivering an impulse to other particles during absorption or reflection. Consequently, electromagnetic radiation results in another form of pressure. It is instructive to rederive the expression for radiation pressure found in Chapter 9 by making use of the pressure integral. Substituting the speed of light for the velocity $v$, using the expression for photon momentum, and using an identity for the distribution function, $n_p \, dp = n_\nu \, d\nu$, the general pressure integral, Eq. (10.9), now describes the effect of radiation, giving

$$P_{\text{rad}} = \frac{1}{3} \int_0^\infty h\nu n_\nu \, d\nu.$$

At this point the problem again becomes finding an appropriate expression for $n_\nu \, d\nu$. Since photons are bosons, the Bose–Einstein distribution function would apply. However, the problem may also be solved by realizing that $n_\nu \, d\nu$ represents the number density of photons having frequencies lying in the range between $\nu$ and $\nu + d\nu$. Multiplying by the energy of each photon in that range would then give the *energy density* over the frequency interval, or

$$P_{\rm rad} = \frac{1}{3}\int_0^\infty u_\nu\,d\nu, \quad (10.23)$$

where $u_\nu\,d\nu = h\nu n_\nu\,d\nu$. But the energy density distribution function is found from the Planck function for blackbody radiation, Eq. (9.4),

$$u_\nu\,d\nu = \frac{8\pi h\nu^3}{c^3}\frac{d\nu}{e^{h\nu/kT}-1}. \quad (10.24)$$

Substituting Eq. (10.24) into Eq. (10.23) and performing the integration leads to

$$P_{\rm rad} = \frac{1}{3}aT^4, \quad (10.25)$$

where $a \equiv 4\sigma/c = 7.56591 \times 10^{-15}$ erg cm$^{-3}$ K$^{-4}$ is the radiation constant.

In many astrophysical situations the pressure due to the radiation can actually exceed the pressure produced by the gas by a significant amount. In fact it is possible that the magnitude of the force due to radiation pressure can become sufficiently great that it surpasses the gravitational force, resulting in an overall expansion of the system.

Combining both the ideal gas and radiation pressure terms, the total pressure becomes

$$P_t = \frac{\rho kT}{\mu m_H} + \frac{1}{3}aT^4. \quad (10.26)$$

---

**Example 10.2** Using the results of Example 10.1, it is possible to estimate the central temperature of the Sun. Neglecting the radiation pressure term, the central temperature is found from the ideal gas law equation of state to be

$$T_c = \frac{P_c \mu m_H}{\rho k}.$$

Using $\bar{\rho}_\odot$, a value of $\mu_i = 0.62$ appropriate for complete ionization,[2] and the estimated value for the central pressure, we find that

$$T_c \sim 1.44 \times 10^7 \text{ K}$$

which is in fairly reasonable agreement with more detailed calculations. One standard solar model gives a central temperature of $1.58 \times 10^7$ K. At this temperature, the pressure due to radiation is $1.57 \times 10^{14}$ dynes cm$^{-2}$, 0.06% of the gas pressure.

---

[2]Since, as we will see in the next chapter, the Sun has already converted a significant amount of its core hydrogen into helium via nuclear reactions, the actual value of $\mu_i$ is closer to 0.852.

## 10.3 Stellar Energy Sources

As we have already seen, the rate of energy output of stars (their luminosities) is very large. However, the question of the source of that energy has not yet been addressed. Clearly, one measure of the lifetime of a star must be related to how long it can sustain its power output.

One likely source of stellar energy is gravitational potential energy. Recall that the gravitational potential energy of a system of two particles is given by Eq. (2.14),

$$U = -G\frac{Mm}{r}.$$

As the distance between $M$ and $m$ diminishes, the gravitational potential energy becomes *more negative*, implying that energy must have been converted to other forms, such as kinetic energy. If a star can manage to convert its gravitational potential energy into heat, and then radiate that heat into space, the star may be able to shine for a significant period of time. However, we must also remember that by the virial theorem (Eq. 2.46) the total energy of a system of particles in equilibrium is one-half of the system's potential energy. Therefore only one-half of the change in gravitational potential energy of a star is actually available to be radiated away; the remaining potential energy supplies the thermal energy that heats the star.

Calculating the gravitational potential energy of a star requires consideration of the interaction between every possible pair of particles. This is not as difficult as might first seem. The gravitational force on a point mass $dm_i$ located outside of a spherically symmetric mass $M_r$ is

$$dF_{g,i} = G\frac{M_r \, dm_i}{r^2}$$

and is directed toward the center of the sphere. This force is one that would exist if all of the mass of the sphere were located at its center, a distance $r$ from the point mass. This immediately implies that the gravitational potential energy of the point mass is

$$dU_{g,i} = -G\frac{M_r \, dm_i}{r}.$$

Rather than considering an individual point mass, if we assume that point masses are distributed uniformly within a shell of thickness $dr$ and mass $dm$ (where $dm$ is the sum of all the point masses $dm_i$), then

$$dm = 4\pi r^2 \rho \, dr,$$

where $\rho$ is the mass density of the shell and $4\pi r^2\, dr$ is its volume. Thus

$$dU_g = -G\frac{M_r 4\pi r^2 \rho}{r}\, dr.$$

Integrating over all mass shells from the center of the star to the surface, its total gravitational potential energy becomes

$$U_g = -4\pi G \int_0^R M_r \rho r\, dr, \qquad (10.27)$$

where $R$ is the radius of the star.

Exact calculation of $U_g$ requires knowledge of how $\rho$, and consequently $M_r$, depends on $r$. Nevertheless, an approximate value can be obtained by assuming that $\rho$ is constant and equal to its average value, or

$$\rho \sim \overline{\rho} = \frac{M}{\frac{4}{3}\pi R^3},$$

$M$ being the total mass of the star. Now we may also approximate $M_r$ as

$$M_r \sim \frac{4}{3}\pi r^3 \overline{\rho}.$$

If we substitute into Eq. (10.27), the total gravitational potential energy becomes

$$U_g \sim -\frac{16\pi^2}{15} G \overline{\rho}^2 R^5 \sim -\frac{3}{5}\frac{GM^2}{R}. \qquad (10.28)$$

Lastly, applying the virial theorem, the total mechanical energy of the star is

$$E \sim -\frac{3}{10}\frac{GM^2}{R}. \qquad (10.29)$$

---

**Example 10.3** If the Sun were originally much larger than it is today, how much energy would have been liberated in its gravitational collapse? Assuming that its original radius was $R_i$, where $R_i \gg 1\, R_\odot$, then the energy radiated away during collapse would be

$$\Delta E_g = -(E_f - E_i) \simeq -E_f \simeq \frac{3}{10}\frac{GM_\odot^2}{R_\odot} \simeq 1.1 \times 10^{48}\ \text{ergs}.$$

Assuming also that the luminosity of the Sun has been roughly constant throughout its lifetime, then it could emit energy at that rate for approximately

$$t_{\rm KH} = \frac{\Delta E_g}{L_\odot} \qquad (10.30)$$

$$\sim 10^7\ \text{yr}.$$

## 10.3 Stellar Energy Sources

$t_{KH}$ is known as the **Kelvin–Helmholtz** time scale. Based on radioactive dating techniques, however, the estimated age of rocks on the Moon's surface is over $4 \times 10^9$ yr. It seems unlikely that the age of the Sun is less than the age of the Moon! Therefore, gravitational potential energy alone cannot account for the Sun's luminosity throughout its lifetime. As we shall see in later chapters, however, gravitational energy can play an important role during some phases of the evolution of stars.

---

Another possible energy source involves chemical processes. However, since chemical reactions are based on the interactions of orbital electrons in atoms, the amount of energy available to be released per atom is not likely to exceed more than one to ten electron volts, typical of the atomic energy levels in hydrogen and helium (see Section 5.3). Given the number of atoms present in a star, the amount of chemical energy available is also far too low to account for the Sun's luminosity over a reasonable period of time (Problem 10.3).

The nuclei of atoms may also be considered as sources of energy. Whereas electron orbits involve energies in the electron volt (eV) range, nuclear processes generally involve energies millions of times larger (MeV). Just as chemical reactions can result in the transformation of atoms into molecules or one kind of molecule into another, nuclear reactions change one type of nucleus into another.

The nucleus of a particular **element** is specified by the number of protons, $Z$, it contains (not to be confused with the mass fraction of metals), with each proton carrying a charge of $+e$. Obviously, in a neutral atom the number of protons must exactly equal the number of orbital electrons. An **isotope** of a given element is identified by the number of neutrons, $N$, in the nucleus, with neutrons being electrically neutral as the name implies. (All isotopes of a given element have the same number of protons.) Collectively, protons and neutrons are referred to as **nucleons**, the number of nucleons in a particular isotope being $A = Z + N$. Since protons and neutrons have very similar masses and greatly exceed the mass of electrons, $A$ is a good indication of the mass of the isotope and is often referred to as the *mass number*.[3] The masses of the proton, neutron, and electron are, respectively,

$$m_p = 1.672623 \times 10^{-24} \text{ g}$$

$$m_n = 1.674929 \times 10^{-24} \text{ g}$$

$$m_e = 9.109390 \times 10^{-28} \text{ g}.$$

---

[3] The quantity $A_j$ defined on page 324 is approximately equal to the mass number.

It is often convenient to express the masses of nuclei in terms of *atomic mass units*; 1 u = $1.660540 \times 10^{-24}$ g, exactly one-twelfth the mass of the isotope carbon-12. The masses of nuclear particles are also frequently expressed in terms of their rest mass energies, in units of MeV. Using Einstein's $E = mc^2$, we find 1 u = 931.49432 MeV/$c^2$. When masses are expressed simply in terms of rest mass energies, as is often the case, the factor of $c^2$ is implicitly assumed.

The simplest isotope of hydrogen is composed of one proton and one electron and has a mass of $m_H$ = 1.007825 u. This mass is actually *less* than the combined masses of the proton and electron taken separately. In fact, the exact mass difference is 13.6 eV if the atom is in its ground state, which is just its ionization potential. Since mass and energy are equivalent, and the total mass–energy of the system must be conserved, any loss in energy when the electron and proton combine to form an atom must come at the expense of a loss in total mass. This is the **binding energy** of the atom. Thus, the binding energy is just the amount of energy that would need to be added to break the nucleus apart into its constituent protons and neutrons.

Energy is also released with an accompanying loss in mass when nucleons are combined to form atomic nuclei. A helium nucleus, composed of two protons and two neutrons, can be formed by a series of nuclear reactions originally involving four hydrogen nuclei (i.e., 4H → He + low mass remnants). Such reactions are known as **fusion** reactions, since lighter particles are "fused" together to form a heavier particle.[4] The total mass of the four hydrogen atoms is 4.031280 u while the mass of one helium atom is $m_{He}$ = 4.002603 u. Neglecting the contribution of the low-mass remnants, the combined mass of the hydrogen atoms *exceeds* the mass of the helium atom by $\Delta m$ = 0.028677 u, or 0.7%. Therefore, the binding energy of the helium nucleus is $E_b = \Delta m c^2$ = 26.71 MeV.

---

**Example 10.4** Is this source of nuclear energy sufficient to power the Sun during its lifetime? For simplicity, assume also that the Sun was originally 100% hydrogen and that only the inner 10% of the Sun's mass can be converted from hydrogen into helium.

Since 0.7% of the mass of hydrogen would be converted to energy in forming a helium nucleus, the amount of nuclear energy available in the Sun would be

$$E_{\text{nuclear}} = 0.1 \times 0.007 \times M_\odot c^2 = 1.3 \times 10^{51} \text{ ergs}.$$

---

[4]A **fission** reaction occurs when a massive nucleus is split into smaller fragments.

## 10.3 Stellar Energy Sources

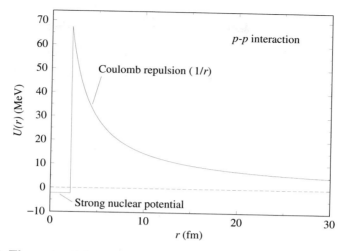

**Figure 10.4** The potential energy curve characteristic of nuclear reactions. The Coulomb repulsion between positive nuclei results in a barrier that is inversely proportional to the separation between nuclei and is proportional to the product of their charges. The nuclear potential well inside the nucleus is due to the attractive strong nuclear force.

This gives a **nuclear time scale** of approximately

$$t_{\text{nuclear}} = \frac{E_{\text{nuclear}}}{L_\odot} \tag{10.31}$$

$$\sim 10^{10} \text{ yr},$$

more than enough time to account for the age of Moon rocks.

---

Apparently, sufficient energy is available in the nuclei of atoms to provide a source for stellar luminosities, but can nuclear reactions actually occur in the interiors of stars? For a reaction to occur, the nuclei of atoms must collide, forming new nuclei in the process. However, all nuclei are positively charged, meaning that a Coulomb potential energy barrier must be overcome before contact can occur. Figure 10.4 shows the characteristic shape of the potential energy curve an atomic nucleus would experience when approaching another nucleus. The curve is composed of two parts: The portion outside of the nucleus is the potential energy that exists between two positively charged nuclei, and the portion inside the nucleus forms a *potential well* governed by the **strong nuclear force** that binds the nucleus together.[5]

If we assume that the energy required to overcome the Coulomb barrier is provided by the thermal energy of the gas, and that all nuclei are moving nonrelativistically, then the temperature $T_{\text{classical}}$ required to overcome the barrier can be estimated. Since all of the particles in the gas are in random motion, it is appropriate to refer to the relative velocity $v$ between two nuclei and their *reduced mass*, $\mu_m$, given by Eq. (2.23) (note that we are not referring here to the mean molecular weight, $\mu$). Equating the initial kinetic energy of the reduced mass to the potential energy of the barrier gives the position of the classical "turn-around point." Now, using Eq. (10.22),

$$\frac{1}{2}\mu_m \overline{v^2} = \frac{3}{2}kT_{\text{classical}} = \frac{Z_1 Z_2 e^2}{r},$$

where $T_{\text{classical}}$ denotes the temperature required for an average particle to overcome the barrier; $Z_1$ and $Z_2$ are the numbers of protons in each nucleus, and $r$ is their distance of separation.[6] Assuming that the radius of a typical nucleus is on the order of 1 fm,[7] the temperature needed to overcome the Coulomb potential energy barrier is approximately

$$T_{\text{classical}} = \frac{2 Z_1 Z_2 e^2}{3kr} \tag{10.32}$$

$$\sim 10^{10} \text{ K}$$

for a collision between two protons ($Z_1 = Z_2 = 1$). However, the central temperature of the Sun is only $1.58 \times 10^7$ K, much lower than required here. Even given the fact that the Maxwell–Boltzmann distribution indicates that a significant number of particles have speeds well in excess of the average speed of particles in the gas, and assuming that our determination of the Sun's central temperature may be somewhat in error, it is unlikely that a sufficient number of particles can overcome the Coulomb barrier to produce the reactions necessary to explain the solar luminosity, assuming classical physics alone.

As was mentioned in Section 5.4, quantum mechanics tells us that it is never possible to know both the position and the momentum of a particle

---

[5]The strong nuclear force is a very short range force that acts between all nucleons within the atom. It is an attractive force that dominates over the Coulomb repulsion between protons. If such a force did not exist, a nucleus would immediately fly apart.

[6]The cgs system of units is assumed in expressing the Coulomb potential energy barrier; $U_e = k_C Z_1 Z_2 e^2 / r$, where $k_C \equiv 1$.

[7]1 fm $\equiv 1 \times 10^{-13}$ cm.

## 10.3 Stellar Energy Sources

to unlimited accuracy. The Heisenberg uncertainty principle states that the uncertainties in position and momentum are related by

$$\Delta x \Delta p_x \geq \frac{\hbar}{2}.$$

The uncertainty in the position of one proton colliding with another may be so large that even though the kinetic energy of the collision is insufficient to overcome the classical Coulomb barrier, one proton might nevertheless find itself within the central potential well defined by the strong force of the other. This quantum mechanical tunneling has no classical counterpart (recall the discussion in Section 5.4). Of course, the greater the ratio of the potential energy barrier height to the particle's kinetic energy or the wider the barrier, the less likely tunneling becomes.

To estimate the effect of tunneling on the temperature necessary to sustain nuclear reactions, assume that a proton must be within approximately one de Broglie wavelength of its target in order to tunnel through the Coulomb barrier. Recalling that the wavelength of a massive particle is given by $\lambda = h/p$ (Eq. 5.17), rewriting the kinetic energy in terms of momentum

$$\frac{1}{2}\mu_m v^2 = \frac{p^2}{2\mu_m},$$

and setting the distance of closest approach equal to one wavelength (where the potential energy barrier height is equal to the original kinetic energy) gives

$$\frac{Z_1 Z_2 e^2}{\lambda} = \frac{p^2}{2\mu_m} = \frac{(h/\lambda)^2}{2\mu_m}.$$

Solving for $\lambda$ and substituting $r = \lambda$ into Eq. (10.32), the quantum mechanical estimate of temperature becomes

$$T_{\text{quantum}} = \frac{4}{3} \frac{\mu_m Z_1^2 Z_2^2 e^4}{k h^2}. \tag{10.33}$$

Again assuming the collision of two protons, $\mu_m = m_p/2$ and $Z_1 = Z_2 = 1$. Substituting, we find that $T_{\text{quantum}} \sim 10^7$ K. In this case, if we assume the effects of quantum mechanics, the temperature required for nuclear reactions is consistent with the estimated central temperature of the Sun.

Now that the possibility of a nuclear energy source has been established, a more detailed description of nuclear *reaction rates* is needed in order to apply them to the development of stellar models. For instance, not all particles in a gas of temperature $T$ will have sufficient kinetic energy and the necessary wavelength to tunnel through the Coulomb barrier successfully. Consequently, the reaction rate per energy interval must be described in terms of the number density of particles having energies within a specific range, and the probability that those particles can actually tunnel through the Coulomb barrier of the target nucleus. The total nuclear reaction rate is then integrated over all possible energies.

First consider the number density of nuclei within a specified energy interval. As we have seen, the Maxwell–Boltzmann distribution (Eq. 8.1) relates the number density of particles with velocities between $v$ and $v+dv$ to the temperature of the gas. Assuming that particles are initially sufficiently far apart that the potential energy may be neglected, the nonrelativistic[8] kinetic energy relation describes the total energy of the particles, or $K = E = \mu_m v^2/2$. Solving for the velocity and substituting, the Maxwell–Boltzmann distribution may be written in terms of the number of particles with kinetic energies between $E$ and $E + dE$ as

$$n_E\, dE = \frac{2n}{\pi^{1/2}} \frac{1}{(kT)^{3/2}} E^{1/2} e^{-E/kT}\, dE. \qquad (10.34)$$

Equation (10.34) gives the number of particles per unit volume that have energies in the range $dE$, but it does not describe the probability that particles will actually interact. To account for this factor, the idea of a **cross section** is introduced. Define the cross section $\sigma(E)$ to be the number of reactions per target nucleus per unit time, divided by the flux of incident particles, or

$$\sigma(E) \equiv \frac{\text{number of reactions/nucleus/time}}{\text{number of incident particles/area/time}}. \qquad (10.35)$$

Although $\sigma(E)$ is strictly a measure of probability, as mentioned in Chapter 9, it can *roughly* be thought of as the cross-sectional area of the target particle; any incoming particle that strikes within that area, centered on the target, will result in a nuclear reaction.

To find the reaction rate in units of reactions volume$^{-1}$ time$^{-1}$, consider the number of particles that will hit a target of cross-sectional area $\sigma(E)$, assuming that all of the incident particles are moving in one direction. Let $x$ denote a target particle and $i$ denote an incident particle. If the number of incident particles per unit volume having energies between $E$ and $E + dE$ is $n_{iE}\, dE$, then the number of reactions, $dN_E$, is the number of particles that can strike $x$ in a time interval $dt$ with a velocity $v(E) = \sqrt{2E/\mu_m}$. The number

---

[8]In astrophysical processes, nuclei are usually nonrelativistic, except in the extreme environment of neutron stars. Because of the much smaller masses of electrons, it cannot be assumed that they are also nonrelativistic, however.

## 10.3 Stellar Energy Sources

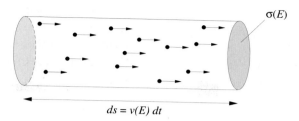

**Figure 10.5** The number of reactions per unit time between particles of type $i$ and a target $x$ of cross section $\sigma(E)$ may be thought of in terms of the number of particles in a cylinder of cross-sectional area $\sigma(E)$ and length $ds = v(E)\,dt$ that will reach the target in a time interval $dt$.

of incident particles is just the number contained within a cylinder of volume $\sigma(E)v(E)\,dt$ (see Fig. 10.5), or

$$dN_E = \sigma(E)v(E)n_{iE}\,dE\,dt.$$

Now, the number of incident particles per unit volume with the appropriate velocity (or kinetic energy) is some fraction of the total number of particles in the sample,

$$n_{iE}\,dE = \frac{n_i}{n}\,n_E\,dE,$$

where $n_i = \int_0^\infty n_{iE}\,dE$, $n = \int_0^\infty n_E\,dE$, and $n_E\,dE$ is given by Eq. (10.34). Therefore, the number of reactions per target nucleus per time interval $dt$ having energies between $E$ and $E + dE$ is

$$\frac{\text{reactions per nucleus}}{\text{time interval}} = \frac{dN_E}{dt} = \sigma(E)v(E)\,\frac{n_i}{n}\,n_E\,dE.$$

Finally, if there are $n_x$ targets per unit volume, the total number of reactions per unit volume per unit time, integrated over all possible energies, is

$$r_{ix} = \int_0^\infty n_x n_i \sigma(E) v(E) \,\frac{n_E}{n}\,dE. \tag{10.36}$$

To evaluate Eq. (10.36) we must know the functional form of $\sigma(E)$. Unfortunately, in general $\sigma(E)$ is quite complicated, changing rapidly with energy, and often only approximate expressions are available. It is also important to compare $\sigma(E)$ with experimental data. However, stellar thermal energies are quite low compared to energies found in laboratory experimentation, and significant extrapolation is usually required to obtain comparison data for stellar nuclear reaction rates.

The process of determining $\sigma(E)$ can be improved somewhat if the terms most strongly dependent on energy are factored out first. We have already suggested that the cross section can be roughly thought of as being a physical area. Moreover, the size of a nucleus, measured in terms of its ability to "touch" target nuclei, is approximately one de Broglie wavelength in radius ($r \sim \lambda$). Combining these ideas, the cross section of the nucleus $\sigma(E)$ should be proportional to

$$\sigma(E) \propto \pi \lambda^2 \propto \pi \left(\frac{h}{p}\right)^2 \propto \frac{1}{E}.$$

To obtain the last expression, we have again used the nonrelativistic relation, $K = E = \mu_m v^2/2 = p^2/2\mu_m$.

We have also mentioned previously that the ability to tunnel through the Coulomb barrier is related to the ratio of the barrier height to the initial kinetic energy of the incoming nucleus, a factor that must be considered in the cross section. If the barrier height $U_c$ is zero, the probability of successfully penetrating it necessarily equals one (100%). As the barrier height increases, the probability of penetration must decrease, asymptotically approaching zero as the potential energy barrier height goes to infinity. In fact, the tunneling probability is exponential in nature. Since $\sigma(E)$ must be related to the tunneling probability, we have

$$\sigma(E) \propto e^{-2\pi^2 U_c/E}. \tag{10.37}$$

The factor of $2\pi^2$ arises from the strict quantum mechanical treatment of the problem. Again assuming that $r \sim \lambda = h/p$, taking the ratio of the barrier potential height $U_c$ to particle kinetic energy $E$ gives

$$\frac{U_c}{E} = \frac{Z_1 Z_2 e^2/r}{\mu_m v^2/2} = \frac{2Z_1 Z_2 e^2}{hv}.$$

After some manipulation, we find that

$$\sigma(E) \propto e^{-bE^{-1/2}}, \tag{10.38}$$

where

$$b \equiv \frac{2^{3/2} \pi^2 \mu_m^{1/2} Z_1 Z_2 e^2}{h}.$$

Clearly, $b$ depends on the composition of the gas.

Combining the previous results[9] and defining $S(E)$ to be some (hopefully)

---

[9]The relative angular momenta of the interacting particles also plays a role in nuclear reaction rates, but it is generally a minor component for reactions of astrophysical significance. Consequently, we will not consider the effects of angular momentum in our present arguments.

## 10.3 Stellar Energy Sources

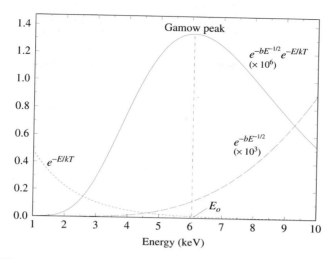

**Figure 10.6** The Gamow peak arises from the contribution of the $e^{-E/kT}$ Maxwell–Boltzmann high-energy tail and the $e^{-bE^{-1/2}}$ Coulomb barrier penetration term. This curve represents the collision of two protons at the central temperature of the Sun. Note that $e^{bE^{-1/2}}$ and $e^{-bE^{-1/2}}e^{-E/kT}$ have been multiplied by $10^3$ and $10^6$, respectively, to more readily illustrate the functional dependence on energy.

slowly varying function of energy, the cross section may now be expressed as

$$\sigma(E) = \frac{S(E)}{E} e^{-bE^{-1/2}}. \tag{10.39}$$

Substituting Eqs. (10.34) and (10.39) into Eq. (10.36) and simplifying, the reaction rate integral becomes

$$r_{ix} = \left(\frac{2}{kT}\right)^{3/2} \frac{n_i n_x}{(\mu_m \pi)^{1/2}} \int_0^\infty S(E) e^{-bE^{-1/2}} e^{-E/kT}\, dE. \tag{10.40}$$

In Eq. (10.40), the term $e^{-E/kT}$ represents the high-energy wing of the Maxwell–Boltzmann distribution and the term $e^{-bE^{-1/2}}$ comes from the penetration probability. As can be seen in Fig. 10.6, the product of these two factors produces a strongly peaked curve, known as the **Gamow peak**, after George Gamow (1904–1968), the physicist who first investigated Coulomb barrier penetration. The top of the curve occurs at the energy

$$E_o = \left(\frac{bkT}{2}\right)^{2/3}. \tag{10.41}$$

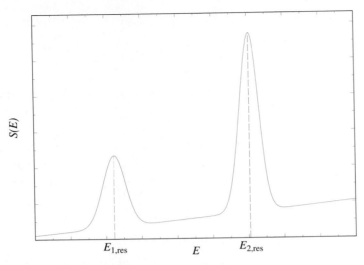

**Figure 10.7** The effect of resonance on $S(E)$.

As a result of the Gamow peak, the greatest contribution to the reaction rate integral comes in a fairly narrow energy band that depends on the temperature and composition of the gas.

Assuming that $S(E)$ is indeed slowly varying across the Gamow peak, it may be approximated by its value at $E_\circ$ [$S(E) \simeq S(E_\circ) =$ constant] and removed from inside of the integral. Also, it is generally much easier to extrapolate laboratory results if they are expressed in terms of $S(E)$.

In some cases, however, $S(E)$ can vary quite rapidly, peaking at specific energies, as in Fig. 10.7. These energies correspond to energy levels within the nucleus, analogous to the orbital energy levels of electrons. It is a *resonance* between the energy of the incoming particle and differences in energy levels within the nucleus that accounts for these strong peaks. A detailed discussion of these resonance peaks is beyond the scope of this book.[10]

Yet another factor influencing reaction rates is **electron screening**. On average, the electrons liberated when atoms are ionized at the high temperatures of stellar interiors produce a "sea" of negative charge that partially hides the target nucleus, reducing its *effective* positive charge. The result of this reduced positive charge is a lower Coulomb barrier to the incoming nucleus and an enhanced reaction rate. By including electron screening, the effective

---

[10] See pp. 348–357 of Clayton (1968) for an excellent and detailed discussion of resonance peaks.

## 10.3 Stellar Energy Sources

Coulomb potential becomes

$$U_{\text{eff}} = \frac{Z_1 Z_2 e^2}{r} + U_s(r),$$

where $U_s(r) < 0$ is the electron screening contribution. Electron screening can be significant, sometimes enhancing the helium-producing reactions by 10% to 50%.

It is often illuminating to write the complicated reaction rate equations in the form of a power law, centered in a particular temperature range. Neglecting the screening factor, in the case of a two-particle interaction, the reaction rate would become

$$r_{ix} \simeq r_\circ X_i X_x \rho^{\alpha'} T^\beta, \tag{10.42}$$

where $r_\circ$ is a constant, $X_i$ and $X_x$ are the mass fractions of the two particles, and $\alpha'$ and $\beta$ are determined from the power law expansion of the reaction rate equations. Usually $\alpha' = 2$ for a two-body collision and $\beta$ can range from near unity to 40 or more.

By combining the reaction rate equation with the amount of energy released per reaction, it is now possible to calculate the amount of energy released per second in each gram of stellar material. If $\mathcal{E}_\circ$ is the amount of energy released per reaction, the amount of energy liberated per gram of material per second becomes

$$\epsilon_{ix} = \left(\frac{\mathcal{E}_\circ}{\rho}\right) r_{ix}, \tag{10.43}$$

or, rewriting in the form of a power law,

$$\epsilon_{ix} = \epsilon'_\circ X_i X_x \rho^\alpha T^\beta, \tag{10.44}$$

where $\alpha = \alpha' - 1$. $\epsilon_{ix}$ has units of ergs g$^{-1}$ s$^{-1}$ and the sum of $\epsilon_{ix}$ for all reactions is the total nuclear energy generation rate. This form of the nuclear energy generation rate will be used later to show the dependence of energy production on temperature and density for several reaction sequences typically operating in stellar interiors.

To determine the luminosity of a star, it is now necessary to consider all of the energy generated by stellar material. The contribution to the total luminosity due to an infinitesimal mass $dm$ is simply

$$dL = \epsilon \, dm,$$

where $\epsilon$ is the *total* energy released per gram per second by all nuclear reactions and by gravity, or $\epsilon = \epsilon_{\text{nuclear}} + \epsilon_{\text{gravity}}$. It is worth noting that $\epsilon_{\text{gravity}}$ could be negative if the star is expanding, a point to be discussed later. For a spherically symmetric star, the mass of a thin shell of thickness $dr$ is just $dm = \rho\, dV = 4\pi r^2 \rho\, dr$. Substituting and dividing by the shell thickness, we have

$$\frac{dL_r}{dr} = 4\pi r^2 \rho \epsilon, \tag{10.45}$$

where $L_r$ is the *interior luminosity* due to all of the energy generated within the star interior to the radius $r$. Equation (10.45) is another of the fundamental stellar structure equations.

The remaining problem in understanding nuclear reactions is the exact sequence of steps by which one element is converted into another. The estimate of the nuclear time scale for the Sun was based on the assumption that four hydrogen nuclei are converted into helium. However, it is highly unlikely that this could occur via a four-body collision (i.e., all nuclei hitting simultaneously). For the process to occur, the final product must be created by a chain of reactions, each involving a much more probable two-body interaction. In fact, we derived the reaction rate equation under the assumption that only two nuclei would collide at any one time.

The process by which a chain of nuclear reactions leads to the final product cannot happen in a completely arbitrary way, however; a series of particle *conservation laws* must be obeyed. In particular, during every reaction it is necessary to conserve electric charge, the number of nucleons, and the number of leptons. The term **lepton** means a "light thing" and includes electrons, positrons, neutrinos, and antineutrinos.

Although antimatter is extremely rare in comparison with matter, it plays an important role in subatomic physics, including nuclear reactions. Antimatter particles are identical to their matter counterparts but have opposite attributes, such as electric charge. Antimatter also has the characteristic (often used in science fiction) that a collision with its matter counterpart results in complete annihilation of both particles, accompanied by the production of energetic photons. For instance

$$e^- + e^+ \rightarrow 2\gamma,$$

where $e^-$, $e^+$, and $\gamma$ denote an electron, positron, and photon, respectively. Note that two photons are required to conserve momentum and energy simultaneously.

Neutrinos and antineutrinos (symbolized by $\nu$ and $\bar{\nu}$, respectively) are an interesting class of particles in their own right and will be discussed often in the remainder of this text. They were originally proposed in 1934 by Enrico

## 10.3 Stellar Energy Sources

Fermi (1901–1954), an Italian physicist, as a way to conserve energy and momentum in certain reaction processes. Neutrinos are electrically neutral and have a very small, if not zero, mass.[11] One of a neutrino's interesting characteristics is its extremely small cross section for interactions with other matter, making it very difficult to detect. Typically $\sigma_\nu \sim 10^{-44}$ cm$^2$, implying that at densities common to stellar interiors, their mean free path—the average distance they would travel before interacting with other particles—is on the order of $10^{20}$ cm or nearly $10^9$ $R_\odot$! After being produced in the interior, neutrinos almost always leave without interacting with the star. One exception to this transparency of stellar material to neutrinos occurs during a supernova explosion with important consequences, as will be discussed in Chapter 13.

Since electrons and positrons have charges equal in magnitude to that of a proton, these leptons can be responsible for a portion of the charge conservation requirement while their total **lepton numbers** are also conserved. Note that in counting the number of leptons involved in a nuclear reaction, it is necessary to treat matter and antimatter differently. Actually the total number of matter leptons *minus* the total number of antimatter leptons must remain a constant.

To assist in counting the number of nucleons and the total electric charge, nuclei will be represented by the symbol

$$^A_Z X,$$

where X is the chemical symbol of the element (H for hydrogen, He for helium, etc.), Z is the number of protons (the total positive charge, in units of $e$), and A is the mass number (the total number of nucleons, protons plus neutrons).

Applying the conservation laws, one chain of reactions that can convert hydrogen into helium is the first **proton–proton chain** (PP I). It involves a reaction sequence that ultimately results in

$$4\,^1_1\text{H} \rightarrow\, ^4_2\text{He} + 2e^+ + 2\nu_e + 2\gamma$$

through the intermediate production of deuterium ($^2_1$H) and helium-3. The entire **PP I** reaction chain is

$$^1_1\text{H} + {}^1_1\text{H} \rightarrow {}^2_1\text{H} + e^+ + \nu_e$$
$$^2_1\text{H} + {}^1_1\text{H} \rightarrow {}^3_2\text{He} + \gamma \qquad (10.46)$$
$$^3_2\text{He} + {}^3_2\text{He} \rightarrow {}^4_2\text{He} + 2\,^1_1\text{H}.$$

---

[11] At the time this text was written, the best estimate for the upper limit to the mass of an electron neutrino was 7.2 eV/$c^2$.

Each step of the PP I chain has its own reaction rate, since different Coulomb barriers and cross sections are involved. The slowest step in the sequence is the initial one, occurring because it involves the *decay* of a proton into a neutron via $p^+ \to n + e^+ + \nu_e$. Such a decay involves the **weak force**, another of the four known forces.[12]

The production of helium-3 nuclei in the PP I chain also provides for the possibility of their interaction directly with helium-4 nuclei, resulting in a second branch of the proton–proton chain. In an environment characteristic of the center of the Sun, 69% of the time a helium-3 interacts with another helium-3 in the PP I chain while 31% of the time the **PP II** chain occurs:

$$\begin{aligned} {}^3_2\text{He} + {}^4_2\text{He} &\to {}^7_4\text{Be} + \gamma \\ {}^7_4\text{Be} + e^- &\to {}^7_3\text{Li} + \nu_e \\ {}^7_3\text{Li} + {}^1_1\text{H} &\to 2\,{}^4_2\text{He}. \end{aligned} \quad (10.47)$$

Yet another branch, the **PP III** chain, is possible because the capture of an electron by the beryllium-7 nucleus in the PP II chain is in competition with the capture of a proton (a proton is captured only 0.3% of the time in the center of the Sun):

$$\begin{aligned} {}^7_4\text{Be} + {}^1_1\text{H} &\to {}^8_5\text{B} + \gamma \\ {}^8_5\text{B} &\to {}^8_4\text{Be} + e^+ + \nu_e \\ {}^8_4\text{Be} &\to 2\,{}^4_2\text{He}. \end{aligned} \quad (10.48)$$

The three branches of the proton–proton (pp) chain, along with their branching ratios, are summarized in Fig. 10.8.

Beginning with Eq. (10.40), the nuclear energy generation rate for the combined pp chain is calculated to be

$$\epsilon_{pp} = 2.38 \times 10^6 \rho X^2 f_{pp} \psi_{pp} C_{pp} T_6^{-2/3} e^{-33.80 T_6^{-1/3}} \text{ ergs g}^{-1} \text{ s}^{-1}, \quad (10.49)$$

where $T_6$ is a dimensionless expression of temperature in units of $10^6$ K (or $T_6 \equiv T/10^6$ K). $f_{pp} = f_{pp}(X, Y, \rho, T) \simeq 1$ is the pp chain screening factor, $\psi_{pp} = \psi_{pp}(X, Y, T) \simeq 1$ is a correction factor that accounts for the simultaneous

---

[12]Each of the four forces has now been mentioned: the gravitational force, which involves all particles with mass–energy; the electromagnetic force associated with photons and electric charge; the strong force that binds nuclei together; and the weak force of radioactive beta (electron/positron) decay.

## 10.3 Stellar Energy Sources

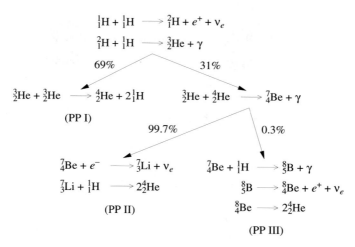

**Figure 10.8** The three branches of the pp chain, along with the branching ratios appropriate for conditions in the core of the Sun.

occurrence of PP I, PP II, and PP III, and $C_{pp} \simeq 1$ involves higher-order correction terms.

When written as a power law (e.g., Eq. 10.44) near $T = 1.5 \times 10^7$ K, the energy generation rate has the form

$$\epsilon_{pp} \simeq \epsilon'_{0,pp} \rho X^2 \psi_{pp} f_{pp} T_6^4, \tag{10.50}$$

where $\epsilon'_{0,pp} = 1.07 \times 10^{-5}$ erg cm$^3$ g$^{-2}$ s$^{-1}$. The power law form of the energy generation rate demonstrates a relatively modest temperature dependence of $T^4$ near $T_6 = 15$.

A second, independent cycle also exists for the production of helium-4 from hydrogen. This cycle was proposed by Hans Bethe in 1938, just six years after the discovery of the neutron. In the **CNO cycle**, carbon, nitrogen, and oxygen are used as catalysts, being consumed and then regenerated during the process. Just as with the pp chain, the CNO cycle has competing branches. The first branch culminates with the production of carbon-12 and helium-4:

$$\begin{aligned}
{}^{12}_{6}\text{C} + {}^{1}_{1}\text{H} &\to {}^{13}_{7}\text{N} + \gamma \\
{}^{13}_{7}\text{N} &\to {}^{13}_{6}\text{C} + e^+ + \nu_e \\
{}^{13}_{6}\text{C} + {}^{1}_{1}\text{H} &\to {}^{14}_{7}\text{N} + \gamma \\
{}^{14}_{7}\text{N} + {}^{1}_{1}\text{H} &\to {}^{15}_{8}\text{O} + \gamma \\
{}^{15}_{8}\text{O} &\to {}^{15}_{7}\text{N} + e^+ + \nu_e \\
{}^{15}_{7}\text{N} + {}^{1}_{1}\text{H} &\to {}^{12}_{6}\text{C} + {}^{4}_{2}\text{He}.
\end{aligned} \tag{10.51}$$

The second branch occurs only about 0.04% of the time, and arises when the final reaction in Eq. (10.51) produces oxygen-16 and a photon, rather than carbon-12 and helium-4:

$$\begin{aligned}{}^{15}_{7}\text{N} + {}^{1}_{1}\text{H} &\rightarrow {}^{16}_{8}\text{O} + \gamma \\ {}^{16}_{8}\text{O} + {}^{1}_{1}\text{H} &\rightarrow {}^{17}_{9}\text{F} + \gamma \\ {}^{17}_{9}\text{F} &\rightarrow {}^{17}_{8}\text{O} + e^{+} + \nu_{e} \\ {}^{17}_{8}\text{O} + {}^{1}_{1}\text{H} &\rightarrow {}^{14}_{7}\text{N} + {}^{4}_{2}\text{He}. \end{aligned} \quad (10.52)$$

The energy generation rate for the CNO cycle is given by

$$\epsilon_{\text{CNO}} = 8.67 \times 10^{27} \rho X X_{\text{CNO}} C_{\text{CNO}} T_6^{-2/3} e^{-152.28 T_6^{-1/3}} \text{ ergs g}^{-1} \text{ s}^{-1}, \quad (10.53)$$

where $X_{\text{CNO}}$ is the total mass fraction of carbon, nitrogen, and oxygen and $C_{\text{CNO}}$ is a higher-order correction term. When written as a power law centered about $T = 1.5 \times 10^7$ K (see Eq. 10.44), this energy equation becomes

$$\epsilon_{\text{CNO}} \simeq \epsilon'_{\text{o,CNO}} \rho X X_{\text{CNO}} T_6^{19.9}, \quad (10.54)$$

where $\epsilon'_{\text{o,CNO}} = 8.24 \times 10^{-24}$ erg cm$^3$ g$^{-2}$ s$^{-1}$. Apparently, the CNO cycle is much more strongly temperature-dependent than is the pp chain. This property implies that low-mass stars, which have smaller central temperatures, are dominated by the pp chains during their "hydrogen burning" evolution while more massive stars, with their higher central temperatures, convert hydrogen to helium by the CNO cycle. The transition in stellar mass between stars dominated by the pp chain and those dominated by the CNO cycle occurs for stars slightly more massive than our Sun. This difference in nuclear reaction processes plays an important role in the structure of stellar interiors, as will be seen in the next section.

When hydrogen is converted into helium by either the pp chain or the CNO cycle, the mean molecular weight $\mu$ of the gas increases. If neither the temperature or density of the gas changes, the ideal gas law predicts that the central pressure would necessarily decrease. As a result, the star would no longer be in hydrostatic equilibrium and would begin to collapse. This collapse has the effect of actually raising both the temperature and the density to compensate for the increase in $\mu$. Equation (10.33) suggests that helium nuclei can overcome their Coulomb repulsion and begin to "burn" when the temperature rises by a factor of approximately 64 above what is necessary to burn hydrogen. This process becomes progressively more likely as the amount of helium increases.

## 10.3 Stellar Energy Sources

The reaction sequence by which helium is converted into carbon is known as the **triple alpha process**. The process takes its name from the historical result that the mysterious alpha particles detected in some types of radioactive decay were shown by Rutherford to be helium-4 nuclei. The triple alpha process is

$$\,^{4}_{2}\text{He} + \,^{4}_{2}\text{He} \rightleftharpoons \,^{8}_{4}\text{Be} \tag{10.55}$$

$$\,^{8}_{4}\text{Be} + \,^{4}_{2}\text{He} \rightarrow \,^{12}_{6}\text{C} + \gamma.$$

In the triple alpha process, the first step produces an unstable beryllium nucleus that will rapidly decay back into two separate helium nuclei if not immediately struck by another alpha particle. As a result, this reaction may be thought of as a three-body interaction and therefore, the reaction rate depends on $(\rho Y)^3$. The nuclear energy generation rate is given by

$$\epsilon_{3\alpha} = 5.09 \times 10^{11} \rho^2 Y^3 T_8^{-3} f_{3\alpha} e^{-44.027 T_8^{-1}} \quad \text{ergs g}^{-1} \text{ s}^{-1}, \tag{10.56}$$

where $T_8 \equiv T/10^8$ K and $f_{3\alpha}$ is the screening factor for the triple alpha process. Written as a power law centered on $T = 10^8$ K (see Eq. 10.44), it demonstrates a very dramatic temperature dependence,

$$\epsilon_{3\alpha} \simeq \epsilon'_{\circ,3\alpha} \rho^2 Y^3 f_{3\alpha} T_8^{41.0}. \tag{10.57}$$

With such a strong dependence, even a small increase in temperature will produce a large increase in the amount of energy generated per second. For instance, an increase of only 10% in temperature raises the energy output rate by more than 50 times!

In the high-temperature environment of helium burning, other competing processes are also at work. After sufficient carbon has been generated by the triple alpha process, it becomes possible for carbon nuclei to capture alpha particles, producing oxygen. Some of the oxygen in turn can capture alpha particles to produce neon.

$$\,^{12}_{6}\text{C} + \,^{4}_{2}\text{He} \rightarrow \,^{16}_{8}\text{O} + \gamma \tag{10.58}$$

$$\,^{16}_{8}\text{O} + \,^{4}_{2}\text{He} \rightarrow \,^{20}_{10}\text{Ne} + \gamma$$

At helium-burning temperatures, the continued capture of alpha particles leading to progressively more massive nuclei quickly becomes prohibitive due to the ever higher Coulomb barrier.

If a star is sufficiently massive, still higher central temperatures can be obtained and many other nuclear products become possible. Examples of available reactions include carbon burning reactions near $6 \times 10^8$ K,

$${}^{12}_{6}\text{C} + {}^{12}_{6}\text{C} \rightarrow \begin{cases} {}^{16}_{8}\text{O} + 2\,{}^{4}_{2}\text{He} & *** \\ {}^{20}_{10}\text{Ne} + {}^{4}_{2}\text{He} & \\ {}^{23}_{11}\text{Na} + p^+ & \\ {}^{23}_{12}\text{Mg} + n & *** \\ {}^{24}_{12}\text{Mg} + \gamma & \end{cases} \qquad (10.59)$$

and oxygen burning reactions near $10^9$ K,

$$ {}^{16}_{8}\text{O} + {}^{16}_{8}\text{O} \rightarrow \begin{cases} {}^{24}_{12}\text{Mg} + 2\,{}^{4}_{2}\text{He} & *** \\ {}^{28}_{14}\text{Si} + {}^{4}_{2}\text{He} & \\ {}^{31}_{15}\text{P} + p^+ & \\ {}^{31}_{16}\text{S} + n & \\ {}^{32}_{16}\text{S} + \gamma & \end{cases} \qquad (10.60)$$

Reactions marked by *** are ones for which energy is absorbed rather than released and are referred to as being **endothermic**; energy-releasing reactions are **exothermic**. In endothermic reactions the product nucleus actually possesses more energy per nucleon than did the nuclei from which it formed. Such reactions occur at the expense of the energy released by exothermic reactions or by gravitational collapse (the virial theorem). In general, endothermic reactions are much less likely to occur than exothermic reactions under conditions that normally prevail in stellar interiors.

A useful quantity in understanding the energy release in nuclear reactions is the binding energy per nucleon, $E_b/A$, where

$$E_b = \Delta m c^2 = (Z m_p + (A - Z) m_n - m_{\text{nucleus}}) c^2.$$

Figure 10.9 shows $E_b/A$ versus the mass number. It is apparent that for relatively small values of $A$ (less than 56), several nuclei have abnormally high values of $E_b/A$ relative to others of similar mass. Among these unusually stable nuclei are ${}^{4}_{2}\text{He}$ and ${}^{16}_{8}\text{O}$, which, along with ${}^{1}_{1}\text{H}$, are the most abundant nuclei in the universe. This unusual stability arises from an inherent shell structure

## 10.3 Stellar Energy Sources

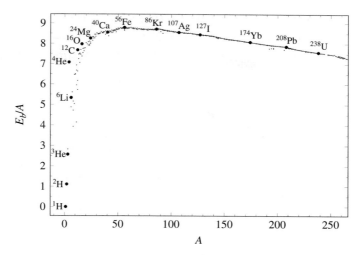

**Figure 10.9** The binding energy per nucleon, $E_b/A$, as a function of mass number, $A$.

of the nucleus, analogous to the shell structure of atomic energy levels that accounts for the chemical nature of elements. These unusually stable nuclei are called *magic nuclei*.

It is believed that shortly after the Big Bang the early universe was composed primarily of hydrogen and helium, with no heavy elements. Today Earth and its inhabitants contain an abundance of heavier metals. The study of stellar **nucleosynthesis** suggests strongly that these heavier nuclei were generated in the interiors of stars. It can be said that we are all "star dust," the product of heavy element generation within previous generations of stars.

Another important feature of Fig. 10.9 is the broad peak around $A = 56$. At the top of the peak is iron, $^{56}_{26}\text{Fe}$, the most stable of all nuclei. As successively more massive nuclei are created in stellar interiors, the iron peak is approached from the left. These fusion reactions result in the liberation of energy.[13] Consequently, the ultimate result of successive chains of nuclear reactions within stars is the production of iron, assuming sufficient energy is available to overcome the Coulomb barrier. If a star is massive enough to create the central temperatures and densities necessary to produce iron, the results are spectacular, as we will see in Chapter 13.

---

[13]Energy is also released when the peak is approached from the right via fission reactions that produce nuclei of smaller mass, again resulting in more stable nuclei. This type of reaction process is important in the fission reactors of nuclear power plants.

Based upon what we have learned in this section about stellar nucleosynthesis, it should come as no surprise that the most abundant nuclear species in the cosmos are, in order, $^{1}_{1}$H, $^{4}_{2}$He, $^{16}_{8}$O, $^{12}_{6}$C, $^{20}_{10}$Ne, $^{14}_{7}$N, and $^{56}_{26}$Fe (recall Table 9.2).[14] The abundances are the result of the dominant nuclear reaction processes that occur in stars, together with the nuclear configurations that result in the most stable nuclei.

## 10.4 Energy Transport and Thermodynamics

One stellar structure equation remains to be developed. We have already related the fundamental quantities $P$, $M$, and $L$ to the independent variable $r$ through differential equations that describe hydrostatic equilibrium, mass conservation, and energy generation, respectively (see Eqs. 10.7, 10.8, and 10.45). However, we have not yet found a differential equation relating the basic parameter of temperature, $T$, to $r$. Moreover, we have not explicitly developed equations that describe the processes by which energy is carried from the deep interior to the surface of the star.

Three different energy transport mechanisms operate in stellar interiors. **Radiation** allows the energy produced by nuclear reactions and gravitation to be carried to the surface via photons, the photons being absorbed and re-emitted in nearly random directions as they encounter matter (recall the discussion in Section 9.3). This suggests that the opacity of the material must play an important role, as one would expect. **Convection** can be a very efficient transport mechanism in many regions of a star, with buoyant, hot mass elements carrying excess energy outward while cool elements fall inward. Finally, **conduction** transports heat via collisions between particles. Although conduction can play an important role in some stellar environments, it is generally insignificant in most stars throughout the majority of their lifetimes and will not be discussed further here.

First consider radiation transport. In Chapter 9 we found that the radiation pressure gradient is given by Eq. (9.25),

$$\frac{dP_\text{rad}}{dr} = -\frac{\overline{\kappa}\rho}{c} F_\text{rad},$$

where $F_\text{rad}$ is the outward radiative flux. However, from Eq. (10.25), the radiation pressure gradient may also be expressed as

$$\frac{dP_\text{rad}}{dr} = \frac{4}{3} aT^3 \frac{dT}{dr}.$$

---

[14]The relative abundances of the elements will be discussed further in Section 13.3; see Fig. 13.24.

## 10.4 Energy Transport and Thermodynamics

Equating the two expressions, we have

$$\frac{dT}{dr} = -\frac{3}{4ac}\frac{\overline{\kappa}\rho}{T^3}F_{\text{rad}}.$$

Finally, if we use the expression for the radiative flux (Eq. 3.2), written in terms of the local radiative luminosity of the star at radius $r$,

$$F_{\text{rad}} = \frac{L_r}{4\pi r^2},$$

the temperature gradient for radiative transport becomes

$$\frac{dT}{dr} = -\frac{3}{4ac}\frac{\overline{\kappa}\rho}{T^3}\frac{L_r}{4\pi r^2}. \tag{10.61}$$

As either the flux or opacity increases, the temperature gradient must become steeper (more negative) if radiation is to transport all of the required luminosity outward. The same situation holds as the density increases or the temperature decreases.

If the temperature gradient becomes too steep, convection can begin to play an important role in the energy transport. Physically, convection involves mass motions; hot parcels of matter move upward as cooler, denser parcels sink. Unfortunately, convection is a much more complex phenomenon than radiation at the macroscopic level. In fact, no truly satisfactory theory yet exists to describe it. *Fluid mechanics*, the field of physics describing the motion of gases and liquids, relies on a complicated set of three-dimensional equations known as the Navier–Stokes equations. However, at present, due in large part to the current limitations in computer power,[15] most stellar structure codes are one-dimensional (i.e., depend on $r$ only). It becomes necessary, therefore, to approximate an explicitly three-dimensional process by a one-dimensional phenomenological theory. To complicate the situation even more, when convection is present in a star, it is generally quite turbulent, requiring a detailed understanding of the amount of viscosity (fluid friction) and heat dissipation involved. Also, a characteristic length scale for convection, typically referred to in terms of the **pressure scale height**, is often comparable to the size of the star. Lastly, the time scale for convection, taken to be the amount of time required for a convective element to travel a characteristic distance, is in some cases approximately equal to the time scale for changes in the structure of the star, implying that convection is strongly coupled to the star's behavior. The impact of these complications on the behavior of the star is not yet fundamentally understood.

---

[15] This limitation is very quickly being overcome with the development of ever faster computers with more memory.

The situation is not completely hopeless, however. Despite the problems encountered in attempting to treat stellar convection exactly, approximate (and even reasonable) results can usually be obtained. To develop an appreciation for the size of a convective region in a star, consider the pressure scale height, $H_P$, defined as

$$\frac{1}{H_P} \equiv -\frac{1}{P}\frac{dP}{dr}. \tag{10.62}$$

If we assume for the moment that $H_P$ is a constant, we can solve for the variation of pressure with radius, giving

$$P = P_\circ e^{-r/H_P}.$$

Obviously, if $r = H_P$, $P = P_\circ e^{-1}$, so that $H_P$ is the distance over which the gas pressure decreases by a factor of $e$. To find a convenient general expression for $H_P$, recall that from the equation for hydrostatic equilibrium (Eq. 10.7), $dP/dr = -\rho g$, where $g = GM_r/r^2$ is the local acceleration of gravity. Substituting into Eq. (10.62), the pressure scale height is simply

$$H_P = \frac{P}{\rho g}. \tag{10.63}$$

---

**Example 10.5** To make an estimate of a typical value for the pressure scale height in the Sun, assume that $\overline{P} = P_c/2$, where $P_c$ is the central pressure, $\overline{\rho}_\odot$ is the average solar density, and

$$\overline{g} = \frac{G(M_\odot/2)}{(R_\odot/2)^2} = 5.5 \times 10^4 \text{ cm s}^{-2}.$$

Then we have

$$H_P \simeq 1.8 \times 10^{10} \text{ cm} \sim R_\odot/4.$$

A detailed calculation shows that $H_P \sim R_\odot/10$ is more typical.

---

Understanding convective heat transport in stars, even in an approximate way, begins with some knowledge of thermodynamics. In the study of heat transport, conservation of energy is expressed by the *first law of thermodynamics*,

$$dU = dQ - dW, \tag{10.64}$$

where the change in the internal energy of a mass element $dU$ is given by the amount of heat *added* to that element, $dQ$, minus the work done *by* that element on its surroundings, $dW$. Throughout our discussion we will assume that these energy changes are measured *per unit mass*.

## 10.4 Energy Transport and Thermodynamics

The internal energy of a system $U$ is a state function, meaning that its value depends only on the present conditions of the gas, not on the history of any changes leading to its current state. Consequently, $dU$ is independent of the actual process involved in the change.[16] Consider an ideal monatomic gas, a gas composed of single particles with no ionization occurring. The total internal energy per unit mass is given by

$$U = (\text{average energy/particle}) \times (\text{number of particles/mass})$$
$$= \overline{K} \times \frac{1}{\overline{m}}$$

where $\overline{m} = \mu m_H$ is the average mass of a single particle in the gas. For an ideal gas, $\overline{K} = 3kT/2$ and the internal energy is given by

$$U = \frac{3}{2}\left(\frac{k}{\mu m_H}\right) T = \frac{3}{2} nRT, \qquad (10.65)$$

where $nR \equiv k/\mu m_H$, $n$ is the number of moles[17] *per unit mass*, and $R = 8.31451 \times 10^7$ ergs mole$^{-1}$ K$^{-1}$ is the universal gas constant. Clearly $U = U(\mu, T)$ is a function of the composition of the gas and its temperature. In this case the internal energy is just the kinetic energy per unit mass.

The change in heat of the mass element $dQ$ is generally expressed in terms of the **specific heat** $C$ of the gas. The specific heat is defined as the amount of heat required to raise the temperature of a unit mass of a material by 1 K, or

$$C_P \equiv \left.\frac{dQ}{dT}\right|_P, \qquad C_V \equiv \left.\frac{dQ}{dT}\right|_V, \qquad (10.66)$$

where $C_P$ and $C_V$ are the specific heats at constant pressure and volume, respectively.

Consider next the amount of work per unit mass, $dW$, done by the gas on its surroundings. Suppose that a cylinder of cross-sectional area $A$ is filled with a gas of mass $m$ and pressure $P$. The gas then exerts a force $F = PA$ on an end of the cylinder. If the end of the cylinder is a piston that moves through a distance $dr$, the work per unit mass performed by the gas may be expressed as

$$dW = \left(\frac{F}{m}\right) dr = \left(\frac{PA}{m}\right) dr = P\,dV,$$

---

[16] On the other hand, neither heat nor work is a state function. The amount of heat added to a system or the amount of work done by a system depends on the ways in which the processes are carried out. $dQ$ and $dW$ are referred to as *inexact differentials*, reflecting this path dependence.

[17] 1 mole = $N_A$ particles, where $N_A = 6.022137 \times 10^{23}$ is Avogadro's number, defined as the number of $^{12}_{6}\text{C}$ atoms required to produce exactly 12 grams of a pure sample; $R = N_A k$.

$V$ being defined as the **specific volume**, the volume per unit mass, or $V = 1/\rho$. The first law of thermodynamics may now be expressed in the useful form

$$dU = dQ - P\, dV. \tag{10.67}$$

At constant volume, $dV = 0$ giving $dU = dU|_V = dQ|_V$, or

$$dU = \left.\frac{dQ}{dT}\right|_V dT = C_V\, dT \tag{10.68}$$

(recall that $dU$ is independent of any specific process). But $dU = (3nR/2)\, dT$ for a monatomic gas. Thus

$$C_V = \frac{3}{2}nR. \tag{10.69}$$

To find $C_P$ for a monatomic gas, note that

$$dU = \left.\frac{dQ}{dT}\right|_P dT - P\left.\frac{\partial V}{\partial T}\right|_P dT.$$

From the ideal gas law (Eq. 10.14),

$$PV = nRT \tag{10.70}$$

or, at constant $P$ and $n$, $P\, dV/dT = nR$. Substituting $dU = C_V\, dT$ and the definition of $C_P$, we arrive at

$$C_P = C_V + nR. \tag{10.71}$$

Equation (10.71) is valid for all situations for which the ideal gas law applies.

Define the parameter $\gamma$ to be the ratio of specific heats, or

$$\gamma \equiv \frac{C_P}{C_V}. \tag{10.72}$$

For a monatomic gas, we see that $\gamma = 5/3$. If ionization is occurring, some of the heat that would normally go into increasing the average kinetic energy of the particles must go into ionizing the atoms instead. Therefore the temperature of the gas, a measure of its internal energy, will not rise as rapidly, implying larger values for the specific heats in a partial ionization zone. As $C_P$ and $C_V$ increase, $\gamma$ approaches unity.[18]

Since the change in internal energy is independent of the process involved, consider the special case of an **adiabatic process** ($dQ = 0$) for which no heat

---

[18]The variation of $\gamma$ also plays an important role in the dynamic stability of stars. This factor will be discussed further in Section 14.3.

## 10.4 Energy Transport and Thermodynamics

flows into or out of the mass element. Then the first law of thermodynamics (Eq. 10.67) becomes

$$dU = -P\,dV.$$

However, from Eq. (10.70),

$$P\,dV + V\,dP = nR\,dT$$

if $n$ is constant. Also, since $dU = C_V\,dT$, we have

$$dT = \frac{dU}{C_V} = -\frac{P\,dV}{C_V}.$$

Combining these results,

$$P\,dV + V\,dP = -\left(\frac{nR}{C_V}\right)P\,dV,$$

which may be rewritten by using Eq. (10.71), to give

$$\gamma\frac{dV}{V} = -\frac{dP}{P}. \tag{10.73}$$

Solving this differential equation leads to the adiabatic gas law,

$$PV^\gamma = K, \tag{10.74}$$

where $K$ is a constant. Using the ideal gas law, a second adiabatic relation may be obtained,

$$P = K'T^{\gamma/(\gamma-1)}, \tag{10.75}$$

where $K'$ is another constant. Because of its special role in Eqs. (10.74) and (10.75), $\gamma$ is often referred to as the "adiabatic gamma," specifying a particularly simple equation of state.

Using the results obtained thus far, it is now possible to calculate a sound speed through the material. The sound speed is related to the compressibility of the gas and its inertia, represented by the density, and is given by

$$v_s = \sqrt{B/\rho},$$

where $B \equiv -V(\partial P/\partial V)_{\text{ad}}$ is the bulk modulus of the gas.[19] The bulk modulus describes how much the volume of the gas will change with changing pressure. From Eq. (10.73), the adiabatic sound speed becomes

$$v_s = \sqrt{\gamma P/\rho}. \tag{10.76}$$

---

[19]Formally, the bulk modulus, and therefore the sound speed, must be defined in terms of a process by which pressure varies with volume. Since sound waves typically propagate through a medium too quickly for a significant amount of heat to enter or leave the system, we usually assume that the process is adiabatic.

**Example 10.6** Assuming a monatomic gas, a typical adiabatic sound speed for the Sun is

$$\overline{v}_s \simeq \left(\frac{5}{3}\frac{\overline{P}}{\overline{\rho}_\odot}\right)^{1/2} \simeq 4 \times 10^7 \text{ cm s}^{-1},$$

where $\overline{P} \sim P_c/2$ was assumed. The amount of time needed for a sound wave to traverse the radius of the Sun would then be

$$t \simeq R_\odot/\overline{v}_s \simeq 29 \text{ minutes.}$$

Returning now to the specific problem of describing convection, we first consider the situation where a hot convective bubble of gas rises and expands *adiabatically*, meaning that the bubble does not exchange heat with its surroundings. After it has traveled some distance, it finally *thermalizes*, giving up any excess heat as it loses its identity and dissolves into the surrounding gas. By differentiating the ideal gas law (Eq. 10.14), an expression involving the bubble's *temperature gradient* (how the bubble's temperature changes with position) is obtained:

$$\frac{dP}{dr} = -\frac{P}{\mu}\frac{d\mu}{dr} + \frac{P}{\rho}\frac{d\rho}{dr} + \frac{P}{T}\frac{dT}{dr}. \tag{10.77}$$

From the adiabatic relationship between pressure and density (Eq. 10.74), and recalling that $V = 1/\rho$ is the specific volume, we have

$$P = K\rho^\gamma. \tag{10.78}$$

Differentiating and rewriting, we obtain

$$\frac{dP}{dr} = \gamma \frac{P}{\rho}\frac{d\rho}{dr}. \tag{10.79}$$

If it is assumed that $\mu$ is a constant, Eqs. (10.77) and (10.79) may be combined to give the adiabatic temperature gradient

$$\left.\frac{dT}{dr}\right|_{\text{ad}} = \left(1 - \frac{1}{\gamma}\right)\frac{T}{P}\frac{dP}{dr}. \tag{10.80}$$

Using Eq. (10.7) and the ideal gas law, we finally obtain

$$\left.\frac{dT}{dr}\right|_{\text{ad}} = -\left(1 - \frac{1}{\gamma}\right)\frac{\mu m_H}{k}\frac{GM_r}{r^2}. \tag{10.81}$$

## 10.4 Energy Transport and Thermodynamics

It is sometimes helpful to express Eq. (10.81) in another, equivalent form. Recalling that $g = GM_r/r^2$, $k/\mu m_H = nR$, $\gamma = C_P/C_V$, and $C_P - C_V = nR$, and that $n$, $C_P$, and $C_V$ are per unit mass, we have

$$\left. \frac{dT}{dr} \right|_{\text{ad}} = -\frac{g}{C_P}. \tag{10.82}$$

This result describes how the temperature of the gas *inside* the bubble changes as the bubble rises and expands adiabatically.

If the star's actual temperature gradient is *steeper* than the adiabatic temperature gradient given in Eq. (10.81), or

$$\left| \frac{dT}{dr} \right|_{\text{act}} > \left| \frac{dT}{dr} \right|_{\text{ad}},$$

the temperature gradient is said to be **superadiabatic**. However, note that $dT/dr < 0$. It will be shown that in the deep interior of a star, if $|dT/dr|_{\text{act}}$ is just *slightly* larger than $|dT/dr|_{\text{ad}}$, this is sufficient to carry all of the luminosity by convection. Consequently, either radiation or convection dominates the energy transport in the deep interiors of stars, while the other energy transport mechanism contributes very little to the total energy outflow. The particular mechanism in operation is determined by the temperature gradient. However, near the surface of the star the situation is much more complicated: Both radiation and convection can carry significant amounts of energy simultaneously.

Just what condition must be met if convection is to dominate over radiation in the deep interior? When will a hot bubble of gas continue to rise? Figure 10.10 shows a convective bubble traveling a distance $dr$ through the surrounding medium. According to Archimedes' principle, if the initial density of the bubble is less than its surroundings ($\rho_i^{(b)} < \rho_i^{(s)}$), it will begin to rise. Now, the buoyant force *per unit volume* exerted on a bubble that is totally submersed in a fluid of density $\rho_i^{(s)}$ is given by

$$f_B = \rho_i^{(s)} g.$$

If we subtract the downward gravitational force per unit volume on the bubble, given by

$$f_g = \rho_i^{(b)} g,$$

the net force per unit volume on the bubble becomes

$$f_{\text{net}} = -g\, \delta\rho, \tag{10.83}$$

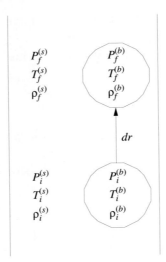

**Figure 10.10** A convective bubble traveling outward a distance $dr$. The initial conditions of the bubble are given by $P_i^{(b)}$, $T_i^{(b)}$, and $\rho_i^{(b)}$, for the pressure, temperature, and density, respectively. Conditions for the surrounding gas are designated by an $(s)$ superscript, initial conditions are indicated by an $i$ subscript, and final conditions are indicated by an $f$ subscript.

where $\delta\rho = \rho_i^{(b)} - \rho_i^{(s)} < 0$ initially. If, after traveling an infinitesimal distance $dr$, the bubble now has a greater density than the surrounding material ($\rho_f^{(b)} > \rho_f^{(s)}$), it will sink again and convection will be prohibited. On the other hand, if $\rho_f^{(b)} < \rho_f^{(s)}$, the bubble will continue to rise and convection will result.

To express this condition in terms of temperature gradients, assume that the gas is initially nearly in *thermal equilibrium*, with $T_i^{(b)} \simeq T_i^{(s)}$ and $\rho_i^{(b)} \simeq \rho_i^{(s)}$. Also assume that the bubble expands adiabatically and that the bubble and surrounding gas pressures are equal at all times, $P_f^{(b)} = P_f^{(s)}$. Now, since it is assumed that the bubble has moved an infinitesimal distance, it is possible to express the final quantities in terms of the initial quantities and their gradients by using a Taylor expansion. To first order

$$\rho_f^{(b)} \simeq \rho_i^{(b)} + \left.\frac{d\rho}{dr}\right|^{(b)} dr$$

and

$$\rho_f^{(s)} \simeq \rho_i^{(s)} + \left.\frac{d\rho}{dr}\right|^{(s)} dr.$$

## 10.4 Energy Transport and Thermodynamics

If the densities inside and outside the bubble remain nearly equal (as is usually the case except near the surfaces of some stars), substituting these results into the convection condition, $\rho_f^{(b)} < \rho_f^{(s)}$, gives

$$\left.\frac{d\rho}{dr}\right|^{(b)} < \left.\frac{d\rho}{dr}\right|^{(s)}. \tag{10.84}$$

We now want to express this solely in terms of quantities for the surroundings. Using Eq. (10.79) for the adiabatically rising bubble to rewrite the left-hand side of 10.84 and using Eq. (10.77) to rewrite the right-hand side (again assuming $d\mu/dr = 0$), we find

$$\frac{1}{\gamma}\frac{\rho_i^{(b)}}{P_i^{(b)}}\left.\frac{dP}{dr}\right|^{(b)} < \frac{\rho_i^{(s)}}{P_i^{(s)}}\left[\left.\frac{dP}{dr}\right|^{(s)} - \frac{P_i^{(s)}}{T_i^{(s)}}\left.\frac{dT}{dr}\right|^{(s)}\right].$$

Recalling that $P^{(b)} = P^{(s)}$ at all times, it is necessary that

$$\left.\frac{dP}{dr}\right|^{(b)} = \left.\frac{dP}{dr}\right|^{(s)} = \frac{dP}{dr},$$

where the superscripts on the pressure gradient are now redundant. Substituting, and canceling equivalent initial conditions,

$$\frac{1}{\gamma}\frac{dP}{dr} < \frac{dP}{dr} - \frac{P_i^{(s)}}{T_i^{(s)}}\left.\frac{dT}{dr}\right|^{(s)}.$$

Dropping subscripts for initial conditions and superscripts designating the surrounding material, we arrive at the requirement

$$\left(\frac{1}{\gamma} - 1\right)\frac{dP}{dr} < -\frac{P}{T}\left.\frac{dT}{dr}\right|_{\text{act}}, \tag{10.85}$$

where the temperature gradient is the *actual* temperature gradient of the surrounding gas. Multiplying by the negative quantity $-T/P$ requires that the direction of the inequality be reversed, giving

$$\left(1 - \frac{1}{\gamma}\right)\frac{T}{P}\frac{dP}{dr} > \left.\frac{dT}{dr}\right|_{\text{act}}.$$

But from Eq. (10.80), we see that the left-hand side of the inequality is just the adiabatic temperature gradient. Thus

$$\left.\frac{dT}{dr}\right|_{\text{ad}} > \left.\frac{dT}{dr}\right|_{\text{act}}.$$

is the condition for the gas bubble to keep rising. Finally, since $dT/dr < 0$ (the temperature decreases as the stellar radius increases), taking the absolute value of the equation again requires that the direction of the inequality be reversed, or

$$\left|\frac{dT}{dr}\right|_{\text{act}} > \left|\frac{dT}{dr}\right|_{\text{ad}}. \tag{10.86}$$

If the actual temperature gradient is superadiabatic, convection will result, assuming that $\mu$ does not vary.

Equation (10.85) may be used to find another useful, although equivalent, condition for convection. Since $dT/dr < 0$ and $1/\gamma - 1 < 0$ (recall that $\gamma > 1$),

$$\frac{T}{P}\left(\frac{dT}{dr}\right)^{-1}\frac{dP}{dr} < -\frac{1}{\frac{1}{\gamma} - 1},$$

which may be simplified to give

$$\frac{T}{P}\frac{dP}{dT} < \frac{\gamma}{\gamma - 1},$$

or, for convection to occur,

$$\frac{d\ln P}{d\ln T} < \frac{\gamma}{\gamma - 1}. \tag{10.87}$$

For an ideal monatomic gas, $\gamma = 5/3$ and convection will occur in some region of a star when $d\ln P/d\ln T < 2.5$. In that case the temperature gradient $(dT/dr)$ is given by Eq. (10.81). When $d\ln P/d\ln T > 2.5$, the region is stable against convection and $dT/dr$ is given by Eq. (10.61).

By comparing Eq. (10.61) for the radiative temperature gradient with either Eq. (10.81) or Eq. (10.82), together with the temperature gradient convection condition, Eq. (10.86), it is possible to develop some understanding of which conditions are likely to lead to convection over radiation. In general, convection will occur when (1) the stellar opacity is large, implying that an unachievably steep temperature gradient ($|dT/dr|_{\text{act}}$) would be necessary for radiative transport, (2) a region exists where ionization is occurring, causing a large specific heat and a low adiabatic temperature gradient ($|dT/dr|_{\text{ad}}$), (3) the local gravitational acceleration is low, as would be the case for very distended stars, again leading to a low adiabatic gradient, and (4) the temperature dependence of the nuclear energy generation rate is large, causing a steep radiative flux gradient and a large temperature gradient. In the atmospheres of many stars, each of the first three conditions could be occurring simultaneously, while the fourth condition would occur only deep in stellar interiors. In particular, the

## 10.4 Energy Transport and Thermodynamics

fourth condition can occur when the highly temperature-dependent CNO cycle or triple alpha processes are occurring.

---

*Optional:* It has already been suggested that the temperature gradient must be only slightly superadiabatic in the deep interior in order for convection to carry all of the energy. We will now justify that assertion.

We begin by returning to the fundamental criterion for convection, $\rho_f^{(b)} < \rho_f^{(s)}$. Since the pressure of the bubble and its surroundings are always equal, the ideal gas law implies that $T_f^{(b)} > T_f^{(s)}$, assuming thermal equilibrium initially. Therefore, the temperature of the surrounding gas must decrease more rapidly with radius, so

$$\left|\frac{dT}{dr}\right|^{(s)} - \left|\frac{dT}{dr}\right|^{(b)} > 0$$

is required for convection. Since the temperature gradients are negative, we have

$$\left.\frac{dT}{dr}\right|^{(b)} - \left.\frac{dT}{dr}\right|^{(s)} > 0.$$

Assuming that the bubble moves adiabatically, and designating the temperature gradient of the surroundings as the actual average temperature gradient of the star, let

$$\left.\frac{dT}{dr}\right|^{(b)} = \left.\frac{dT}{dr}\right|_{ad} \quad \text{and} \quad \left.\frac{dT}{dr}\right|^{(s)} = \left.\frac{dT}{dr}\right|_{act}.$$

After traveling a distance $dr$, the temperature of the bubble will exceed the temperature of the surrounding gas by[20]

$$\delta T = \left(\left.\frac{dT}{dr}\right|_{ad} - \left.\frac{dT}{dr}\right|_{act}\right) dr = \delta\left(\frac{dT}{dr}\right) dr. \tag{10.88}$$

We use $\delta$ here to indicate the difference between the value of a quantity associated with the bubble and the same quantity associated with the surroundings, both determined at a specified radius $r$, just as was done for Eq. (10.83).

Now, assume that a hot, rising bubble travels some distance $\ell = \alpha H_P$ before dissipating, at which point it thermalizes with its surroundings, giving up its excess heat at constant pressure (since $P^{(b)} = P^{(s)}$ at all times). The distance $\ell$ is called the **mixing length**, $H_P$ is the pressure scale height (see

---

[20] In some texts, $\delta\left(\frac{dT}{dr}\right) \equiv \Delta \nabla T$.

Eq. 10.63), and $\alpha \equiv \ell/H_P$ is a *free parameter*, generally assumed to be of order unity. (From comparisons of numerical stellar models with observations, values of $0.5 < \alpha < 3$ are typical.)

After traveling one mixing length, the excess heat flow *per unit volume* from the bubble into its surroundings is just

$$\delta q = (C_P \, \delta T) \rho,$$

where $\delta T$ is calculated from Eq. (10.88) by substituting $\ell$ for $dr$. Multiplying by the average velocity $\overline{v}_c$ of the convective bubble, we obtain the convective flux (the amount of energy per unit area per unit time carried by a bubble),

$$F_c = \delta q \, \overline{v}_c = (C_P \, \delta T) \rho \overline{v}_c. \tag{10.89}$$

Note that $\rho \overline{v}$ is a *mass flux*, or the amount of mass per second that crosses a unit area oriented perpendicular to the direction of the flow. Mass flux is a quantity that is often encountered in fluid mechanics.

$\overline{v}$ may be found from the net force *per unit volume*, $f_{\text{net}}$, acting on the bubble. Using the ideal gas law, we can write

$$\delta P = \frac{P}{\rho} \delta \rho + \frac{P}{T} \delta T,$$

assuming constant $\mu$. Since the pressure is always equal between the bubble and its surroundings, $\delta P \equiv P^{(b)} - P^{(s)} = 0$. Thus

$$\delta \rho = -\frac{\rho}{T} \delta T.$$

From Eq. (10.83),

$$f_{\text{net}} = \frac{\rho g}{T} \delta T. \tag{10.90}$$

However, it is assumed that the initial temperature difference between the bubble and its surroundings is zero, or $\delta T_i = 0$. Consequently the buoyant force must also be zero initially. Since $f_{\text{net}}$ increases linearly with $\delta T$, we may take an average over the distance $\ell$ between the initial and final positions, or

$$\langle f_{\text{net}} \rangle = \frac{1}{2} \frac{\rho g}{T} \delta T_f.$$

Neglecting viscous forces, the work done per unit volume by the buoyant force over the distance $\ell$ goes into the kinetic energy of the bubble, or

$$\frac{1}{2} \rho v_f^2 = \langle f_{\text{net}} \rangle \ell.$$

## 10.4 Energy Transport and Thermodynamics

Choosing an *average* kinetic energy over one mixing length leads to some average value of $v^2$, namely $\beta v^2$, where $\beta$ has a value in the range $0 < \beta < 1$. Now the average convective bubble velocity becomes

$$\bar{v}_c = \left(\frac{2\beta \langle f_{\text{net}} \rangle \ell}{\rho}\right)^{1/2}.$$

Substituting the net force per unit volume, using Eq. (10.88) with $dr = \ell$, and rearranging, we have

$$\bar{v}_c = \left(\frac{\beta g}{T}\right)^{1/2} \left[\delta\left(\frac{dT}{dr}\right)\right]^{1/2} \ell$$

$$= \beta^{1/2} \left(\frac{T}{g}\right)^{1/2} \left(\frac{k}{\mu m_H}\right) \left[\delta\left(\frac{dT}{dr}\right)\right]^{1/2} \alpha, \tag{10.91}$$

where the last equation was obtained by replacing the mixing length with $\alpha H_P$, and using Eq. (10.63) together with the ideal gas law.

After some manipulation, Eqs. (10.89) and (10.91) finally give

$$F_c = \rho C_P \left(\frac{k}{\mu m_H}\right)^2 \left(\frac{T}{g}\right)^{3/2} \beta^{1/2} \left[\delta\left(\frac{dT}{dr}\right)\right]^{3/2} \alpha^2. \tag{10.92}$$

Fortunately $F_c$ is not very sensitive to $\beta$, but it does depend strongly on $\alpha$ and $\delta(dT/dr)$.

The derivation leading to the prescription for the convective flux given by Eq. (10.92) is known as the **mixing length theory**. Although basically a phenomenological theory, containing arbitrary constants, the mixing length theory is generally quite successful in predicting the results of observations.

To evaluate $F_c$, it is still necessary to know the difference between the temperature gradients of the bubble and its surroundings. Suppose, for simplicity, that *all* of the flux is carried by convection, so that

$$F_c = \frac{L_r}{4\pi r^2},$$

where $L_r$ is the interior luminosity. This will allow us to estimate the difference in temperature gradients needed for this special case. Solving Eq. (10.92) for the temperature gradient difference gives

$$\delta\left(\frac{dT}{dr}\right) = \left\{\frac{L_r}{4\pi r^2} \frac{1}{\rho C_P \alpha^2} \left(\frac{\mu m_H}{k}\right)^2 \left(\frac{g}{T}\right)^{3/2} \beta^{-1/2}\right\}^{2/3}. \tag{10.93}$$

Dividing Eq. (10.93) by Eq. (10.82) gives an estimate of how superadiabatic the actual temperature gradient must be to carry all of the flux by convection alone:

$$\frac{\delta(dT/dr)}{|dT/dr|_{\text{ad}}} = \left(\frac{L_r}{4\pi r^2}\right)^{2/3} C_P^{1/3} \rho^{-2/3} \alpha^{-4/3} \left(\frac{\mu m_H}{k}\right)^{4/3} \frac{1}{T} \beta^{-1/3}. \quad (10.94)$$

---

**Example 10.7** Using values typical of the base of the Sun's convection zone, assuming a monatomic gas throughout, and assuming $\alpha = \beta = 1$, it is possible to estimate a characteristic adiabatic temperature gradient, the degree to which the actual gradient is superadiabatic, and the convective bubble velocity.

Assume that $M_r \simeq 1$ M$_\odot$, $L_r \simeq 1$ L$_\odot$, $r = 0.75$ R$_\odot$, $C_P = 5nR/2$, $P = 3 \times 10^{13}$ dynes cm$^{-2}$, $\rho = 0.1$ g cm$^{-3}$, $\mu = 0.6$, and $T = 1.8 \times 10^6$ K. Then

$$\left|\frac{dT}{dr}\right|_{\text{ad}} \sim 1.4 \times 10^{-4} \text{ K cm}^{-1}$$

and

$$\delta\left(\frac{dT}{dr}\right) \sim 8 \times 10^{-11} \text{ K cm}^{-1}.$$

The relative amount by which the actual temperature gradient is superadiabatic is then

$$\frac{\delta(dT/dr)}{|dT/dr|_{\text{ad}}} \sim 6 \times 10^{-7}.$$

For parameters appropriate for the deep interior, convection is certainly adequately approximated by the adiabatic temperature gradient.

The convective velocity needed to carry all of the convective flux is found from Eq. (10.91),

$$\overline{v}_c \sim 7 \times 10^3 \text{ cm s}^{-1} \sim 10^{-4} v_s,$$

where $v_s$ is the local solar sound speed.

---

Near the surface of a star, where the presence of ionization results in a larger value for $C_P$ and where $\rho$ and $T$ get much smaller, the ratio of the superadiabatic excess to the adiabatic gradient can become significantly larger, with the convective velocity possibly approaching the sound speed. In this situation, a detailed study of the relative amounts of convective and radiative flux must be considered. This will not be discussed further here.

Although the mixing length theory is adequate for many problems, it is inherently incomplete. For instance, $\alpha$ and $\beta$ are *free parameters* that must be chosen for a particular problem; they may even vary throughout the star. There are also stellar conditions for which the mixing length theory is unsatisfactory, such as during stellar pulsations, when the outer layers of the star are oscillating with periods comparable to the time scale for convection, given by $t_c = \ell/\overline{v}_c$. In such cases, rapid changes in the physical conditions in the star directly couple to the driving of the convective bubbles, which in turn alters the structure of the star. Unfortunately, at present, no theory exists to completely describe this behavior. Much work remains to be done in understanding the important details of stellar convection.

## 10.5 Stellar Model Building

We have now derived all of the fundamental differential equations of stellar structure. These equations, together with a set of relations describing the physical properties of the stellar material, may be solved to obtain a theoretical stellar model.

For convenience, the basic *time-independent* (static) stellar structure equations are summarized:

$$\frac{dP}{dr} = -G\frac{M_r \rho}{r^2} \tag{10.7}$$

$$\frac{dM_r}{dr} = 4\pi r^2 \rho \tag{10.8}$$

$$\frac{dL_r}{dr} = 4\pi r^2 \rho \epsilon \tag{10.45}$$

$$\frac{dT}{dr} = -\frac{3}{4ac}\frac{\overline{\kappa}\rho}{T^3}\frac{L_r}{4\pi r^2} \quad \text{(radiation)} \tag{10.61}$$

$$\frac{dT}{dr} = -\left(1 - \frac{1}{\gamma}\right)\frac{\mu m_H}{k}\frac{GM_r}{r^2} \quad \text{(adiabatic convection)}, \tag{10.81}$$

where $\epsilon = \epsilon_{\text{nuclear}}$. The last equation assumes that the convective temperature gradient is purely adiabatic and is applied when

$$\frac{d\ln P}{d\ln T} < \frac{\gamma}{\gamma - 1}. \tag{10.87}$$

If the structure of the stellar model is changing, then the contribution of gravitational energy must be included, $\epsilon = \epsilon_{\text{nuclear}} + \epsilon_{\text{gravity}}$. The introduction of the gravitational energy term adds a time dependence to the equations that is not present in the purely static case. This can be seen by realizing that the virial theorem requires that gravitational collapse convert one-half of the potential energy that is lost into heat. The *rate* of energy production (per unit mass) by gravity is then $dQ/dt$. Therefore $\epsilon_{\text{gravity}} = -dQ/dt$, the minus sign indicating the heat liberated *from* the material.

As a note of interest, it is often useful to express the gravitational energy generation rate in terms of the change in the **entropy** per unit mass (the specific entropy), defined by

$$dS \equiv \frac{dQ}{T}.$$

Then the energy generation rate is seen to be due to the change in entropy of the material, or

$$\epsilon_{\text{gravity}} = -T\frac{dS}{dt}. \tag{10.95}$$

If the star is collapsing, $\epsilon_{\text{gravity}}$ will be positive; if it is expanding, $\epsilon_{\text{gravity}}$ will be negative. Thus, as the star contracts, its entropy decreases. This is not a violation of the second law of thermodynamics, which states that the entropy of a *closed* system must always remain the same (reversible process) or increase (irreversible process). Since a star is not a closed system, its entropy may decrease locally while the entropy of the remainder of the universe increases by a greater amount. The entropy is carried out of the star by photons.

When changes in the structure of the star are sufficiently rapid that accelerations can no longer be neglected, Eq. (10.7) must be replaced by the exact expression, Eq. (10.6). Such a situation can occur during a supernova explosion or during stellar pulsations. The effect of the acceleration term in stellar pulsations will be discussed in Chapter 14.

The basic stellar structure equations [Eqs. (10.7), (10.8), (10.45), (10.61), and (10.81)] require information concerning the physical properties of the matter from which the star is made. The required conditions are the **equations of state** of the material and are collectively referred to as **constitutive relations**. Specifically, it is necessary to obtain relationships for the pressure, the opacity, and the energy generation rate, in terms of fundamental characteristics of the material: the density, temperature, and composition. In general,

$$P = P(\rho, T, \text{composition})$$

$$\overline{\kappa} = \overline{\kappa}(\rho, T, \text{composition})$$

$$\epsilon = \epsilon(\rho, T, \text{composition})$$

The pressure equation of state can be quite complex in the deep interiors of certain classes of stars, where the density and temperature can become extremely high. However, in most situations, the ideal gas law, combined with

## 10.5 Stellar Model Building

the expression for radiation pressure, is a good first approximation, particularly when the variation in the mean molecular weight with composition and ionization is properly calculated. Equation (10.26) includes the contributions to pressure due to the ideal gas law and radiation.

The opacity of the stellar material cannot be expressed exactly by a single formula. It is actually calculated explicitly for various compositions at specific densities and temperatures and presented in tabular form. Stellar structure codes either interpolate in a density–temperature grid to obtain the opacity for the specified conditions, or alternatively, a "fitting function" can be generated based on the tabulated values. A similar situation also occurs for accurate calculations of the pressure equation of state. Although no accurate fitting function can be constructed to account for bound–bound opacities, approximate expressions for bound–free, free–free, and electron scattering opacities were presented in Section 9.2 [see Eqs. (9.19), (9.20), and (9.21)].

To calculate the nuclear energy generation rate, we can use formulas such as those presented in Section 10.3 for the pp chain (Eq. 10.49) and the CNO cycle (Eq. 10.53).

The actual solution of the stellar structure equations and the constitutive relations requires appropriate **boundary conditions** that specify physical constraints to the mathematical equations. Boundary conditions essentially play the role of defining limits of integration. The central boundary conditions are fairly obvious, namely that the interior mass and luminosity must approach zero at the center of the star, or

$$\left. \begin{array}{rcl} M_r & \to & 0 \\ L_r & \to & 0 \end{array} \right\} \quad \text{as } r \to 0. \quad (10.96)$$

This simply means that the star is physically realistic and does not contain a hole, a core of negative luminosity, or central points of infinite $\rho$ or $\epsilon$!

A second set of boundary conditions are required at the surface of the star. The simplest set of assumptions is that the temperature, pressure, and density all approach zero at some surface value for the star's radius, $R_\star$, or

$$\left. \begin{array}{rcl} T & \to & 0 \\ P & \to & 0 \\ \rho & \to & 0 \end{array} \right\} \quad \text{as } r \to R_\star. \quad (10.97)$$

Strictly, the conditions of Eqs. (10.97) will never be obtained in a real star. Therefore it often necessary to use more sophisticated surface boundary conditions, such as when the star being modeled has an extended atmosphere or is losing mass, as most stars do.

Finally, given the basic stellar structure equations, constitutive relations, and boundary conditions, we can now specify the type of star to be modeled. As can be seen by examination of Eq. (10.7), the pressure gradient at a given radius is dependent on the interior mass and the density. Similarly, the radiative temperature gradient (Eq. 10.45) depends on the local temperature, density, opacity, and interior luminosity, while the luminosity gradient is a function of the density and energy generation rate. The density, opacity, and energy generation rate in turn depend explicitly on the pressure, temperature, and composition at that location. If the interior mass at the surface of the star (i.e., the entire stellar mass) is specified, along with the composition, surface radius, and luminosity, application of the surface boundary conditions allows for a determination of the pressure, interior mass, temperature, and interior luminosity at an infinitesimal distance $dr$ below the surface of the star.[21] Continuing this *numerical integration* of the stellar structure equations to the center of the star must result in agreement with the central boundary conditions. Since the values of the various gradients are directly related to the composition of the star, it is not possible to specify any arbitrary combination of surface radius and luminosity after the mass and composition have been selected. Similarly, if the mass, radius, and luminosity are specified, a unique composition structure is required. This set of constraints is known as the **Vogt–Russell theorem**:

> The mass and composition of a star uniquely determine its radius, luminosity, and internal structure, as well as its subsequent evolution.

*The dependence of a star's evolution on mass and composition is a consequence of the change in composition due to nuclear burning.* The statement of the Vogt–Russell "theorem" given here is somewhat misleading since there are other parameters that can influence stellar interiors, such as magnetic fields and rotation. However, these parameters are assumed to have little effect in most stars and will not be discussed further.[22]

---

[21]It is also necessary to specify the average density over that distance. Since $\rho$ is assumed to be zero at the surface, and since it depends explicitly on the pressure and temperature, which are also assumed to be zero at the surface and are initially unknown below the surface, an immediate difficulty arises; the left-hand sides of Eqs. (10.7) and (10.61) are *zero*, so $P$ and $T$ never increase from their surface values! One solution to this initial step problem is outlined in Appendix H. More sophisticated solutions require an iterative procedure, continually correcting previous estimates until a self-consistent answer is obtained to within some specified level of accuracy.

[22]Even without the complications of magnetic fields and rotation, the Vogt–Russell "theorem" can be violated in certain special circumstances. However, an actual star (as opposed

## 10.5 Stellar Model Building

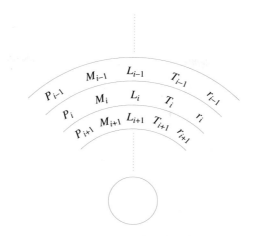

**Figure 10.11** Zoning in a numerical stellar model. The star is assumed to be constructed of spherically symmetric mass shells, with the physical parameters associated with each zone being specified by the stellar structure equations, the constitutive relations, the boundary conditions, and the star's mass and composition.

With the exception of very special solutions to the stellar structure equations, the system of differential equations, along with their constitutive relations, cannot be solved analytically.[23] Instead, as already mentioned, it is necessary to integrate the system of equations numerically. This is accomplished by approximating the differential equations by *difference equations*, by replacing $dP/dr$ by $\Delta P/\Delta r$, for instance. The star is then imagined to be constructed of spherically symmetric **shells** as in Fig. 10.11 and the "integration" is carried out from some initial radius in finite steps by specifying some increment $\delta r$.[24] It is then possible to increment each of the fundamental physi-

---

to a theoretical model) would likely select one unique structure as a consequence of its evolutionary history. In this sense, the Vogt–Russell "theorem" should more appropriately be considered as a general rule rather than as a rigorous law.

[23]**Polytropes** assume that the pressure equation of state may be written in the form $P = K\rho^{(n+1)/n}$, where $K$ is a constant and $n$ is known as the *polytropic index*. For example, $n = 1.5$ corresponds to an adiabatic gas with $\gamma = 5/3$, appropriate for purely convective models, while $n = 3$ corresponds to a star in radiative equilibrium. Only models corresponding to $n = 0, 1,$ or $5$ have analytic solutions; all other choices of $n$ require numerical techniques to solve the differential equations. There are many excellent discussions of polytropes available for the interested reader, including Clayton (1968).

[24]Codes that treat the radius as an independent variable are called **Eulerian** codes. **Lagrangian** codes treat the mass as an independent variable. In the Lagrangian formulation the differential equations are rewritten through the use of Eq. (10.8); the hydrostatic equilibrium equation can be written in the form $dP/dM$, for instance.

cal parameters through successive applications of the difference equations. For instance, if the pressure in zone $i$ is given by $P_i$, then the pressure in the next deepest zone, $P_{i+1}$, is found from

$$P_{i+1} = P_i + \frac{\Delta P}{\Delta r}\delta r,$$

where $\delta r$ is negative.

The numerical integration of the stellar structure equations may be carried out from the surface toward the center, from the center toward the surface, or, as is often done, in both directions simultaneously. If the integration is carried out in both directions, the solutions will meet at some *fitting point* where the variables must vary smoothly from one solution to the other. This last approach is frequently taken because the most important physical processes in the outer layers of stars generally differ from those in the deep interiors. The transfer of radiation through optically thin zones and the ionization of hydrogen and helium occur close to the surface while nuclear reactions occur near the center. By integrating in both directions, these processes may be decoupled somewhat, simplifying the problem.

Simultaneously matching the surface and central boundary conditions for a desired stellar model usually requires several iterations before a satisfactory solution is obtained. If the surface-to-center and center-to-surface integrations do not agree at the fitting point, the starting conditions must be changed. This is accomplished in a series of attempts, called *iterations*, where the initial conditions of the next integration are estimated from the outcome of the previous integration. A process of successive iterations is also necessary if the star is integrated from the surface to the center or from the center to the surface; in these cases the fitting points are simply the center and surface, respectively.

A very simple stellar structure code (called STATSTAR), written in FORTRAN, is given in Appendix H. STATSTAR integrates the stellar structure equations developed in this chapter in their time-independent form from the outside of the star to the center using the appropriate constitutive relations; it also assumes a constant (or homogeneous) composition throughout. So that the basic elements of stellar model building can be more easily understood, many of the sophisticated numerical techniques present in research codes have been neglected, as have the detailed calculations of the pressure equation of state and the opacity. The complex formalism of the mixing length theory has also been left out in favor of the simplifying assumption of adiabatic convection. Despite these approximations, very reasonable models may be obtained for stars lying on the main sequence of the H–R diagram. An example of a 1 $M_\odot$ star is given in Appendix I to help the student interpret the output of STATSTAR.

## 10.6 The Main Sequence

The analysis of stellar spectra tells us that the atmospheres of the vast majority of all stars are composed primarily of hydrogen, usually in the range of 70% by mass ($X \sim 0.7$), while the amount of metals varies from near zero to approximately 3% ($0 < Z < 0.03$). Assuming that the initial composition of a star is **homogeneous** (meaning that the composition is the same throughout), then the first set of nuclear fusion reactions ought to be those that convert hydrogen into helium, either the pp chain or the CNO cycle. Recall that these reactions occur at the lowest temperatures because the associated Coulomb barrier is lower than for the burning of more massive nuclei. Consequently, the structure of the star ought to be strongly influenced by hydrogen nuclear burning deep within its interior.

Because of the predominance of hydrogen that initially exists in the core, and since hydrogen burning is a relatively slow process, the interior composition and structure of the star will slowly change. As we saw in Example 10.4, a rough estimate of the hydrogen burning lifetime of the Sun is 10 billion years. Of course, the surface conditions will not be completely static. By the Vogt–Russell theorem, any change in composition or mass requires a readjustment of the effective temperature and luminosity; *the observational characteristic of the star must change as a consequence of the central nuclear reactions.* As long as changes in the core are slow, so are the evolutionary changes in the observed surface features.[25] Consequently, it is expected that the observational characteristics of stars should change very little during their core hydrogen-burning lifetimes.

Since most stars have similar compositions, there ought to be a smooth change in the structures of stars with mass. Recall from Examples 10.1 and 10.2 that as the mass increases, the central pressure and the central temperature should increase. Therefore, for stars of low mass, the pp chain will dominate since less energy is required to initiate these reactions than the reactions of the CNO cycle. For high-mass stars, the CNO cycle will likely dominate because of its very strong temperature dependence. This also implies that the total luminosity of a star should increase with mass because the nuclear reaction rates increase with $T$.

---

[25]Some short-period surface changes can occur that are essentially decoupled from the long-term variations in the core. Stellar pulsations require specific conditions to exist, but their time scales are usually much shorter than the nuclear time scale. These oscillations will be discussed in Chapter 14.

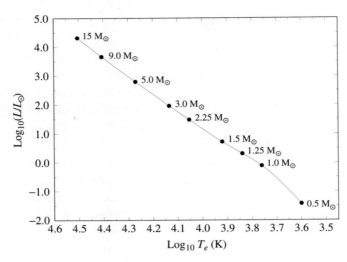

**Figure 10.12** The locations of stellar models on a theoretical H–R diagram. The models were computed using the stellar structure equations and constitutive relations. (Data from Iben, *Annu. Rev. Astron. Astrophys.*, 5, 571, 1967.)

At some point, as progressively less massive stars are considered, the central temperature will diminish to the point where stable nuclear reactions are no longer possible. This has been shown to occur at approximately 0.08 $M_\odot$. At the other extreme, stars with masses greater than approximately 90 $M_\odot$ become subject to thermal oscillations in their centers that produce dramatic variations in the nuclear energy generation rates over time scales as short as 8 hours, prohibiting the formation of stable stars.[26] As a result, core hydrogen burning in main-sequence stars is limited to objects whose masses are between these limits.

From theoretical models that are computed in the mass range of hydrogen burning, it is possible to obtain a numerical relationship between $M$ and $L$ that agrees well with the observational mass–luminosity relation shown in Fig. 7.7. It is also possible to locate each of the models on a theoretical H–R diagram (see Fig. 10.12). By comparison with Fig. 8.11, it can be seen that stars undergoing hydrogen burning in their cores lie along the observational main sequence!

The range in main-sequence luminosities is from near $5 \times 10^{-4}$ $L_\odot$ to approximately $1 \times 10^6$ $L_\odot$, a variation of over nine orders of magnitude, while the masses change by only three orders of magnitude. Because of the enormous

---

[26]This so-called $\epsilon$ *pulsation mechanism* will be discussed in Section 14.2.

## 10.6 The Main Sequence

rate of energy output from upper main-sequence stars, they consume their core hydrogen in a much shorter period of time than do stars on the lower end of the main sequence. As a result, main-sequence lifetimes decrease with increasing luminosity. Estimates of the range of main-sequence lifetimes are left as an exercise.

Effective temperatures are much less dependent on stellar mass. From approximately 2700 K for 0.08 $M_\odot$ stars to near 53,000 K for 90 $M_\odot$ stars, the increase in effective temperature amounts to only a factor of about 20. However, this variation is still large enough to change dramatically the stellar spectrum that is observed, since the dissociation energies of molecules and the ionization potentials of most elements lie within this range, as was demonstrated in Chapter 8. Consequently, by comparison with theoretical models, it is possible to correlate main-sequence masses with observed spectra.

The interior structure of stars along the main sequence also varies with mass, primarily in the location of convection zones. In the upper portion of the main sequence, where energy generation is due to the strongly temperature-dependent CNO cycle, convection is dominant in the core. This occurs because the rate of energy generation changes quickly with radius and radiation is not efficient enough to transport all of the energy being released in nuclear reactions. Outside of the hydrogen burning core, radiation is again capable of handling the flux and convection ceases. As the stellar mass decreases, so does the central temperature and the energy output of the CNO cycle until, near 1.2 $M_\odot$, the pp chain begins to dominate and the core becomes radiative. Meanwhile, near the surface of the star, as the effective temperature decreases with decreasing mass, the opacity increases, in part because of the location of the zone of hydrogen ionization (recall Fig. 9.10). The increase in opacity makes convection more efficient than radiation near the surfaces of stars with masses less than approximately 1.3 $M_\odot$. This has the effect of creating convection zones near the surfaces of these stars. As we continue to move down the main sequence, the bottom of the convection zone lowers until the entire star becomes convective near 0.3 $M_\odot$.

Through the use of the fundamental physical principles developed thus far in this text, we have been able to build realistic models of main-sequence stars and develop an understanding of their interiors. However, other stars remain on the observational H–R diagram that do not lie along the main sequence (see Fig. 8.11). By considering the changes in stellar structure that occur because of changes in composition due to nuclear burning (the Vogt–Russell theorem), it will become possible to explain their existence as well. The evolution of stars is discussed in Chapters 12 and 13.

## Suggested Readings

GENERAL

Kippenhahn, Rudolf, *100 Billion Suns*, Basic Books, New York, 1983.

TECHNICAL

Bahcall, John N., *Neutrino Astrophysics*, Cambridge University Press, Cambridge, 1989.

Bahcall, John N., and Ulrich, Roger K., "Solar Models, Neutrino Experiments, and Helioseismology," *Reviews of Modern Physics*, 60, 297, 1988.

Barnes, C. A., Clayton, D. D., and Schramm, D. N. (eds.), *Essays in Nuclear Astrophysics*, Cambridge University Press, Cambridge, 1982.

Bowers, Richard L., and Deeming, Terry, *Astrophysics I: Stars*, Jones and Bartlett Publishers, Boston, 1984.

Clayton, Donald D., *Principles of Stellar Evolution and Nucleosynthesis*, McGraw-Hill, New York, 1968.

Cox, J. P., and Giuli, R. T., *Principles of Stellar Structure*, Gordon and Breach, New York, 1968.

Fowler, William A., Caughlan, Georgeanne R., and Zimmerman, Barbara A., "Thermonuclear Reaction Rates, I," *Annual Review of Astronomy and Astrophysics*, 5, 525, 1967.

Fowler, William A., Caughlan, Georgeanne R., and Zimmerman, Barbara A., "Thermonuclear Reaction Rates, II," *Annual Review of Astronomy and Astrophysics*, 13, 69, 1975.

Hansen, C. J., and Kawaler, S. D., *Stellar Interiors: Physical Principles, Structure, and Evolution*, Springer-Verlag, New York, 1994.

Harris, Michael J., Fowler, William A., Caughlan, Georgeanne R., and Zimmerman, Barbara A., "Thermonuclear Reaction Rates, III," *Annual Review of Astronomy and Astrophysics*, 21, 165, 1983.

Iben, Icko, Jr., "Stellar Evolution Within and Off the Main Sequence," *Annual Review of Astronomy and Astrophysics*, 5, 571, 1967.

Kippenhahn, Rudolf, and Weigert, Alfred, *Stellar Structure and Evolution*, Springer-Verlag, Berlin, 1990.

Liebert, James, and Probst, Ronald G., "Very Low Mass Stars," *Annual Review of Astronomy and Astrophysics, 25*, 473, 1987.

Novotny, Eva, *Introduction to Stellar Atmospheres and Interiors*, Oxford University Press, New York, 1973.

# Problems

10.1 Show that the equation for hydrostatic equilibrium, Eq. (10.7), can also be written in terms of the optical depth $\tau$, as
$$\frac{dP}{d\tau} = \frac{g}{\kappa}.$$
This form of the equation is often useful in building model stellar atmospheres.

10.2 Prove that the gravitational force on a point mass located anywhere inside a hollow, spherically symmetric shell is zero. Assume that the mass of the shell is $M$ and has a constant density $\rho$. Assume also that the radius of the inside surface of the shell is $r_1$ and that the radius of the outside surface is $r_2$. The mass of the point is $m$.

10.3 Assuming that 10 eV could be released by every atom in the Sun through chemical reactions, estimate how long the Sun could shine at its current rate through chemical processes alone. For simplicity, assume that the Sun is composed entirely of hydrogen. Is it possible that the Sun's energy is entirely chemical? Why or why not?

10.4 (a) What temperature would be required for two protons to collide if quantum mechanical tunneling is neglected? Assume that nuclei having velocities ten times the root-mean-square (rms) value for the Maxwell–Boltzmann distribution can overcome the Coulomb barrier. Compare your answer with the estimated central temperature of the Sun.

(b) Using Eq. (8.1), calculate the ratio of the number of protons having velocities ten times the rms value to those moving at the rms velocity.

(c) Assuming (incorrectly) that the Sun is pure hydrogen, estimate the number of hydrogen nuclei in the Sun. Could there be enough protons moving with a speed ten times the rms value to account for the Sun's luminosity?

**10.5** Derive the ideal gas law, Eq. (10.11). Begin with the pressure integral (Eq. 10.10) and the Maxwell–Boltzmann velocity distribution function (Eq. 8.1).

**10.6** Derive Eq. (10.34) from Eq. (8.1).

**10.7** Show that the form of the Coulomb potential barrier penetration probability given by Eq. (10.38) follows directly from Eq. (10.37).

**10.8** Prove that the energy corresponding to the Gamow peak is given by Eq. (10.41).

**10.9** Calculate the ratio of the energy generation rate for the pp chain to the energy generation rate for the CNO cycle given conditions characteristic of the center of the present-day (evolved) Sun, namely $T = 1.58 \times 10^7$ K, $\rho = 162$ g cm$^{-3}$, $X = 0.34$, and $X_{\text{CNO}} = 0.013$.[27] Assume that the pp chain screening factor is unity ($f_{pp} = 1$) and that the pp chain branching factor is unity ($\psi_{pp} = 1$).

**10.10** Beginning with Eq. (10.56) and writing the energy generation rate in the form
$$\epsilon(T) = \epsilon'' T_8^\alpha,$$
show that the temperature dependence for the triple alpha process, given by Eq. (10.57), is correct. $\epsilon''$ is a function that is independent of temperature.

*Hint:* First take the natural logarithm of both sides of Eq. (10.56) and then differentiate with respect to $\ln T_8$. Follow the same procedure with your power law form of the equation and compare the results. You may want to make use of the relation
$$\frac{d \ln \epsilon}{d \ln T_8} = \frac{d \ln \epsilon}{\frac{1}{T_8} dT_8} = T_8 \frac{d \ln \epsilon}{dT_8}.$$

**10.11** The $Q$ value of a reaction is the amount of energy released (or absorbed) during the reaction. Calculate the $Q$ value for each step of the PP I reaction chain (Eq. 10.46). Express your answers in MeV. The masses of $^2_1$H and $^3_2$He are 2.0141 u and 3.0160 u, respectively.

---

[27]The interior values assumed here are taken from the standard solar model of Guzik (private communication), 1994; see Section 11.1.

# Problems

**10.12** Calculate the amount of energy released or absorbed in the following reactions (express your answers in MeV):

(a) $^{12}_{6}\text{C} + ^{12}_{6}\text{C} \rightarrow ^{24}_{12}\text{Mg} + \gamma$

(b) $^{12}_{6}\text{C} + ^{12}_{6}\text{C} \rightarrow ^{16}_{8}\text{O} + 2\,^{4}_{2}\text{He}$

(c) $^{19}_{9}\text{F} + ^{1}_{1}\text{H} \rightarrow ^{16}_{8}\text{O} + ^{4}_{2}\text{He}$

The mass of $^{12}_{6}\text{C}$ is 12.0000 u, by definition, and the masses of $^{16}_{8}\text{O}$, $^{19}_{9}\text{F}$, and $^{24}_{12}\text{Mg}$ are 15.99491 u, 18.99840 u, and 23.98504 u, respectively. Are these reactions exothermic or endothermic?

**10.13** Complete the following reaction sequences. Be sure to include any necessary leptons.

(a) $^{27}_{14}\text{Si} \rightarrow ^{?}_{13}\text{Al} + e^{+} + \;?$

(b) $^{?}_{13}\text{Al} + ^{1}_{1}\text{H} \rightarrow ^{24}_{12}\text{Mg} + ^{4}_{?}?$

(c) $^{35}_{17}\text{Cl} + ^{1}_{1}\text{H} \rightarrow ^{36}_{18}\text{Ar} + \;?$

**10.14** Prove that Eq. (10.75) follows from Eq. (10.74).

**10.15** Estimate the hydrogen burning lifetimes of stars on the lower and upper ends of the main sequence. The lower end of the main sequence[28] occurs near 0.085 $M_{\odot}$, with $\log_{10} T_e = 3.438$ and $\log_{10}(L/L_{\odot}) = -3.297$, while the upper end of the main sequence[29] occurs at approximately 90 $M_{\odot}$ with $\log_{10} T_e = 4.722$ and $\log_{10}(L/L_{\odot}) = 6.045$. Assume that the 0.085 $M_{\odot}$ star is entirely convective so that, through convective mixing, all of its hydrogen becomes available for burning rather than just the inner 10%.

**10.16** Using the information given in Problem 10.15, calculate the radii of a 0.085 $M_{\odot}$ star and a 90 $M_{\odot}$ star. What is the ratio of their radii?

**10.17** Verify that the basic stellar structure equations [Eqs. (10.7), (10.8), (10.45), (10.61)] are satisfied by the 1 $M_{\odot}$ STATSTAR model found in Appendix I. This may be done by selecting two adjacent zones and

---

[28] Data from Grossman, Hays, and Graboske, *Astron. Astrophys.*, 30, 95, 1974.
[29] Data from Cahn, Cox, and Ostlie, in *Lecture Notes in Physics: Stellar Pulsation*, Arthur N. Cox, Warren M. Sparks, and Sumner G. Starrfield (eds.), Springer-Verlag, Berlin, 51, 1987.

numerically computing the derivatives on the left-hand sides of the equations, for example

$$\frac{dP}{dr} \simeq \frac{P_{i+1} - P_i}{r_{i+1} - r_i},$$

and comparing your results with results obtained from the right-hand sides using average values of quantities for the two zones [e.g., $M_r = (M_i + M_{i+1})/2$]. Carry out your calculations for the two shells at $r = 1.27 \times 10^{10}$ cm and $r = 1.34 \times 10^{10}$ cm and then compare your results for the right- and left-hand sides of each equation by determining relative errors. Note that the model in Appendix I assumes complete ionization everywhere and has the uniform composition $X = 0.7$, $Y = 0.292$, $Z = 0.008$. Your results on the right- and left-hand sides will not agree exactly because STATSTAR uses a Runge–Kutta numerical algorithm that carries out intermediate steps not shown in Appendix I.

10.18 **Computer Problem** Appendix I contains an example of a theoretical 1.0 $M_\odot$ main-sequence star produced by the stellar structure code STATSTAR, found in Appendix H. Using STATSTAR, build a second main-sequence star with a mass of 0.75 $M_\odot$ that has a homogeneous composition of $X = 0.7$, $Y = 0.292$, and $Z = 0.008$. For these values, the model's luminosity and effective temperature are 0.1877 $L_\odot$ and 3839.1 K, respectively. Compare the central temperatures, pressures, densities, and energy generation rates between the 1.0 $M_\odot$ and 0.75 $M_\odot$ models. Explain the differences in the central conditions of the two models.

10.19 **Computer Problem** Use the stellar structure code STATSTAR found in Appendix H, together with the theoretical STATSTAR H–R diagram and mass–effective temperature data provided in Appendix I, to calculate a homogeneous, main-sequence model having the composition $X = 0.7$, $Y = 0.292$, and $Z = 0.008$. (*Note:* It may be more illustrative to assign each student in the class a different mass for this problem so that the results can be compared.)

(a) After obtaining a satisfactory model, plot $P$ versus $r$, $M_r$ versus $r$, $L_r$ versus $r$, and $T$ versus $r$.

(b) At what temperature has $L_r$ reached approximately 99% of its surface value? 50% of its surface value? Is the temperature associated with 50% of the total luminosity consistent with the rough estimate found in Eq. (10.33)? Why or why not?

(c) What is the value of $M_r/M_\star$ for the two temperatures found in part (b)? $M_\star$ is the total mass of the stellar model.

(d) If each student in the class calculated a different mass, compare the changes in the following quantities with mass:

    (i)   The central temperature.
    (ii)  The central density.
    (iii) The central energy generation rate.
    (iv) The extent of the central convection zone with mass fraction and radius.
    (v)  The effective temperature.
    (vi) The luminosity.

(e) If each student in the class calculated a different mass,

    (i)   Plot each model on a graph of luminosity versus mass (i.e., plot $L_\star/L_\odot$ versus $M_\star/M_\odot$).
    (ii)  Plot $\log_{10}(L_\star/L_\odot)$ versus $\log_{10}(M_\star/M_\odot)$ for each stellar model.
    (iii) Using an approximate power law relation of the form

$$L_\star/L_\odot = (M_\star/M_\odot)^\alpha,$$

find an appropriate value for $\alpha$. $\alpha$ may differ for different compositions or vary somewhat with mass. This is known as the mass–luminosity relation (see Fig. 7.7).

10.20 **Computer Problem** Repeat Problem 10.19 using the same mass but a different composition; assume $X = 0.7$, $Y = 0.290$, $Z = 0.010$.

(a) For a given mass, which model has the largest central temperature? the largest central density?

(b) Referring to the appropriate stellar structure equations and constitutive relations, explain your results in part (a).

(c) Which model has the largest energy generation rate at the center? Why?

(d) How do you account for the differences in effective temperature and luminosity between your two models?

# Chapter 11

# The Sun

## 11.1 The Solar Interior

Over the last few chapters we have investigated the theoretical foundations of stellar structure, treating the star as being comprised of an atmosphere and an interior. The distinction between the two regions is fairly nebulous. Loosely, the atmosphere is considered to be that region where the optical depth is less than unity and the simple approximation of photons *diffusing* through optically thick material is not justified (see Eq. 9.25). Instead, atomic line absorption and emission must be considered in detail in the stellar atmosphere. On the other hand, nuclear reaction processes deep in the stellar interior play a crucial role in the star's energy output and its inevitable evolution.

Due to its proximity to us, the star for which we have the greatest amount of observational data is our Sun. This wealth of information provides us with an excellent test of the theory of stellar atmospheres and interiors. Although most of the fundamental predictions of standard computer models of the Sun are in excellent agreement with observations, some discrepancies still remain. The resolution of these differences will certainly shed new light on a number of important astrophysical problems and could very well have a significant impact on other more fundamental areas of physics.

Based on its observed luminosity and effective temperature, our Sun is classified as a typical main-sequence star of spectral class G2 with a surface composition of $X = 0.73$ and $Z = 0.02$ (the mass fractions of hydrogen and metals, respectively). To understand how it has evolved to this point, recall that according to the Vogt–Russell theorem the mass and composition of a star dictate its internal structure. Neglecting any mass loss that may have occurred in the past, our Sun has been converting hydrogen to helium via the

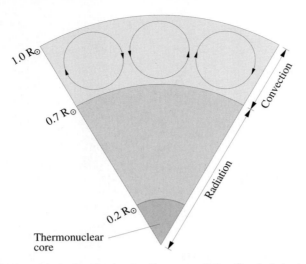

**Figure 11.1** A schematic diagram of the Sun's interior.

pp chain during most of its lifetime, thereby changing its composition and its structure. By comparing the results of radioactive dating tests of Moon rocks and meteorites with stellar evolution calculations, the current age of the Sun is determined to be approximately $4.52 \times 10^9$ yr.

From this information, a **standard solar model** may be constructed for the present-day Sun using the physical principles discussed in preceding chapters. A schematic diagram of such a model is shown in Fig. 11.1.[1] According to one evolutionary sequence leading to a present-day standard model, during its lifetime, the mass fraction of hydrogen ($X$) in the Sun's center has decreased from its initial value of 0.71 to 0.34 while the central mass fraction of helium ($Y$) has increased accordingly, from 0.27 to 0.64.[2] Furthermore, due to diffusive settling of elements heavier than hydrogen, the mass fraction of hydrogen near the surface may have increased by approximately 0.03 while the mass fraction of helium has decreased by a similar amount. The change in the Sun's central composition since it began nuclear burning has had a direct influence on its observable luminosity and radius, as shown in Fig. 11.2; the luminosity has increased by 40% while the size of the star has increased by more than 10%.

---

[1] *Nonstandard* solar models invoke exotic or hypothetical physical mechanisms in an attempt to explain lingering discrepancies between the standard models and certain observations, such as the flux of solar neutrinos; see below.

[2] The data quoted here and in the following discussion are from the standard solar model of Joyce Guzik, private communication, 1994.

## 11.1 The Solar Interior

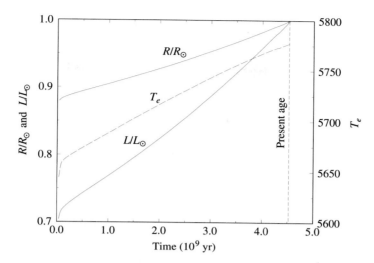

**Figure 11.2** The evolution of the Sun from its birth to the present. As a result of changes in its internal composition, the Sun has become larger and brighter. The solid lines indicate its radius and luminosity while the dashed line represents its effective temperature. The radius and luminosity curves are relative to present-day values. (Data from Guzik, private communication, 1994.)

From a theoretical standpoint, it is not at all clear how this change in solar energy output has altered Earth during its history, primarily because of uncertainties in the behavior of the terrestrial environment. Understanding the complex interaction between the Sun and Earth involves the detailed calculation of convection in Earth's atmosphere, as well as the effects of the atmosphere's time-varying composition and the nature of the continually changing reflectivity of Earth's surface.[3]

Table 11.1 gives the values of the central temperature, pressure, density, and composition for a standard solar model of Guzik.[4] Because of the Sun's past evolution, its composition is no longer homogeneous (constant throughout) but instead shows the influence of ongoing nucleosynthesis, surface convection, and elemental diffusion (settling of heavier elements). The composition structure of the Sun is shown in Fig. 11.3 for $^1_1$H, $^3_2$He, and $^4_2$He. Since the

---

[3]The ratio of the amount of reflected sunlight to incident sunlight is known as the **albedo**. Earth's albedo is affected by, among other things, the amount of surface water and ice.

[4]Various researchers find slightly different values for central parameters, depending on assumptions about composition, opacities, convection, and so on. For instance, typical values of the central density range from approximately 150 g cm$^{-3}$ to 160 g cm$^{-3}$ while central temperatures range from $1.56 \times 10^7$ K to $1.58 \times 10^7$ K.

| | |
|---|---|
| Temperature | $1.58 \times 10^7$ K |
| Pressure | $2.50 \times 10^{17}$ dynes cm$^{-2}$ |
| Density | $1.62 \times 10^2$ g cm$^{-3}$ |
| X | 0.336 |
| Y | 0.643 |

**Table 11.1** Central Conditions in the Sun. (Data from Guzik, private communication, 1994.)

Sun's primary energy production mechanism is the pp chain, $^3_2$He is an intermediate species in the reaction sequence. During the conversion of hydrogen to helium, $^3_2$He is produced and then destroyed again [see Eqs. (10.46), (10.47), and (10.48)]. At the top of the hydrogen burning region, where the temperature is lower, $^3_2$He is relatively more abundant because it is produced more easily than it is destroyed.[5] At greater depths, the higher temperatures allow the helium–helium interaction to proceed more rapidly and the $^3_2$He abundance again decreases (the temperature structure of the Sun is shown in Fig. 11.4). The slight ramp in the $^1_1$H and $^4_2$He curves near 0.7 R$_\odot$ reflects evolutionary changes in the position of the base of the surface convection zone, combined with the effects of elemental diffusion. Within the convection zone, turbulence results in essentially complete mixing and a homogeneous composition.

The largest contribution to the energy production in the Sun occurs at approximately one-tenth of the solar radius, as can be seen in the luminosity curve (Fig. 11.5) and the curve of its derivative with respect to radius (Fig. 11.6). If this result seems unexpected, consider that the mass conservation equation (Eq. 10.8)

$$\frac{dM_r}{dr} = 4\pi r^2 \rho$$

gives

$$dM_r = 4\pi r^2 \rho \, dr = \rho \, dV,$$

indicating that the amount of mass within a certain radius interval increases with radius simply because the volume of a spherical shell, $dV = 4\pi r^2 \, dr$, increases with $r$ for a fixed choice of $dr$. Of course, the mass contained in the shell also depends on the density of the gas. Consequently, even if the amount of energy liberated per gram of material ($\epsilon$) decreases steadily from the center outward, the largest contribution to the total luminosity will occur, not at the

---

[5] Recall that much higher temperatures are required for helium–helium compared to proton–proton interactions.

## 11.1 The Solar Interior

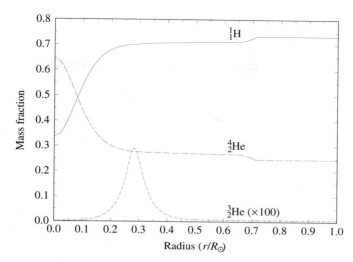

**Figure 11.3** The abundances of $^1_1$H, $^3_2$He, and $^4_2$He as a function of radius for the Sun. (Data from Guzik, private communication, 1994.)

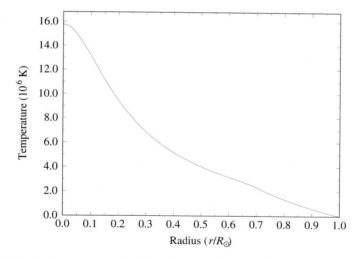

**Figure 11.4** The temperature in the solar interior as a function of radius. (Data from Guzik, private communication, 1994.)

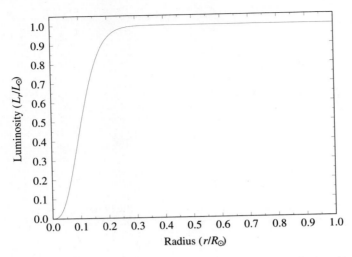

**Figure 11.5** The interior luminosity of the Sun as a function of radius. (Data from Guzik, private communication, 1994.)

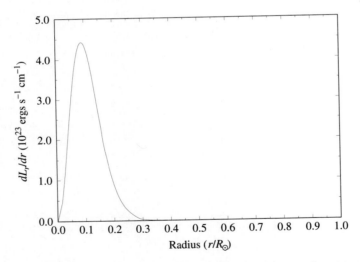

**Figure 11.6** The derivative of the Sun's interior luminosity with respect to radius, showing the location of the greatest contribution to the energy output. (Data from Guzik, private communication, 1994.)

## 11.1 The Solar Interior

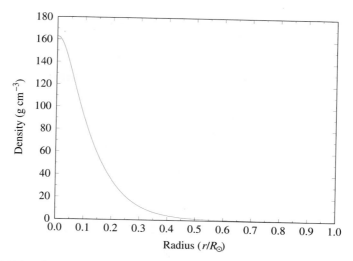

**Figure 11.7** The density structure of the Sun as a function of radius. (Data from Guzik, private communication, 1994.)

center, but in a shell that contains a significant amount of mass. In the case of the middle-aged Sun, the decrease in the amount of available fuel (hydrogen) at its center will also influence the location of the peak in the energy production region (see Eq. 10.49).

Figures 11.7 and 11.8 show just how rapidly the density and pressure change with radius in the Sun. These variations are forced on the solar structure by the condition of hydrostatic equilibrium (Eq. 10.7), the ideal gas law (Eq. 10.14), and the composition structure of the star. Of course, boundary conditions applied to the stellar structure equations require that both $\rho$ and $P$ become negligible at the surface (Eq. 10.97).

Figure 11.9 shows the interior mass ($M_r$) as a function of radius. Notice that 90% of the mass of the star is located within one-half of its radius. This should not come as a complete surprise, since Fig. 11.7 indicates how the density increases significantly as the center of the Sun is approached. Integration of the density function over the volume of the star from the center outward [i.e., the integration of Eq. (10.8)] yields the interior mass function.

The question remains as to how the energy generated in the interior is transported outward. Recall that in Chapter 10 we determined a criterion for the onset of convection in stellar interiors, namely that the temperature gradient become superadiabatic (Eq. 10.86),

$$\left|\frac{dT}{dr}\right|_{\text{act}} > \left|\frac{dT}{dr}\right|_{\text{ad}}.$$

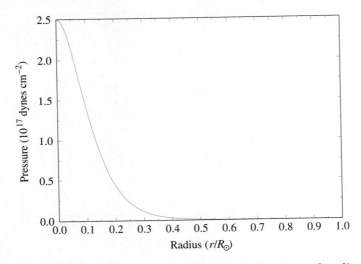

**Figure 11.8** The interior pressure in the Sun as a function of radius. (Data from Guzik, private communication, 1994.)

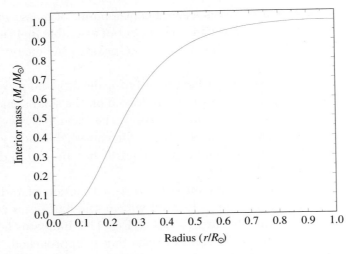

**Figure 11.9** The interior mass as a function of radius for the Sun. (Data from Guzik, private communication, 1994.)

## 11.1 The Solar Interior

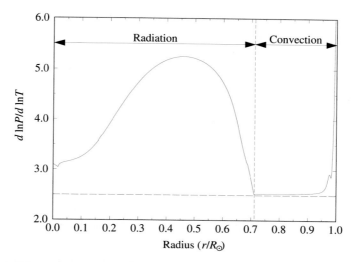

**Figure 11.10** The convection condition $d\ln P/d\ln T$ plotted versus $r/R_\odot$. The dashed horizontal line represents the boundary between adiabatic convection and radiation for an ideal monatomic gas. The onset of convection does not exactly agree with the ideal adiabatic case because of the incorporation of a sophisticated equation of state and a more detailed treatment of convection physics. The rapid rise in $d\ln P/d\ln T$ near the surface is associated with the highly superadiabatic nature of convection in that region (i.e., the adiabatic approximation that convection occurs when $d\ln P/d\ln T < 2.5$ is invalid near the surface of the Sun). (Data from Guzik, private communication, 1994.)

Under the simplifying assumption of an ideal monatomic gas, this condition becomes (Eq. 10.87)

$$\frac{d\ln P}{d\ln T} < 2.5.$$

$d\ln P/d\ln T$ is plotted versus $r/R_\odot$ in Fig. 11.10. As can be seen, the Sun is purely radiative below approximately $r/R_\odot = 0.71$ and becomes convective above that point. Physically this occurs because the opacity in the outer portion of the Sun becomes large enough to inhibit the transport of energy by radiation; recall that the radiative temperature gradient is proportional to the opacity (see Eq. 10.61). When the temperature gradient becomes too large, convection becomes the more efficient means of energy transport. In the region of convective energy transport, $d\ln P/d\ln T \simeq 2.5$, which is characteristic of the nearly adiabatic temperature gradient of most convection zones. The rapid rise in $d\ln P/d\ln T$ above 0.95 $R_\odot$ is due to the significant departure of the actual temperature gradient from an adiabatic one. In this case convection must be described by a more detailed treatment, such as the mixing length theory discussed at the end of Section 10.4.

Notice also that a minimum in $d\ln P/d\ln T$ occurs near the core of the Sun. Although the Sun remains purely radiative in the center, the large amount of energy that must be transported outward pushes the temperature gradient in the direction of becoming superadiabatic. We will see in Chapter 13 that stars only slightly more massive than the Sun are convective in their centers because of the stronger temperature dependence of the CNO cycle as compared to the pp chain.

Clearly, an enormous amount of information is available regarding the solar interior, derived from the direct and careful application of the stellar structure equations and the fundamental physical principles described in the last three chapters. A very complete and reasonable model of the Sun can be produced that would be consistent with evolutionary time scales and would fit the global characteristics of the star, specifically its mass, luminosity, radius, effective temperature, surface composition, and, as we will see in the next section, its observed surface convection zone. However, there are several aspects of standard solar models of the interior that do not completely agree with observations and do not as yet have satisfactory explanations: the number of neutrinos that are detected coming from the Sun's core, the observed surface abundance of lithium, and the frequencies of the oscillations the Sun experiences. The lithium problem will be discussed in Chapter 13 and solar oscillations will be considered in Chapter 14. These problems, taken together, provide important information about the Sun's interior and may indicate the need for revisions in our understanding of some fundamental physical processes.

The **solar neutrino problem** has been around for some two decades, ever since Raymond Davis began measuring the neutrino flux from the Sun using a detector located almost one mile below ground in the Homestake Gold Mine in Lead, South Dakota (Fig. 11.11). Because of the very low cross section of neutrino interactions with other matter, neutrinos can easily travel completely through Earth while other particles originating from space cannot. As a result, the underground detector is assured of measuring what it is designed to measure—neutrinos created eight minutes earlier in the solar core.

The Davis neutrino detector contains 100,000 gallons of cleaning fluid, $C_2Cl_4$ (perchloroethylene). One isotope of chlorine ($^{37}_{17}Cl$) is capable of interacting with neutrinos of sufficient energy to produce a radioactive isotope of argon that has a half-life of 35 days,

$$^{37}_{17}Cl + \nu_e \rightleftharpoons {}^{37}_{18}Ar + e^-. \tag{11.1}$$

The threshold energy for this reaction, 0.814 MeV, is less than the energies

## 11.1 The Solar Interior

**Figure 11.11** The Davis solar neutrino detector. The tank is located 4900 feet below ground in the Homestake Gold Mine in Lead, South Dakota, and is filled with 100,000 gallons of $C_2Cl_4$. (Courtesy of Brookhaven National Laboratory.)

of the neutrinos produced in every step of the pp chain except the crucial first one, $^1_1H + ^1_1H \rightarrow ^2_1H + e^+ + \nu_e$ [see Eqs. (10.46), (10.47), and (10.48)]. However, the reaction that accounts for 77% of the neutrinos detected in the Davis experiment is the decay of $^8_5B$ in the PP III chain,

$$^8_5B \rightarrow ^8_4Be + e^+ + \nu_e. \tag{11.2}$$

Unfortunately, this reaction is very rare, accounting only for one pp chain termination in 5000.

Once every few months Davis and his collaborators carefully purge the accumulated argon from the tank and count the atoms produced. The counting rate is measured in terms of the **solar neutrino unit**, or SNU (1 SNU $\equiv 10^{-36}$ reactions per target atom per second). With approximately $2.2 \times 10^{30}$ atoms of $^{37}_{17}Cl$ atoms in the tank, if only one argon atom is produced each day, this rate would correspond to 5.35 SNU.

Results of the Davis experiment are shown in Fig. 11.12. Standard solar models predict that the outcome of the experiment should yield a capture rate of 7.9 SNU while the actual data give $2.23 \pm 0.26$ SNU; only one argon atom is produced every two to three days! This discrepancy is the solar neutrino problem.

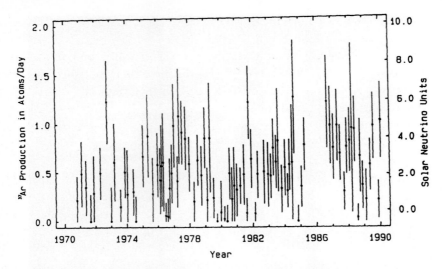

**Figure 11.12** Results of the Davis solar neutrino experiment from 1970 to 1990. The uncertainties in the experimental data are shown by vertical error bars associated with each run. The dashed horizontal line is the predicted solar neutrino capture rate for the $^{37}_{17}$Cl detector. (Figure from Davis and Cox, *Solar Interior and Atmosphere*, Cox, Livingston, and Matthews (eds.), The University of Arizona Press, Tucson, 1991.)

Other neutrino experiments, fundamentally different from the $^{37}_{17}$Cl experiment, have confirmed the existence of the solar neutrino problem. Japan's underground Kamiokande II detector uses 3000 tons of water and detects the **Cerenkov light** that is produced when neutrinos scatter electrons, causing the electrons to move at speeds greater than the speed of light in water.[6] The number of neutrinos detected by Kamiokande II remains nearly constant at about half the number expected from standard solar models. The Soviet–American Gallium Experiment (SAGE), located at the Baksan Neutrino Laboratory (inside a mountain in the Caucasus), and GALLEX (at the Gran Sasso underground laboratory in Italy) are designed to measure the low-energy pp chain neutrinos that dominate the Sun's neutrino flux. SAGE and GALLEX make their detections via a reaction that converts gallium into germanium,

$$\nu_e + {}^{71}_{31}\text{Ga} \rightarrow {}^{71}_{32}\text{Ge} + e^-.$$

After considering the expected number of background counts from sources

---

[6]Note that this *does not* violate Einstein's special theory of relativity since the special theory applies to the speed of light *in a vacuum*. The speed of light in any other medium is always less than the speed of light in a vacuum.

## 11.1 The Solar Interior

other than the Sun, both experiments confirm the deficit of neutrinos first established by the Davis detector.

The significance of the solar neutrino problem cannot be overstated. Researchers have been trying to resolve the disagreement between experiment and theory ever since the problem first appeared. Given the small numbers of neutrinos that are detected, one obvious solution (at least from a theorist's point of view!) is that there is something wrong with the experiments or the analysis of the resulting data. However, all attempts to find errors that would produce systematically low counts have failed.

Can the standard models be fundamentally flawed in some way? Do we really understand the solar interior well enough to predict neutrino fluxes accurately? Assuming that our knowledge of stellar nucleosynthesis is essentially correct, if the temperature of the solar interior was approximately 10% lower than theory predicts, the number of solar neutrinos emitted from the core would come into agreement with observation. However, using conventional physics, the condition of hydrostatic equilibrium requires a specific temperature in the core. As we will learn in the following chapters, other aspects of stellar structure theory agree very well with the observational data available for a wide variety of stars. Consequently, significant constraints on possible solutions to the solar neutrino problem exist.

The search for a theoretical resolution to the problems of the solar neutrinos and oscillations has considered two general approaches: Either some fundamental physical process operating in the standard models is incorrect, or something happens to the neutrinos on their way from the Sun's core to Earth. The first of these possibilities has resulted in the reexamination of a host of features of the solar model, including nuclear reaction rates, the opacity of stellar material, the evolution of the Sun up to its present state, and variations in the composition of the solar interior. Unfortunately, these models do not appear to satisfy all of the observational constraints simultaneously. For example, enhancing the helium abundance in the core may improve the agreement between theory and the observations with regard to solar oscillations, but it actually makes the solar neutrino problem worse.

A radical and controversial solution to the problem involves hypothetical particles called **WIMPs** (weakly interacting massive particles). Perhaps WIMPs are responsible for transporting some of the energy near the center of the Sun to other regions of the star, thereby cooling the core enough to lower the neutrino flux. Although this idea of creating a new particle to explain the solar neutrino problem may seem excessive, it does have some other interesting consequences. The theoretical frequencies in solar oscillations predicted by a WIMP model appear to be in better agreement with observations than are the frequencies predicted by the standard models. Also, another long-standing problem of astrophysics, which deals with identifying the unseen mass (referred to as **dark matter**) responsible for certain characteristics of galactic rotational velocity curves and the gravitational interactions between galaxies, may be solved by the existence of WIMPs.

An elegant solution to the solar neutrino problem that has been proposed suggests that the standard solar models are essentially correct but that the neutrinos produced in the Sun's core actually change before they reach Earth. The **Mikheyev–Smirnov–Wolfenstein** (or **MSW**) **effect** involves the transformation of neutrinos from one type to another. (This idea is an extension of the *electroweak theory* of particle physics that combines the electromagnetic theory with the theory of weak interactions governing some types of radioactive decay.) The neutrinos produced in the various branches of the pp chain are all electron neutrinos ($\nu_e$); however, two other *flavors* of neutrinos also exist—the muon neutrino ($\nu_\mu$) and the tau neutrino ($\nu_\tau$). The MSW effect suggests that neutrinos oscillate between being electron neutrinos, muon neutrinos, or tau neutrinos during their passage through the Sun. The neutrino oscillations are caused by interactions with electrons as the neutrinos travel toward the surface. Because the Davis, Kamiokande II, and gallium detectors (SAGE and GALLEX) have different threshold energies and they are sensitive only to the electron neutrino, their results seem to be consistent with the MSW theory.

One testable consequence of the MSW effect is that neutrinos would necessarily have mass, with $m(\nu_e) \ll m(\nu_\mu) \ll m(\nu_\tau)$. The required mass of the electron neutrino is well within the upper mass limit of approximately 7.2 eV established by experiment. Moreover, even though the current electroweak theory does not require masses for the neutrinos, most extensions of the electroweak theory that also attempt to include the strong nuclear force do require nonzero neutrino masses. Such extended theories, known as **grand unified theories** (**GUTs**), are currently the focus of intense research by high-energy (particle) physicists.

Whatever the solution to the solar neutrino problem, it is likely to play a significant role in further advancing our understanding of the most fundamental problems of physics.

## 11.2 The Solar Atmosphere

When we observe the Sun visually, it appears as though there is a very abrupt and clear edge to this hot, gaseous ball (Fig. 11.13). Of course, an actual "surface" does not exist; rather, what we are seeing is a region where the solar

## 11.2 The Solar Atmosphere

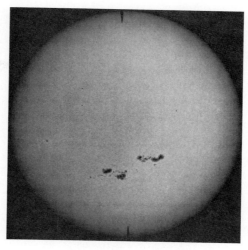

**Figure 11.13** The solar disk appears sharp because of the rapid increase in optical depth with distance through the photosphere; however, note also the effect of limb darkening. Sunspots are visible on the surface of the disk, as well. (Courtesy of The Observatories of the Carnegie Institution of Washington.)

atmosphere is *optically thin* and photons originating from that level travel unimpeded through space. Even this region is not clearly defined, however, since some photons can always escape when the optical depth is somewhat greater than unity while others may be absorbed when the optical depth is less than unity; but the odds of a photon leaving the solar atmosphere diminish rapidly as the optical depth increases. Consequently, the Sun's atmosphere changes from being optically thin to optically thick in only 500 km. This relatively small distance (about 0.07% of the Sun's radius) is what gives the "edge" of the Sun its sharp appearance.

The region where the observed optical photons originate is known as the solar **photosphere**. Within this region the temperature of the gas decreases from a value of 6500 K at the base of the photosphere (defined to be where the optical depth at 5000 Å is unity) to a minimum of 4400 K about 500 km higher up, the location of the top of the photosphere. Above this point, the temperature begins to rise again. The approximate thicknesses of the various components of the Sun's atmosphere to be discussed in this section, including the photosphere and chromosphere, are depicted in Fig. 11.14.

As was discussed in Section 9.3, the solar flux is emitted (on average) from an optical depth of $\tau = 2/3$ (the Eddington approximation). This leads to the identification of the effective temperature with the temperature of the gas

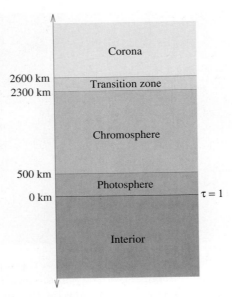

**Figure 11.14** The thicknesses of the components of the Sun's atmosphere. The base of the photosphere is located at $\tau = 1$.

at this depth, or $T_e = T_{\tau=2/3}$. Using the Sun's luminosity, determined from the solar constant (see Example 3.2), together with its radius, found from the angular size of the Sun, the effective temperature is calculated from Eq. (3.17) to be $T_e = 5770$ K.

Of course, optical depth is a function not only of the distance that a photon must travel to the surface of the Sun, but also of the wavelength-dependent opacity of the solar material (Eq. 9.15). Consequently, photons can originate from or be absorbed at different physical depths in the atmosphere, depending upon their wavelengths. Since a spectral line is not infinitesimally thin, but actually covers a range of wavelengths, even different parts of the same line are formed at different levels of the atmosphere. Thus solar observations with high wavelength resolution can be used to probe the atmosphere at various depths, providing a wealth of information about its structure.

When the base of the photosphere is observed, it appears as a patchwork of bright and dark regions that are constantly changing, with individual regions appearing and then disappearing. With a spatial extent of roughly 700 km, the characteristic lifetime for one of these regions is five to ten minutes. This patchwork structure is known as **granulation** and is the top of the convection zone protruding into the base of the photosphere. Figure 11.15 shows an image of the solar granulation present at the base of the photosphere. Significantly

## 11.2 The Solar Atmosphere

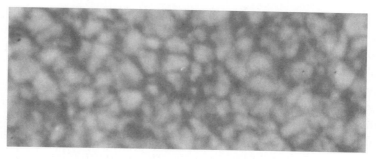

**Figure 11.15** Granulation at the base of the photosphere is due to the rising and falling gas bubbles produced by the underlying convection zone. (Courtesy of W. Livingston and the National Optical Astronomy Observatories.)

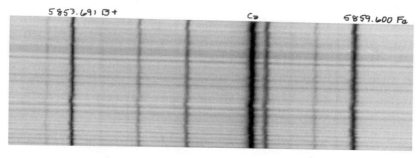

**Figure 11.16** A spectrum of a portion of the photospheric granulation showing absorption lines that indicate the presence of radial motions. Wiggles to the left are toward shorter wavelengths and are blueshifted while wiggles to the right are redshifted. (Courtesy of W. Livingston and the National Optical Astronomy Observatories.)

smaller features, if they exist, are not yet observable due to our resolution limit of about $1''$ for ground-based observations through Earth's turbulent atmosphere (see Chapter 6).

If a high-resolution spectrum of solar granulation spanning a number of convection cells is made, the absorption lines appear to have wiggles in them (Fig. 11.16). This occurs because some parts of the region are Doppler blueshifted while others are redshifted. Using Eq. (4.39), we find that radial velocities of 0.4 km s$^{-1}$ are common; brighter regions produce the blueshifted sections of the lines while darker regions produce the redshifted sections. Thus the bright cells are the vertically rising hot convective bubbles carrying energy from the solar interior. When those bubbles reach the optically thin photosphere, the energy is released via photons and the resulting cooler, darker gas sinks back into the interior. The lifetime of a typical granule is the amount of time needed for a convective eddy to rise and fall the distance of one mixing length. Solar granulation provides us with a visual verification of the results of the stellar structure equations applied to our Sun.

Clearly, the structure at the top of the convection zone is quite complex. Unfortunately, no theoretical description of solar convection is at present sufficiently detailed to reproduce all of these features. However, the relatively crude treatment of convection used in stellar structure codes can still provide us with some important information and produce reasonable models of global behavior.[7]

Recall that the convection zone is driven by an increase in opacity near the surface of the Sun. The Sun also radiates predominantly as a blackbody in the visible and infrared portions of the spectrum (note the relatively smooth features of the solar spectrum depicted in Fig. 9.5). This second observation suggests that there exists a source of opacity that is basically continuous across wavelength. Both phenomena are due in part to the existence of a unique ion in the photosphere—the $H^-$ ion, a hydrogen atom that possesses an extra electron. Because of the partial shielding that the nucleus provides, a second electron can be loosely bound to the atom on the side of the ion opposite that of the first electron. In this position the second electron is closer to the positively charged nucleus than it is to the negatively charged electron. Therefore, according to Coulomb's law (Eq. 5.9), the net force on the extra electron is attractive.

The binding energy of the $H^-$ ion is only 0.75 eV, compared with the 13.6 eV required to ionize the ground state hydrogen atom. As a result, any photon with energy in excess of the ionization energy can be absorbed by a $H^-$ ion, liberating the extra electron; the remaining energy becomes kinetic energy. Conversely, an electron captured by a hydrogen atom to form $H^-$ will release a photon corresponding to the kinetic energy lost by the electron together with the ion's binding energy,

$$H + e^- \rightleftharpoons H^- + \gamma.$$

Since 0.75 eV corresponds roughly to a photon of wavelength 17,000 Å, $H^-$ is capable of contributing to the continuum opacity in the infrared and at shorter wavelengths.

Using the Saha equation (Eq. 8.6), we can determine the ratio of the number

---

[7]At the time this text was written, research was underway to simulate the full three-dimensional process of solar convection. These simulations are very computationally intensive. Solar convection is one of the so-called Grand Challenge problems in computational astrophysics.

## 11.2 The Solar Atmosphere

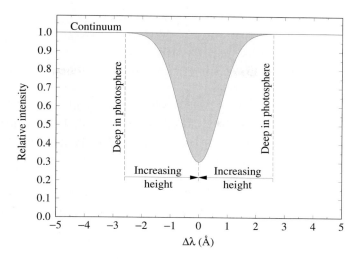

**Figure 11.17** The relationship between absorption line strength and depth in the photosphere for a typical spectral line.

of H$^-$ ions to neutral hydrogen atoms. It is left as an exercise to show that in the Sun's photosphere, only about one in $10^7$ hydrogen atoms actually forms an H$^-$ ion. The importance of H$^-$ in the Sun is due to the fact that even though the abundance of the ion is quite low, neutral hydrogen is not capable of contributing significantly to the continuum.

Fraunhofer absorption lines are also produced in the photosphere (see Section 5.1). According to Kirchhoff's laws, the absorption lines must be produced where the gas is cooler than the bulk of the continuum-forming region. Line formation must also occur between the observer and the region where much of the continuum is produced. In reality, the Fraunhofer lines are formed in the same layers where H$^-$ produces the continuum. However, the darkest part of the line (its center) originates from regions higher in the photosphere, where the gas is cooler. The reason that you don't see as deeply into the Sun at wavelengths where the opacity is greatest (at the central wavelength) is due to the existence of more intervening material between the source of most of the continuum light deep within the photosphere and the absorbing region higher up in the atmosphere. Moving away from the central wavelength toward the wing of the line implies that absorption is occurring at progressively deeper levels. At wavelengths sufficiently far from the central peak, the edge of the line merges with the continuum being produced at the base of the photosphere. This effect is illustrated in Fig. 11.17 (recall the discussion at the beginning of Section 9.4).

Photospheric absorption lines may also be used to measure the rotation rate of the Sun. By measuring Doppler shifts at the solar limb, we find that the Sun rotates *differentially* (i.e., the rate of rotation depends on the latitude being observed). At the equator the rotation period is approximately 26 days, increasing to 37 days at the poles.

The **chromosphere**, with an intensity that is only about $10^{-4}$ of the value for the photosphere, is that portion of the solar atmosphere that lies just above the photosphere and extends upward for approximately 2000 km. Analysis of the light produced in the chromosphere indicates that the gas density drops by nearly a factor of $10^4$ and that the temperature begins to *increase* with increasing altitude, from 4400 K to about 25,000 K. Reference to the Boltzmann and Saha equations [Eqs. (8.4) and (8.6), respectively] shows that lines that are not produced at the lower temperatures and higher densities of the photosphere can form in the environment of the chromosphere. For instance, along with the hydrogen Balmer lines, the lines of He II, Fe II, Si II, Cr II, and Ca II (in particular, the Ca II H and K lines, 3968 Å and 3933 Å) can appear in the spectrum.

Although the Fraunhofer lines appear as absorption lines in the visible and near ultraviolet portions of the spectrum, they begin to appear as emission lines at shorter (and much longer) wavelengths. Again Kirchhoff's laws offer an explanation, suggesting that a hot, low-density gas must be responsible. Because the interior of the Sun is optically thick below the base of the photosphere, the area of emission line production must occur elsewhere. With the peak of the blackbody spectrum near 5000 Å, the strength of the ultraviolet continuum decreases rapidly at shorter and longer wavelengths (Fig. 3.8). As a result, emission lines produced outside of the visible portion of the spectrum are not overwhelmed by the blackbody radiation.

Visible wavelength emission lines that are not normally seen against the bright solar disk can be observed for a few seconds at the beginning and end of a total eclipse of the Sun; this phenomenon is referred to as a **flash spectrum**. During this period, the portion of the Sun that is still visible takes on a reddish hue because of the dominance of the Balmer $H_\alpha$ emission line, a line that is normally observed only as an absorption line in the Sun's atmosphere.

Using filters that restrict observations to the wavelengths of the emission lines produced in the chromosphere (particularly $H_\alpha$), it is possible to see a great deal of structure in this portion of the atmosphere. **Supergranulation** becomes evident on scales of 30,000 km, showing the continued effects of the underlying convection zone. Doppler studies again reveal convective velocities on the order of 0.4 km s$^{-1}$, with gas rising in the centers of the supergranules and sinking at their edges. Also present are vertical filaments of gas, known as

## 11.2 The Solar Atmosphere

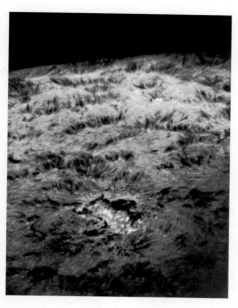

**Figure 11.18** Spicules in the chromosphere of the Sun. Observations were made using the H$_\alpha$ emission line. (Courtesy of the National Optical Astronomy Observatories/National Solar Observatory.)

**spicules**, extending upward from the chromosphere for 10,000 km (Fig. 11.18). An individual spicule may have a lifetime of only 15 minutes, but it is estimated that 30,000 spicules exist at any given moment, covering a few percent of the surface of the Sun. Doppler studies show that mass motions are present in spicules, with material moving outward at approximately 20 km s$^{-1}$.

Above the chromosphere, the temperature rises very rapidly in only a few hundred kilometers (see Fig. 11.19), reaching more than $10^6$ K before the temperature gradient flattens. This **transition region** may be selectively observed at various altitudes in the ultraviolet and extreme ultraviolet parts of the electromagnetic spectrum. For instance, the 1216 Å Lyman $\alpha$ emission line of hydrogen ($n = 2 \rightarrow n = 1$) is produced at the top of the chromosphere at 20,000 K, the C III 977 Å line originates at a level where the temperature is 90,000 K, the 1032 Å line of O VI occurs at 300,000 K, and Mg X creates a 625 Å line at $1.4 \times 10^6$ K. Figure 11.20 shows that the features evident in the chromosphere continue throughout the transition region, until structures begin to change significantly at the level of Mg X formation.

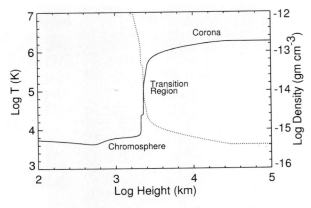

**Figure 11.19** The temperature (solid line) and number density structure (dotted line) of the upper atmosphere of the Sun. (Figure from Mariska, *The Solar Transition Region*, Cambridge University Press, Cambridge, 1992.)

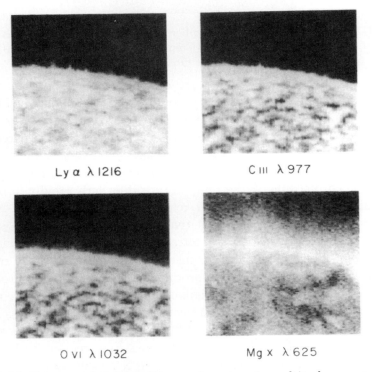

**Figure 11.20** Features of the transition region at various altitudes, ranging from the top of the chromosphere to the corona. (Courtesy of G. L. Withbroe.)

## 11.2 The Solar Atmosphere

**Figure 11.21** (a) The quiet solar corona seen during a total solar eclipse in 1954. The shape of the corona is elongated along the Sun's equator. (Courtesy of J. D. R. Bahng and K. L. Hallam.) (b) The active corona tends to have a very complex structure. This image of the July 11, 1991 eclipse is a composite of five photographs that was processed electronically. (Courtesy of S. Albers.)

When the Moon fully occults the photosphere during a total solar eclipse, the radiation from the faint **corona** becomes visible (Fig. 11.21). The corona, located above the transition region, extends out into space for several solar radii and has an energy output nearly $10^6$ times less intense than that of the photosphere. Because the density of the corona is so low (typically $10^5$ particles cm$^{-3}$, compared with $10^{19}$ particles cm$^{-3}$ at sea level in Earth's atmosphere), it is essentially transparent to most electromagnetic radiation (except long radio wavelengths) and is not in local thermodynamic equilibrium. For gases that are not in LTE, a unique temperature is not strictly definable (see Section 9.2). However, the temperatures obtained by considering thermal motions, ionization levels, and radio emissions do give reasonably consistent results. For instance, the presence of Fe XIV lines indicate temperatures in excess of $2 \times 10^6$ K, as do line widths produced by thermal Doppler broadening.

Based on the radiation coming from the corona, three distinct structural components can be identified:

- The **K corona** (from *Kontinuierlich*, the German word for "continuous") produces the continuous white light emission that results from photospheric radiation scattered by free electrons. Contributions to the coronal light due to the K corona primarily occur between 1 and 2.3 R$_\odot$

from the center of the Sun. The spectral lines evident in the photosphere are essentially blended by the large Doppler shifts that are caused by the high thermal velocities of the electrons.

- The **F corona** (for Fraunhofer) comes from the scattering of photospheric light by dust grains that are located beyond 2.3 $R_\odot$. Because dust grains are much more massive and slower than electrons, Doppler broadening is minimal and the Fraunhofer lines are still detectable. The F corona actually merges with the **zodiacal light**, the faint glow found along the ecliptic that is a reflection of the Sun's light from interplanetary dust.

- The **E corona** is the source of the emission lines that are produced by the highly ionized atoms located throughout the corona; the E corona overlaps the K and F coronas. Since the temperatures are extremely high in the corona, the exponential term in the Saha equation encourages ionization because thermal energies are comparable to ionization potentials. The very low number densities also encourage ionization since the chance of recombination is greatly reduced.

The low number densities allow so-called **forbidden transitions** to occur, producing spectral lines that are generally seen only in astrophysical environments where gases are extremely thin. Forbidden transitions occur from atomic energy levels that are **metastable**; electrons do not readily make transitions from metastable states to lower energy states without assistance. While **allowed transitions** occur on time scales on the order of $10^{-8}$ s, spontaneous forbidden transitions may require one second or longer. In gases at higher densities, electrons are able to escape from metastable states through collisions with other atoms or ions, but in the corona these collisions are rare. Consequently, given enough time, some electrons will be able to make spontaneous transitions from metastable states to lower energy states, accompanied by the emission of photons.

Since the blackbody continuum emission from the Sun decreases like $\lambda^{-2}$ for sufficiently long wavelengths (see Eq. 3.20), the amount of photospheric radio emission is negligible. The solar corona, however, *is* a source of radio-wavelength radiation. Some radio emission arises from free–free transitions of electrons that pass near atoms and ions. During these close encounters, photons may be emitted as the electrons' energies are decreased slightly. From the conservation of energy, the greater the change in the energy of an electron, the more energetic the resulting photon and the shorter its wavelength. Clearly, the closer an electron comes to an ion, the more likely it is that the electron's energy will change appreciably. Since more frequent and closer encounters are

## 11.2 The Solar Atmosphere

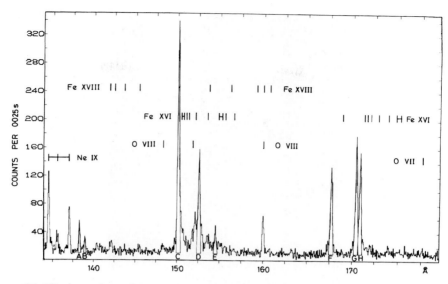

**Figure 11.22** A section of the x-ray emission spectrum of the solar corona. (Figure from Parkinson, *Astron. Astrophys.*, *24*, 215, 1973.)

expected if the number density is larger, shorter wavelength radio emissions should be observed nearer the Sun. Radio wavelengths of 1 to 20 cm are observed from the chromosphere through the lower corona, while longer wavelength radiation originates from the outer corona. It is important to note that synchrotron radiation by relativistic electrons also contributes to the observed radio emission from the solar corona (recall the discussion of the headlight effect in Example 4.3).

Photospheric emissions are negligible in the x-ray wavelength range as well. In this case the blackbody continuum decreases very rapidly, dropping off like $\lambda^{-5} e^{-hc/\lambda kT}$. Consequently, any emission in x-ray wavelengths from the corona will completely overwhelm the output from the photosphere. In fact, because of the high temperatures of the corona, its x-ray spectrum is very rich in emission lines. This is due to the high degree of ionization that exists for all of the elements present, together with the ability of the corona to excite a large number of atomic transitions. Given the many electrons that are present in heavy elements such as iron, and the vast number of available energy levels, each such element is capable of producing an extensive emission spectrum. Figure 11.22 shows a sample of the lines that are observed in one portion of the x-ray wavelength band, along with the ions responsible for their production.

An image of the x-ray Sun is shown in Fig. 11.23. This fascinating picture indicates that x-ray emission is not uniform, that x-ray dark regions called

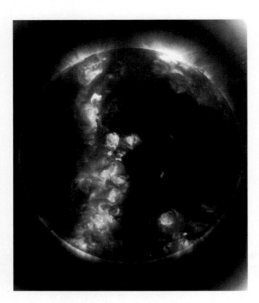

**Figure 11.23** An x-ray image of the Sun obtained September 11, 1989 by the Normal Incidence X-ray Telescope. A solar flare is visible in the upper right corner of the image. (Courtesy of the Smithsonian Astrophysical Observatory and IBM T.J. Watson Research Center.)

**coronal holes** exist. Moreover, even in the coronal holes, localized bright spots of enhanced x-ray emission appear and disappear on a time scale of several hours. Smaller features are also apparent within the regions of generally bright x-ray emission.

The weaker x-ray emission coming from coronal holes is characteristic of the lower densities and temperatures that exist in those regions, as compared to the rest of the corona. The explanation for the existence of coronal holes is tied to the Sun's magnetic field and the generation of the **solar wind**, a continuous stream of ions and electrons escaping from the Sun and moving through interplanetary space.

Just like the magnetic field that is produced by a current loop, the magnetic field of the Sun is generally that of a dipole, at least on a global scale (Fig. 11.24). Although its value can differ significantly in localized regions (as we will see in the next section), the strength of the field is typically a few gauss near the surface.[8] Coronal holes correspond to those parts of the magnetic field where the field lines are *open* while the x-ray bright regions are associated with *closed* field lines; open field lines extend out to great distances from the Sun

---

[8] The magnetic field near the surface of Earth is approximately 0.6 G.

## 11.2 The Solar Atmosphere

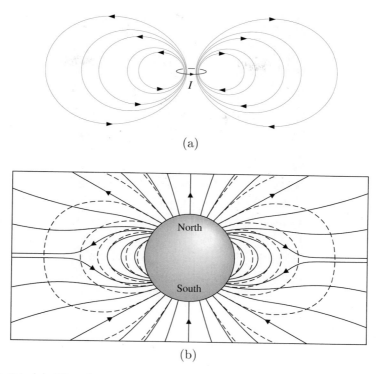

**Figure 11.24** (a) The characteristic dipole magnetic field of a current loop. (b) The global magnetic field of the Sun. The dashed lines show the field of a perfect magnetic dipole.

while closed lines form loops that return to the Sun.

The Lorentz force equation (in cgs units),

$$\mathbf{F} = q\left(\mathbf{E} + \frac{\mathbf{v}}{c} \times \mathbf{B}\right), \tag{11.3}$$

describing the force exerted on a charged particle of velocity **v** in an electric field **E** and a magnetic field **B** states that the force due to the magnetic field is always perpendicular to both the direction of the velocity vector and the field (the cross product). Providing that electric fields are negligible, charged particles are forced to spiral around magnetic field lines and cannot actually cross them except by collisions (Fig. 11.25). This implies that closed magnetic field lines tend to trap charged particles, not allowing them to escape. In regions of open field lines, however, particles can actually follow the lines out away from the Sun. Consequently, the solar wind originates from the regions of open magnetic field lines, namely the coronal holes. The details observed

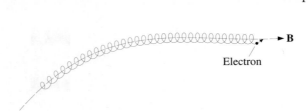

**Figure 11.25** A charged particle is forced to spiral around a magnetic field line because the Lorentz force is mutually perpendicular to both the velocity of the particle and the direction of the magnetic field.

in the x-ray-bright regions, as well as the localized bright spots in the coronal holes, are due to the higher densities of the electrons and ions that are trapped in large and small magnetic field loops.

The existence of ongoing mass loss from the Sun was deduced long before it was ever detected directly. The tails of comets are generally composed of two parts, a curved dust tail and a straight ion tail, both of which are always pointed away from the Sun (Fig. 11.26). The force exerted on dust grains by radiation pressure is sufficient to push the dust tail back; the curvature of the tail is due to the different orbital speeds of the individual dust grains, which, according to Kepler's third law, is a function of their varying distances from the Sun. However, the ion tail cannot be explained by radiation pressure; the interaction between photons and the ions is not efficient enough. On the other hand, the electric force between the ions of the solar wind and the ions in the comet can account for the direction of the ion tail. This interaction allows momentum to be transferred to the cometary ions, driving them straight away from the Sun.

The **aurora borealis** and the **aurora australis** (the northern and southern lights, respectively) are also products of the solar wind. As the ions from the Sun interact with Earth's magnetic field, they become trapped in it. Bouncing back and forth between the north magnetic pole and the south magnetic pole, these ions form the **Van Allen radiation belts**. Ions that are sufficiently energetic will collide with the atoms in Earth's upper atmosphere, causing the atmospheric atoms to become excited or ionized. The resulting deexcitations or recombinations emit the photons that produce the spectacular light displays observed from high northern and southern latitudes.

Using rockets and satellites, characteristics of the solar wind can be measured as it passes near Earth. At a distance of 1 AU from the Sun, the solar wind velocity ranges from approximately 200 km s$^{-1}$ to 700 km s$^{-1}$, with a typical density of 7 ions cm$^{-3}$ and characteristic kinetic temperatures of $4 \times 10^4$ K for protons and $10^5$ K for electrons. Although the wind is composed primarily

## 11.2 The Solar Atmosphere

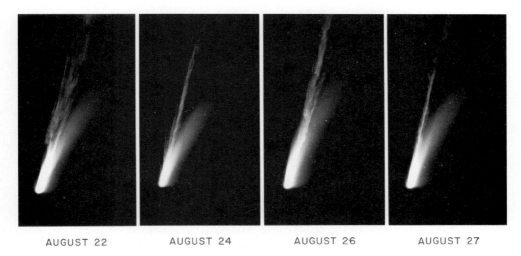

**Figure 11.26** Comet Mrkos in 1957. The dust tail of a comet is curved and its ion tail is straight. (Courtesy of Palomar/Caltech.)

of protons and electrons, heavier ions are present as well.

---

**Example 11.1** The mass loss rate of the Sun may be estimated from the data given above. We know that all of the mass leaving the Sun must also pass through a sphere of radius 1 AU centered on the Sun; otherwise it would collect at some location in space. If we further assume (for simplicity) that the mass loss rate is spherically symmetric, then the amount of mass crossing a spherical surface of radius $r$ in an amount of time $t$ is just the mass density of the gas multiplied by the volume of the shell of gas that can travel across the sphere during that time interval, or

$$dM = \rho\, dV = (nm_H)(4\pi r^2\, v\, dt),$$

where $n$ is the number density of ions (mostly protons), $m_H$ is the mass of a proton, and $dV = A\, dr \simeq 4\pi r^2\, v\, dt$ is the volume of a shell that crosses a spherical surface in an amount of time $dt$. Dividing both sides by $dt$, we obtain the mass loss rate,

$$\frac{dM}{dt} = 4\pi r^2\, n m_H v = 4\pi r^2\, \rho v. \tag{11.4}$$

By convention, stellar mass loss rates are generally given in *solar masses per year* and symbolized by $\dot{M} \equiv dM/dt$. Using $v = 500$ km s$^{-1}$, $r = 1$ AU, and $n = 7$ protons cm$^{-3}$, we find that

$$\dot{M}_\odot \simeq 3 \times 10^{-14}\ M_\odot\ \text{yr}^{-1}.$$

At this rate it would require more than $10^{13}$ yr before the entire mass of the Sun is dissipated. However, the interior structure of the Sun is changing much more rapidly than this, so the effect of the present-day solar wind on the evolution of the Sun is minimal.[9]

We now consider how the expansion of the solar corona produces the solar wind. This is a result of the corona's high temperature, together with the high thermal conductivity of the ionized gas, referred to as a **plasma**. The ability of the plasma to conduct heat implies that the corona is almost isothermal (see Fig. 11.19).

In 1958 Eugene Parker developed an approximately isothermal model of the solar wind that has been successful in describing many of its basic features. To see why the solar wind is inevitable, begin by considering the condition of hydrostatic equilibrium, Eq. (10.7). If the mass of the corona is insignificant compared to the total mass of the Sun, then $M_r \simeq M_\odot$ in that region and the hydrostatic equilibrium equation becomes

$$\frac{dP}{dr} = -\frac{GM_\odot \rho}{r^2}. \tag{11.5}$$

Next, assuming for simplicity that the gas is completely ionized and composed entirely of hydrogen, the number density of protons is given by

$$n \simeq \frac{\rho}{m_p}$$

since $m_p \simeq m_H$. From the ideal gas law (Eq. 10.14), the pressure of the gas may be written as (taking $\mu = 1/2$ for ionized hydrogen)

$$P = 2nkT.$$

Substituting expressions for the pressure and density into Eq. (11.5), the hydrostatic equilibrium equation becomes

$$\frac{d}{dr}(2nkT) = -\frac{GM_\odot n m_p}{r^2}. \tag{11.6}$$

---

[9]In 1992 both *Voyagers I* and *II* detected radio noise at frequencies of 1.8 to 3.5 kHz originating from the outer reaches of the solar system. It is believed that the signals were produced where particles from the solar wind collided with the interstellar medium (dust and gas located between the stars; see Section 12.1). If so, this would represent the first detection of the **heliopause**, the outer limit of the Sun's electromagnetic influence. The distance from the Sun to the heliopause is estimated to be between 80 and 170 AU.

## 11.2 The Solar Atmosphere

Making the assumption that the gas is isothermal, Eq. (11.6) can be integrated directly to give an expression for the number density (and therefore the pressure) as a function of radius. It is left as an exercise to show that

$$n(r) = n_0 e^{-\lambda(1-r_0/r)}, \qquad (11.7)$$

where

$$\lambda \equiv \frac{GM_\odot m_p}{2kTr_0}$$

and $n = n_0$ at some radius $r = r_0$. Note that $\lambda$ is approximately the ratio of a proton's gravitational potential energy and its kinetic energy at a distance $r_0$ from the center of the Sun. We now see that the pressure structure is just

$$P(r) = P_0 e^{-\lambda(1-r_0/r)}, \qquad (11.8)$$

where $P_0 = 2n_0 kT$.

An immediate consequence of Eq. (11.8) is that in our isothermal approximation the pressure *does not* approach zero as $r$ goes to infinity. To estimate the limiting values of $n(r)$ and $P(r)$, let $T = 1.5 \times 10^6$ K and $n_0 = 3 \times 10^7$ cm$^{-3}$ at about $r_0 = 1.4$ R$_\odot$, values typical of the inner corona. Then $\lambda \simeq 5.5$, $n(\infty) \simeq 1.2 \times 10^5$ cm$^{-3}$, and $P(\infty) \simeq 5 \times 10^{-5}$ dyne cm$^{-2}$. However, as we will see in Section 12.1, with the exception of localized clouds of material, the actual densities and pressures of interstellar dust and gas are much lower than those just derived.

Given the inconsistency that exists between the isothermal, hydrostatic solution to the structure of the corona and the conditions in interstellar space, at least one of the assumptions made in the derivation must be incorrect. Although the choice that the corona is approximately isothermal is not completely valid, it is roughly consistent with observations. Recall that near Earth ($r \sim 215$ R$_\odot$), the solar wind is characterized by temperatures on the order of $10^5$ K, indicating that the temperature of the gas is not decreasing rapidly with distance. It can be shown that solutions that allow for a realistically varying temperature structure still do not eliminate the problem of a predicted gas pressure significantly in excess of the interstellar value. Apparently, it is the assumption that the corona is in hydrostatic equilibrium that is wrong. Since $P(\infty)$ greatly exceeds the pressures in interstellar space, material must be expanding outward from the Sun, implying the existence of the solar wind.

Given the existence of a continuous mass outflow, if we are to develop an understanding of the structure of the solar atmosphere, the simple approximation of hydrostatic equilibrium must be replaced by a set of **hydrodynamic**

**equations** that describe the flow. In particular, by writing

$$\frac{d^2r}{dt^2} = \frac{dv}{dt} = \frac{dv}{dr}\frac{dr}{dt} = v\frac{dv}{dr},$$

Eq. (10.6) becomes

$$\rho v \frac{dv}{dr} = -\frac{dP}{dr} - G\frac{M_r \rho}{r^2}, \qquad (11.9)$$

where $v$ is the velocity of the flow. With the introduction of a new variable (velocity), another expression that describes the conservation of mass flow across boundaries must also be included, specifically

$$4\pi r^2 \rho v = \text{constant},$$

which is just the relationship that was used in Example 11.1 to estimate the Sun's mass loss rate. This expression immediately implies that

$$\frac{d(\rho v r^2)}{dr} = 0. \qquad (11.10)$$

At the top of the convection zone, the motion of the hot, rising gas and the return flow of the cool gas sets up longitudinal waves (pressure waves) that propagate outward through the photosphere and into the chromosphere. The outward flux of wave energy, $F_E$, is governed by the expression

$$F_E = \frac{1}{2}\rho v_w^2 v_s, \qquad (11.11)$$

where $v_s$ is the local sound speed and $v_w$ is the velocity amplitude of the oscillatory wave motion for individual particles being driven about their equilibrium positions by the "piston" of the convection zone. From Eq. (10.76),

$$v_s = \sqrt{\gamma P/\rho}.$$

Since, according to the ideal gas law, $P = \rho kT/\mu m_H$, the sound speed may also be written as

$$v_s = \sqrt{\frac{\gamma kT}{\mu m_H}} \propto \sqrt{T}$$

for fixed $\gamma$ and $\mu$.

When the wave is first generated at the top of the convection zone, $v_w < v_s$. However, the density of the gas that these waves travel through decreases significantly with altitude, dropping by four orders of magnitude in approximately 1000 km. If we assume that very little mechanical energy is lost in moving

## 11.2 The Solar Atmosphere

through the photosphere (i.e., $4\pi r^2 F_E$ is approximately constant) and that $v_s$ remains essentially unchanged since the temperature varies less than one order of magnitude across the photosphere and chromosphere, the rapid decrease in density means that $v_w$ must increase significantly (approximately two orders of magnitude). As a result, the wave motion quickly becomes *supersonic* ($v_w > v_s$) as particles in the wave try to travel through the medium faster than the local speed of sound. The result is that the wave develops into a **shock wave**, much like the shock waves that produce sonic booms behind supersonic aircraft.

A shock wave is characterized by a very steep density change over a short distance, called the **shock front**. As a shock moves through a gas it produces a great deal of heating via collisions, leaving the gas behind the shock highly ionized. This heating comes at the expense of the mechanical energy of the shock, and the shock quickly dissipates. Thus the gas in the chromosphere and above is effectively heated by the mass motions created in the convection zone.

Our discussion of the hydrodynamic equations has failed to account for the influence of the Sun's magnetic field. It is believed that the temperature structure throughout the outer solar atmosphere, including the very steep *positive* temperature gradient in the transition region, is due at least in part to the presence of the magnetic field, coupled with mass motions produced by the convection zone. **Magnetohydrodynamics** (usually mercifully shortened to **MHD**) is the study of the interactions between magnetic fields and plasmas. Owing to the great complexity of the problem, a complete solution to the set of MHD equations applied to the outer atmosphere of the Sun does not yet exist. However, some aspects of the solution can be described.

The presence of the magnetic field allows for the generation of a second kind of wave motion. These waves may be thought of as transverse waves that propagate along the magnetic field lines, due to the restoring force of their *tension*. To understand the origin of this restoring force, recall that establishing a magnetic field (which is always generated by moving electric charges, or currents) requires that an amount of energy be expended. The energy used to establish the field can be thought of as being stored within the magnetic field itself; thus the space containing the magnetic field also contains a magnetic energy density. The value of the magnetic energy density is given (in units of ergs cm$^{-3}$) by

$$u_m = \frac{B^2}{8\pi}. \tag{11.12}$$

If a volume $V$ of plasma containing a number of magnetic field lines is compressed in a direction perpendicular to the lines, the density of field lines necessarily increases.[10] But the density of field lines is just a description of the strength of the magnetic field itself, so the energy density of the magnetic field also increases during compression. An amount of mechanical work must therefore have been done in compressing the field lines in the gas. Since work is given by $W = \int P\, dV$, the compression of the plasma must imply the existence of a magnetic pressure. It can be shown that the magnetic pressure is numerically equal to the magnetic energy density, or

$$P_m = \frac{B^2}{8\pi}. \tag{11.13}$$

When a magnetic field line gets displaced by some amount perpendicular to the direction of the line, a *magnetic pressure gradient* becomes established; the pressure in the direction of the displacement increases because the number density of field lines increases, while the pressure in the opposite direction decreases. This pressure change then tends to push the line back again, restoring the original density of field lines. This process may be thought of as being analogous to the oscillations that occur in a string when a portion of the string is displaced; it is the tension in the string that pulls it back when it is plucked. The "tension" that restores the position of the magnetic field line is just the magnetic pressure gradient.

As with the traveling motion of a wave on a string, a disturbance in the magnetic field line can also propagate down the line. This transverse MHD wave is called an **Alfvén wave**. The speed of propagation of the Alfvén wave may be estimated by making a comparison with the sound speed in a gas. Since the adiabatic sound speed is given by

$$v_s = \sqrt{\frac{\gamma P_g}{\rho}},$$

where $\gamma$ is of order unity, by analogy, the *Alfvén speed* should be approximately

$$v_m \sim \sqrt{\frac{P_m}{\rho}} = \frac{B}{\sqrt{8\pi\rho}}.$$

A more careful treatment gives the result,

$$v_m = \frac{B}{\sqrt{4\pi\rho}}. \tag{11.14}$$

---

[10]Recall that if the electric field is negligible, charged particles must spiral around field lines, so that if the charged particles are pushed, they drag the field lines with them; the field lines are said to be "frozen in" the plasma.

## 11.2 The Solar Atmosphere

---

**Example 11.2** Using Eqs. (10.76) and (11.14), the sound speed and Alfvén speed may be compared for the photosphere. The gas pressure at the base of the photosphere is roughly $5 \times 10^4$ dynes cm$^{-2}$, with a density of $2.5 \times 10^{-7}$ g cm$^{-3}$. Assuming $\gamma = 5/3$ (for an ideal monatomic gas),

$$v_s \simeq 10^6 \text{ cm s}^{-1}.$$

Notice that this speed is significantly lower than was found using global solar values in Example 10.6; apparently, the sound speed is much larger in the Sun's interior.

Taking a typical surface magnetic field strength to be 2 G, the magnetic pressure is (from Eq. 11.13)

$$P_m = 0.2 \text{ dyne cm}^{-2},$$

and the Alfvén speed is

$$v_m = 1 \times 10^3 \text{ cm s}^{-1}.$$

The magnetic pressure may generally be neglected in photospheric hydrostatic considerations, since it is smaller than the gas pressure by five orders of magnitude. We will see in the next section, however, that much larger magnetic field strengths can exist in localized regions on the Sun's surface.

---

Since Alfvén waves can propagate along magnetic field lines, they may also transport energy outward. Now, according to Maxwell's equations, a time-varying magnetic field produces an electric field, which in turn creates electrical currents in the highly conductive plasma. This implies that some resistive Joule heating will occur in the ionized gas, causing the temperature to rise. Thus MHD waves can also contribute to the temperature structure of the upper solar atmosphere.

Because of the Sun's rotation, its open magnetic field lines are dragged along through interplanetary space (Fig. 11.27). Since the solar wind is forced to move with the field lines, a torque is produced that actually slows the Sun's rotation. Said another way, the solar wind is transferring angular momentum away from the Sun. As a result, the Sun's rotation rate will decrease significantly over its lifetime. Interestingly, the differential rotation present in the photosphere is not manifested in the corona. Apparently, the magnetic field, which so strongly influences the structure of the corona, does not exhibit differential rotation at this height. Actually, the corona rotates *more rapidly* than does the photosphere. This may be an indication that the solar interior,

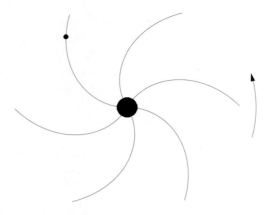

**Figure 11.27** The drag of the solar magnetic field creates a spiral pattern in interplanetary space and causes angular momentum to be transferred away from the Sun.

where the magnetic field lines are anchored, also rotates more rapidly than the photosphere.

Although this chapter is devoted to our Sun, the most thoroughly studied of all stars, the outer atmospheres of other stars can be investigated as well. For instance, observations indicate that the rotation rates of solar-type stars seem to decrease with age. Furthermore, late main-sequence stars, with their convective envelopes, generally have much slower rotation rates than stars on the upper end of the main sequence. Perhaps winds are transferring angular momentum away from these stars.

Satellites such as IUE and Einstein have also provided us with valuable UV and x-ray observations of other stars. It appears that stars along the main sequence that are cooler than spectral class F have emission lines in the ultraviolet that are similar to those observed coming from the Sun's chromosphere and transition region, and x-ray observations indicate corona-like emissions as well. These stars are also those for which stellar structure calculations indicate that surface convection zones should exist. Apparently the same mechanisms that are in operation in the Sun, heating its outer atmosphere, are also in operation in other stars.

## 11.3 The Solar Cycle

Some of the most fascinating and complex features of the solar atmosphere are transient in nature, including its **sunspots** (recall Fig. 11.13). It was Galileo who made the first telescopic observations of sunspots. Sunspots are

## 11.3 The Solar Cycle

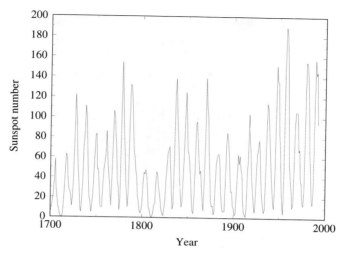

**Figure 11.28** The number of sunspots between 1700 and 1992 indicates an 11-year periodicity. (Data from the National Center for Atmospheric Research.)

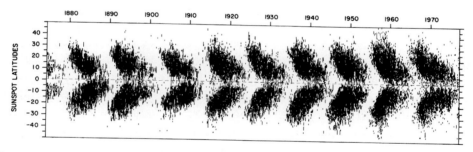

**Figure 11.29** The butterfly diagram, showing sunspot latitudes with time. (Adapted from a figure courtesy of J. A. Eddy, High Altitude Observatory.)

even visible occasionally with the unaided eye, but making such observations is *strongly discouraged* because of the potential for eye damage.

Reliable observations made over the past two centuries indicate that the number of sunspots is approximately periodic, going from minimum to maximum and back to minimum again nearly every 11 years (Fig. 11.28). The average latitude of sunspot formation is also periodic, again over an 11-year cycle. A plot of sunspot location as a function of time is shown in Fig. 11.29. Because of its winglike appearance, Fig. 11.29 has come to be known as the **butterfly diagram**. Individual sunspots are short-lived features, typically surviving no more than a month or so. During its lifetime, a sunspot will

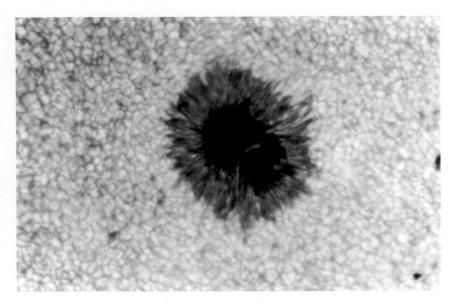

**Figure 11.30** A typical sunspot, showing the presence of a dark umbra and a penumbra, with its filamentary structure. (Courtesy of W. Livingston and the National Optical Astronomy Observatories.)

remain at a constant latitude, although succeeding sunspots tend to form at progressively lower latitudes, moving from the poles toward the equator. As the last sunspots of one cycle vanish near the Sun's equator, a new cycle begins near $\pm 40°$ (north and south) of the equator. The largest number of spots (sunspot maximum) typically occurs at intermediate latitudes.

The key to understanding sunspots lies in their strong magnetic fields. A typical sunspot is shown in Fig. 11.30. The darkest portion of the sunspot is known as the **umbra** and may measure as much as 30,000 km in diameter. (For reference, the diameter of Earth is 12,756 km.) The umbra is usually surrounded by a filamentlike structure, called the **penumbra**, whose mere appearance suggests the presence of magnetic lines of force. The existence of a strong magnetic field can be verified by observing individual spectral lines produced within the spot. As was discussed in Section 5.4, the strength and polarity of magnetic fields can be measured by observing the Zeeman effect, the splitting of spectral lines that results from removing the degeneracy inherent in atomic energy levels. The amount of splitting is proportional to the strength of the magnetic field and the polarization of the light corresponds to the direction of the field. Figure 11.31 shows an example of the splitting of a spectral line measured across a sunspot. Magnetic field strengths of several

## 11.3 The Solar Cycle

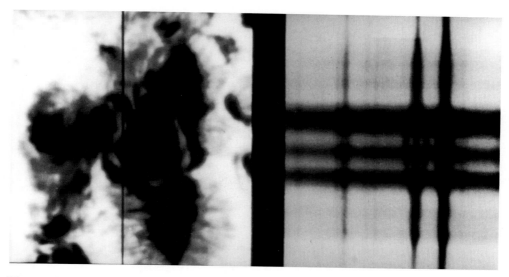

**Figure 11.31** The Zeeman splitting of the Fe 5250.2 Å spectral line due to the presence of a strong magnetic field in a sunspot. The spectrograph slit was aligned vertically across a sunspot, resulting in a wavelength dependence that runs from left to right in the image. The slit extended beyond the image of the sunspot. (Courtesy of the National Optical Astronomy Observatories/National Solar Observatory.)

thousand gauss have been measured in the centers of umbral regions, with field strengths decreasing across penumbral regions. Furthermore, polarization measurements indicate that the direction of a typical umbral magnetic field is vertical, becoming horizontal across the penumbra.

Sunspots are generally located in groups. Typically, a dominant sunspot leads in the direction of rotation, and one or more sunspots follow. During an 11-year cycle, the lead sunspot will always have the same polarity in one hemisphere—say, a north pole in the geographic northern hemisphere—while the lead sunspot in the other hemisphere will have the opposite polarity (e.g., a south pole in the geographic southern hemisphere); trailing sunspots have the opposite polarity. Even when a large collection of trailing spots exist, creating a tangled magnetic field pattern, a basically *bipolar* field is present. During the next 11-year cycle, polarities will be reversed; the sunspot with a magnetic south polarity will lead in the northern hemisphere, and vice versa in the southern hemisphere. Accompanying this local polarity reversal is a global polarity reversal, the overall dipole field of the Sun will change so that the magnetic north pole of the Sun will switch from the geographic north

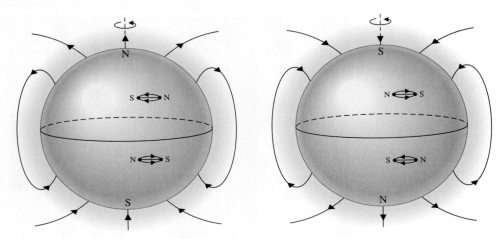

**Figure 11.32** The global magnetic field orientation of the Sun, along with the magnetic polarity of sunspots during successive 11-year periods.

pole to the geographic south pole. Polarity reversal always occurs during sunspot minimum, when the first sunspots are beginning to form at the highest latitudes. For this reason, the Sun is said to have a *22-year cycle*. This important magnetic behavior is illustrated in Fig. 11.32.

The dark appearance of sunspots is due to their significantly lower temperatures. In the central portion of the umbra the temperature may be as low as 3900 K, compared with the Sun's effective temperature of 5770 K. From Eq. (3.18), this implies a surface bolometric flux that is a factor of $(5770/3900)^4 = 4.8$ lower than that of the surrounding photosphere.[11] Observations obtained from the Solar Maximum Mission satellite (SMM) have shown that this decrease in surface flux affects the overall energy output of the Sun. When a number of large sunspots exist, the solar luminosity is depressed by roughly 0.1%. Since convection is the principal energy transport mechanism just below the photosphere, and since strong magnetic fields inhibit motion through the "freezing in" of field lines in a plasma, it is likely that the mass motion of convective bubbles is inhibited in sunspots, thereby decreasing the flow of energy through the sunspots.

Along with luminosity variations on a time scale of months (the typical lifetime of an individual sunspot), the Sun's luminosity seems to experience variability on a much longer time scale, as does the number of sunspots. For

---

[11] A 3900-K blackbody is very bright of course. However, when seen through a filter dark enough to make viewing the rest of the 5770-K photosphere comfortable, the sunspot appears dark.

## 11.3 The Solar Cycle

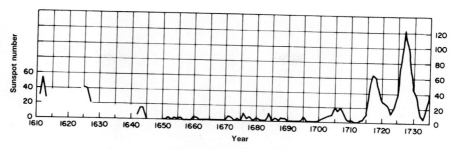

**Figure 11.33** An unusually small number of sunspots were observed between 1645 and 1715 (the Maunder minimum). (Adapted from a figure courtesy of J. A. Eddy, High Altitude Observatory.)

instance, very few sunspots were observed between 1645 and 1715; this time interval has come to be called the **Maunder minimum** (see Fig. 11.33). Surprisingly, during this period the average temperature in Europe was significantly lower, consistent with the solar luminosity being a few tenths of a percent less than it is today. John Eddy has proposed that there is a very long term periodicity on which the solar cycle is superimposed. This long-period variation goes through grand sunspot maxima and minima that may last for centuries. Evidence in support of this suggestion is found on Earth in the relative numbers of atmospheric carbon dioxide molecules that contain radioactive carbon atoms ($^{14}_{6}C$), as preserved in the 7000-year-long record of tree rings. The importance of $^{14}_{6}C$ in long-term sunspot studies lies in an inverse correlation between sunspots and the amount of $^{14}_{6}C$ present in Earth's atmosphere. $^{14}_{6}C$ is a radioactive isotope of carbon that is produced when extremely energetic charged particles from space, called **cosmic rays**, collide with atmospheric nitrogen. Cosmic rays are affected by the charged particles in the solar wind, which in turn is affected by solar activity. During the Maunder minimum the amount of atmospheric $^{14}_{6}C$ increased significantly and was incorporated into the rings of living trees. The amount of $^{14}_{6}C$ also seems to correlate well with the advance and retreat of glaciers over the past 5000 years.

With lower temperatures in sunspots, the gas pressure is necessarily lower than in the surrounding material (see Eq. 10.14). However, the gravitational force is essentially the same. From these considerations alone, it seems as though the gas within a sunspot ought to sink into the interior of the star, an effect that is not observed. Without the benefit of a sufficiently large gas pressure gradient to support a sunspot, another component to the pressure must exist. As we have already seen in the last section, a magnetic field is accompanied by a pressure term. It is this extra magnetic pressure that provides

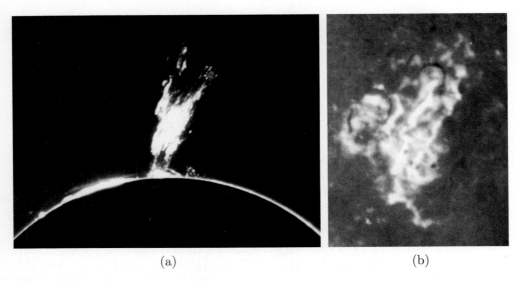

**Figure 11.34** (a) A solar flare ejection seen at the limb of the Sun. (Courtesy of M. McCabe, Haleakala Observatory, Institute for Astronomy, University of Hawaii.) (b) A two-ribbon flare seen in $H_\alpha$ on October 19, 1989. (Courtesy of the National Optical Astronomy Observatories.)

the support necessary to keep a sunspot from sinking or being compressed by the surrounding gas pressure.

A variety of other phenomena are also associated with sunspot activity. **Plages** (from the French word for *beach*) are chromospheric regions of bright $H_\alpha$ emission located near active sunspots. They usually form before the sunspots appear and disappear after the sunspots vanish from a particular area. Plages have higher densities than the surrounding gas and are products of the magnetic fields. Apparently the cause of the decreased brightness of sunspots does not play an important role in plages.

**Solar flares** are eruptive events that release between $10^{29}$ and $10^{32}$ ergs of energy over time intervals ranging from a few minutes to more than an hour.[12] The physical dimensions of a flare are enormous, with a large flare reaching 100,000 km in length (see Fig. 11.34a). During an eruption, the hydrogen Balmer line, $H_\alpha$, appears locally in emission rather than in absorption, as is usually the case, implying that photon production occurs above much of the absorbing material. When observed in $H_\alpha$, a flare is often seen on the disk as two ribbons of light (Fig. 11.34b). Along with $H_\alpha$, other types of electromagnetic radiation are produced that can range from kilometer-wavelength

---

[12] For comparison, a one megaton bomb releases approximately $10^{23}$ ergs.

## 11.3 The Solar Cycle

*nonthermal* radio waves (synchrotron radiation; see Section 4.3) to very short wavelength nuclear gamma-ray emission lines. Charged particles are also accelerated to high speeds, many escaping into interplanetary space as *solar cosmic rays*. In the largest flares the ejected charged particles, mostly protons and helium nuclei, may reach Earth in 30 minutes, disrupting some communications and posing a very serious threat to any unprotected astronauts. Shock waves are also generated and can occasionally propagate several astronomical units before dissipating.

The answer to the question of what powers solar flares lies in the location of the flare eruption. Flares develop in regions where the magnetic field intensity is great, namely in sunspot groups. From the discussion of the previous section, the creation of magnetic fields results in energy being stored in those magnetic fields (Eq. 11.12). If a magnetic field disturbance could quickly release the stored energy, a flare might develop. It is left as an exercise to show that both the amount of energy stored in the magnetic field and the time scale involved in perturbing it via Alfvén waves are consistent with the creation of a solar flare. However, details of the energy conversion, such as particle acceleration, are still a matter of active research.

A model of a solar flare is illustrated in Fig. 11.35. It appears that the general mechanism of a solar flare involves the *reconnection* of magnetic field lines. A disturbance in magnetic field loops (perhaps due to the Sun's convection zone) causes the creation of a sheet of current in the highly conducting plasma (recall Lenz's law). The finite resistance in the plasma results in Joule heating of the gas, causing temperatures to reach $10^7$ K. Such high temperatures can produce x-ray and gamma-ray emission. Radio wavelength radiation is generated by the synchrotron process of charged particles spiraling around the magnetic field lines. $H_\alpha$ emission at the base of the magnetic field lines (the two $H_\alpha$ ribbons) is produced by recombining electrons and protons that are accelerated away from the reconnection point, toward the photosphere. Particles accelerated away from the reconnection point and away from the Sun may escape entirely, producing solar cosmic rays.

**Solar prominences** are also related to the Sun's magnetic field. *Quiescent* prominences are curtains of ionized gas that reach well into the corona. The material in the prominence has collected along the magnetic field lines of an active region, with the result that the gas is cooler and more dense than the surrounding coronal gas. This causes the gas to "rain" back down into the chromosphere. When viewed in $H_\alpha$ at the limb of the Sun, quiescent prominences appear as bright structures against the thin corona. When viewed in the continuum against the solar disk, however, a quiescent prominence appears

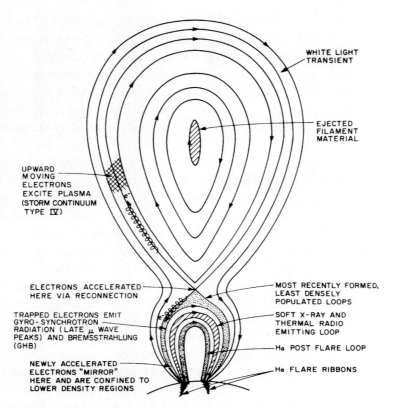

**Figure 11.35** A model of a solar flare. Note the two $H_\alpha$ flare ribbons. Electrons are also accelerated along the magnetic field lines, producing synchrotron radiation. (Figure from Cliver et al., *Ap. J.*, *305*, 920, 1986.)

as a dark *filament*, absorbing the light emitted from below. An example of a quiescent *hedgerow* prominence is shown in Fig. 11.36(a).

An *active* solar prominence (Fig. 11.36b) may exist for only a few hours. Again material is falling down along magnetic field lines toward the magnetic poles anchored in an active sunspot region. Although active solar prominences are closely tied to flares, quiescent prominences are generally decoupled from these explosive phenomena. Occasionally, however, a quiescent prominence can be affected by a flare as well, causing the prominence to erupt and eject its material into interplanetary space.

Yet another feature of the solar cycle involves the shape of the corona itself. During a period of little solar activity, when there are few sunspots and few, if any, flares or prominences, the *quiet* solar corona is generally more extended at the equator than at the poles. Near sunspot maximum, the *active* corona is

## 11.3 The Solar Cycle

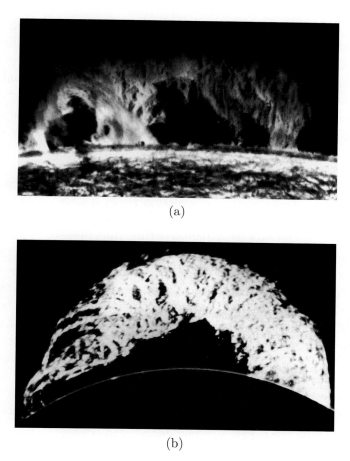

**Figure 11.36** (a) A quiescent hedgerow prominence. (Courtesy of Big Bear Solar Observatory, California Institute of Technology.) (b) An active loop prominence seen in June 1946. (Courtesy of the High Altitude Observatory.)

more complex in shape. Examples of the shape of the corona during sunspot minimum and maximum are seen in Figs. 11.21(a) and 11.21(b), respectively. The changing shape of the corona, as with other solar activity, is due to the dynamic structure of the Sun's magnetic field.

To understand all the various phenomena associated with the 22-year solar cycle, it is clear that we must understand the generation of the Sun's magnetic field, along with its interaction with the ionized gases in the solar interior and atmosphere. A **magnetic dynamo** model describing many of the components of the solar cycle was first proposed by Horace Babcock in 1961. Despite its general success in describing the major features of the solar cycle, the model is as yet unable to provide adequate explanations of many of the important details of solar activity. Any complete picture of the solar cycle will require a full treatment of the MHD equations in the solar environment, including differential rotation, a more rapidly rotating core, convection, solar oscillations, heating of the upper atmosphere, and mass loss. Of course, not all of these processes are likely to play equally important roles in the study of the solar cycle, but it is important to understand the degree to which each of them contributes to the particular phenomenon under investigation.

The major components incorporated into the Babcock dynamo model are the Sun's differential rotation, its convection zone, and its magnetic field. The generation of the Sun's magnetic field, as with all magnetic fields, is due to electrical currents. For the Sun, it is believed that the currents move through the highly conducting plasma in the convection zone. Consequently, the magnetic field exists predominantly in the outer 30% of the Sun's radius.

As depicted in Fig. 11.37, because the magnetic field lines are "frozen into" the gas, the differential rotation of the Sun drags the lines along, converting a *poloidal* field (essentially a simple magnetic dipole) to one that has a significant *toroidal* component (field lines that are wrapped around the Sun). The turbulent convection zone then has the effect of twisting the lines, creating regions of intense magnetic fields, called magnetic *ropes*. The buoyancy produced by magnetic pressure (Eq. 11.13) causes the ropes to rise to the surface, appearing as sunspot groups. The polarity of the sunspots is due to the direction of the magnetic field along the ropes; consequently, every lead spot in one hemisphere will have the same polarity while the lead spots in the other hemisphere will have the opposite polarity.

Initially, the little twisting that does develop occurs at higher latitudes; during sunspot minimum. As the differential rotation continues to drag the field lines along and convective turbulence ties them in knots, more sunspots develop at intermediate latitudes, producing a sunspot maximum. It would seem that ultimately the greatest amount of twisting and the largest number of sunspots should develop near the equator. However, sunspots from the two hemispheres tend to cancel out near the equator since the polarities of their leading spots are opposed. As a result, the number of sunspots appearing near the equator is small. Finally, the cancellation of magnetic fields near the equator causes the poloidal field to be reestablished, but with its original polarity reversed. This process takes approximately 11 years. The entire procedure repeats continuously, with the polarity of the magnetic field returning to its original orientation every other cycle. Hence, the entire solar cycle is actually 22 years long when magnetic field polarities are considered.

## 11.3 The Solar Cycle

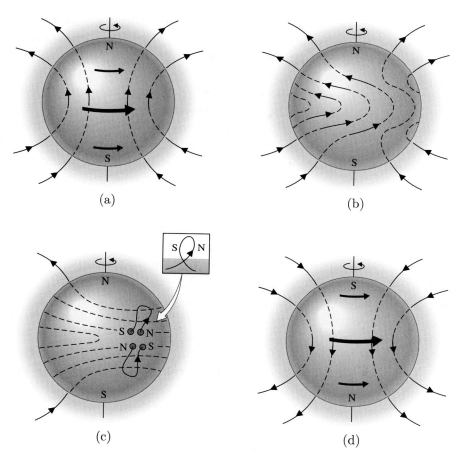

**Figure 11.37** The magnetic dynamo. (a) The solar magnetic field is initially a poloidal field. (b) Differential rotation drags the "frozen-in" magnetic field lines around the Sun, converting the poloidal field into a toroidal field. (c) Turbulent convection twists the field lines into magnetic ropes, causing them to rise to the surface as sunspots, the polarity of the lead spots corresponding to the original polarity of the poloidal field. (d) As the cycle progresses, successive sunspot groups migrate toward the equator where magnetic field reconnection reestablishes the poloidal field, but with the original polarity reversed.

As we have already seen, details related to specific phenomena, such as the cause of the decreased flux coming from sunspots or the exact process of flare generation, are as yet not well understood. The same situation also holds for the more fundamental magnetic dynamo itself. Although the preceding discussion describes the behavior of the solar cycle in an approximate way, even such basic results as the time scales involved have not yet been accurately modeled. A successful magnetic dynamo model must not only produce the general location and numbers of sunspots and flares, but it must also do so with the observed 22-year periodicity. Moreover, the dynamo model must replicate the much slower variation that appears to be responsible for the Maunder minimum.

Fortunately, some evidence does exist that the basic ideas behind the solar cycle are correct. Observations of other cool main-sequence stars indicate that they possess activity cycles much like the solar cycle. It was pointed out in the last section that late main-sequence stars exhibit observational characteristics consistent with the existence of hot coronae. It was also mentioned that angular momentum is apparently lost via stellar winds. Both agree with the theoretical onset of surface convection in low-mass stars, the major component of the dynamo theory.

Other forms of magnetic activity have also been seen in some stars. Observations indicate the existence of **flare stars**, main-sequence stars of class M that demonstrate occasional, rapid fluctuations in brightness. If flares the size of those on the Sun were to occur on the much dimmer M stars, the flares would contribute significantly to the total luminosity of those stars, producing the short-term changes that are observed.

**Starspots** are even observed to exist on stars other than the Sun. Starspots are revealed by their effect on the luminosity of a star. Two classes of stars, RS Canum Venaticorum and BY Draconis stars,[13] show significant long-term variations that are attributed to starspots covering appreciable fractions of their surface. For example, Fig. 11.38 shows a variation of over 0.6 magnitude in the B band for the BY Draconis star, BD + 26°730.

Magnetic fields have also been detected directly on several cool main-sequence stars by measuring Zeeman-broadened spectral lines. Analysis of the data indicates field strengths of several thousand gauss over significant fractions

---

[13]Classes of stars that show light variations, *variable stars*, are usually named after the first star discovered that exhibits the specific characteristics. RS CVn and BY Dra are main-sequence stars of spectral class F–G and K–M, respectively. The letters RS and BY indicate that these are variable stars; Canum Venaticorum and Draconis are the constellations in which the stars are located.

## 11.3 The Solar Cycle

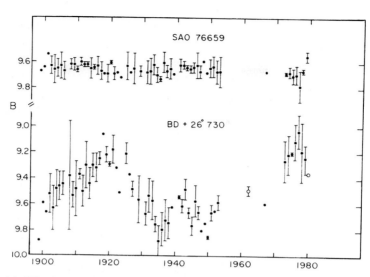

**Figure 11.38** The light curve of BD+26°730, a BY Dra star. SAO 76659 is a nearby reference star. (Figure from Hartmann et al., *Ap. J.*, *249*, 662, 1981.)

of the stellar surfaces. The existence of the strong fields correlates with their observed luminosity variations.

One more possible effect of the solar cycle is worth mentioning. Davis has suggested that the data of his neutrino experiment (Section 11.1) are consistent with an 11-year periodicity in the number of neutrinos detected (Fig. 11.39). If this suggestion proves to be correct, it would seem to indicate that the neutrinos produced in the Sun's core are altered as they pass through the time-varying magnetic field near its surface. Such an interaction would require that neutrinos possess fairly large magnetic dipole moments and masses, although at present, no laboratory experiments have been successful in definitively measuring a nonzero value for either quantity. Because the Davis experiment is sensitive only to electron neutrinos with spins oriented in the same direction as those produced by the $^{8}_{5}$B PP III reaction, a neutrino spin-flip might explain the solar neutrino problem. It should be pointed out, however, that neutrino observations still cover only a small number of 11-year sunspot number cycles, making a periodicity in neutrino detections very difficult to establish (in fact, some researchers question the statistical significance of the anticorrelation with sunspot number). Data covering a significantly longer time span are required before any final conclusions can be drawn concerning the effect of the solar cycle on neutrino counts.

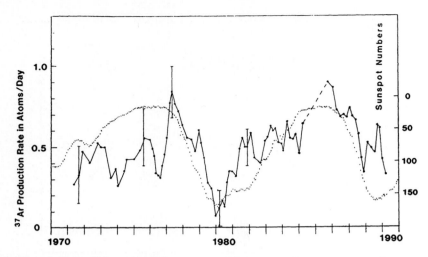

**Figure 11.39** The possible inverse correlation between sunspot number and the number of neutrinos detected in the Davis $^{37}_{17}$Cl experiment. The monthly averages of sunspot numbers are indicated by the small dots. Note that the sunspot count is inverted; sunspot maxima occurred in 1980 and 1991. (Figure from Davis and Cox, *Solar Interior and Atmosphere*, Cox, Livingston, and Matthews (eds.), The University of Arizona Press, Tucson, 1991.)

From our discussion in this chapter it should be clear that astrophysics has had a great deal of success in explaining many of the features of our Sun. The stellar structure equations describe the major aspects of the solar interior, and much of the Sun's complex atmosphere is also understood. But many other important questions remain to be answered, such as the solar neutrino problem and the details of the solar cycle, not to mention the interaction between the Sun and Earth's climate. Much exciting and challenging work remains to be done before we can feel confident that we understand the star that is closest to us.

# Suggested Readings

## General

Frazier, Kendrick, *Our Turbulent Sun*, Prentice-Hall, Englewood Cliffs, NJ, 1982.

Giovanelli, Ronald, *Secrets of the Sun*, Cambridge University Press, Cambridge, 1984.

Newkirk, Gordon Jr., and Frazier, Kendrick, "The Solar Cycle," *Physics Today*, April 1982.

Schwarzschild, Bertram, "Solar Neutrino Update: Three Detectors Tell Three Stories," *Physics Today*, October 1990.

## Technical

Abazov, A. I., et al., "Search for Neutrinos from the Sun Using the Reaction $^{71}Ga(\nu_e, e^-)^{71}Ge$," *Physical Review Letters*, *67*, 3332, 1991.

Bahcall, John N., *Neutrino Astrophysics*, Cambridge University Press, Cambridge, 1989.

Bahcall, John N., and Ulrich, Roger K., "Solar Models, Neutrino Experiments, and Helioseismology," *Reviews of Modern Physics*, *60*, 297, 1988.

Bai, T., and Sturrock, P. A., "Classification of Solar Flares," *Annual Review of Astronomy and Astrophysics*, *27*, 421, 1989.

Baliunas, Sallie, and Vaughan, Arthur H., "Stellar Activity Cycles," *Annual Review of Astronomy and Astrophysics*, *23*, 379, 1985.

Böhm-Vitense, Erika, *Introduction to Stellar Astrophysics, Volume I: Basic Stellar Observations and Data*, Cambridge University Press, Cambridge, 1989.

Böhm-Vitense, Erika, *Introduction to Stellar Astrophysics, Volume II: Stellar Atmospheres*, Cambridge University Press, Cambridge, 1989.

Cox, A. N., Livingston, W. C., and Matthews, M. S. (eds.), *Solar Interior and Atmosphere*, University of Arizona Press, Tucson, 1991.

Faulkner, John, and Gilliland, Ronald L., "Weakly Interacting Massive Particles and the Solar Neutrino Flux," *The Astrophysical Journal*, *299*, 994, 1985.

Foukal, Peter V., *Solar Astrophysics*, John Wiley and Sons, New York, 1990.

Griffiths, David J., *Introduction to Electrodynamics*, Second Edition, Prentice-Hall, Englewood Cliffs, NJ, 1989.

Hathaway, David H., and Wilson, Robert M., "Solar Rotation and the Sunspot Cycle," *The Astrophysical Journal, 357*, 271, 1990.

Mariska, John T., "The Quiet Solar Transition Region," *Annual Review of Astronomy and Astrophysics, 24*, 23, 1986.

Moore, Ronald and Rabin, Douglas, "Sunspots," *Annual Review of Astronomy and Astrophysics, 23*, 239, 1985.

Parker, E. N., "Dynamics of Interplanetary Gas and Magnetic Fields," *The Astrophysical Journal, 128*, 664, 1958.

Zirin, Harold, *Astrophysics of the Sun*, Cambridge University Press, Cambridge, 1988.

# Problems

11.1 Using Fig. 11.2, verify that the change in the Sun's effective temperature over the past 4.5 billion years is consistent with the variations in its radius and luminosity.

11.2 (a) At what rate is the Sun's mass decreasing due to nuclear reactions? Express your answer in solar masses per year.

  (b) Compare your answer to part (a) with the mass loss rate due to the solar wind.

  (c) Assuming that the solar wind mass loss rate remains constant, would either mass loss process significantly affect the total mass of the Sun over its entire main-sequence lifetime?

11.3 Using the Saha equation, calculate the ratio of the number of $H^-$ ions to neutral hydrogen atoms in the Sun's photosphere. Take the temperature of the gas to be the effective temperature, and assume that the electron pressure is 15 dyne cm$^{-2}$. Note that the Pauli exclusion principle requires that only one state can exist for the ion because its two electrons must have opposite spins.

11.4 The Paschen series of hydrogen ($n = 3$) can contribute to the visible continuum for the Sun since the series limit occurs at 8208 Å. However, it is the contribution from the $H^-$ ion that dominates the formation of the continuum. Using the results of Problem 11.3, along with the Boltzmann equation, estimate the ratio of the number of H$^-$ ions to hydrogen atoms in the $n = 3$ state.

11.5 (a) Using Eq. (9.58) and neglecting turbulence, estimate the full width at half-maximum of the hydrogen H$_\alpha$ absorption line due to random thermal motions in the Sun's photosphere. Assume that the temperature is the Sun's effective temperature.

(b) Using H$_\alpha$ redshift data for solar granulation, estimate the full width at half-maximum when convective turbulent motions are included with thermal motions.

(c) What is the ratio of $v_{\text{turb}}^2$ to $2kT/m$?

(d) Determine the relative change in the full width at half-maximum due to Doppler broadening when turbulence is included. Does turbulence make a significant contribution to $(\Delta \lambda)_{1/2}$ in the solar photosphere?

11.6 Estimate the thermally Doppler-broadened line widths for the hydrogen Lyman $\alpha$, C III, O VI, and Mg X lines given on page 401; use the temperatures provided. Take the masses of H, C, O, and Mg to be 1 u, 12 u, 16 u, and 24 u, respectively.

11.7 (a) Using Eq. (3.20), show that in the Sun's photosphere

$$\ln (B_a/B_b) \approx 11.5 + \frac{hc}{kT} \left( \frac{1}{\lambda_b} - \frac{1}{\lambda_a} \right)$$

where $B_a/B_b$ is the ratio of the amount of blackbody radiation emitted at $\lambda_a = 100$ Å to the amount emitted at $\lambda_b = 1000$ Å, centered in a wavelength band 1 Å wide.

(b) What is the value of this expression for the case where the temperature is taken to be the effective temperature of the Sun?

(c) Writing the ratio in the form $B_a/B_b = 10^x$, determine the value of $x$.

11.8 Suppose that you are attempting to make observations through an optically thick gas that has a constant density and temperature. Assume that the density and temperature of the gas are $2.5 \times 10^{-7}$ g cm$^{-3}$ and 5770 K, respectively, typical of the values found at the base of the Sun's photosphere. If the opacity of the gas at one wavelength ($\lambda_1$) is $\kappa_{\lambda_1} = 0.26$ cm$^2$ g$^{-1}$ and the opacity at another wavelength ($\lambda_2$) is

$\kappa_{\lambda 2} = 0.30$ cm$^2$ g$^{-1}$, calculate the distance into the gas where the optical depth equals 2/3 for each wavelength. At which wavelength can you see farther into the gas? How much farther? This effect allows astronomers to probe the Sun's atmosphere at different depths (see Fig. 11.17).

11.9 (a) Using the data given in Example 11.2, estimate the pressure scale height at the base of the photosphere.

(b) Assuming that the mixing length to pressure scale height ratio is 2.2, use the measured Doppler velocity of solar granulation to estimate the amount of time required for a convective bubble to travel one mixing length. Compare this value to the characteristic lifetime of a granule.

11.10 Show that Eq. (11.7) follows directly from Eq. (11.6).

11.11 Calculate the magnetic pressure in the center of the umbra of a large sunspot. Assume that the magnetic field strength is 2000 G. Compare your answer with a typical value for the gas pressure at the base of the photosphere.

11.12 Assume that a large solar flare erupts in a region where the magnetic field strength is 300 G and it releases $10^{32}$ ergs in one hour.

(a) What was the magnetic energy density in that region before the eruption began?

(b) What minimum volume would be required to supply the magnetic energy necessary to fuel the flare?

(c) Assuming for simplicity that the volume involved in supplying the energy for the flare eruption was a cube, compare the length of one side of the cube with the typical size of a large flare.

(d) How long would it take an Alfvén wave to travel the length of the flare?

(e) What can you conclude about the assumption that magnetic energy is the source of solar flares, given the physical dimensions and time scales involved?

11.13 (a) Calculate the frequency shift produced by the normal Zeeman effect in the center of a sunspot that has a magnetic field strength of 3000 G.

(b) By what fraction would the wavelength of one component of the 6302.5 Å Fe I spectral line change due to a magnetic field of 3000 G?

11.14 Argue from Eq. (11.12) and the work integral that magnetic pressure is given by Eq. (11.13).

# Chapter 12

# The Process of Star Formation

## 12.1  Interstellar Dust and Gas

When we look into the heavens, it appears as though the stars are unchanging, pointlike sources of light that shine steadily. On casual inspection, even our own Sun appears constant. But, as we have seen in the last chapter, this is not the case; sunspots come and go, flares erupt, the corona changes shape, and even the Sun's luminosity appears to be fluctuating very slightly over long periods of time.

In fact all stars change. Usually the changes are so gradual and over such long time intervals when measured in human terms that we do not notice them. Occasionally, however, the changes are extremely rapid and dramatic, as in the case of a supernova explosion. By invoking our understanding of the physics of stellar interiors and atmospheres developed thus far, we can now examine this process of how stars evolve during their lives.

In some sense the evolution of a star is cyclic. It is born out of gas and dust that exists between the stars, known as the **interstellar medium** (ISM). During its lifetime, depending on the star's total mass, much of that material may be returned to the ISM through stellar winds and explosive events. Subsequent generations of stars can then form from this processed material. As a result, to understand the evolution of a star, it is important to study the nature of the ISM.

On a dark night some of the dust clouds that populate our Galaxy can be seen in the band of stars that is the disk of the Milky Way Galaxy (see Fig. 12.1). It is not that these dark regions are devoid of stars, but rather

**Figure 12.1** Dust clouds obscure the stars located behind them in the disk of the Milky Way. (Courtesy of Palomar/Caltech.)

that the stars located behind intervening dust clouds are obscured. This obscuration, referred to as **interstellar extinction**, is due to the scattering and absorption of the starlight (see Fig. 12.2).

Given the effect that extinction can have on the apparent magnitude of a star, the distance modulus equation (Eq. 3.6) must be modified appropriately. In a given wavelength band centered on $\lambda$, we now have

$$m_\lambda = M_\lambda + 5\log_{10} d - 5 + a_\lambda, \qquad (12.1)$$

where $d$ is in pc and $a_\lambda$ represents the number of magnitudes of absorption or scattering present along the line of sight. If $a_\lambda$ is large enough, a star that would otherwise be visible to the naked eye or through a telescope could no longer be detected. This is the reason for the dark bands running through the Milky Way.

Clearly $a_\lambda$ must be related to the optical depth of the material, measured back along the line of sight. From Eq. (9.16), the fractional change in the intensity of the light is given by

$$I_\lambda/I_{\lambda,\circ} = e^{-\tau_\lambda}.$$

Combining this with Eq. (3.4), we can now relate the optical depth to the

## 12.1 Interstellar Dust and Gas

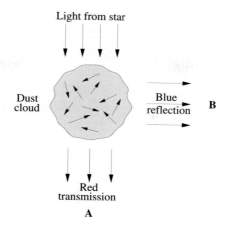

**Figure 12.2** A dust cloud can either scatter or absorb light that passes through it. The amount of scattering and absorption depends on the number density of particles, the wavelength of the light, and the thickness of the cloud. Since shorter wavelengths are affected more significantly than longer ones, a star lying behind the cloud appears reddened to observer A. Observer B sees the scattered shorter wavelengths as a blue reflection nebula.

change in apparent magnitude due to extinction, giving

$$m_\lambda - m_{\lambda,o} = -2.5 \log_{10}\left(e^{-\tau_\lambda}\right) = 2.5\tau_\lambda \log_{10} e = 1.086\tau_\lambda.$$

But the change in apparent magnitude is just $a_\lambda$, so

$$a_\lambda = 1.086\tau_\lambda. \tag{12.2}$$

*The change in magnitude due to extinction is approximately equal to the optical depth along the line of sight.*

From the expression for the optical depth given by Eq. (9.15) and the discussion at the end of Example 9.2 (page 265), the optical depth through the cloud is given by

$$\tau_\lambda = \int_0^s n(s)\sigma_\lambda\, ds, \tag{12.3}$$

where $n(s)$ is the number density of scattering particles and $\sigma_\lambda$ is the scattering cross section. If $\sigma_\lambda$ is constant along the line of sight, then

$$\tau_\lambda = \sigma_\lambda \int_0^s n(s)\, ds = \sigma_\lambda N_d,$$

where $N_d$, the *column density*, is the number of scattering dust particles in a thin cylinder with a cross section of 1 cm$^2$ stretching from the observer to the star. Thus we see that the amount of extinction depends on the amount of interstellar dust that the light passes through, as one would expect.

If we assume for simplicity, as was first done by G. von Mie in 1908, that the dust particles are spherical and each has a radius $a$, then the geometrical cross section that a particle presents to a passing photon is just $\sigma_g = \pi a^2$. We may now define the dimensionless *extinction coefficient* $Q_\lambda$ to be

$$Q_\lambda = \frac{\sigma_\lambda}{\sigma_g},$$

where $Q_\lambda$ depends on the composition of the dust grain.

Mie was able to show that when the wavelength of the light is on the order of the size of the dust grains, then $Q_\lambda \sim a/\lambda$ (implying that $\sigma_\lambda \propto \lambda^{-1}$). In the limit that $\lambda$ becomes very large relative to $a$, $Q_\lambda$ goes to zero, and at the other extreme, if $\lambda$ becomes very small relative to $a$, $Q_\lambda$ approaches a constant, independent of $\lambda$. This behavior can be understood by analogy to waves on the surface of a lake. If the wavelength of the waves is much larger than an object in their way, such as a grain of sand, the waves pass by almost completely unaffected ($Q_\lambda \sim 0$). On the other hand, if the waves are much smaller than the obstructing object—for instance, an island—they are simply blocked; the only waves that continue on are those that miss the island altogether. Similarly, at sufficiently short wavelengths, the only light we detect passing through the dust cloud is the light that travels between the particles.

Combining the ideas already discussed, it is clear that the amount of extinction, as measured by $a_\lambda$, must be wavelength-dependent. Since the longer-wavelength red light is not scattered as strongly as blue light, the starlight passing through intervening dust clouds becomes reddened as the blue light is removed. This **interstellar reddening** causes stars to appear redder than their effective temperatures would otherwise imply. Fortunately, it is possible to detect this change by carefully analyzing the absorption and emission lines in the star's spectrum.

Much of the incident blue light is scattered out of its original path and can leave the cloud in virtually any direction. As a result, looking at the cloud in a direction other than along the line of sight to a bright star behind the cloud, an observer will see a blue **reflection nebula** (recall Fig. 12.2). This process is analogous to Rayleigh scattering, which produces a blue sky on Earth. The difference between Mie scattering and Rayleigh scattering is that the sizes of the scattering molecules of the latter are much smaller than the wavelength of visible light, leading to $\sigma_\lambda \propto \lambda^{-4}$.

## 12.1 Interstellar Dust and Gas

**Example 12.1** A certain star, located 0.8 kpc from Earth, is found to be dimmer than expected at 5500 Å by $a_V = 1.1$ magnitudes, where $a_V$ is the amount of extinction as measured through the *visual wavelength* filter (see Section 3.6). If $Q_{5500} = 1.5$ and the dust grains are assumed to be spherical with radii of 0.2 $\mu$m, estimate the average density ($\bar{n}$) of material between the star and Earth.

From Eq. (12.2), the optical depth along the line of sight is nearly equal to the amount of extinction in magnitudes, or $\tau_{5500} \simeq 1$. Also,

$$\sigma_{5500} = \pi a^2 Q_{5500} \simeq 2 \times 10^{-9} \text{ cm}^2.$$

Now the column density of the dust along the line of sight is

$$N_d = \frac{\tau_{5500}}{\sigma_{5500}} \simeq 5 \times 10^8 \text{ cm}^{-2}.$$

Finally, since $N_d = \int_0^s n(s)\, ds = \bar{n} \times 0.8$ kpc, we have

$$\bar{n} = \frac{N_d}{0.8 \text{ kpc}} = 2 \times 10^{-13} \text{ cm}^{-3}.$$

Number densities of this magnitude are typical of the plane of the Milky Way Galaxy.

---

Predictions of the Mie theory work well for longer wavelengths, typically from the infrared into the visible wavelength region. However, at ultraviolet wavelengths significant deviations become apparent, as can be seen by considering the ratio of $a_\lambda$, the extinction in a wavelength band centered at $\lambda$, to the extinction in some reference wavelength band, such as $a_V$. This ratio is generally plotted versus reciprocal wavelength $\lambda^{-1}$, as in Fig. 12.3.[1]

At longer wavelengths (the left side of the graph) the data agree well with the Mie theory. For wavelengths shorter than the blue wavelength band ($B$), however, the curves begin to diverge significantly, deviating from the expected relation, $a_\lambda/a_V \propto \lambda^{-1}$. Particularly evident is the "bump" in the ultraviolet at 2175 Å or 4.6 $\mu$m$^{-1}$. At even shorter wavelengths, the extinction curve tends to rise sharply as the wavelength decreases.

The existence of the "bump" in Fig. 12.3 gives us some hint of the composition of the dust. Graphite, a well-ordered form of carbon, interacts strongly with light near 2175 Å. Although it is unclear how carbon can organize into large graphite particles in the interstellar medium, the strength of the "bump,"

---

[1] Alternatively, **color excesses** may be plotted, such as $(a_\lambda - a_V)/(a_B - a_V)$.

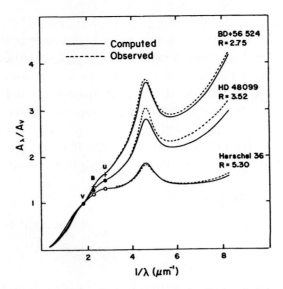

**Figure 12.3** Interstellar extinction curves along the lines of sight to three stars. The dashed lines represent the observational data and the solid lines are theoretical fits. The $U$, $B$, and $V$ wavelength bands are indicated for reference. (Figure adapted from Mathis, *Annu. Rev. Astron. Astrophys.*, **28**, 37, 1990. Reproduced with permission from the *Annual Review of Astronomy and Astrophysics*, Volume 28, ©1990 by Annual Reviews Inc.)

the abundance of carbon, and the existence of the 1275 Å resonance have led most researchers to suggest that graphite must be a major component of interstellar dust.

Apparently interstellar dust is composed of other particles as well, as evidenced by the existence of dark absorption bands at wavelengths of 9.7 $\mu$m and 18 $\mu$m in the near-infrared. These features are believed to be the result of the stretching in silicates of the Si-O chemical bond and the bending of Si-O-Si bonds, respectively. Just as transitions between atomic energy levels are quantized, so are the energies associated with chemical bonds. In the case of chemical bonds, however, the energy levels tend to be grouped in closely spaced bands, producing characteristic broad features in the spectrum of the light. The existence of these absorption bands involving silicon indicates that silicate grains are also present in the dust clouds and the diffuse dust of the ISM.

Yet another series of molecular bands have been observed in the light from diffuse dust, in this case in emission rather than in absorption. The so-called *unidentified infrared emission bands* exist in the wavelength range between

## 12.1 Interstellar Dust and Gas

3.3 $\mu$m and 12 $\mu$m. These bands appear to be due to vibrations in C-C and C-H bonds. As a result, astronomers have been led to suggest the existence of planar molecules with organic benzene ringlike structures known as **polycyclic aromatic hydrocarbons** (PAHs).

An important characteristic of the light from interstellar dust is that it tends to be slightly polarized. The amount of polarization is typically a few percent and depends on wavelength. This necessarily implies that the dust grains cannot be perfectly spherical. Furthermore, they must be at least somewhat aligned along a unique direction since the electric field vectors of the radiation are preferentially oriented in a particular direction. The most likely way to establish such an alignment is for the grains to interact with a weak magnetic field. Because less energy is required, the particles tend to rotate with their long axes perpendicular to the direction of the magnetic field.

All of these observations give us some clues as to the composition of the dust in the ISM. However, as yet no consensus exists regarding the details relating to the range of sizes of the grains or their overall structure. One possibility is that the dust in the ISM is composed of both graphite and silicate grains ranging in size from about 0.25 $\mu$m down to several angstroms, the size of PAHs. By combining the contributions from each of these components, it appears that many of the features of the interstellar extinction curve can be reproduced.

Although dust produces most of the obscuration that is readily noticeable, the dominant component of the ISM is hydrogen gas in its various forms; neutral hydrogen (H I), ionized hydrogen (H II), and molecular hydrogen ($H_2$). Hydrogen comprises approximately 70% of the matter in the ISM, and helium makes up most of the remaining mass; metals, such as carbon and silicon, account for only a few percent of the total.

Most hydrogen in diffuse interstellar hydrogen clouds is in the form of H I in the ground state. As a result, the H I is generally incapable of producing emission lines by downward transitions of electrons from one orbit to another. It is also difficult to observe H I in absorption, since UV wavelength photons are required to lift the electrons out of the ground state.[2]

Fortunately, it is still possible to identify neutral hydrogen in the ISM. This is done by detecting the unique radio-wavelength **21-cm line**. The 21-cm line is produced by the reversal of the spin of the electron relative to the proton in the atom's nucleus. Recall (Section 5.4) that both electrons and protons possess an inherent spin angular momentum, with the $z$-component

---

[2]Orbiting observatories, such as the Hubble Space Telescope and the International Ultraviolet Explorer, have detected absorption lines produced by cold clouds of H I with strong UV sources lying behind them.

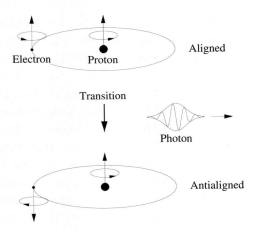

**Figure 12.4** When the spins of the electron and proton in a hydrogen atom go from being aligned to antialigned, a 21-cm wavelength photon is emitted.

of the spin angular momentum vector having one of two possible orientations (corresponding to the two allowed values of the quantum number, $m_s = \pm\frac{1}{2}$). Because these particles are also electrically charged, their intrinsic spins endow them with dipole magnetic fields, much like those of bar magnets. If the spins of the electron and proton are aligned (e.g., both spin axes are in the same direction), the atom has slightly more energy than if they are antialigned (see Fig. 12.4). As a result, if the electron's spin "flips" from being aligned with the proton to being antialigned, energy must be lost from the atom. If the spin flip is not due to a collision with another atom, then a photon is emitted. Of course, a photon can also be absorbed, exciting a hydrogen atom into aligning its electron and proton spins. The wavelength of the photon associated with the spin flip is 21.1 cm, corresponding to a frequency of 1420 MHz.

The emission of a 21-cm photon from an individual hydrogen atom is extremely rare. Once in the excited state, on average several million years can pass before that atom will emit a photon. Competing with this spontaneous emission are collisions between hydrogen atoms that may result in either excitation or de-excitation. In the ISM, collisions occur on time scales of hundreds of years. Although this is far shorter than the spontaneous emission time scale, statistically some atoms are still able to make the necessary spontaneous transition. On the other hand, the best vacuums produced in Earth-based laboratories have densities much greater than those found in the ISM, meaning that collision rates are significantly higher and essentially all of the atoms in the laboratory are de-excited before they can emit 21-cm radiation. As a result, the only place from which 21-cm radiation can be originating is space. The existence of 21-cm radiation was predicted in the early 1940s and first detected in 1951. Since then it has become an important tool in mapping the location and density of H I, measuring radial velocities using the Doppler effect, and estimating magnetic fields using the Zeeman effect. The 21-cm radiation is particularly valuable in determining the structure and kinematic properties of galaxies, including our own.

Although H I is quite abundant, the rarity of 21-cm emission (or absorption) from individual atoms means that the center of this line can remain optically thin over large interstellar distances. Assuming that the line profile is a Gaussian, like the shape of the Doppler line profile shown in Fig. 9.19, the optical depth of the line center is given by

$$\tau_H = 5.2 \times 10^{-19} \frac{N_H}{T \Delta v}, \qquad (12.4)$$

where $N_H$ is the column density of H I (in units of cm$^{-2}$), $T$ is the temperature of the gas (in kelvin), and $\Delta v$ is the full width of the line at half maximum. However, since the line width is due primarily to the Doppler effect (Eq. 4.30), rather than expressing the width in wavelength units, $\Delta v$ is expressed in units of velocity; typically $\Delta v \sim 10$ km s$^{-1}$. As long as the line is optically thin (i.e., on the linear part of the curve of growth; Fig. 9.21), the optical depth is proportional to the neutral hydrogen column density. Studies of **diffuse H I clouds** indicate temperatures of 30 to 80 K, number densities in the range $100 - 800$ cm$^{-3}$, and masses on the order of $1 - 100$ M$_\odot$.

Comparing $\tau_H$ with $a_V$ along the same line of sight shows that $N_H$ is generally proportional to $N_d$ when $a_V < 1$. This observation suggests that dust and gas are distributed together throughout the ISM. However, when $a_V > 1$, this correlation breaks down; the column density of H I no longer increases as rapidly as the column density of dust. Apparently, other physical processes are involved when the dust becomes optically thick.

Optically thick dust clouds shield hydrogen from sources of ultraviolet radiation. One consequence of this shielding is that molecular hydrogen can exist without the threat of undergoing dissociation by UV photon absorption. Dust can also enhance the H$_2$ formation rate beyond what would be expected by random collisions of hydrogen atoms. This enhancement occurs for two reasons: (1) A dust grain (or possibly a PAH) can provide a *site* for the hydrogen atoms to meet, and (2) the dust provides a *sink* for the binding energy that must be liberated if a stable molecule is to form. The liberated energy goes into heating the grain and ejecting the H$_2$ molecule from the formation site. Since the structure of H$_2$ differs greatly from that of atomic hydrogen, the H$_2$

molecule does not emit 21-cm radiation and thus explains why $N_H$ and $a_V$ are poorly correlated in **molecular clouds** for which $a_V > 1$.

If the column density of atomic hydrogen is sufficiently large ($N_H$ on the order of $10^{21}$ cm$^{-2}$), it can also shield H$_2$ from UV photodissociation. Consequently, molecular clouds are surrounded by shells of H I.

Unfortunately, H$_2$ is very difficult to observe directly because the molecule does not have any emission or absorption lines in the visible or radio portions of the electromagnetic spectrum. As a result, it becomes necessary to use other molecules as *tracers* of H$_2$. Because of its relative abundance (approximately $10^{-4}$ that of H$_2$), the most commonly investigated tracer is carbon monoxide, CO, although other molecules have also been used, including CH, OH, CS, and C$_3$H$_2$. During collisions the tracer molecules become excited (or de-excited) and spontaneous transitions from excited states result in the emission of photons in wavelength regions that are more easily observed than those associated with H$_2$, such as the 2.6 mm transition of CO. Since collision rates depend on both the gas temperature (or kinetic energy) and the number densities of the species, molecular tracers can provide information about the environment within a molecular cloud.[3]

The results of these studies show that conditions within molecular clouds can vary widely. In clouds where the hydrogen gas is primarily atomic and $a_V \simeq 1 - 5$, molecular hydrogen may be found in regions of higher column density. Such clouds are referred to as **translucent molecular clouds**. Conditions in translucent clouds are typical of diffuse H I clouds but with somewhat higher masses; they have temperatures of 15 – 50 K, $n \sim 500 - 5000$ cm$^{-3}$, $m \sim 3 - 100$ M$_\odot$, and they measure several parsecs across. Both H I clouds and translucent molecular clouds tend to be irregularly shaped.

**Giant molecular clouds** are enormous complexes of dust and gas where temperatures are typically $T \sim 20$ K, number densities are in the range $n \sim 100 - 300$ cm$^{-3}$, masses may reach $10^6$ M$_\odot$, and typical sizes are on the order of 50 pc. Residing within GMC complexes are hot, dense cores with characteristic dimensions of $r \sim 0.05 - 1$ pc, where $a_V \sim 50 - 1000$, $T \sim 100 - 200$ K, $n \sim 10^7 - 10^9$ cm$^{-3}$, and $m \sim 10 - 1000$ M$_\odot$. Thousands of GMCs are known to exist in our Galaxy, mostly in its spiral arms.

At the other extreme are the small, dense, almost spherical clouds known as **Bok globules**.[4] These globules are characterized by large visual extinctions ($a_V \sim 10$), low temperatures ($T \sim 10$ K), relatively large number densities

---

[3] An estimate of atomic and molecular collision rates can be made in a way completely analogous to the approach used to obtain the nuclear reaction rate equation (Eq. 10.36).

[4] Bok globules are named after Bart Bok, who first studied these objects in the 1940s.

($n > 10^4$ cm$^{-3}$), low masses ($m \sim 1 - 1000$ M$_\odot$), and small sizes ($r \sim 1$ pc).

Infrared surveys of Bok globules have revealed that many, perhaps most, of these objects harbor young stars in their centers. Apparently Bok globules are active sites of star formation. Furthermore, the physical association between GMCs and very young O and B main-sequence stars suggests that star formation occurs in these regions as well. The process by which stars form out of the ISM will be considered in the next section.

Before leaving our discussion of clouds, it is worth mentioning the existence of other molecules in the ISM. To date, radio observations have revealed the presence of more than 50 molecules, ranging in complexity from diatomic molecules such as H$_2$ and CO to fairly long organic strings, including HC$_{11}$N. Such complex molecules cannot be expected to form as a result of random collisions in the low densities of the ISM. Rather, it is believed that dust serves the role of providing sites where atoms can congregate to form these molecules.

## 12.2 The Formation of Protostars

Over the past three decades, our understanding of stellar evolution has developed significantly, reaching the point where much of the life history of a star is well determined. This success has been due to advances in observational techniques, improvements in our knowledge of the physical processes important in stars, and increases in computational power. In the remainder of this chapter and in Chapter 13 we will present an overview of the lives of stars, leaving detailed discussions of some special phases of evolution until later, specifically stellar pulsation and compact objects (stellar corpses).

Despite many successes, important questions remain concerning how stars change during their lifetimes. One area where the picture is far from complete is in the earliest stage of evolution, the formation of pre-nuclear-burning objects known as **protostars** from interstellar molecular clouds.

If molecular clouds are the sites of star formation, what conditions must exist for collapse to occur? Sir James Jeans (1877–1946) first investigated this problem in 1902 by considering the effects of small deviations from hydrostatic equilibrium. Although several simplifying assumptions are made in the analysis, such as neglecting effects due to rotation and galactic magnetic fields, it provides important insights into the development of protostars.

The virial theorem (Eq. 2.45),

$$2K + U = 0,$$

describes the condition of equilibrium for a stable, gravitationally bound, system.[5] We have already seen that the virial theorem arises naturally in the discussion of orbital motion, and we have also invoked it in estimating the amount of gravitational energy contained within a star (Eq. 10.28). The virial theorem may also be used to estimate the necessary conditions for protostellar collapse.

If twice the total internal kinetic energy of a molecular cloud ($2K$) exceeds the absolute value of the gravitational potential energy ($|U|$), the force due to the gas pressure will dominate over the force of gravity and the cloud will expand. On the other hand, if the internal kinetic energy is too low, the cloud will collapse. The boundary between these two cases describes the critical condition for stability.

From Eq. (10.28), the gravitational potential energy of a spherical cloud of constant density is approximately

$$U \sim -\frac{3}{5}\frac{GM_c^2}{R_c},$$

where $M_c$ and $R_c$ are the mass and radius of the cloud, respectively. We may also estimate the cloud's internal kinetic energy, given by

$$K = \frac{3}{2}NkT,$$

where $N$ is the total number of particles. But $N$ is just

$$N = \frac{M_c}{\mu m_H},$$

where $\mu$ is the mean molecular weight. Now, by the virial theorem, the condition for collapse ($2K < |U|$) becomes

$$\frac{3M_c kT}{\mu m_H} < \frac{3}{5}\frac{GM_c^2}{R_c}. \tag{12.5}$$

The radius may be replaced by using the initial mass density of the cloud, $\rho_0$, assumed here to be constant throughout the cloud,

$$R_c = \left(\frac{3M_c}{4\pi\rho_0}\right)^{1/3}. \tag{12.6}$$

After substitution into Eq. (12.5), we may solve for the minimum mass necessary to initiate the spontaneous collapse of the cloud. This condition is known as the **Jeans criterion**,

$$M_c > M_J,$$

---

[5]We have implicitly assumed that the kinetic and potential energy terms are averaged over time, as in Section 2.4.

## 12.2 The Formation of Protostars

where

$$M_J \simeq \left(\frac{5kT}{G\mu m_H}\right)^{3/2} \left(\frac{3}{4\pi\rho_0}\right)^{1/2}. \tag{12.7}$$

The critical value $M_J$ is called the **Jeans mass**. Using Eq. (12.6), the Jeans criterion may also be expressed in terms of the minimum radius necessary to collapse a cloud of density $\rho_0$.

$$R_c > R_J,$$

where

$$R_J \simeq \left(\frac{15kT}{4\pi G\mu m_H \rho_0}\right)^{1/2} \tag{12.8}$$

is the **Jeans length**.

---

**Example 12.2** For a typical diffuse hydrogen cloud, $T = 50$ K and $n = 500$ cm$^{-3}$. If we assume that the cloud is entirely composed of H I, $\rho_0 = m_H n_H = 8.4 \times 10^{-22}$ g cm$^{-3}$. Taking $\mu = 1$ and using Eq. (12.7), the minimum mass necessary to cause the cloud to collapse spontaneously is approximately $M_J \sim 1500$ $M_\odot$. However, this value significantly exceeds the estimated $1 - 100$ $M_\odot$ believed to be contained in H I clouds. Hence diffuse hydrogen clouds are stable against gravitational collapse.

On the other hand, for a dense core of a giant molecular cloud, typical temperatures and number densities are $T = 150$ K and $n = 10^8$ cm$^{-3}$, implying that $\rho_0 = m_H n_H = 2 \times 10^{-16}$ g cm$^{-3}$. In this case the Jeans mass is $M_J \sim 17$ $M_\odot$, well within the observationally determined range of $10 - 1000$ $M_\odot$ for these objects. Apparently the cores of GMCs are unstable to gravitational collapse, consistent with their being sites of star formation.

---

In the case that the Jeans criterion for gravitational collapse has been satisfied, the collapsing molecular cloud is essentially in free-fall during the first part of its evolution; any existing pressure gradients are too small to influence the motion appreciably. Furthermore, throughout the free-fall phase, the temperature of the gas remains nearly constant (i.e., the collapse is said to be *isothermal*). This is true as long as the cloud remains optically thin and the gravitational potential energy released during the collapse can be efficiently radiated away. Therefore, the spherically symmetric hydrodynamic equation (Eq. 10.6) can be used to describe the contraction if we assume that $|dP/dr| \ll GM_r\rho/r^2$. After canceling the density on both sides of the expression, we have

$$\frac{d^2r}{dt^2} = -G\frac{M_r}{r^2}. \tag{12.9}$$

Of course, the right-hand side of Eq. (12.9) is just the local acceleration of gravity at a distance $r$ from the center of a spherical cloud. As usual, the mass of the sphere interior to the radius $r$ is denoted by $M_r$.

To describe the behavior of the surface of a sphere of radius $r$ within the collapsing cloud as a function of time, Eq. (12.9) must be integrated over time. Since we are interested only in the surface that encloses $M_r$, the mass interior to $r$ will remain a constant during that collapse. As a result, we may replace $M_r$ by the product of the initial density $\rho_0$ and the initial spherical volume, $4\pi r_0^3/3$. Then, if we multiply both sides of Eq. (12.9) by the velocity of the surface of the sphere, we arrive at the expression

$$\frac{dr}{dt}\frac{d^2r}{dt^2} = -\left(\frac{4\pi}{3}G\rho_0 r_0^3\right)\frac{1}{r^2}\frac{dr}{dt},$$

which can be integrated to give

$$\frac{1}{2}\left(\frac{dr}{dt}\right)^2 = \left(\frac{4\pi}{3}G\rho_0 r_0^3\right)\frac{1}{r} + C_1.$$

The integration constant, $C_1$, can be evaluated by requiring that the velocity of the sphere's surface be zero at the beginning of the collapse, or $dr/dt = 0$ when $r = r_0$. This gives

$$C_1 = -\frac{4\pi}{3}G\rho_0 r_0^2.$$

Substituting and solving for the velocity at the surface, we have

$$\frac{dr}{dt} = -\left[\frac{8\pi}{3}G\rho_0 r_0^2\left(\frac{r_0}{r} - 1\right)\right]^{1/2}. \tag{12.10}$$

Note that the negative root was chosen because the cloud is collapsing.

To integrate Eq. (12.10) so that we can obtain an expression for the position as a function of time, we make the substitutions

$$\theta \equiv \frac{r}{r_0}$$

and

$$K \equiv \left(\frac{8\pi}{3}G\rho_0\right)^{1/2},$$

which leads to the differential equation

$$\frac{d\theta}{dt} = -K\left(\frac{1}{\theta} - 1\right)^{1/2}. \tag{12.11}$$

## 12.2 The Formation of Protostars

Making yet another substitution,

$$\theta \equiv \cos^2 \xi, \qquad (12.12)$$

and after some manipulation, Eq. (12.11) becomes

$$\cos^2 \xi \frac{d\xi}{dt} = \frac{K}{2}. \qquad (12.13)$$

Equation (12.13) may now be integrated directly to yield

$$\frac{\xi}{2} + \frac{1}{4}\sin 2\xi = \frac{K}{2}t + C_2. \qquad (12.14)$$

Lastly, the integration constant, $C_2$, must be evaluated. Doing so requires that $r = r_0$ when $t = 0$, which implies that $\theta = 1$, or $\xi = 0$ at the beginning of the collapse. Therefore $C_2 = 0$.

We have finally arrived at the equation of motion for the gravitational collapse of the cloud, given in parameterized form by

$$\xi + \frac{1}{2}\sin 2\xi = Kt. \qquad (12.15)$$

Our task now is to extract the behavior of the collapsing cloud from this equation. From Eq. (12.15), it is possible to calculate the **free-fall time scale** for a cloud that has satisfied the Jeans criterion. Let $t = t_\text{ff}$ when the radius of the collapsing sphere reaches zero ($\theta = 0$, $\xi = \pi/2$).[6] Then

$$t_\text{ff} = \frac{\pi}{2K}.$$

Substituting the value for $K$, we have

$$t_\text{ff} = \left(\frac{3\pi}{32}\frac{1}{G\rho_0}\right)^{1/2}. \qquad (12.16)$$

The reader should notice that the free-fall time is actually independent of the initial radius of the sphere. Consequently, as long as the original density of the spherical molecular cloud was uniform, all parts of the cloud will take the same amount of time to collapse and the density will increase at the same rate everywhere. This behavior is known as a **homologous collapse**. If the cloud is somewhat centrally condensed when the collapse begins, however, the free-fall time for material near the center will be shorter than for material farther out. Thus, as the collapse progresses, the density near the center will increase more rapidly than in other regions.

---

[6]This is obviously an unphysical final condition, since it implies infinite density. If $r_0 \gg r_\text{final}$, however, then $r_\text{final} \simeq 0$ is a reasonable approximation for our purposes here.

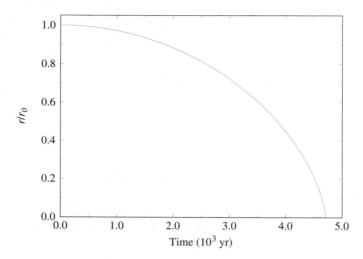

**Figure 12.5** The ratio of the radius relative to its initial value as a function of time for the homologous collapse of a molecular cloud. The collapse is assumed to be isothermal, beginning with a density of $\rho_0 = 2 \times 10^{-16}$ g cm$^{-3}$.

---

**Example 12.3** Using data given in Example 12.2 for a dense core of a giant molecular cloud, we may estimate the amount of time required for the collapse. Assuming $\rho_0 = 2 \times 10^{-16}$ g cm$^{-3}$, Eq. (12.16) gives

$$t_{\rm ff} = 4700 \text{ yr}.$$

To investigate the actual behavior of the collapse in our simplified model, we must first solve Eq. (12.15) for $\xi$, given a value for $t$, and then use Eq. (12.12) to find $\theta = r/r_0$. However, Eq. (12.15) cannot be solved explicitly, so numerical techniques must be employed. The numerical solution of the homologous collapse of the molecular cloud is shown in Figs. 12.5 and 12.6. Notice that the collapse is quite slow initially and accelerates quickly as $t_{\rm ff}$ is approached. At the same time, the density increases very rapidly during the final stages of collapse.

---

Since the masses of fairly large molecular clouds could exceed the Jeans limit, from Eq. (12.7) our simple analysis seems to imply that stars can form with very large masses, possibly up to the initial mass of the cloud. However, observations show that this does not happen. Furthermore, it appears that stars frequently (perhaps even preferentially) tend to form in groups, ranging from binary star systems to clusters that contain hundreds of thousands of members (see Section 13.4).

## 12.2 The Formation of Protostars

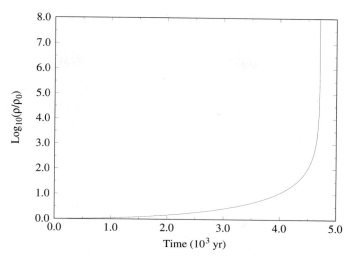

**Figure 12.6** The ratio of the cloud's density relative to its initial value as a function of time for the isothermal, homologous collapse of a molecular cloud with an initial density of $\rho_0 = 2 \times 10^{-16}$ g cm$^{-3}$.

The process of **fragmentation** that segments a collapsing cloud can be understood by referring to the equation for the Jeans mass (Eq. 12.7). An important consequence of the collapse of a molecular cloud is that the density of the cloud increases by many orders of magnitude during free-fall (Fig. 12.6). Consequently, since $T$ remains nearly constant throughout much of the collapse, it appears that the Jeans mass must decrease. Given any initial inhomogeneities in density, after collapse has begun, sections of the cloud will independently satisfy the Jeans mass limit and begin to collapse locally, producing smaller features within the original cloud. This cascading collapse could lead to the formation of large numbers of smaller objects.

But what stops the fragmentation process? Since we observe a galaxy filled with stars that have masses on the order of the mass of the Sun, the cascading fragmentation of the cloud cannot proceed without interruption. The answer to the question lies in our implicit assumption that the collapse is isothermal, which in turn implies that the only term that changes in Eq. (12.7) is the density. Clearly this cannot be the case since stars have temperatures much higher than $10 - 100$ K. If the energy that is released during a gravitational collapse (as required by the virial theorem) is radiated away efficiently, the temperature can remain nearly constant. At the other extreme, if the energy cannot be transported out of the cloud at all (an *adiabatic* collapse), then the temperature must rise. Of course, the real situation must be somewhere between these two limits, but by considering each of these special cases carefully we can begin to understand some of the important features of the problem.

If the collapse changes from being essentially isothermal to adiabatic, the associated temperature rise would begin to affect the value of the Jeans mass. In Chapter 10 we saw that for an adiabatic process the pressure of the gas is related to its density by $\gamma$, the ratio of specific heats (Eq. 10.78). Using the ideal gas law (Eq. 10.14), an adiabatic relation between density and temperature can be obtained,

$$T = K'' \rho^{\gamma-1}, \tag{12.17}$$

where $K''$ is a constant. Substituting this expression for temperature into Eq. (12.7), we find that for an adiabatic collapse the dependence of the Jeans mass on density becomes

$$M_J \propto \rho^{(3\gamma-4)/2}.$$

For atomic hydrogen $\gamma = 5/3$, giving $M_J \propto \rho^{1/2}$; the Jeans mass *increases* with increasing density for a perfectly adiabatic collapse of a cloud. This behavior means that the collapse results in a minimum value for the mass of the fragments produced. The minimum mass depends on the point when the collapse goes from being predominantly isothermal to adiabatic.

Of course, this transition is not instantaneous or even complete. However, it is possible to make a crude order of magnitude estimate of the lower mass limit of the fragments. As we have already mentioned, according to the virial theorem, energy must be liberated during the collapse of the cloud. From Eq. (10.28) and the discussion contained in Example 10.3, the energy released is roughly

$$\Delta E_g \simeq \frac{3}{10} \frac{GM_J^2}{R_J}$$

for a spherical cloud just satisfying the Jeans criterion at some point during the collapse. Averaged over the free-fall time, the luminosity due to gravity is given by

$$L_{\text{ff}} \simeq \frac{\Delta E_g}{t_{\text{ff}}} \sim G^{3/2} \left(\frac{M_J}{R_J}\right)^{5/2},$$

where we have made use of Eq. (12.16) and neglected terms of order unity.

If the cloud were optically thick and in thermodynamic equilibrium, the energy would be emitted as blackbody radiation. During collapse, however, the process of releasing the energy is less efficient than for an ideal blackbody. Following Eq. (3.17), we may express the radiated luminosity as

$$L_{\text{rad}} = 4\pi R^2 e\sigma T^4,$$

## 12.2 The Formation of Protostars

where an efficiency factor, $0 < e < 1$, has been introduced to indicate the deviation from thermodynamic equilibrium. If the collapse is perfectly isothermal and escaping radiation does not interact at all with overlying infalling material, $e \sim 0$. If, on the other hand, energy emitted by some of parts of the cloud is absorbed and then re-emitted by other parts of the cloud, thermodynamic equilibrium would more nearly apply and $e$ would be closer to unity.

Equating the two expressions for the cloud's luminosity,

$$L_{\text{ff}} = L_{\text{rad}},$$

and rearranging, we have

$$M_J^{5/2} = \frac{4\pi}{G^{3/2}} R_J^{9/2} e\sigma T^4.$$

Making use of Eq. (12.6) to eliminate the radius, and then using Eq. (12.7) to write the density in terms of the Jeans mass, we arrive at an estimate of when adiabatic effects become important, expressed in terms of the minimum obtainable Jeans mass,

$$M_{J_{\min}} = 0.03 \left( \frac{T^{1/4}}{e^{1/2} \mu^{9/4}} \right) \, M_\odot, \qquad (12.18)$$

where $T$ is expressed in kelvin. If we take $\mu \sim 1$, $e \sim 0.1$, and $T \sim 1000$ K at the time when adiabatic effects may start to become significant, $M_J \sim 0.5 \, M_\odot$. This is just the right order of magnitude to explain what is observed; fragmentation ceases when the segments of the original cloud begin to reach the range of solar mass objects. The estimate is relatively insensitive to other reasonable choices for $T$, $e$, and $\mu$.

We have, of course, left out a number of important features in our calculations. For instance, we have freely used the Jeans criterion during each point in the collapse of the cloud to discuss the process of fragmentation. This cannot be correct, since our estimate of the Jeans criterion was based on a perturbation of a static cloud; no consideration was made of the initial velocity of the cloud's outer layers. We have also neglected the details of radiation transport through the cloud, as well as vaporization of the dust grains, dissociation of molecules, and ionization of the atoms. Nevertheless, it is worth noting that as unsophisticated as the preceding analysis was, it did illustrate important aspects of the fundamental problem. Such preliminary approaches to understanding complex physical systems are powerful tools in our study of nature.[7]

Perhaps just as important to the questions of stability and the rate of collapse are the possible effects of rotation (angular momentum), the deviation from spherical symmetry, or the presence of magnetic fields. An appreciable amount of angular momentum present in the original cloud is likely to result in a disklike structure for at least a part of the original material, since collapse will proceed at a more rapid rate along the axis of rotation when compared with the equator. Also, the interaction between magnetic fields and charged particles could significantly influence the evolution of the cloud.

To investigate the nature of the gravitational collapse of a cloud in more detail, we must solve the hydrodynamic equations numerically.[8] These calculations exhibit many of the characteristics that were illustrated by our crude analytical studies, although other important aspects of the collapse become apparent that were not contained in the physics that has already been discussed.

Consider a spherical cloud of approximately 1 $M_\odot$ and solar composition that exceeds the Jeans criterion. Initially the cloud is optically thin, and the early stages of the free-fall collapse are nearly isothermal. Owing to an initial slight increase in density toward the center of the cloud, the free-fall time scale is shorter near the center and the density increases more rapidly there. When the density of the material near the core reaches approximately $10^{-13}$ g cm$^{-3}$, the region becomes optically thick and the collapse becomes more adiabatic. The opacity of the cloud at this point is primarily due to the presence of dust.

The increased pressure that occurs when the collapse becomes adiabatic substantially slows the rate of collapse near the core. At this point the central region is nearly in hydrostatic equilibrium with a radius of approximately 5 AU. It is this central object that is referred to as a protostar.

One observable consequence of the cloud becoming optically thick is that the gravitational potential energy being released during the collapse is converted into heat and then radiated away in the infrared as blackbody radiation. By computing the rate of energy release (the luminosity) and the radius of the cloud where the optical depth is $\tau = 2/3$, the effective temperature may be determined using Eq. (3.17). It then becomes possible to plot the location of the cloud on the H–R diagram as a function of time. Curves that depict the

---

[7]This type of approach is sometimes called a "back-of-the-envelope" calculation because of the relatively small space required to carry out the estimate. Extensive use of back-of-the-envelope calculations are made throughout this book to illustrate the effects of key physical processes.

[8]Some of the first calculations of protostellar collapse were performed by Richard Larson in 1969.

## 12.2 The Formation of Protostars

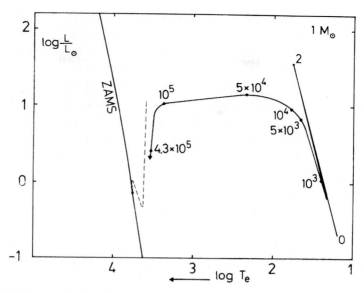

**Figure 12.7** A theoretical evolutionary track of the gravitational collapse of a 1 $M_\odot$ cloud through the protostar phase (times are labeled in years since the development of a hydrostatic core). The dashed line shows the quasi-hydrostatic evolution of a pre-main-sequence star, beginning on the Hayashi track. Also shown is a portion of the zero-age main sequence (ZAMS). Both the Hayashi track and the ZAMS are discussed in Section 12.3. (Figure from Appenzeller and Tscharnuter, *Astron. Astrophys.*, 40, 397, 1975.)

life histories of stars on the H–R diagram are known as **evolutionary tracks**. Figure 12.7 shows a theoretical evolutionary track of a 1 $M_\odot$ cloud during its collapse. As the collapse continues to accelerate, the luminosity of the cloud increases significantly while its effective temperature also increases.

Above the developing protostellar core, material is still in free-fall. When the infalling material meets the nearly hydrostatic core, a shock wave develops where the speed of the material exceeds the local sound speed (the material is supersonic). It is at this shock front that the infalling material loses a significant fraction of its kinetic energy in the form of heat that "powers" the cloud and produces much of its luminosity.

When the temperature reaches approximately 1000 K, the dust begins to vaporize and the opacity drops. This means that the radius where $\tau = 2/3$ is substantially reduced, approaching the surface of the hydrostatic core. Since the luminosity remains high during this phase, a corresponding increase in the effective temperature must occur.

As the overlying material continues to fall onto the hydrostatic core, the temperature of the core slowly increases. Eventually the temperature becomes high enough (approximately 2000 K) to cause the molecular hydrogen to dissociate into individual atoms. This process absorbs energy that would otherwise provide a pressure gradient sufficient to maintain hydrostatic equilibrium. As a result, the core becomes dynamically unstable and a second collapse occurs.[9] After the core radius has decreased to a value about 30% larger than the present size of the Sun, hydrostatic equilibrium is re-established. At this point, the core mass is still much less than its final value.

After the core collapse, a second shock front is established as the envelope continues to accrete infalling material. With only a finite amount of mass available from the original cloud, the accretion rate, and therefore the luminosity, must eventually decrease.

The theoretical scenario just described leads to the possibility of observational verification. Since it is expected that the collapse should occur deep within a molecular cloud, the protostar itself would likely be shielded from direct view by a cocoon of dust. Consequently, any observational evidence of the collapse would be in the form of small infrared sources embedded within molecular clouds. The detection of protostellar collapse is made more difficult by the relatively small value for the free-fall time, meaning that protostars are fairly short-lived objects. The search for protostars has been underway for some twenty years and a number of strong candidates have been identified, including the Bok globule, B335, in the constellation of Aquilae, and numerous objects in the Orion Nebula.

## 12.3 Pre-Main-Sequence Evolution

As we discovered in the last section, the collapse of a molecular cloud is characterized by the free-fall time scale, given by Eq. (12.16). With the formation of a quasi-static protostar, the rate of evolution becomes controlled by the rate at which the star can thermally adjust to the collapse. This is just the Kelvin–Helmholtz time scale discussed in Example 10.3 (see Eq. 10.30); the gravitational potential energy liberated by the collapse is released over time and is the source of the object's luminosity. Since $t_{KH} \gg t_{ff}$, protostellar evolution proceeds at a much slower rate than free-fall collapse. For instance, a 1 $M_\odot$ star requires on the order of $10^7$ yr to contract quasi-statically to its main-sequence structure.

---

[9]Dynamical instabilities will be discussed further in Chapter 14 in connection with pulsating stars.

## 12.3 Pre-Main-Sequence Evolution

With the steadily increasing effective temperature of the protostar, the opacity of the outer layers becomes dominated by the H$^-$ ion, the extra electrons coming from the partial ionization of some of the heavier elements in the gas that have lower ionization potentials. As with the envelope of the main-sequence Sun, this large opacity contribution causes the envelope of a contracting protostar to become convective. In fact, in some cases the convection zone extends all the way to the center of the star. In 1961, C. Hayashi demonstrated that, because of the constraints convection puts on the structure of a star, a deep convective envelope limits its quasi-static evolutionary path to a line that is nearly vertical in the H–R diagram. Consequently, as the protostar collapse slows, its luminosity decreases while its effective temperature increases slightly. It is this evolution along the **Hayashi track** that is shown as the last phase of evolution in Fig. 12.7.

The Hayashi track actually represents a boundary between "allowed" hydrostatic stellar models and those that are "forbidden." To the right of the Hayashi track, there is no mechanism that can adequately transport the luminosity out of the star at those low effective temperatures; hence no stable stars can exist there. To the left of the Hayashi track, convection and/or radiation is responsible for the necessary energy transport. Note that this distinction between allowed and forbidden models is not in conflict with the free-fall evolution of collapsing gas clouds found to the right of the Hayashi track, since those objects are far from being in hydrostatic equilibrium.

In 1965, before detailed protostellar collapse calculations were performed, Icko Iben, Jr. computed the final stages of collapse onto the main sequence for stars of a variety of masses, starting his models on the Hayashi track. All of those models neglected the effects of rotation, magnetic fields, and mass loss. The **pre-main-sequence evolutionary tracks** for a sequence of masses are shown in Fig. 12.8 with the times of various points during the evolution given in Table 12.1.

Following the results of Iben and others, consider the pre-main-sequence evolution of a 1 $M_\odot$ star, beginning on the Hayashi track. With the high H$^-$ opacity near the surface, the star is completely convective during approximately the first one million years of the collapse (near point 2 in Fig. 12.8). During this early period of collapse, the first stage of nuclear burning occurs in the star's center (about point 1). The reaction involved is *deuterium burning*, the second step of the PP I chain (Eq. 10.46). This reaction is favored over the first step because it has a fairly large cross section, $\sigma(E)$, at low temperatures. Since $^2_1$H is not very abundant, however, the nuclear reactions have little effect on the overall collapse; they simply slow the rate of collapse slightly.

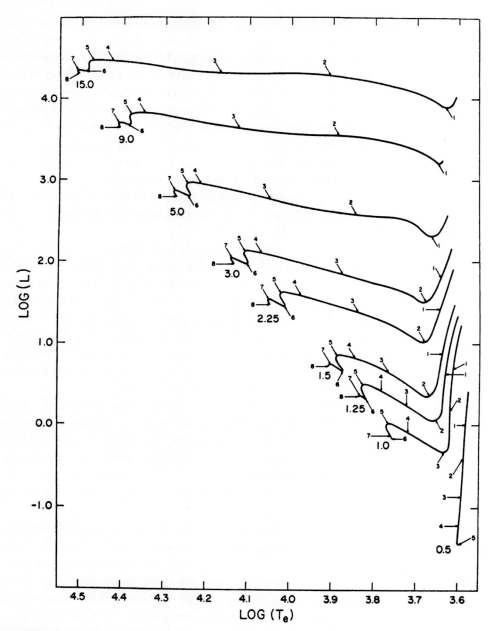

**Figure 12.8** Theoretical pre-main-sequence evolutionary tracks for stars of various masses. The mass of each model (in solar mass units) is indicated beside its evolutionary track. (Figure from Iben, *Ap. J.*, *141*, 993, 1965.)

## 12.3 Pre-Main-Sequence Evolution

| Mass $(M_\odot)$ | Age of Model at Points (in years) | | | |
|---|---|---|---|---|
| | 1 | 2 | 3 | 4 |
| 15.0 | 6.740 (2) | 3.766 (3) | 9.350 (3) | 2.203 (4) |
| 9.0 | 1.443 (3) | 1.473 (4) | 3.645 (4) | 6.987 (4) |
| 5.0 | 2.936 (4) | 1.069 (5) | 2.001 (5) | 2.860 (5) |
| 3.0 | 3.420 (4) | 2.078 (5) | 7.633 (5) | 1.135 (6) |
| 2.25 | 7.862 (4) | 5.940 (5) | 1.883 (6) | 2.505 (6) |
| 1.5 | 2.347 (5) | 2.363 (6) | 5.801 (6) | 7.584 (6) |
| 1.25 | 4.508 (5) | 3.957 (6) | 8.800 (6) | 1.155 (7) |
| 1.0 | 1.189 (5) | 1.058 (6) | 8.910 (6) | 1.821 (7) |
| 0.5 | 3.195 (5) | 1.786 (6) | 8.711 (6) | 3.092 (7) |

| Mass $(M_\odot)$ | Age of Model at Points (in years) | | | |
|---|---|---|---|---|
| | 5 | 6 | 7 | 8 |
| 15.0 | 2.657 (4) | 3.984 (4) | 4.585 (7) | 6.170 (8) |
| 9.0 | 7.922 (4) | 1.019 (5) | 1.195 (5) | 1.505 (5) |
| 5.0 | 3.137 (4) | 3.880 (5) | 4.559 (5) | 5.759 (5) |
| 3.0 | 1.250 (6) | 1.465 (6) | 1.741 (6) | 2.514 (6) |
| 2.25 | 2.818 (6) | 3.319 (6) | 3.993 (6) | 5.855 (6) |
| 1.5 | 8.620 (6) | 1.043 (7) | 1.339 (7) | 1.821 (7) |
| 1.25 | 1.404 (7) | 1.755 (7) | 2.796 (7) | 2.945 (7) |
| 1.0 | 2.529 (7) | 3.418 (7) | 5.016 (7) | |
| 0.5 | 1.550 (8) | | | |

**Table 12.1** Pre-Main-Sequence Evolutionary Times. The times represent the time that has elapsed (in years) since the initial model (powers of 10 are given in parentheses). The points correspond to the labels in Fig. 12.8. (Data from Iben, *Ap. J.*, *141*, 993, 1965.)

As the central temperature continues to rise, increasing levels of ionization decrease the opacity in that region (see Fig. 9.10) and a radiative core develops, progressively encompassing more and more of the star's mass. By point 3, the radiative core allows energy to escape into the convective envelope more readily, causing the luminosity of the star to increase again. Also, as required by Eq. (3.17), the effective temperature continues to increase, since the star is still shrinking.

At about the time that the luminosity begins to increase again, the temperature near the center has become high enough for nuclear reactions to begin in earnest, although not yet at their equilibrium rates. Initially, the first two steps of the PP I chain [the conversion of $^1_1$H to $^3_2$He; Eq. (10.46)] and the CNO reactions that turn $^{12}_{6}$C into $^{14}_{7}$N [Eqs. (10.51) and (10.52)] dominate the nuclear energy production. With time, these reactions provide an increasingly larger fraction of the luminosity, while the energy production due to gravitational collapse and the virial theorem make less of a contribution to $L$.

Due to the onset of the highly temperature-dependent CNO reactions, a steep temperature gradient is established in the core and some convection again develops in that region. At point 5, the rate of nuclear energy production has become so great that the central core is forced to expand somewhat, causing the gravitational energy term in Eq. (10.45) to become negative [recall that $\epsilon = \epsilon_{\text{nuclear}} + \epsilon_{\text{gravity}}$; see Eq. (10.95)]. This effect is apparent at the surface as the total luminosity decreases toward its main-sequence value, accompanied by a decrease in the effective temperature.

When the $^{12}_{6}$C is finally exhausted, the core completes its readjustment to nuclear burning, reaching a sufficiently high temperature for the remainder of the PP I chain to become important. At the same time, with the establishment of a stable energy source, the gravitational energy term becomes insignificant and the star finally settles onto the main sequence. It is worth noting that the time required for a 1 $M_\odot$ star to reach the main sequence (point 7), according to the detailed numerical model just described, is not very different from the crude estimate of the Kelvin–Helmholtz time scale performed in Example 10.3.

For stars with masses lower than our Sun's, the evolution is somewhat different. In particular, for the 0.5 $M_\odot$ star shown in Fig. 12.8, the upward branch is missing just before the main sequence. This happens because the central temperature never gets hot enough to burn $^{12}_{6}$C efficiently (recall that our estimates of the central pressure and temperature of the Sun in Examples 10.1 and 10.2 were roughly proportional to the mass of the star). In fact, as was mentioned in Section 10.6, if the mass of the collapsing protostar is less than approximately 0.08 $M_\odot$, the core never gets hot enough to sustain any nuclear reactions, in which case the stable hydrogen-burning main sequence is never reached. This explains the lower end of the main sequence.

Another important difference exists between solar-mass stars and stars of lower mass that can reach the main sequence: Temperatures remain cool enough and the opacity stays sufficiently high in low-mass stars that a radiative core never develops. Consequently, these stars remain fully convective all the way to the main sequence.

For massive stars, the central temperature quickly becomes high enough to burn $^{12}_{6}$C as well as convert $^{1}_{1}$H into $^{3}_{2}$He. This means that these stars leave the Hayashi track at higher luminosities and evolve nearly horizontally across the H–R diagram. Because of the much larger central temperatures, the full

## 12.3 Pre-Main-Sequence Evolution

CNO cycle becomes the dominant mechanism for hydrogen burning in these main-sequence stars. Since the CNO cycle is so strongly temperature-dependent, the core remains convective even after the main sequence is reached.

We mentioned in Section 10.6 that the upper mass limit for stable stars on the main sequence is approximately 90 $M_\odot$ and is due to the onset of pulsations deep in the core, the so-called $\epsilon$ mechanism of stellar pulsation. Other factors also affect the stability of very massive stars, including their extremely high luminosities.

As can be seen by Eq. (10.26), if the temperature is sufficiently high and the gas density is low enough, it is possible for radiation pressure to dominate over the gas pressure in certain regions of the star, a situation that can occur in the outer layers of very massive stars. In this case the pressure gradient is approximately given by Eq. (9.25). Combined with the relationship between radiant flux and luminosity (Eq. 3.2), the pressure gradient near the surface may be written as

$$\frac{dP}{dr} \simeq -\frac{\overline{\kappa}\rho}{c} \frac{L}{4\pi r^2}.$$

But hydrostatic equilibrium (Eq. 10.7) demands that the pressure gradient near the star's surface must also be given by

$$\frac{dP}{dr} = -G\frac{M\rho}{r^2},$$

where $M$ is the star's mass. Combining, and solving for the luminosity, we have

$$L_{\text{Ed}} = \frac{4\pi G c}{\overline{\kappa}} M. \tag{12.19}$$

$L_{\text{Ed}}$ is the maximum radiative luminosity that a star can have and still remain in hydrostatic equilibrium. If the luminosity exceeds $L_{\text{Ed}}$, mass loss must occur, driven by radiation pressure. This luminosity maximum, known as the **Eddington limit**, appears in a number of areas of astrophysics, including late stages of stellar evolution, novae, and the structure of accretion disks.

For our purposes, it is possible to make an estimate of the Eddington luminosity for stars on the upper end of the main sequence. The effective temperatures of these massive stars are in the range of 50,000 K, high enough that most of the hydrogen is ionized in their photospheres. Therefore the major contribution to the opacity is from electron scattering and we can replace $\overline{\kappa}$ by Eq. (9.21). For $X = 0.7$, Eq. (12.19) becomes

$$L_{\text{Ed}} \simeq 1.5 \times 10^{38} \frac{M}{M_\odot} \text{ ergs s}^{-1}, \tag{12.20}$$

or
$$\frac{L_{\text{Ed}}}{L_\odot} \simeq 3.8 \times 10^4 \, \frac{M}{M_\odot}.$$

For a 90 $M_\odot$ star, $L_{\text{Ed}} \simeq 3.5 \times 10^6$ $L_\odot$, roughly three times the expected main-sequence value.

The fairly close correspondence between the theoretical and Eddington luminosities implies that the envelopes of massive main-sequence stars are loosely bound at best. In fact, observations of the few stars with masses estimated to be near 100 $M_\odot$ indicate that they are suffering from large amounts of mass loss and exhibit variability in their luminosities.

The diagonal line in the H–R diagram where stars of various masses first reach the main sequence and begin equilibrium hydrogen burning is known as the **zero-age main sequence** (ZAMS). Inspection of Table 12.1 shows that the amount of time required for stars to collapse onto the ZAMS is inversely related to mass; a 0.5 $M_\odot$ star takes over 100 million years to reach the ZAMS while a 15 $M_\odot$ star makes it to the ZAMS in only 60,000 years!

Generally, more low-mass than high-mass stars form when an interstellar cloud fragments. This implies that the number of stars that form per mass interval per unit volume (or per unit area in the Milky Way's disk) is strongly mass-dependent. This functional dependence is known as the **initial mass function**. One theoretical estimate of the IMF is shown in Fig. 12.9. As a consequence of the process of fragmentation, most stars form with relatively low mass. Given the disparity in the numbers of stars formed in different mass ranges, combined with the very different rates of evolution, it is not surprising that massive stars are extremely rare, while low-mass stars are found in abundance.[10]

When hot, massive stars reach the ZAMS with O or B spectral types, the bulk of their radiation is emitted in the ultraviolet portion of the electromagnetic spectrum. If the photons that are produced have energies in excess of 13.6 eV, they can ionize the ground-state hydrogen gas (H I) in the ISM that still surrounds the newly formed star. Of course, if these **H II regions** are in equilibrium (meaning that their volumes remain constant), the rate of ionization must equal the rate of recombination; photons must be absorbed and ions must be produced at the same rate that free electrons and protons recombine to form neutral hydrogen atoms. When recombination occurs, the electron does not necessarily fall directly to the ground state, but it can cascade downward, producing a number of lower-energy photons, many of which will be in

---

[10] However, recall also our estimate of the minimum fragmentation mass, based on the Jeans mass (Eq. 12.18).

## 12.3 Pre-Main-Sequence Evolution

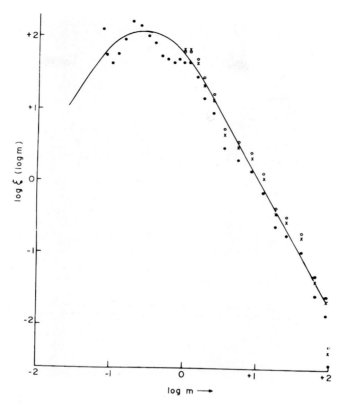

**Figure 12.9** The initial mass function, $\xi$, shows the number of stars per unit area of the Milky Way's disk per unit interval of logarithmic mass that is produced in different mass intervals. Masses are in solar units. (Figure adapted from Rana, *Astron. Astrophys.*, *184*, 104, 1987.)

the visible portion of the spectrum. The dominant visible wavelength photon produced in this way results from the transition between $n = 3$ and $n = 2$, the red line of the Balmer series ($H_\alpha$). Consequently, because of this energy cascade, H II regions appear to fluoresce in red light.

These nebulae are considered by some to be among the most beautiful objects in the night sky. One of the more famous H II regions is the Orion nebula (M42),[11] found in the sword of the Orion constellation. M42 is part of the Orion A complex (see Fig. 12.10), which also contains a giant molecular cloud (OMC 1) and a very young cluster of stars (the Trapezium cluster). The first protostar candidates were discovered in this region.

---

[11]M42 is the entry number in the well-known Messier catalog, a popular collection of observing objects for amateur astronomers. The Messier catalog is listed in Appendix F.

**Figure 12.10** The H II region in Orion A is associated with a young OB association, the Trapezium cluster, and a giant molecular cloud. The Orion complex is 450 pc away. (Courtesy of the National Optical Astronomy Observatories.)

The size of an H II region can be estimated by considering the requirement of equilibrium. Let $N$ be the number of photons *per second* produced by the O or B star that are sufficiently energetic to ionize hydrogen from the ground state ($\lambda < 912$ Å). Assuming that all of the energetic photons are ultimately absorbed by the hydrogen in the H II region, the rate of photon creation must equal the rate of recombination. If this equilibrium condition did not develop, the size of the region would continue to grow as the photons traveled ever farther before encountering un-ionized gas.

Next, let $\alpha n_e n_H$ be the number of recombinations *per unit volume per second*, where $\alpha$ is a quantum-mechanical recombination coefficient that describes the likelihood that an electron and a proton can form a hydrogen atom, given their number densities (obviously, the more electrons and protons that are present, the greater the chance of recombination, hence the

## 12.3 Pre-Main-Sequence Evolution

product $n_e n_H$).[12] At about 8000 K, a temperature characteristic of H II regions, $\alpha = 3.1 \times 10^{-13}$ cm$^3$ s$^{-1}$. If we assume that the gas is composed entirely of hydrogen and is electrically neutral, then for every ion produced, one electron must have been liberated, or $n_e = n_H$. With this equality, the expression for the recombination rate can be multiplied by the volume of the H II region, assumed here to be spherical, and then set equal to the number of ionizing photons produced per second. Finally, solving for the radius of the H II region gives

$$r_S \simeq \left(\frac{3N}{4\pi\alpha}\right)^{1/3} n_H^{-2/3}. \tag{12.21}$$

$r_S$ is called the **Strömgren radius**, after Bengt Strömgren, the astrophysicist who first carried out the analysis in the late 1930s. H II regions are sometimes referred to as Strömgren spheres.

---

**Example 12.4** From Appendix E, the effective temperature and luminosity of an O6 star are $T_e \simeq 45{,}000$ K and $L \simeq 1.3 \times 10^5$ L$_\odot$, respectively. According to Wien's law (Eq. 3.15), the peak wavelength of the blackbody spectrum is given by

$$\lambda_{\max} = \frac{0.29 \text{ cm K}}{T_e} = 640 \text{ Å}.$$

Since this is significantly shorter than the 912 Å limit necessary to produce ionization from the hydrogen ground state, it can be assumed that most of the photons created by an O6 star are capable of causing ionization.

The energy of one 640 Å photon can be calculated from Eq. (5.3),

$$E_\gamma = \frac{hc}{\lambda} = 19 \text{ eV}.$$

Now, assuming for simplicity that all of the emitted photons have the same (peak) wavelength, the total number of photons produced by the star per second is just

$$N \simeq L/E_\gamma \simeq 1.6 \times 10^{49} \text{ photons s}^{-1}.$$

Lastly, taking $n_H \sim 5000$ cm$^{-3}$ to be a typical value for Orion A's H II region, we find

$$r_S \simeq 0.3 \text{ pc}.$$

Actually the H II region in Orion A is produced by the ionizing radiation of a number of O and B stars.

---

[12] Note that this expression is somewhat analogous to Eq. (10.36), the generalized nuclear reaction rate equation.

As we saw from Table 12.1, the massive O and B stars will form first from a collapsing cloud that is undergoing fragmentation. As a massive star forms, the protostar will first appear as an infrared source embedded inside the molecular cloud. With the rising temperature, first the dust will vaporize, then the molecules will dissociate, and finally, as the star reaches the main sequence, the gas immediately surrounding it will ionize, resulting in the creation of an H II region inside of an existing H I region.

Now, because of the star's high luminosity, radiation pressure will begin to drive significant amounts of mass loss, which then tends to disperse the remainder of the cloud. If several O and B stars form at the same time, it may be that much of the mass that has not yet become gravitationally bound to more slowly forming low-mass protostars will be driven away, halting any further star formation. Moreover, if the cloud was originally marginally bound (near the limit of the Jeans criterion), the loss of mass will diminish the potential energy term in the virial theorem, with the result that the newly formed cluster of stars and protostars will become unbound (i.e., the stars will tend to drift apart).

Groups of stars that are dominated by O and B main-sequence stars are referred to as **OB associations**. Studies of their individual kinematic velocities and masses generally lead to the conclusion that they cannot remain gravitationally bound to one another as permanent stellar clusters. One such example is the Trapezium cluster in the Orion A complex, believed to be less than 10 million years old. It is currently densely populated with stars ($> 2 \times 10^3$ pc$^{-3}$), most of which have masses in the range of 0.5 to 2.0 M$_\odot$. Doppler shift measurements of the radial velocities of $^{13}$CO show that the gas in the vicinity is very turbulent. Furthermore, radial velocity studies of the stars themselves show high velocities relative to the center of mass of the cluster. Apparently, the nearby O and B stars are dispersing the gas, and the cluster is becoming unbound.

**T Tauri stars** are an important class of low-mass pre-main-sequence objects that represent a transition between stars that are still shrouded in dust (IR sources) and main-sequence stars. T Tauri stars, named after the first star of their class (located in the constellation of Taurus), are characterized by unusual spectral features and by large and fairly rapid irregular variations in luminosity, with time scales on the order of days. The positions of T Tauri stars on the H–R diagram are shown in Fig. 12.11; theoretical pre-main-sequence evolutionary tracks are also included. The masses of T Tauri stars appear to range from 0.5 to 3 M$_\odot$.

Many T Tauri stars exhibit strong emission lines from hydrogen (the Balmer series), from Ca II (the H and K lines), and from iron, as well as absorption

## 12.3 Pre-Main-Sequence Evolution

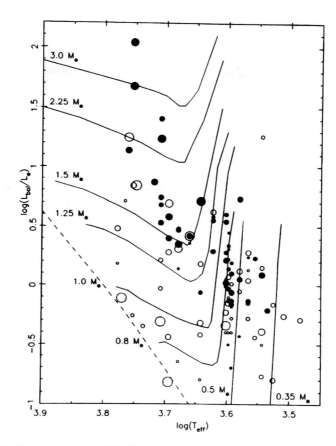

**Figure 12.11** The positions of T Tauri stars on the H–R diagram. The sizes of the circles indicate the rate of rotation. Stars with strong emission lines are indicated by filled circles and weak emission line stars are represented by open circles. Theoretical pre-main-sequence evolutionary tracks are also included. (Figure from Bertout, *Annu. Rev. Astron. Astrophys.*, **27**, 351, 1989. Reproduced with permission from the *Annual Review of Astronomy and Astrophysics*, Volume 27, ©1989 by Annual Reviews Inc.)

lines of lithium. Interestingly, forbidden lines of [O I] and [S II] are also present in the spectra of many T Tauri stars.[13] The existence of forbidden lines in a spectrum is an indication of extremely low gas densities. (Note that, to distinguish them from "allowed" lines, forbidden lines are usually indicated by square brackets, e.g., [O I].)

---

[13] Recall the discussion of allowed and forbidden transitions in the Sun's corona (Section 11.2).

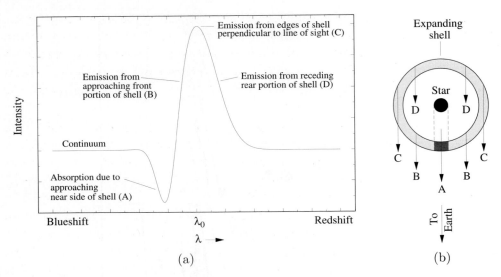

**Figure 12.12** (a) A spectral line exhibiting a P Cygni profile is characterized by a broad emission peak with a superimposed blueshifted absorption trough. (b) A P Cygni profile is produced by an expanding mass shell. The emission peak is due to the outward movement of material perpendicular to the line of sight, while the blueshifted absorption feature is caused by the approaching matter in the shaded region, intercepting photons coming from the central star.

Not only can information be gleaned from stellar spectra by determining which lines are present and with what strengths, but information is also contained in the *shapes* of those lines as a function of wavelength.[14] An important example is found in the shapes of some of the lines in T Tauri stars. The $H_\alpha$ line often exhibits the characteristic shape shown in Fig. 12.12(a). Superimposed on a rather broad emission peak is an absorption trough at the short wavelength edge of the line. This unique line shape is known as a **P Cygni profile**, after the first star observed to have emission lines with blueshifted absorption components. The interpretation given for the existence of P Cygni profiles in a star's spectrum is that the star is experiencing significant mass loss. Recall from Kirchhoff's laws (Section 5.1) that emission lines are produced by a hot, diffuse gas when there is little intervening material between the source and the observer. In this case the emission source is that portion of the expanding shell of the T Tauri star that is moving nearly perpendicular to the line of sight, as illustrated by the geometry shown in Fig. 12.12(b). Absorption lines are the result of light passing through a cooler, diffuse gas;

---

[14]Line profiles were first discussed in Section 9.4.

## 12.3 Pre-Main-Sequence Evolution

the shaded portion of the expanding shell absorbs the photons emitted by the hotter star behind it. Since the shaded part of the shell (A) is moving toward the observer, the absorption is blueshifted relative to the emission component (typically by 80 km s$^{-1}$ for T Tauri stars). The mass loss rates of T Tauri stars appear to be in the range of $\dot{M} = 10^{-8}$ to $10^{-7}$ M$_\odot$ yr$^{-1}$, although the estimates are still quite uncertain.[15]

In some extreme cases, line profiles of T Tauri stars have gone from P Cygni profiles to *inverse* P Cygni profiles (redshifted absorption) on time scales of days, indicating mass accretion rather than mass loss. Mass accretion rates appear to be on the same order as mass loss rates. Apparently the environment around a T Tauri star is very unstable.

Along with expanding shells, mass loss also occurs via the production of **jets** of gas that have been ejected in rather narrow beams in opposite directions.[16] **Herbig–Haro objects**, first discovered in the vicinity of the Orion nebula in the early 1950s by George Herbig and Guillermo Haro, are apparently associated with the jets produced by young protostars, such as T Tauri stars. As the jets expand supersonically into the interstellar medium, collisions excite the gas, resulting in bright objects with emission-line spectra. Figure 12.13(a) shows a Hubble Space Telescope image of the Herbig–Haro objects HH 1 and HH 2, which were created by material ejected at speeds of several hundred kilometers per second from a star enshrouded in a cocoon of dust. The jets associated with another Herbig–Haro object, HH 47, are shown in Fig. 12.13(b).

Continuous emission is also observed in some Herbig–Haro objects and is due to the reflection of light from the parent star. Apparent in Fig. 12.14 is a **circumstellar accretion disk** around HH 30.[17] The surfaces of the disk are illuminated by the central star, which is again hidden from view behind the dust in the disk. Also apparent are jets originating from deep within the accretion disk, possibly from the central star itself. These accretion disks seem to be responsible for many of the characteristics associated with the protostellar objects, including emission lines, mass loss, jets, and perhaps even some of the luminosity variations. Unfortunately, details concerning the physical processes involved are not yet well understood. A model of the production of Herbig–Haro objects like HH 1 and HH 2 is shown in Fig. 12.15.

---

[15]These values are much higher than the Sun's current rate of mass loss ($10^{-14}$ M$_\odot$ yr$^{-1}$; see Example 11.1).

[16]As we shall see in later chapters, astrophysical jets occur in a variety of phenomena over enormous ranges of energy and physical size.

[17]Accretion disks will be discussed in detail in Chapter 17.

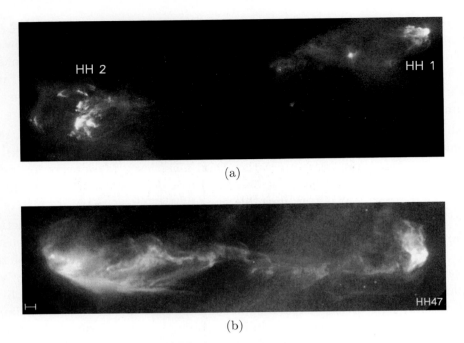

**Figure 12.13** (a) The Herbig–Haro objects HH 1 and HH 2 are located just south of the Orion nebula and are moving away from a young protostar hidden inside a dust cloud near the center of the image. [Courtesy of J. Hester (Arizona State University), the WF/PC 2 Investigation Definition Team, and NASA.] (b) A jet associated with HH 47. The scale in the lower left is 1000 AU. (Courtesy of J. Morse/STScI, and NASA.)

Observations have revealed that other young stars also possess circumstellar disks of material orbiting around them. Two well-known examples are Vega and $\beta$ Pictoris. An infrared image of $\beta$ Pic and its disk is shown in Fig. 12.16. $\beta$ Pic has also been observed in the ultraviolet lines of Fe II by the Goddard High-Resolution Spectrograph onboard the Hubble Space Telescope. It appears that clumps of material are falling from the disk into the star at the rate of 2 or 3 per week. Larger objects may be forming in the disk as well, possibly protoplanets. An artist's conception of the $\beta$ Pic system is shown in Fig. 12.17.

Shortly after the December 1993 refurbishment mission, the Hubble Space Telescope made observations of the Orion Nebula using WF/PC 2. The images shown in Fig. 12.18 were obtained using the emission lines of $H_\alpha$, [N II], and [O III]. Analysis of the data has revealed that 56 of the 110 stars brighter than $V = 21$ mag are surrounded by disks of circumstellar dust and gas. The circumstellar disks, termed **proplyds**, appear to be protoplanetary disks associated

## 12.3 Pre-Main-Sequence Evolution

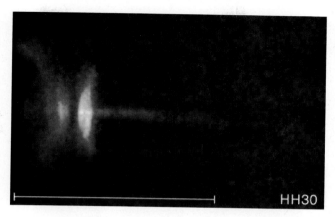

**Figure 12.14** The circumstellar disk and jets of the protostellar object, HH 30. The central star is obscured by dust in the plane of the disk. The scale in the lower left is 1000 AU. [Courtesy of C. Burrows (STScI and ESA), the WF/PC 2 Investigation Definition Team, and NASA.]

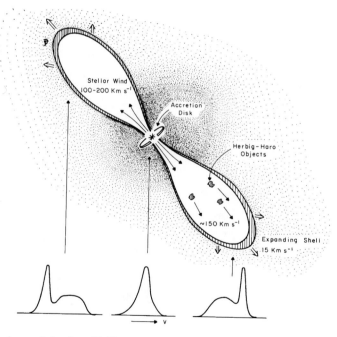

**Figure 12.15** A model of a T Tauri star with an accretion disk. The disk powers and collimates jets that expand into the interstellar medium, producing Herbig–Haro objects. (Figure from Snell, Loren, and Plambeck, *Ap. J. Lett.*, **239**, L17, 1980.)

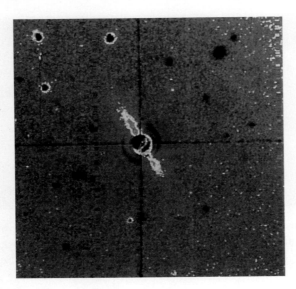

**Figure 12.16** An infrared image of $\beta$ Pictoris, showing its circumstellar disk. (Courtesy of NASA/JPL.)

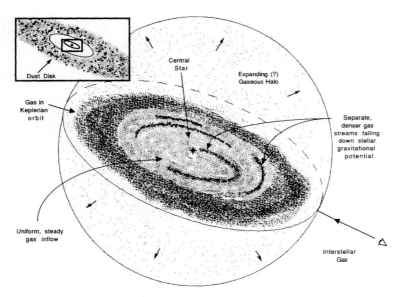

**Figure 12.17** An artist's conception of the $\beta$ Pictoris system. Clumps of material appear to be falling into the star at the rate of 2 or 3 clumps per week. Some matter may also be leaving the system as an expanding halo. (Figure adapted from Boggess et al., *Ap. J. Lett.*, *377*, L49, 1991.)

## 12.3 Pre-Main-Sequence Evolution

with young stars that are less than 1 million years old. Based on observations of the ionized material in the proplyds, the disks seem to have masses much greater than $2 \times 10^{28}$ g (for reference, the mass of Earth is $5.974 \times 10^{27}$ g).

Apparently, disk formation is fairly common during the collapse of protostellar clouds. Undoubtedly this is due to the spin-up of the cloud as required by the conservation of angular momentum. As the radius of the protostar decreases, so does its moment of inertia. This implies that in the absence of external torques, the protostar's angular velocity must increase. It is left as an exercise (Problem 12.13) to show that by including a centripetal acceleration term in Eq. (12.9) and requiring conservation of angular momentum, the collapse perpendicular to the axis of rotation can be halted before the collapse along the axis, resulting in disk formation.

A problem immediately arises when the effect of angular momentum is included in the collapse. Conservation of angular momentum arguments lead us to expect that all main-sequence stars ought to be rotating very rapidly, at rates close to breakup. However, observations show that this is not generally the case. Apparently the angular momentum is transferred away from the collapsing star. One suggestion (discussed in Section 11.2) is that magnetic fields, anchored to convection zones within the stars and coupled to ionized stellar winds, slow the rotation by applying torques. Evidence in support of this idea exists in the form of apparent solarlike coronal activity in the outer atmospheres of many T Tauri stars.

Along with the problems associated with rotation and magnetic fields, mass loss may also play an important role in the evolution of pre-main-sequence stars. If mass loss rates are sufficiently high, the quasi-static evolutionary tracks shown in Fig. 12.8 could be modified. Although these problems are being investigated, much work remains to be done before we can hope to understand all of the details of pre-main-sequence evolution.

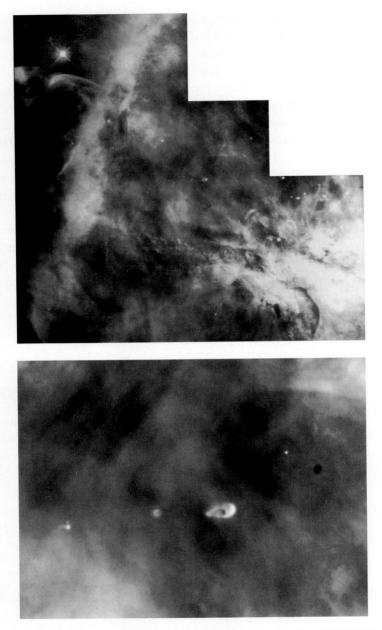

**Figure 12.18** Images of the Orion Nebula (M42) obtained using WF/PC 2 on the Hubble Space Telescope. The lower image is an enlarged view of the central region of the upper image. Note that numerous proplyds are visible in the field of view of the camera. (Courtesy of C. Robert O'Dell/Rice University and NASA.)

# Suggested Readings

## General

Boss, Alan P., "Collapse and Formation of Stars," *Scientific American*, January 1985.

Knapp, Gillian, "The Stuff Between the Stars," *Sky and Telescope*, May 1995.

O'Dell, C. Robert, "Exploring the Orion Nebula," *Sky and Telescope*, December 1994.

Schild, Rudolph E., "A Star Is Born," *Sky and Telescope*, December 1990.

## Technical

Aller, Lawrence H., *Atoms, Stars, and Nebulae*, Third Edition, Cambridge University Press, Cambridge, 1991.

Bertout, Claude, "T Tauri Stars: Wild As Dust," *Annual Review of Astronomy and Astrophysics*, *27*, 351, 1989.

Böhm-Vitense, Erika, *Introduction to Stellar Astrophysics, Volume I: Basic Stellar Observations and Data*, Cambridge University Press, Cambridge, 1989.

Dickey, John M., and Lockman, Felix J., "H I in the Galaxy," *Annual Review of Astronomy and Astrophysics*, *28*, 215, 1990.

Dyson, J. E., and Williams, D. A., *Physics of the Interstellar Medium*, Halsted Press, New York, 1980.

Iben, Icko Jr., "Stellar Evolution. I. The Approach to the Main Sequence," *The Astrophysical Journal*, *141*, 993, 1965.

Kippenhahn, Rudolf, and Weigert, Alfred, *Stellar Structure and Evolution*, Springer-Verlag, Berlin, 1990.

Larson, Richard B., "Numerical Calculations of the Dynamics of a Collapsing Protostar," *Monthly Notices of the Royal Astronomical Society*, *145*, 271, 1969.

Levy, Eugene H., and Lunine Jonathan I. (eds.), *Protostars and Planets III*, The University of Arizona Press, Tucson, 1993.

Mathis, John S., "Interstellar Dust and Extinctions," *Annual Review of Astronomy and Astrophysics*, *28*, 37, 1990.

O'Dell, C. R., and Wen, Zheng, "Postrefurbishment Mission Hubble Space Telescope Images of the Core of the Orion Nebula: Proplyds, Herbig-Haro Objects, and Measurements of a Circumstellar Disk," *The Astrophysical Journal*, *436*, 194, 1994.

Osterbrock, Donald E., *Astrophysics of Gaseous Nebulae and Active Galactic Nuclei*, University Science Books, Mill Valley, CA, 1989.

Puget, J. L., and Léger, A., "A New Component of the Interstellar Matter: Small Grains and Large Aromatic Molecules," *Annual Review of Astronomy and Astrophysics*, *27*, 161, 1989.

Shu, Frank H., Adams, Fred C., and Lizano, Susana, "Star Formation in Molecular Clouds: Observation and Theory," *Annual Review of Astronomy and Astrophysics*, *25*, 23, 1987.

Stahler, Steven W., "Understanding Young Stars: A History," *Publications of the Astronomical Society of the Pacific*, *100*, 1474, 1988.

Zhou, Shudong, Evans, Neal J. II, Kömpe, Carsten, and Walmsley, C. M., "Evidence for Protostellar Collapse in B335," *The Astrophysical Journal*, *404*, 232, 1993.

# Problems

12.1 In a certain part of the North American Nebula, the amount of interstellar extinction in the visual wavelength band is 1.1 magnitudes. The thickness of the nebula is estimated to be 20 pc and it is located 700 pc from Earth. Suppose that a B spectral class main-sequence star is observed in the direction of the nebula, and that the absolute visual magnitude of the star is known to be $M_V = -1.1$ from spectroscopic data. Neglect any other sources of extinction between the observer and the nebula.

(a) Find the apparent visual magnitude of the star if it is lying just in front of the nebula.

(b) Find the apparent visual magnitude of the star if it is lying just behind the nebula.

(c) Without taking the existence of the nebula into consideration, based on its apparent magnitude, how far away does the star in part (b) appear to be? What would the percentage error be in determining the distance if interstellar extinction were neglected?

## Problems

**12.2** Estimate the temperature of a dust grain that is located 100 AU from a newly formed F0 main-sequence star. *Hint:* Assume that the dust grain is in thermal equilibrium—meaning that the amount of energy absorbed by the grain in a given time interval must equal the amount of energy radiated away during the same interval of time. Assume also that the dust grain is spherically symmetric and emits and absorbs radiation as a perfect blackbody. You may want to refer to Appendix E for the effective temperature and radius of an F0 main-sequence star.

**12.3** The Boltzmann factor, $e^{-(E_2-E_1)/kT}$, helps determine the relative populations of energy levels (see Section 8.1). Using the Boltzmann factor, estimate the temperature required for a hydrogen atom's electron and proton to go from being antialigned to aligned. Are the temperatures in H I clouds sufficient to produce this low-energy excited state?

**12.4** An H I cloud produces a 21-cm line with an optical depth at its center of $\tau_H = 0.5$ (the line is optically thin). The temperature of the gas is 100 K, the line's full width at half-maximum is 10 km s$^{-1}$, and the average atomic number density of the cloud is estimated to be 10 cm$^{-3}$. From this information and Eq. (12.4), find the thickness of the cloud. Express your answer in pc.

**12.5** Using an approach analogous to the development of Eq. (10.36) for nuclear reaction rates, make a crude estimate of the number of random collisions per cubic centimeter per second between CO and H$_2$ molecules in a giant molecular cloud that has a temperature of 15 K and a number density of $n_{H_2} = 10^2$ cm$^{-3}$. Assume (incorrectly) that the molecules are spherical in shape with radii of approximately 1 Å, the characteristic size of an atom.

**12.6** The rotational kinetic energy of a molecule is given by

$$E_{\rm rot} = \frac{1}{2}I\omega^2 = \frac{L^2}{2I},$$

where $L$ is the molecule's angular momentum and $I$ is its moment of inertia. The angular momentum is restricted by quantum mechanics to the discrete values

$$L = \sqrt{\ell(\ell+1)}\hbar$$

where $\ell = 0, 1, 2, \ldots$.

(a) For a diatomic molecule,
$$I = m_1 r_1^2 + m_2 r_2^2,$$
where $m_1$ and $m_2$ are the masses of the individual atoms and $r_1$ and $r_2$ are their separations from the center of mass of the molecule. Using the ideas developed in Section 2.3, show that $I$ may be written as
$$I = \mu r^2,$$
where $\mu$ is the reduced mass and $r$ is the separation between the atoms in the molecule.

(b) The separation between the C and O atoms in CO is approximately 1.2 Å, and the atomic masses of $^{12}$C, $^{13}$C, and $^{16}$O are 12.000 u, 13.003 u, and 15.995 u, respectively. Calculate the moments of inertia for $^{12}$CO and $^{13}$CO.

(c) What is the wavelength of the photon that is emitted by $^{12}$CO during a transition between the rotational angular momentum states $\ell = 3$ and $\ell = 2$? To which part of the electromagnetic spectrum does this correspond?

(d) Repeat part (c) for $^{13}$CO. How do astronomers distinguish between different isotopes in the interstellar medium?

12.7 Calculate the Jeans length for the giant molecular cloud in Example 12.2.

12.8 (a) By using the ideal gas law, calculate $|dP/dr| \approx |\Delta P/\Delta r| \sim P_c/R_J$ at the beginning of the collapse of a giant molecular cloud, where $P_c$ is an approximate value for the central pressure of the cloud. Assume that $P = 0$ at the edge of the molecular cloud and take its mass and radius to be the Jeans values found in Example 12.2 and in Problem 12.7. You should also assume the cloud temperature and density given in Example 12.2.

(b) Show that, given the accuracy of our crude estimates, $|dP/dr|$ found in part (a) is comparable to (i.e., within an order of magnitude of) $GM_r\rho/r^2$, as required for quasi-hydrostatic equilibrium.

(c) As long as the collapse remains isothermal, show that the contribution of $dP/dr$ in Eq. (10.6) continues to decrease relative to $GM_r\rho/r^2$, supporting the assumption made in Eq. (12.9) that $dP/dr$ can be neglected once free-fall collapse begins.

**12.9** Assuming that the free-fall acceleration of the surface of a collapsing cloud remains constant during the entire collapse, derive an expression for the free-fall time. Show that your answer only differs from Eq. (12.16) by a term of order unity.

**12.10** Using Eq. (10.76), estimate the sound speed of the giant molecular cloud discussed in Examples 12.2 and 12.3. Use this speed to find the amount of time required for a sound wave to cross the cloud, $t_{\text{sound}} = 2R_J/v_{\text{sound}}$, and compare your answer to the estimate of the free-fall time found in Example 12.3. Why would you expect the two values to be approximately the same?

**12.11** Using the information contained in the text, derive Eq. (12.18).

**12.12** Estimate the gravitational energy per unit volume in the giant molecular cloud in Example 12.2 and compare that with the magnetic energy density that would be contained in the cloud if it has a magnetic field of uniform strength, $B = 10 \ \mu G$. [*Hint:* Refer to Eq. (11.12).] Could magnetic fields play a significant role in the collapse of a cloud?

**12.13** (a) Beginning with Eq. (12.9), adding a centripetal acceleration term, and using conservation of angular momentum, show that the collapse of a cloud will stop in the plane perpendicular to its axis of rotation when the radius reaches

$$r_f = \frac{\omega_0^2 r_0^4}{2GM_r}$$

where $M_r$ is the interior mass, and $\omega_0$ and $r_0$ are the original angular velocity and radius of the surface of the cloud, respectively. Assume that the initial radial velocity of the cloud is zero and that $r_f \ll r_0$. You may also assume (incorrectly) that the cloud rotates as a rigid body during the entire collapse. *Hint:* Recall from the discussion leading to Eq. (11.9) that $d^2r/dt^2 = v_r \, dv_r/dr$. (Since no centripetal acceleration term exists for collapse along the rotation axis, disk formation is a consequence of the original angular momentum of the cloud.)

(b) Assume that the original cloud had a mass of 1 $M_\odot$ and an initial radius of 0.5 pc. If collapse is halted at approximately 100 AU, find the initial angular velocity of the cloud.

(c) What was the original rotational velocity (in cm s$^{-1}$) of the edge of the cloud?

(d) Assuming that the moment of inertia is approximately that of a uniform solid sphere, $I_\text{sphere} = \frac{2}{5}Mr^2$, when the collapse begins and a uniform disk, $I_\text{disk} = \frac{1}{2}Mr^2$, when it stops, determine the rotational velocity at 100 AU.

(e) After the collapse has stopped, calculate the time required for a piece of mass to make one complete revolution around the central protostar. Compare your answer with the orbital period at 100 AU expected from Kepler's third law. Why would you not expect the two periods to be identical?

12.14 Estimate the Eddington luminosity of a 0.085 $M_\odot$ star and compare your answer to the main-sequence luminosity given in Problem 10.15. Assume $\overline{\kappa} = 0.01$ cm$^2$ g$^{-1}$. Is radiation pressure likely to be significant in the stability of a low-mass main-sequence star?

12.15 Assuming a mass loss rate of $10^{-7}$ $M_\odot$ yr$^{-1}$ and a stellar wind velocity of 80 km s$^{-1}$ from a T Tauri star, estimate the mass density of the wind at a distance of 100 AU from the star. (*Hint:* Refer to Example 11.1.) Compare your answer with the density of the giant molecular cloud in Example 12.2.

# Chapter 13

# POST-MAIN-SEQUENCE STELLAR EVOLUTION

## 13.1 Evolution on the Main Sequence

In Section 10.6 we learned that the existence of the main sequence is due to the nuclear reactions that convert hydrogen into helium in the cores of stars. The evolutionary process of protostellar collapse to the zero-age main sequence was discussed in Chapter 12. In this chapter we will follow the lives of stars as they age, beginning on the main sequence. This evolutionary process is an inevitable consequence of the relentless force of gravity and the change in chemical composition due to nuclear reactions.

To maintain their luminosities, stars must tap sources of energy contained within, either nuclear or gravitational.[1] Pre-main-sequence evolution is characterized by two basic time scales, the free-fall time scale (Eq. 12.16) and the Kelvin–Helmholtz (or thermal readjustment) time scale (Eq. 10.30). Main-sequence and post-main-sequence evolution are also governed by a third time scale, the time scale of nuclear reactions (Eq. 10.31). As we saw in Example 10.4, the nuclear time scale is on the order of $10^{10}$ years for the Sun, much longer than the Kelvin–Helmholtz time scale of roughly $10^7$ years, estimated in Example 10.3. It is the difference in time scales for the various phases of evolution of individual stars that explains why approximately 80% to 90% of all stars in the solar neighborhood are observed to be main-sequence stars (see Section 8.2); we are more likely to find stars on the main sequence simply because that stage of evolution requires the most time, while later stages of

---

[1]We have already seen in Problem 10.3 that chemical energy cannot play a significant role in the energy budgets of stars.

evolution proceed more rapidly. However, as a star switches from one nuclear source to the next, gravitational energy can play a major role and the Kelvin–Helmholtz time scale will again become important.

Careful study of the main sequence of an observational H–R diagram such as Fig. 8.11 or the observational mass–luminosity relation (Fig. 7.7) reveals that these curves are not simply thin lines but have finite widths. The widths of the main sequence and the mass–luminosity relation are due to a number of factors, including observational errors, differing chemical compositions of the individual stars in the study, and varying stages of evolution on the main sequence.

In this section, the evolution of stars on the main sequence will be considered. Although all stars on the main sequence are converting hydrogen into helium and, as a result, share similar evolutionary characteristics, differences do exist. For instance, as was mentioned in Section 10.6, ZAMS stars with masses less than about 1.2 $M_\odot$ have radiative cores while more massive stars have convective cores.

First consider a star typical of those at the low-mass end of the main sequence, such as the Sun. As was mentioned in the discussion of Fig. 11.2, the Sun's luminosity, radius, and temperature have all increased steadily since it reached the ZAMS some 4.5 billion years ago. This evolution occurs because, as the pp chain converts hydrogen into helium, the mean molecular weight $\mu$ of the core increases (Eq. 10.21). According to the ideal gas law (Eq. 10.14), unless the density and/or temperature of the core also increases, there will be insufficient gas pressure to support the overlying layers of the star. As a result, the core must be compressed. While the density of the core increases, gravitational potential energy is released, and, as required by the virial theorem (Section 2.4), half of the energy is radiated away and half of the energy goes into increasing the thermal energy and hence the temperature of the gas.[2] Now, since the pp chain nuclear reaction rate goes as $\rho X^2 T_6^4$ (see Eq. 10.50), the increased temperature and density more than offset the decrease in the mass fraction of hydrogen, and the luminosity of the star slowly increases, along with its radius and effective temperature.

Main-sequence and post-main-sequence evolutionary tracks of stars of various masses, as computed in the pioneering study by Icko Iben, Jr., are shown in Fig. 13.1, and the amount of time required to evolve between points indicated on the figure are given in Table 13.1.[3] The locus of points labeled 1 represents

---

[2]This temperature increase means that the region of the star that is hot enough to undergo nuclear reactions increases slightly during the main-sequence phase of evolution.

[3]Note that in these models, the solar luminosity was taken to be $L_\odot = 3.86 \times 10^{33}$ ergs s$^{-1}$, rather than the value adopted elsewhere in this book.

13.1 Evolution on the Main Sequence

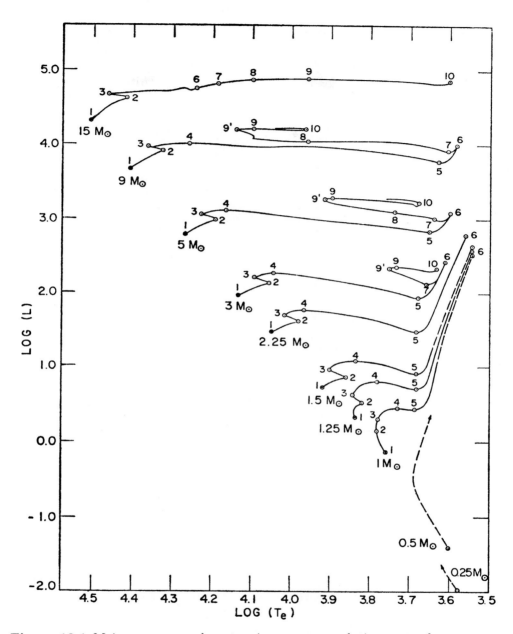

**Figure 13.1** Main-sequence and post-main-sequence evolutionary tracks of stars with an initial composition of $X = 0.708$, $Y = 0.272$, and $Z = 0.020$. (Figure from Iben, *Annu. Rev. Astron. Astrophys.*, 5, 571, 1967. Reproduced with permission from the *Annual Review of Astronomy and Astrophysics*, Volume 5, ©1967 by Annual Reviews Inc.)

| Mass | Time Interval Between Points (in years) | | | | |
|---|---|---|---|---|---|
| (M☉) | 1 – 2 | 2 – 3 | 3 – 4 | 4 – 5 | 5 – 6 |
| 15 | 1.010 (7) | 2.270 (5) | ← 7.55 (4) → | | |
| 9 | 2.144 (7) | 6.053 (5) | 9.113 (4) | 1.477 (5) | 6.552(4) |
| 5 | 6.547 (7) | 2.173 (6) | 1.372 (6) | 7.532 (5) | 4.857 (5) |
| 3 | 2.212 (8) | 1.042 (7) | 1.033 (7) | 4.505 (6) | 4.238 (6) |
| 2.25 | 4.802 (8) | 1.647 (7) | 3.696 (7) | 1.310 (7) | 3.829 (7) |
| 1.5 | 1.553 (9) | 8.10 (7) | 3.490 (8) | 1.049 (8) | ≥2 (8) |
| 1.25 | 2.803 (9) | 1.824 (8) | 1.045 (9) | 1.463 (8) | ≥4 (8) |
| 1.0 | 7 (9) | 2 (9) | 1.20 (9) | 1.57 (9) | ≥1 (9) |

| Mass | Time Interval Between Points (in years) | | | |
|---|---|---|---|---|
| (M☉) | 6 – 7 | 7 – 8 | 8 – 9 | 9 – 10 |
| 15 | 7.17 (5) | 6.20 (5) | 1.9 (5) | 3.5 (4) |
| 9 | 4.90 (5) | 9.50 (4) | 3.28 (6) | 1.55 (5) |
| 5 | 6.05 (6) | 1.02 (6) | 9.00 (6) | 9.30 (5) |
| 3 | 2.51 (7) | 4.08 (7) | 6.00 (6) | |

**Table 13.1** The Time Intervals Between Points in Fig. 13.1, Measured in Years (powers of 10 are given in parentheses). (Data from Iben, *Annu. Rev. Astron. Astrophys.*, 5, 571, 1967.)

the theoretical ZAMS, with the present-day Sun located between points 1 and 2 on the 1 M☉ track.[4]

Standard solar models were discussed in some detail in Section 11.1, with Figs. 11.3 – 11.10 showing the internal structure of one such model. In particular, Fig. 11.3 illustrates the partial depletion of hydrogen in the core, together with the accompanying increase in the amount of helium. The internal structure of a 1 M☉ star is also shown in Fig. 13.2, approximately 4.3 billion years after reaching the ZAMS (the location of the model is between points 1 and 2 in Fig. 13.1). Along with radius, density, temperature, and luminosity, the figure illustrates the mass fractions of the species $^{1}_{1}H$, $^{3}_{2}He$, $^{12}_{6}C$, $^{14}_{7}N$, as functions of interior mass. As the star's evolution on the main sequence continues, eventually the hydrogen at its center will be completely depleted. Such a situation is illustrated in Fig. 13.3 for a 1 M☉ star 9.2 billion years after arriving on the ZAMS; this model corresponds to point 3 in Fig. 13.1.

---

[4]More recent calculations of evolutionary tracks include significant refinements in input physics, but the results of these early models remain essentially correct and serve to illustrate many of the important features of stellar evolution. For examples of newer results, the interested reader is directed to the extensive grid of models published in a series of papers by Schaerer et al. (e.g., *Astron. Astrophys. Suppl.*, *102*, 339, 1993).

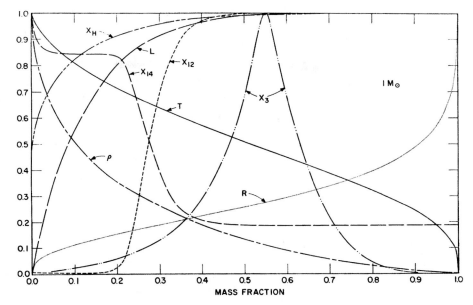

**Figure 13.2** The interior structure of a 1 $M_\odot$ star, $4.26990 \times 10^9$ years after reaching the ZAMS. The model is located between points 1 and 2 in Fig. 13.1. The maximum ordinate values of the parameters are $R = 0.96830$ $R_\odot$, $\rho = 159.93$ g cm$^{-3}$, $T = 15.910 \times 10^6$ K, $L = 1.0575$ $L_\odot$, $X_H = 0.708$, $X_3 = 4.20 \times 10^{-3}$, $X_{12} = 3.61 \times 10^{-3}$, and $X_{14} = 6.40 \times 10^{-3}$. (Figure from Iben, *Ap. J.*, *147*, 624, 1967.)

With the depletion of hydrogen in the core, the generation of energy via the pp chain must stop. However, by now the core temperature has increased to the point that nuclear fusion continues to generate energy in a thick hydrogen-burning shell around a small, predominantly helium core. This effect can be seen in the luminosity curve in Fig. 13.3. Note that the luminosity remains close to zero throughout the inner 3% of the star's mass. At the same time, the temperature is nearly constant over the same region. That the helium core must be isothermal when the luminosity gradient is zero can be seen from the radiative temperature gradient, given by Eq. (10.61). Since $L_r \simeq 0$ over a finite region, $dT/dr \simeq 0$ and $T$ is nearly constant. For an isothermal core to support the material above it in hydrostatic equilibrium, the required pressure gradient must be the result of a continuous increase in density as the center of the star is approached.

At this point, the luminosity being generated in the thick shell actually exceeds what was produced by the core during the phase of core hydrogen-burning. As a result, the evolutionary track continues to rise beyond point 3

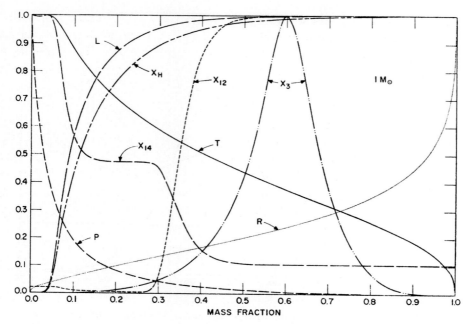

**Figure 13.3** The interior structure of a 1 $M_\odot$ star, $9.20150 \times 10^9$ years after reaching the ZAMS. The model is located near point 3 in Fig. 13.1. The maximum ordinate values of the parameters are $R = 1.2681$ $R_\odot$, $P = 13.146 \times 10^{17}$ dynes cm$^{-2}$, $T = 19.097 \times 10^6$ K, $L = 2.1283$ $L_\odot$, $X_H = 0.708$, $X_3 = 5.15 \times 10^{-3}$, $X_{12} = 3.61 \times 10^{-3}$, and $X_{14} = 1.15 \times 10^{-2}$. The radius of the star is 1.3526 $R_\odot$. (Figure from Iben, *Ap. J.*, *147*, 624, 1967.)

in Fig. 13.1, although not all of the energy generated reaches the surface; some of it goes into a slow expansion of the envelope. Consequently, the effective temperature begins to decrease slightly and the evolutionary track bends to the right. As the hydrogen-burning shell continues to consume its nuclear fuel, the isothermal helium core grows in mass while the star moves farther to the red in the H–R diagram. The locus of points during this redward phase of evolution is known as the **subgiant branch** (SGB).

This phase of evolution ends when the mass of the isothermal core has become too great and the core is no longer capable of supporting the material above it. The maximum fraction of a star's mass that can exist in an isothermal core and still support the overlying layers was first estimated by Schönberg and Chandrasekhar in 1942; it is given by

$$\left(\frac{M_{ic}}{M}\right)_{SC} \simeq 0.37 \left(\frac{\mu_e}{\mu_{ic}}\right)^2. \tag{13.1}$$

## 13.1 Evolution on the Main Sequence

The existence of the **Schönberg–Chandrasekhar limit** is another consequence of the virial theorem. Based on the physical tools we have developed so far, an approximate form of this result can be obtained. The analysis is presented in the optional material at the end of this section (beginning on the next page).

The maximum fraction of the mass of a star that can be contained in an isothermal core and still maintain hydrostatic equilibrium is a function of the mean molecular weights of the core and the envelope. When the mass of the isothermal helium core exceeds this limit, the core collapses on a Kelvin–Helmholtz time scale and the star evolves very rapidly relative to the nuclear time scale of main-sequence evolution. This occurs at the points labeled 4 in Fig. 13.1. For stars below about 1.2 $M_\odot$, this defines the end of the main-sequence phase.[5]

---

**Example 13.1** If a star is formed with the initial composition $X = 0.708$, $Y = 0.272$, and $Z = 0.020$, and if complete ionization is assumed at the core-envelope boundary, we find from Eq. (10.21) that $\mu_e \simeq 0.61$. Assuming that all of the hydrogen has been converted into helium in the isothermal core, $\mu_{ic} \simeq 1.34$. Therefore, from Eq. (13.1), the Schönberg–Chandrasekhar limit is

$$\left(\frac{M_{ic}}{M}\right)_{\text{SC}} \simeq 0.08.$$

The isothermal core will collapse if its mass exceeds 8% of the star's total mass.

---

The mass of an isothermal core can be increased beyond the Schönberg–Chandrasekhar limit if an additional source of pressure can be found to supplement the ideal gas pressure. This can occur if the electrons in the gas start to become **degenerate**. When the density of a gas becomes sufficiently high, the electrons in the gas are forced to occupy the lowest available energy levels. Since electrons are fermions and obey the Pauli exclusion principle (see Section 5.4), they cannot all occupy the same quantum state. Consequently, the electrons must be stacked into progressively higher energy states, beginning with the ground state. In the case of *complete degeneracy*, the pressure of the gas is due to the nonthermal motions of the electrons and, as a result, becomes independent of the temperature of the gas.

If the electrons are nonrelativistic, the pressure of a completely degenerate electron gas is given by

$$P_e = K\rho^{5/3}, \tag{13.2}$$

---

[5]What happens next is the subject of Section 13.2.

where $K$ is a constant. If the degeneracy is only partial, some temperature dependence remains.[6] The isothermal core of a 1 $M_\odot$ star between points 3 and 4 in Fig. 13.1 is partially degenerate; consequently, the core mass can reach approximately 13% of the entire mass of the star before it begins to collapse. Less massive stars exhibit even higher levels of degeneracy on the main sequence and therefore may not exceed the Schönberg–Chandrasekhar limit at all before the next stage of nuclear burning commences.

The evolution of more massive stars on the main sequence is similar to that of their lower-mass cousins with one important difference, the existence of a convective core. The convection zone has the effect of continually mixing the material, keeping the core composition nearly homogeneous.[7] For a 5 $M_\odot$ star, the central convection zone decreases somewhat in mass during core hydrogen-burning, leaving behind a slight composition gradient. As we move up the main sequence, as the star evolves the convection zone in the core retreats more rapidly with increasing stellar mass, disappearing entirely before the hydrogen is exhausted for those stars with masses greater than about 10 $M_\odot$.

When the mass fraction of hydrogen reaches about $X = 0.05$ in the core of a 5 $M_\odot$ star (point 2 in Fig. 13.1), the entire star begins to contract. With the release of some gravitational potential energy, the luminosity increases slightly. Since the radius decreases, the effective temperature must also increase. For massive stars, this stage of overall contraction is defined to be the end of the main-sequence phase of evolution.

---

*Optional:* To estimate the Schönberg–Chandrasekhar limit, begin by dividing the equation of hydrostatic equilibrium (Eq. 10.7) by the equation of mass conservation (Eq. 10.8). This gives

$$\frac{dP}{dM_r} = -\frac{GM_r}{4\pi r^4}, \qquad (13.3)$$

which is just the condition of hydrostatic equilibrium, written with the interior mass as the independent variable.[8] Rewriting, Eq. (13.3) may be expressed as

$$4\pi r^3 \frac{dP}{dM_r} = -\frac{GM_r}{r}. \qquad (13.4)$$

---

[6]The physics of degenerate gases will be discussed in more detail in Section 15.3.

[7]The time scale for convection, defined by the amount of time it takes a convective element to travel one mixing length (see Section 10.4), is much shorter than the nuclear time scale.

[8]This is the Lagrangian form of the condition for hydrostatic equilibrium.

## 13.1 Evolution on the Main Sequence

The left-hand side is just

$$4\pi r^3 \frac{dP}{dM_r} = \frac{d(4\pi r^3 P)}{dM_r} - 12\pi r^2 P \frac{dr}{dM_r} = \frac{d(4\pi r^3 P)}{dM_r} - \frac{3P}{\rho},$$

where Eq. (10.8) was used to obtain the last expression. Substituting back into Eq. (13.4) and integrating over the mass ($M_{ic}$) of the isothermal core, we have

$$\int_0^{M_{ic}} \frac{d(4\pi r^3 P)}{dM_r} dM_r - \int_0^{M_{ic}} \frac{3P}{\rho} dM_r = -\int_0^{M_{ic}} \frac{GM_r}{r} dM_r. \qquad (13.5)$$

To evaluate Eq. (13.5), we will consider each term separately. The first term on the left-hand side is just

$$\int_0^{M_{ic}} \frac{d(4\pi r^3 P)}{dM_r} dM_r = 4\pi R_{ic}^3 P_{ic},$$

where $R_{ic}$ and $P_{ic}$ are the radius and the gas pressure at the *surface* of the isothermal core, respectively (note that $r = 0$ at $M_r = 0$).

The second term on the left-hand side of Eq. (13.5) can also be evaluated quickly by realizing that, from the ideal gas law,

$$\frac{P}{\rho} = \frac{kT_{ic}}{\mu_{ic} m_H},$$

where $T_{ic}$ and $\mu_{ic}$ are the temperature and mean molecular weight throughout the isothermal core, respectively.[9] Thus

$$\int_0^{M_{ic}} \frac{3P}{\rho} dM_r = \frac{3M_{ic} kT_{ic}}{\mu_{ic} m_H} = 3N_{ic} kT_{ic} = 2K_{ic},$$

where

$$N_{ic} \equiv \frac{M_{ic}}{\mu_{ic} m_H}$$

is the number of gas particles in the core and

$$K_{ic} = \frac{3}{2} N_{ic} kT_{ic}$$

is the total thermal energy of the core, assuming an ideal monatomic gas.

---

[9]The core is actually supported in part by electron degeneracy pressure, meaning that the ideal gas law is not strictly valid. For our purposes here, however, the assumption of an ideal gas gives reasonable results.

The right-hand side of Eq. (13.5) is simply the gravitational potential energy of the core, or
$$-\int_0^{M_{ic}} \frac{GM_r}{r} dM_r = U_{ic}.$$
Substituting each term into Eq. (13.5), we find
$$4\pi R_{ic}^3 P_{ic} - 2K_{ic} = U_{ic}. \tag{13.6}$$

This expression should be compared with the virial theorem in the form of Eq. (2.45). If we had integrated from the center of the star to the surface, where $P \simeq 0$, we would have obtained our original version of the theorem. The difference lies in the nonzero pressure boundary condition. Thus Eq. (13.6) is a generalized form of the virial theorem for stellar interiors in hydrostatic equilibrium.

Next, from Eq. (10.28), the gravitational potential energy of the core may be approximated as
$$U_{ic} \sim -\frac{3}{5}\frac{GM_{ic}^2}{R_{ic}}.$$
Furthermore, the internal thermal energy of the core is just
$$K_{ic} = \frac{3M_{ic}kT_{ic}}{2\mu_{ic}m_H}.$$

Introducing these expressions into Eq. (13.6) and solving for the pressure at the surface of the isothermal core, we have
$$P_{ic} = \frac{3}{4\pi R_{ic}^3}\left(\frac{M_{ic}kT_{ic}}{\mu_{ic}m_H} - \frac{1}{5}\frac{GM_{ic}^2}{R_{ic}}\right). \tag{13.7}$$

Notice that there are two competing terms in Eq. (13.7); the first term is due to the thermal energy in the core and the second is due to gravitational effects. For specific values of $T_{ic}$ and $R_{ic}$, as the core mass increases, the thermal energy tends to increase the pressure at the surface of the core while the gravitational term tends to decrease it. For some value of $M_{ic}$, $P_{ic}$ is maximized, meaning that there exists an upper limit on how much pressure the isothermal core can exert in order to support the overlying envelope.

To determine when $P_{ic}$ is a maximum, we must differentiate Eq. (13.7) with respect to $M_{ic}$ and set the derivative equal to zero. It is left as an exercise to show that the radius of the isothermal core for which $P_{ic}$ is a maximum is given by
$$R_{ic} = \frac{2}{5}\frac{GM_{ic}\mu_{ic}m_H}{kT_{ic}} \tag{13.8}$$

## 13.1 Evolution on the Main Sequence

and that the maximum value of the surface pressure that can be produced by an isothermal core is given by

$$P_{ic,\max} = \frac{375}{64\pi} \frac{1}{G^3 M_{ic}^2} \left( \frac{kT_{ic}}{\mu_{ic} m_H} \right)^4. \tag{13.9}$$

The important feature of Eq. (13.9) is that, as the core mass increases, the maximum pressure at the surface of the core *decreases*. At some point, it may no longer be possible for the core to support the overlying layers of the star's envelope. Clearly this critical condition must be related to the mass contained in the envelope and therefore to the total mass of the star.

To estimate the maximum mass that can be supported by the isothermal core, we must determine the pressure exerted on the core by the overlying envelope. In hydrostatic equilibrium, this pressure must not exceed the maximum possible pressure due to the isothermal core. To estimate the envelope pressure, we will start again with Eq. (13.3), this time integrating from the surface of the star to the surface of the isothermal core. Assuming for simplicity that the pressure at the surface of the star is zero,

$$P_{ic,\text{env}} = \int_0^{P_{ic,\text{env}}} dP$$

$$= -\int_M^{M_{ic}} \frac{GM_r}{4\pi r^4} dM_r$$

$$\simeq -\frac{G}{8\pi \langle r^4 \rangle} \left( M_{ic}^2 - M^2 \right),$$

where $M$ is the total mass of the star and $\langle r^4 \rangle$ is some average value of $r^4$ between the surface of the star of radius $R$ and the surface of the core. Assuming that $M_{ic}^2 \ll M^2$, and making the crude approximation that $\langle r^4 \rangle \sim R^4/2$, we have

$$P_{ic,\text{env}} \sim \frac{G}{4\pi} \frac{M^2}{R^4} \tag{13.10}$$

for the pressure at the core's surface due to the weight of the envelope.

The quantity $R^4$ can be written in terms of the mass of the star and the temperature of the isothermal core through the use of the ideal gas law,

$$T_{ic} = \frac{P_{ic,\text{env}} \mu_e m_H}{\rho_{ic,\text{env}} k}, \tag{13.11}$$

where $\mu_e$ is the mean molecular weight of the envelope and $\rho_{ic,\text{env}}$ is the gas density at the core-envelope interface. Making the rough estimate that

$$\rho_{ic,\text{env}} \sim \frac{M}{4\pi R^3/3},$$

using Eq. (13.10), and solving for $R$, Eq. (13.11) gives

$$R \sim \frac{1}{3} \frac{GM}{T_{ic}} \frac{\mu_e m_H}{k}.$$

Substituting our solution for the radius of the envelope back into Eq. (13.10), we arrive at an expression for the pressure at core–envelope interface due to the overlying envelope,

$$P_{ic,\text{env}} \sim \frac{81}{4\pi} \frac{1}{G^3 M^2} \left( \frac{kT_{ic}}{\mu_e m_H} \right)^4. \qquad (13.12)$$

Note that $P_{ic,\text{env}}$ is independent of the mass of the isothermal core.

Finally, to estimate the Schönberg–Chandrasekhar limit, we set the maximum pressure of the isothermal core (Eq. 13.9) equal to the pressure needed to support the overlying envelope (Eq. 13.12). This immediately simplifies to give

$$\frac{M_{ic}}{M} \sim 0.54 \left( \frac{\mu_e}{\mu_{ic}} \right)^2.$$

Our result is only slightly larger than the one obtained originally by Schönberg and Chandrasekhar (Eq. 13.1).

## 13.2 Late Stages of Stellar Evolution

Following the completion of the main-sequence phase of stellar evolution, a complicated sequence of evolutionary stages occurs that may involve nuclear burning in the cores of stars together with nuclear burning in concentric mass shells. At various times, core burning and/or nuclear burning in a mass shell may cease, accompanied by a readjustment of the structure of the star. This readjustment may involve expansion or contraction of the core or envelope and the development of extended convection zones. As the final stages of evolution are approached, extensive mass loss from the surface also plays a critical role in determining the star's ultimate fate.

As an example of post-main-sequence stellar evolution, we will consider the ongoing changes in the 5 $M_\odot$ star discussed at the end of the last section. A detailed depiction of its evolution is shown in Fig. 13.4. Note that the numbers marked on the evolutionary track in that figure *do not* necessarily correspond to those used in Fig. 13.1, even though both diagrams illustrate many of the same features.

Figure 13.5 illustrates the changes in the star's internal structure with time. The numbers labeled on the top of Fig. 13.5 correspond to the points in

## 13.2 Late Stages of Stellar Evolution

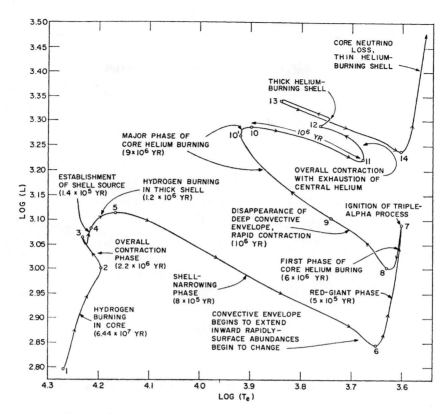

**Figure 13.4** The evolution of a 5 $M_\odot$ star from the zero-age main sequence to the asymptotic giant branch. The luminosity is given in solar units. (Figure from Iben, *Annu. Rev. Astron. Astrophys.*, 5, 571, 1967. Reproduced with permission from the *Annual Review of Astronomy and Astrophysics*, Volume 5, ©1967 by Annual Reviews Inc.)

Fig. 13.4. At a given time, a vertical slice in the diagram shows the structure of the star as a function of interior mass. (The center of the star is at the bottom of the figure.)

The end of the main-sequence phase of evolution at point 2 in Fig. 13.4 (or Fig. 13.1) corresponds to the stage of overall contraction that begins as a result of the near depletion of hydrogen fuel in the nuclear-burning core. By the time the star reaches point 3, the mass fraction of hydrogen has been reduced to $X = 0.01$ in the core (see Fig. 13.6) and hydrogen burning begins in a thick shell immediately surrounding the core. Because the ignition of the shell is quite rapid, the overlying envelope is forced to expand slightly, absorbing some of the energy released by the shell. As a result, the luminosity decreases

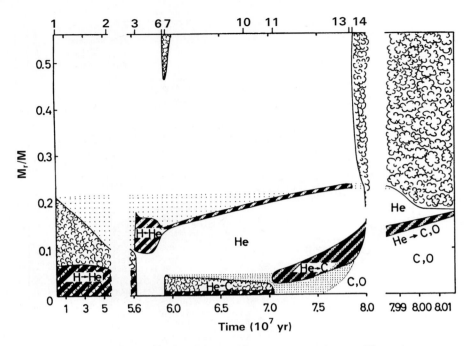

**Figure 13.5** The structure of a 5 $M_\odot$ star as a function of time. Note that the time axis is not linear. Cloudy regions represent convection zones, dots show regions of changing composition, and stripes designate regions where a significant amount of energy is being generated by nuclear reactions. Although the numbers on the top of the diagram correspond to those in Fig. 13.4, this evolutionary model is based on calculations carried out by another research group. (Figure adapted from Kippenhahn, Thomas, and Weigert, *Z. Astrophys.*, *61*, 241, 1965.)

momentarily, as can be seen in Fig. 13.4. A sketch of the star's structure at this point is given in Fig. 13.7.

As the shell continues to consume the hydrogen that is available at the base of the envelope, the size of the helium core steadily increases and, as with the 1 $M_\odot$ star, it becomes nearly isothermal. At point 5 in Fig. 13.4, the Schönberg–Chandrasekhar limit is reached and the core begins to collapse, causing the evolution to proceed on the much faster Kelvin–Helmholtz time scale.

As the core collapses, a nonzero temperature gradient is reestablished because of the release of gravitational potential energy. At the same time, the temperature and density of the hydrogen-burning shell increase, and, although the shell begins to narrow significantly, the rate at which energy is generated

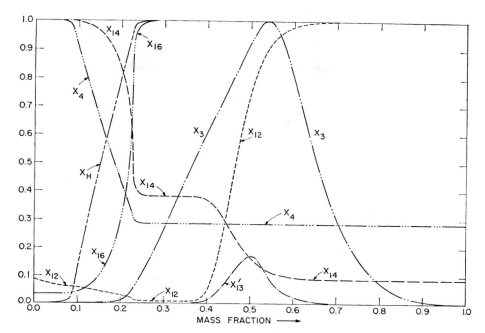

**Figure 13.6** The chemical composition as a function of interior mass fraction for a 5 $M_\odot$ star during the phase of overall contraction, labeled as point 3 in Fig. 13.4. The maximum mass fractions of the indicated species are $X_H < 0.708$, $X_3 < 1.296 \times 10^{-4}$, $X_4 < 0.9762$, $X_{12} < 3.61 \times 10^{-3}$, $X'_{13} < 3.61 \times 10^{-3}$ ($^{13}_{6}C$), $X_{14} < 0.0145$, and $X_{16} < 0.01080$ ($^{16}_{8}O$). (Figure from Iben, *Ap. J.*, *143*, 483, 1966.)

by the shell increases rapidly. This forces the envelope of the star to expand, absorbing much of the shell's energy before it reaches the surface. As a result, the surface luminosity of the star decreases, just as it did following point 3.

With the expansion of the envelope, the star's effective temperature drops, the photospheric opacity increases (due to the additional contribution of the $H^-$ ion), and a convection zone begins to develop near the surface. As the evolution continues toward point 6 in Fig. 13.4, the base of this convection zone extends down into regions where the chemical composition has been modified by nuclear processes (see Fig. 13.5). In particular, because of its rather large nuclear reaction cross section, lithium burns via collisions with protons at temperatures greater than about $2.7 \times 10^6$ K. This means that because of the star's evolution to this point, lithium has become nearly depleted over the inner 98% of its mass. At the same time, nuclear processing has increased the mass fraction of $^3_2He$ over the middle third of the star (Fig. 13.6) as well as altered the

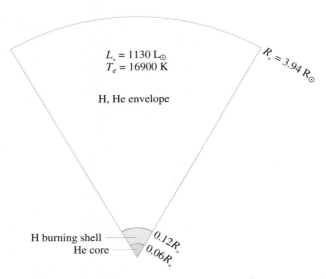

**Figure 13.7** A 5 $M_\odot$ star approaching the red giant branch with a helium core and a hydrogen-burning shell shortly after point 3 in Fig. 13.4. (Data from Iben, *Ap. J.*, *143*, 483, 1966.)

abundance ratios of the various species in the CNO cycle. When the surface convection zone encounters these chemically modified regions, the processed material will become mixed with the material above it. The result is a change in the observable composition of the photosphere; the amount of lithium at the surface will decrease, the amount of $^3_2$He will increase, and abundance ratios such as $X'_{13}/X_{12}$ will be modified. This is referred to as the **first dredge-up** phase. These observable changes in surface composition provide an important test of stellar evolution theory.

As the surface of the envelope continues to cool, recombination removes free electrons from the gas, and the contribution of the H$^-$ ion to the opacity decreases, allowing more energy to escape from the envelope. At the same time, the core continues to contract and the energy production of the narrowing hydrogen-burning shell increases still further. This now causes the luminosity of the star to increase and the radius to expand. By this time the surface of the star has cooled to the point that it has reached the Hayashi track (the path in the H–R diagram that the fully convective pre-main-sequence star followed as it approached the main sequence during its phase of quasi-hydrostatic contraction). The subsequent evolution proceeds nearly vertically in the H–R diagram. It is this phase of the star's evolution that first carries it into the red giant region, along the **red giant branch** (RGB); see Fig. 8.11. During

## 13.2 Late Stages of Stellar Evolution

the rise up the RGB, the still-deepening convective envelope transports $^{12}_{6}$C inward and $^{14}_{7}$N outward, decreasing the observable ratio, $X_{12}/X_{14}$.

At point 7 in Fig. 13.4, the central temperature ($1.3 \times 10^8$ K) and density (7700 g cm$^{-3}$) have finally become high enough that quantum-mechanical tunneling through the Coulomb barrier acting between $^{4}_{2}$He nuclei becomes effective and the triple alpha process begins. Some of the resulting $^{12}_{6}$C is further processed into $^{16}_{8}$O.

With the onset of a new and strongly temperature-dependent source of energy [see Eqs. (10.55) and (10.57)], the core expands. Although the hydrogen-burning shell remains the dominant source of the star's luminosity, the expansion of the core causes the rate of energy output of the shell to decrease somewhat, resulting in a decrease in the luminosity of the star. At the same time, the envelope contracts and the effective temperature begins to increase again.

An interesting difference arises at this point between the evolution of stars with masses greater than about 2 M$_\odot$ (as discussed above) and those that have masses less than 2 M$_\odot$. For stars of lower mass, as the helium core continues to collapse during evolution up to the tip of the red giant branch, the core becomes strongly electron-degenerate. When the temperature and density become high enough to initiate the triple alpha process (approximately $10^8$ K and $10^4$ g cm$^{-3}$, respectively), the ensuing energy release is almost explosive.[10] The luminosity generated by the core reaches $10^{11}$ L$_\odot$, comparable to that of an entire galaxy! However, this tremendous energy release lasts for only a few seconds, and most of the energy never even reaches the surface. Instead, it is absorbed by the overlying layers of the envelope, possibly causing some mass to be lost from the surface of the star. This short-lived phase of evolution of low-mass stars is referred to as the **helium core flash**. The origin of the explosive energy release is in the very weak temperature dependence of electron degeneracy pressure and the strong temperature dependence of the triple alpha process. The energy generated must first go into "lifting" the degeneracy. Only after this occurs can the energy go into thermal (kinetic) energy required to expand the core, which decreases the density and lowers the temperature, throttling the reaction rate.

As the envelope of the 5 M$_\odot$ model contracts beyond point 7, the increasing compression of the hydrogen-burning shell causes the energy output of the shell, and the overall energy output of the star, to rise. With the increased effective

---

[10]Significant neutrino losses from the core of the star prior to reaching the tip of the RGB may result in a negative temperature gradient near the center (a temperature inversion). This implies that the ignition of the triple alpha process is likely to occur in a shell around the center of the star.

temperature, the deep convection zone in the envelope retreats toward the surface, while at the same time, a convective core develops. The convective core is due to the high temperature sensitivity of the triple alpha process (just as the convective cores of upper main-sequence stars arise because of the temperature dependence of the CNO cycle).

When the star's evolution reaches point 10′, the mean molecular weight in the core has increased to the point that the core begins to contract, accompanied by the expansion and cooling of the envelope. The resulting drop in the star's luminosity is due once again to the required increase in the gravitational potential energy of the envelope. At point 11, the helium in the core is exhausted and the entire star contracts, exactly like the overall contraction associated with the exhaustion of hydrogen in the core at the end of the main-sequence phase (between points 2 and 3).

The generally horizontal evolution between points 7 and 11 in Fig. 13.4 (between points 6 and 10 in Fig. 13.1), known as the **horizontal branch** (HB), is characterized by a helium-burning core and a hydrogen-burning shell. During their passage along the horizontal branch, many stars develop instabilities in their outer envelopes, leading to periodic pulsations that are readily observable as variations in luminosity, temperature, radius, and surface radial velocity. These pulsations provide yet another test of stellar structure theory.[11]

With the increase of the core's temperature associated with its contraction, a thick helium-burning shell develops near point 12 in Fig. 13.4. As the core continues to contract, the helium-burning shell narrows and strengthens, forcing the material above it to expand and cool. The result is a temporary turn-off of the hydrogen-burning shell.

Along with the contraction of the helium-exhausted core, neutrino production increases to the point that the core actually cools somewhat; it is refrigerated because of the energy that is carried away by the easily escaping neutrinos! As a consequence of the increasing central density and decreasing temperature, electron degeneracy pressure becomes an important component of the total pressure in the carbon–oxygen core.

The remainder of the evolution illustrated in Fig. 13.4 (beyond point 13) is very similar to the evolution between points 5 and 7. The expanding envelope initially absorbs much of the energy produced by the helium-burning shell. As the effective temperature continues to decrease, the convective envelope deepens again, this time extending downward to the chemical discontinuity between the hydrogen-rich outer layers and the helium-rich region above the helium-burning shell. The mixing that results during this **second dredge-up**

---

[11]Pulsating variable stars will be discussed in greater detail in Chapter 14.

## 13.2 Late Stages of Stellar Evolution

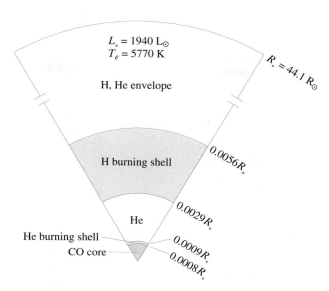

**Figure 13.8** A 5 $M_\odot$ star on (or approaching) the asymptotic giant branch with a carbon–oxygen core and hydrogen and helium-burning shells following point 11 in Fig. 13.4. Note that relative to the surface radius, the scale of the shells and core have been increased by a factor of 100 for clarity. (Data from Iben, *Ap. J.*, *143*, 483, 1966.)

phase increases the helium and nitrogen content of the envelope. (The increase in nitrogen is due to the previous conversion of carbon and oxygen into nitrogen in the intershell region.)

When the redward evolution reaches the Hayashi track, the evolutionary track bends upward along a path referred to as the **asymptotic giant branch** (AGB).[12] At this point in its evolution the core temperature is approximately $2 \times 10^8$ K and its density is on the order of $10^6$ g cm$^{-3}$. A diagram of an AGB star with two shell sources is shown in Fig. 13.8. Note that the diagram is not to scale; in order to visualize the structure from the hydrogen-burning shell, inward, that region was enlarged by a factor of 100 relative to the surface of the star.

During the AGB phase, the dormant hydrogen-burning shell eventually re-ignites and the narrowing helium-burning shell below it begins to turn on and off periodically. These periodic **helium shell flashes** occur because the hydrogen-burning shell is dumping helium onto the helium layer below. As the mass of the helium layer increases, it becomes slightly degenerate. Then, when

---
[12]The AGB gets its name because the evolutionary track approaches the line of the RGB asymptotically from the left.

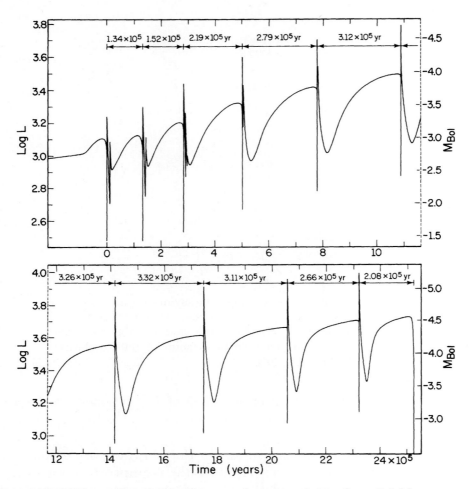

**Figure 13.9** The surface luminosity as a function of time for a 0.6 $M_\odot$ stellar model that is undergoing helium shell flashes. (Figure from Iben, *Ap. J.*, *260*, 821, 1982.)

the temperature of the helium shell increases sufficiently, a helium shell flash occurs, analogous to the earlier helium core flashes of low-mass stars (although much less energetic). This drives the hydrogen-burning shell outward, causing it to cool and turn off for a time. Eventually the burning in the helium shell diminishes, the hydrogen-burning shell recovers, and the process repeats. The period between pulses is a function of the mass of the star, ranging from thousands of years for stars near 5 $M_\odot$ to hundreds of thousands of years for low-mass stars (0.6 $M_\odot$), with the pulse amplitude growing with each successive event; see Fig. 13.9.

## 13.2 Late Stages of Stellar Evolution

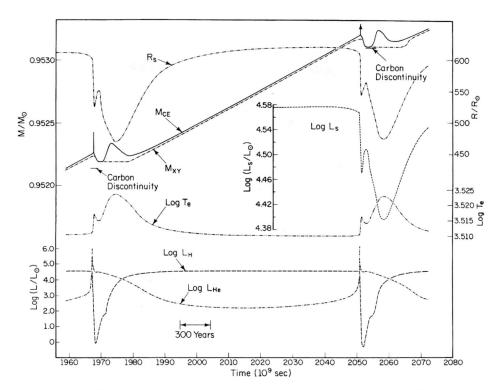

**Figure 13.10** Time-dependent changes in the properties of a 7 $M_\odot$ AGB star produced by helium shell flashes. The quantities shown are the surface radius ($R_S$), the interior mass fractions of the base of the convective envelope ($M_{CE}$) and the hydrogen–helium discontinuity ($M_{XY}$), the star's luminosity and effective temperature ($L_s$ and $T_e$, respectively), and the luminosities of the hydrogen and helium-burning shells ($L_H$ and $L_{He}$, respectively). (Figure from Iben, *Ap. J.*, **196**, 525, 1975.)

Details of thermal pulses for a 7 $M_\odot$ star are shown in Fig. 13.10. Following a helium shell flash, the luminosity arising from the hydrogen-burning shell drops appreciably while the energy output from the helium-burning shell increases. Since, after its re-ignition, the hydrogen-burning shell once again dominates the total energy output of the star, the surface luminosity drops, eventually relaxing back to its preflash value. Simultaneously, the radius of the star decreases and the effective temperature increases for a period of time following a helium shell flash.

A class of pulsating variable stars known as **long period variables** (LPVs) are AGB stars. (LPVs have pulsation periods of 100 to 700 days and include the subclass of **Mira variable stars**.) It has been suggested that the structural changes arising from shell flashes could cause observable changes in the periods of these stars, providing another possible test of stellar evolution theory. In fact, several Miras (e.g., W Dra, R Aql, and R Hya) have been observed to be undergoing period changes.

Because of the sudden increase in energy flux from the helium-burning shell during a flash episode, a convection zone is established between it and the hydrogen-burning shell. At the same time, the depth of the envelope convection zone increases with pulse strength. For stars that are massive enough ($M > 2$ $M_\odot$), the convection zones will merge and eventually extend down into regions where carbon has been synthesized, carrying freshly processed material to the surface. This **third dredge-up** phase appears to explain the difference between oxygen-rich giants ($X_O > X_C$) and carbon-rich giants ($X_C > X_O$), called **carbon stars**, that have been observed spectroscopically.

AGB stars are known to lose mass at a rapid rate, sometimes as high as $\dot{M} \sim 10^{-4}$ $M_\odot$ yr$^{-1}$. The effective temperatures of these stars are also quite cool (around 3000 K). As a result, it is expected that dust grains should exist in the matter that is being expelled. Since silicate grains tend to form in an environment rich in oxygen, and graphite grains will form in a carbon-rich environment, the composition of the ISM may be related to the relative numbers of carbon and oxygen-rich stars. Observations of ultraviolet extinction curves in the Milky Way and the Large and Small Magellanic Clouds[13] support the idea that mass loss from these stars does, in fact, help enrich the ISM.

As evolution up the AGB continues, what happens next is strongly dependent on the original mass of the star and the amount of mass loss experienced by that star during its lifetime. Although the results of ongoing observational and theoretical studies are still inconclusive, it appears that the final evolutionary behavior of stars can be separated into two basic groups: those with ZAMS masses above about 8 $M_\odot$ and those with masses below this value.[14] The distinction between the two mass regimes is based on whether or not the central carbon–oxygen core will undergo further nuclear burning. In the remainder of this section we will consider the final evolution of stars with initial masses less than 8 $M_\odot$, leaving the ultimate evolution of more massive stars to the next section.

As stars with initial masses below 8 $M_\odot$ continue to evolve up the AGB, the helium-burning shell converts more and more of the helium into carbon and then into oxygen, increasing the mass of the carbon-oxygen core. At the

---

[13]The LMC and the SMC are small satellite galaxies of the Milky Way, visible in the southern hemisphere.

[14]Some researchers suggest that the dividing line may be closer to 9 $M_\odot$.

## 13.2 Late Stages of Stellar Evolution

same time, the core continues to contract slowly, causing its central density to increase. Depending on the star's mass, neutrino energy losses may decrease the central temperature somewhat during this phase. In any event, the densities in the core become large enough that electron degeneracy pressure begins to dominate. This situation is very similar to the development of an electron-degenerate helium core in a low-mass star during its rise up the red giant branch.

For stars with ZAMS masses less than about 4 $M_\odot$, the carbon–oxygen core will never become large enough and hot enough to ignite nuclear burning. However, *if the important contribution of mass loss is ignored* for stars between 4 $M_\odot$ and 8 $M_\odot$, theory suggests that the C–O core could reach a sufficiently large mass that it is no longer capable of remaining in hydrostatic equilibrium, even with the assistance of pressure from the degenerate electron gas. The outcome of this situation is catastrophic core collapse. The maximum value of 1.4 $M_\odot$ for a completely degenerate core is known as the **Chandrasekhar limit**.[15]

When the collapse of the C–O core begins, the temperature and density rise to such high levels that the rate of energy production is enormous, and just as in the case of the helium core flash of low-mass stars at the tip of the red giant branch, the reaction rate is not slowed immediately by core expansion. However, in this instance so much energy is released in such a short period of time that, unlike the helium core flash, the **carbon–oxygen core flash** is truly explosive, destroying the entire star in a spectacular visual display, known as a **supernova**.[16]

Whether the situation described in the preceding two paragraphs actually occurs in nature depends heavily on just how much mass loss the star experiences during its lifetime. Our current understanding of the situation suggests that the type of supernova explosion described here will not happen because the star loses enough mass to bring it below the value required to reach the Chandrasekhar limit in the core.

As has already been mentioned, observations of AGB stars do show enormous mass loss rates. Unfortunately, our understanding of the mechanism(s) that cause this mass loss is poor. Some astronomers have suggested that the mass loss may be linked to the helium shell flashes, or perhaps to the periodic envelope pulsations of LPVs. Other proposed mechanisms stem from the high

---

[15]The Chandrasekhar limit plays a critical role in the formation of the final products of stellar evolution, namely white dwarfs, neutron stars, and black holes. The physics of the Chandrasekhar limit will be discussed in some detail in Section 15.4.

[16]This hypothetical type of supernova explosion has been termed a Type I$\frac{1}{2}$. Type II supernovae are the subject of the next section, and Type I will be discussed in Chapter 17.

luminosities and low surface gravities of these stars. Whatever the cause, its influence on the evolution of AGB stars is significant.

One might expect that the rate of mass loss would actually accelerate with time since the luminosity and radius are increasing during continued evolution up the AGB. At the same time, the continually decreasing mass of the star implies that the surface gravity is decreasing as well, meaning that the surface material is less tightly bound. Consequently, mass loss becomes progressively more important as AGB evolution continues.

It has been suggested that a **superwind** develops near the end of the mass loss phase ($\dot{M} \sim 10^{-4}$ $M_\odot$ yr$^{-1}$). Whether shell flashes, envelope pulsations, or some other mechanism is the reason, high mass loss rates seem to be responsible for the existence of a class of objects known as **OH/IR sources**. These objects appear to be stars shrouded in optically thick dust clouds that radiate their energy primarily in the infrared part of the electromagnetic spectrum.

The OH part of the OH/IR designation is due to the detection of OH molecules, which can be seen via their **maser emission**.[17] A maser is the molecular analog of a laser; electrons are "pumped up" from a lower energy level into a higher, long-lived metastable energy state. The electron then makes a downward transition back to a lower state when it is stimulated by a photon with an energy equal to the difference in energies between the two states. The original photon and the emitted photon will travel in the same direction and will be in phase with each other, hence the amplification of radiation. A schematic energy level diagram of a hypothetical three-level maser is depicted in Fig. 13.11.

As the cloud around the OH/IR source continues to expand, it eventually becomes optically thin, exposing the central star, which characteristically exhibits the spectrum of an F or G supergiant. During the ensuing final phase of mass loss, the remainder of the star's envelope is expelled, revealing the cinders produced by its long history of nuclear reactions. With only a very thin layer of material remaining above them, the hydrogen- and helium-burning shells are extinguished and the luminosity of the star drops rapidly. The hot central object, now revealed, will cool to become a **white dwarf star**, which is essentially the old red supergiant's degenerate carbon–oxygen core, surrounded by a thin layer of residual hydrogen and helium. This important class of stars, the end products of the evolution of stars with initial main-sequence masses less than 8 $M_\odot$, will be discussed in Chapter 15.

---

[17]The term *maser* is an acronym for *m*icrowave *a*mplification by *s*timulated *e*mission of *r*adiation.

## 13.2 Late Stages of Stellar Evolution

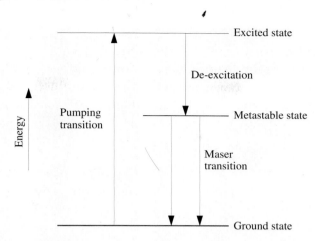

**Figure 13.11** A schematic diagram of a hypothetical three-level maser. The intermediate energy level is a relatively long-lived metastable state. A transition from the metastable state to the lowest energy level can occur through stimulated emission by a photon of energy equal to the energy difference between the two states.

The expanding shell of gas around a white dwarf progenitor is called a **planetary nebula**. Examples of planetary nebulae are shown in Fig. 13.12. This class of beautiful, glowing clouds of gas was given its name in the nineteenth century because, when viewed through a small telescope, they look somewhat like giant gaseous planets. A planetary nebula owes its appearance to the ultraviolet light emitted by the hot, condensed central star. The ultraviolet photons are absorbed by the gas in the nebula, causing the atoms to become excited or ionized. When the electrons cascade back down to lower energy levels, photons are emitted whose wavelengths are in the visible portion of the electromagnetic spectrum. As a result, the cloud appears to glow in visible light.[18] Many planetaries, like the Ring Nebula in Fig. 13.12(a), look like they have a ringlike structure rather than a spherically symmetric shell because of the increase in optical depth when looking tangent to the edge of the expanding shell, rather than perpendicular to the shell's surface.[19] The blueish-green coloration of many planetary nebulae is due to the 5006.8 Å and 4958.9 Å forbidden lines of [O III] (forbidden lines of [O II] and [Ne III] are also common) and the reddish coloration comes from ionized hydrogen and nitrogen. Characteristic temperatures of these objects are in the range of the ionization temperature of hydrogen, $10^4$ K.

---

[18]This process is reminiscent of the creation of H II regions around newly formed O and B main-sequence stars, discussed in Section 12.3.

[19]The same effect occurs when looking at soap bubbles.

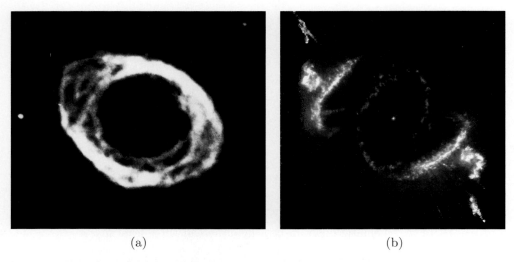

(a) (b)

**Figure 13.12** (a) The Ring Nebula in Lyra, a typical planetary nebula, is approximately 1200 pc from Earth. (Courtesy of the National Optical Astronomy Observatories.) (b) NGC 6543 (the "Cat's Eye") is in Draco, 900 pc away. This image was produced by taking the ratio of emission from the 6584 Å line of [N III] and $H_\alpha$ (the data were obtained by the Hubble Space Telescope's WF/PC 2). The complex structure may be due to high-speed jets and the presence of a companion star, making NGC 6543 part of a binary star system. (Courtesy of J. P. Harrington, K. J. Borkowski, and NASA.)

The expansion velocities of planetaries, as measured by Doppler-shifted spectral lines, show that the gas is moving away from the central stars with typical speeds of between 10 and 30 km s$^{-1}$. Combined with radii of around 0.3 pc, their estimated ages are on the order of 10,000 years. After only about 50,000 years, a planetary nebula will dissipate into the ISM. Compared with the entire lifetime of a star, the phase of planetary nebula ejection is fleeting indeed.

Despite their short lifetimes, roughly 1600 planetary nebulae are known to exist in the Milky Way Galaxy. Given the fact that we are unable to observe the entire Galaxy from Earth, it is estimated that the number of planetaries is probably close to 50,000. If, on average, each planetary contains about 0.5 $M_\odot$ of material, the ISM is being enriched at the rate of several solar masses per year through this process.

The last stages of evolution of a 0.6 $M_\odot$ model, from the AGB through planetary nebula ejection, is depicted in Fig. 13.13. The lower right-hand portion of the track (labeled E-AGB) represents the early evolution up the

## 13.2 Late Stages of Stellar Evolution

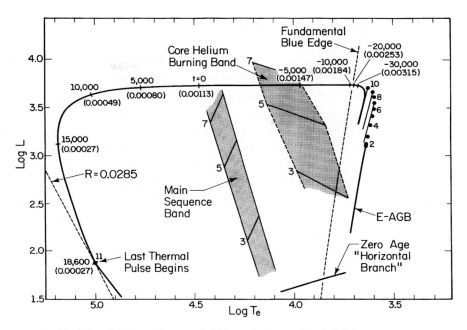

**Figure 13.13** The AGB and post-AGB evolution of a 0.6 $M_\odot$ star undergoing mass loss. The initial composition of the model is $X = 0.749$, $Y = 0.25$, and $Z = 0.001$. The main-sequence and horizontal branches of 3, 5, and 7 $M_\odot$ stars are shown for reference. Details of the figure are discussed in the body of the text. (Figure adapted from Iben, *Ap. J.*, *260*, 821, 1982.)

asymptotic giant branch before the onset of helium shell flashes. The position of the star on the H–R diagram at the start of each flash episode is indicated by a number next to the evolutionary track (eleven pulses in all), with the resulting excursions in luminosity and effective temperature indicated for pulses 7, 9, and 10. It is after the tenth pulse that the star leaves the AGB, ejecting its envelope as a planetary nebula during the nearly constant luminosity path across the H–R diagram. The amount of mass remaining in the hydrogen-rich envelope is indicated in parentheses along the evolutionary track (in $M_\odot$). Also indicated is the amount of time before (negative) or after (positive) the point when the star's effective temperature was 30,000 K (the time is measured in years). Following the eleventh helium shell flash, the star finally loses the last remnants of its envelope and becomes a white dwarf of radius 0.0285 $R_\odot$.[20]

---

[20]The line labeled "Fundamental Blue Edge" corresponds to the high-temperature limit for fundamental mode pulsations of a class of variable stars known as **RR Lyraes**. This important class of objects will be discussed extensively in Chapter 14.

## 13.3 The Fate of Massive Stars

The AGB and post-AGB evolution of stars more massive than about 8 $M_\odot$ is decidedly different from what was described in the previous section. As the helium-burning shell continues to add ash to the carbon–oxygen core, and as the core continues to contract, it eventually ignites in carbon-burning, generating a variety of by-products, such as $^{16}_{8}O$, $^{20}_{10}Ne$, $^{23}_{11}Na$, $^{23}_{12}Mg$, and $^{24}_{12}Mg$ [see Eqs. (10.58) and (10.59)]. What follows is a succession of nuclear reaction sequences, the exact details of which depend sensitively on the mass of the star.

Assuming that each reaction sequence reaches equilibrium, an "onionlike" shell structure develops in the interior of the star. Following carbon burning, the oxygen in the resulting neon–oxygen core will ignite (Eq. 10.60), producing a new core composition dominated by $^{28}_{14}Si$. Finally, at temperatures near $3 \times 10^9$ K, silicon burning can commence through a series of reactions such as

$$^{28}_{14}Si + ^{4}_{2}He \rightleftharpoons ^{32}_{16}S + \gamma$$
$$^{32}_{16}S + ^{4}_{2}He \rightleftharpoons ^{36}_{18}Ar + \gamma \qquad (13.13)$$
$$\vdots$$
$$^{52}_{24}Cr + ^{4}_{2}He \rightleftharpoons ^{56}_{28}Ni + \gamma.$$

Silicon burning produces a host of nuclei centered near the $^{56}_{26}Fe$ peak of the binding energy per nucleon curve, shown in Fig. 10.9, the most abundant of which are probably $^{54}_{26}Fe$, $^{56}_{26}Fe$, and $^{56}_{28}Ni$. Any further reactions that produce nuclei more massive than $^{56}_{26}Fe$ are endothermic and cannot contribute to the luminosity of the star. Grouping all of the products together, silicon burning is said to produce an *iron core*.

A sketch of the onionlike interior structure of a massive star following silicon burning is given in Fig. 13.14. Figure 13.15 shows the detailed structure of a 15 $M_\odot$ star with an iron core. Evident are the nuclear burning shells (spikes in $\epsilon_{\text{nuc}}$), convection zones, and the extensive nuclear processing of elements.

Because carbon, oxygen, and silicon burning produce nuclei with masses progressively nearer the *iron peak* of the binding energy curve, less and less energy is generated per gram of fuel. As a result, the time scale for each succeeding reaction sequence becomes shorter (recall Example 10.3). For example, for a 20 $M_\odot$ star, its main-sequence lifetime (core hydrogen burning) is roughly $10^7$ years, core helium burning requires $10^6$ years, carbon burning lasts 300 years, oxygen burning takes roughly 200 days, and silicon burning is completed in only two days!

## 13.3 The Fate of Massive Stars

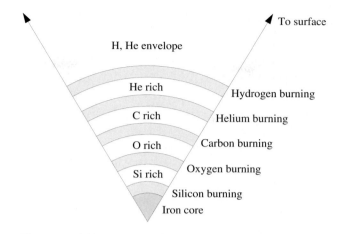

**Figure 13.14** The onionlike interior of a massive star that has evolved through core silicon burning. Inert regions of processed material are sandwiched between the nuclear burning shells. The drawing is not to scale.

At the very high temperatures now present in the core, the photons possess enough energy to destroy heavy nuclei (note the reverse arrows in the silicon-burning sequence), a process known as **photodisintegration**. Particularly important are the photodisintegration of $^{56}_{26}$Fe and $^{4}_{2}$He:

$$^{56}_{26}\text{Fe} + \gamma \rightarrow 13\,^{4}_{2}\text{He} + 4n \tag{13.14}$$

$$^{4}_{2}\text{He} + \gamma \rightarrow 2p^{+} + 2n. \tag{13.15}$$

When the mass of the contracting iron core has become large enough and the temperature sufficiently high, photodisintegration can, in a very short period of time, undo what the star has been trying to do its entire life, namely produce elements more massive than hydrogen and helium. Of course, this process of stripping iron down to individual protons and neutrons is highly endothermic; thermal energy is removed from the gas that would otherwise have resulted in the pressure necessary to support the core of the star. The core masses for which this process occurs vary from 1.3 $M_\odot$ for a 10 $M_\odot$ ZAMS star to 2.5 $M_\odot$ for a 50 $M_\odot$ star.

Under the extreme conditions that now exist (e.g., $T_c \sim 8 \times 10^9$ K and $\rho_c \sim 10^{10}$ g cm$^{-3}$ for a 15 $M_\odot$ star), the free electrons that had assisted in supporting the star through degeneracy pressure are captured by heavy nuclei and by the protons that were produced through photodisintegration; for instance,

$$p^{+} + e^{-} \rightarrow n + \nu_e. \tag{13.16}$$

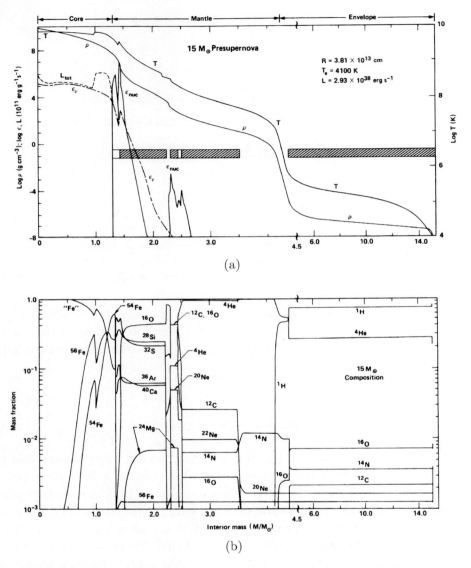

**Figure 13.15** The interior structure of an evolved 15 $M_\odot$ star just before the collapse of the iron core. (a) Temperature ($T$), density ($\rho$), the energy from nuclear reactions and losses from neutrinos ($\epsilon_{\text{nuc}}$ and $\epsilon_\nu$, respectively), and the total luminosity ($L_{\text{tot}}$) are shown. Convection zones are indicated by hatched bars. (b) The composition structure reveals extensive nuclear processing. (Figures from Woosley and Weaver, *Annu. Rev. Astron. Astrophys.*, 24, 205, 1986. Reproduced with permission from the *Annual Review of Astronomy and Astrophysics*, Volume 24, ©1986 by Annual Reviews Inc.)

## 13.3 The Fate of Massive Stars

The amount of energy that escapes the star in the form of neutrinos becomes enormous; during silicon-burning the photon luminosity of a 20 $M_\odot$ stellar model is $4.4 \times 10^{38}$ ergs s$^{-1}$ while the neutrino luminosity is $3.1 \times 10^{45}$ ergs s$^{-1}$.

Through the photodisintegration of iron, combined with electron capture by protons and heavy nuclei, most of the core's support in the form of electron degeneracy pressure is suddenly gone and the core begins to collapse extremely rapidly. In the inner portion of the core, the collapse is homologous and the velocity of the collapse is proportional to the distance away from the center of the star [recall the discussion of a homologous free-fall collapse during the formation of a protostar and Eq. (12.16)]. At the radius where the velocity exceeds the local sound speed, the collapse can no longer remain homologous and the inner core decouples from the now supersonic outer core, which is left behind and nearly in free-fall. During the collapse, speeds can reach almost 70,000 km s$^{-1}$ in the outer core, and within about one second a volume the size of Earth has been compressed down to a radius of 50 km!

Since mechanical information will only propagate through the star at the speed of sound and because the core collapse proceeds so quickly, there is not enough time for the outer layers to learn about what has happened inside. The outer layers, including the oxygen, carbon, and helium shells, as well as the outer envelope, are left in the precarious position of being almost suspended above the catastrophically collapsing core.

The homologous collapse of the inner core continues until the density there exceeds about $8 \times 10^{14}$ g cm$^{-3}$, roughly three times the density of an atomic nucleus. At that point, the nuclear material that now makes up the inner core stiffens because the strong force (usually attractive) suddenly becomes repulsive. This is a consequence of the Pauli exclusion principle applied to neutrons.[21] The result is that the inner core rebounds somewhat, sending pressure waves outward into the infalling material from the outer core. When the velocity of the pressure waves reach the sound speed, they build into a shock wave that begins to move outward.

As the shock wave encounters the infalling outer iron core, the high temperatures that result cause further photodisintegration, robbing the shock of much of its energy. For every 0.1 $M_\odot$ of iron that is broken down into protons and neutrons, the shock loses $1.7 \times 10^{51}$ ergs. If the remainder of the iron core is not too massive (the entire initial iron core should not exceed about 1.2 $M_\odot$), the shock will fight its way through the rest of the outer core and collide with the remainder of the nuclear-processed material and the outer envelope. Once the shock forms above the surface of the inner core, the time

---

[21] Neutrons, along with electrons and protons, are fermions.

required to penetrate the outer core is only 20 milliseconds. This process is known as a *prompt hydrodynamic explosion*.

If the initial iron core is too large, computer simulations indicate that the shock stalls, becoming nearly stationary, with infalling material accreting onto it. In this case the shock becomes an *accretion shock*, somewhat akin to the situation during protostellar collapse, discussed in Section 12.2. However, below the shock, a *neutrinosphere* develops from the processes of photodisintegration and electron capture. Since the overlying material is now so dense that even neutrinos cannot easily penetrate it, some of the neutrino energy ($\sim 5\%$) is deposited in the matter just behind the shock. This additional energy heats the material and allows the shock to resume its march toward the surface. The temporary stalling of the shock front is called a *delayed explosion mechanism*. The success of delayed explosion models seems to hinge on two- or three-dimensional simulations, which allow for hot, rising plumes of gas to mix with colder, infalling gas. Older one-dimensional calculations that were necessitated by limits in computing power had an annoying tendency to stall out completely and never explode.

If the initial mass of the star on the main sequence was not too large (perhaps $M_{\text{ZAMS}} < 25 \, M_\odot$), the remnant in the inner core will stabilize and become a **neutron star**, supported by degenerate neutron pressure. However, if the initial stellar mass is much larger, even the pressure of neutron degeneracy cannot support the remnant against the pull of gravity, and the final collapse will be complete, producing a **black hole**.[22] In either case, the creation of these exotic objects is accompanied by a tremendous production of neutrinos, the majority of which escape into space with a total energy on the order of the binding energy of a neutron star, approximately $3 \times 10^{53}$ ergs. This represents roughly 100 times more energy than the Sun will produce over its entire main-sequence lifetime!

Meanwhile, the shock is still working its way toward the surface, driving the hydrogen-rich envelope and the remainder of the nuclear-processed matter in front of it. The total kinetic energy in the expanding material is on the order of $10^{51}$ ergs, roughly 1% of the energy liberated in neutrinos. Finally, when the material becomes optically thin at a radius of about $10^{15}$ cm, a tremendous optical display results, releasing approximately $10^{49}$ ergs of energy in the form of photons, with a peak luminosity of nearly $10^{43}$ ergs s$^{-1}$, or roughly $10^9 \, L_\odot$, which is capable of competing with the brightness of an entire galaxy.

---

[22]We leave the detailed discussion of neutron stars and black holes to Chapters 15 and 16, respectively.

## 13.3 The Fate of Massive Stars

The events just described—the catastrophic collapse of an iron core, the generation of a shock wave, and the ensuing ejection of the star's envelope—are believed to be the mechanism that creates a **Type II supernova**. Observationally, Type II supernovae are characterized by a rapid rise in luminosity, reaching a limiting absolute bolometric magnitude of about $-18$, followed by a steady decrease, dropping six to eight magnitudes in a year. Their spectra also exhibit lines associated with hydrogen and heavier elements. Furthermore, P Cygni profiles are common in many lines (indicating rapid expansion; see Section 12.3).

Another class of objects, known as **Type I supernovae**, do not contain prominent hydrogen lines in their spectra. As will be discussed in Section 17.4, a subclass of Type I's, known as Type Ia's, involve the explosion of a carbon–oxygen white dwarf in a close binary system, while Types Ib and Ic involve mechanisms similar to Type II's. Current estimates suggest that Type II events occur approximately once every 44 years in the Milky Way Galaxy, while Type I's occur once every 36 years. However, because we are unable to see most of the Galaxy from our location, naked-eye supernova events are extremely rare. It is believed that SN 1006 and the supernovae detected by Tycho (SN 1572) and Kepler (SN 1604) were probably Type I's. On the other hand, the Crab supernova (SN 1054) and the detection of a supernova in the Large Magellanic Cloud (SN 1987A) were Type II's.[23] In the remainder of this section we will discuss only the observations of Type II's, postponing further discussion of Type I supernovae until Chapter 17.

The light curves of Type II supernovae can be classified as either Type II-L (linear) or Type II-P (plateau). Composite B (blue) magnitude light curves of each type are shown in Fig. 13.16. A temporary but clear plateau exists between about 30 and 80 days after maximum light for Type II-P supernovae; no such detectable plateau exists for Type II-L objects.

The source of the plateau in Type II-P light curves is the **radioactive decay** of the large amount of $^{56}_{28}$Ni that was produced by the shock front during its march through the star (the half-life of $^{56}_{28}$Ni is $\tau_{1/2} = 6.1$ days). It is expected that the explosive nucleosynthesis of the supernova shock should have produced significant amounts of other radioactive isotopes as well, such as $^{57}_{27}$Co ($\tau_{1/2} = 271$ days), $^{22}_{11}$Na ($\tau_{1/2} = 2.6$ yr), and $^{44}_{22}$Ti ($\tau_{1/2} \simeq 47$ yr). If the isotopes are present in sufficient quantities, each in turn may contribute to the overall light curve, causing the slope of the curve to change.

---

[23]Supernovae are named for the year in which they occurred, and in what order they were detected. For instance, SN 1987A was the first supernova discovered in 1987.

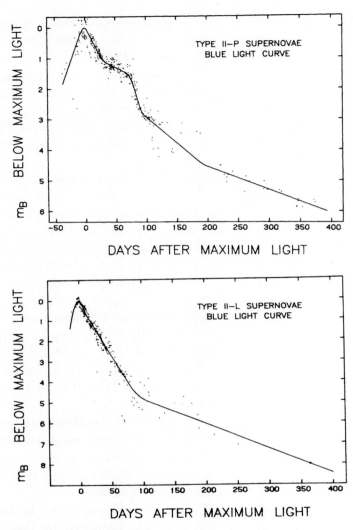

**Figure 13.16** The characteristic shapes of Type II-P and Type II-L light curves. These are composite light curves, based on the observations of many supernovae. (Figures from Doggett and Branch, *Astron. J.*, *90*, 2303, 1985.)

## 13.3 The Fate of Massive Stars

The $^{56}_{28}\text{Ni}$ is transformed into $^{56}_{27}\text{Co}$ through the *beta-decay* reaction[24]

$$^{56}_{28}\text{Ni} \rightarrow {}^{56}_{27}\text{Co} + e^+ + \nu_e + \gamma. \tag{13.17}$$

The energy released by the decay is deposited into the optically thick expanding shell, which is then radiated away from the supernova remnant's photosphere. This "holds up" the light curve for a time, resulting in the observed plateau. Eventually the expanding gas cloud will become optically thin, exposing the central product of the explosion, the neutron star or black hole.

$^{56}_{27}\text{Co}$, the product of the radioactive decay of $^{56}_{28}\text{Ni}$, is itself radioactive, with a longer half-life of 77.7 days:

$$^{56}_{27}\text{Co} \rightarrow {}^{56}_{26}\text{Fe} + e^+ + \nu_e + \gamma. \tag{13.18}$$

This implies that as the luminosity of the supernova diminishes over time, it should be possible to detect the contribution to the light being made by $^{56}_{27}\text{Co}$.

Since radioactive decay is a statistical process, the rate of decay must be proportional to the number of atoms remaining in the sample, or

$$\frac{dN}{dt} = -\lambda N, \tag{13.19}$$

where $\lambda$ is a constant. It is left as an exercise to show that Eq. (13.19) can be integrated to give

$$N(t) = N_0 e^{-\lambda t}, \tag{13.20}$$

where $N_0$ is the original number of radioactive atoms in the sample (see Fig. 13.17), and

$$\lambda = \frac{\ln 2}{\tau_{1/2}}.$$

Since the rate at which decay energy is being deposited into the supernova remnant must be proportional to $dN/dt$, the slope of the bolometric light curve is given by

$$\frac{d \log_{10} L}{dt} = -0.434\lambda \tag{13.21}$$

or

$$\frac{dM_{\text{bol}}}{dt} = 1.086\lambda. \tag{13.22}$$

Therefore, by measuring the slope of the light curve, we can verify the presence of large quantities of a specific radioactive isotope, like $^{56}_{27}\text{Co}$.

There are now many examples of **supernova remnants**, including the Crab Nebula, located in the constellation of Taurus (Fig. 13.18a). The very

---

[24]Electrons and positrons are also known as $\beta$ particles.

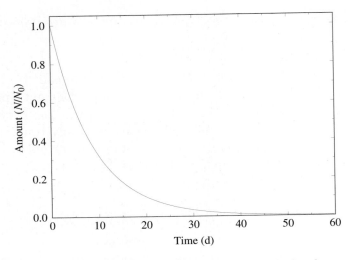

**Figure 13.17** The radioactive decay of $^{56}_{28}$Ni, with a half-life of $\tau_{1/2}=$ 6.1 days. There is a 50% chance that any given $^{56}_{28}$Ni atom will decay during a time interval of 6.1 days. If the original sample is entirely composed of $^{56}_{28}$Ni, after $n$ successive half-lives the fraction of Ni atoms remaining is $2^{-n}$.

rapid brightening that followed the explosion is believed to have been observed on July 4, 1054 A.D. by Yang Wei-T'e, a court astrologer during China's Sung dynasty. There is also evidence, based on a rock painting in Chaco Canyon, New Mexico, that the supernova was observed by Anasazi Indians as well. Furthermore, although this is the subject of some debate, it seems that Europeans also witnessed the event.

Today, more than 940 years since the supernova explosion occurred, the Crab is still expanding at a rate of nearly 1450 km s$^{-1}$ and it has a luminosity of $8 \times 10^4$ L$_\odot$. Much of the radiation being emitted is in the form of highly polarized synchrotron radiation (see Section 4.3), indicating the presence of relativistic electrons that are spiraling around magnetic field lines. The ongoing source of the electrons and the continued high luminosity so long after the explosion remained major puzzles in astronomy until the discovery of a **pulsar** (a rapidly spinning neutron star) at the center of the Crab SNR. Pulsars will be discussed extensively in Chapter 15.

A second example of a supernova remnant is shown in Fig. 13.18(b). The image is of a small portion of the 15,000 year old Cygnus Loop nebula, located 800 pc from Earth in the constellation of Cygnus. The remnant is expanding from left to right in the image, producing shock fronts several astronomical units wide as the debris from the supernova explosion encounters material in

## 13.3 The Fate of Massive Stars

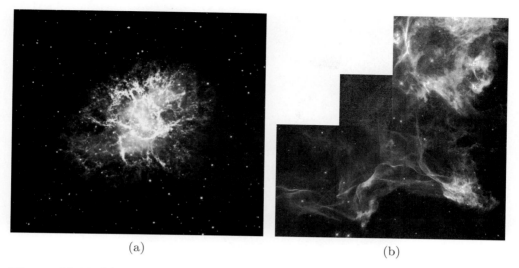

(a) (b)

**Figure 13.18** (a) The Crab supernova remnant, located 2000 pc away in the constellation of Taurus. The remnant is the result of a Type II supernova, that was observed for the first time on July 4, 1054 A.D. (Courtesy of Palomar/Caltech.) (b) An HST WF/PC 2 image of a portion of the Cygnus Loop, 800 pc away. (Courtesy of J. Hester/Arizona State University and NASA.)

the interstellar medium. The shocks excite and ionize the ISM, causing the observed emission.

On February 24.23 UT, 1987, Ian Shelton, using a 10-inch astrograph at the Las Campanas Observatory in Chile, detected SN 1987A just southwest of 30 Doradus, in the Large Magellanic Cloud; the supernova is shown in Fig. 13.19. It was the first time since the development of modern instruments that a supernova had been seen so close to Earth (the distance to the LMC is 50 kpc). The excitement of the astronomical community world-wide was immediate and intense; the chance to observe a Type II supernova from such a close vantage point provided an ideal opportunity to test our theory of the final stages of evolution of massive stars.

Almost right away it was realized that SN 1987A was unusual when compared with other, more distant Type II's that had been observed. This was most evident in the rather slow rise to maximum light (taking 80 days), which peaked only at an absolute bolometric magnitude of $-15.5$ (recall that a typical Type II reaches $-18$). The light curve through day 1444 after the outburst is shown in Fig. 13.20.

The mystery of the subluminous nature of SN 1987A was solved when the

**Figure 13.19** A portion of the Large Magellanic Cloud showing SN 1987A on the lower right-hand side of the photograph. The Tarantula Nebula, an immense H II region, appears on the left-hand side of the photograph. (Courtesy of the European Southern Observatory, ©ESO.)

identity of its progenitor was established. The star that blew up was a fairly well-studied 12th-magnitude *blue* supergiant (spectral class B3 I), known as Sanduleak −69 202.[25] Since what exploded was a much smaller blue supergiant, rather than a red supergiant (as is usually assumed to be the case), the star was more dense. As a result, before the thermal energy produced by the shock could diffuse out and escape as light, it was converted into the mechanical energy required to lift the envelope of the star out of the deeper potential well of a blue supergiant. Measurements of $H_\alpha$ lines indicate that some of the outer hydrogen envelope was ejected at speeds near 30,000 km s$^{-1}$, or $0.1c$!

The decay of the 0.075 M$_\odot$ of $^{56}_{28}$Ni that was produced by the shock occurred while the time scale required for energy to be radiated away was still quite long. Consequently, the added decay energy produced a bump on the light curve near maximum light. By the time the resulting $^{56}_{27}$Co began to decay, this diffusion time scale had become sufficiently short that the decrease in the luminosity of the remnant began to track closely the rate of decay of cobalt-56. Subsequently, the next important radioactive isotope, $^{57}_{27}$Co, began to play an important role in the development of the light curve. The expected contributions of the various

---

[25]Sk −69 202 gets its name from being the 202nd entry in the −69° declination band of the Sanduleak catalog of stars in the Magellanic Clouds.

## 13.3 The Fate of Massive Stars

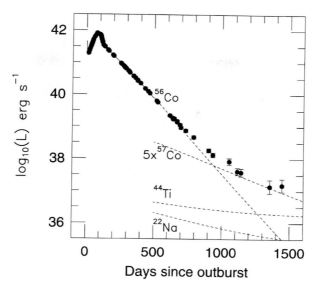

**Figure 13.20** The bolometric light curve of SN 1987A through the first 1444 days after the explosion. The dashed lines show the contributions expected from the radioactive isotopes produced by the shock wave. The initial masses are estimated to be $^{56}_{28}$Ni (and later $^{56}_{27}$Co), 0.075 M$_\odot$; $^{57}_{27}$Co, 0.009 M$_\odot$ (five times the solar abundance); $^{44}_{22}$Ti, $1 \times 10^{-4}$ M$_\odot$; and $^{22}_{11}$Na, $2 \times 10^{-6}$ M$_\odot$. (Figure from Suntzeff et al., *Ap. J. Lett.*, *384*, L33, 1992.)

radioactive isotopes to the light curve of SN 1987A are shown in Fig. 13.20, the slopes in the light curve being related to the half-lives of the isotopes through Eq. (13.21).

For the first time, astronomers were able to directly measure the x-ray and gamma-ray emission lines produced by radioactive decay. In particular, the 847 keV and 1238 keV lines of $^{56}_{27}$Co were detected by a number of experiments, confirming the presence of this isotope. Doppler shift measurements indicate that the heavier isotopes in the remnant are expanding at several thousand kilometers per second.

The available observations of Sk −69 202, together with theoretical evolutionary models, suggest that the progenitor of SN 1987A had a mass of around 20 M$_\odot$ when it was on the main sequence and that it lost perhaps a few solar masses before its iron core collapsed (estimated to be between 1.4 and 1.6 M$_\odot$). Although it was apparently a red supergiant for between several hundred thousand and one million years, it evolved to the blue just 40,000 years before the explosion. Supporting this hypothesis is the observation that hydrogen was more abundant in the envelope of Sk −69 202 than was helium, suggesting

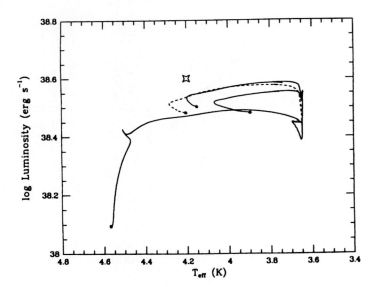

**Figure 13.21** Three theoretical evolutionary tracks of a star with an initial main-sequence mass of 18 $M_\odot$. The leftmost and rightmost solid lines represent stars with final masses of 16.2 $M_\odot$ and 14.7 $M_\odot$, respectively. The evolutionary track represented by the dashed line incorporated artificial convective mixing of helium up into the envelope. The position of Sk −69 202 is indicated by the star. (Figure from Arnett et al., *Annu. Rev. Astron. Astrophys.*, *27*, 629, 1989. Reproduced with permission from the *Annual Review of Astronomy and Astrophysics*, Volume 27, ©1989 by Annual Reviews Inc.)

that the star had not suffered extensive amounts of mass loss. Whether and when a massive star evolves from being a red supergiant to a blue supergiant before exploding depends sensitively on the mass of the star (it cannot be much more than about 20 $M_\odot$), its composition (it must be metal poor, as are the stars of the LMC), the rate of mass loss (which must be low), and the treatment of convection (always a major uncertainty in theoretical stellar models). A set of theoretical evolutionary tracks for a star with an initial mass of 18 $M_\odot$ are shown in Fig. 13.21.

Although mass loss prior to the explosion could not have been excessive, the progenitor did lose some mass. The Hubble Space Telescope has recorded three rings around SN 1987A (Fig. 13.22). The innermost ring measures 0.42 pc in diameter and lies in a plane that contains the center of the supernova explosion. It glows in visible light due to emissions from O III energized by radiation from the supernova and appears elongated because it is inclined relative to our line

## 13.3 The Fate of Massive Stars

**Figure 13.22** Rings around SN 1987A, detected by the Hubble Space Telescope's WF/PC 2. The diameter of the inner ring is 0.42 pc. (Courtesy of Dr. Christopher Burrows, ESA/STScI and NASA.)

of sight. The material making up the central ring was ejected by a stellar wind prior to the explosion of SN 1987A.

The two larger rings are not in planes containing the central explosion but lie in front of and behind the star. One explanation for these fascinating and unexpected features is that Sk −69 202 resided near a companion star, possibly a neutron star or a black hole (see Chapters 15 and 16, respectively). As this companion source wobbles, narrow jets of radiation from the source "paint" the rings on an hourglass-shaped, bipolar distribution of mass that was ejected from Sk −69 202. It is in the denser equatorial plane of the bipolar mass distribution that the central ring is located. In support of this hypothesis, researchers believe that they may have identified the source of these beams of radiation about 0.1 pc from the center of the supernova explosion, consistent with the fact that the larger rings appear to be offset from the explosion's center. Opponents of this model believe that the explanation is simply too complicated since it requires two sources of high-energy radiation, one to explain the central ring and another to explain the larger ones.

In the summer of 1990, fluctuating radio emissions were finally detected from the supernova. Although radio wavelength energy was detected during the first days following the explosion, SN 1987A had remained radio quiet since that time. Apparently the shock wave, still propagating outward at a speed close to $0.1c$, collided with clumps of material lost from Sk −69 202 prior to the supernova event. At that speed, the shock wave should encounter the central ring of gas sometime between 1997 and 2004, causing it to glow more brightly.

Arguably the most exciting early observations of SN1987A were based on its neutrinos, representing the first time that neutrinos had been detected from an astronomical source other than the Sun. The measurement of the neutrino burst confirmed the basic theory of Type II supernovae and amounts to our "seeing" the formation of a neutron star out of the collapsed iron core.

The arrival of the neutrino burst was recorded over a period of $12\frac{1}{2}$ seconds, beginning at February 23.316 UT, 1987, three hours *before* the arrival of the photons at February 23.443 UT. Twelve events were recorded at Japan's Kamiokande II Cerenkov detector, and at the same time, eight events were detected by the underground IMB[26] Cerenkov detector near Fairport, Ohio.[27] Figure 13.23 shows the early light curve of SN 1987A, together with the time marks of the neutrino detections. Assuming that the exploding star became optically thin to neutrinos before the shock wave reached the surface, and assuming further that the neutrinos traveled faster than the shock while still inside the star, the neutrinos began their trip to Earth ahead of the photons. Given the fact that the neutrinos arrived ahead of the light, their velocity through space must have been very near the speed of light (within one part in $10^8$). This observation, together with the absence of any significant dispersion in the arrival time of neutrinos of different energies (i.e., higher-energy neutrinos did not arrive any earlier than lower-energy ones), suggests that the rest mass of electron neutrinos must be quite small. The upper limit on the electron neutrino, based on data from SN 1987A, is $m_e \leq 16$ eV, consistent with the results of laboratory experiments that place the upper limit at 7.2 eV. Although SN 1987A has presented some interesting twists in our study of stellar evolution, it has also confirmed or clarified important aspects of the theory.

---

[26] IMB stands for the consortium that operates the observatory: University of California at Irvine, University of Michigan, and Brookhaven National Laboratory.

[27] Two other observatories made claims that they detected the arrival of neutrinos from SN 1987A as well: the liquid scintillator detector in Mont Blanc and the Cherenkov Baksan detector in the Caucasus mountains. The Mont Blanc experiment recorded five events about 4.7 hours before the simultaneous detections of Kamiokande II and IMB, while the five events reported by the Baksan experiment were roughly simultaneous with Kamiokande II and IMB. It is now believed that the results of both the Mont Blanc and Baksan detectors are unreliable because the energies of the neutrinos were very close to the thresholds of the experiments, leading to confusion with background events. Furthermore, the Mont Blanc arrival time was much too early to be reconcilable with the other measurements. The famous Davis solar neutrino detector (discussed extensively in Section 11.1) did not measure any neutrinos from SN 1987A; the solar neutrino background was much *larger* than the neutrino count from the supernova!

## 13.3 The Fate of Massive Stars

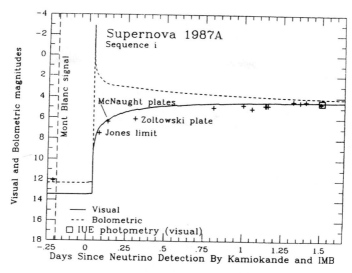

**Figure 13.23** The early light curve of SN 1987A, showing the elapsed time since the detection of neutrinos by Kamiokande II and IMB. The solid line is the apparent visual magnitude and the dashed line is the estimated apparent bolometric magnitude of the supernova. The signal recorded by the Mont Blanc neutrino detector is also included, although it is inconsistent with the other neutrino measurements. (Figure from Arnett et al., *Annu. Rev. Astron. Astrophys.*, *27*, 629, 1989. Reproduced with permission from the *Annual Review of Astronomy and Astrophysics*, Volume 27, ©1989 by Annual Reviews Inc.)

Another important success of stellar evolution theory is the ability to explain most of the observed abundance ratios of the elements. For instance, the chemical composition of the Sun's photosphere is shown in Fig. 13.24, with all values normalized to $10^{12}$ for hydrogen. (A portion of this data was also presented in Table 9.2.) By far the most abundant element in the universe is hydrogen, with helium being less abundant by about a factor of ten. It is believed that hydrogen is primordial, having been synthesized immediately following the Big Bang that began the universe. Much of the present-day helium was also produced directly from the Big Bang, while the remainder was generated from hydrogen burning in stellar interiors.

Relative to hydrogen and helium, lithium, beryllium, and boron are very underabundant. There are two reasons for this: They are not prominent endproducts of nuclear reaction chains, and they can be destroyed by collisions with protons. For lithium this occurs at temperatures greater than about $2.7 \times 10^6$ K, while for beryllium the required temperature is $3.5 \times 10^6$ K. It is the

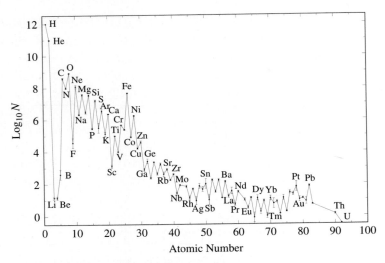

**Figure 13.24** The relative abundances of elements in the Sun's photosphere. All abundances are normalized relative to $10^{12}$ hydrogen atoms. (Data from Grevesse and Anders, *Solar Interior and Atmosphere*, Cox, Livingston, and Matthews (eds.), University of Arizona Press, Tucson, 1991.)

Sun's surface convection zone that is responsible for transporting the surface lithium, beryllium, and boron into the interior. When the present-day solar composition is compared with the abundances of meteorites (which should be similar to the Sun's primordial composition), we find that the relative abundances of beryllium are comparable but that the Sun's surface composition of lithium is smaller than the meteorites' lithium abundance by a factor of about 100. This suggests that lithium has been destroyed in the Sun since the star's formation, but that beryllium has not been appreciably depleted. Apparently the base of the solar convection zone extends down sufficiently far to burn lithium but not far enough to burn beryllium. However, combining stellar structure theory, including the mixing-length theory of convection, with the analysis of solar oscillations (see Section 14.5) indicates that the base of the convection zone extends down to $2.3 \times 10^6$ K, not far enough to burn lithium adequately. The disagreement of standard models with the observations is known as the **solar lithium problem**.[28]

---

[28] Perhaps because of their momenta, descending convective bubbles *overshoot* the bottom of the zone of convective instability, causing lithium to be transported deeper than the standard models suggest. Additional effects may also exist due to diffusion and the interaction of convection with rotation.

## 13.3 The Fate of Massive Stars

Peaks occur in Fig. 13.24 for elements such as carbon, nitrogen, oxygen, neon, and so on because they are created as a consequence of a star's evolutionary march toward the iron peak. Type II supernovae are also responsible for the generation of significant quantities of oxygen, while Type I supernovae are responsible for the creation of most of the iron observed in the cosmos.

When nuclei having progressively higher values of $Z$ (the number of protons) form via stellar nucleosynthesis, it becomes increasingly difficult for other charged particles such as protons, alpha particles, and so on, to react with them. The cause is the existence of a high Coulomb potential barrier. However, the same limitation does not exist when neutrons collide with these nuclei. Consequently, nuclear reactions involving neutrons can occur even at relatively low temperatures, assuming of course that free neutrons are present in the gas. The reactions with neutrons

$$^{A}_{Z}X + n \rightarrow {}^{A+1}_{Z}X + \gamma$$

result in more massive nuclei that are either stable or unstable against the beta-decay reaction,

$$^{A+1}_{Z}X \rightarrow {}^{A+1}_{Z+1}X + e^- + \overline{\nu}_e + \gamma.$$

If the beta-decay half-life is short compared to the time scale for neutron capture, the neutron-capture reaction is said to be a *slow process* or an **s-process** reaction. s-process reactions tend to yield stable nuclei, either directly or secondarily via beta decay. On the other hand, if the half-life for the beta-decay reaction is long compared with the time scale for neutron capture, the neutron-capture reaction is termed a *rapid process* or **r-process** and results in neutron-rich nuclei. s-process reactions tend to occur in normal phases of stellar evolution, while r-processes can occur during a supernova when a large flux of neutrinos exists. Although neither process plays significant roles in energy production, they do account for the abundance ratios of nuclei with $A > 60$.

Despite the successes of stellar evolution theory, interesting and complicated problems remain to be solved. Particularly intriguing are the **Wolf–Rayet stars** (WR). These very massive stars are probably post-main-sequence objects, although they are found to be near the main sequence and are losing mass at very high rates, as evidenced by their P Cygni profiles. They also frequently exhibit strong emission lines of nitrogen, oxygen, or carbon (WN, WO, or WC stars, respectively). Moreover, only very weak hydrogen lines appear in their spectra (if they are seen at all). Current research suggests that mass loss can strip these stars of their envelopes, exposing nuclear processed material.

**Figure 13.25** $\eta$ Carinae is estimated to have a mass of 150 $M_\odot$ and is rapidly losing mass. This image was obtained by WF/PC 2 on board the Hubble Space Telescope. (Courtesy of J. Hester/Arizona State University and NASA.)

Another curious example is $\eta$ Carinae (Fig. 13.25), a seventh magnitude star believed to have a mass of roughly 150 $M_\odot$ and a luminosity of $6 \times 10^6$ $L_\odot$. This star is losing mass at the prodigious rate of $3 \times 10^{-4}$ $M_\odot$ yr$^{-1}$ with wind speeds on the order of 450 km s$^{-1}$. Prior to the 1830s, $\eta$ Carinae was a second magnitude star before brightening significantly to $-1$ mag, making it the second brightest star in the sky. At the time of brightening, it underwent a major episode of mass loss, ejecting roughly 3 $M_\odot$ of very dusty and metal-rich material. Its lower visual magnitude today is due to the shielding caused by the dust, which absorbs radiation from the star and re-emits the energy in the infrared.

Numerous models have been proposed to explain $\eta$ Carinae, including its being a massive pre-main-sequence star, a slow supernova, a binary or multiple star system, or a post-main-sequence object! The most likely explanation seems to be that $\eta$ Carinae is an atypical example of a **luminous blue variable** (LBV), a post-main-sequence object in a stage of evolution just prior to becoming a Wolf–Rayet star. If so, questions remain regarding how this star was able to form with such a high mass and what is driving its enormous mass loss rate. Given its strange behavior, it does appear that $\eta$ Carinae is at the edge of stability, as expected of such high-mass stars.

In the next section we will look at yet another important test of stellar evolution, the stellar clusters.

## 13.4 Stellar Clusters

Over the past two chapters we have seen a story develop that depicts the lives of stars. They are formed from the ISM, only to return most of that material to the ISM through stellar winds, by the ejection of planetary nebulae, or via supernova explosions. The matter that is given back, however, has been enriched with heavier elements that were produced through the various sequences of nuclear reactions governing a star's life. As a result, when the next generation of stars is formed, it possesses higher concentrations of these heavy elements than did its ancestors. This cyclic process of star formation, death, and rebirth is evident in the variations in composition between stars.

It is generally believed that the universe began with the Big Bang some 10 to 20 billion years ago and that hydrogen and helium were essentially the only elements produced by the nucleosynthesis that occurred during the initial fireball. Consequently, the first stars to form were extremely *metal poor*, having very low values for $Z$. Each succeeding generation of star production resulted in higher and higher proportions of heavier elements, leading to *metal-rich* stars for which $Z \sim 0.03$. Metal-poor stars are referred to as **Population II** while metal-rich stars are called **Population I**.

The classifications of Population II and Population I are due originally to their identifications with kinematically distinct groups of stars within our Galaxy. Population I stars have velocities relative to the Sun that are low compared to Population II stars. Furthermore, Population I stars are found predominantly in the disk of the Milky Way, while Population II stars can be found well above or below the disk. It was only later that astronomers realized that these two groups of stars differed chemically as well. Not only do populations tell us something about evolution, but the kinematic characteristics, positions, and compositions of Population I and Population II stars also provide us with a great deal of information about the formation and evolution of the Milky Way Galaxy.

Recall from Section 12.2 that during the collapse of a molecular cloud, a process of cascading fragmentation results. This leads to the creation of **stellar clusters**, ranging in size from tens of stars to hundreds of thousands of stars. Every member of a given cluster is formed from the same cloud, at the same time, and all with essentially identical compositions. Thus, excluding such effects as rotation, magnetic fields, or membership in a binary star system, the Vogt–Russell theorem suggests that the differences in evolutionary states between the various stars in the cluster are due solely to their initial masses.

Extreme Population II clusters formed when the Galaxy was very young,

**Figure 13.26** (a) M13, the great globular cluster in Hercules, is located approximately 7000 pc from Earth. (From the Digitized Sky Survey at STScI. Courtesy of Palomar/Caltech, the National Geographic Society, and the Space Telescope Science Institute.) (b) The Pleiades is a galactic cluster found in the constellation of Taurus, at a distance of 130 pc. (Courtesy of the National Optical Astronomical Observatories.)

making them some of the oldest objects in the Milky Way. They also contain the largest number of members. Figure 13.26(a) shows M13, one such **globular cluster**, located in the constellation of Hercules. Population I clusters, like the Pleiades (Fig. 13.26b), tend to be smaller and younger. These smaller clusters are called **galactic clusters**, also known as **open clusters**.

The H–R diagrams of clusters can be constructed in a self-consistent way without knowledge of the exact distances to them. Since the dimensions of a typical cluster are small relative to its distance from Earth, little error is introduced by assuming that each member of the cluster has the same distance modulus. As a result, plotting the apparent magnitude, rather than the absolute magnitude only amounts to shifting the position of each star in the diagram vertically by the same amount. By matching the observational main sequence of the cluster to a main sequence calibrated in absolute magnitude, the distance modulus of the cluster can be determined, giving the cluster's distance from the observer. This method of distance determination is known as **main-sequence fitting**.

Rather than attempting to determine the effective temperatures of every member of a cluster by undertaking a detailed spectral line analysis of each star (which would be a major project for a globular cluster, even assuming that

## 13.4 Stellar Clusters

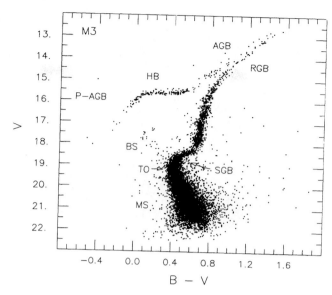

**Figure 13.27** A color–magnitude diagram for M3, an old globular cluster. The major phases of stellar evolution are indicated: main sequence (MS); blue stragglers (BS); the main-sequence turn-off point (TO); the subgiant branch of hydrogen shell burning (SGB); the red giant branch along the Hayashi track, prior to helium core burning (RGB); the horizontal branch during helium core burning (HB); the asymptotic giant branch during hydrogen and helium shell burning (AGB); post-AGB evolution proceeding to the white dwarf phase (P-AGB). (Figure from Renzini and Fusi Pecci, *Annu. Rev. Astron. Astrophys.*, **26**, 199, 1988. Reproduced with permission from the *Annual Review of Astronomy and Astrophysics*, Volume 26, ©1988 by Annual Reviews Inc.)

the stars are bright enough to get good spectra), it is much faster to determine their color indices $(B-V)$. With knowledge of the apparent magnitude and the color index of each star, a **color–magnitude diagram** can be constructed. Color–magnitude diagrams for M3 (a globular cluster) and h and $\chi$ Persei (a *double* galactic cluster) are shown in Figs. 13.27 and 13.28, respectively.

Clusters, and their associated color–magnitude diagrams, offer nearly ideal tests of many aspects of stellar evolution theory. By computing the evolutionary tracks of stars of various masses, all having the same composition as the cluster, it is possible to plot the position of each model on the H–R diagram when the models reach the age of the cluster (the curve connecting these positions is known as an **isochrone**). The relative number of stars at each location on the isochrone depends on the number of stars in each mass range

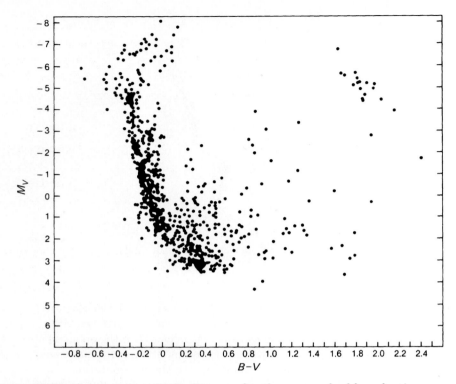

**Figure 13.28** A color–magnitude diagram for the young double galactic cluster, h and χ Persei. Note that the most massive stars are pulling away from the main sequence while the low-mass stars in the middle of the diagram are still contracting onto the main sequence. Red giants are present in the upper right-hand corner of the diagram. (Figure adapted from Wildey, *Ap. J. Suppl.*, *8*, 439, 1964.)

within the cluster (the initial mass function; see Fig. 12.9), combined with the different rates of evolution during each phase. Therefore, star counts in a color–magnitude diagram can shed light on the time scales involved in stellar evolution.

As the theoretical cluster "ages," beginning with the initial collapse of the molecular cloud, the most massive and least abundant stars will arrive on the main sequence first, evolving rapidly. Before the lowest-mass stars have even reached the main sequence, the most massive ones have already evolved into the red giant region, perhaps even undergoing supernova explosions. These disparate rates of evolution can be seen by comparing Figs. 12.8 and 13.1 for pre-main-sequence and post-main-sequence evolution, respectively, together with their associated tables.

## 13.4 Stellar Clusters

The color–magnitude diagram of h and $\chi$ Persei illustrates such a situation (Fig. 13.28); apparent are red giants, together with low-mass pre-main-sequence stars. Also evident in the diagram is the complete absence of stars between the massive ones that are just leaving the main sequence and the few in the red giant region. It is unlikely that this represents an incomplete survey, since these stars are the brightest members of the cluster. Rather, it points out the very rapid evolution that occurs just after leaving the main sequence. This feature, known as the **Hertzsprung gap**, is a common characteristic of the color–magnitude diagrams of young, galactic clusters. The existence of the Hertzsprung gap is due to evolution on a Kelvin–Helmholtz time scale, following the point when the hydrogen-depleted core exceeds the Schönberg–Chandrasekhar limit.

Since core hydrogen-burning lifetimes are inversely related to mass, continued evolution of the cluster means that the main-sequence **turn-off point**, defined as the point where stars in the cluster are currently leaving the main sequence, becomes redder and less luminous with time. It then becomes possible to estimate the age of a cluster by the location of the uppermost point of its main sequence. A composite color–magnitude diagram of a number of clusters is shown in Fig. 13.29. Labeled vertically on the right-hand side is the age of the cluster corresponding to the location of the main-sequence turn-off point.

Notice in Fig. 13.29 that the cluster M67 does not show the existence of the Hertzsprung gap; the same can be said of M3 (Fig. 13.27). Recall that below about 1.25 $M_\odot$, the rapid contraction phase related to the Schönberg–Chandrasekhar limit is much less pronounced. As a result, color–magnitude diagrams of old globular clusters with turn-off points near 1 $M_\odot$ have continuous distributions of stars leading to the red giant region.

Close inspection of Fig. 13.27 shows that a relatively small number of stars exists on the asymptotic giant branch and only a few stars are to be found in the region labeled P-AGB (post-asymptotic giant branch). This is just a consequence of the very rapid pace of evolution during this phase of heavy mass loss that leads directly to the formation of white dwarfs.

It should be pointed out that a group of stars, known as **blue stragglers**, can be found above the turn-off point of M3. Although our understanding of these stars is incomplete, it appears that their tardiness in leaving the main sequence is due to some unusual aspect of their evolution. The most likely scenarios to date appear to be mass exchange with a binary star companion,[29] some process of internal chemical mixing that has provided more hydrogen fuel

---

[29]Mass exchange between close binaries is the subject of Chapter 17.

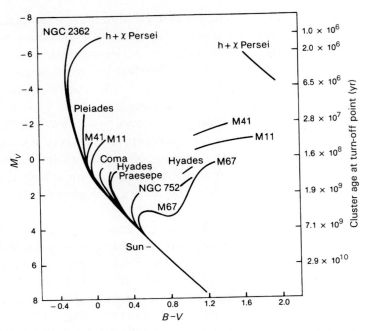

**Figure 13.29** A composite color–magnitude diagram for a set of Population I galactic clusters. The absolute visual magnitude is indicated on the left-hand vertical axis and the age of the cluster, based on the location of its turn-off point, is labeled on the right-hand side. (Figure adapted from an original diagram by A. Sandage.)

in the core, extending the star's main-sequence lifetime, or a merger with a close companion.

The successful comparisons between theory and observation that are provided by stellar clusters give strong support to the idea that our picture of stellar evolution is fairly complete, although perhaps in need of some fine-tuning. Continued refinements in stellar opacities, revisions in nuclear reaction cross sections, or much-needed improvements in the treatment of convection will likely lead to even better agreement with observations. However, much fundamental work remains to be done as well, such as developing a better understanding of the effects of mass loss, rotation, magnetic fields, and the presence of a close companion.

# Suggested Readings

## General

Balick, B., et al., "The Shaping of Planetary Nebulae," *Sky and Telescope*, February 1987.

Harpaz, Amos, "The Formation of a Planetary Nebula," *The Physics Teacher*, May 1991.

Kaler, J., "Planetary Nebulae and Stellar Evolution," *Mercury*, July/August 1981.

Kippenhahn, Rudolf, *100 Billion Suns*, Basic Books, New York, 1983.

Lattimer, J., and Burrows, A., "Neutrinos from Supernova 1987A," *Sky and Telescope*, October 1988.

Meadows, A. J., *Stellar Evolution*, Second Edition, Pergamon Press, Oxford, 1978.

Payne-Gaposchkin, Cecilia, *Stars and Clusters*, Harvard University Press, Cambridge, MA, 1979.

Woosley, S., and Weaver, T., "The Great Supernova of 1987," *Scientific American*, August 1989.

## Technical

Aller, Lawrence H., *Atoms, Stars, and Nebulae*, Third Edition, Cambridge University Press, Cambridge, 1991.

Arnett, W. David, Bahcall, John N., Kirshner, Robert P., and Woosley, Stanford E., "Supernova 1987A," *Annual Review of Astronomy and Astrophysics*, 27, 629, 1989.

Bowen, G. H., and Willson, L. A., "From Wind to Superwind: The Evolution of Mass-Loss Rates for Mira Models," *The Astrophysical Journal Letters*, 375, L53, 1991.

Chiosi, Cesare, Bertelli, Gianpaolo, and Bressan, Alessandro, "New Developments in Understanding the H–R Diagram," *Annual Review of Astronomy and Astrophysics*, 30, 235, 1992.

Chiosi, Cesare, and Maeder, André, "The Evolution of Massive Stars with Mass Loss," *Annual Review of Astronomy and Astrophysics*, 24, 329, 1986.

Dupree, A. K., "Mass Loss From Cool Stars," *Annual Review of Astronomy and Astrophysics*, *24*, 377, 1986.

Hanes, Dave, and Madore, Barry (eds.), *Globular Clusters*, Cambridge University Press, Cambridge, 1980.

Hansen, C. J., and Kawaler, S. D., *Stellar Interiors: Physical Principles, Structure, and Evolution*, Springer-Verlag, New York, 1994.

Harpaz, Amos, *Stellar Evolution*, A K Peters, Wellesley, MA, 1994.

Iben, Icko Jr., "Stellar Evolution Within and Off the Main Sequence," *Annual Review of Astronomy and Astrophysics*, *5*, 571, 1967.

Iben, Icko Jr., "Single and Binary Star Evolution," *The Astrophysical Journal Supplement Series*, *76*, 55, 1991.

Iben, Icko Jr., and Renzini, Alvio, "Asymptotic Giant Branch Evolution and Beyond," *Annual Review of Astronomy and Astrophysics*, *21*, 271, 1983.

Kippenhahn, Rudolf, and Weigert, Alfred, *Stellar Structure and Evolution*, Springer-Verlag, Berlin, 1990.

Schaerer, D., et al., "Grids of Stellar Models. IV. From 0.8 to 120 $M_\odot$ at Z=0.040," *Astron. Astrophys. Suppl.*, *102*, 339, 1993.

Shklovskii, Iosif S., *Stars: Their Birth, Life, and Death*, W. H. Freeman and Company, San Francisco, 1978.

Suntzeff, Nicholas B., et al., "The Energy Sources Powering the Late-Type Bolometric Evolution of SN 1987A," *The Astrophysical Journal Letters*, *384*, L33, 1992.

Woosley, S. E., and Weaver, Thomas A., "The Physics of Supernova Explosions," *Annual Review of Astronomy and Astrophysics*, *24*, 205, 1986.

# Problems

13.1 (a) Beginning with Eq. (13.7), show that the radius of the isothermal core for which the gas pressure is a maximum is given by Eq. (13.8). Recall that this solution assumes that the gas in the core is ideal and monatomic.

(b) From your results in part (a), show that the maximum pressure at the surface of the isothermal core is given by Eq. (13.9).

## Problems

**13.2** During the first dredge-up phase of a 5 $M_\odot$ star, would you expect the composition ratio $X'_{13}/X_{12}$ to increase or decrease? Explain your reasoning. *Hint:* You may find Fig. 13.6 helpful.

**13.3** Use Eq. (10.33) to show that the ignition of the triple alpha process at the tip of the red giant branch ought to occur at more than $10^8$ K.

**13.4** In an attempt to identify the important components of AGB mass loss, various researchers have proposed parameterizations of the mass loss rate that are based on fitting observed rates for a specified set of stars with some general equation that includes measurable quantities associated with the stars in the sample. One of the most popular, developed by D. Reimers, is given by

$$\dot{M} = -4 \times 10^{-13} \eta \frac{L}{gR} \; M_\odot \; \text{yr}^{-1}, \quad (13.23)$$

where $L$, $g$, and $R$ are the luminosity, surface gravity, and radius of the star, respectively (all in solar units; $g_\odot = 2.74 \times 10^4$ cm s$^{-2}$). $\eta$ is a *free parameter* whose value is expected to be near unity. Note that the minus sign has been explicitly included here, indicating that the mass of the star is decreasing.

(a) Explain qualitatively why $L$, $g$, and $R$ enter Eq. (13.23) in the way they do.

(b) Estimate the mass loss rate of a 1 $M_\odot$ AGB star that has a luminosity of 7000 $L_\odot$ and a temperature of 3000 K.

**13.5** (a) Show that the Reimers mass loss rate, given by Eq. (13.23) in Problem 13.4, can also be written in the form

$$\dot{M} = -4 \times 10^{-13} \eta \frac{LR}{M},$$

where $L$, $R$, and $M$ are all in solar units.

(b) Assuming (incorrectly) that $L$, $R$, and $\eta$ do not change with time, derive an expression for the mass of the star as a function of time. Let $M = M_0$ when the mass loss phase begins.

(c) Using $L = 7000$ $L_\odot$, $R = 310$ $R_\odot$, $M_0 = 1$ $M_\odot$, and $\eta = 1$, make a graph of the star's mass as a function of time.

(d) How long would it take for a star with an initial mass of 1 $M_\odot$ to be reduced to the mass of the degenerate carbon–oxygen core (0.6 $M_\odot$)?

13.6 The Helix Nebula is a planetary nebula with an angular diameter of 15' that is located approximately 120 pc from Earth.

(a) Calculate the diameter of the nebula.

(b) Assuming that the nebula is expanding away from the central star at a constant velocity of 20 km s$^{-1}$, estimate its age.

13.7 Using Eq. (12.16), make a crude estimate of the amount of time required for the homologous collapse of the inner portion of the iron core of a massive star, marking the beginning of a Type II supernova.

13.8 (a) Show that the amount of radioactive material remaining in an initially pure sample is given by Eq. (13.20).

(b) Prove that
$$\lambda = \frac{\ln 2}{\tau_{1/2}}.$$

13.9 (a) The angular size of the Crab SNR is $4' \times 2'$ and its distance from Earth is approximately 2000 pc (see Fig. 13.18a). Estimate the linear dimensions of the nebula.

(b) Using the measured expansion rate of the Crab and ignoring any accelerations since the time of the supernova explosion, estimate the age of the nebula.

13.10 Taking the distance to the Crab to be 2000 pc, and assuming that the absolute bolometric magnitude at maximum brightness was characteristic of a Type II supernova, estimate its peak apparent magnitude. Compare this to the maximum brightness of the planet Venus ($m \simeq -4$), which is sometimes visible in the daytime.

13.11 (a) Assuming that the light curve of a supernova is dominated by the energy released in the radioactive decay of an isotope that has a decay constant of $\lambda$, show that the slope of the light curve is given by Eq. (13.21).

(b) Prove that Eq. (13.22) follows from Eq. (13.21).

13.12 The energy released during the decay of one $^{56}_{27}$Co atom is 3.72 MeV. If 0.075 $M_\odot$ of cobalt was produced by the decay of $^{56}_{28}$Ni following the explosion of SN 1987A, estimate the amount of energy released per second through the radioactive decay of cobalt

(a) just after the formation of the cobalt.

(b) one year after the explosion.

(c) Compare your answers with the light curve of SN 1987A given in Fig. 13.20.

13.13 The neutrino flux from SN 1987A was estimated to be $1.3 \times 10^{10}$ cm$^{-2}$ at the location of Earth. If the average energy per neutrino was approximately 4.2 MeV, estimate the amount of energy released via neutrinos during the supernova explosion.

13.14 Using Eq. (10.28), estimate the gravitational binding energy of a neutron star with a mass 1.4 M$_\odot$ and a radius of 10 km. Compare your answer with the amount of energy released in neutrinos during the collapse of the iron core of Sk −69 202 (the progenitor of SN 1987A).

13.15 Estimate the Eddington limit for $\eta$ Car and compare your answer with the luminosity of that star. Is your answer consistent with its behavior? Why or why not?

13.16 An old version of stellar evolution, popular at the beginning of the twentieth century, maintained that stars begin their lives as large, cool spheres of gas, like the giant stars on the H–R diagram. They then contract and heat up under the pull of their own gravity to become hot, bright blue O stars. For the remainder of their lives they lose energy, becoming dimmer and redder with age. As they slowly move down the main sequence, they eventually end up as cool, dim red M stars. Explain how observations of stellar clusters, plotted on an H–R diagram, contradict this idea.

13.17 (a) Show that $\log_{10}(L_V/L_B)$ + constant is, to within a constant, equivalent to the color index, $B - V$.

(b) Estimating best-fit curves through the data given in Fig. 13.30 on the next page, trace the two color–magnitude diagrams, placing them on a single diagram. Note that the abscissas have been normalized so that the lowest-luminosity stars of both clusters are located at the same positions on their respective diagrams.

(c) Given that 47 Tuc is relatively metal-rich for a globular cluster ($Z/Z_\odot = 0.17$, where $Z_\odot$ is the solar value) and M15 is metal-poor ($Z/Z_\odot = 0.0060$), explain the difference in colors between the two clusters. *Hint:* You may wish to refer back to the discussion in Example 9.10 (Section 9.4).

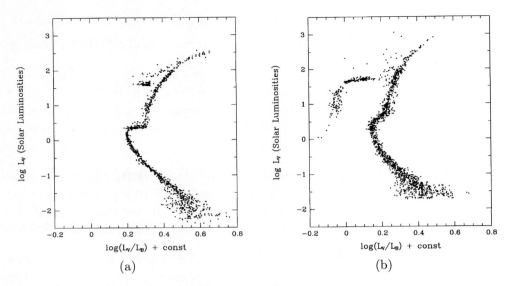

**Figure 13.30** (a) A color–magnitude diagram for 47 Tuc, a relatively metal-rich globular cluster with $Z/Z_\odot = 0.17$. (Data from Hesser et al., *Publ. Astron. Soc. Pac.*, *99*, 739, 1987; figure courtesy of William E. Harris.) (b) A color–magnitude diagram for M15, a metal-poor globular cluster with $Z/Z_\odot = 0.0060$. (Data from Durrell and Harris, *Astron. J.*, *105*, 1420, 1993; figure courtesy of William E. Harris.)

13.18  Using the technique of main-sequence fitting, estimate the distance to M3; refer to Figs. 13.27 and 13.29.

# Chapter 14

# STELLAR PULSATION

## 14.1 Observations of Pulsating Stars

In August of 1595, a Lutheran pastor and amateur astronomer named David Fabricius observed the star $o$ Ceti. As he watched over a period of months, the brightness of this second magnitude star in the constellation Cetus (the Sea Monster) slowly faded. By October, the star had vanished from the sky. Several more months passed as the star eventually recovered and returned to its former brilliance. In honor of this miraculous event, $o$ Ceti was named Mira, meaning "wonderful."

Mira continued its rhythmic dimming and brightening, and by 1660 the 11-month period of its cycle was established. The regular changes in brightness were mistakenly attributed to dark "blotches" on the surface of a rotating star. Supposedly, Mira would appear fainter when these dark areas were turned toward Earth.

Figure 14.1 shows the *light curve* of Mira for a 42-year interval. Today astronomers recognize that the changes in Mira's brightness are due not to dark spots on its surface, but to the fact that Mira is a **pulsating star**, a star that dims and brightens as its surface expands and contracts. Mira is the prototype of the **long-period variables**, stars that have somewhat irregular light curves and pulsation periods between 100 and 700 days.

Nearly two centuries elapsed before another pulsating star was discovered. In 1784 John Goodricke of York, England, found that the brightness of the star $\delta$ Cephei varies regularly with a period of 5 days, 8 hours, 48 minutes. This discovery cost Goodricke his life; he contracted pneumonia while observing $\delta$ Cephei and died at the age of 21. The light curve of $\delta$ Cephei, shown in Fig. 14.5, is less spectacular than that of $o$ Ceti. It varies by less than one

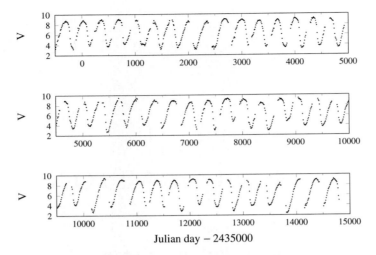

**Figure 14.1** The light curve of Mira from 1953 to March 1995. The time is measured in **Julian days**; Julian day 0 started at noon (UT) on January 1, 4713 B.C., and JD 2435000 is September 14, 1954. (Courtesy of Janet A. Mattei, AAVSO Director.)

magnitude in brightness and never fades from view. Nevertheless, pulsating stars similar to $\delta$ Cephei, called **classical Cepheids**, are vitally important to astronomy.

Today, some 20,000 pulsating stars have been cataloged by astronomers. One woman, Henrietta Swan Leavitt (1868–1921; see Fig. 14.2), discovered more than 10% of these stars while working as a "computer" for Edward Charles Pickering at Harvard University. Her tedious task was to compare two photographs of the same field of stars taken at different times, and detect any star that varied in brightness. Eventually she discovered 2400 classical Cepheids, most of them located in the Small Magellanic Cloud, with periods between 1 and 50 days. Leavitt took advantage of this opportunity to investigate the nature of the classical Cepheids in the Small Magellanic Cloud. Noticing that the more luminous Cepheids took longer to go through their pulsation cycles, she plotted the apparent magnitudes of these stars against their pulsation periods. The resulting graph, shown in Fig. 14.3 demonstrated that the apparent magnitudes of classical Cepheids are closely correlated with their periods, with an intrinsic uncertainty of only $\Delta m \approx \pm 0.5$ at a given period.

Because all of the stars in the Small Magellanic Cloud are roughly the same distance from us (about 60 kpc), the differences in their apparent magnitudes must be the same as the differences in their absolute magnitudes [c.f. Eq. (3.6) for the distance modulus]. Thus the observed differences in these stars'

## 14.1 Observations of Pulsating Stars

**Figure 14.2** Henrietta Swan Leavitt (1868–1921). (Courtesy of Harvard College Observatory.)

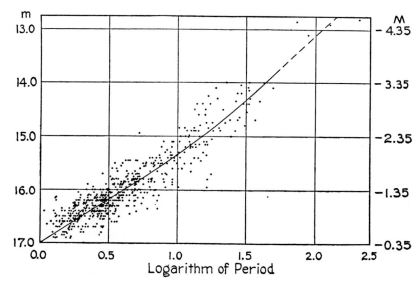

**Figure 14.3** Classical Cepheids in the Small Magellanic Cloud, with the period in units of days. (Figure from Shapley, *Galaxies*, Harvard University Press, Cambridge, MA, 1961.)

apparent brightnesses must reflect intrinsic differences in their luminosities. Astronomers were excited at the prospect of determining the absolute magnitude or luminosity of a distant Cepheid simply by timing its pulsation, because knowing both a star's apparent and absolute magnitudes allows the distance of the star to be easily determined from the distance modulus, Eq. (3.6). This would permit the measurement of large distances in the universe, far beyond the limited range of parallax techniques. The only stumbling block was the calibration of Leavitt's relation. An independent distance to a single Cepheid had to be obtained to measure its absolute magnitude and luminosity. Once this difficult chore was accomplished,[1] the resulting **period–luminosity relation** could be used to measure the distance to any Cepheid. This relation is described by

$$\log_{10} \frac{\langle L \rangle}{L_\odot} = 1.15 \log_{10} \Pi^d + 2.47, \qquad (14.1)$$

where $\langle L \rangle$ is the star's average luminosity and $\Pi^d$ is its pulsation period in units of days. In terms of the average absolute visual magnitude, $M_{\langle V \rangle}$, the relation is

$$M_{\langle V \rangle} = -2.80 \log_{10} \Pi^d - 1.43, \qquad (14.2)$$

and is shown in Fig. 14.4.

Classical Cepheids provide astronomy with its third dimension and supply the foundation for the measurement of extragalactic distances. Because Cepheids are supergiant stars (luminosity class Ib), about fifty times the Sun's size and thousands of times more luminous, they can be seen over intergalactic distances. They serve as "standard candles," beacons scattered throughout the night sky that serve as mileposts for astronomical surveys of the universe.

The important use of Cepheids as cosmic distance indicators does not in any way require an understanding of the physical reasons for their light variations. In fact, the observed changes in brightness were once thought to be caused by tidal effects in the atmospheres of binary stars. However, in 1914 the American astronomer Harlow Shapley (1885–1972) argued that the binary theory was fatally flawed because the size of the star would exceed the size of the orbit for some variables. Shapley advanced an alternative idea, that the observed variations in the brightness and temperature of classical Cepheids

---

[1]The nearest classical Cepheid is Polaris, some 200 pc away. In the early twentieth century, this distance was too great to be reliably measured by stellar parallax. In 1913, Ejnar Hertzsprung succeeded in using the longer baseline provided by the Sun's motion through space together with statistical methods to find the average distance to Cepheids having a specified period. The measurement of the absolute magnitude of a Cepheid is also complicated by the dimming effect of interstellar extinction; recall Eq. (12.1).

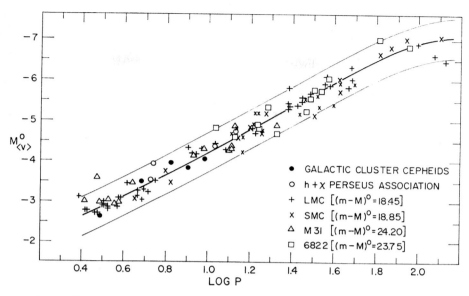

**Figure 14.4** The period–luminosity relation for classical Cepheids. (Figure from Sandage and Tammann, *Ap. J.*, *151*, 531, 1968.)

were caused by the radial pulsation of single stars. He proposed that these stars were rhythmically "breathing" in and out, becoming alternately brighter and dimmer in the process. Four years later Sir Arthur Stanley Eddington provided a firm theoretical framework for the pulsation hypothesis, which received strong support from the observed correlations among the variations in brightness, temperature, and surface velocity throughout the pulsation cycle. Figure 14.5 shows the measured changes in magnitude, temperature, radius, and radial velocity for $\delta$ Cephei.[2] The change in brightness is primarily due to the 1000–1500 K variation in $\delta$ Cephei's surface temperature; the accompanying change in size makes a lesser contribution to the luminosity. Although the total excursion of $\delta$ Cephei's surface from its equilibrium radius is large in absolute terms (a bit more than the diameter of the Sun), it is still only about 5% to 10% of the size of this supergiant star. The spectral type of $\delta$ Cephei changes continuously throughout the cycle, varying between F5 (hottest) and G2 (coolest). A careful examination of Fig. 14.5 reveals that the magnitude and radial velocity curves are mirror images of one another. The star is thus brightest when its surface is expanding outward most rapidly, *after* it has passed through its minimum radius. Later in this chapter we will see that the

---

[2]$\delta$ Cephei itself is moving through space with a radial velocity of $-16$ km s$^{-1}$ (approaching), so its surface velocity varies about this value with an amplitude of 19 km s$^{-1}$.

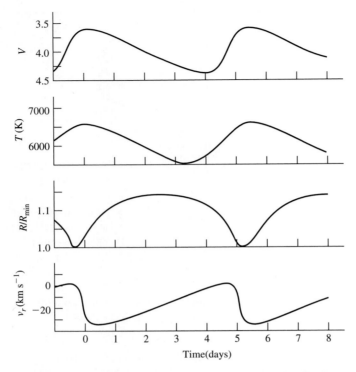

**Figure 14.5** Observed pulsation properties of $\delta$ Cephei.

explanation of this **phase lag** of maximum luminosity behind minimum radius has its origin in the mechanism that maintains the oscillations.

The Milky Way Galaxy is estimated to contain several million pulsating stars. Considering that our Galaxy consists of several hundred billion stars, this implies that stellar pulsation must be a transient phenomenon. The positions of the pulsating variables on the H–R diagram (see Figs. 14.6 and 8.15) confirms this conclusion. Rather than being located on the main sequence, where stars spend most of their lives, the majority of pulsating stars occupy a narrow (about 600–1100 K wide), nearly vertical **instability strip** on the right-hand side of the H–R diagram. Theoretical evolutionary tracks for stars of various masses are also shown in Fig. 14.6. As stars evolve along these tracks, they begin to pulsate as they enter the instability strip and cease their oscillations upon leaving. Of course, evolutionary time scales are far too long to observe the onset and cessation of a single star's oscillations, but several stars have been caught in the final phase of their pulsational history.[3]

---

[3]The amplitude of the oscillations of Polaris, a classical Cepheid, has declined sharply

## 14.1 Observations of Pulsating Stars

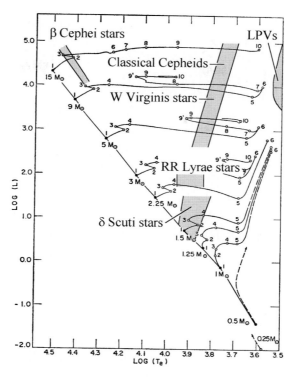

**Figure 14.6** Pulsating stars on the H–R diagram. The evolutionary tracks are incomplete, and those of the lower-mass stars extend into the LPV (long-period variable) region. (The evolutionary tracks are from Iben, *Annu. Rev. Astron. Astrophys.*, 5, 571, 1967. Reproduced with permission from the *Annual Review of Astronomy and Astrophysics*, Volume 5, ©1967 by Annual Reviews Inc.)

Astronomers have divided pulsating stars into several classes. Some of these are listed in Table 14.1, adapted from Cox (1980). The W Virginis stars are metal-deficient (Population II) Cepheids and are about four times less luminous than classical Cepheids with the same period. Their period–luminosity relation is thus lower than and parallel to the one shown for the classical Cepheids in Fig. 14.4. RR Lyrae stars, also Population II, are horizontal-branch stars found in globular clusters. Because all RR Lyrae stars have nearly the same luminosity, they are also useful yardsticks for distance measurements. The $\delta$ Scuti variables are evolved F stars found near the main sequence of the H–R

---

even though it is well within the instability strip. The reason for its diminished motion is not yet understood. As of this writing, Polaris continues to pulsate with a period of 3.97 days, brightening and dimming by about 0.03 magnitude.

| Type | Range of Periods | Population Type | Radial or Nonradial |
|---|---|---|---|
| Long-Period Variables | 100–700 days | I,II | R |
| Classical Cepheids | 1–50 days | I | R |
| W Virginis stars | 2–45 days | II | R |
| RR Lyrae stars | 1.5–24 hours | II | R |
| δ Scuti stars | 1–3 hours | I | R,NR |
| β Cephei stars | 3–7 hours | I | R,NR |
| ZZ Ceti stars | 100–1000 seconds | I | NR |

**Table 14.1** Pulsating Stars.

diagram. They exhibit both radial and nonradial oscillations; the latter is a more complicated motion that will be discussed in Section 14.4. Below the main sequence (not shown in Fig. 14.6; however, see Fig. 15.4) are the pulsating white dwarfs, called ZZ Ceti stars. All of the types of stars listed thus far lie within the instability strip, and they share a common mechanism that drives the oscillations. The long-period variables such as Mira and the β Cephei stars are located outside of the instability strip occupied by the classical Cepheids and RR Lyrae stars. Their unusual positions on the H–R diagram will be discussed in the next section.

## 14.2 The Physics of Stellar Pulsation

Geologists and geophysicists have obtained a wealth of information about Earth's interior from their study of the seismic waves produced by earthquakes and other sources. In the same manner, astrophysicists model the pulsational properties of stars to understand better their internal structure. By numerically calculating an evolutionary sequence of stellar models and then comparing the pulsational characteristics (periods, amplitudes, and details of the light and radial velocity curves) of the models with those actually observed, astronomers can further test their theories of stellar structure and evolution and obtain a detailed view of the interior of a star.[4]

The radial oscillations of a pulsating star are the result of sound waves resonating in the star's interior. A rough estimate of the pulsation period, $\Pi$, may be easily obtained by considering how long it would take a sound wave to cross the diameter of a model star of radius $R$ and constant density $\rho$. The

---
[4]Several other ways of testing the ideas of stellar structure and evolution were discussed in Chapter 13.

## 14.2 The Physics of Stellar Pulsation

adiabatic sound speed is given by Eq. (10.76),

$$v_s = \sqrt{\frac{\gamma P}{\rho}}.$$

The pressure may be found from Eq. (10.7) for hydrostatic equilibrium, using the (unrealistic) assumption of constant density. Thus

$$\frac{dP}{dr} = -\frac{GM_r\rho}{r^2} = -\frac{G(\frac{4}{3}\pi r^3\rho)\rho}{r^2} = -\frac{4}{3}\pi G\rho^2 r.$$

This is readily integrated using the boundary condition that $P = 0$ at the surface to obtain the pressure as a function of $r$,

$$P(r) = \frac{2}{3}\pi G\rho^2 \left(R^2 - r^2\right). \tag{14.3}$$

Thus the pulsation period is roughly

$$\Pi \approx 2\int_0^R \frac{dr}{v_s} \approx 2\int_0^R \frac{dr}{\sqrt{\frac{2}{3}\gamma\pi G\rho (R^2 - r^2)}},$$

or

$$\Pi \approx \sqrt{\frac{3\pi}{2\gamma G\rho}}. \tag{14.4}$$

Qualitatively, this shows that the pulsation period of a star is inversely proportional to the square root of its mean density. Referring to Fig. 14.6 and Table 14.1, this **period–mean density relation** explains why the pulsation period decreases as you move down the instability strip from the very tenuous supergiants to the very dense white dwarfs.[5] The tight period–luminosity relation discovered by Leavitt exists because the instability strip is roughly parallel to the luminosity axis of the H–R diagram.[6] The quantitative agreement of Eq. (14.4) with the observed periods of Cepheids is not too bad, considering its crude derivation. If we take $M = 5$ $M_\odot$ and $R = 50$ $R_\odot$ for a typical Cepheid, then $\Pi \approx 10$ days. This falls nicely within the range of periods measured for the classical Cepheids.

The sound waves involved in the **radial modes** of stellar pulsation are essentially *standing waves*, similar to the standing waves that occur in an organ pipe that is open at one end; see Fig. 14.7. Both the star and the organ pipe can sustain several modes of oscillation. The standing wave for

---
[5]The pulsating white dwarfs exhibit nonradial oscillations, and their periods are longer than predicted by the period–mean density relation.

[6]The finite width of the instability strip is reflected in the ±0.5 magnitude uncertainty in the period–luminosity relation.

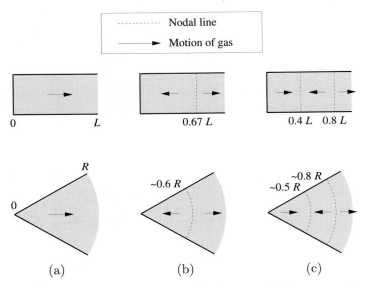

**Figure 14.7** Standing sound waves in an organ pipe and in a star for (a) the fundamental mode, (b) the first overtone, and (c) the second overtone.

each mode has a *node* at one end (the star's center, the pipe's closed end) where the gases do not move, and an *antinode* at the other end (the star's surface, the pipe's open end). For the *fundamental mode*, the gases move in the same direction at every point in the star or pipe. There is a single node between the center and the surface for the *first overtone* mode,[7] with the gases moving in opposite directions on either side of the node, and two nodes for the *second overtone* mode. Figure 14.8 shows the fractional displacement, $\delta r/R$, of the stellar material from its equilibrium position for several radial modes of a 12 $M_\odot$ main-sequence star model. Note that $\delta r/R$ has been arbitrarily scaled to unity at the stellar surface. For radial modes, the motion of the stellar material occurs primarily in the surface regions.

The vast majority of the classical Cepheids and W Virginis stars pulsate in the fundamental mode. The RR Lyrae variables pulsate in either the fundamental or first overtone modes, with a few oscillating simultaneously in both. The long-period variables such as Mira are probably also fundamental mode pulsators, although this has been the subject of considerable debate in the past.

---

[7]Some texts use the unfortunate term "first harmonic" for the first overtone.

## 14.2 The Physics of Stellar Pulsation

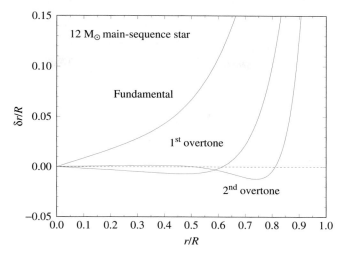

**Figure 14.8** Radial modes for a pulsating star. The waveform for each mode has been arbitrarily scaled so that $\delta r/R = 1$ at the surface of the star. Actually, $\delta r/R \approx 0.05$–$0.10$ at the surface of a classical Cepheid.

To explain the mechanism that powers these standing sound waves, Eddington proposed that pulsating stars are thermodynamic heat engines. The gases comprising the layers of the star do $PdV$ work as they expand and contract throughout the pulsation cycle. If the integral $\oint P\,dV > 0$ for the cycle, a layer does net positive work on its surroundings and contributes to driving the oscillations; if $\oint P\,dV < 0$, the net work done by the layer is negative and tends to dampen the oscillations. Figures 14.9 and 14.10 show $P$–$V$ diagrams for a driving layer and a damping layer, respectively, in a numerical calculation of the oscillation of an RR Lyrae star. If the total work (found by adding up the contributions of all the layers of the star) is positive, the oscillations will grow in amplitude. Similarly, the oscillations will decay if the total work is negative. These changes in the pulsation amplitude continue until an equilibrium value is reached, when the total work done by all the layers is zero.

As for any heat engine, the net work done by each layer of the star during one cycle is the difference between the heat flowing into the gas and the heat leaving the gas. For driving, the heat must enter the layer during the high-temperature part of the cycle and leave during the low-temperature part. Just as the spark plug of an automobile engine fires at the end of the compression stroke, the driving layers of a pulsating star must absorb heat around the time of their maximum compression. In this case the maximum pressure will occur *after* maximum compression, and the oscillations will be amplified.

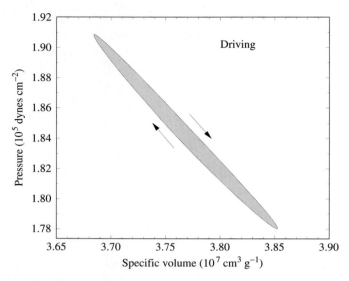

**Figure 14.9** $P$–$V$ diagram for a driving layer of an RR Lyrae star model.

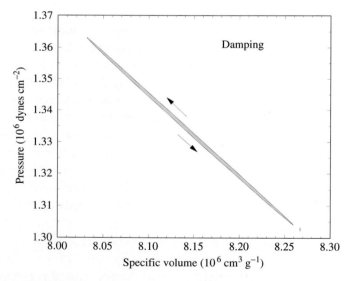

**Figure 14.10** $P$–$V$ diagram for a damping layer of an RR Lyrae star model.

## 14.2 The Physics of Stellar Pulsation

In what region of the star can this driving take place? An obvious possibility was first considered by Eddington: When the center of the star is compressed, its temperature rises and so increases the rate at which thermonuclear energy is generated. However, recall from Fig. 14.8 that the displacement $\delta r/R$ has a node at the center of the star. The pulsation amplitude is very small near the center. Although this energy mechanism (called the **$\epsilon$-mechanism**) does in fact operate in the core of a star, it is usually not enough to drive the star's pulsation. However, as mentioned in Section 10.6, variations in the nuclear energy generation rate ($\epsilon$) produce oscillations that may prevent the formation of stars with masses greater than approximately 90 $M_\odot$.

Eddington then suggested an alternative, a *valve mechanism*. If a layer of the star became more opaque upon compression, it could "dam up" the energy flowing toward the surface and push the surface layers upward. Then, as this expanding layer became more transparent, the trapped heat could escape and the layer would fall back down to begin the cycle anew. In Eddington's own words, "To apply this method we must make the star more heat-tight when compressed than when expanded; in other words, *the opacity must increase with compression.*"

In most regions of the star, however, the opacity actually *decreases* with compression. Recall from Section 9.2 that for a Kramers law, the opacity $\kappa$ depends on the density and temperature of the stellar material as $\kappa \propto \rho/T^{3.5}$. As the layers of a star are compressed, their density and temperature both increase. But because the opacity is more sensitive to the temperature than to the density, the opacity of the gases usually decreases upon compression. It takes special circumstances to overcome the damping effect of most stellar layers, which explains why stellar pulsation is observed for only one of every $10^5$ stars.

The special conditions responsible for exciting and maintaining the stellar oscillations were first identified by the Russian astronomer S. A. Zhevakin and verified in detailed calculations by a German and two Americans, Rudolph Kippenhahn, Norman Baker, and John Cox. They found that the regions of a star where Eddington's valve mechanism can successfully operate are its *partial ionization zones* (c.f. Section 8.1). In these layers of the star where the gases are partially ionized, part of the work done on the gases as they are compressed produces further ionization rather than raising the temperature of the gas.[8] With a smaller temperature rise, the increase in density with compression produces a corresponding increase in the Kramers opacity; see

---

[8]As discussed in Section 10.4, this causes the specific heats $C_P$ and $C_V$ to have larger values in a partial ionization zone.

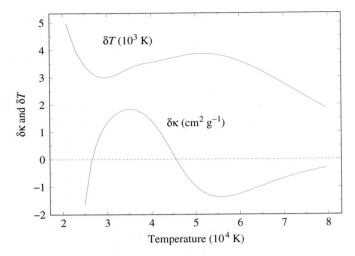

**Figure 14.11** Variations in the temperature and opacity throughout an RR Lyrae star model at the time of maximum compression. In the He II partial ionization zone ($T \approx 40{,}000$ K), $\delta\kappa > 0$ and $\delta T$ is reduced. These are the $\kappa$- and $\gamma$-mechanisms that drive the star's oscillations.

Fig. 14.11. Similarly, during expansion, the temperature does not decrease as much as expected since the ions now recombine with electrons and release energy. Again, the density term in the Kramers law dominates, and the opacity decreases with decreasing density during the expansion. This layer of the star can thus absorb heat during compression, be pushed outward to release the heat during expansion, and fall back down again to begin another cycle. Astronomers refer to this opacity mechanism as the **$\kappa$-mechanism**. In a partial ionization zone, the $\kappa$-mechanism is reinforced by the tendency of heat to flow into the zone during compression simply because its temperature has increased less than the adjacent stellar layers. This effect is called the **$\gamma$-mechanism**, after the smaller ratio of specific heats caused by the increased values of $C_P$ and $C_V$. Partial ionization zones are the pistons that drive the oscillations of stars; they modulate the flow of energy through the layers of the star and are the direct cause of stellar pulsation.

In most stars there are two main ionization zones. The first is a broad zone where both the ionization of neutral hydrogen (H I→H II) and the first ionization of helium (He I→He II) occur in layers with a characteristic temperature of 1 to 1.5 $\times 10^4$ K. These layers are collectively referred to as the *hydrogen partial ionization zone*. The second, deeper zone involves the second ionization of helium (He II→He III), which occurs at a characteristic temperature of $4 \times 10^4$ K, and is called the *He II partial ionization zone*. The location of

## 14.2 The Physics of Stellar Pulsation

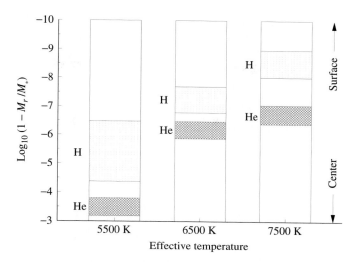

**Figure 14.12** Hydrogen and helium ionization zones in stars of different temperatures. For each point in the star, the vertical axis displays the logarithm of the fraction of the star's mass that lies *above* that point.

these ionization zones within the star determines its pulsational properties. As shown in Fig. 14.12, if the star is too hot (7500 K), the ionization zones will be located very near the surface. At this position, the density is quite low and there is not enough mass available to drive the oscillations effectively. This accounts for the blue (hot) edge of the instability strip on the H–R diagram. In a cooler star (6500 K), the characteristic temperatures of the ionization zones are found deeper in the star. There is more mass for the ionization zone "piston" to push around, and the first overtone mode may be excited.[9] In a still cooler star (5500 K), the ionization zones occur deep enough to drive the fundamental mode of pulsation. However, if a star's surface temperature is too low, the onset of efficient convection in its outer layers may dampen the oscillations. Because the transport of energy by convection is more effective when the star is compressed, the convecting stellar material may lose heat at minimum radius. This could overcome the damming up of heat by the ionization zones, and so quench the pulsation of the star. The red (cool) edge of the instability strip is believed to be the result of the damping effect of convection.[10]

---

[9]Whether a mode is actually excited depends on whether the positive work generated within the ionization zones is sufficient to overcome the damping (negative work) of the other layers of the star.

[10]Much work remains to be done on the effect of convection on stellar pulsation, although some results have been obtained for RR Lyrae and ZZ Ceti stars. Progress has been hampered by the present lack of a fundamental theory of time-dependent convection.

Detailed numerical calculations of the pulsation of model stars produce an instability strip that is in good agreement with its observed location on the H–R diagram. These computations show that it is the *He II ionization zone* that is primarily responsible for driving the oscillations of stars within the instability strip.[11] If the effect of the helium ionization zone is artificially removed, the model stars will not pulsate.

The hydrogen ionization zone plays a more subtle role. As a star pulsates, the hydrogen ionization zone moves toward or away from the surface as the zone expands and contracts in response to the changing temperature of the stellar gases. It happens that the star is brightest when the *least mass* lies between the hydrogen ionization zone and the surface. As a star oscillates, the location of an ionization zone changes both with respect to its radial position, $r$, and mass interior to $r$, $M_r$. The luminosity incident on the *bottom* of the hydrogen ionization zone is indeed a maximum at minimum radius, but this merely propels the zone outward (through mass) most rapidly at that instant. The emergent luminosity is thus greatest *after* minimum radius, when the zone is nearest the surface. This delaying action of the hydrogen partial ionization zone produces the phase lag observed for classical Cepheids and RR Lyrae stars.

The mechanism responsible for the pulsation of stars outside the instability strip is not as well understood. The long-period variables are red supergiants (AGB stars) with huge, diffuse convective envelopes surrounding a compact core. Their spectra are dominated by molecular absorption lines and emission lines that reveal the existence of atmospheric shock waves and significant mass loss. While we understand that the *hydrogen* ionization zone drives the pulsation of a long-period variable star, many details remain to be explained, such as how its oscillations interact with its outer atmosphere. Equally puzzling are the $\beta$ Cephei variables, the pulsating blue-white stars found near the upper end of the main sequence of the H–R diagram. Since their discovery in 1902, the mechanism responsible for the pulsation of the $\beta$ Cephei stars has eluded astronomers. These stars were considered too hot to be powered by their ionization zones, which are found very close to the surface. However, improvements in the calculated values of stellar opacities indicate that the $\kappa$-mechanism may be strong enough to drive the oscillations of the $\beta$ Cephei stars. Much remains to be learned about the structure and evolution of stars from the theory of stellar pulsation.

---

[11] In Section 15.2, we will see that the ZZ Ceti stars are driven by the hydrogen ionization zone.

## 14.3 Modeling Stellar Pulsation

In Section 10.5, the construction of a stellar model in hydrostatic equilibrium was described. The star was considered to be divided into a number of concentric mass shells. The differential equations of static stellar structure were then applied to each mass shell, and the system of equations was solved on a computer subject to certain boundary conditions at the center and surface of the stellar model.

Because a pulsating star is *not* in hydrostatic equilibrium, the stellar structure equations collected at the beginning of Section 10.5 cannot be used in their present form. Instead, a more general set of equations is employed that takes the oscillation of the mass shells into account. For example, Newton's second law (Eq. 10.6),

$$\rho \frac{d^2 r}{dt^2} = -G \frac{M_r \rho}{r^2} - \frac{dP}{dr}, \tag{14.5}$$

must be used instead of Eq. (10.7) for hydrostatic equilibrium. Once the differential equations describing the nonequilibrium mechanical and thermal behavior of a star have been assembled, along with the appropriate constitutive relations, they may be replaced by difference equations as described in Section 10.5 and solved numerically. In essence, the model star is mathematically displaced from its equilibrium configuration and then "released" to begin its oscillation. The mass shells expand and contract, pushing against each other as they move. If conditions are right, the ionization zones in the model star will drive the oscillations, and the pulsation amplitude will slowly increase; otherwise the amplitude will decay away. Computer programs that carry out these calculations have been very successful at modeling the details observed in the light and radial velocity curves of Cepheid variables.

The main advantage of the preceding approach is that it is a **nonlinear** calculation, capable of modeling the complexities of large pulsation amplitudes and reproducing the nonsinusoidal shape of actual light curves. The disadvantage lies in the computer resources required: This process uses a large amount of CPU time and memory. Many oscillations must be calculated before the model settles down into a well-behaved periodic motion, and even more periods may be required for the model to reach its *limit cycle*, when the pulsation amplitude has reached its final value.[12]

An alternative to the nonlinear approach is to **linearize** the differential equations by considering only small-amplitude oscillations. This is done by

---

[12] Some models may never attain a truly periodic solution. Instead, they exhibit chaotic behavior, as observed in some real stars.

writing every variable in the differential equations as an equilibrium value (found in the static model of the star) plus a small change due to the pulsation. For example, the pressure $P$ would be written as $P = P_\circ + \delta P$, where $P_\circ$ is the value of the pressure in a mass shell of the equilibrium model, and $\delta P$ is the small change in pressure that occurs as that mass shell moves in the oscillating model star. Thus $\delta P$ is a function of time, but $P_\circ$ is constant. When the variables written in this manner are inserted into the differential equations, the terms containing only equilibrium quantities cancel and terms that involve powers of the deltas higher than the first, such as $(\delta P)^2$, may be discarded because they are negligibly small. The resulting linearized differential equations and their associated boundary conditions, also linearized, are similar to the equations for a wave on a string or in an organ pipe. Only certain standing waves with specific periods are permitted, and so the pulsation modes of the star are cleanly identified. The equations are still sufficiently complicated that a computer solution is required, but the time involved is much less than that required for a nonlinear calculation. The penalties for adopting the linearized approach are that the motion of the star is forced to be sinusoidal (as it must be for small amplitudes of oscillation), and the limiting value of the pulsation amplitude cannot be determined. Modeling the complexities of the full nonlinear behavior of the stellar model is thus sacrificed.

---

**Example 14.1** In this example, we consider an unrealistic, but very instructive, model of a pulsating star called a **one-zone model**; see Fig. 14.13. It consists of a central point mass equal to the entire mass of the star, $M$, surrounded by a single thin spherical shell of mass $m$ and radius $R$ that represents the surface layer of the star. The interior of the shell is filled with a massless gas of pressure $P$ whose sole function is to support the shell against the gravitational pull of the central mass $M$. Newton's second law (Eq. 14.5) applied to this shell is

$$m\frac{d^2 R}{dt^2} = -\frac{GMm}{R^2} + 4\pi R^2 P. \qquad (14.6)$$

For the equilibrium model, the left-hand side of this equation is zero, so

$$\frac{GMm}{R_\circ^2} = 4\pi R_\circ^2 P_\circ. \qquad (14.7)$$

The linearization is accomplished by writing the star's radius and pressure as

$$R = R_\circ + \delta R \quad \text{and} \quad P = P_\circ + \delta P,$$

## 14.3 Modeling Stellar Pulsation

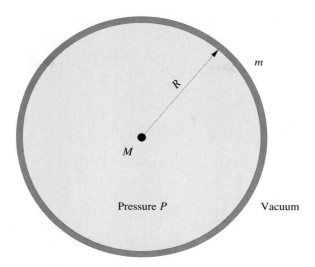

**Figure 14.13** One-zone model of a pulsating star.

and inserting these expressions into Eq. (14.6), giving

$$m\frac{d^2(R_\circ + \delta R)}{dt^2} = -\frac{GMm}{(R_\circ + \delta R)^2} + 4\pi(R_\circ + \delta R)^2(P_\circ + \delta P). \tag{14.8}$$

Using the first-order approximation

$$\frac{1}{(R_\circ + \delta R)^2} \approx \frac{1}{R_\circ^2}\left(1 - 2\frac{\delta R}{R_\circ}\right)$$

and keeping only those terms involving the first powers of the deltas results in

$$m\frac{d^2(\delta R)}{dt^2} = -\frac{GMm}{R_\circ^2} + \frac{2GMm}{R_\circ^3}\delta R$$
$$+ 4\pi R_\circ^2 P_\circ + 8\pi R_\circ P_\circ \delta R + 4\pi R_\circ^2 \delta P, \tag{14.9}$$

where $d^2 R_\circ/dt^2 = 0$ has been used for the equilibrium model. The first and third terms on the right-hand side cancel (see Eq. 14.7), leaving

$$m\frac{d^2(\delta R)}{dt^2} = \frac{2GMm}{R_\circ^3}\delta R + 8\pi R_\circ P_\circ \delta R + 4\pi R_\circ^2 \delta P. \tag{14.10}$$

This is the linearized version of Newton's second law for our one-zone model.

To reduce the two variables, $\delta R$ and $\delta P$, to one, we now assume that the oscillations are *adiabatic*. In this case, the pressure and volume of the model are related by the adiabatic relation $PV^\gamma = $ constant, where $\gamma$ is the ratio of specific heats of the gas. Since the volume of the one-zone model is just $\frac{4}{3}\pi R^3$, the adiabatic relation says that $PR^{3\gamma} = $ constant. It is left as a problem to show that the linearized version of this expression is

$$\frac{\delta P}{P_\circ} = -3\gamma \frac{\delta R}{R_\circ}. \tag{14.11}$$

Using this equation, $\delta P$ can be eliminated from Eq. (14.10). In addition, Eq. (14.7) can be used to replace $8\pi R_\circ P_\circ$ with $2GMm/R_\circ^3$. As a result, the mass $m$ of the shell cancels, leaving the linearized equation for $\delta R$:

$$\frac{d^2(\delta R)}{dt^2} = -(3\gamma - 4)\frac{GM}{R_\circ^3}\delta R. \tag{14.12}$$

If $\gamma > 4/3$ (so the right-hand side of the equation is negative), this is just the familiar equation for simple harmonic motion. It has the solution $\delta R = A\sin(\omega t)$, where $A$ is the pulsation amplitude and $\omega$ is the angular pulsation frequency. Inserting this expression for $\delta R$ into Eq. (14.12) results in

$$\omega^2 = (3\gamma - 4)\frac{GM}{R_\circ^3}. \tag{14.13}$$

Finally, the pulsation period of the one-zone model is just $\Pi = 2\pi/\omega$, or

$$\Pi = \frac{2\pi}{\sqrt{\frac{4}{3}\pi G \rho_\circ (3\gamma - 4)}}, \tag{14.14}$$

where $\rho_\circ = M/\frac{4}{3}\pi R_\circ^3$ is the average density of the equilibrium model. For an ideal monatomic gas (appropriate for hot stellar gases), $\gamma = 5/3$. Except for factors of order unity, this is the same as our earlier period estimate (Eq. 14.4) obtained from the time for a sound wave to cross the diameter of a star.

---

In Example 14.1, the approximations that the pulsation of the one-zone model was *linear* and *adiabatic* were used to simplify the calculation. Note that the pulsation amplitude, $A$, canceled in this example. The inability to calculate the amplitude of the oscillations is an inherent drawback of the linearized approach to pulsation.

Because no heat is allowed to enter or leave the layers of a stellar model in an adiabatic analysis, the amplitude (whatever it may be) of the oscillation remains constant. However, astronomers need to know which modes will grow and which will decay away. This calculation must include the physics involved in Eddington's valve mechanism. The equations describing the transfer of heat and radiation through the stellar layers (similar to those discussed in Section 10.4) must be incorporated in such a *nonadiabatic* computation. These nonadiabatic expressions may also be linearized and solved to obtain the periods and growth rates of the individual modes. However, a more complicated and costly *nonlinear nonadiabatic* calculation is needed to reproduce the complicated light and radial velocity curves that are observed for some variable stars. The computer problem at the end of the chapter asks you to carry out a nonlinear (but still adiabatic) calculation of the pulsation of this one-zone model.

Equation (14.12) provides a very important insight into the **dynamical stability** of a star. If $\gamma < 4/3$, then the right-hand side of Eq. (14.12) is positive. The solution is now $\delta R = A e^{-\kappa t}$, where $\kappa^2$ is the same as $\omega^2$ in Eq. (14.13). Instead of pulsating, the star *collapses* if $\gamma < 4/3$. The increase in gas pressure is not enough to overcome the inward pull of gravity and push the mass shell back out again, resulting in a *dynamically unstable* model. This instability caused by a reduction in the value of $\gamma$ will be seen again in Section 15.4, where the effect of relativity on white dwarf stars is described.[13]

## 14.4 Nonradial Stellar Pulsation

As some types of stars pulsate, their surfaces do not move uniformly in and out in a simple "breathing" motion. Instead, such a star executes a more complicated type of **nonradial** motion in which some regions of its surface expand while other areas contract. Figure 14.14 shows the angular patterns for several nonradial modes. If the stellar surface is moving outward within the lighter regions, then it is moving inward within the shaded areas. Scalar quantities such as the change in pressure ($\delta P$) follow the same pattern, having positive values in some areas and negative values in others. These patterns are described by two integers, $\ell$ and $m$.[14] There are $\ell$ *nodal circles* (where $\delta r = 0$), with $|m|$ of these circles passing through the poles of the star and the rest parallel to the star's equator. If $\ell = m = 0$, then the pulsation is purely radial.

---

[13] For the case of *nonadiabatic* oscillations, the time-dependence of the pulsation is usually taken to be (the real part of) $e^{i\sigma t}$, where $\sigma$ is the complex frequency $\sigma = \omega + i\kappa$. In this expression, $\omega$ is the usual pulsation frequency, while $\kappa$ is a *stability coefficient*. The pulsation amplitude is then proportional to $e^{-\kappa t}$, and $1/\kappa$ is the characteristic time for the growth or decay of the oscillations.

[14] The reader may recognize that these patterns correspond to the real parts of the spherical harmonic functions, $Y_\ell^m(\theta, \phi)$, where $\ell$ is a nonnegative integer and $m$ is equal to any of the $2\ell + 1$ integers between $-\ell$ and $+\ell$.

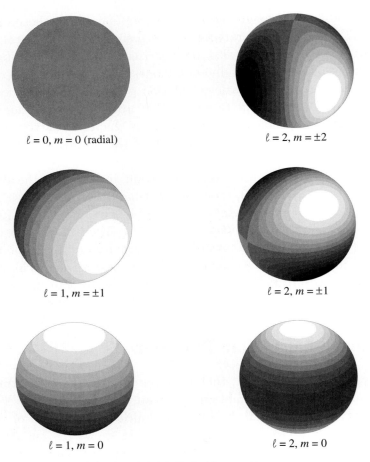

**Figure 14.14** Pulsation patterns.

The patterns for nonzero $m$ represent *traveling waves* that move across the star parallel to its equator. (Imagine these patterns on a beach ball, with the ball slowly spinning about the vertical axis.) The time required for the waves to travel around the star is $|m|$ times the star's pulsation period. Note that the star itself may not be rotating at all. Just as water waves may travel across the surface of a lake without the water itself making the trip, these traveling waves are disturbances that pass through the stellar gases.[15]

In Section 14.2, the radial pulsation of stars was attributed to standing sound waves in the stellar interior. For the case of nonradial oscillations, the sound waves can propagate horizontally as well as radially to produce

---

[15] Observations of nonradially pulsating stars are considered in Problem 14.10.

## 14.4 Nonradial Stellar Pulsation

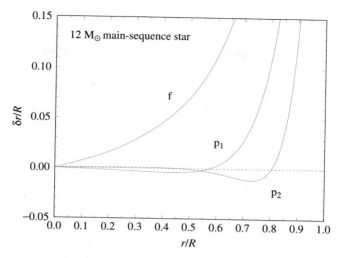

**Figure 14.15** Nonradial p-modes with $\ell = 2$. The waveforms have been arbitrarily scaled so that $\delta r/R = 1$ at the star's surface.

waves that travel around the star. Because *pressure* provides the restoring force for sound waves, these nonradial oscillations are called **p-modes**. A complete description of a p-mode requires a specification of its radial and angular nodes. For example, a $p_2$ mode may be thought of as the nonradial analog of a radial second overtone mode. The $p_2$ mode with $\ell = 4$ and $m = -3$ has two radial nodes between the center and the surface, and its angular pattern has four nodal lines, three through the poles and one parallel to the equator. Figure 14.15 shows several p-modes for a 12 $M_\odot$ main-sequence star model; the reader may note the similarities between this figure and Fig. 14.8, with most of the motion occurring near the stellar surface. The **f-mode** shown in Fig. 14.15 is the nonradial analog of the fundamental radial mode.

An estimate of the angular frequency of a p-mode may be obtained from the time for a sound wave to travel one horizontal wavelength, from one angular nodal line to the next. This horizontal wavelength is given by the expression

$$\lambda_h = \frac{2\pi r}{\sqrt{\ell(\ell+1)}}, \tag{14.15}$$

where $r$ is the radial distance from the center of the star. The **acoustic frequency** at this depth in the star is then defined as

$$S_\ell = \frac{2\pi}{\text{time for sound to travel } \lambda_h},$$

which can be written as

$$S_\ell = 2\pi \left[ \frac{v_s}{2\pi r / \sqrt{\ell(\ell+1)}} \right]$$

$$= \sqrt{\frac{\gamma P}{\rho}} \frac{\sqrt{\ell(\ell+1)}}{r}, \tag{14.16}$$

where $v_s$ is the adiabatic sound speed given by Eq. (10.76). Because the speed of sound is proportional to the square root of the temperature,[16] the critical acoustic frequency is large in the deep interior of the star and decreases with increasing $r$. The frequency of a p-mode is determined by the average value of $S_\ell$ in the regions of the star where the oscillations are most energetic.

In the absence of rotation, the pulsation period depends only on the number of radial nodes and the integer $\ell$. The period is independent of $m$ because with no rotation there are no well-defined poles or equator; thus $m$ has no physical significance. On the other hand, if the star is rotating, the rotation itself defines the poles and equator, and the pulsation frequencies for modes with different values of $m$ become separated or *split* as the traveling waves move either with or against the rotation.[17] The amount by which the pulsation frequencies are split depends on the angular rotation frequency, $\Omega$, of the star, with the rotationally produced shift in frequency proportional to the product $m\Omega$ (for the simple case of uniform rotation). As will be discussed later, this frequency splitting provides a powerful probe for measuring the rotation of the Sun's interior.

Just as pressure supplies the restoring force for the compression and expansion of the p-mode sound waves, *gravity* is the source of the restoring force for another class of nonradial oscillations called **g-modes**. The g-modes are produced by *internal gravity waves*. These waves involve a "sloshing" back and forth of the stellar gases, which is ultimately connected to the *buoyancy* of stellar material.[18]

To gain a better understanding of this oscillatory motion for g-modes, consider a small bubble of stellar material that is displaced upward from its equilibrium position in the star by an amount $dr$, as shown in Fig. 10.10. We will assume that this motion occurs

1. slowly enough that the pressure within the bubble, $P^{(b)}$, is always equal to the pressure of its surroundings, $P^{(s)}$; and

---

[16] According to the ideal gas law, Eq. (10.14), $P/\rho \propto T$.

[17] The sign of $m$ determines the direction in which the waves move around the star.

[18] Because "sloshing" cannot occur for purely radial motion, there is no radial analog for the g-modes.

## 14.4 Nonradial Stellar Pulsation

2. rapidly enough that there is no heat exchanged between the bubble and its surroundings.

The second assumption means that the expansion and compression of the gas bubble is *adiabatic*. If the density of the displaced bubble is greater than the density of its new surroundings, the bubble will fall back to its original position.[19] The net restoring force *per unit volume* on the bubble in its final position is the difference between the upward buoyant force (given by Archimedes' law) and the downward gravitational force:

$$f_{\text{net}} = (\rho_f^{(s)} - \rho_f^{(b)}) g,$$

where $g = GM_r/r^2$ is the local value of the gravitational acceleration. Using a Taylor expansion for the densities about their initial positions results in

$$f_{\text{net}} = \left[ \left( \rho_i^{(s)} + \frac{d\rho^{(s)}}{dr} dr \right) - \left( \rho_i^{(b)} + \frac{d\rho^{(b)}}{dr} dr \right) \right] g.$$

The initial densities of the bubble and its surroundings are the same, so these terms cancel, leaving

$$f_{\text{net}} = \left( \frac{d\rho^{(s)}}{dr} - \frac{d\rho^{(b)}}{dr} \right) g\, dr.$$

Because the motion of the bubble is adiabatic, Eq. (10.79) can be used to replace $d\rho^{(b)}/dr$:

$$f_{\text{net}} = \left( \frac{d\rho^{(s)}}{dr} - \frac{\rho_i^{(b)}}{\gamma P_i^{(b)}} \frac{dP^{(b)}}{dr} \right) g\, dr.$$

Looking at this equation, all of the "b" superscripts may be changed to "s" because the initial densities are equal, and according to the first assumption given, the pressures inside and outside the bubble are *always* the same. Thus all quantities in this equation refer to the stellar material surrounding the bubble. With that understanding the subscripts may be dropped completely, resulting in

$$f_{\text{net}} = \left( \frac{1}{\rho} \frac{d\rho}{dr} - \frac{1}{\gamma P} \frac{dP}{dr} \right) \rho g\, dr.$$

For convenience, the term in parentheses is defined as

$$A \equiv \frac{1}{\rho} \frac{d\rho}{dr} - \frac{1}{\gamma P} \frac{dP}{dr}. \tag{14.17}$$

---

[19]This is just a reexamination of the problem of convection, last seen in Section 10.4, from another perspective.

Thus the net force per unit volume acting on the bubble is

$$f_{\text{net}} = \rho A g \, dr. \tag{14.18}$$

If $A > 0$, the net force on the displaced bubble has the same sign as $dr$, and so the bubble will continue to move away from its equilibrium position. This is the condition necessary for *convection* to occur, and it is equivalent to the other requirements previously found for convective instability, such as Eq. (10.86). However, if $A < 0$, then the net force on the bubble will be in a direction opposite to the displacement, and so the bubble will be pushed back toward its equilibrium position. In this case, Eq. (14.18) has the form of Hooke's law, with the restoring force proportional to the displacement. Thus if $A < 0$, the bubble will oscillate about its equilibrium position with simple harmonic motion.

Dividing the force per unit volume, $f_{\text{net}}$, by the mass per unit volume, $\rho$, gives the force per unit mass or acceleration, $a = f_{\text{net}}/\rho = Ag\,dr$. Because the acceleration is simply related to the displacement for simple harmonic motion,[20] we have

$$a = -N^2 \, dr = Ag \, dr,$$

where $N$ is the angular frequency of the bubble about its equilibrium position, called the **buoyancy frequency** or the **Brunt–Väisälä frequency**,

$$N = \sqrt{-Ag} = \sqrt{\left(\frac{1}{\gamma P}\frac{dP}{dr} - \frac{1}{\rho}\frac{d\rho}{dr}\right)g}. \tag{14.19}$$

The buoyancy frequency is zero at the center of star (where $g = 0$) and at the edges of a convection zone (where $A = 0$). Recall that $A < 0$ where there is no convection, so $N$ is larger in regions that are more stable against convection. Inside a convection zone, where $A > 0$, the buoyancy frequency is not defined.

The "sloshing" effect of neighboring regions of the star produces the internal gravity waves that are responsible for the g-modes of a nonradially pulsating star. The frequency of a g-mode is determined by the average value of $N$ in the regions of the star where the oscillations are most energetic. Figure 14.16 shows several g-modes for the same stellar model that was used for Fig. 14.15. A comparison of these two figures reveals significant differences between these classes of modes, making them very useful to astronomers attempting to study the interior of the Sun and other stars. Most importantly, notice the difference

---

[20] For example, recall that $F = ma = -kx$ for a spring. The acceleration is $a = -\omega^2 x$, where $\omega = \sqrt{k/m}$ is the angular frequency of the spring's motion.

## 14.5 Helioseismology

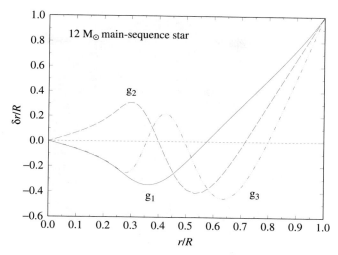

**Figure 14.16** Nonradial g-modes with $\ell = 2$. The waveforms have been arbitrarily scaled so that $\delta r/R = 1$ at the star's surface.

in the vertical scales of the two figures. The g-modes involve significant movement of the stellar material deep within the star, while the p-mode's motions are confined near the stellar surface. Thus g-modes provide a view into the very heart of a star, while p-modes allow a diagnosis of the conditions in its surface layers.

## 14.5 Helioseismology

All of the ideas of nonradial pulsation come into play in the science of **helioseismology**, the study of the oscillations of the Sun first observed in 1962 by American astronomers Robert Leighton, Robert Noyes, and George Simon. A typical solar oscillation mode has a very low amplitude, with a surface velocity of only 10 cm s$^{-1}$ or less,[21] and a luminosity variation $\delta L/L_\odot$ of only $10^{-6}$. With an incoherent superposition of roughly *ten million* modes rippling through its surface and interior, our star is "ringing" like a bell.

The oscillations observed on the Sun fall into two general categories:

- Modes with periods between three and eight minutes and very short horizontal wavelengths ($\ell$ ranging from 0 to 1000 or more). These so-called *five-minute oscillations* have been identified as p-modes.

---

[21]These incredibly precise velocity measurements are made by carefully observing the Doppler shift of a spectral absorption line such as Fe I (5576.099 Å) through a narrow slit that follows the rotating solar surface.

**Figure 14.17** Five-minute $p_{15}$ mode with $\ell = 20$ and $m = 16$. The solar convection zone is the stippled region, where the p-modes are found. (Courtesy of National Optical Astronomy Observatories.)

- Modes with longer periods of about 160 minutes. The observations of these modes are still controversial; they may be g-modes with $\ell \approx 1\text{--}4$.

The five-minute p-modes are concentrated below the photosphere within the Sun's convection zone; Fig. 14.17 shows a typical p-mode. The g-modes are found deep in the solar interior, below the convection zone. By studying these oscillations, astronomers have been able to gain new insights into the structure of the Sun in these regions.

Figure 14.18 shows the relative power contained in the solar p-modes. This information can also be plotted in another manner, as shown in Fig. 14.19, with $\ell$ on the horizontal axis and the pulsation frequency on the vertical axis. Circles show the observed frequencies, and each continuous ridge corresponds to a different p-mode ($p_1$, $p_2$, $p_3$, etc.). The superimposed lines are the *theoretical* frequencies calculated for a solar model. All of the observed five-minute modes have been identified in this way. The fit is certainly impressive but not quite exact. A solar model must be carefully tuned to obtain the best agreement between the theoretical and observed p-mode frequencies. This procedure has the potential for revealing much about the depth of the solar convection zone and about the rotation and composition of the outer layers of the Sun.

As of this writing, it appears that the convection zone extends down to a temperature of about $2.3 \times 10^6$ K and occupies roughly the outer 30% of the

## 14.5 Helioseismology

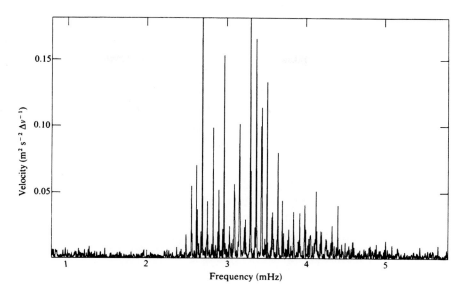

**Figure 14.18** Relative power of solar p-modes; a period of five minutes corresponds to a frequency of 3.33 mHz. (Figure from Grec, Fossat, and Pomerantz, *Nature*, *288*, 541, 1980.)

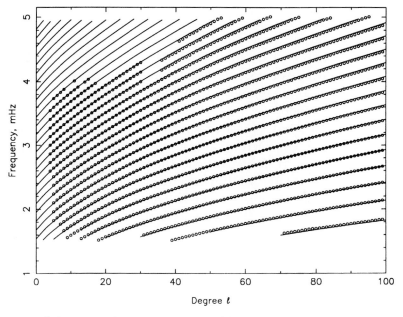

**Figure 14.19** Solar p-modes: observations (circles) and theory (lines). (Figure from Libbrecht, *Space Sci. Rev.*, *47*, 275, 1988.)

Sun's interior. The rotational splitting observed for p-mode frequencies indicates that differential rotation observed at the Sun's surface decreases slightly down through the convection zone.[22] Below the convection zone, the rotation rates may change more rapidly, with the equatorial and polar rotation rates converging toward comparable values at $r/R_\odot \approx 0.5$. Because a change in the rotation rate with depth is needed to convert the Sun's magnetic field from a poloidal to a toroidal geometry (as discussed in Section 11.3), these results indicate that the Sun's magnetic dynamo may be seated at the base of the convection zone.

The abundance of helium in the outer layers of the Sun can also be inferred from a comparison of the observed and theoretical p-mode ridges in Fig. 14.19. Although a definitive answer has not yet been obtained, the results thus far are consistent with the accepted value of $Y = 0.27$ for the mass fraction of helium.

Astronomers have experienced more difficulty in their attempts to use the solar g-modes as a probe of the Sun's interior. Because the g-modes dwell beneath the convection zone, their amplitudes are significantly diminished at the Sun's surface. To date, a definite identification of the g-modes has not been made, and even the existence of the most prominent g-mode is doubted by some astronomers. Its 160-minute period is exactly one-ninth of a solar day, raising the suspicion that it is an observational artifact produced by the Sun's influence on Earth's atmosphere.[23] Nevertheless, the potential rewards of using these oscillations to learn more about the core of the Sun, with possible implications for testing some of the solutions of the solar neutrino problem discussed in Section 11.1, compel astronomers to apply their observational ingenuity to these g-modes.

The question of the mechanism responsible for driving the solar oscillations has not yet been conclusively answered. Our main-sequence Sun is not a normal pulsating star. It lies far beyond the red edge of the instability strip on the H–R diagram (see Fig. 14.6) where turbulent convection overcomes the tendency of the ionization zones to absorb heat at maximum compression. Eddington's valve mechanism thus cannot be responsible for the solar oscillations. The time scale for convection near the top of the convection zone is a

---

[22]Those p-modes with shorter horizontal wavelengths (larger $\ell$) penetrate less deeply into the convection zone, so the difference in rotational frequency splitting with $\ell$ reveals the depth dependence of the rotation. The variation in the rotation with the distance from the solar equator comes from the dependence of the rotational frequency splitting on $m$.

[23]The European SOHO (*So*lar and *H*eliospheric *O*bservatory) spacecraft planned for the late 1990s will be devoted to solar oscillations, making its measurements far above the contaminating effects of Earth's atmosphere.

few minutes, and it is strongly suspected that the p-modes are driven by tapping into the turbulent energy of the convection zone itself, where the p-modes are confined. The driving mechanism responsible for the g-modes remains as mysterious as the modes themselves.

## Suggested Readings

GENERAL

Aller, Lawrence H., *Atoms, Stars, and Nebulae*, Revised Edition, Harvard University Press, Cambridge, MA, 1971.

Giovanelli, Ronald, *Secrets of the Sun*, Cambridge University Press, Cambridge, 1984.

Kaler, James B., *Stars and Their Spectra*, Cambridge University Press, Cambridge, 1989.

Leibacher, John W., et al., "Helioseismology," *Scientific American*, September 1985.

Percy, John R., "Pulsating Stars," *Scientific American*, June 1975.

Wentzel, Donat G., "The Solar Chimes: Searching for Oscillations Inside the Sun," *Mercury*, May/June 1991.

TECHNICAL

Brown, Timothy M., et al., "Inferring the Sun's Internal Angular Velocity from Observed p-Mode Frequency Splittings," *The Astrophysical Journal*, *343*, 526, 1989.

Clayton, Donald D., *Principles of Stellar Evolution and Nucleosynthesis*, McGraw-Hill, New York, 1968.

Cox, John P., *The Theory of Stellar Pulsation*, Princeton University Press, Princeton, NJ, 1980.

Freedman, Wendy L., et al., "Distance to the Virgo Cluster Galaxy M100 from Hubble Space Telescope Observations of Cepheids," *Nature*, *371*, 757, 1994.

Hansen, C. J., and Kawaler, S. D., *Stellar Interiors: Physical Principles, Structure, and Evolution*, Springer-Verlag, New York, 1994.

Leibacher, John W., and Stein, Robert F., "Oscillations and Pulsations," *The Sun as a Star*, NASA SP–450, 1981.

# Problems

**14.1** Use the light curve for Mira, Fig. 14.1, to estimate the ratio of Mira's luminosity at visible wavelengths, when it is brightest to when it is dimmest. For what fraction of its pulsation cycle is Mira visible to the naked eye?

**14.2** If the intrinsic uncertainty in the period–luminosity relation is $\Delta M \approx 0.5$ magnitude, find the resulting fractional uncertainty in the calculated distance to a classical Cepheid. (Astronomers can substantially decrease this uncertainty by also taking account of the color of the Cepheid.)

**14.3** The most distant classical Cepheids discovered to date were found by the Hubble Space Telescope in the galaxy denoted M100. (M100 is a member of the Virgo cluster, a rich cluster of galaxies.) Figure 14.20 shows the period–luminosity relation for these Cepheids. Use the two Cepheids nearest the figure's best-fit line to estimate the distance to M100. Compare your result to the distance of $17.1 \pm 1.8$ Mpc obtained by Wendy Freedman and her colleagues. The reader is referred to Freedman et al. (1994) for more information on the discovery and importance of these remote pulsating stars.

**14.4** Make a graph similar to Fig. 14.4 showing the period–luminosity relation for both the classical Cepheids and W Virginis stars.

**14.5** Assuming (incorrectly) that the oscillations of $\delta$ Cephei are sinusoidal, calculate the greatest excursion of its surface from its equilibrium position.

**14.6** Use Eq. (14.4) to estimate the pulsation period the Sun would have if it were to oscillate radially.

**14.7** Derive Eq. (14.11) by linearizing the adiabatic relation

$$PV^\gamma = \text{constant}.$$

**14.8** (a) Linearize the Stefan–Boltzmann equation in the form of Eq. (3.17) to show that

$$\frac{\delta L}{L_\circ} = 2\frac{\delta R}{R_\circ} + 4\frac{\delta T}{T_\circ}.$$

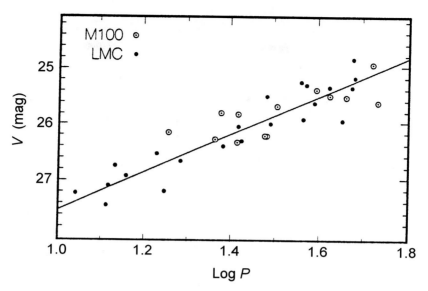

**Figure 14.20** A composite period–luminosity relation for Problem 14.3. The white circles denote Cepheids in M100, and the black circles show nearby Cepheids found in the Large Magellanic Cloud (a small galaxy that neighbors our Milky Way Galaxy). The average visual magnitudes of the LMC Cepheids have been increased by the same amount to match those of the M100 variables. The required increase in $V$ for a best fit is then used to find the relative distances to the LMC and M100. (Adapted from Freedman et al., *Nature*, *371*, 757, 1994.)

(b) Linearize the adiabatic relation $TV^{\gamma-1} =$ constant, and so find a relation between $\delta L/L_\circ$ and $\delta R/R_\circ$ for a spherical blackbody model star composed of an ideal monatomic gas.

14.9 Consider a general potential energy function, $U(r)$, for a force $\mathbf{F} = -(dU/dr)\,\hat{\mathbf{r}}$ on a particle of mass $m$. Assume that the origin ($r = 0$) is a point of stable equilibrium. By expanding $U(r)$ in a Taylor series about the origin, show that if it is displaced slightly from the origin and then released, the particle will undergo simple harmonic motion about the origin. This explains why the linearization procedure of Section 14.3 is guaranteed to result in sinusoidal oscillations.

14.10 Figure 14.21 shows a view of a nonradially pulsating ($\ell = 2, m = -2$), rotating star from above the star's north pole. Astronomers on Earth view the star, looking down on its equator. Assuming that a spectral absorption line appears as in Fig. 9.18 when the bottom of Fig. 14.21

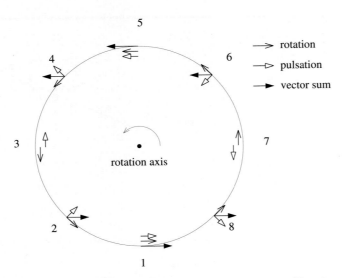

**Figure 14.21** Surface velocities for a rotating, pulsating star ($\ell = 2, m = -2$) for Problem 14.10. The arrows indicate the surface velocities due to rotation alone, pulsation alone, and their vector sum.

is facing Earth, sketch the changes in the appearance of the line profile due to Doppler shifts caused by the total surface velocity as the star rotates. (Don't worry about the timing; just sketch the spectral line as seen from the eight different points of view shown that are directly over the star's equator.) Assume that the equivalent width of the line does not change. You may wish to compare your line profiles with those actually observed for a nonradially pulsating star such as the $\beta$ Cephei star 12 Lacertae; see Smith, *Ap. J.*, *240*, 149, 1980. For convenience, the magnitudes of the rotation and pulsation velocities are assumed to be equal.

14.11 Show that Eq. (10.86), the condition for convection to occur, is the same as the requirement that $A > 0$, where $A$ is given by Eq. (14.17). Assume that the mean molecular weight, $\mu$, does not vary.

14.12 In a convection zone, the time scale for convection (see Section 10.4) is related to the value of $A$ (Eq. 14.17) by $t_c \simeq 2\sqrt{2/Ag}$. Table 14.2 shows the values of the pressure and density at two points near the top of the Sun's convection zone. Use these values and $\gamma = 5/3$ to obtain an estimate of the time scale for convection near the top of the

| $r$ (cm) | $P$ (dyne cm$^{-2}$) | $\rho$ (g cm$^{-3}$) |
|---|---|---|
| $6.959318 \times 10^{10}$ | $9.2860 \times 10^{4}$ | $2.2291 \times 10^{-7}$ |
| $6.959366 \times 10^{10}$ | $8.9957 \times 10^{4}$ | $2.1925 \times 10^{-7}$ |

**Table 14.2** Data from a Standard Solar Model for Problem 14.12. (Data from Guzik, private communication, 1994.)

Sun's convection zone. How does your answer compare with the range of periods observed for the Sun's p-modes?

14.13 **Computer Problem** In this problem you will carry out a nonlinear calculation of the radial pulsation of the one-zone model described in Example 14.1. The equations that describe the oscillation of this model star are Newton's second law for the forces on the shell,

$$m\frac{dv}{dt} = -\frac{GMm}{R^2} + 4\pi R^2 P, \qquad (14.20)$$

and the definition of the velocity, $v$, of the mass shell,

$$v = \frac{dR}{dt}. \qquad (14.21)$$

As in Example 14.1, we assume that the expansion and contraction of the gas is adiabatic:

$$P_i V_i^\gamma = P_f V_f^\gamma, \qquad (14.22)$$

where the "initial" and "final" subscripts refer to any two instants during the pulsation cycle.

(a) Explain in words the meaning of each term in Eq. (14.20). Also, use Eq. (14.22) to show that

$$P_i R_i^{3\gamma} = P_f R_f^{3\gamma}. \qquad (14.23)$$

(b) You will not be taking derivatives. Instead, you will take the difference between the initial and final values of the radius $R$ and radial velocity $v$ of the shell divided by the time interval $\Delta t$ separating the initial and final values. That is, you will use $(v_f - v_i)/\Delta t$ instead of $dv/dt$, and $(R_f - R_i)/\Delta t$ instead of $dR/dt$ in Eqs. (14.20) and (14.21). A careful analysis shows that you should use $R = R_i$ and $P = P_i$ on the right-hand side of Eq. (14.20), and use $v = v_f$ on the left-hand side of Eq. (14.21). Make these substitutions in Eqs. (14.20) and (14.21), and show that you can write

$$v_f = v_i + \left(\frac{4\pi R_i^2 P_i}{m} - \frac{GM}{R_i^2}\right)\Delta t \qquad (14.24)$$

and

$$R_f = R_i + v_f \Delta t. \qquad (14.25)$$

(c) Now you are ready to calculate the oscillation of the model star. The mass of a typical classical Cepheid is $M = 1 \times 10^{34}$ g (5 $M_\odot$), and the mass of the surface layers may be arbitrarily assigned $m = 1 \times 10^{29}$ g. For starting values at time $t = 0$, take

$$R_i = 1.7 \times 10^{12} \text{ cm}$$
$$v_i = 0 \text{ cm s}^{-1}$$
$$P_i = 5.6 \times 10^5 \text{ dyne cm}^{-2}$$

and use a time interval of $\Delta t = 10^4$ s. Take the ratio of specific heats to be $\gamma = 5/3$ for an ideal monatomic gas. Use Eq. (14.24) to calculate the final velocity $v_f$ at the end of one time interval (at time $t = 1 \times 10^4$ s); then use Eq. (14.25) to calculate the final radius $R_f$ and Eq. (14.23) to calculate the final pressure $P_f$. Now take these final values to be your new initial values, and find new values for $R$, $v$, and $P$ after two time intervals (at time $t = 2 \times 10^4$ s). Continue to find $R$, $v$, and $P$ for 150 time intervals, until $t = 1.5 \times 10^6$ s. Make three graphs of your results: $R$ vs. $t$, $v$ vs. $t$, and $P$ vs. $t$. Plot the time on the horizontal axis.

(d) From your graphs, measure the period $\Pi$ of the oscillation (both in seconds and in days) and the equilibrium radius, $R_o$, of the model star. Compare this value of the period with that obtained from Eq. (14.14). Also compare your results with the period and radial velocity observed for $\delta$ Cephei.

# Chapter 15

# The Degenerate Remnants of Stars

## 15.1 The Discovery of Sirius B

In 1838 Friedrich Wilhelm Bessel used the technique of stellar parallax to find the distance to the star 61 Cygni. Following this first successful measurement of a stellar distance, Bessel applied his talents to another likely candidate: Sirius, the brightest appearing star in the sky. Its parallax angle of $p'' = 0.377''$ corresponds to a distance of only 2.65 pc, or 8.65 ly. Sirius's brilliance in the night sky is in part due to its proximity to Earth. As he followed the star's path through the heavens, Bessel found that it deviated slightly from a straight line. After ten years of precise observations, Bessel concluded in 1844 that Sirius is actually a binary star system. Although unable to detect the companion of the brighter star, he deduced that its orbital period was about 50 years (the modern value is 49.9 years) and predicted its position. The search was on for the unseen "Pup," the faint companion of the luminous "Dog Star."

The telescopes of Bessel's time were incapable of finding the Pup so close to the glare of its bright counterpart, and following Bessel's death in 1846 the enthusiasm for the quest waned. Finally in 1862, the son of the prominent American lensmaker Alvan Clark tested his father's new 18-inch refractor (3 inches larger than any previous instrument) on Sirius, and he promptly discovered the Pup at its predicted position. The dominant Sirius A was found to be nearly one thousand times brighter than the Pup, now called Sirius B; see Fig. 15.1. The details of their orbits about their center of mass (see Fig. 15.2 and Problem 7.4) revealed that Sirius A and Sirius B have masses of about 2.3 $M_\odot$ and

**Figure 15.1** The white dwarf, Sirius B, beside the overexposed image of Sirius A. (Courtesy of Lick Observatory.)

1.0 $M_\odot$, respectively. A more recent determination for the mass of Sirius B is $1.053 \pm 0.028$ $M_\odot$, and it is this value that will be used below.

Clark's discovery of Sirius B was made near the opportune time of apastron, when the two stars were most widely separated (by just $10''$). The great difference in their luminosities ($L_A = 23.5$ $L_\odot$ and $L_B = 0.03$ $L_\odot$) makes observations at other times much more difficult. When the next apastron arrived 50 years later, spectroscopists had developed the tools to measure the stars' surface temperatures. From the Pup's faint appearance, astronomers expected it to be cool and red. They were startled when Walter Adams, working at Mt. Wilson Observatory in 1915, discovered that to the contrary, Sirius B is a hot, blue-white star that emits much of its energy in the ultraviolet. A modern value of the temperature of Sirius B is 27,000 K, much hotter than Sirius A's 9910 K.

The implications for the star's physical characteristics were astounding. Using the Stefan–Boltzmann law, Eq. (3.17), to calculate the size of Sirius B results in a radius of only $5.5 \times 10^8$ cm $\approx 0.008$ $R_\odot$. Sirius B has the mass of the Sun confined within a volume smaller than Earth! The average density of Sirius B is $3.0 \times 10^6$ g cm$^{-3}$, and the acceleration due to gravity at its surface is about $4.6 \times 10^8$ cm s$^{-2}$. On Earth, the pull of gravity on a teaspoon of white-dwarf material would be 14.5 billion dynes (over 16 tons), and on the surface of the white dwarf it would weigh 470,000 times more. This fierce

## 15.2 White Dwarfs

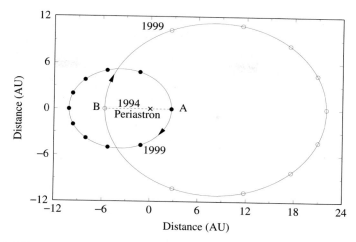

**Figure 15.2** The orbits of Sirius A and Sirius B. The center of mass of the system is marked with an "×."

gravity reveals itself in the spectrum of Sirius B; it produces an immense pressure near the surface that results in very broad hydrogen absorption lines; see Fig. 8.14.[1] Other than these lines, its spectrum is a featureless continuum.

Astronomers first reacted to the discovery of Sirius B by dismissing the results, calling them "absurd." However, the calculations are so simple and straightforward that this attitude soon changed to the one expressed by Eddington in 1922: "Strange objects, which persist in showing a type of spectrum entirely out of keeping with their luminosity, may ultimately teach us more than a host which radiate according to rule." Like all sciences, astronomy advances most rapidly when confronted with exceptions to its theories.

## 15.2 White Dwarfs

Obviously Sirius B is not a normal star. It is a **white dwarf**, a class of stars that have approximately the mass of the Sun and the size of Earth. Although as many as one-quarter of the stars in the vicinity of the Sun may be white dwarfs, the average characteristics of these faint stars have been difficult to determine because a complete sample has been obtained only within 10 pc of the Sun.

Figures 8.13 and 8.15 show that the white dwarfs occupy a narrow sliver of the H–R diagram that is roughly parallel to and below the main sequence.

---

[1] Recall the discussion of pressure broadening in Section 9.4.

Although white dwarfs are typically whiter than normal stars, the name itself is something of a misnomer since they come in all colors, with surface temperatures ranging from less than 5000 K to more than 80,000 K. Their spectral type, D (for "dwarf"), has several subdivisions. The largest group (about two-thirds of the total number, including Sirius B), called **DA white dwarfs**, display only pressure-broadened hydrogen absorption lines in their spectra. Hydrogen lines are absent from the **DB white dwarfs** (8%), which show only helium absorption lines, and the **DC white dwarfs** (14%) show no lines at all—only a continuum devoid of features.

It is instructive to estimate the conditions at the center of a white dwarf of mass $M_{\rm wd}$ and radius $R_{\rm wd}$, using the values for Sirius B given in the preceding section. Equation (14.3) with $r = 0$ shows that the central pressure is roughly[2]

$$P_c \approx \frac{2}{3}\pi G \rho^2 R_{\rm wd}^2 \approx 3.8 \times 10^{23} \text{ dyne cm}^{-2}, \tag{15.1}$$

about 1.5 million times larger than the pressure at the center of the Sun. A crude estimate of the central temperature may be obtained from Eq. (10.61) for the radiative temperature gradient,[3]

$$\frac{dT}{dr} = -\frac{3}{4ac}\frac{\overline{\kappa}\rho}{T^3}\frac{L_r}{4\pi r^2}$$

or

$$\frac{T_{\rm wd} - T_c}{R_{\rm wd} - 0} = -\frac{3}{4ac}\frac{\overline{\kappa}\rho}{T_{\rm wd}^3}\frac{L_{\rm wd}}{4\pi R_{\rm wd}^2}.$$

Assuming that the surface temperature, $T_{\rm wd}$, is much smaller than the central temperature and using $\overline{\kappa} = 0.2$ cm$^2$ g$^{-1}$ for electron scattering [Eq. (9.21) with $X = 0$] gives

$$T_c \approx \left[\frac{3\overline{\kappa}\rho}{4ac}\frac{L_{\rm wd}}{4\pi R_{\rm wd}}\right]^{1/4} \approx 7.6 \times 10^7 \text{ K}.$$

Thus the central temperature of a white dwarf is several times $10^7$ K.

These estimated values for a white dwarf lead directly to a surprising conclusion. Although hydrogen makes up roughly 70% of the visible mass of the universe, it cannot be present in appreciable amounts below the surface layers of a white dwarf. Otherwise, the dependence of the nuclear energy generation rates on density and temperature [see Eq. (10.49) for the pp chain and

---

[2]Remember that Eq. (14.3) was obtained for the unrealistic assumption of constant density.

[3]As will be discussed later in Section 15.5, the assumption of a radiative temperature gradient is incorrect because the energy is actually carried outward by electron conduction. However, Eq. (10.61) is sufficient for the purpose of this estimation.

## 15.2 White Dwarfs

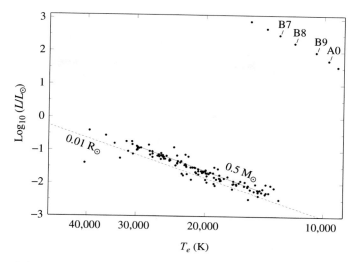

**Figure 15.3** DA white dwarfs on an H–R diagram. A line marks the location of the 0.50 M$_\odot$ white dwarfs, and a portion of the main sequence is at the upper right. (Data from Bergeron, Saffer, and Liebert, *Ap. J.*, *394*, 228, 1992.)

Eq. (10.53) for the CNO cycle] would produce white dwarf luminosities several orders of magnitude larger than those actually observed. Similar reasoning applied to other reaction sequences implies that thermonuclear reactions are not involved in producing the energy radiated by white dwarfs and that their centers must therefore consist of particles that are incapable of fusion at these densities and temperatures.

As was discussed in Section 13.2, white dwarfs are manufactured in the cores of low and intermediate-mass stars (those with an initial mass below 8 or 9 M$_\odot$ on the main sequence) near the end of their lives on the asymptotic giant branch of the H–R diagram. Because any star with a helium core mass exceeding about 0.5 M$_\odot$ will undergo fusion, most white dwarfs consist primarily of completely ionized carbon and oxygen nuclei.[4] As the aging giant expels its surface layers as a planetary nebula, the core is exposed as a white dwarf progenitor. The distribution of DA white dwarf masses is sharply peaked at 0.56 M$_\odot$, with some 80 percent lying between 0.42 M$_\odot$ and 0.70 M$_\odot$; see Fig. 15.3. The much larger main-sequence masses quoted earlier imply high mass-loss rates while on the asymptotic giant branch, involving thermal pulses and perhaps a superwind phase as well.

---

[4]Low-mass helium white dwarfs may also exist, and rare oxygen–neon–magnesium white dwarfs have been detected in a few novae.

The exceptionally strong pull of the white dwarf's gravity is responsible for the characteristic hydrogen spectrum of DA white dwarfs. Heavier nuclei are pulled below as the lighter hydrogen rises to the surface, resulting in a thin outer layer of hydrogen covering a layer of helium on top of the carbon–oxygen core.[5] This vertical stratification of nuclei according to their mass takes only 100 years or so in the hot atmosphere of the star. The origin of the non-DA (e.g., DB and DC) white dwarfs is not yet clear. Efficient mass-loss may occur on the asymptotic giant branch associated with the thermal pulse or superwind phases, stripping the white dwarf of nearly all of its hydrogen. Alternately, a single white dwarf may be transformed between the DA and non-DA spectral types by convective mixing in its surface layers.[6] For example, the helium convection zone's penetration into a thin hydrogen layer above could change a DA into a DB white dwarf.

White dwarfs with surface temperatures of $T_e \approx 12,000$ K lie within the instability strip of the H–R diagram and pulsate with periods between 100 and 1000 s; see Fig. 8.15 and Table 14.1. These **ZZ Ceti** variables, named after the prototype discovered in 1968 by Arlo Landolt, are variable DA white dwarfs; hence they are also known as **DAV stars**. The pulsation periods correspond to nonradial g-modes that resonate within the white dwarf's surface layers of hydrogen and helium.[7] Because these g-modes involve almost perfectly horizontal displacements, the radii of these compact pulsators hardly change. Their brightness variations (typically a few tenths of a magnitude) are due to temperature variations on the stars' surfaces. Since most stars will end their lives as white dwarfs, these must be the most common type of variable star in the universe, although only about thirty have been detected.

Successful numerical calculations of pulsating white dwarf models were carried out by American astronomer Don Winget and others. They were able to demonstrate that it is the *hydrogen* partial ionization zone that is responsible for driving the oscillations of the ZZ Ceti stars, as mentioned in Section 14.2. These computations also confirmed the elemental stratification of white dwarf envelopes. Winget and his colleagues went on to predict that hotter DB white dwarfs should also exhibit g-mode oscillations driven by the *helium* partial ionization zone. Within a year's time, this prediction was confirmed when the first

---

[5]Estimates of the relative masses of the hydrogen and helium layers range from $m(H)/m(He) \approx 10^{-2}$ to $10^{-11}$ for DA white dwarfs.

[6]As will be seen in Section 15.5, steep temperature gradients produce convection zones in the white dwarf's surface layers.

[7]The nonradial pulsation of stars was discussed in Section 14.4. Unlike the g-modes of normal stars, shown in Fig. 14.16, the g-modes of white dwarfs are confined to their surface layers.

## 15.3 The Physics of Degenerate Matter

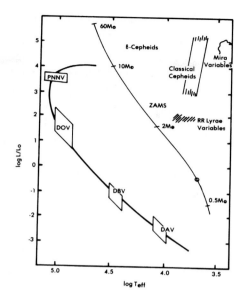

**Figure 15.4** Compact pulsators on the H–R diagram. (Figure from Winget, *Advances in Helio- and Asteroseismology*, Christensen-Dalsgaard and Frandsen (eds.), Reidel, Dordrecht, 1988.)

**DBV** star ($T_e \approx 27{,}000$ K) was discovered by Winget and his collaborators.[8] The location of the DAV and DBV stars on the H–R diagram is shown in Fig. 15.4, along with the very hot DOV and PNNV ($T_e \approx 10^5$ K) variables that are associated with the birth of white dwarfs. ("PNN" stands for planetary nebula nuclei and the DO spectral type marks the transition to the white dwarf stage.) All of these stars have multiple periods, simultaneously displaying at least 3, and as many as 125, different frequencies. Astronomers are deciphering the data to obtain a detailed look at the structure of white dwarfs.

## 15.3 The Physics of Degenerate Matter

We now delve below the surface to ask, What can support a white dwarf against the relentless pull of its gravity? It is easy to show (Problem 15.4) that normal gas and radiation pressure are completely inadequate. The answer was discovered in 1926 by the American physicist R. H. Fowler, who applied the new idea of the Pauli exclusion principle (recall Section 5.4) to the electrons within the

---

[8] Readers interested in this unique prediction and subsequent discovery of a new type of star are referred to Winget et al. (1982a,b).

white dwarf. The qualitative argument that follows elucidates the fundamental physics of the **electron degeneracy pressure** described by Fowler.

Any system—whether an atom of hydrogen, an oven filled with blackbody photons, or a box filled with gas particles—consists of quantum states that are identified by a set of quantum numbers. Just as the oven is filled with standing waves of electromagnetic radiation that are described by three quantum numbers (specifying the number of photons of wavelength $\lambda$ traveling in the $x$-, $y$-, and $z$-directions), a box of gas particles is filled with standing de Broglie waves that are also described by three quantum numbers (specifying the particle's component of momentum in each of three directions). If the gas particles are fermions (such as electrons), then the Pauli exclusion principle allows at most one fermion in each quantum state because no two fermions can have the same set of quantum numbers.

In an everyday gas at standard temperature and pressure, only one of every $10^7$ quantum states is occupied by a gas particle, and the limitations imposed by the Pauli exclusion become insignificant. Ordinary gas has a *thermal* pressure that is related to its temperature by the ideal gas law. However, as energy is removed from the gas and its temperature falls, an increasingly large fraction of the particles are forced into the lower-energy states. If the gas particles are fermions, only one particle is allowed in each state; thus all the particles cannot crowd into the ground state. Instead, as the temperature of the gas is lowered, the fermions will fill up the lowest available unoccupied states starting with the ground state, and then successively occupy the excited states with the lowest energy. Even in the limit $T \to 0$ K, the vigorous motion of the fermions in excited states produces a pressure in the fermion gas. At zero temperature, *all* of the lower energy states and *none* of the higher energy states are occupied. Such a fermion gas is said to be completely **degenerate**.

As shown in Fig. 15.5, at $T = 0$ K the energy dividing the occupied from the vacant states is called the **Fermi energy**,

$$\varepsilon_F = \frac{\hbar^2}{2m} \left(3\pi^2 n\right)^{2/3}, \tag{15.2}$$

where $m$ is the mass of the fermion and $n$ is the number of fermions per unit volume. The average energy per fermion at zero temperature is $\frac{3}{5}\varepsilon_F$.

At any temperature above absolute zero, some of the states with an energy less than $\varepsilon_F$ will become vacant as fermions use their thermal energy to occupy other, more energetic states. Although the degeneracy will not be precisely complete when $T > 0$ K, the assumption of complete degeneracy is a good approximation at the densities encountered in the interior of a white dwarf. All but the most energetic particles will have an energy less than the Fermi

## 15.3 The Physics of Degenerate Matter

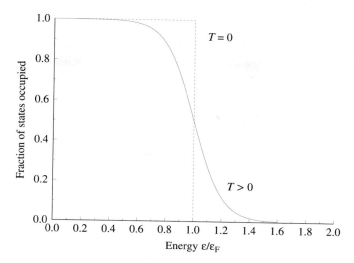

**Figure 15.5** Fraction of states of energy $\varepsilon$ occupied by fermions.

energy. To understand how the degree of degeneracy depends on both the temperature and density of the white dwarf, we first express the Fermi energy in terms of the density of the electron gas. For full ionization, the number of electrons per unit volume is

$$n_e = \left(\frac{\#\text{electrons}}{\text{nucleon}}\right)\left(\frac{\#\text{nucleons}}{\text{volume}}\right) = \left(\frac{Z}{A}\right)\frac{\rho}{m_H}, \qquad (15.3)$$

where $Z$ and $A$ are the number of protons and nucleons, respectively, in the white dwarf's nuclei, and $m_H$ is the mass of a hydrogen atom.[9] Thus the Fermi energy is proportional to the 2/3 power of the density,

$$\varepsilon_F = \frac{\hbar^2}{2m_e}\left[3\pi^2\left(\frac{Z}{A}\right)\frac{\rho}{m_H}\right]^{2/3}. \qquad (15.4)$$

Now compare the Fermi energy with the average thermal energy of an electron, $\frac{3}{2}kT$ [where $k$ is Boltzmann's constant; see Eq. (10.22)]. In rough terms, if $\frac{3}{2}kT < \varepsilon_F$, then an average electron will be unable to make a transition to an unoccupied state, and the electron gas will be degenerate. That is, for a degenerate gas,

$$\frac{3}{2}kT < \frac{\hbar^2}{2m_e}\left[3\pi^2\left(\frac{Z}{A}\right)\frac{\rho}{m_H}\right]^{2/3},$$

---

[9] The hydrogen mass is adopted as a representative mass of the proton and neutron.

or

$$\frac{T}{\rho^{2/3}} < \frac{\hbar^2}{3m_e k} \left[ \frac{3\pi^2}{m_H} \left( \frac{Z}{A} \right) \right]^{2/3} = 1.3 \times 10^5 \text{ K cm}^2 \text{ g}^{-2/3}$$

for $Z/A = 0.5$. Defining

$$\mathcal{D} \equiv 1.3 \times 10^5 \text{ K cm}^2 \text{ g}^{-2/3},$$

the condition for degeneracy is

$$\frac{T}{\rho^{2/3}} < \mathcal{D}. \tag{15.5}$$

The smaller the value of $T/\rho^{2/3}$, the more degenerate is the gas.

---

**Example 15.1** How important is electron degeneracy at the centers of the Sun and Sirius B? At the center of one standard solar model, $T_c = 1.58 \times 10^7$ K and $\rho_c = 162$ g cm$^{-3}$. Then

$$\frac{T_c}{\rho_c^{2/3}} = 5.3 \times 10^5 > 1.3 \times 10^5 = \mathcal{D}.$$

In the Sun, electron degeneracy is quite weak and plays a very minor role, supplying only a few tenths of a percent of the central pressure. However, as the Sun continues to evolve, electron degeneracy will become increasingly important (Fig. 15.6). As described in Section 13.2, the Sun will develop a degenerate helium core while on the red giant branch of the H–R diagram, leading eventually to a core helium flash. Later, on the asymptotic giant branch, the progenitor of a carbon–oxygen white dwarf will form in the core to be revealed with the ejection of the Sun's surface layers as a planetary nebula.

For Sirius B, the values of the density and central temperature estimated above lead to

$$\frac{T_c}{\rho_c^{2/3}} = 3.6 \times 10^3 \ll 1.3 \times 10^5 = \mathcal{D},$$

and so complete degeneracy is a valid assumption for Sirius B.

---

We now estimate the electron degeneracy pressure by combining two key ideas of quantum mechanics:

1. The Pauli exclusion principle, which allows at most one electron in each quantum state; and

## 15.3 The Physics of Degenerate Matter

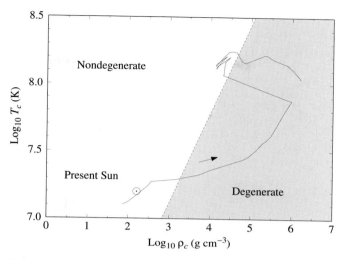

**Figure 15.6** Degeneracy in the Sun's center as it evolves. (Data from Mazzitelli and D'Antonia, *Ap. J.*, *311*, 762, 1986.)

2. Heisenberg's uncertainty principle in the form of Eq. (5.18),

$$\Delta x \, \Delta p \approx \hbar,$$

which requires that an electron confined to a small volume of space have a correspondingly high uncertainty in its momentum. Because the minimum value of the electron's momentum, $p_{\min}$, is approximately $\Delta p$, more closely confined electrons will have greater momenta.

Making the unrealistic assumption that all of the electrons have the same momentum, $p$, Eq. (10.9) for the pressure integral becomes

$$P \approx \frac{1}{3} n_e p v, \tag{15.6}$$

where $n_e$ is the total electron number density.

In a completely degenerate electron gas, the electrons are packed as tightly as possible, with a separation between neighboring electrons of about $n_e^{-1/3}$. However, to satisfy the Pauli exclusion principle, the electrons must maintain their identities as different particles. That is, the uncertainty in their positions cannot be larger than their physical separation. Identifying $\Delta x \approx n_e^{-1/3}$ for the limiting case of complete degeneracy, we can use Heisenberg's uncertainty relation to estimate the momentum of an electron,

$$p \approx \Delta p \approx \frac{\hbar}{\Delta x} \approx \hbar n_e^{1/3} \tag{15.7}$$

(see Example 5.5). Using Eq. (15.3) for the electron number density with full ionization gives

$$p \approx \hbar \left[ \left(\frac{Z}{A}\right) \frac{\rho}{m_H} \right]^{1/3}.$$

For nonrelativistic electrons, the speed is

$$v = \frac{p}{m_e}$$

$$\approx \frac{\hbar}{m_e} n_e^{1/3} \tag{15.8}$$

$$\approx \frac{\hbar}{m_e} \left[ \left(\frac{Z}{A}\right) \frac{\rho}{m_H} \right]^{1/3}. \tag{15.9}$$

Inserting Eqs. (15.3), (15.7), and (15.9) into Eq. (15.6) for the electron degeneracy pressure results in

$$P \approx \frac{1}{3} \frac{\hbar^2}{m_e} \left[ \left(\frac{Z}{A}\right) \frac{\rho}{m_H} \right]^{5/3}. \tag{15.10}$$

This is roughly a factor of five smaller than the exact expression for $P$,

$$P = \frac{(3\pi^2)^{2/3}}{5} \frac{\hbar^2}{m_e} n_e^{5/3}$$

$$= \frac{(3\pi^2)^{2/3}}{5} \frac{\hbar^2}{m_e} \left[ \left(\frac{Z}{A}\right) \frac{\rho}{m_H} \right]^{5/3}, \tag{15.11}$$

the pressure due to a completely degenerate, nonrelativistic electron gas.

Using $Z/A = 0.5$ for a carbon–oxygen white dwarf, Eq. (15.11) shows that the electron degeneracy pressure available to support a white dwarf such as Sirius B is about $1.9 \times 10^{23}$ dyne cm$^{-2}$, within a factor of two of the estimate of the central pressure made previously. *Electron degeneracy pressure is responsible for maintaining hydrostatic equilibrium in a white dwarf.*

## 15.4 The Chandrasekhar Limit

The relation between the radius, $R_{\rm wd}$, of a white dwarf and its mass, $M_{\rm wd}$, may be found by setting the estimate of the central pressure, Eq. (15.1), equal to the electron degeneracy pressure, Eq. (15.11):

$$\frac{2}{3} \pi G \rho^2 R_{\rm wd}^2 = \frac{(3\pi^2)^{2/3}}{5} \frac{\hbar^2}{m_e} \left[ \left(\frac{Z}{A}\right) \frac{\rho}{m_H} \right]^{5/3}.$$

## 15.4 The Chandrasekhar Limit

Using $\rho = M_{\text{wd}}/\frac{4}{3}\pi R_{\text{wd}}^3$ (assuming constant density), this leads to an estimate of the radius of the white dwarf,

$$R_{\text{wd}} \approx \frac{(18\pi)^{2/3}}{10} \frac{\hbar^2}{Gm_e M_{\text{wd}}^{1/3}} \left[\left(\frac{Z}{A}\right) \frac{1}{m_H}\right]^{5/3}. \tag{15.12}$$

For a 1 $M_\odot$ carbon–oxygen white dwarf, $R \approx 2.9 \times 10^8$ cm, too small by roughly a factor of two but an acceptable estimate. More important is the surprising implication that $M_{\text{wd}} R_{\text{wd}}^3 = $ constant, or

$$M_{\text{wd}} V_{\text{wd}} = \text{constant}. \tag{15.13}$$

The volume of a white dwarf is inversely proportional to its mass, so more massive white dwarfs are actually *smaller*. This **mass–volume relation** is a result of the star deriving its support from electron degeneracy pressure. The electrons must be more closely confined to generate the larger degeneracy pressure required to support a more massive star. In fact, the mass–volume relation implies that $\rho \propto M_{\text{wd}}^2$.

According to the mass–volume relation, piling more and more mass onto a white dwarf would eventually result in shrinking the star down to zero volume as its mass becomes infinite. However, if the density exceeds about $10^6$ g cm$^{-3}$, there is a departure from this relation. To see why this is so, use Eq. (15.9) to estimate the speed of the electrons in Sirius B:

$$v \approx \frac{\hbar}{m_e} \left[\left(\frac{Z}{A}\right) \frac{\rho}{m_H}\right]^{1/3} = 1.1 \times 10^{10} \text{ cm s}^{-1},$$

over one-third the speed of light! If the mass–volume relation were correct, white dwarfs a bit more massive than Sirius B would be so small and dense that their electrons would exceed the limiting value of the speed of light. This impossibility points out the dangers of ignoring the effects of relativity in our expression for the electron speed (Eq. 15.9) and pressure (Eq. 15.10).[10] Because the electrons are moving more slowly than the nonrelativistic Eq. (15.9) would indicate, there is less electron pressure available to support the star. Thus a massive white dwarf is *smaller* than predicted by the mass–volume relation. Indeed, zero volume occurs for a finite value of the mass. *There is therefore a maximum mass for white dwarfs*, a limit to the amount of matter that can be supported by electron degeneracy pressure. This amazing conclusion, the result

---

[10] It is left as an exercise to show that relativistic effects must be included for densities greater than $10^6$ g cm$^{-3}$.

of correctly including relativity in the physics of white dwarfs, was announced in 1931 by the brilliant Indian physicist S. Chandrasekhar.

To appreciate the effect of relativity on the stability of a white dwarf, note that Eq. (15.11) (which is valid only for approximately $\rho < 10^6$ g cm$^{-3}$) is of the form $P = K\rho^{5/3}$, where $K$ is a constant. Comparing this with Eq. (10.78) shows that the value of the ratio of specific heats is $\gamma = 5/3$ in the nonrelativistic limit. As was discussed in Section 14.3, this means that the white dwarf is dynamically stable. If it suffers a small perturbation, it will return to its equilibrium structure instead of collapsing. However, in the extreme relativistic limit, the electron speed $v = c$ must be used instead of Eq. (15.9) to find the electron degeneracy pressure. The result is

$$P = \frac{(3\pi^2)^{1/3}}{4} \hbar c \left[ \left(\frac{Z}{A}\right) \frac{\rho}{m_H} \right]^{4/3}. \tag{15.14}$$

In this limit $\gamma = 4/3$, which corresponds to *dynamical instability*. The smallest departure from equilibrium will cause the white dwarf to collapse as electron degeneracy pressure fails.[11] As was explained in Section 13.3, approaching the Chandrasekhar limit leads to the collapse of the degenerate core in an aging supergiant, resulting in a Type II supernova.

An approximate value for the maximum white-dwarf mass may be obtained by setting the estimate of the central pressure, Eq. (15.1) with $\rho = M_{wd}/\frac{4}{3}\pi R_{wd}^3$, equal to Eq. (15.14) with $Z/A = 0.5$. The radius of the white dwarf cancels, leaving

$$M_{Ch} \approx \frac{3\sqrt{2\pi}}{8} \left(\frac{\hbar c}{G}\right)^{3/2} \left[ \left(\frac{Z}{A}\right) \frac{1}{m_H} \right]^2 = 8.7 \times 10^{32} \text{ g} = 0.44 \text{ M}_\odot$$

for the greatest possible mass. This formula is truly remarkable. It contains three fundamental constants—$\hbar$, $c$, and $G$—representing the combined effects of quantum mechanics, relativity, and Newtonian gravitation on the structure of a white dwarf. A precise derivation with $Z/A = 0.5$ results in a value of $M_{Ch} = 1.44$ M$_\odot$, called the **Chandrasekhar limit**. Figure 15.7 shows the mass–radius relation for white dwarfs.[12] No white dwarf has been discovered with a mass exceeding the Chandrasekhar limit.[13]

---

[11] In fact, the strong gravity of the white dwarf, as described by Einstein's general theory of relativity (see Section 16.1), acts to raise the critical value of $\gamma$ for dynamical instability slightly above 4/3.

[12] Figure 15.7 does not include complications such as the electrostatic attraction between the nuclei and electrons in a white dwarf, thus tending to reduce the radius slightly.

[13] It is natural to wonder about the outcome of sneaking up on the Chandrasekhar limit by

## 15.4 The Chandrasekhar Limit

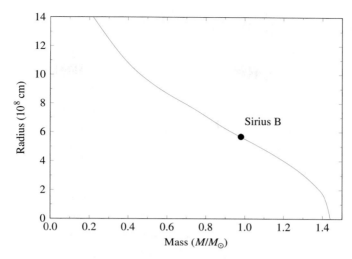

**Figure 15.7** Radii of white dwarfs of $M_{\text{wd}} \leq M_{Ch}$ at $T = 0$ K.

It is important to emphasize that neither the nonrelativistic or relativistic formulas for the electron degeneracy pressure [Eqs. (15.11) and (15.14), respectively] contains the temperature. Unlike the gas pressure of the ideal gas law and the expression for radiation pressure, the pressure of a completely degenerate electron gas is independent of its temperature. This has the effect of decoupling the mechanical structure of the star from its thermal properties.

We have already seen one implication of this decoupling in Section 13.2, where the helium core flash was described as the result of the independence of the mechanical and thermal behavior of the degenerate helium core of a low-mass star. When helium burning begins in the core, it proceeds without an accompanying increase in pressure that would normally expand the core and therefore restrain the rising temperature. The resulting rapid rise in temperature leads to a runaway production of nuclear energy—the helium flash—which lasts until the temperature becomes sufficiently high to remove the degeneracy of the core, allowing it to expand. On the other hand, a star may have so little mass that its core temperature never becomes high enough to initiate the helium burning. The result in this case is the formation of a helium white dwarf.

---

adding just a bit more mass to white dwarf with very nearly 1.44 $M_\odot$. This will be considered in Section 17.4, where Type I supernovae are discussed. A Type I supernova occurs when a white dwarf pulls gas from a giant companion star in a binary system.

## 15.5 The Cooling of White Dwarfs

Most stars end their lives as white dwarfs. These glowing embers scattered throughout space are a galaxy's memory of its past glory. Because no fusion occurs in their interiors, white dwarfs simply cool off at an essentially constant radius as they slowly deplete their supply of thermal energy (recall Fig. 15.3). Much effort has been directed at understanding the rate at which a white dwarf cools so its lifetime and the time of its birth may be calculated. Just as paleontologists can read the history of Earth's life in the fossil record, astronomers may be able to recover the history of star formation in our Galaxy by studying the statistics of white-dwarf temperatures. This section will be devoted to a discussion of the principles involved in this stellar archaeology.

First we must ask how energy is transported outward from the interior of a white dwarf. In an ordinary star, photons travel much farther than atoms do before suffering a collision that robs them of energy (recall Examples 9.1 and 9.2). As a result, photons are normally a more efficient carrier of energy to the stellar surface. In a white dwarf, however, the degenerate electrons can travel long distances before losing energy in a collision with a nucleus, since the vast majority of the lower-energy electron states are already occupied. Thus, in a white dwarf, energy is carried by **electron conduction** rather than by radiation. This is so efficient that the interior of a white dwarf is nearly isothermal, with the temperature dropping significantly only in the nondegenerate surface layers. Figure 15.8 shows that a white dwarf consists of a nearly constant-temperature interior surrounded by a thin nondegenerate envelope that transfers heat less efficiently, allowing the energy to leak out slowly. The steep temperature gradient near the surface creates convection zones that may alter the appearance of the white dwarf's spectrum as it cools (as described in Section 15.2).

The structure of the (nondegenerate) surface layers of a star is described at the beginning of Appendix H. For a white dwarf of surface luminosity $L_{\rm wd}$ and mass $M_{\rm wd}$, Eq. (H.1) for the pressure $P$ as a function of the temperature $T$ in the envelope is[14]

$$P = \left( \frac{4}{17} \frac{16\pi ac}{3} \frac{GM_{\rm wd}}{L_{\rm wd}} \frac{k}{\kappa_\circ \mu m_H} \right)^{1/2} T^{17/4}, \qquad (15.15)$$

where $\kappa_\circ$ [called "$A$" in Eq. (H.1)] is the coefficient of the bound–free Kramers

---

[14]Equation (15.15) assumes that the envelope is in radiative equilibrium, with the energy carried outward by photons. Even when convection occurs in the surface layers of a white dwarf, it is not expected to have a large effect on the cooling.

## 15.5 The Cooling of White Dwarfs

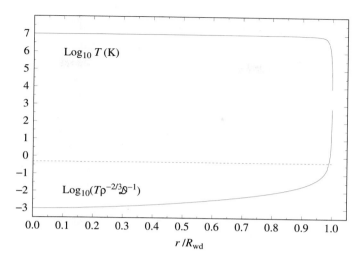

**Figure 15.8** Temperature and degree of degeneracy in the interior of a white-dwarf model.

opacity law in Eq. (9.19),

$$\kappa_\circ = 4.34 \times 10^{25} \, Z(1+X) \text{ cm}^2 \text{ g}^{-1}.$$

Using the ideal gas law, Eq. (10.14), to replace the pressure results in a relation between the density and the temperature,

$$\rho = \left( \frac{4}{17} \frac{16\pi ac}{3} \frac{GM_{\text{wd}}}{L_{\text{wd}}} \frac{\mu m_H}{\kappa_\circ k} \right)^{1/2} T^{13/4}. \tag{15.16}$$

The transition between the nondegenerate surface layers and the degenerate isothermal interior of temperature $T_c$ is described by setting the two sides of Eq. (15.5) equal to each other. Using this to replace the density results in an expression for the luminosity at the white dwarf's surface in terms of its interior temperature,

$$L_{\text{wd}} = \frac{4D^3}{17} \frac{16\pi ac}{3} \frac{Gm_H}{\kappa_\circ k} \mu M_{\text{wd}} T_c^{7/2}$$

$$= C T_c^{7/2}, \tag{15.17}$$

where

$$C \equiv \frac{4D^3}{17} \frac{16\pi ac}{3} \frac{Gm_H}{\kappa_\circ k} \mu M_{\text{wd}}$$

$$= 7.3 \times 10^5 \left( \frac{M_{\text{wd}}}{M_\odot} \right) \frac{\mu}{Z(1+X)}$$

in cgs units. Note that the luminosity is proportional to $T_c^{7/2}$ (the *interior* temperature), while it varies as the fourth power of the *effective* temperature according to the Stefan–Boltzmann law, Eq. (3.17). The surface of a white dwarf cools more slowly than its isothermal interior as its thermal energy leaks into space.

---

**Example 15.2** Equation (15.17) can be used to estimate the interior temperature of a 1 $M_\odot$ white dwarf with $L_{\rm wd} = 0.03\ L_\odot$. Arbitrarily assuming values of $X = 0$, $Y = 0.9$, $Z = 0.1$ for the nondegenerate envelope (so $\mu \simeq 1.4$) results in[15]

$$T_c = \left[ \frac{L_{\rm wd}}{7.3 \times 10^4} \left( \frac{M_\odot}{M_{\rm wd}} \right) \frac{Z(1+X)}{\mu} \right]^{2/7} = 2.8 \times 10^7\ {\rm K}.$$

Equating the two sides of the degeneracy condition, Eq. (15.5), shows that the density at the base of the nondegenerate envelope is about

$$\rho = \left( \frac{T_c}{\mathcal{D}} \right)^{3/2} = 3.1 \times 10^3\ {\rm g\ cm}^{-3}.$$

This result is several orders of magnitude less than the average density of a 1 $M_\odot$ white dwarf such as Sirius B and verifies that the envelope is indeed thin, contributing very little to the star's total mass.

---

A white dwarf's thermal energy resides primarily in the kinetic energy of its nuclei; the degenerate electrons cannot give up a significant amount of energy because nearly all of the lower energy states are already occupied. The total number of nuclei in the white dwarf is equal to the star's mass, $M_{\rm wd}$, divided by the mass of a nucleus, $Am_H$. Furthermore, since the average thermal energy of a nucleus is $\frac{3}{2}kT$, the thermal energy available for radiation is

$$U = \frac{M_{\rm wd}}{Am_H} \frac{3}{2} kT_c. \qquad (15.18)$$

If we use the value of $T_c$ from Example 15.2 and $A = 12$ for carbon, Eq. (15.18) gives approximately $6.0 \times 10^{47}$ ergs. A crude estimate of the characteristic time scale for cooling, $\tau_{\rm cool}$, can be obtained simply by dividing the thermal energy by the luminosity. Thus

$$\tau_{\rm cool} = \frac{U}{L_{\rm wd}} = \frac{3}{2} \frac{M_{\rm wd} k}{Am_H C T_c^{5/2}}, \qquad (15.19)$$

---

[15] Because the amount of hydrogen is quite small even in a DA white dwarf, this composition is a reasonable choice for both types DA and DB.

## 15.5 The Cooling of White Dwarfs

which is about $5.2 \times 10^{15}$ s $\approx$ 170 million years. This is an underestimate, because the cooling time scale increases as $T_c$ decreases. The more detailed calculation that follows shows that a white dwarf spends most of its life cooling slowly with a low temperature and luminosity.

The depletion of the internal energy provides the luminosity, so Eqs. (15.17) and (15.18) give

$$-\frac{dU}{dt} = L_{\text{wd}}$$

or

$$-\frac{d}{dt}\left(\frac{M_{\text{wd}}}{Am_H}\frac{3}{2}kT_c\right) = CT_c^{7/2}.$$

If the initial temperature of the interior is $T_\circ$ when $t = 0$, then this expression may be integrated to obtain the core temperature as a function of time,

$$T_c(t) = T_\circ\left(1 + \frac{5}{3}\frac{Am_H CT_\circ^{5/2}}{M_{\text{wd}}k}t\right)^{-2/5} = T_\circ\left(1 + \frac{5}{2}\frac{t}{\tau_\circ}\right)^{-2/5}, \quad (15.20)$$

where $\tau_\circ$ is the time scale for cooling at the initial temperature of $T_\circ$; that is, $\tau_\circ = \tau_{\text{cool}}$ at time $t_\circ$. Inserting this into Eq. (15.17) shows that at first the luminosity of the white dwarf declines sharply from its initial value of $L_\circ = CT_\circ^{7/2}$ and then dims much more gradually as time passes:

$$L_{\text{wd}} = L_\circ\left(1 + \frac{5}{3}\frac{Am_H C^{2/7}L_\circ^{5/7}}{M_{\text{wd}}k}t\right)^{-7/5} = L_\circ\left(1 + \frac{5}{2}\frac{t}{\tau_\circ}\right)^{-7/5}. \quad (15.21)$$

The solid line in Fig. 15.9 shows the decline in the luminosity of a pure carbon 0.6 $M_\odot$ white dwarf calculated from Eq. (15.21). The dashed line is a curve obtained for a sequence of more realistic white dwarf models[16] that include thin surface layers of hydrogen and helium overlying the carbon core. The insulating effect of these layers slows the cooling by about 15%. Also included are some of the intriguing phenomena that occur as the white dwarf's internal temperature drops.

As a white dwarf cools, it crystallizes in a gradual process that starts at the center and moves outward. The upturned "knee" in the dashed curve in Fig. 15.9 at about $L_{\text{wd}}/L_\odot \approx 10^{-4}$ occurs when the cooling nuclei begin settling into a crystalline lattice. The regular crystal structure is maintained by the mutual electrostatic repulsion of the nuclei; it minimizes their energy as they vibrate about their average position in the lattice. As the nuclei undergo this phase change, they release their latent heat (about $kT$ per nucleus), slowing

---
[16]The reader is referred to Winget et al. (1987) for details of this and other cooling curves.

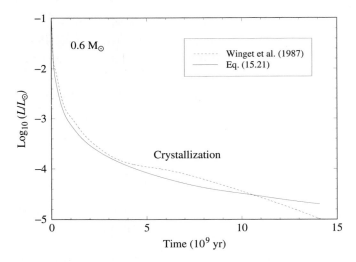

**Figure 15.9** Theoretical cooling curves for 0.6 M$_\odot$ white-dwarf models. [The solid line is from Eq. (15.21), and the dashed line is from Winget et al., *Ap. J. Lett.*, *315*, L77, 1987.]

the star's cooling and producing the knee in the cooling curve. Later, as the white dwarf's temperature continues to drop, the crystalline lattice actually accelerates the cooling as the coherent vibration of the regularly spaced nuclei promotes further energy loss. This is reflected in the subsequent downturn in the cooling curve. Thus the ultimate monument to the lives of most stars will be a "diamond in the sky," a cold, dark, Earth-size sphere of crystallized carbon and oxygen floating through the depths of space.[17]

Despite the large uncertainties in the measurement of surface temperatures,[18] it is possible to observe the cooling of a pulsating white dwarf. As the star's temperature declines, its period $\Pi$ slowly changes according to $d\Pi/dt \propto T^{-1}$ (approximately). Extremely precise measurements of a rapidly cooling DOV star yield a period derivative of $\Pi/|d\Pi/dt| = 1.4 \times 10^6$ years, in excellent agreement with the theoretical value. Measuring period changes for the more slowly cooling DBV and DAV stars will be even more difficult, although preliminary results are encouraging. Astronomers are determined to fulfill the promise of testing and calibrating the theoretical cooling curves for white dwarfs.

This interest in an accurate calculation of the decline in a white dwarf's

---

[17] Unlike a terrestrial diamond, the white dwarf's nuclei will be arrayed in a body-centered cubic lattice like that of metallic sodium.

[18] For Sirius B, effective temperatures ranging from 27,000 K to 32,000 K are often quoted.

## 15.5 The Cooling of White Dwarfs

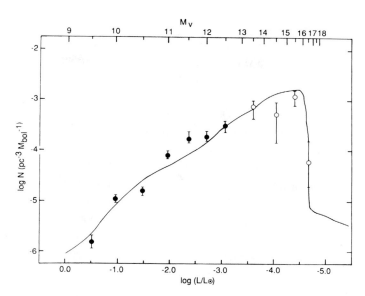

**Figure 15.10** Observed and theoretical distribution of white dwarf luminosities. (Figure from Winget et al., *Ap. J. Lett.*, *315*, L77, 1987.)

temperature reflects the hope of using these fossil stars as a tool for uncovering the history of star formation in our Galaxy. Figure 15.10, from Winget et al. (1987), illustrates how this might be accomplished. The circles (both open and filled) in the figure are the observed number of white dwarfs per cubic parsec with the absolute visual magnitude given at the top of the figure. The dramatically sudden drop in the population of white dwarfs with $L_{\rm wd}/L_\odot < -4.5$ is inconsistent with the assumption that stars have been forming in our Galaxy throughout the infinite past. Instead, this decline can best be explained if the first white dwarfs were formed and began cooling $9.0 \pm 1.8$ billion years ago. Figure 15.10 shows the theoretically expected distribution of white dwarf luminosities based on this cooling time, calculated using theoretical cooling curves similar to the one shown in Fig. 15.9 together with the observed distribution of white-dwarf masses. Furthermore, adding the time spent in the pre-white-dwarf stages of stellar evolution implies that star formation in the disk of our Galaxy began $9.3 \pm 2.0$ billion years ago. This time is about 6 billion years shorter than the age determined for the Milky Way's globular clusters (which formed at an earlier epoch) and may conflict with estimates of the age of the Galactic disk obtained by other methods. However, such a simple result, depending only on the observed shortage of faint white dwarfs and the calculated cooling times, is worthy of serious consideration.

## 15.6 Neutron Stars

Two years after James Chadwick discovered the neutron in 1932, a German astronomer and a Swiss astrophysicist, Walter Baade (1893–1960) and Fritz Zwicky (1898–1974) of Mount Wilson Observatory, proposed the existence of **neutron stars**. These two astronomers, who also coined the term "supernova," went on to suggest that "supernovae represent the transitions from ordinary stars into neutron stars, which in their final stages consist of extremely closely packed neutrons." Because neutron stars are formed when the degenerate core of an aging supergiant star nears the Chandrasekhar limit and collapses, we take $M_{Ch}$ (rounded to two figures) for a typical neutron star mass. A 1.4 solar-mass neutron star would consist of $1.4 \, M_\odot / m_n \approx 10^{57}$ neutrons—in effect, a huge nucleus with a mass number of $A \approx 10^{57}$ that is held together by gravity and supported by **neutron degeneracy pressure**.[19] It is left as an exercise to show that

$$R_{ns} \approx \frac{(18\pi)^{2/3}}{10} \frac{\hbar^2}{G M_{ns}^{1/3}} \left(\frac{1}{m_H}\right)^{8/3} \tag{15.22}$$

is the expression for the estimated neutron star radius that is analogous to Eq. (15.12) for a white dwarf. For $M_{ns} = 1.4 \, M_\odot$, this yields a value of $4.4 \times 10^5$ cm. As was found with Eq. (15.12) for white dwarfs, this estimate is too small by a factor of about 3. That is, the actual radius of a 1.4 $M_\odot$ neutron star lies roughly between 10 and 15 km; we will adopt a value of 10 km for the radius. As will be seen, there are many uncertainties involved in the construction of a model neutron star.

This incredibly compact stellar remnant would have an average density of $6.65 \times 10^{14}$ g cm$^{-3}$, greater than the typical density of an atomic nucleus, $\rho_{nuc} \approx 2.3 \times 10^{14}$ g cm$^{-3}$. In some sense, the neutrons in a neutron star must be "touching" one another. At the density of a neutron star, all of Earth's 5.5 billion inhabitants could be crowded into a cube 1 cm on each side.[20]

The pull of gravity at the surface of a neutron star is fierce. For a 1.4 $M_\odot$ neutron star with a radius of 10 km, $g = 1.86 \times 10^{14}$ cm s$^{-2}$, 190 billion times stronger than the acceleration of gravity at Earth's surface. An object dropped from a height of one meter would arrive at the star's surface with a speed of $1.93 \times 10^8$ cm s$^{-1}$ (about 4.3 million mph).

---

[19] Like electrons, neutrons are also fermions and so are subject to the Pauli exclusion principle.

[20] Astronomer Frank Shu has commented that this shows "how much of humanity is empty space"!

## 15.6 Neutron Stars

**Example 15.3** The inadequacy of using Newtonian mechanics to describe neutron stars can be demonstrated by calculating the escape velocity at the surface. Using Eq. (2.17), we find

$$v_{\text{esc}} = \sqrt{2GM_{\text{ns}}/R_{\text{ns}}} = 1.93 \times 10^{10} \text{ cm s}^{-1} = 0.643c.$$

The ratio of the (Newtonian) gravitational potential energy to the rest energy of an object of mass $m$ at the star's surface is

$$\frac{GM_{\text{ns}}m/R_{\text{ns}}}{mc^2} = 0.207.$$

Clearly, the effects of relativity must be included for an accurate description of a neutron star. This applies not only to Einstein's theory of special relativity, described in Chapter 4, but also to his theory of gravity, called the *general theory of relativity*, which will be considered in Section 16.1. Nevertheless, we will use both relativistic formulas and the more familiar Newtonian physics to reach qualitatively correct conclusions about neutron stars.

To appreciate the exotic nature of the material comprising a neutron star and the difficulties involved in calculating the equation of state, imagine compressing the mixture of iron nuclei and degenerate electrons that make up an iron white dwarf at the center of a massive supergiant star.[21] Specifically, we are interested in the equilibrium configuration of $10^{57}$ nucleons (protons and neutrons), together with enough free electrons to provide zero net charge. The equilibrium arrangement is the one that involves the least energy.

To begin with, at low densities the nucleons are found in iron nuclei. This is the outcome of the minimum-energy compromise between the repulsive Coulomb force between the protons and the attractive nuclear force between all of the nucleons. However, as mentioned in the previous section, when $\rho \approx 10^6$ g cm$^{-3}$ the electrons become relativistic. Soon thereafter, the minimum energy arrangement of protons and neutrons changes because the energetic electrons can convert protons in the iron nuclei into neutrons by the process of electron capture (Eq. 13.16),

$$p^+ + e^- \rightarrow n + \nu_e.$$

---

[21] Because the mechanical and thermal properties of degenerate matter are independent of one another, we will assume for convenience that $T = 0$ K. The iron nuclei are then arranged in a crystalline lattice.

Because the neutron mass is slightly greater than the sum of the proton and electron masses ($m_n c^2 - m_p c^2 - m_e c^2 = 0.78$ MeV), the electron must supply the kinetic energy to make up the difference.

---

**Example 15.4** We will obtain an estimate of the density at which the process of electron capture begins for a simple mixture of hydrogen nuclei (protons) and relativistic degenerate electrons,

$$p^+ + e^- \rightarrow n + \nu_e.$$

In the limiting case when the neutrino carries away no energy, we can equate the relativistic expression for the electron kinetic energy, Eq. (4.45), with the difference between the neutron rest energy and combined proton and electron rest energies and write

$$m_e c^2 \left( \frac{1}{\sqrt{1 - v^2/c^2}} - 1 \right) = (m_n - m_p - m_e) c^2,$$

or

$$\left( \frac{m_e}{m_n - m_p} \right)^2 = 1 - \frac{v^2}{c^2}.$$

Although Eq. (15.9) for the electron speed is strictly valid only for nonrelativistic electrons, it is accurate enough to be used in this estimate. Inserting this expression for $v$ leads to

$$\left( \frac{m_e}{m_n - m_p} \right)^2 \approx 1 - \frac{\hbar^2}{m_e^2 c^2} \left[ \left( \frac{Z}{A} \right) \frac{\rho}{m_H} \right]^{2/3}.$$

Solving for $\rho$ shows that the density at which electron capture begins is approximately

$$\rho \approx \frac{A m_H}{Z} \left( \frac{m_e c}{\hbar} \right)^3 \left[ 1 - \left( \frac{m_e}{m_n - m_p} \right)^2 \right]^{3/2}$$

$$\approx 2.3 \times 10^7 \text{ g cm}^{-3},$$

using $A/Z = 1$ for hydrogen. This is in reasonable agreement with the actual value of $\rho = 1.2 \times 10^7$ g cm$^{-3}$.

---

## 15.6 Neutron Stars

Free protons were considered in Example 15.4 to avoid the complications that arise when they are bound in heavy nuclei. A careful calculation that takes into account the surrounding nuclei and relativistic degenerate electrons as well as the complexities of nuclear physics reveals that the density must exceed $10^9$ g cm$^{-3}$ for the protons in $^{56}_{26}$Fe nuclei to capture electrons. At still higher densities, the most stable arrangement of nucleons is one where the neutrons and protons are found in a lattice of increasingly neutron-rich nuclei. This process is known as **neutronization**, and produces a sequence of nuclei such as $^{56}_{26}$Fe, $^{62}_{28}$Ni, $^{64}_{28}$Ni, $^{66}_{28}$Ni, $^{86}_{36}$Kr, ..., $^{118}_{36}$Kr. Ordinarily, these supernumerary neutrons would revert to protons via the standard $\beta$-decay process,

$$n \rightarrow p^+ + e^- + \overline{\nu}_e.$$

However, under the conditions of complete electron degeneracy, there are no vacant states available for an emitted electron to occupy, and so the neutrons cannot decay back into protons.[22]

When the density reaches about $4 \times 10^{11}$ g cm$^{-3}$, the minimum-energy arrangement is one in which some of the neutrons are found *outside* the nuclei. The appearance of these free neutrons is called **neutron drip** and marks the start of a three-component mixture of a lattice of neutron-rich nuclei, nonrelativistic degenerate free neutrons, and relativistic degenerate electrons.

The fluid of free neutrons has the striking property that it has no viscosity. This occurs because a spontaneous pairing of the degenerate neutrons has taken place. The resulting combination of two fermions (the neutrons) is a boson (recall Section 5.4), and so is not subject to the restrictions of the Pauli exclusion principle. Because degenerate bosons can *all* crowd into the lowest energy state, the fluid of paired neutrons can lose no energy. It is a **superfluid** that flows without resistance. Any whirlpools or vortices in the fluid will continue to spin forever without stopping.

As the density increases further, the number of free neutrons increases as the number of electrons declines. The neutron degeneracy pressure exceeds the electron degeneracy pressure when the density reaches roughly $4 \times 10^{12}$ g cm$^{-3}$. As the density approaches $\rho_{\rm nuc}$, the nuclei effectively dissolve as the distinction between neutrons inside and outside of nuclei becomes meaningless. This results in a fluid mixture of free neutrons, protons, and electrons dominated by neutron degeneracy pressure, with both the neutrons and protons paired to form superfluids. The fluid of pairs of positively charged protons is also **superconducting**, with zero electrical resistance. As the density increases

---

[22] An *isolated* neutron will decay into a proton in about 11 minutes, the half-life for that process.

| Transition density ($g\ cm^{-3}$) | Composition | Degeneracy pressure |
|---|---|---|
| | iron nuclei, nonrelativistic free electrons | electron |
| $\approx 1 \times 10^6$ | electrons become relativistic | |
| | iron nuclei, relativistic free electrons | electron |
| $\approx 1 \times 10^9$ | neutronization | |
| | neutron-rich nuclei, relativistic free electrons | electron |
| $\approx 4 \times 10^{11}$ | neutron drip | |
| | neutron-rich nuclei, free neutrons, relativistic free electrons | electron |
| $\approx 4 \times 10^{12}$ | neutron degeneracy pressure dominates | |
| | neutron-rich nuclei, superfluid free neutrons, relativistic free electrons | neutron |
| $\approx 2 \times 10^{14}$ | nuclei dissolve | |
| | superfluid free neutrons, superconducting free protons, relativistic free electrons | neutron |
| $\approx 4 \times 10^{14}$ | pion production | |
| | superfluid free neutrons, superconducting free protons, relativistic free electrons, other elementary particles (pions, ...?) | neutron |

**Table 15.1** Composition of Neutron Star Material.

further, the ratio of neutrons:protons:electrons approaches a limiting value of 8:1:1, as determined by the balance between the competing processes of electron capture and $\beta$-decay inhibited by the presence of degenerate electrons.

The properties of the neutron star material when $\rho > \rho_{\rm nuc}$ is still poorly understood. A complete theoretical description of the behavior of a sea of free neutrons interacting via the strong nuclear force in the presence of protons and electrons is not yet available, and there is little experimental data on the behavior of matter in this density range. A further complication is the appearance of exotic elementary particles such as *pions* ($\pi$) produced by the

## 15.6 Neutron Stars

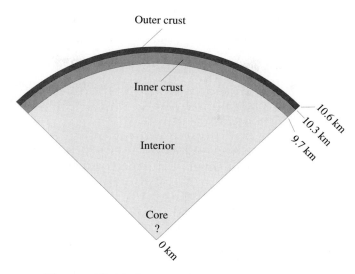

**Figure 15.11** A 1.4 $M_\odot$ neutron star model.

decay of a neutron into a proton and a negatively charged pion, $n \to p^+ + \pi^-$, which occurs spontaneously in neutron stars when $\rho > 2\rho_{\text{nuc}}$.[23] Nevertheless, these are the values of the density encountered in the interiors of neutron stars, and the difficulties mentioned are the primary reasons for the uncertainty in the structure calculated for model neutron stars.

Table 15.1 summarizes the composition of the neutron star material at various densities. After an equation of state that relates the density and pressure has been obtained, a model of the star can be calculated by numerically integrating general-relativistic versions of the stellar structure equations collected at the beginning of Section 10.5. The first quantitative model of a neutron star was calculated by J. Robert Oppenheimer and G. M. Volkoff at Berkeley in 1939. Figure 15.11 shows the result of a recent calculation of a 1.4 $M_\odot$ neutron star model. Although the details are sensitive to the equation of state used, this model displays some typical features.

1. The outer crust consists of heavy nuclei, either in the form of a fluid "ocean" or a solid lattice, and relativistic degenerate electrons. Nearest the surface, the nuclei are probably $^{56}_{26}$Fe. At greater depth and density,

---

[23]The $\pi^-$ is a negatively charged particle that is 273 times more massive than the electron. It mediates the strong nuclear force that holds an atomic nucleus together. (The strong force between nucleons was described in Section 10.3.) Pions have been produced and studied in high-energy accelerator laboratories.

increasingly neutron-rich nuclei are encountered until neutron drip begins at the bottom of the outer crust (where $\rho \approx 4 \times 10^{11}$ g cm$^{-3}$).

2. The inner crust consists of a three-part mixture of a lattice of nuclei such as $^{118}_{36}$Kr, a superfluid of free neutrons, and relativistic degenerate electrons. The bottom of the inner crust occurs where $\rho \approx \rho_{\text{nuc}}$, and the nuclei dissolve.

3. The interior of the neutron star consists primarily of superfluid neutrons, with a smaller number of superfluid, superconducting protons and relativistic degenerate electrons.

4. There may or may not be a solid core consisting of pions or other elementary particles. The density at the center of a 1.4 M$_\odot$ neutron star is about $10^{15}$ g cm$^{-3}$.

Like white dwarfs, neutron stars obey a mass–volume relation,

$$M_{\text{ns}} V_{\text{ns}} = \text{constant},$$

so neutron stars become smaller and more dense with increasing mass. However, this mass–volume relation fails for more massive neutron stars because there is a point beyond which neutron degeneracy pressure can no longer support the star. Hence, there is a maximum mass for neutron stars, analogous to the Chandrasekhar mass for white dwarfs. As might be expected, the value of this maximum mass is different for different choices of the equation of state. However, a very general argument involving the general theory of relativity shows that the maximum mass possible for a neutron star cannot exceed about 3 M$_\odot$. If a neutron star is to remain dynamically stable and resist collapsing, it must be able to respond to a small disturbance in its structure by rapidly adjusting its pressure to compensate. However, there is a limit to how quickly such an adjustment can be made because these changes are conveyed by sound waves that must move more slowly than light. If a neutron star's mass exceeds 3 M$_\odot$, it cannot generate pressure quickly enough to avoid collapsing. The result, as will be discussed in Section 16.3, is a black hole.

Several properties of neutron stars were anticipated before they were observed. For example, neutron stars must rotate very rapidly. If the iron white dwarf core of the presupernova supergiant star were rotating even slowly, the decrease in radius is so great that the conservation of angular momentum guarantees the formation of a rapidly rotating neutron star.

The magnitude of the collapse can be found from Eqs. (15.12) and (15.22) for the estimated radii of a white dwarf and neutron star. Although the leading

## 15.6 Neutron Stars

constants in both expressions are spurious (a by-product of the approximations made), the *ratio* of the radii is more accurate:

$$\frac{R_{wd}}{R_{ns}} \approx \frac{m_n}{m_e} \left(\frac{Z}{A}\right)^{5/3} = 512, \qquad (15.23)$$

where $Z/A = 26/56$ for iron has been used for the white dwarf. Now apply the conservation of angular momentum to the collapsing core (which is assumed to lose no mass, so $M_{wd} = M_{ns}$). Treating each star as a sphere with a moment of inertia of the form $I = CMR^2$, we have[24]

$$I_i \omega_i = I_f \omega_f$$
$$CM_i R_i^2 \omega_i = CM_f R_f^2 \omega_f$$
$$\omega_f = \omega_i \left(\frac{R_i}{R_f}\right)^2.$$

In terms of the rotation period $P$, this is

$$P_f = P_i \left(\frac{R_f}{R_i}\right)^2. \qquad (15.24)$$

For the specific case of an iron white dwarf collapsing to form a neutron star, Eq. (15.23) shows that

$$P_{ns} \approx 3.8 \times 10^{-6} \, P_{wd}. \qquad (15.25)$$

The question of how fast the white dwarf core may be rotating is difficult to answer. As a star evolves, its contracting core is not completely isolated from the surrounding envelope, so one cannot use the simple approach to angular momentum conservation described above.[25] For the purposes of estimation we will take $P_{wd} = 1350$ s, the rotation period observed for the white dwarf 40 Eridani B (shown in the H–R diagrams of Figs. 8.10 and 8.15). Inserting this into Eq. (15.25) results in a rotation period of about $5 \times 10^{-3}$ s. Thus neutron stars will be rotating very rapidly when they are formed, with rotation periods on the order of a few milliseconds.

---

[24] The constant $C$ is determined by the distribution of mass inside the star. For example, $C = 2/5$ for a uniform sphere. We assume that the white dwarf and neutron star have about the same value of $C$.

[25] The core and envelope may exchange angular momentum by magnetic fields or rotational mixing via the very slow *meridional currents* that generally circulate upward at the poles and downward at the equator of a rotating star.

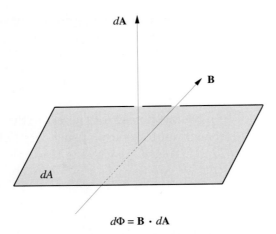

**Figure 15.12** Magnetic flux through an element of surface area $d\mathbf{A}$.

Another property predicted for neutron stars is that they should have extremely strong magnetic fields. The "freezing in" of magnetic field lines in a conducting fluid or gas (mentioned in Section 11.3 in connection with sunspots) implies that the *magnetic flux* through the surface of a white dwarf will be conserved as it collapses to form a neutron star. The flux of a magnetic field through a surface $\mathcal{S}$ is defined as the surface integral

$$\Phi \equiv \int_{\mathcal{S}} \mathbf{B} \cdot d\mathbf{A},$$

where $\mathbf{B}$ is the magnetic field vector (see Fig. 15.12). In approximate terms, if we ignore the geometry of the magnetic field, this means that the product of the magnetic field strength and the area of the star's surface remains constant. Thus

$$B_i 4\pi R_i^2 = B_f 4\pi R_f^2.$$

It is not at all clear what should be considered as the value of the initial magnetic field of an iron white dwarf at the center of a presupernova star. As an extreme case, we can use the largest observed white-dwarf magnetic field of $B \approx 5 \times 10^8$ G (which is large compared to a typical white-dwarf magnetic field of perhaps $10^5$ G, and huge compared with the Sun's global 2 G field). Then, using Eq. (15.23), the magnetic field of the neutron star would be

$$B_{\text{ns}} \approx B_{\text{wd}} \left(\frac{R_{\text{wd}}}{R_{\text{ns}}}\right)^2 = 1.3 \times 10^{14} \text{ G}.$$

This shows that neutron stars could be formed with extremely strong magnetic

## 15.6 Neutron Stars

fields, although this particular estimate must be viewed as an upper limit rather than a typical value.

The final property of neutron stars is the most obvious. They are extremely hot when they were forged in the "fires" of a supernova, with $T \sim 10^{11}$ K. During the first day, the neutron star cools by emitting neutrinos via the so-called **URCA process**,[26]

$$n \to p^+ + e^- + \overline{\nu}_e$$
$$p^+ + e^- \to n + \nu_e.$$

As the nucleons shuttle between being in the form of neutrons and protons, large numbers of neutrinos and antineutrinos are produced that fly unhindered into space, carrying away energy and thus cooling the neutron star. This process can continue only as long as the nucleons are not degenerate, and it is suppressed after the protons and neutrons settle into the lowest unoccupied energy states. This degeneracy occurs about one day after the formation of the neutron star, when its internal temperature has dropped to about $10^9$ K. Other neutrino-emitting processes continue to dominate the cooling for approximately the first thousand years, after which photons emitted from the star's surface take over. The neutron star is a few hundred years old when its internal temperature has declined to $10^8$ K, with a surface temperature of several million K. By now the cooling rate has slowed considerably, and the surface temperature will hover around $10^6$ K for the next ten thousand years or so as the neutron star cools at an essentially constant radius.

It is interesting to calculate the blackbody luminosity of a 1.4 M$_\odot$ neutron star with a surface temperature of $T = 10^6$ K. From the Stefan–Boltzmann law, Eq. (3.17),

$$L = 4\pi R^2 \sigma T_e^4 = 7.13 \times 10^{32} \text{ erg s}^{-1}.$$

Although this is comparable to the luminosity of the Sun, the radiation is primarily in the form of x-rays since according to Wien's displacement law, Eq. (3.19),

$$\lambda_{\max} = \frac{(5000 \text{ Å})(5800 \text{ K})}{T} = 29 \text{ Å}.$$

Astronomers held little hope of ever observing such an exotic object, barely the size of San Diego, California.

---

[26]The URCA process, which efficiently removes energy from a hot neutron star, is named for the Casino de URCA in Rio de Janeiro in remembrance of the efficiency with which it removed money from an unlucky physicist. The casino was closed by Brazil in 1955.

## 15.7 Pulsars

Jocelyn Bell spent two years setting up a forest of 2048 radio dipole antennas over four and a half acres of English countryside. She and her Ph.D. thesis advisor, Anthony Hewish, were using this radio telescope, tuned to a frequency of 81.5 MHz, to study the scintillation ("flickering") that is observed when the radio waves from distant sources known as quasars passed through the solar wind. In July 1967, Bell was puzzled to find a bit of "scruff" that reappeared every 400 feet or so on the rolls of her strip chart recorder; see Fig. 15.13. Careful measurements showed that this quarter inch of ink reappeared every 23 hours and 56 minutes, indicating that its source passed over her fixed array of antennae once every sidereal day. Bell concluded that the source was out among the stars rather than within the solar system. To better resolve the signal, she used a faster recorder and discovered that the scruff consisted of a series of regularly spaced radio pulses 1.337 s apart (the pulse **period**, $P$). Such a precise celestial clock was unheard of, and Bell and Hewish considered the possibility that these might be signals from an extraterrestrial civilization. If this were true, she felt annoyed that the aliens had chosen such an inconvenient time to make contact. She recalled, "I was now two and a half years through a three year studentship and here was some silly lot of Little Green Men using *my* telescope and *my* frequency to signal to planet Earth." When Bell found another bit of scruff, coming from another part of the sky, her relief was palpable. She wrote, "It was highly unlikely that two lots of Little Green Men could choose the same unusual frequency and unlikely technique to signal to the same inconspicuous planet Earth!"

Hewish, Bell, and their colleagues announced the discovery of these mysterious **pulsars**,[27] and several more were quickly found by other radio observatories. Today some 550 pulsars are known, and each is designated by a "PSR" prefix (for *P*ulsating *S*ource of *R*adio) followed by its right ascension ($\alpha$) and declination ($\delta$). For example, the source of Bell's scruff is PSR 1919+21, identifying its position as $\alpha = 19^{\text{h}}19^{\text{m}}$ and $\delta = +21°$.

All known pulsars share the following characteristics, which are crucial clues to their physical nature:

- Most pulsars have periods between 0.25 s and 2 s, with an average time between pulses of about 0.79 s (see Fig. 15.14). The pulsar with the

---

[27]See Hewish et al. (1968) for details of the discovery of pulsars. The term *pulsar* was coined by the science correspondent for the London *Daily Telegraph*.

## 15.7 Pulsars

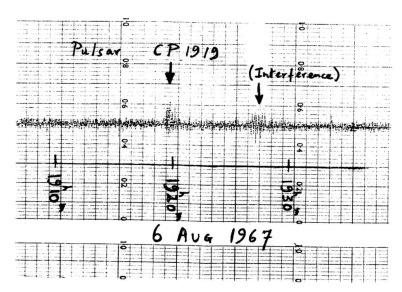

**Figure 15.13** Discovery of the first pulsar, PSR 1919+21 ("CP" stands for Cambridge Pulsar). (Figure from Lyne and Graham-Smith, *Pulsar Astronomy*, ©Cambridge University Press, New York, 1990. Reprinted with the permission of Cambridge University Press.)

longest period is PSR 1845−19 ($P = 4.308$ s); the fastest pulsar is PSR 1937+214 ($P = 0.0016$ s).[28]

- Pulsars have extremely well-defined pulse periods and would make exceptionally accurate clocks. For example, the period of PSR 1937+214 has been determined to be $P = 0.00155780644887275$ s, a measurement that challenges the accuracy of the best atomic clocks.

- The periods of all pulsars increase very gradually as the pulses slow down, the rate of increase given by the period derivative $\dot{P} \equiv dP/dt$.[29] Typically, $\dot{P} \approx 10^{-15}$, and the *characteristic lifetime* (the time for the pulses to cease if $\dot{P}$ were constant) is $P/\dot{P} \approx$ a few $10^7$ years. The value of $\dot{P}$ for PSR 1937+214 is unusually small, $\dot{P} = 1.051054 \times 10^{-19}$. This corresponds to a characteristic lifetime of $P/\dot{P} = 1.48 \times 10^{16}$ s, or about 470 million years.

---

[28] The declination coordinate, $\delta$, takes on values between $\pm 90°$. The "214" for the fastest pulsar contains an unwritten decimal point, and should be interpreted as $\delta = 21.4°$.

[29] Note that $\dot{P}$ is measured in terms of seconds of period change per second and so is unitless.

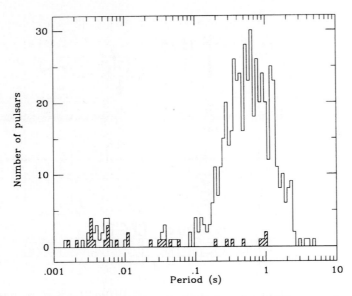

**Figure 15.14** The distribution of periods for 558 pulsars. Binary pulsars are indicated by the shaded area, and the millisecond pulsars are on the left. The average period is about 0.8 s. (Figure from Taylor, Manchester, and Lyne, *Ap. J. Suppl.*, *88*, 529, 1993.)

These characteristics allowed astronomers to deduce the basic components of pulsars. In the paper announcing their discovery, Hewish, Bell, and their co-authors suggested that an oscillating neutron star might be involved, but American astronomer Thomas Gold quickly and convincingly argued instead that pulsars are rapidly rotating neutron stars.

There are three obvious ways of obtaining rapid regular pulses in astronomy:

1. **Binary stars.** If the orbital periods of a binary star system are to fall in the range of the observed pulsar periods, then extremely compact stars must be involved—either white dwarfs or neutron stars. The general form of Kepler's third law, Eq. (2.35), shows that if two 1 $M_\odot$ stars were to orbit each other every 0.79 s (the average pulsar period), then their separation would be only $1.6 \times 10^8$ cm. This is much less than the $5.5 \times 10^8$ cm radius of Sirius B, and the separation would be even smaller for more rapid pulsars. This eliminates even the smallest, most massive white dwarfs from consideration.

   Neutron stars are so small that two of them could orbit each other with a period in agreement with those observed for pulsars. However, this

## 15.7 Pulsars

possibility is ruled out by Einstein's general theory of relativity. As the two neutron stars rapidly move through space and time, gravitational waves are generated that carry energy away from the binary system. As the neutron stars slowly spiral closer together, their orbital period *decreases*, according to Kepler's third law. This contradicts the observed *increase* in the periods of the pulsars and so eliminates binary neutron stars as a source of the radio pulses.[30]

2. **Pulsating stars**. As discussed in Section 15.2, white dwarfs oscillate with periods between 100 and 1000 s. The periods of these nonradial g-modes are much longer than the observed pulsar periods. Of course, it might be imagined that a radial oscillation is involved with the pulsars. However, the period for the radial fundamental mode is a few seconds, too long to explain the faster pulses.

A similar argument eliminates neutron star oscillations. Neutron stars are about $10^8$ times more dense than white dwarfs. According to the period–mean density relation for stellar pulsation (recall Section 14.2), the period of oscillation is proportional to $1/\sqrt{\rho}$. This implies that neutron stars should vibrate approximately $10^4$ times more rapidly than white dwarfs, with a radial fundamental mode period around $10^{-4}$ s and nonradial g-modes between $10^{-2}$ s and $10^{-1}$ s. These periods fall below the range of the slower pulsars.

3. **Rotating stars**. The enormous angular momentum of a rapidly rotating compact star would guarantee its precise clocklike behavior. But how fast can a star spin? Its angular velocity, $\omega$, is limited by the ability of gravity to supply the centripetal force that keeps the star from flying apart. This constraint is most severe at the star's equator, where the stellar material moves most rapidly. Ignore the inevitable equatorial bulging caused by rotation and assume that the star remains circular with radius $R$ and mass $M$. Then the maximum angular velocity may be found by equating the centripetal and gravitational accelerations at the equator,

$$\omega_{\max}^2 R = G \frac{M}{R^2},$$

so that the minimum rotation period is $P_{\min} = 2\pi/\omega_{\max}$, or

$$P_{\min} = 2\pi \sqrt{\frac{R^3}{GM}}. \tag{15.26}$$

---

[30]Gravitational waves will be described in more detail in Section 17.5, as will the binary system of two neutron stars in which these waves have been indirectly detected.

For Sirius B, $P_{\min} \approx 7$ s, which is much too long. However, for a 1.4 $M_\odot$ neutron star, $P_{\min} \approx 5 \times 10^{-4}$ s. Because this is a *minimum* rotation time, it can accommodate the complete range of periods observed for pulsars.

Only one alternative has emerged unscathed from this process of elimination, namely, that pulsars are rapidly rotating neutron stars. This conclusion was strengthened by the discovery in 1968 of pulsars associated with the Vela and Crab supernovae remnants.[31] In addition, the Crab pulsar PSR 0531−21 has a very short pulse period of only 0.0333 s. No white dwarf could rotate 30 times per second without disintegrating, and the last doubts about the identity of pulsars were laid to rest. Until the discovery of the millisecond pulsars ($P \approx 10$ ms or less) in 1982, the Crab pulsar held the title of the fastest known pulsar (see Fig. 15.14). The Vela and Crab pulsars not only produce radio bursts, they pulse in other regions of the electromagnetic spectrum ranging from radio to gamma rays, including visible flashes as shown in Fig. 15.15. These young pulsars (and a few others) also display *glitches* when their periods abruptly *decrease* by a tiny amount ($|\Delta P|/P \approx 10^{-6}$ to $10^{-8}$); see Fig. 15.16. These sudden spinups are separated by uneven intervals of several years.

The nearest pulsar yet detected is only some 90 pc away. Nicknamed "Geminga," it was well known as a strong source of gamma rays for 17 years before its identity as a pulsar was established in 1992.[32] With a period of 0.237 s, Geminga pulses in both gamma and x-rays (but not at radio wavelengths) and may display glitches. In visible light, its absolute magnitude is fainter than +23.

The supernova explanation for the origin of pulsars is consistent with the fact that only 1% of known pulsars belong to a binary system, while at least half of the stars in the sky are found in multiple star systems. Pulsars also move much faster through space than do normal stars, as though the explosion's violence flung them through space while it disrupted any nearby companion.

---

**Example 15.5** Observations of the Crab Nebula, the remnant of the A.D. 1054 supernova, clearly reveal its intimate connection with the pulsar at its center. As shown on the front cover of this text and in Fig. 15.15, the expanding nebula produces a ghostly glow surrounding gaseous filaments that wind throughout it.

---

[31] Only three pulsars are associated with supernova remnants, a number roughly consistent with the long lifetime ($\approx 10^7$ years) of a pulsar and the shorter lifetime ($\approx 10^5$ years) of the glowing gases of the nebula. The Crab supernova was recorded by Chinese astrologers on July 4, 1054 A.D., while the Vela supernova occurred some 30,000 to 50,000 years ago.

[32] "Geminga" means "does not exist" in Milanese dialect, accurately reflecting its long-mysterious nature.

## 15.7 Pulsars

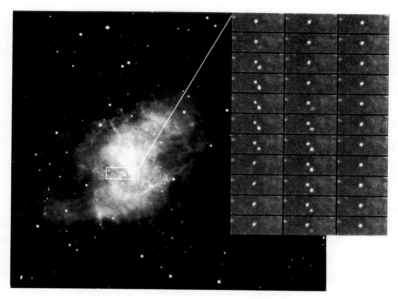

**Figure 15.15** A sequence of images showing the flashes at visible wavelengths from the Crab pulsar, located at the center of the Crab Nebula (left). (Courtesy of National Optical Astronomy Observatories.)

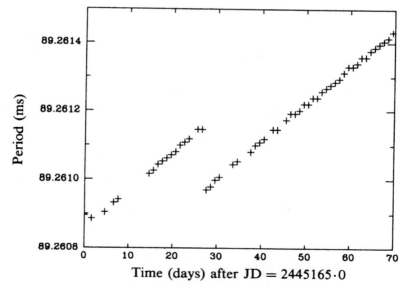

**Figure 15.16** A glitch in the Vela pulsar. (Figure from McCulloch et al., *Aust. J. Phys.*, *40*, 725, 1987.)

Interestingly, if the present rate of expansion is extrapolated backward in time, the nebula converges to a point about 90 years *after* the supernova explosion was observed. Obviously the nebula must have been expanding more slowly in the past than it is now, which implies that the expansion is actually accelerating.

In 1953, the Russian astronomer I. Shklovsky proposed that the white light is **synchrotron radiation** produced when relativistic electrons spiral along magnetic field lines. From the equation for the magnetic force on a moving charge $q$,

$$\mathbf{F}_m = \frac{q}{c} \mathbf{v} \times \mathbf{B}$$

(in cgs units), the component of an electron's velocity $\mathbf{v}$ perpendicular to the field lines produces a circular motion around the lines, while the component of the velocity along the lines is not affected; see Fig. 15.17. As they follow the curved field lines, the relativistic electrons accelerate and emit electromagnetic radiation. It is called *synchrotron radiation* if the circular motion around the field lines dominates, and *curvature radiation* if the motion is primarily along the field lines. In both cases, the shape of the continuous spectrum produced depends on the energy distribution of the emitting electrons, and so is easily distinguished from the spectrum of blackbody radiation.[33] Either way, the radiation is strongly linearly polarized, in the plane of the circular motion for synchrotron radiation, and in the plane of the curving magnetic field line for curvature radiation. As a test of his theory, Shklovsky predicted that the white light from the Crab Nebula would be found to be strongly linearly polarized. His prediction was subsequently confirmed as the light from some emitting regions of the nebula was measured to be 60% linearly polarized.

The identification of the white glow as synchrotron radiation raised new questions. It implied that magnetic fields of $10^{-3}$ G must permeate the Crab Nebula. This was puzzling because, according to theoretical estimates, long ago the expansion of the nebula should have weakened the magnetic field far below this value. Furthermore, the electrons should have radiated away all of their energy after only 100 years. It is clear that the production of synchrotron radiation today requires both a replenishment of the magnetic field and a continuous injection of new energetic electrons. The total power needed for the accelerating expansion of the nebula, the relativistic electrons, and the magnetic field is calculated to be about $5 \times 10^{38}$ erg s$^{-1}$, or more than $10^5$ L$_\odot$.

The energy source is the rotating neutron star at the heart of the Crab Nebula. It acts as a huge flywheel and stores an immense amount of rotational

---

[33]Both synchrotron and curvature radiation are sometimes called *nonthermal* to distinguish them from the thermal origin of blackbody radiation.

## 15.7 Pulsars

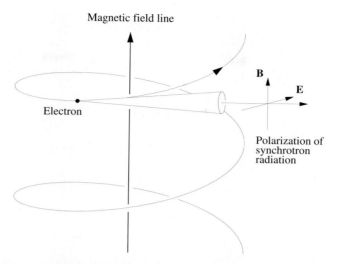

**Figure 15.17** Synchrotron radiation emitted by a relativistic electron as it spirals around a magnetic field line.

kinetic energy. As the star slows down, its energy supply decreases. To calculate the rate of energy loss, first write the rotational kinetic in terms of the period and moment of inertia of the neutron star,

$$K = \frac{1}{2}I\omega^2 = \frac{2\pi^2 I}{P^2}.$$

Then the rate at which the rotating neutron star is losing energy is

$$\frac{dK}{dt} = -\frac{4\pi^2 I \dot{P}}{P^3}. \tag{15.27}$$

Assuming that the neutron star is a uniform sphere with $R = 10^6$ cm and $M = 1.4$ M$_\odot$, its moment of inertia is approximately

$$I = \frac{2}{5}MR^2 = 1.1 \times 10^{45} \text{ g cm}^2.$$

Inserting $P = 0.0333$ s and $\dot{P} = 4.21 \times 10^{-13}$ for the Crab pulsar gives $dK/dt \approx 5.0 \times 10^{38}$ erg s$^{-1}$. Remarkably, this is exactly the energy required to power the Crab Nebula. The slowing down of the neutron star flywheel has enabled the nebula to continue shining and expanding for more than 900 years.

It is important to realize that this energy is not transported to the nebula by the pulse itself. The radio luminosity of the Crab's pulse is about $10^{31}$ erg s$^{-1}$, 200 million times smaller than the rate at which energy is delivered to the

**Figure 15.18** An HST image of the immediate surroundings of the Crab pulsar. (Figure from Hester et al., *Ap. J.*, *448*, 240, 1995.)

nebula. (For older pulsars, the radio pulse luminosity is typically $10^{-5}$ of the spin-down rate of energy loss.) Thus the pulse process, whatever it may be, is a minor component of the total energy-loss mechanism.

Figure 15.18 shows an HST view of the immediate environment of the Crab pulsar. The ringlike halo seen on the west side of the pulsar is a glowing torus of gas; it may be the result of a polar jet from the pulsar forcing its way through the surrounding nebula. Just to the east of the pulsar, about 1500 AU away, is a bright knot of emission from shocked material in the jet, perhaps due to an instability in the jet itself. Another knot is seen at a distance of 9060 AU.

Before describing the details of a model pulsar, it is worth taking a closer look at the pulses themselves. As can be seen in Fig. 15.19, the pulses are brief, received over a small fraction (typically from 1% to 5%) of the pulse period. Generally, they are received at radio wave frequencies between roughly 20 MHz and 10 GHz.[34]

---

[34] 1 gigahertz (GHz) = $10^3$ megahertz (MHz).

## 15.7 Pulsars

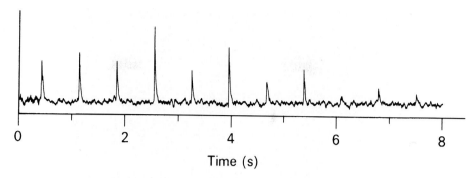

**Figure 15.19** Pulses from PSR 0329+54 with a period of 0.714 s. (Figure from Manchester and Taylor, *Pulsars*, W. H. Freeman and Co., New York, 1977.)

As the pulses travel through interstellar space, the time-varying electric field of the radio waves causes the electrons that are encountered along the way to vibrate. This process slows the radio waves below the speed of light in a vacuum, $c$, with a greater retardation at lower frequencies. Thus a sharp pulse emitted at the neutron star, with all frequencies peaking at the same time, is gradually drawn out or *dispersed* as it travels to Earth (see Fig. 15.20). Because more distant pulsars exhibit a greater pulse dispersion, these time delays can be used to measure the distances to pulsars. The results show that the known pulsars are concentrated within the plane of our Milky Way Galaxy (Fig. 15.21) at typical distances of hundreds to thousands of parsecs.

Figure 15.22 shows that there is a substantial variation in the shape of the individual pulses received from a pulsar. Although a typical pulse consists of a number of brief *subpulses*, the *integrated pulse profile*, an average built up by adding together a train of 100 or more pulses, is remarkably stable. Some pulsars have more than one average pulse profile, and abruptly switch back and forth between them (Fig. 15.23). The subpulses may appear at random times in the "window" of the main pulse, or they may march across in a phenomenon known as *drifting subpulses*, as shown in Fig. 15.24. For about 30% of all known pulsars, the individual pulses may simply disappear or *null*, only to reappear up to 100 periods later. Drifting subpulses may even emerge from a nulling event in step with those that entered the null! Finally, the radio waves of many pulsars are strongly linearly polarized (up to 100%), a feature that indicates the presence of a strong magnetic field.

The basic pulsar model, shown in Fig. 15.25, consists of a rapidly rotating neutron star with a strong dipole magnetic field (two poles, north and south) that is inclined to the rotation axis at an angle $\theta$. As explained in the previous

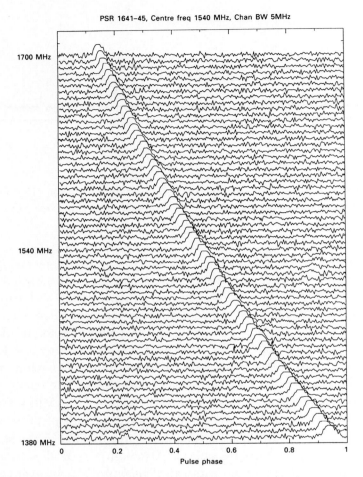

**Figure 15.20** Dispersion of the pulse from PSR 1641−45. (Figure from Lyne and Graham-Smith, *Pulsar Astronomy*, ©Cambridge University Press, New York, 1990. Reprinted with the permission of Cambridge University Press.)

section, both of these features arise naturally following the collapse of the core of a supergiant star.[35]

First, we need to obtain a measure of the strength of the pulsar's magnetic field. As the pulsar rotates, the magnetic field at any point in space will change rapidly. According to Faraday's law, this will induce an electric field at that

---

[35]It is likely that the millisecond pulsars have a different origin in a close binary system; more than half of the known millisecond pulsars belong to binaries. For this reason, they will be discussed separately in Section 17.5.

## 15.7 Pulsars

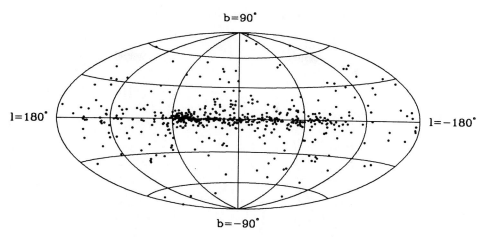

**Figure 15.21** Distribution of 558 pulsars in galactic coordinates, with the center of the Milky Way in the middle. The clump of pulsars at 60° is a selection effect due to the fixed orientation of the Arecibo radio telescope. (Figure from Taylor, Manchester, and Lyne, *Ap. J. Suppl.*, *88*, 529, 1993.)

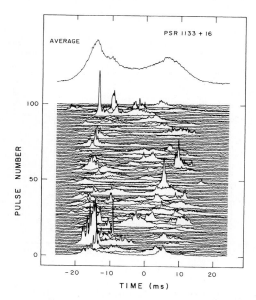

**Figure 15.22** The average of 500 pulses (top) and a series of 100 consecutive pulses (below) for PSR 1133+16. (Figure from Cordes, *Space Sci. Review*, *24*, 567, 1979.)

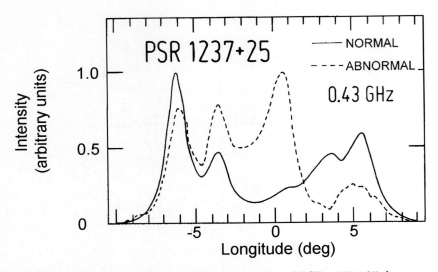

**Figure 15.23** Changes in the integrated pulse profile of PSR 1237+25 due to mode switching. This pulsar displays five distinct subpulses, the most for any known pulsar. (Figure adapted from Bartel et al., *Ap. J.*, *258*, 776, 1982.)

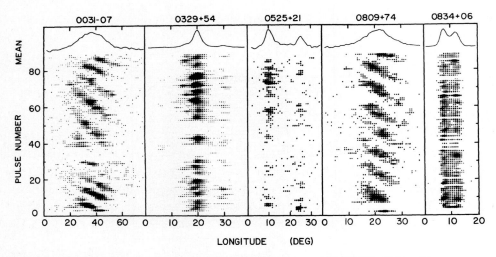

**Figure 15.24** Drifting subpulses for two pulsars; note that PSR 0031−07 also nulls. (Figure from Taylor et al., *Ap. J.*, *195*, 513, 1975.)

## 15.7 Pulsars

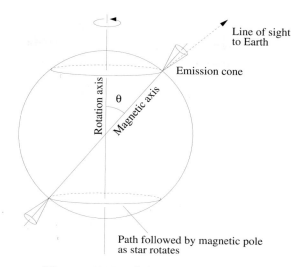

**Figure 15.25** A basic pulsar model.

point. Far from the star (near the **light cylinder** defined in Fig. 15.26) the time-varying electric and magnetic fields form an electromagnetic wave that carries energy away from the star. For this particular situation, the radiation is called *magnetic dipole radiation*. Although it is beyond the scope of this book to consider the model in detail, we note that the energy per second emitted by the rotating magnetic dipole is (in cgs units)

$$\frac{dE}{dt} = -\frac{16\pi^4 B^2 R^6 \sin^2\theta}{6c^3 P^4}, \tag{15.28}$$

where $B$ is the field strength at the magnetic pole of the star of radius $R$. The minus sign indicates that the neutron star is drained of energy, causing its rotation period, $P$, to increase. Note that the factor of $1/P^4$ means that the neutron star will lose energy much more quickly at smaller periods. Since the average pulsar period is 0.79 s, most pulsars are born spinning considerably faster than their current rates, with typical initial periods of a few milliseconds.

Assuming that all of the rotational kinetic energy lost by the star is carried away by magnetic dipole radiation, then $dE/dt = dK/dt$. Using Eqs. (15.27) and (15.28), this is

$$-\frac{16\pi^4 B^2 R^6 \sin^2\theta}{6c^3 P^4} = -\frac{4\pi^2 I \dot{P}}{P^3}. \tag{15.29}$$

This can be easily solved for the magnetic field at the pole of the neutron star,

$$B = \frac{1}{2\pi R^3 \sin\theta}\sqrt{6c^3 I P \dot{P}}. \tag{15.30}$$

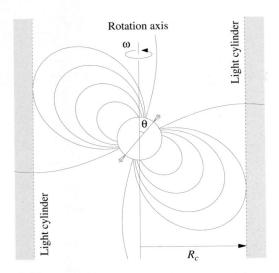

**Figure 15.26** The light cylinder around a rotating neutron star. The cylinder's radius $R_c$ is where a point co-rotating with the neutron star would move at the speed of light: $R_c = c/\omega = cP/2\pi$.

---

**Example 15.6** We will estimate the magnetic field strength at the poles of the Crab pulsar (PSR 0531−21), with $P = 0.0333$ s and $\dot{P} = 4.21 \times 10^{-13}$. Assuming that $\theta = 90°$, Eq. (15.30) then gives a value of $8.0 \times 10^{12}$ G. As we have seen, the Crab pulsar is interacting with the dust and gas in the surrounding nebula, so there are other torques that contribute to slowing down the pulsar's spin. This value of $B$ is therefore an overestimate; the accepted value of the Crab pulsar's magnetic field is $4 \times 10^{12}$ G.[36] Values of $B$ around $10^{12}$ G are typical for most pulsars.

However, repeating the calculation for PSR 1937+214 (the fastest known millisecond pulsar) with $P = 0.00156$ s, $\dot{P} = 1.05 \times 10^{-19}$, and assuming the same value for the moment of inertia, the magnetic field strength is found to be only $B = 8.6 \times 10^8$ G. This much smaller value distinguishes the millisecond pulsars and provides another hint that these fastest pulsars may have a different origin or environment.

---

Developing a detailed model of the pulsar's emission mechanism has been an exercise in frustration because almost every observation is open to more than

---

[36] The suggestion that the Crab Nebula is powered by the magnetic dipole radiation from a rotating neutron star was made by an Italian astronomer, Franco Pacini, in 1967, a year *before* the discovery of pulsars!

one interpretation. The emission of radiation is the most poorly understood aspect of pulsars, and at present there is agreement only on the most general features of how a neutron star manages to produce radio waves. The following discussion summarizes a popular model of the pulse process. The reader should keep in mind, however, that there is as yet no general consensus on whether the object being discussed actually occurs in Nature or only in the minds of astronomers!

It is at least certain that the rapidly changing magnetic field near the rotating pulsar induces a huge electric field at the surface. The electric field of about $2 \times 10^8$ statvolts cm$^{-1}$ ($6.3 \times 10^{10}$ V m$^{-1}$ in SI units) easily overcomes the pull of gravity on charged particles in the neutron star's crust. For example, the electric force on a proton is about 300 million times stronger than the force of gravity, and the ratio of the electric force on an electron to the gravitational force is even more overwhelming. Depending on the direction of the electric field, either electrons or ions will be continuously ripped from the neutron star's polar regions. This creates a **magnetosphere** of charged particles surrounding the pulsar that is dragged around with the pulsar's rotation. However, the speed of the co-rotating particles cannot exceed the speed of light, so at the light cylinder the charged particles are spun away, carrying the magnetic field with them in a pulsar "wind." Such a wind may be responsible for the replenishment of the Crab Nebula's magnetic field and the continual delivery of relativistic particles needed to keep the nebula shining.

The charged particles ejected from the vicinity of the pulsar's magnetic poles are quickly accelerated to relativistic speeds by the induced electric field. As the electrons follow the curved magnetic field lines, they emit curvature radiation in the form of energetic gamma-ray photons. This radiation is emitted in a narrow beam in the instantaneous direction of motion of the electron, a consequence of the relativistic headlight effect discussed in Section 4.3. Each gamma-ray photon has so much energy that it can spontaneously convert this energy into an electron–positron pair via Einstein's relation $E = mc^2$. (This process, described by $\gamma \to e^- + e^+$, is just the inverse of the annihilation process mentioned in Section 10.3 for the Sun's interior.) The electrons and positrons are accelerated and in turn emit their own gamma rays, which create more electron–positron pairs, and so on. A cascade of pair-production is thus initiated near the magnetic poles of the neutron star. Coherent beams of curvature radiation emitted by bunches of these particles may be responsible for the individual subpulses that contribute to the integrated pulse profile.

As these particles continue to curve along the magnetic field lines, they emit a continuous spectrum of curvature radiation in the forward direction, producing a narrow cone of radio waves radiating from the magnetic polar regions.[37] As the neutron star rotates, these radio waves sweep through space in a way reminiscent of the light from a rotating lighthouse beacon. If the beam happens to fall on a radio telescope on a blue-green planet in a distant solar system, the astronomers there will detect a regular series of brief radio pulses.

As the pulsar ages and slows down, the structure of the underlying neutron star must adapt to the reduced rotational stresses. As a consequence, perhaps the crust settles a fraction of a millimeter and the star spins faster as a result of its decreased moment of inertia, or perhaps the superfluid vortices in the neutron star's core become momentarily "unpinned" from the underside of the solid crust where they are normally attached, giving the crust a sudden jolt. Either possibility could produce a small but abrupt increase in the rotation speed, and the astronomers on Earth would record a glitch for the pulsar.

The question of a pulsar's final fate, as its period increases beyond several seconds, has several possible answers. It may be that the neutron star's magnetic field, originally produced by the collapse of the presupernova star's degenerate stellar core, decays with a characteristic time of 9 million years or so. Then at some future time when the pulsar's period has been reduced to several seconds, its magnetic field may no longer be strong enough to sustain the pulse mechanism and the pulsar turns off. On the other hand, it may be that the magnetic field does not decay appreciably but is maintained by a dynamolike mechanism involving the differential rotation of the crust and core of the neutron star. However, rotation itself is an essential ingredient of any pulsar emission mechanism. As a pulsar ages and slows down, its beam will become weaker even if the magnetic field does not decay. In this case, the radio pulses may become too faint to be detected as the pulsar simply fades below the sensitivity of radio telescopes. The time scale for the decay of a neutron star's magnetic field is a matter of considerable debate, and both scenarios are consistent with the observations.

The preceding sketch reflects the current state of uncertainty about the true nature of pulsars. There are few objects in astronomy that offer such a wealth of intriguing observational detail and yet are so lacking in a consistent theoretical description. Regardless of whether the basic picture outlined is vindicated or whether it is supplanted by another view (perhaps involving a disk of material surrounding the neutron star), pulsar theorists continue to take advantage of this unique natural laboratory for studying matter under the most extreme conditions.

---

[37]The visible, x-ray, and gamma-ray pulses received from the Crab, Vela, Circinus, and Geminga pulsars may originate farther out in the pulsar's magnetosphere.

# Suggested Readings

### General

Burnell, Jocelyn Bell, "The Discovery of Pulsars," *Serendipitous Discoveries in Radio Astronomy*, National Radio Astronomy Observatory, Green Bank, WV, 1983.

Graham-Smith, F., "Pulsars Today," *Sky and Telescope*, September 1990.

Greenstein, George, *Frozen Star*, Freundlich Books, New York, 1983.

Kawaler, Stephen D., and Winget, Donald E., "White Dwarfs: Fossil Stars," *Sky and Telescope*, August 1987.

Mitton, Simon, *The Crab Nebula*, Charles Scribner's Sons, New York, 1978.

Nather, R. Edward, and Winget, Donald E., "Taking the Pulse of White Dwarfs," *Sky and Telescope*, April 1992.

Trimble, Virginia, "White Dwarfs: The Once and Future Suns," *Sky and Telescope*, October 1986.

Van Horn, Hugh M., "The Physics of White Dwarfs," *Physics Today*, January 1979.

Winget, Donald E., "ZZ Ceti Stars: Variable White Dwarfs," *Sky and Telescope*, September 1982.

### Technical

D'Antonia, Francesca, and Mazzitelli, Italo, "Cooling of White Dwarfs," *Annual Review of Astronomy and Astrophysics*, *28*, 139, 1990.

Gold, T., "Rotating Neutron Stars as the Origin of the Pulsating Radio Sources," *Nature*, *218*, 731, 1968.

Hewish, A., et al., "Observations of a Rapidly Pulsating Radio Source," *Nature*, *217*, 709, 1968.

Liebert, James, "White Dwarf Stars," *Annual Review of Astronomy and Astrophysics*, *18*, 363, 1980.

Lyne, A. G., and Graham-Smith, F., *Pulsar Astronomy*, Cambridge University Press, Cambridge, 1990.

Manchester, Joseph H., and Taylor, Richard N., *Pulsars*, W. H. Freeman and Company, San Francisco, CA, 1977.

Michel, F. Curtis, *Theory of Neutron Star Magnetospheres*, The University of Chicago Press, Chicago, 1991.

Pacini, F., "Energy Emission from a Neutron Star," *Nature*, *216*, 567, 1967.

Shapiro, Stuart L., and Teukolsky, Saul A., *Black Holes, White Dwarfs, and Neutron Stars*, John Wiley and Sons, New York, 1983.

Winget, D. E., et al., "An Independent Method for Determining the Age of the Universe," *The Astrophysical Journal Letters*, *315*, L77, 1987.

Winget, D. E., et al., "Hydrogen-Driving and the Blue Edge of Compositionally Stratified ZZ Ceti Star Models," *The Astrophysical Journal Letters*, *252*, L65, 1982a.

Winget, Donald E., et al., "Photometric Observations of GD 358: DB White Dwarfs Do Pulsate," *The Astrophysical Journal Letters*, *262*, L11, 1982b.

# Problems

15.1 The most easily observed white dwarf in the sky is in the constellation of Eridanus (the River Eridanus). Three stars comprise the 40 Eridani system: 40 Eri A is a 4th-magnitude star similar to the Sun; 40 Eri B is a 10th-magnitude white dwarf; and 40 Eri C is an 11th-magnitude red M5 star. This problem deals only with the latter two stars, which are widely separated from 40 Eri A by 400 AU.

(a) The period of the 40 Eri B and C system is 247.9 years. The system's measured trigonometric parallax is $0.201''$ and the true angular extent of the semimajor axis of the reduced mass is $6.89''$. The ratio of the distances of 40 Eri B and C from the center of mass is $a_B/a_C = 0.37$. Find the mass of 40 Eri B and C in terms of the mass of the Sun.

(b) The absolute bolometric magnitude of 40 Eri B is 9.6. Determine its luminosity in terms of the luminosity of the Sun.

(c) The effective temperature of 40 Eri B is 16,900 K. Calculate its radius, and compare your answer to the radii of the Sun, Earth, and Sirius B.

(d) Calculate the average density of 40 Eri B, and compare your result with the average density of Sirius B. Which is more dense, and why?

# Problems

(e) Calculate the product of the mass and volume of both 40 Eri B and Sirius B. Is there a departure from the mass–volume relation? What might be the cause?

15.2 The helium absorption lines seen in the spectra of DB white dwarfs are formed by excited He I atoms with one electron in the lowest ($n = 1$) orbital and the other in an $n = 2$ orbital. White dwarfs of spectral type DB are not observed with temperatures below about 11,000 K. Using what you know about spectral line formation, give a *qualitative* explanation of why the helium lines would not be seen at lower temperatures. As a DB white dwarf cools below 12,000 K, into what spectral type would it change?

15.3 Deduce a rough upper limit for $X$, the mass fraction of hydrogen, in the interior of a white dwarf. *Hint*: Use the mass and average density for Sirius B in the equations for the nuclear energy generation rate, and take $T = 10^7$ K for the central temperature. Set $\psi_{pp}$ and $f_{pp} = 1$ in Eq. (10.50) for the pp chain, and $X_{CNO} = 1$ in Eq. (10.54) for the CNO cycle.

15.4 Estimate the ideal gas pressure and the radiation pressure at the center of Sirius B, using $3 \times 10^7$ K for the central temperature. Compare these values with the estimated central pressure, Eq. (15.1).

15.5 By equating the pressure of an ideal gas of electrons to the pressure of a degenerate electron gas, determine a condition for the electrons to be degenerate, and compare it with the condition of Eq. (15.5). Use the exact expression (Eq. 15.11) for the electron degeneracy pressure.

15.6 In the extreme relativistic limit, the electron speed $v = c$ must be used instead of Eq. (15.9) to find the electron degeneracy pressure. Use this to repeat the derivation of Eq. (15.10) and find

$$P \approx \frac{\hbar c}{3} \left[ \left(\frac{Z}{A}\right) \frac{\rho}{m_H} \right]^{4/3}.$$

15.7 (a) At what speed do relativistic effects become important at a level of 10%? In other words, for what value of $v$ does the Lorentz factor, $\gamma$, become equal to 1.1?

(b) Estimate the density of the white dwarf for which the speed of a degenerate electron is equal to the value found in part (a).

(c) Use the mass–volume relation to find the approximate mass of a white dwarf with this average density. This is roughly the mass where white dwarfs depart from the mass–volume relation.

15.8 Crystallization will occur in a cooling white dwarf when the electrostatic potential energy between neighboring nuclei, $Z^2 e^2/r$ in cgs units, dominates the characteristic thermal energy $kT$. The ratio of the two is defined to be $\Gamma$,

$$\Gamma = \frac{Z^2 e^2}{rkT}.$$

In this expression, the distance $r$ between neighboring nuclei is customarily (and somewhat awkwardly) defined to be the radius of a sphere whose volume is equal to the volume per nucleus. Specifically, since the average volume per nucleus is $Am_H/\rho$, $r$ is found from

$$\frac{4}{3}\pi r^3 = \frac{Am_H}{\rho}.$$

(a) Calculate the value of the average separation $r$ for a 0.6 $M_\odot$ pure carbon white dwarf of radius 0.012 $R_\odot$.

(b) Much effort has been spent on precise numerical calculations of $\Gamma$ to obtain increasingly realistic cooling curves. The results indicate a value of about $\Gamma = 160$ for the onset of crystallization. Estimate the interior temperature, $T_c$, at which this occurs.

(c) Estimate the luminosity of a pure carbon white dwarf with this interior temperature. Assume a composition like that of Example 15.2 for the nondegenerate envelope.

(d) For roughly how many years could the white dwarf sustain the luminosity found in part (c), using just the latent heat of $kT$ per nucleus released upon crystallization? Compare this amount of time (when the white dwarf cools more slowly) with Fig. 15.9.

15.9 In the *liquid-drop model* of an atomic nucleus, a nucleus with mass number $A$ has a radius of $r_\circ A^{1/3}$, where $r_\circ = 1.2 \times 10^{-13}$ cm. Find the density of this nuclear model.

15.10 If our Moon were as dense as a neutron star, what would its diameter be?

15.11 (a) Consider two point masses, each having mass $m$, that are separated vertically by a distance of 1 cm just above the surface of a neutron star of radius $R$ and mass $M$. Using Newton's law of gravity (Eq. 2.11), find an expression for the ratio of the gravitational force on the lower mass to that on the upper mass, and evaluate this expression for $R = 10$ km, $M = 1.4$ $M_\odot$, and $m = 1$ g.

(b) An iron cube 1 cm on each side is held just above the surface of the neutron star described in (a) above. The density of iron is 7.86 g cm$^{-3}$. If iron experiences a stress (force per cross-sectional area) of $4.2 \times 10^8$ dyne cm$^{-2}$, it will be permanently stretched; if the stress reaches $1.5 \times 10^9$ dyne cm$^{-2}$, the iron will rupture. What will happen to the iron cube? (*Hint*: Imagine concentrating half of the cube's mass on each of its top and bottom surfaces.) What would happen to an iron meteoroid falling toward the surface of a neutron star?

15.12 Estimate the neutron degeneracy pressure at the center of a 1.4 $M_\odot$ neutron star (take the central density to be $1.5 \times 10^{15}$ g cm$^{-3}$), and compare this with the estimated pressure at the center of Sirius B.

15.13 (a) At a density just below neutron drip, assume that all of the neutrons are in heavy neutron-rich nuclei such as $^{118}_{36}$Kr. Estimate the pressure due to relativistic degenerate electrons.

(b) At a density just above neutron drip, assume (*wrongly!*) that all of the neutrons are free (and not in nuclei). Estimate the speed of the degenerate neutrons and the pressure they would produce.

15.14 Suppose that the Sun were to collapse down to the size of a neutron star (10 km radius).

(a) Assuming that no mass is lost in the collapse, find the rotation period of the neutron star.

(b) Find the magnetic field strength of the neutron star.

Even though our Sun will not end its life as a neutron star, this shows that the conservation of angular momentum and magnetic flux can easily produce pulsarlike rotation speeds and magnetic fields.

15.15 (a) Use Eq. (14.14) with $\gamma = 5/3$ to calculate the fundamental radial pulsation period for a one-zone model of a pulsating white dwarf (use the values for Sirius B) and a 1.4 $M_\odot$ neutron star. Compare these to the observed range of pulsar periods.

(b) Use Eq. (15.26) to calculate the minimum rotation period for the same stars, and compare them to the range of pulsar periods.

(c) Give an explanation for the similarity of your results.

15.16 (a) Determine the minimum rotation period for a 1.4 M$_\odot$ neutron star (the fastest it can spin without flying apart). For convenience, assume that the star remains spherical with a radius of 10 km.

(b) Newton studied the equatorial bulge of a homogeneous fluid body of mass $M$ that is *slowly* rotating with angular velocity $\Omega$. He proved that the difference between its equatorial radius ($E$) and its polar radius ($P$) is related to its average radius ($R$) by

$$\frac{E - P}{R} = \frac{5\Omega^2 R^3}{4GM}.$$

Use this to estimate the equatorial and polar radii for a 1.4 M$_\odot$ neutron star rotating with twice the minimum rotation period you found in part (a).

15.17 If you measured the period of PRS 1937+214 and obtained the value on page 609, about how long would you have to wait before the last digit changed from a "5" to a "6"?

15.18 Consider a pulsar that has a period $P_\circ$ and period derivative $\dot{P}_\circ$ at $t = 0$. Assume that the product $P\dot{P}$ remains constant for the pulsar (c.f. Eq. 15.29).

(a) Integrate to obtain an expression for the pulsar's period $P$ at time $t$.

(b) Imagine that you have constructed a clock that would keep time by counting the radio pulses received from this pulsar. Suppose you also have a *perfect* clock ($\dot{P} = 0$) that is initially synchronized with the pulsar clock when they both read zero. Show that when the perfect clock displays the characteristic lifetime $P_\circ/\dot{P}_\circ$, the time displayed by the pulsar clock is $(\sqrt{3} - 1)P_\circ/\dot{P}_\circ$.

15.19 During a glitch, the period of the Crab pulsar decreased by $|\Delta P| \approx 10^{-8} P$. If the increased rotation was due to an overall contraction of the neutron star, find the change in the star's radius. Assume that the pulsar is a rotating sphere of uniform density with an initial radius of 10 km.

## Problems

**15.20** The Geminga pulsar has a period of $P = 0.237$ s and a period derivative of $\dot{P} = 1.1 \times 10^{-14}$. Assuming that $\theta = 90°$, estimate the magnetic field strength at the pulsar's poles.

**15.21** (a) Find the radii of the light cylinders for the Crab pulsar and for the slowest pulsar PSR 1845−19. Compare these values to the radius of a 1.4 $M_\odot$ neutron star.

(b) The strength of a magnetic dipole is proportional to $1/r^3$. Determine the ratio of the magnetic field strengths at the light cylinder for the Crab pulsar and for PSR 1845−19.

**15.22** (a) Integrate Eq. (15.29) to obtain an expression for a pulsar's period $P$ at time $t$ if its initial period was $P_\circ$ at time $t = 0$.

(b) Assuming that the pulsar has had time to slow down enough that $P_\circ \ll P$, show that the age $t$ of the pulsar is given approximately by

$$t = \frac{P}{2\dot{P}},$$

where $\dot{P}$ is the period derivative at time $t$.

(c) Evaluate this age for the case of the Crab pulsar, using the values found in Example 15.5. Compare your answer with the known age.

**15.23** One way of qualitatively understanding the flow of charged particles into a pulsar's magnetosphere is to imagine a charged particle of mass $m$ and charge $e$ (the fundamental unit of charge) at the equator of the neutron star. Assume for convenience that the star's rotation carries it perpendicular to the pulsar's magnetic field. The moving charge experiences a magnetic Lorentz force of $F_m = evB/c$ (in cgs units), and a gravitational force, $F_g$. Show that the ratio of these forces is

$$\frac{F_m}{F_g} = \frac{2\pi eBR}{Pcmg},$$

where $R$ is the star's radius and $g$ is the acceleration due to gravity at the surface. Evaluate this ratio for the case of a proton at the surface of the Crab pulsar, using a magnetic field strength of $10^{12}$ G.

**15.24** Find the minimum photon energy required for the creation of an electron–positron pair via the pair-production process $\gamma \rightarrow e^- + e^+$. What is the wavelength of this photon? In what region of the electromagnetic spectrum is this wavelength found?

15.25 A subpulse involves a very narrow radio beam with a width between 1° and 3°. Use Eq. (4.43) for the headlight effect to calculate the minimum speed of the electrons responsible for a 1° subpulse.

*Chapter 16*

# BLACK HOLES

## 16.1 The General Theory of Relativity

Gravity, the weakest of the four forces of nature, is nonetheless the dominant force in sculpting the universe on the largest scale. Newton's law of universal gravitation,

$$F = G\frac{Mm}{r^2}, \tag{16.1}$$

remained an unquestioned cornerstone of astronomers' understanding of heavenly motion until the beginning of the twentieth century. Its application had explained the motions of the known planets and had accurately predicted the existence and position of the planet Neptune in 1846. The sole blemish on Newtonian gravitation was the inexplicably large rate of shift in the orientation of Mercury's orbit.

The gravitational influences of the other planets cause the major axis of Mercury's elliptical orbit to slowly swing around the Sun in a counterclockwise direction relative to the fixed stars; see Fig. 16.1. The angular position at which perihelion occurs shifts at a rate of 574″ per century.[1] However, Newton's law of gravity was unable to explain 43″ per century of this shift, an inconsistency that led some mid-nineteenth century physicists to suggest that Eq. (16.1) should be modified from an exact inverse-square law. Others thought that an unseen planet, nicknamed Vulcan, might occupy an orbit inside Mercury's.

Between the years 1907 and 1915 Albert Einstein developed a new theory of gravity, his **general theory of relativity**. In addition to resolving the

---

[1]The value of 1.5° per century encountered in some texts includes the very large effect of the precession of Earth's rotation axis on the celestial coordinate system, described in Section 1.3.

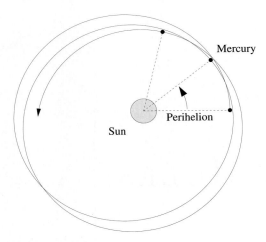

**Figure 16.1** Perihelion shift of Mercury's orbit; the shift per orbit has been exaggerated by a factor of $10^5$.

mystery of Mercury's orbit, it predicted many new phenomena that were later confirmed by experiment. In this and the next section we will describe just enough of the physical content of general relativity to provide the background needed for future discussions of black holes and cosmology. Einstein's view of the universe provides an exhilarating challenge to the imaginations of all students of astrophysics. Before embarking on our study of general relativity, it will be helpful to take an advanced look at this new gravitational landscape.

The general theory of relativity is fundamentally a geometric description of how distances (intervals) in spacetime are measured in the presence of mass. For the moment, the effects on space and time will be considered separately, although the reader should always keep in mind that relativity deals with a unified spacetime. Near a massive object, both space and time must be described in a new way.

Distances between points in the space surrounding a massive object are altered in a way that can be interpreted as space becoming *curved* through a fourth spatial dimension perpendicular to *all* of the usual three directions. The human mind balks at picturing this situation, but an analogy is easily found. Imagine four people holding the corners of a rubber sheet, stretching it tight and flat. This represents the flatness of empty space that exists in the absence of mass. Also imagine that a polar coordinate system has been painted on the sheet, with evenly spaced concentric circles spreading out from its center. Now lay a heavy bowling ball (representing the Sun) at the center of the sheet, and watch the indentation of the sheet as it curves down and stretches in response

## 16.1 The General Theory of Relativity

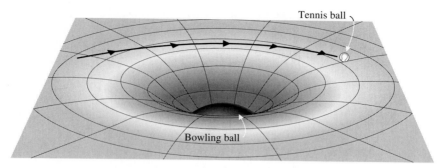

**Figure 16.2** Rubber sheet analogy for curved space around the Sun.

to the ball's weight, as pictured in Fig. 16.2. Closer to the ball, the sheet's curvature increases and the distance between points on the circles is stretched more. Just as the sheet curves in a third direction perpendicular to its original flat two-dimensional plane, the space surrounding a massive object may be thought of as curving in a fourth spatial dimension perpendicular to the usual three of "flat space."[2] The fact that mass has an effect on the surrounding space is the first essential element of general relativity. The curvature of space is just one aspect of the effect of mass on spacetime. In the language of unified spacetime, *mass acts on spacetime, telling it how to curve.*

Now imagine rolling a tennis ball, representing a planet, across the sheet. As it passes near the bowling ball, the tennis ball's path is curved. If the ball were rolled in just the right way under ideal conditions, it could even "orbit" the more massive bowling ball. In a similar manner, a planet orbits the Sun as it responds to the curved spacetime around it. Thus *curved spacetime acts on mass, telling it how to move.*

The passage of a ray of light near the Sun can be represented by rolling a ping-pong ball very rapidly past the bowling ball. Although the analogy with a massless photon is strained, it is reasonable to expect that as the photon moves through the curved space surrounding the Sun, its path will be deflected from a straight line. The bend of the photon's trajectory is small because the photon's speed carries it quickly through the curved space; see Fig. 16.3. In general relativity, gravity is the result of objects moving through curved spacetime, and everything that passes through, even massless particles such as photons, is affected.

Figure 16.3 hints at another aspect of general relativity. Nothing can move

---

[2]It is important to note that this fourth spatial dimension has nothing at all to do with the role played by time as a fourth nonspatial *coordinate* in the theory of relativity.

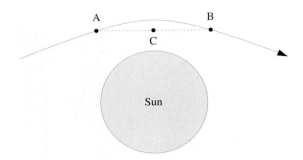

**Figure 16.3** A photon's path around the Sun is shown by the solid line. The bend in the photon's trajectory is greatly exaggerated.

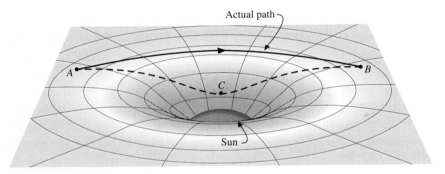

**Figure 16.4** Comparison of two photon paths through curved space between points $A$ and $B$.

between two points in space faster than light.[3] This means that light must always follow the quickest route between any two points. In flat, empty space, this path is a straight line, but what is the quickest route through curved space? Suppose we use a series of mirrors to force the light beam to travel between points $A$ and $B$ by the apparent "shortcut" indicated by the dotted lines in Figs. 16.3 and 16.4. Would the light taking the dotted path outrace the beam free to follow its natural route through curved space? The answer is no—the curved beam would win the race. This result seems to imply that the beam following the dotted line would slow down along the way. However, this inference can't be correct because, according to the postulates of relativity (page 97), every observer, including one at point $C$, measures the same value for the speed of light. There are just two possible answers. The distance along the dotted line might actually be *longer* than the light beam's natural path, and/or time might run more slowly along the dotted path; either would

---

[3]Throughout this chapter, light is assumed to be traveling in a vacuum.

## 16.1 The General Theory of Relativity

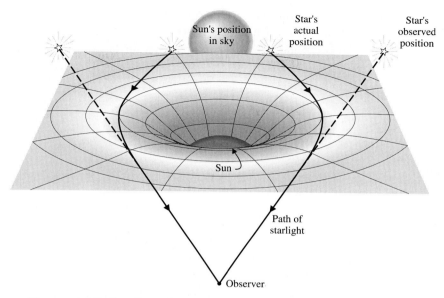

**Figure 16.5** Bending of starlight measured during a solar eclipse.

retard the beam's passage. In fact, according to general relativity these effects contribute equally to delaying the light beam's trip from $A$ to $B$ along the dotted line. The curving light beam actually does travel the *shorter* path. If two space travelers were to lay meter sticks end-to-end along the two paths, the dotted path would require a greater number of meter sticks because it penetrates farther into curved space, as shown in Fig. 16.4. In addition, the curvature of space involves a concomitant slowing down of time, so clocks placed along the dotted path would actually *run more slowly*. This is the final essential feature of general relativity: *Time runs more slowly in curved spacetime.*

It is important to note that all of the foregoing ideas have been tested experimentally many times, and in every case the results agree with general relativity. As soon as Einstein completed his theory, he applied it to the problem of Mercury's unexplained residual perihelion shift of $43''$ per century. Einstein wrote that his heart raced when his calculations exactly explained the discrepancy in terms of the planet's passage through the curved space near the Sun, saying that, "For a few days, I was beside myself with joyous excitement." Another triumph came in 1919 when the curving path of starlight passing near the Sun was first measured during a total solar eclipse by Arthur Stanley Eddington. As shown in Fig. 16.5, the apparent positions of stars close to the Sun's eclipsed edge were shifted from their actual positions by a small angle. Einstein's theory predicted that this angular deflection would be 1.75″, in good agreement with Eddington's observations. General relativity has been tested continuously ever since. For instance, the superior conjunction of Mars that occurred in 1976 led to a spectacular confirmation of Einstein's theory. Radio signals beamed to Earth from the Viking spacecraft on Mars' surface were delayed as they traveled deep into the curved space surrounding the Sun. The time delay agreed with the predictions of general relativity to within 0.1%.

It is now time to retrace our steps and discover how Einstein came to his revolutionary understanding of gravity as geometry. One of the postulates of special relativity states that the laws of physics are the same in all inertial reference frames. Accelerating frames of reference are not inertial frames, because they introduce fictitious forces that depend on the acceleration. For example, an apple at rest on the seat of a car will not remain at rest if the car suddenly brakes to a halt. However, the acceleration produced by the force of gravity has a unique aspect. This may be clearly seen by noting a fundamental difference between Newton's law of gravity and Coulomb's law for the electrical force (Eq. 5.9).

Consider two objects, one of mass $m$ and charge $q$, and the other of mass $M$ and charge $Q$, separated by a distance $r$. If at least one of the charges is zero, then the magnitude of the acceleration of mass $m$ due to the gravitational force is

$$ma = G\frac{mM}{r^2}.$$

On the other hand, if neither of the charges is zero, then the acceleration due to the electrical force is

$$ma = \frac{qQ}{r^2}.$$

The mass $m$ on the left-hand sides is an *inertial mass* and measures the object's resistance to being accelerated (its inertia). On the right-hand sides, the masses $m$ and $M$ and charges $q$ and $Q$ are numbers that couple the masses or charges to their respective forces and determine the strength of these forces. The mystery is the appearance of $m$ on both sides of the gravitational formula. Why should a quantity that measures an object's inertia (which exists even in the complete absence of gravity) be the same as the "gravitational charge" that determines the force of gravity? The answer is that the equation's notation is flawed, and the formula should be properly written as

$$m_i a = G\frac{m_g M_g}{r^2}$$

or

$$a = G\frac{M_g}{r^2}\frac{m_g}{m_i}$$

## 16.1 The General Theory of Relativity

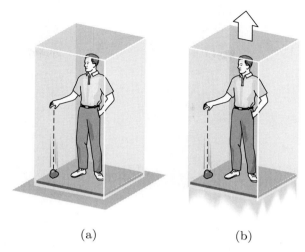

**Figure 16.6** Gravity is equivalent to an accelerating laboratory: (a) a laboratory on Earth, and (b) a laboratory accelerating in space.

to clearly distinguish between the inertial and gravitational mass of each object. Yet it is an experimental fact, tested to a precision of 1 part in $10^{12}$, that $m_g/m_i$ is a constant; for convenience, this constant is chosen to be unity so the two types of mass will be numerically equal.[4] The proportionality of the inertial and gravitational masses means that at a given location, all objects experience the *same* gravitational acceleration.

This distinctive aspect of gravity, that every object falls with the same acceleration, has been known since the time of Galileo. It presented Einstein with both a problem and an opportunity to extend his theory of special relativity. He realized that if an entire laboratory were in free-fall, with all of its contents falling together, there would then be no way to detect its acceleration. In such a freely falling laboratory, it would be impossible to experimentally determine whether the laboratory were floating in space, far from any massive object, or if it were falling freely in a gravitational field. Similarly, an observer watching an apple falling with an acceleration $g$ toward the floor of a laboratory would be unable to tell whether the laboratory were on Earth or far out in space, accelerating at a rate $g$ in the direction of the ceiling, as illustrated in Fig. 16.6. This posed a serious problem for the theory of special relativity, which requires that inertial reference frames have a constant velocity. Because

---

[4] If the gravitational mass were chosen to be *twice* the inertial mass, for example, the laws of physics would be unchanged except the gravitational constant $G$ would be assigned a new value only one-fourth as large.

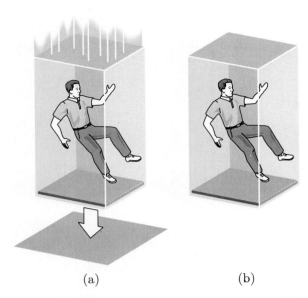

(a)          (b)

**Figure 16.7** Gravity abolished in a freely falling laboratory: (a) a laboratory in free-fall, and (b) a laboratory floating in space.

gravity is equivalent to an accelerating laboratory, an inertial reference frame cannot even be defined in the presence of gravity. Einstein had to find a way to remove gravity from the laboratory.

In 1907, Einstein had "the happiest thought of my life."

> I was sitting in a chair in the patent office at Bern when all of a sudden a thought occurred to me: "If a person falls freely he will not feel his own weight." I was startled. This simple thought made a deep impression on me. It impelled me toward a theory of gravitation.

The way to eliminate gravity in a laboratory is to surrender to it by entering into a state of free-fall; see Fig. 16.7.[5] However, there was an obstacle to applying this to special relativity because its inertial reference frames are *infinite* collections of meter sticks and synchronized clocks (recall Section 4.1). It would be impossible to eliminate gravity everywhere in an infinite, freely

---

[5] "Free-fall" means that there are no nongravitational forces accelerating the laboratory. In his meditation on general relativity, *A Journey into Gravity and Spacetime* (see Suggested Readings), John A. Wheeler prefers the term "free-float." Since gravity has been abolished, why should falling even be mentioned? The reader is also urged to browse through the pages of *Gravitation* by Misner, Thorne, and Wheeler (1973) for additional insights into general relativity.

## 16.1 The General Theory of Relativity

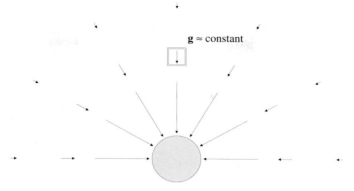

**Figure 16.8** A local inertial reference frame, with **g** ≈ constant inside.

falling reference frame, because different points would have to be falling at different rates in different directions (toward the center of Earth, for example). Einstein realized that he would have to use *local* reference frames, just small enough that the acceleration due to gravity would be essentially constant (in both magnitude and direction) everywhere inside (see Fig. 16.8). Gravity would then be abolished inside a local, freely falling reference frame.

In 1907 Einstein adopted this as the cornerstone of his theory of gravity, calling it the principle of equivalence.

> **The Principle of Equivalence:** All local, freely falling, nonrotating laboratories are fully equivalent for the performance of all physical experiments.

The restriction to nonrotating labs is necessary to eliminate the fictitious forces associated with rotation, such as the Coriolis and centrifugal forces. We will call these local, freely falling, nonrotating laboratories **local inertial reference frames**.

Note that special relativity is incorporated into the principle of equivalence. For example, measurements made from two local inertial frames in relative motion are related by the Lorentz transformations (Eqs. 4.16–4.19) using the *instantaneous* value of the relative velocity between the two frames. Thus general relativity is in fact an extension of the theory of special relativity.

We now move on to two simple thought experiments involving the equivalence principle that demonstrate the curvature of spacetime. For the first experiment, imagine a laboratory suspended above the ground by a cable (see Fig. 16.9a). Let a photon of light leave a horizontal flashlight at the same instant the cable holding the lab is severed (Fig. 16.9b). Gravity has been

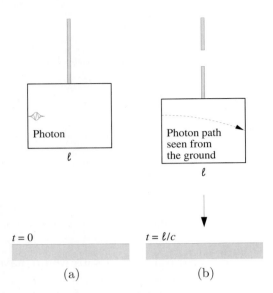

**Figure 16.9** The equivalence principle for a horizontally traveling photon. The photon (a) leaves the left wall at $t = 0$, and (b) arrives at the right wall at $t = \ell/c$.

abolished from this freely falling lab, so it is now a local inertial reference frame. According to the equivalence principle, an observer falling with the lab will measure the light's path across the room as a straight horizontal line, in agreement with all of the laws of physics. But another observer on the ground sees a lab that is falling under the influence of gravity. Because the photon maintains a constant height above the lab's floor, the ground observer must measure a photon that falls with the lab, following a curved path. This displays the spacetime curvature represented by the rubber sheet analogy. The curved path taken by the photon is the quickest route possible through the curved spacetime surrounding Earth.

The angle of deflection, $\phi$, of the photon is very slight, as the following bit of geometry shows. Although the photon does not follow a circular path, we will use the *best-fitting circle* of radius $r_c$ to the actual path measured by the ground observer. Referring to Fig. 16.10, the center of the best-fitting circle is at point $O$, and the arc of the circle subtends an angle $\phi$ (exaggerated in the figure) between the radii $OA$ and $OB$. If the width of the lab is $\ell$, then the photon crosses the lab in time $t = \ell/c$. (The difference between the length of the arc and the width of the lab is negligible.) In this amount of time, the lab falls a distance $d = \frac{1}{2}gt^2$. Because triangles $ABC$ and $OBD$ are similar (each

## 16.1 The General Theory of Relativity

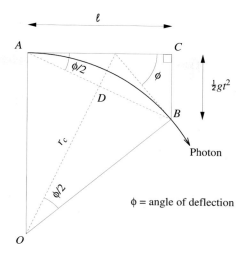

**Figure 16.10** Geometry for the radius of curvature, $r$, and angular deflection, $\phi$.

containing a right angle and another angle $\phi/2$),

$$\overline{BC}/\overline{AC} = \overline{BD}/\overline{OD}$$

$$\left(\frac{1}{2}gt^2\right)\bigg/\ell = \left[\frac{\ell}{2\cos(\phi/2)}\right]\bigg/\overline{OD}.$$

In fact, $\phi$ is so small that we can set $\cos(\phi/2) \simeq 1$ and the distance $\overline{OD} \simeq r_c$. Then, using $t = \ell/c$ and $g = 980$ cm s$^{-2}$ for the acceleration of gravity near the surface of Earth, we find

$$r_c = \frac{c^2}{g} \tag{16.2}$$

$$= 9.17 \times 10^{17} \text{ cm},$$

for the radius of curvature of the photon's path, which is nearly a light-year!

Of course, the angular deflection $\phi$ depends on the width $\ell$ of the lab. For example, if $\ell = 10$ m, then

$$\phi = \frac{\ell}{r_c} = 1.09 \times 10^{-15} \text{ rad},$$

or only $2.25 \times 10^{-10}$ arcsecond. The large radius of the photon's path indicates that spacetime near Earth is only slightly curved. Nonetheless, the curvature

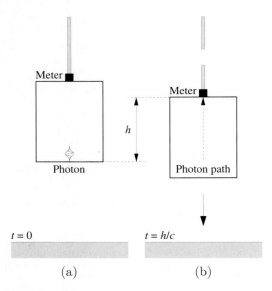

**Figure 16.11** Equivalence principle for a vertically traveling light. The photon (a) leaves the floor at $t = 0$, and (b) arrives at the ceiling at $t = h/c$.

is great enough to produce the circular orbits of satellites, which move slowly through the curved spacetime (slowly, that is, compared to the speed of light).

Our second thought experiment also begins with the laboratory suspended above the ground by a cable. This time, monochromatic light of frequency $\nu_\circ$ leaves a vertical flashlight on the floor at the same instant the cable holding the lab is severed. The freely falling lab is again a local inertial frame where gravity has been abolished, and so the equivalence principle requires that a frequency meter in the lab's ceiling record the *same* frequency, $\nu_\circ$, for the light that it receives. But an observer on the ground sees a lab that is falling under the influence of gravity. As shown in Fig. 16.11, if the light has traveled upward a height $h$ toward the meter in time $t = h/c$, then the meter has gained a downward speed toward the light of $v = gt = gh/c$ since the cable was released. Accordingly, we would expect that from the point of view of the ground observer, the meter should have measured a blueshifted frequency greater than $\nu_\circ$ by an amount given by Eq. (4.33). For the slow free-fall speeds involved here, this expected *increase* in frequency is

$$\frac{\Delta \nu}{\nu_\circ} = \frac{v}{c} = \frac{gh}{c^2}.$$

But in fact, the meter recorded *no change* in frequency. Therefore there must be another effect of the light's upward journey through the curved spacetime

## 16.1 The General Theory of Relativity

around Earth that exactly compensates for this blueshift. This is a **gravitational redshift** that tends to *decrease* the frequency of the light as it travels upward a distance $h$, given by

$$\frac{\Delta \nu}{\nu_\circ} = -\frac{v}{c} = -\frac{gh}{c^2}. \tag{16.3}$$

An outside observer, not in free-fall inside the lab, would measure only this gravitational redshift. If the light were traveling downward, a corresponding blueshift would be measured. It is left as an exercise to show that this formula remains valid even if the light is traveling at an angle to the vertical, as long as $h$ is taken to be the *vertical* distance covered by the light.

---

**Example 16.1** In 1960, a test of the gravitational redshift formula was carried out at Harvard University. A gamma ray was emitted by an unstable isotope of iron, $^{57}_{26}\text{Fe}$, at the bottom of a 22.6 m tall tower, and received at the top of the tower.[6] Using this value for $h$, the expected decrease in frequency of the gamma ray due to the gravitational redshift is

$$\frac{\Delta \nu}{\nu_\circ} = -\frac{gh}{c^2} = -2.46 \times 10^{-15}, \tag{16.4}$$

in excellent agreement with the experimental result of $\Delta\nu/\nu_\circ = -(2.57 \pm 0.26) \times 10^{-15}$. More precise experiments carried out since that time have obtained agreement to within 0.007%.

---

An approximate expression for the total gravitational redshift for a beam of light that escapes out to infinity can be calculated by integrating Eq. (16.3) from an initial position $r_\circ$ to infinity, using $g = GM/r^2$ (Newtonian gravity) and $h = dr$ for a spherical mass, $M$, located at the origin. Some care must be taken when carrying out the integration, because Eq. (16.4) was derived using a *local* inertial reference frame. By integrating, we are really adding up the redshifts obtained for a chain of *different* frames. The radial coordinate $r$ can be used to measure distances for these frames only if spacetime is nearly flat [that is, if the radius of curvature given by Eq. (16.2) is very large compared with $r_\circ$]. In this case, the "stretching" of distances seen previously in the rubber sheet analogy is not too severe, and we can integrate

$$\int_{\nu_\circ}^{\nu_\infty} \frac{d\nu}{\nu} \simeq -\int_{r_\circ}^{\infty} \frac{GM}{r^2 c^2}\, dr,$$

---

[6]Actually, the experiment was performed with both upward and downward traveling gamma rays, providing tests of both the gravitational redshift and blueshift.

where $\nu_o$ and $\nu_\infty$ are the frequencies at $r_o$ and infinity, respectively. The result is

$$\ln\left(\frac{\nu_\infty}{\nu_o}\right) \simeq -\frac{GM}{r_o c^2},$$

which is valid when gravity is weak ($r_o/r_c = GM/r_o c^2 \ll 1$). This can be rewritten as

$$\frac{\nu_\infty}{\nu_o} \simeq e^{-GM/r_o c^2}. \tag{16.5}$$

Because the exponent is $\ll 1$, we use $e^{-x} \simeq 1-x$ to get

$$\frac{\nu_\infty}{\nu_o} \simeq 1 - \frac{GM}{r_o c^2}. \tag{16.6}$$

This approximation shows the first-order correction to the frequency of the photon.

The exact result for the gravitational redshift, valid even for a strong gravitational field, is

$$\frac{\nu_\infty}{\nu_o} = \left(1 - \frac{2GM}{r_o c^2}\right)^{1/2}. \tag{16.7}$$

When gravity is weak and the exponent is $\ll 1$, we use $(1-x)^{1/2} \simeq 1 - x/2$ to recover Eq. (16.6),

$$\frac{\nu_\infty}{\nu_o} = \left(1 - \frac{2GM}{r_o c^2}\right)^{1/2} \simeq 1 - \frac{GM}{r_o c^2}.$$

The gravitational redshift can be incorporated into the redshift parameter (Eq. 4.34), giving

$$z = \frac{\lambda_\infty - \lambda_o}{\lambda_o} = \frac{\nu_o}{\nu_\infty} - 1$$

$$= \left(1 - \frac{2GM}{r_o c^2}\right)^{-1/2} - 1 \tag{16.8}$$

$$\simeq \frac{GM}{r_o c^2}, \tag{16.9}$$

where Eq. (16.9) is valid only for a weak gravitational field.

To understand the origin of the gravitational redshift, imagine a clock that is constructed to tick once with each vibration of a monochromatic light wave. The time between ticks is then equal to the period of the oscillation of the wave, $\Delta t = 1/\nu$. Then according to Eq. (16.7), as seen from an infinite distance, the

## 16.1 The General Theory of Relativity

gravitational redshift implies that the clock at $r_o$ will be observed to run more slowly than an identical clock at $r = \infty$. If an amount of time $\Delta t_o$ passes at position $r_o$ outside a spherical mass, $M$, then the time $\Delta t_\infty$ at $r = \infty$ is

$$\frac{\Delta t_o}{\Delta t_\infty} = \frac{\nu_\infty}{\nu_o} = \left(1 - \frac{2GM}{r_o c^2}\right)^{1/2} \tag{16.10}$$

$$\simeq 1 - \frac{GM}{r_o c^2}, \tag{16.11}$$

where Eq. (16.11) holds for a weak field. We must conclude that *time passes more slowly as the surrounding spacetime becomes more curved*, an effect called **gravitational time dilation**. The gravitational redshift is therefore a consequence of time running at a slower rate near a massive object.

In other words, suppose two perfect, identical clocks are initially standing side by side, equally distant from a spherical mass. They are synchronized, and then one is slowly lowered below the other and then raised back to its original level. All observers will agree that when the clocks are again side by side, the clock that was lowered will be running behind the other because time in its vicinity passed more slowly while it was deeper in the mass's gravitational field.

---

**Example 16.2** The white dwarf Sirius B has a radius of $R = 5.5 \times 10^8$ cm and a mass of $M = 2.1 \times 10^{33}$ g. The radius of curvature of the path of a horizontally traveling light beam near the surface of Sirius B is given by Eq. (16.2),

$$r_c = \frac{c^2}{g} = \frac{R^2 c^2}{GM} = 1.9 \times 10^{12} \text{ cm.}$$

The fact that $GM/Rc^2 = R/r_c \ll 1$ indicates that the curvature of spacetime is not severe. Even at the surface of a white dwarf, gravity is considered relatively weak.

From Eq. (16.9), the gravitational redshift suffered by a photon emitted at the star's surface is

$$z \simeq \frac{GM}{Rc^2} = 2.8 \times 10^{-4}.$$

This is in excellent agreement with the measured gravitational redshift for Sirius B of $(3.0 \pm 0.5) \times 10^{-4}$.

To compare the rate at which time passes at the surface of Sirius B with the rate at a great distance, suppose that exactly one hour is measured by a distant clock. The time recorded by a clock at the surface of Sirius B would be *less* than one hour by an amount found using Eq. (16.11):

$$\Delta t_\infty - \Delta t_\circ = \Delta t_\infty \left(1 - \frac{\Delta t_\circ}{\Delta t_\infty}\right) \simeq (3600 \text{ s}) \left(\frac{GM}{Rc^2}\right) = 1.0 \text{ s}.$$

The clock at the surface of Sirius B runs more slowly by about one second per hour compared to an identical clock far out in space.

---

The preceding experimental results (results obtained from tests of the equivalence principle) confirm the curvature of spacetime. In Section 16.2, we will learn that a freely falling particle takes the straightest possible path through curved spacetime.

## 16.2 Intervals and Geodesics

We now consider the united concepts of space and time as expressed in *spacetime*, with four coordinates $(x, y, z, t)$ specifying each *event*.[7] Einstein's crowning achievement was the deduction of his *field equations* for calculating the geometry of spacetime produced by a given distribution of mass and energy. His equations have the form

$$\mathcal{G} = -\frac{8\pi G}{c^4} \mathcal{T}. \tag{16.12}$$

On the right is the *stress–energy tensor*, $\mathcal{T}$, which evaluates the effect of a given distribution of mass and energy on the curvature of spacetime, as described mathematically by the Einstein tensor, $\mathcal{G}$ (for $\mathcal{G}$ravity), on the left.[8] The appearance of Newton's gravitational constant, $G$, and the speed of light symbolizes the extension of relativity theory to include gravity. It is far beyond the scope of this book to delve further into this fascinating equation. We will be content merely to describe the curvature of spacetime around a spherical object of mass $M$ and radius $R$, then demonstrate how an object moves through the curved spacetime it encounters.

Figure 16.12 shows three examples of some paths traced out in spacetime. In these *spacetime diagrams*, time is represented on the vertical axis, while space is depicted by the horizontal $x$–$y$ plane. The third spatial dimension, $z$, cannot be shown, so this figure deals only with motion that occurs in a plane.

---

[7]Nothing special (in fact, nothing at all) need happen at an event. Recall from Section 4.1 that an event is simply a location in spacetime identified by $(x, y, z, t)$.

[8]Note that $E_{\text{rest}} = mc^2$ implies that both mass and energy contribute to the curvature of spacetime.

## 16.2 Intervals and Geodesics

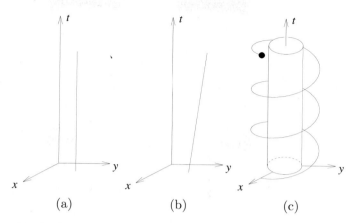

**Figure 16.12** Worldlines for (a) a man at rest, (b) a woman running with constant velocity, and (c) a satellite orbiting Earth.

The path followed by an object as it moves through spacetime is called its **worldline**. Our task will be to calculate the worldline of a freely falling object in response to the local curvature of spacetime. The spatial components of such a worldline describe the trajectory of a baseball arcing toward an outfielder, a planet orbiting the Sun, or a photon attempting to escape from a black hole.

The worldlines of photons in flat spacetime point the way to an understanding of the geometry of spacetime. Suppose a flashbulb is set off at the origin at time $t = 0$; call this event $A$. As shown in Fig. 16.13, the worldlines of photons traveling in the $x$–$y$ plane form a **light cone** that represents a widening series of horizontal circular slices through the expanding spherical wavefront of light. The graph's axes are scaled so that the straight worldlines of light rays make $45°$ angles with the time axis.

A massive object initially at event $A$ must travel slower than light, so the angle between its worldline and the time axis must be less than $45°$. Therefore the region inside the light cone represents the possible *future* of event $A$. It consists of all of the events that can possibly be reached by a traveler initially at event $A$ and therefore all of the events that the traveler could ever influence in a causal way.

Extending the diverging photon worldlines back through the origin generates a lower light cone. Within this lower light cone is the possible *past* of event $A$, the collection of all events from which a traveler could have arrived just as the bulb flashed. In other words, the possible past consists of the locations in space and time of every event that could possibly have caused the flashbulb to go off.

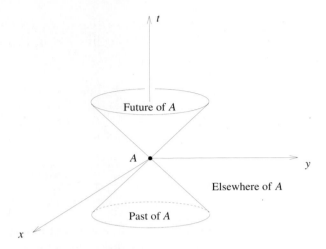

**Figure 16.13** Light cones generated by horizontally traveling photons leaving the origin at time $t = 0$.

Outside the future and past light cones is an unknowable *elsewhere*, that part of spacetime of which a traveler at event $A$ can have no knowledge and over which he or she can have no influence. It may come as a surprise to realize that vast regions of spacetime are hidden from us. You just can't get there from here.

In principle, every event in spacetime has a pair of light cones extending from it. The light cone divides spacetime into that event's future, past, and elsewhere. For any event in the past to have possibly influenced you, that event must lie within your past light cone, just as any event that you can ever possibly affect must lie within your future light cone. Your entire future worldline, your *destiny*, must therefore lie within your future light cone at every instant. Light cones act as spacetime horizons, separating the knowable from the unknowable.

Measuring the progress of an object as it moves along its worldline involves defining a "distance" for spacetime. Consider the familiar case of purely spatial distances. If two points have Cartesian coordinates

$$(x_1, y_1, z_1) \quad \text{and} \quad (x_2, y_2, z_2),$$

then the distance $\Delta \ell$ measured along the straight line between the two points in flat space is defined by

$$(\Delta \ell)^2 = (x_2 - x_1)^2 + (y_2 - y_1)^2 + (z_2 - z_1)^2.$$

## 16.2 Intervals and Geodesics

The analogous measure of "distance" in spacetime is called the **spacetime interval** (or simply *interval* for short), first encountered in Problem 4.11. Let two events $A$ and $B$ have spacetime coordinates

$$(x_A, y_A, z_A, t_A) \quad \text{and} \quad (x_B, y_B, z_B, t_B),$$

measured by an observer in an inertial reference frame, $S$. Then the interval $\Delta s$ measured along the straight worldline between the two events in flat spacetime is defined by

$$(\Delta s)^2 = [c(t_B - t_A)]^2 - (x_B - x_A)^2 - (y_B - y_A)^2 - (z_B - z_A)^2. \qquad (16.13)$$

In words,

$$(\text{interval})^2 = (\text{distance traveled by light in time } |t_B - t_A|)^2$$
$$- (\text{distance between events } A \text{ and } B)^2.$$

This definition of the interval is very useful because, as shown in Problem 4.11, $(\Delta s)^2$ is *invariant* under a Lorentz transformation (Eqs. 4.16–4.19). An observer in another inertial reference frame, $S'$, will measure the same value for the interval between events $A$ and $B$; that is, $\Delta s = \Delta s'$.

Note that $(\Delta s)^2$ may be positive, negative, or zero. The sign tells us whether light has enough time to travel between the two events. If $(\Delta s)^2 > 0$, then the interval is *timelike* and light has more than enough time to travel between events $A$ and $B$. An inertial reference frame $S$ can therefore be chosen that moves along the straight worldline connecting events $A$ and $B$ so that the two events happen at the *same location* in $S$ (at the origin, for example); see Fig. 16.14. Because the two events occur at the same place in $S$, the time measured between the two events is $\Delta s/c$. By definition, the time between two events that occur at the same location is the **proper time**, $\Delta \tau$, where

$$\Delta \tau \equiv \frac{\Delta s}{c} \qquad (16.14)$$

(recall Section 4.3). The proper time is just the elapsed time recorded by a watch moving along the worldline from $A$ to $B$. An observer in any inertial reference frame can use the interval to calculate the proper time between two events that are separated by a timelike interval.

If $(\Delta s)^2 = 0$, then the interval is *lightlike* or *null*. In this case, light has exactly enough time to travel between events $A$ and $B$. Only light can make the journey from one event to the other, and the proper time measured along a null interval is zero.

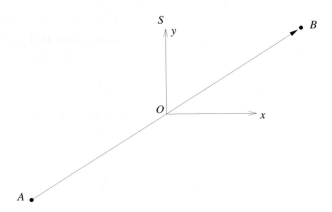

**Figure 16.14** An inertial reference frame $S$ moving along the timelike worldline connecting events $A$ and $B$. Both events occur at the origin of $S$.

Finally, if $(\Delta s)^2 < 0$, then the interval is *spacelike*; light does not have enough time to travel between events $A$ and $B$. No observer could travel between the two events because speeds greater than $c$ would be required. The lack of absolute simultaneity in this situation, however, means that there are inertial reference frames in which the two events occur in the opposite temporal order, or even at the *same time*. By definition, the distance measured between two events $A$ and $B$ in a reference frame for which they occur simultaneously ($t_A = t_B$) is the **proper distance** separating them,[9]

$$\Delta \mathcal{L} = \sqrt{-(\Delta s)^2}. \tag{16.15}$$

If a straight rod were connected between the locations of the two events, this would be the *rest length* of the rod. An observer in any inertial reference frame can use this to calculate the proper distance between two events that are separated by a spacelike interval.[10]

The interval is clearly related to the light cones discussed in the foregoing paragraphs. Let event $A$ be a flashbulb set off at the origin at time $t = 0$. The surfaces of the light cones, where the photons are at any time $t$, are the locations of all events $B$ that are connected to $A$ by a null interval. The events within the future and past light cones are connected to $A$ by a timelike interval, and the events that occur elsewhere are connected to $A$ by a spacelike interval.

---

[9] In Section 4.3, when the emphasis was on length rather than distance, this was called the *proper length*. The terms may be used interchangeably, depending on the context.

[10] For both proper time and proper distance, the term *proper* has the connotation of "measured by an observer who is right there, moving along with the clock or the rod."

## 16.2 Intervals and Geodesics

Returning to three-dimensional space for a moment, it is obvious that a path connecting two points in space doesn't have to be straight. Two points can be connected by infinitely many curved lines. To measure the distance along a curved path, $\mathcal{P}$, from one point to the other, we use a differential distance formula called a *metric*,

$$(d\ell)^2 = (dx)^2 + (dy)^2 + (dz)^2.$$

Then $d\ell$ may be integrated along the path $\mathcal{P}$ (a *line integral*) to calculate the total distance between the two points,

$$\Delta \ell = \int_1^2 \sqrt{(d\ell)^2} = \int_1^2 \sqrt{(dx)^2 + (dy)^2 + (dz)^2} \qquad \text{along } \mathcal{P}.$$

The distance between two points thus depends on the path connecting them. Of course, the *shortest* distance between two points in flat space is measured along a straight line. In fact, we can *define* the "straightest possible line" between two points as the path for which $\Delta \ell$ is a *minimum*.

Similarly, a worldline between two events in spacetime is not required to be straight; the two events can be connected by infinitely many curved worldlines. To measure the interval along a curved worldline, $\mathcal{W}$, connecting two events in spacetime with no mass present, we use the **metric for flat spacetime**,

$$(ds)^2 = (c\,dt)^2 - (dx)^2 - (dy)^2 - (dz)^2.$$

Then $ds$ is integrated to determine the total interval along the worldline $\mathcal{W}$,

$$\Delta s = \int_A^B \sqrt{(ds)^2} = \int_A^B \sqrt{(c\,dt)^2 - (dx)^2 - (dy)^2 - (dz)^2} \qquad \text{along } \mathcal{W}.$$

The interval is still related to the proper time measured along the worldline by Eq. (16.14). *The interval measured along any timelike worldline divided by the speed of light is always the proper time measured by a watch moving along that worldline.* The proper time is zero along a null worldline, and undefined for a spacelike worldline.

In flat spacetime, the interval measured along a straight timelike worldline between two events is a *maximum*. Any other worldline between the same two events will not be straight, and will have a smaller interval. For a massless particle such as a photon, all worldlines have a null interval (so $\int \sqrt{(ds)^2} = 0$).

The maximal character of the interval of a straight worldline in flat spacetime is easily demonstrated. Figure 16.15 is a spacetime diagram showing two events, $A$ and $B$, that occur at times $t_A$ and $t_B$. The events are observed from

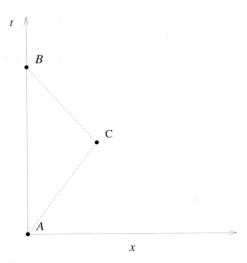

**Figure 16.15** Worldlines connecting events $A$ and $B$.

a inertial reference $S$ frame that moves from $A$ to $B$, chosen so the two events occur at the origin of $S$. The interval measured along the straight worldline connecting $A$ and $B$ is

$$\Delta s(A \to B) = \int_A^B \sqrt{(ds)^2} = \int_A^B \sqrt{(c\,dt)^2 - (dx)^2 - (dy)^2 - (dz)^2}$$

$$= \int_{t_A}^{t_B} c\,dt = c(t_B - t_A).$$

Now consider the interval measured along another worldline connecting $A$ and $B$ that includes event $C$, which occurs at $(x, y, z, t) = (x_C, 0, 0, t_C)$. In this case,

$$\Delta s(A \to C \to B) = \int_A^C \sqrt{(ds)^2} + \int_C^B \sqrt{(ds)^2}$$

$$= \int_A^C \sqrt{(c\,dt)^2 - (dx)^2 - (dy)^2 - (dz)^2}$$

$$+ \int_C^B \sqrt{(c\,dt)^2 - (dx)^2 - (dy)^2 - (dz)^2}.$$

Using $dx/dt = v_{AC}$ for the constant velocity along worldline $A \to C$ in the first integral, and $dx/dt = v_{CB}$ for the constant velocity along $C \to B$ in

## 16.2 Intervals and Geodesics

the second integral, leads to

$$\Delta s(A \to C \to B) = (t_C - t_A)\sqrt{c^2 - v_{AC}^2} + (t_B - t_C)\sqrt{c^2 - v_{CB}^2}$$
$$< c(t_C - t_A) + c(t_B - t_C)$$
$$= \Delta s(A \to B).$$

Thus the straight worldline has the longer interval. Any worldline connecting event $A$ and $B$ can be represented as a series of short segments, so we can conclude that the interval $\Delta s$ is indeed a maximum for the straight worldline.

In a spacetime that is curved by the presence of mass, the situation is slightly more complicated. Even the "straightest possible worldline" will be curved. These straightest possible worldlines are called **geodesics**. In flat spacetime a geodesic is a straight worldline.

In curved spacetime, a timelike geodesic between two events has either a *maximum* or a *minimum* interval. In other words, the value of $\Delta s$ along a timelike geodesic is an *extremum*, either a maximum or a minimum, when compared with the intervals of nearby worldlines between the same two events.[11] In the situations we will encounter in this chapter, the intervals of timelike geodesics will be maxima. A massless particle such as a photon follows a *null geodesic*, with $\int \sqrt{(ds)^2} = 0$.[12] Einstein's key realization was that *the paths followed by freely falling objects through spacetime are geodesics.*

We are now prepared to deal with the effect of mass on the geometry of spacetime, based on the three fundamental features of general relativity.

- Mass acts on spacetime, telling it how to curve.

- Spacetime in turn acts on mass, telling it how to move.

- Any freely falling particle (including a photon) follows the straightest possible worldline, a geodesic, through spacetime. For a massive particle, the geodesic has a *maximum* or a *minimum* interval, while for light, the geodesic has a *null* interval.

---

[11] In fact, a calculation of the intervals of nearby worldlines would show that the interval of a timelike geodesic corresponds to a maximum, minimum, or an inflection point. The reader is referred to Section 13.4 of Misner, Thorne, and Wheeler (1973) for an interesting discussion of geodesics as worldlines of extremal proper time.

[12] The extremal principle for intervals cannot be directly applied to find the straightest possible worldline for a photon, since its interval is always null. However, the straightest possible worldline for a massless particle is the same as that for a massive particle in the limit of a vanishingly small mass as its velocity $v \to c$.

These components of the theory will allow us to describe the curvature of spacetime around a massive spherical object and to determine how another object will move in response, whether it is a satellite orbiting Earth or a photon orbiting a black hole. For situations with spherical symmetry, it will be more convenient to use the familiar spherical coordinates $(r, \theta, \phi)$ instead of Cartesian coordinates. The metric between two nearby points in flat space is then

$$(d\ell)^2 = (dr)^2 + (r\,d\theta)^2 + (r\sin\theta\,d\phi)^2, \tag{16.16}$$

and the corresponding expression for the flat spacetime metric is

$$(ds)^2 = (c\,dt)^2 - (dr)^2 - (r\,d\theta)^2 - (r\sin\theta\,d\phi)^2. \tag{16.17}$$

Of course, spacetime will not be flat in the vicinity of a massive object. The specific situation to be investigated here is the motion of a particle through the curved spacetime produced by a massive sphere. It could be a planet, a star, or a black hole. The first task is to calculate how this massive object acts on spacetime, telling it how to curve. This requires a description of the metric for this curved spacetime that will replace Eq. (16.17) for a flat spacetime.

Before presenting this metric, it should be emphasized that the variables $r$, $\theta$, $\phi$, and $t$ that appear in the expression for the metric are the *coordinates* used by an observer at rest a great ($\simeq$ infinite) distance from the origin. In the absence of a central mass at the origin, $r$ would be the distance from the origin, and differences in $r$ would measure the distance between points on a radial line. The time $t$ measured by clocks scattered throughout the coordinate system would remain synchronized, advancing everywhere at the same rate.

Now we place a sphere of mass $M$ and radius $R$ (which will be called a "planet") at the origin of our coordinate system. Some care must be taken in laying out the radial coordinate. The origin (which is inside the sphere) should not be used as a point of reference, and so we will avoid defining $r$ as "the distance from the origin." Instead, imagine a series of nested concentric spheres centered at the origin. The surface area of a sphere can be measured without approaching the origin, so the coordinate $r$ will be defined by the surface of that sphere having an area $4\pi r^2$. With this careful approach, we will find that these coordinates can be used with the metric for curved spacetime to measure distances in space and the passage of time near this massive sphere. As an object moves through this curved spacetime, its **coordinate speed** is just the rate at which its spatial coordinates change.

At a large distance ($r \simeq \infty$) from the massive sphere, spacetime is essentially flat, and the gravitational time dilation of a photon received from the planet is given by Eq. (16.10). From this, it might be expected that the factor

## 16.2 Intervals and Geodesics

$\sqrt{1 - 2GM/rc^2}$ would play a role in the metric for the spacetime surrounding the planet. Furthermore, recall from Section 16.1 that the stretching of space and the slowing down of time contribute equally to delaying a light beam's passage through curved spacetime. This provides a hint that the same factor will be involved in the metric's radial term. The angular terms are the same as those in Eq. (16.17) for flat spacetime.

These effects are indeed present in the metric that describes the curved spacetime surrounding a spherical mass, $M$. In 1916, just two months after Einstein published his general theory of relativity, the German astronomer Karl Schwarzschild (1873–1916) solved Einstein's field equations to obtain what is now called the **Schwarzschild metric**:

$$(ds)^2 = \left(c\,dt\sqrt{1 - 2GM/rc^2}\right)^2 - \left(\frac{dr}{\sqrt{1 - 2GM/rc^2}}\right)^2$$
$$-(r\,d\theta)^2 - (r\sin\theta\,d\phi)^2. \tag{16.18}$$

There is no other, easier way to obtain the Schwarzschild metric, so we must be content with the foregoing heuristic description of its terms.

It is important to realize that the Schwarzschild metric is a *vacuum solution* of Einstein's field equations.[13] That is, it is valid only in the empty space *outside* the object. The mathematical form of the metric is different in the object's interior, which is occupied by matter.

The Schwarzschild metric contains all of the effects considered in the last section. The "curvature of space" resides in the radial term. The radial distance measured simultaneously ($dt = 0$) between two nearby points on the same radial line ($d\theta = d\phi = 0$) is just the proper distance, Eq. (16.15),

$$d\mathcal{L} = \sqrt{-(ds)^2} = \frac{dr}{\sqrt{1 - 2GM/rc^2}}. \tag{16.19}$$

Thus the spatial distance $d\mathcal{L}$ between two points on the same radial line is *greater* than the coordinate difference $dr$. This is precisely what is represented by the stretched grid lines in the rubber sheet analogy of the previous section. The factor of $1/\sqrt{1 - 2GM/rc^2}$ must be included in any calculation of spatial distances. This is analogous to using a topographic map when planning a hike up a steep trail. The additional information provided by the map's elevation contour lines must be included in any calculation of the actual hiking distance, which is always greater than the difference in map coordinates; see Fig. 16.16.

---

[13]In fact, the Schwarzschild metric is the *only* spherically symmetric vacuum solution of Einstein's field equations.

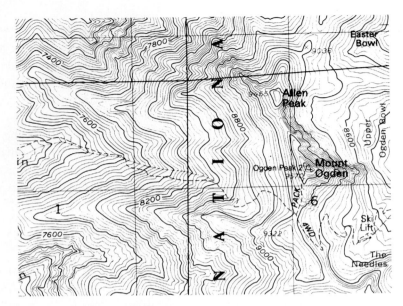

**Figure 16.16** Topographical map with elevation contour lines. The shortest distance between two points on the map may not be a straight line. (Courtesy of USGS.)

The Schwarzschild metric also incorporates time dilation and the gravitational redshift (two aspects of the same effect). If a clock is at rest at the radial coordinate $r$, then the proper time $d\tau$ it records (Eq. 16.14) is related to the time $dt$ that elapses at an infinite distance by

$$d\tau = \frac{ds}{c} = dt\sqrt{1 - \frac{2GM}{rc^2}}, \tag{16.20}$$

which is, of course, just Eq. (16.10). Since $d\tau < dt$, this shows that time passes more slowly closer to the planet.

Having finally learned how a spherical object of any mass acts on spacetime, telling it how to curve, we are now ready to calculate how curved spacetime acts on a particle, telling it how to move. The rest of this section will be devoted to using general relativity to find the motion of a satellite about the planet. All that is needed is the rule that it will follow the straightest possible worldline, the worldline with an *extremal* interval.[14]

At this point, the reader may be fondly recalling the simplicity of Newtonian gravity. According to Newton, the motion of a satellite in a circular orbit

---

[14] It is assumed that the satellite's mass $m$ is small enough that its effect on the surrounding spacetime is negligible.

## 16.2 Intervals and Geodesics

around Earth is found by simply equating the centripetal and gravitational accelerations. That is,

$$\frac{v^2}{r} = \frac{GM}{r^2},$$

where $v$ is the orbital speed. This immediately results in

$$v = \sqrt{\frac{GM}{r}}.$$

Einstein and Newton must agree in the limiting case of weak gravity, so this result must be concealed within the Schwarzschild metric for curved spacetime.[15] It can be found by using the Schwarzschild metric to find the straightest possible worldline for the satellite's circular orbit.

Powerful tools are available for calculating the worldline with the maximum interval between two fixed events, and employing them would ignite some mathematical fireworks. The orbit of the satellite would emerge along with the laws of the conservation of energy, momentum, and angular momentum because they are built into Einstein's field equations. Instead we will follow a quieter and quicker path and assume from the beginning that satellite travels above Earth's equator ($\theta = 90°$) in a circular orbit with a specified angular speed $\omega = v/r$. Inserting these choices, along with $dr = 0$, $d\theta = 0$, and $d\phi = \omega\, dt$, into the Schwarzschild metric gives

$$(ds)^2 = \left[ \left( c\sqrt{1 - 2GM/rc^2} \right)^2 - r^2\omega^2 \right] dt^2$$

$$= \left( c^2 - \frac{2GM}{r} - r^2\omega^2 \right) dt^2.$$

Integrating, the spacetime interval for one orbit is just

$$\Delta s = \int_0^{2\pi/\omega} \sqrt{c^2 - \frac{2GM}{r} - r^2\omega^2}\, dt. \tag{16.21}$$

When finding the value of $r$ for which the interval is an extremum, we must be certain that the endpoints of the satellite's worldline remain fixed. That is, the satellite's orbit must always begin and end at the same position, $r_o$, for all of the worldlines. To accommodate orbits of different radii, we start the satellite at $r_o$ and then move it (at nearly the speed of light) radially outward

---

[15]To avoid succumbing to Newtonian nostalgia, the reader should remember that when Einstein and Newton disagree, nature sides with Einstein.

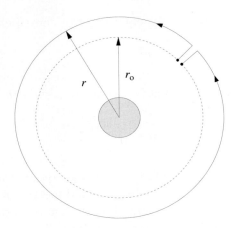

**Figure 16.17** The "orbit" of a satellite, showing the radial motions used to keep the endpoints of the satellite's worldline fixed. The net effect is a circular orbit.

or inward to the radius $r$ of its actual orbit. At the end of the orbit, the satellite returns just as rapidly to its starting point at $r_o$. The "orbit" then appears as shown in Fig. 16.17. Fortunately, the quick radial excursions at the beginning and the end of the orbit can be made with negligible contribution to the integral for the spacetime interval. (At almost the speed of light, the contribution is nearly null.) The net effect is that of a purely circular motion, and so Eq. (16.21) can be used to evaluate the interval.

In Eq. (16.21), the limits of integration are constant and the only variable is $r$. The value of the radial coordinate $r$ for the orbit actually followed by the satellite must be the one for which $\Delta s$ is an extremum. This value may be found by taking the derivative of $\Delta s$ with respect to $r$ and setting it equal to zero:

$$\frac{d}{dr}(\Delta s) = \frac{d}{dr}\left(\int_0^{2\pi/\omega} \sqrt{c^2 - \frac{2GM}{r} - r^2\omega^2}\, dt\right) = 0.$$

The derivative may be taken inside the integral to obtain

$$\frac{d}{dr}\sqrt{c^2 - \frac{2GM}{r} - r^2\omega^2} = 0,$$

implying

$$\frac{2GM}{r^2} - 2r\omega^2 = 0.$$

## 16.3 Black Holes

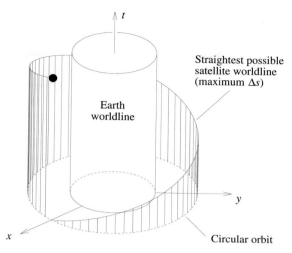

**Figure 16.18** The straightest possible worldline through curved spacetime and its projection onto the orbital plane of the satellite.

Thus, as promised,

$$v = r\omega = \sqrt{\frac{GM}{r}} \qquad (16.22)$$

is the coordinate speed of the satellite for a circular orbit. [By coordinate speed, we simply mean that $v = r\, d\phi/dt$ is speed of the satellite measured in the $(r, \theta, \phi, t)$ coordinate system used by a distant observer.] Figure 16.18 illustrates how this straightest possible worldline through curved spacetime is projected onto the orbital plane, resulting in the satellite's circular orbit around Earth. In fact, this result is valid even for the very large spacetime curvature encountered around a black hole.

## 16.3  Black Holes

In 1783 George Michell, an English clergyman and amateur astronomer, considered the implications of Newton's corpuscular theory of light. If light were indeed a stream of particles, then it should be influenced by gravity. In particular, he conjectured that the gravity of a star 500 times larger than the Sun, but with the Sun's average density, would be sufficiently strong that even light could not escape from it. As the reader may verify using Eq. (2.17), the escape velocity of Michell's star would be the speed of light. Naively setting the Newtonian formula for the escape velocity equal to $c$ shows that $R = 2GM/c^2$ is the radius of a star whose escape velocity equals the speed of light. In terms of the mass of the Sun, $R = 2.95(M/M_\odot)$ km. Even if this Newtonian derivation were correct, the resulting radius of such a star seemed unrealistically small and so it held little interest for astronomers until the middle of the twentieth century. In 1939 American physicists J. Robert Oppenheimer and Hartland Snyder described the ultimate gravitational collapse of a massive star that had exhausted its sources of nuclear fusion. It was earlier that year that Oppenheimer and G. M. Volkoff had calculated the first models of neutron stars (see Section 15.6). We have seen that a neutron star cannot be more massive than about 3 $M_\odot$. Oppenheimer and Snyder pursued the question of the fate of a degenerate star that might exceed this limit and surrender completely to the force of gravity.

For the simplest case of a nonrotating star, the answer lies in the Schwarzschild metric, Eq. (16.18),

$$(ds)^2 = \left(c\,dt\sqrt{1 - 2GM/rc^2}\right)^2 - \left(\frac{dr}{\sqrt{1 - 2GM/rc^2}}\right)^2 - (r\,d\theta)^2 - (r\sin\theta\,d\phi)^2.$$

When the radial coordinate of the star's surface has collapsed to

$$R_S = 2GM/c^2, \tag{16.23}$$

called the **Schwarzschild radius**, the square roots in the metric go to zero. The resulting behavior of space and time at $r = R_S$ is remarkable. For example, according to Eq. (16.14), the proper time measured by a clock at the Schwarzschild radius is $d\tau = 0$. Time has slowed to a complete stop, as measured from our vantage point at rest a great distance away.[16] From our viewpoint, *nothing ever happens at the Schwarzschild radius!*

This behavior is quite curious; does it imply that even light is frozen in time? The speed of light measured by an observer suspended above the collapsed star must always be $c$. But from far away, we can determine that light is delayed as it moves through curved spacetime. (Recall the time delay of radio signals from the Viking lander on Mars described in Section 16.1.) The apparent speed of light, the rate at which the spatial coordinates of a photon change, is called the *coordinate speed of light*. Starting with the Schwarzschild metric with $ds = 0$ for light,

$$0 = \left(c\,dt\sqrt{1 - 2GM/rc^2}\right)^2 - \left(\frac{dr}{\sqrt{1 - 2GM/rc^2}}\right)^2 - (r\,d\theta)^2 - (r\sin\theta\,d\phi)^2,$$

---

[16] The reader should recall that the spacetime coordinates $(r, \theta, \phi, t)$ in the Schwarzschild metric were established for use by an observer at rest at $r \simeq \infty$.

## 16.3 Black Holes

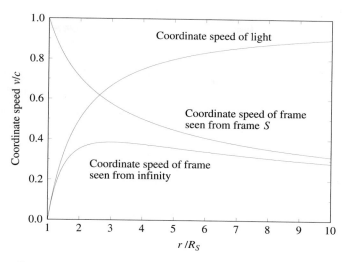

**Figure 16.19** Coordinate speed of light, and coordinate speeds of a freely falling frame $S$ seen by an observer at rest at infinity and by an observer in the frame $S$. The radial coordinates are in terms of $R_S$ for a 10 $M_\odot$ black hole having a Schwarzschild radius of $\approx$ 30 km.

we can calculate the coordinate speed of a vertically traveling photon. Inserting $d\theta = d\phi = 0$ shows that, in general, the coordinate speed of light in the radial direction is

$$\frac{dr}{dt} = c\left(1 - \frac{2GM}{rc^2}\right) = c\left(1 - \frac{R_S}{r}\right). \tag{16.24}$$

When $r \gg R_S$, $dr/dt \simeq c$, as expected in flat spacetime. However, at $r = R_S$, $dr/dt = 0$ (see Fig. 16.19). Light is indeed frozen in time at the Schwarzschild radius. The spherical surface at $r = R_S$ acts as a barrier and prevents our receiving any information from within. For this reason, a star that has collapsed down within the Schwarzschild radius is called a **black hole**.[17] It is enclosed by the **event horizon**, the spherical surface at $r = R_S$. Note that the event horizon is a mathematical surface and need not coincide with any physical surface.

Although the interior of a black hole, inside the event horizon, is a region that is forever hidden from us on the outside, its properties may still be calculated. A nonrotating black hole has a particularly simple structure. At the center is the **singularity**, a point of zero volume and infinite density where all of the black hole's mass is located. Spacetime is infinitely curved at the

---

[17]The term "black hole" is the 1968 invention of the American theoretical physicist John A. Wheeler.

singularity.[18] Cloaking the central singularity is the event horizon, so the singularity can never be observed. In fact, there is a hypothesis dubbed the "Law of Cosmic Censorship" that forbids a *naked singularity* from appearing unclothed (without an associated event horizon).

An object as bizarre as a black hole deserves closer scrutiny. Imagine an attempt to investigate the black hole by starting at a safe distance and reflecting a radio wave from an object at the event horizon. How much time will it take for a radio photon (or any photon) to reach the event horizon from a radial coordinate $r \gg R_S$ and then return? Since the round trip is symmetric, it is necessary only to find the time for either the journey in or out and then double the answer. It is easiest to integrate the coordinate speed of light in the radial direction, Eq. (16.24), between two arbitrary values of $r_1$ and $r_2$ to obtain the general answer,

$$\Delta t = \int_{r_1}^{r_2} \frac{dr}{dr/dt} = \int_{r_1}^{r_2} \frac{dr}{c(1 - R_S/r)}$$
$$= \frac{r_2 - r_1}{c} + \frac{R_S}{c} \ln \left( \frac{r_2 - R_S}{r_1 - R_S} \right),$$

assuming that $r_1 < r_2$. Inserting $r_1 = R_S$ for the photon's destination, we find that $\Delta t = \infty$. According to the distant observer, the radio photon will *never* reach the event horizon. Instead, according to gravitational time dilation, the photon's coordinate velocity will slow down until it finally stops at the event horizon in the infinite future. In fact, any object falling toward the event horizon will suffer the same fate. Seen from the outside, even the surface of the star that collapsed to form the event horizon would be frozen, and so a black hole is in this sense a *frozen star*.

A brave (and *indestructible*) astronomer decides to test this remarkable conclusion. Starting from rest at a great distance, she volunteers to fall freely toward a 10 $M_\odot$ black hole ($R_S \simeq 30$ km). We remain behind to watch her local inertial frame $S$ as it falls with coordinate speed $dr/dt$ all the way to the event horizon. She gradually accelerates as she monitors her watch and shines a monochromatic flashlight back in our direction once every second. As her fall progresses, the light signals arrive farther and farther apart for several reasons: Subsequent signals must travel a longer distance as she accelerates, and her proper time $\tau$ is running more slowly than our coordinate time $t$ due to her

---

[18] The black hole's singularity is a real physical entity. It is not a mathematical artifact, as is the *mathematical* singularity exhibited by the Schwarzschild metric at the event horizon (where $1/\sqrt{1 - 2GM/rc^2} \to \infty$). Choosing another coordinate system would remove the divergence at the event horizon, so that divergence has no physical significance.

## 16.3 Black Holes

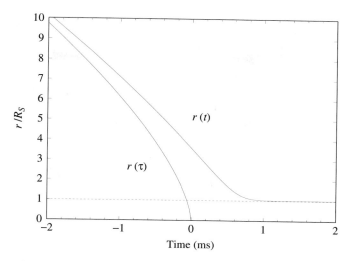

**Figure 16.20** Coordinate $r(t)$ of a freely falling frame $S$ according to an observer at rest at infinity, and $r(\tau)$ according to an observer in the frame $S$. The radial coordinates are in terms of $R_S$ for a 10 $M_\odot$ black hole.

location (gravitational time dilation) and her motion (special relativity time dilation). Furthermore, the coordinate speed of light becomes slower as she approaches the black hole, so the signals travel back to us more slowly. The frequency of the light waves we receive is also increasingly redshifted. This is caused both by her acceleration away from us and the gravitational redshift. The light becomes dimmer as well, as the rate at which her flashlight emits photons decreases (seen from our vantage point) and the energy per photon ($hc/\lambda$) also declines. Then when she is about $2R_S$ from the event horizon, the time between her signals begins to increase without limit as the strength of the signals decreases. The light is redshifted and dimmed into invisibility as time dilation brings her coordinate speed to zero (see Figs. 16.19 and 16.20). She is frozen in time, held for eternity like a fly caught in amber. Our successors could watch for millennia while stars are born, evolve, and die without receiving a single photon from her.

How does all of this appear to the brave astronomer, freely falling toward the black hole? Because gravity has been abolished in her local inertial frame, initially she does not notice her approach to the black hole. She monitors her watch (which displays her proper time, $\tau$), and she turns on her flashlight once per second. However, as she draws closer, she begins to feel as though she is being stretched in the radial direction and compressed in the perpendicular directions; see Fig. 16.21. The gravitational pull on her feet (nearer the black

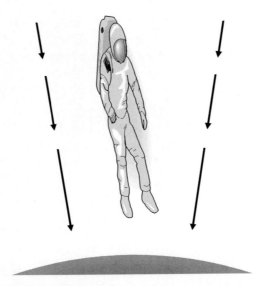

**Figure 16.21** Tidal forces near a black hole.

hole) is stronger than on her head, and the variation in the direction of gravity from side to side produces a compression that is even more severe. These differential *tidal forces* increase in strength as she falls. In other words, the size of her local inertial frame (where gravity has been abolished) becomes increasingly smaller as the spatial variation in the gravitational acceleration vector, **g**, increases. Were she not indestructible, our astronomer would be torn apart by these tidal forces while still several hundred kilometers from the black hole.[19]

In just two milliseconds (proper time), she falls the final few hundred kilometers to the event horizon and crosses it. Her proper time continues normally, and she encounters no frozen stellar surface since it has fallen through long ago.[20] However, once inside the event horizon, her fate is sealed. It is impossible for any particle to be at rest when $r < R_S$, as can be easily seen from the Schwarzschild metric (Eq. 16.18). Using $dr = d\theta = d\phi = 0$ for an object at rest, the interval is given by

$$(ds)^2 = (c\,dt)^2 \left(1 - \frac{R_S}{r}\right) < 0$$

when $r < R_S$. This is a *spacelike* interval, which is not permitted for particles. Therefore it is impossible to remain at rest where $r < R_S$. Within the event

---

[19] You may recall Problem 15.11 concerning the stretching of an iron cube near a neutron star.

[20] The presence or absence of a frozen stellar surface at the event horizon makes no real difference; the Schwarzschild metric specifies the same spacetime curvature outside.

## 16.3 Black Holes

horizon of a nonrotating black hole, all worldlines converge at the singularity. Even photons are pulled in toward the center. This means that the astronomer never has an opportunity to glimpse the singularity because no photons can reach her from there. She can, however, see the light that falls in behind her from events in the outside universe, but she does not see the entire history of the universe as it unfolds. Although the elapsed coordinate time in the outside world does become infinite, the light from all of these events does not have time to reach the astronomer. Instead, these events occur in her "elsewhere." Just $6.6 \times 10^{-5}$ s of proper time after passing the event horizon, she is inexorably drawn to the singularity.[21]

Black holes may be formed in several ways. The collapse of the center of a sufficiently massive supergiant star may result in the formation of a central black hole. A neutron star in a close binary system may gravitationally strip enough mass from its companion that the neutron star's self-gravity exceeds the ability of the degeneracy pressure to support it, again resulting in a black hole. Black holes may also have been manufactured in the earliest instants of the universe. Presumably, these **primordial black holes** would have been formed with a wide range of masses, both much greater ($\sim 10^5$ $M_\odot$) and much less ($\sim 10^{-5}$ g) than the lower limit of 3 $M_\odot$ for black holes formed by stellar collapse. The only criterion for a black hole is that its entire mass must lie within the Schwarzschild radius, so the Schwarzschild metric is valid at the event horizon.

---

**Example 16.3** If Earth could somehow (miraculously) be compressed sufficiently to become a black hole, its radius would only be $R_S = 2GM_\oplus/c^2 = 0.887$ cm. Although a primordial black hole could be this size, it is almost impossible to imagine packing Earth's entire mass into so small a ball. In reality, the cubic centimeter of material at Earth's center contains only 13 g, so nothing special happens 0.887 cm from the center.[22]

---

Whatever the formation process, it is certain to be very complicated. The collapse of a star will rarely be symmetrical. Detailed calculations have demonstrated, however, that any irregularities are radiated away by gravitational waves (see Section 17.5). As a result, once the surface of the collapsing star reaches the event horizon, the exterior spacetime horizon is spherically symmetric and described by the Schwarzschild metric.

---

[21] A thorough description of the final view of the falling astronomer may be found in Rothman et al. (1985).

[22] The reader is reminded that the Schwarzschild metric is valid only outside matter. It does not describe the spacetime inside Earth.

Another complication is the fact that all stars rotate, and therefore so will the resulting black hole. Remarkably, however, any black hole can be completely described by just three numbers: its mass, angular momentum, and electric charge.[23] Black holes have no other attributes or adornments, a condition commonly expressed by saying that "a black hole has no hair."[24]

There is a firm upper limit for a rotating black hole's angular momentum. The maximum value of the angular momentum for a black hole of mass $M$ is

$$L_{\max} = \frac{GM^2}{c}. \qquad (16.25)$$

If the angular momentum of a rotating black hole were to exceed this limit, there would be no event horizon and a naked singularity would appear, in violation of the Law of Cosmic Censorship.

---

**Example 16.4** The maximum angular momentum for a solar-mass black hole is

$$L_{\max} = \frac{GM_\odot^2}{c} = 8.81 \times 10^{48} \text{ g cm}^2 \text{ s}^{-1}.$$

By comparison, the angular momentum of the Sun (assuming uniform rotation) is $1.63 \times 10^{48}$ g cm$^2$ s$^{-1}$, about 18% of $L_{\max}$. We should expect that many stars will have angular momenta that are comparable to $L_{\max}$, and so vigorous (if not maximal) rotation ought to be common for stellar-mass black holes.

---

The structure of a maximally rotating black hole is shown in Fig. 16.22.[25] The rotation has distorted the central singularity from a point into a flat ring, and the event horizon has assumed the shape of an ellipsoid. The figure also shows additional features caused by the rotation. As a massive object spins, it induces a rotation in the surrounding spacetime, a phenomenon known as *frame dragging*. To gain some insight into this effect, recall the behavior of a pendulum swinging at the north pole of Earth. As Earth rotates, the plane of the pendulum's swing remains fixed with respect to the distant stars. The stars define a nonrotating frame of reference for the universe, and it is relative to this frame that the pendulum's swing remains planar. However, the rotating

---

[23]If magnetic monopoles exist, the "magnetic charge" would also be required for a complete specification. However, both magnetic and electric charge can be safely ignored because stars should be very nearly neutral.

[24]The "no hair" theorem actually applies only to the universe outside the event horizon. Inside, the spacetime geometry is complicated by the mass distribution of the collapsed star.

[25]The *Kerr metric* for a rotating black hole was derived from Einstein's field equations by a New Zealand mathematician, Roy Kerr, in 1963.

## 16.3 Black Holes

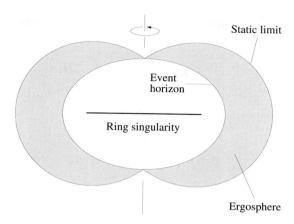

**Figure 16.22** The structure of a maximally rotating black hole, with the ring singularity seen edge-on. The location of the event horizon at the equator is $r = \frac{1}{2}R_S = GM/c^2$.

spacetime close to a massive spinning object produces a local deviation from the nonrotating frame that describes the universe at large. Near a rotating black hole, frame dragging is so severe that there is a nonspherical region outside the event horizon called the **ergosphere** where any particle *must* move in the same direction that the black hole rotates; see Fig. 16.22. Spacetime within the ergosphere is rotating so rapidly that a particle would have to travel faster than the speed of light to remain at the same angular coordinate (e.g., at the same value of $\phi$ in the coordinate system used by a distant observer). The outer boundary of the ergosphere is called the **static limit**, so named because once beyond this boundary a particle can remain at the same coordinate as the effect of frame dragging diminishes.

Even Earth's rotation produces very weak frame dragging. Scheduled to be launched into a polar orbit in 1999, the Stanford Gravity Probe B experiment will attempt to detect Earth's frame dragging by measuring the precession of four gyroscopes made of precisely shaped spheres of fused quartz 3.8 cm in diameter. Although the expected precession rate is only $0.042''$ yr$^{-1}$, the effect is cumulative. By comparing the changes that occur in the gyroscopes' different initial orientations, the frame dragging should be measurable.

At this point, the reader should be warned that the previous descriptions of a black hole's structure inside the event horizon, such as Fig. 16.22, are based on vacuum solutions to Einstein's field equations. These solutions were obtained by ignoring the effects of the mass of the collapsing star, so the vacuum solutions do not describe the interior of a real black hole. Furthermore,

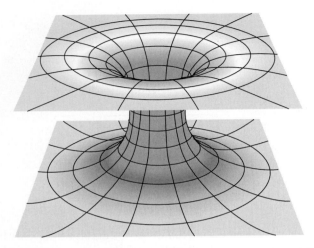

**Figure 16.23** Depiction of a Schwarzschild throat connecting two different regions of spacetime. Any attempted passage of matter or energy through the throat would cause it to collapse.

the present laws of physics, including general relativity, break down under the extreme conditions found very near the center. The details of the singularity cannot be fully described until a theory of quantum gravity is found. The presence of a singularity is assured, however. In 1965 an English mathematician, Roger Penrose, proved conclusively that *every* complete gravitational collapse must form a singularity.

The possibility of using a black hole as a tunnel connecting one location in spacetime with another (perhaps in a different universe) has inspired both physicists and science fiction writers. Most conjectures of spacetime tunnels are based on vacuum solutions to Einstein's field equations and as such don't apply to the interiors of real black holes. Still, they have become part of the popular culture and we will consider them briefly here. Figure 16.23 depicts a spacetime tunnel called a **Schwarzschild throat** (also known as an *Einstein–Rosen bridge*), which uses the Schwarzschild geometry of a nonrotating black hole to connect two regions of spacetime. The width of the throat is a minimum at the event horizon, and the "mouths" may be interpreted as opening onto two different locations in spacetime. It is tempting to imagine this as a tunnel, and writers of speculative fiction have dreamed of *white holes* pouring out mass or serving as passageways for starships and cute Disney robots. However, it appears that any attempt to send a tiny amount of matter or energy (even a stray photon) through the throat would cause it to collapse. For a real nonrotating black hole, all worldlines end at the inescapable singularity, where

## 16.3 Black Holes

spacetime is infinitely curved. There is simply no way to bypass the singularity.

The story is somewhat different for a rotating black hole. Although spacetime is still infinitely curved at the ring singularity, all worldlines need not converge there. In fact, it is difficult for an infalling object to hit the singularity in a rotating black hole. Theorists have calculated worldlines for vacuum solutions that miss the singularity and emerge in the spacetime of another universe. However, as for nonrotating black holes, any attempt to pass the smallest amount of matter or energy along such a route would cause the passageway to collapse, thereby pinching it off. In summary, it seems extremely unlikely that black holes can provide a stable passageway for any matter or energy, even for idealized cases. For more realistic situations, any voyager attempting a trip through a black hole would end up being torn apart by the singularity.

One final possibility is that of a **wormhole**, a hypothetical tunnel between two points in spacetime separated by an arbitrarily great distance.[26] We will briefly consider nonrotating, spherically symmetric wormholes. They are described by *nonvacuum* solutions to Einstein's field equation. In other words, a wormhole must be threaded by some sort of *exotic material* whose tension prevents the collapse of the wormhole. There is no known mechanism that would allow a wormhole to arise naturally; it would have to be constructed by an incredibly advanced civilization. However, the theoretical possibilities alone are fascinating. These solutions to Einstein's field equations have no event horizon (permitting two-way trips through the wormhole) and involve survivable tidal forces. Journey times from one end through to the other can be less than one year (traveler's proper time), although the ends of the wormhole may be separated by interstellar or intergalactic distances.

The catch, of course, is the problematic existence of the exotic material needed to stabilize the wormhole. The unusual nature of the exotic material becomes apparent if we consider two light rays that converge on the wormhole and enter it, only to diverge when they exit the other end. This implies that the exotic material must be capable of gravitationally defocusing light, an "antigravity" effect involving the gravitational repulsion of the light by the material through which the rays pass. Exotic material meeting this requirement would have a negative energy density ($\rho c^2 < 0$), at least as experienced by the light rays. Although a negative energy density arises in certain quantum situations, it may or may not be allowed physically on macroscopic scales. We will leave wormholes as a fascinating possibility and abandon the discussion at this point,

---

[26]The term *wormhole* recalls the holes eaten in some ancient map books by worms, providing a symbolic shortcut between the distant locations portrayed on the maps.

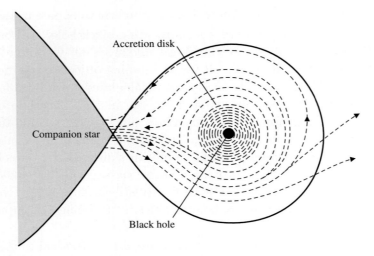

**Figure 16.24** Gas pulled from a companion star forms an x-ray emitting disk around a black hole.

recalling Einstein's remark that "all our thinking is of the nature of a free play with concepts."[27]

The reader may feel as though much of this section were borrowed from the pages of a science fiction novel. Extraordinary claims require extraordinary proof, and proof of the mere existence of black holes has been difficult to obtain. The problem lies in detecting an object only a few tens of kilometers across that emits no radiation directly. The best hope of astronomers has been to find a black hole in a close binary system. If the black hole in such a system is able to pull gas from the envelope of the normal companion star, the angular momentum of their orbital motion would cause a disk of gas to form around the black hole (see Fig. 16.24).[28] As the gas spirals down toward the event horizon, it is compressed and heated to millions of kelvins and emits x-rays. Only the gravity of a neutron star or a black hole can produce x-rays in a close binary system, and in fact the compact object in most x-ray binaries is believed to be a neutron star. However, if an x-ray binary can be found in which the mass of the compact object exceeds 3 $M_\odot$, then a strong case can be made that the compact object is a black hole. The first black hole to be tentatively identified in this way is Cygnus X-1, near the bright star $\eta$ Cygni in the middle of the swan's neck. Another promising candidate is LMC X-3, an

---

[27] The reader is referred to Morris and Thorne (1988) and Thorne (1994) for more details and further speculations concerning wormholes.

[28] Accretion disks in a close binary system will be discussed in more detail in Section 17.2.

## 16.3 Black Holes

x-ray binary in the Large Magellanic Cloud. The most compelling case to date involves the x-ray binary A0620−00.[29] The orbital velocities along the line of sight of *both* components of A0620−00 have been determined using Doppler shifted spectral lines. A simple application of Kepler's laws shows that the mass of the compact object must be *at least* $3.82 \pm 0.24$ $M_\odot$, well above the 3 $M_\odot$ upper limit for a neutron star.[30] As more evidence accumulates, it seems that astronomers have finally found the extraordinary proof required for the existence of a black hole.

The black holes of classical general relativity last forever. There is a very general result due to Stephen Hawking, stating that the surface area of a black hole's event horizon can never decrease. If a black hole coalesces with any other object, the result is an even larger black hole.[31] In 1974, however, Hawking discovered a loophole in this law when he combined quantum mechanics with the theory of black holes and found that black holes can slowly *evaporate*. The key to this process is pair-production, the formation of a particle–antiparticle pair just outside the event horizon of a black hole. Ordinarily the particles quickly recombine and disappear, but if one of the particles falls into the event horizon while its partner escapes, as shown in Fig. 16.25, this disappearing act may be thwarted. The black hole's gravitational energy was used to produce the two particles, and so the escaping particle has carried away some of the black hole's mass. The net effect as seen by an observer at a great distance is the emission of particles, known as **Hawking radiation**, by the black hole, accompanied by a reduction in the black hole's mass.

The rate at which energy is carried away by particles in this manner is inversely proportional to the square of the black hole's mass, or $1/M^2$. For solar-mass black holes, the emitted particles are photons and the rate of emission is minuscule. As the black hole's mass declines, however, the rate of emission increases. The final stages of a black hole's evaporation proceeds extremely rapidly, releasing a burst of all types of elementary particles. This tremendous explosion most probably leaves behind only an empty region of flat spacetime.

The lifetime of a primordial black hole prior to its evaporation, $t_{\text{evap}}$, is

---

[29] A0620−00 is on the border of the constellations Monoceros and Orion, about one-third of the way along a line from Betelgeuse to Sirius.

[30] See Haswell and Shafter (1990) for details concerning this black hole binary. This and other black hole candidates will be discussed further in Section 17.5.

[31] There is evidence for the presence of extremely massive black holes, containing between roughly $10^5$ and $10^9$ $M_\odot$, located in the central regions of galaxies.

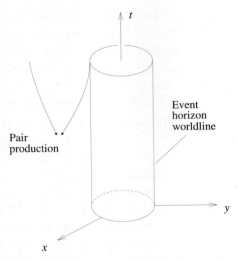

**Figure 16.25** Spacetime diagram showing particle–antiparticle pairs created near the event horizon of a black hole.

quite long,

$$t_{\text{evap}} \approx \left(\frac{M}{M_\odot}\right)^3 \times 10^{66} \text{ yr.}$$

Since the estimated age of the universe is approximately 10 to 20 billion years, this process is of no consequence for black holes formed by a collapsing star. However, a primordial black hole with a mass of roughly $5 \times 10^{14}$ g would evaporate in about 15 billion years. Thus primordial black holes with this mass should be in the final, explosive stage of evaporation right now and could possibly be detected. The final burst of Hawking radiation is thought to release high-energy ($\approx 100$ MeV) gamma rays at a rate of $10^{20}$ erg s$^{-1}$, together with electrons, positrons, and many other particles. The subsequent decay of these particles should produce additional gamma rays that would be observable by Earth-orbiting satellites. To date, measurements of the cosmic gamma-ray background at this energy have not detected anything that can be identified with the demise of a nearby primordial black hole. Although there is as yet no positive evidence that primordial black holes exist, this negative result is still important. It implies that on average there cannot be more than 200 primordial black holes with this mass in every cubic light-year of space.

# Suggested Readings

### GENERAL

Davies, Paul, "Wormholes and Time Machines," *Sky and Telescope*, January 1992.

Hawking, Stephen W., "The Quantum Mechanics of Black Holes," *Scientific American*, January 1977.

Hawking, Stephen W., *A Brief History of Time*, Bantam Books, Toronto, 1988.

Luminet, Jean-Pierre, *Black Holes*, Cambridge University Press, Cambridge, 1992.

McClintock, Jeffrey, "Do Black Holes Exist?" *Sky and Telescope*, January 1988.

Morris, Michael S., and Thorne, Kip S., "Wormholes in Spacetime and Their Use for Interstellar Travel: A Tool for Teaching General Relativity," *American Journal of Physics*, 56, 395, 1988.

Rothman, Tony, et al., *Frontiers of Modern Physics*, Dover Publications, New York, 1985.

Shapiro, Stuart L., and Teukolsky, Saul A., *Black Holes, White Dwarfs, and Neutron Stars*, John Wiley and Sons, New York, 1983.

Thorne, Kip S., *Black Holes and Time Warps: Einstein's Outrageous Legacy*, W. W. Norton and Co., New York, 1994.

Wald, Robert M., *Space, Time, and Gravity*, The University of Chicago Press, Chicago, 1977.

Wheeler, John A., *A Journey into Gravity and Spacetime*, Scientific American Library, New York, 1990.

Will, Clifford, *Was Einstein Right?*, Basic Books, New York, 1986.

### TECHNICAL

Bekenstein, Jacob D., "Black Hole Thermodynamics," *Physics Today*, January 1980.

Berry, Michael, *Principles of Cosmology and Gravitation*, Cambridge University Press, Cambridge, 1976.

Haswell, Carole A., and Shafter, Allen W., "A Detection of Orbital Radial Velocity Variations of the Primary Component the Black Hole Binary A0620−00 (= V616 Monocerotis)," *The Astrophysical Journal Letters*, *359*, L47, 1990.

Misner, Charles W., Thorne, Kip S., and Wheeler, John A., *Gravitation*, W. H. Freeman and Co., San Francisco, 1973.

Ruffini, Remo, and Wheeler, John A., "Introducing the Black Hole," *Physics Today*, January 1991.

Taylor, Edwin F., and Wheeler, John Archibald, *Scouting Black Holes*, preprint, 1995. (For current version, contact Edwin F. Taylor, 22 Hopkins Road, Arlington, MA 02174 USA, or email eftaylor@mit.edu.)

# Problems

16.1 In the rubber sheet analogy of Section 16.1, a keen eye would notice that the tennis ball also depresses the sheet slightly, and so the soccer ball constantly tilts slightly toward the tennis ball as they orbit each other. Qualitatively compare this with the motion of two stars in a binary orbit.

16.2 Show that Eq. (16.3) for the gravitational redshift remains valid even if the light travels upward at an angle $\theta$ measured from the vertical as long as $h$ is taken to be the *vertical distance* traveled by the light pulse.

16.3 A photon near the surface of Earth travels a horizontal distance of 1 km. How far does the photon "fall" in this time?

16.4 Leadville, Colorado, is at an altitude of 3.1 km above sea level. If a person there lives for 75 years (as measured by an observer at a great distance from Earth), how much longer would gravitational time dilation have allowed that person to live if he or she had moved at birth from Leadville to a city at sea level?

16.5 (a) Estimate the radius of curvature of a horizontally traveling photon at the surface of a 1.4 $M_\odot$ neutron star, and compare the result with the 10 km radius of the star. Can general relativity be neglected when studying neutron stars?

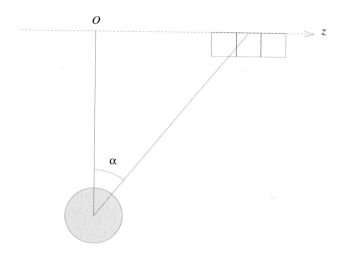

**Figure 16.26** Local inertial frames for measuring the deflection of light near the Sun (Problem 16.6).

(b) If one hour passes at the surface of the neutron star, how much time passes at a great distance? Compare the times obtained from the exact and approximate expressions, Eqs. (16.10) and (16.11), respectively.

16.6 Imagine a series of rectangular local inertial reference frames suspended by cables in a line near the Sun's surface, as shown in Fig. 16.26. The frames are carefully lined up so the tops and sides of neighboring frames are parallel, and the tops of the frames lie along the $z$-axis. A photon travels unhindered through the frames. As the photon enters each frame, the frame is released from rest and falls freely toward the center of the Sun.

(a) Show that in passing through the frame located at angle $\alpha$ (shown in the figure), the angular deflection of the photon's path is

$$d\phi = \frac{g_\circ \cos^3 \alpha}{c^2} \, dz,$$

where $dz$ is the width of the reference frame and $g_\circ$ is the Newtonian gravitational acceleration at the point of closest approach, $O$. The angular deflection is small, so assume that the photon is initially traveling in the $z$-direction as it enters the frame. (*Hint*: The width of the frame in the $z$-direction is $dz$, so the time for the photon to cross the frame can be taken to be $dz/c$.)

(b) Integrate the result you found in part (a) from $\alpha = -\pi/2$ to $+\pi/2$ and so find the total angular deflection of the photon as it passes through the curved spacetime near the Sun.

(c) Your answer (which is also the answer obtained by Einstein in 1911 before he arrived at his field equations) is only half the correct value of 1.75″. Can you *qualitatively* account for the missing factor of two?

16.7 Assume that you are at the origin of a laboratory reference system at time $t = 0$ when you start your clock (event $A$). Determine whether the following events are within the future lightcone or past lightcone of event $A$, or elsewhere.

(a) A flashbulb goes off 7 m away at time $t = 0$.
(b) A flashbulb goes off 7 m away at time $t = 2$ s.
(c) A flashbulb goes off 70 km away at time $t = 2$ s.
(d) A flashbulb goes off 700,000 km away at time $t = 2$ s.
(e) A supernova explodes 180,000 ly away at time $t = -5.7 \times 10^{12}$ s.
(f) A supernova explodes 180,000 ly away at time $t = 5.7 \times 10^{12}$ s.
(g) A supernova explodes 180,000 ly away at time $t = -5.6 \times 10^{12}$ s.
(h) A supernova explodes 180,000 ly away at time $t = 5.6 \times 10^{12}$ s.

For items (e) and (g), could an observer in another reference frame moving relative to yours measure that the supernova exploded *after* event $A$? For items (f) and (h), could an observer in another frame measure that the supernova exploded *before* event $A$?

16.8 $\tau$ Ceti is the closest single star that is similar to the Sun. At time $t = 0$, Alice leaves Earth in her starship and travels at a speed of $0.95c$ to $\tau$ Ceti, 11.7 ly away as measured by astronomers on Earth. Her twin brother, Bob, remains at home, at $x = 0$.

(a) According to Bob, what is the interval between Alice's leaving Earth and arriving at $\tau$ Ceti?

(b) According to Alice, what is the interval between her leaving Earth and arriving at $\tau$ Ceti?

(c) Upon arriving at $\tau$ Ceti, Alice immediately turns around and returns to Earth at a speed of $0.95c$. (Assume that the actual turnaround takes negligible time.) What was the proper time for Alice during her round trip to $\tau$ Ceti?

(d) When she and Bob meet on her return to Earth, how much younger will Alice be than her brother?

16.9 Consider a spherical blackbody of constant temperature and mass $M$ whose surface lies at radial coordinate $r = R$. An observer located at the surface of the sphere and a distant observer both measure the blackbody radiation given off by the sphere.

(a) If the observer at the surface of the sphere measures the luminosity of the blackbody to be $L$, use the gravitational time dilation formula, Eq. (16.10), to show that the observer at infinity measures

$$L_\infty = L\left(1 - \frac{2GM}{Rc^2}\right). \tag{16.26}$$

(b) Both observers use Wien's law, Eq. (3.15), to determine the blackbody's temperature. Show that

$$T_\infty = T\sqrt{1 - \frac{2GM}{Rc^2}}. \tag{16.27}$$

(c) Both observers use the Stefan–Boltzmann law, Eq. (3.17), to determine the radius of the spherical blackbody. Show that

$$R_\infty = \frac{R}{\sqrt{1 - 2GM/Rc^2}}. \tag{16.28}$$

Thus using the Stefan–Boltzmann law without including the effects of general relativity will lead to an *overestimate* of the size of a compact blackbody.

16.10 In 1792 the French mathematician Simon-Pierre de Laplace (1749–1827) wrote that a hypothetical star, "of the same density as Earth, and whose diameter would be two hundred and fifty times larger than the Sun, would not, in consequence of its attraction, allow any of its rays to arrive at us." Use Newtonian mechanics to calculate the escape velocity of Laplace's star.

16.11 Qualitatively describe the effects on the orbits of the planets if the Sun were suddenly to become a black hole.

16.12 Consider four black holes with masses of $10^{15}$ g, 10 $M_\odot$, $10^5$ $M_\odot$, and $10^9$ $M_\odot$.

(a) Calculate the Schwarzschild radius for each.

(b) Calculate the average density, defined by $\rho = M/\left(\frac{4}{3}\pi R_S^3\right)$, for each.

16.13 (a) Show that the proper distance from the event horizon to a radial coordinate $r$ is given by

$$\Delta \mathcal{L} = r\sqrt{1 - \frac{R_S}{r}} + \frac{R_S}{2} \ln\left(\frac{1 + \sqrt{1 - R_S/r}}{1 - \sqrt{1 - R_S/r}}\right).$$

This illustrates the danger of interpreting $r$ as a distance instead of a coordinate. *Hint*: Integrate Eq. (16.19).

(b) Make a graph of $\Delta \mathcal{L}$ as a function of $r$ for values of $r$ between $r = R_S$ and $r = 10 R_S$.

(c) Show for large values of $r$ that

$$\Delta \mathcal{L} \simeq r.$$

Thus, far from the black hole, the radial coordinate $r$ can be treated as a distance.

16.14 Verify that the area of the event horizon of a black hole is $4\pi R_S^2$. (*Hint*: Remember that the radial coordinate $r$ is *not* the distance to the center. Use the Schwarzschild metric as your starting point.)

16.15 Equation (16.22) describes the coordinate speed of a massive particle orbiting a nonrotating black hole. However, it can be shown that the orbit is not stable unless $r \geq 3 R_S$; any disturbance will cause a particle in a smaller orbit to spiral down to the event horizon.

(a) Find the coordinate speed of a particle in the smallest stable orbit around a 10 $M_\odot$ black hole.

(b) Find the orbital period (in coordinate time $t$) for this smallest stable orbit around a 10 $M_\odot$ black hole.

16.16 (a) Find an expression for the coordinate speed of light in the $\phi$-direction.

(b) Consider Eq. (16.22) in the limit that the particle's mass goes to zero and its speed approaches that of light. Use your result for part (a) to show that $r = 1.5 R_S$ for the circular orbit of a photon around a black hole.

(c) Find the orbital period (in coordinate time $t$) for this orbit around a 10 $M_\odot$ black hole.

(d) If a flashlight were beamed in the $\phi$-direction at $r = 1.5 R_S$, what would happen? (The surface at $r = 1.5 R_S$ is called the **photon sphere**.)

16.17 Use Eq. (16.25) to compare the maximum angular momentum of a 1.4 $M_\odot$ black hole with the angular momentum of the fastest known pulsar, which rotates with a period of 0.00156 s. Assume that the pulsar is a 1.4 $M_\odot$ uniform sphere of radius 10 km.

16.18 An electron is a pointlike particle of zero radius, so it is natural to wonder whether an electron could be a black hole. However, a black hole of mass $M$ cannot have an arbitrary amount of angular momentum $L$ and charge $Q$. These values must satisfy an inequality,

$$\left(\frac{GM}{c}\right)^2 \geq G\left(\frac{Q}{c}\right)^2 + \left(\frac{L}{M}\right)^2.$$

If this inequality were violated, the singularity would be found *outside* the event horizon, in violation of the Law of Cosmic Censorship. Use $\hbar/2$ for the electron's angular momentum to determine whether or not an electron is a black hole.

16.19 (a) The angular rotation rate, $\Omega$, at which spacetime is dragged around a rotating mass must be proportional to its angular momentum $L$. The expression for $\Omega$ may also contain the constants $G$ and $c$, together with the radial coordinate $r$. Show on purely dimensional grounds that

$$\Omega = \text{constant} \times \frac{GL}{r^3 c^2},$$

where the constant (which you need not determine) is of order unity.

(b) Evaluate this for Earth, assuming that it is a uniformly rotating sphere. Set the leading constant equal to one, and express your answer in arcseconds per year. How much time would it take for a pendulum at the north pole to rotate *once* relative to the distant stars because of frame dragging?

(c) Repeat part (b) for the fastest known pulsar, expressing $\Omega$ in revolutions per second.

16.20 (a) Use dimensional arguments to combine the fundamental constants $\hbar$, $c$, and $G$ into an expression that has units of mass. Evaluate your result, which is an estimate of the least massive primordial black hole formed in the first instant after the Big Bang. What is the mass in grams?

(b) What is the Schwarzschild radius for such a black hole?

(c) How long would it take light to travel this distance?

(d) What is the lifetime of this black hole before its evaporation?

16.21 In the x-ray binary system A0620−00, the radial orbital velocities for the normal star and the compact object are $v_{sr} = 457$ km s$^{-1}$ and $v_{cr} = 43$ km s$^{-1}$, respectively. The orbital period is 0.3226 days.

(a) Calculate the mass function [the right-hand side of Eq. (7.8)],

$$\frac{m_c^3}{(m_s + m_c)^2} \sin^3 i,$$

where $m_s$ is the mass of the normal star, $m_c$ is the mass of its compact companion, and $i$ is the angle of inclination of the orbit. What does this result say about the mass of the compact object? (Note that the value of $v_{cr}$ was not needed to obtain this result.)

(b) Now use the value of the orbital radial velocity of the compact object to determine its mass, assuming $i = 90°$. What does this result say about the mass of the compact object?

(c) The x-rays are not eclipsed in this system, so the angle of inclination must be less than approximately 85°. Suppose that the angle of inclination were 45°; what would the mass of the compact object be then?

This simple calculation, based only on the dynamics of the binary system, provides the best evidence to date for the existence of a black hole.

# Chapter 17

# Close Binary Star Systems

## 17.1 Gravity in a Close Binary Star System

As explained in Chapter 7, at least half of all "stars" in the sky are actually multiple systems, consisting of two (or more) stars in orbit about their common center of mass. In most of these systems the stars are sufficiently far apart that they have a negligible impact on one another. They evolve essentially independently, living out their lives in isolation except for the gentle grip of gravity that binds them together.

If the stars are very close, with a separation roughly equal to the diameter of the larger star, then one or both stars may have their outer layers gravitationally deformed into a teardrop shape. As a star rotates through the tidal bulge raised by its partner's gravitational pull, it is forced to pulsate. These oscillations are damped by the mechanisms discussed in Section 14.2. Orbital and rotational energy are dissipated in this way until the system reaches the state of minimum energy for its (constant) angular momentum, resulting in synchronous rotation and circular orbits. Thereafter the same side of each star always faces the other as the system rotates rigidly in space and no further energy can be lost by tidally driven oscillations.[1] The distorted star may even lose some of its photospheric gases to its companion. The spilling of gas from one star onto another can lead to some spectacular celestial fireworks, the subject of this chapter.

To understand how gravity operates in a close binary star system, consider

---

[1] If one of the stars is a compact object such as a white dwarf or a neutron star, its spin may not be synchronized.

two stars in a circular orbit in the $x$–$y$ plane with angular velocity $\omega = v_1/r_1 = v_2/r_2$. Here, $v_1$ and $r_1$ are the orbital speed of star 1 and its distance from the center of mass of the system, and similarly for star 2. It is useful to choose a corotating coordinate system that follows the rotation of the two stars about their center of mass. If the center of mass is at the origin, then the stars will be at rest in this rotating reference frame, with their mutual gravitational attraction balanced by the outward "push" of a centrifugal force.[2] The centrifugal force vector on a mass $m$ in this frame a distance $r$ from the origin is then

$$\mathbf{F}_c = m\omega^2 r\,\hat{\mathbf{r}}, \tag{17.1}$$

in the outward radial direction.

It is usually easier to work with the gravitational potential energy, given by Eq. (2.14),

$$U_g = -G\frac{Mm}{r},$$

instead of the gravitational force.[3] To do this in a rotating coordinate system, a fictitious "centrifugal potential energy" must be included in the potential energy through the use of Eq. (2.13),

$$U_f - U_i = \Delta U_c = -\int_{\mathbf{r}_i}^{\mathbf{r}_f} \mathbf{F}_c \cdot d\mathbf{r}.$$

Here, $\mathbf{F}_c$ is the centrifugal force vector, $\mathbf{r}_i$ and $\mathbf{r}_f$ are the initial and final position vectors, respectively, and $d\mathbf{r}$ is the infinitesimal change in the position vector (recall Fig. 2.9). The change in centrifugal potential energy is thus

$$\Delta U_c = -\int_{r_i}^{r_f} m\omega^2 r\,dr = -\frac{1}{2}m\omega^2(r_f^2 - r_i^2).$$

Realizing that only *changes* in potential energy are physically meaningful, $U_c = 0$ at $r = 0$ can arbitrarily be chosen to give the final result for the centrifugal potential energy,

$$U_c = -\frac{1}{2}m\omega^2 r^2. \tag{17.2}$$

Figure 17.1 shows a corotating coordinate system in which two stars with masses $M_1$ and $M_2$ are separated by a distance $a$. The stars are located on

---

[2]The centrifugal force is an *inertial force* (as opposed to a physical force) that must be included when describing motion in a rotating coordinate system. There is another inertial force, called the *Coriolis force*, that will be neglected in what follows.

[3]Most stars can be treated as point masses in what follows because the mass is concentrated at their centers, allowing their teardrop shapes to be neglected.

## 17.1 Gravity in a Close Binary Star System

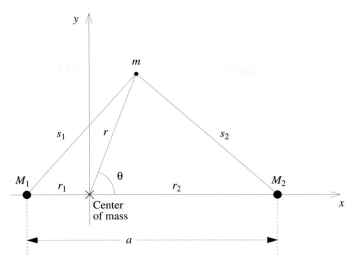

**Figure 17.1** Corotating coordinates for a binary star system.

the $x$-axis at distances $r_1$ and $r_2$, respectively, from the center of mass, which is placed at the origin. Thus

$$r_1 + r_2 = a \quad \text{and} \quad M_1 r_1 = M_2 r_2. \tag{17.3}$$

Including the centrifugal term, the effective potential energy for a small test mass $m$ located in the plane of the orbit (the $x$–$y$ plane) is

$$U = -G\left(\frac{M_1 m}{s_1} + \frac{M_2 m}{s_2}\right) - \frac{1}{2}m\omega^2 r^2.$$

For convenience, the effective potential energy can be divided by $m$ to obtain the **effective gravitational potential** $\Phi$,

$$\Phi = -G\left(\frac{M_1}{s_1} + \frac{M_2}{s_2}\right) - \frac{1}{2}\omega^2 r^2. \tag{17.4}$$

This is just the effective potential energy *per unit mass*, in units of ergs g$^{-1}$. From the law of cosines, the distances $s_1$ and $s_2$ are given by

$$\left.\begin{array}{l} s_1^2 = r_1^2 + r^2 + 2r_1 r \cos\theta \\ s_2^2 = r_2^2 + r^2 - 2r_2 r \cos\theta. \end{array}\right\} \tag{17.5}$$

The angular frequency of the orbit, $\omega$, comes from Kepler's third law for the orbital period $P$, Eq. (2.35),

$$\omega^2 = \left(\frac{2\pi}{P}\right)^2 = \frac{G(M_1 + M_2)}{a^3}. \tag{17.6}$$

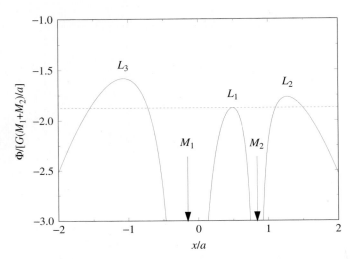

**Figure 17.2** The effective gravitational potential $\Phi$ for two stars of mass $M_1 = 0.85$ $M_\odot$, $M_2 = 0.17$ $M_\odot$ on the $x$-axis. The stars are separated by a distance $a = 5 \times 10^{10}$ cm $= 0.718$ $R_\odot$, with their center of mass located at the origin. The $x$-axis is in units of $a$, and $\Phi$ is expressed in units of $G(M_1 + M_2)/a = 2.71 \times 10^{15}$ ergs g$^{-1}$. (In fact, the figure is the same for any $M_2/M_1 = 0.2$.) The dashed line is the value of $\Phi$ at the inner Lagrangian point. If the total energy per gram of a particle exceeds this value of $\Phi$, it can flow through the inner Lagrangian point between the two stars.

Equations (17.3)–(17.6) can be used to evaluate the effective gravitational potential $\Phi$ at every point in the orbital plane of a binary star system. For example, Fig. 17.2 shows the value of $\Phi$ along the $x$-axis. The significance of this graph becomes clear when the $x$-component of the force on a small test mass $m$, initially at rest on the $x$-axis, is written as

$$F_x = -\frac{dU}{dx} = -m\frac{d\Phi}{dx} \tag{17.7}$$

(recall Eq. 2.15). The three "hilltops" labeled $L_1$, $L_2$, and $L_3$ are **Lagrangian points**, where there is no force on the test mass ($d\Phi/dx = 0$). At these three equilibrium points, the gravitational forces on $m$ due to $M_1$ and $M_2$ are balanced by the centrifugal force.[4] These equilibrium points are *unstable* because they are *local maxima* of $\Phi$; if the test mass is displaced slightly, the minus sign in Eq. (17.7) indicates that it will accelerate "downhill," away from

---

[4] From an inertial (nonrotating) frame of reference, this motion would be described by saying that gravitational forces of $M_1$ and $M_2$ produce the inward centripetal acceleration of the test mass as it orbits the center of mass of the system.

## 17.1 Gravity in a Close Binary Star System

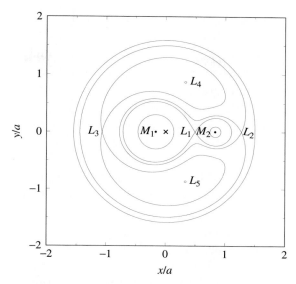

**Figure 17.3** Equipotentials for $M_1 = 0.85$ $M_\odot$, $M_2 = 0.17$ $M_\odot$, and $a = 5 \times 10^{10}$ cm $= 0.718$ $R_\odot$. The axes are in units of $a$, with the system's center of mass (the "×") at the origin. Starting at the top of the figure and moving down toward the center of mass, the values of $\Phi$ (in units of $G(M_1 + M_2)/a = 2.71 \times 10^{15}$ ergs g$^{-1}$) for the equipotential curves are $\Phi = -1.875, -1.768, -1.583, -1.583, -1.768$ (the "dumbbell"), $-1.875$ (the Roche lobe), and $-3$ (the spheres). $L_4$ and $L_5$ are local maxima, with $\Phi = -1.431$.

its equilibrium position. The *inner Lagrangian point*, $L_1$, plays a central role in close binary systems. Approximate expressions for the distances from $L_1$ to $M_1$ and $M_2$, denoted respectively by $\ell_1$ and $\ell_2$, are

$$\ell_1 = a\left[0.500 - 0.227 \log_{10}\left(\frac{M_2}{M_1}\right)\right] \tag{17.8}$$

$$\ell_2 = a\left[0.500 + 0.227 \log_{10}\left(\frac{M_2}{M_1}\right)\right]. \tag{17.9}$$

Points in space that share the same value of $\Phi$ form an **equipotential surface**. Figure 17.3 shows *equipotential contours* that outline the intersection of several equipotential surfaces with the plane of the orbit. Very close to either of the masses $M_1$ or $M_2$, the equipotential surfaces are nearly spherical and centered on each mass. Farther away, the combined gravitational influence of $M_1$ and $M_2$ distorts the equipotential surfaces into teardrop shapes until they finally touch at the inner Lagrangian point. At even greater distances, the equipotential surfaces assume a "dumbbell" shape surrounding both masses.[5]

These equipotential surfaces are *level surfaces* for binary stars. In a binary system, as one of the stars evolves, it will expand to fill successively larger equipotential surfaces (somewhat like inflating a balloon). To see this, consider that the effective gravity at each point is always *perpendicular* to the equipotential surface there.[6] Hydrostatic equilibrium guarantees that the pressure is constant along a surface of constant $\Phi$; there is no component of gravity parallel to an equipotential surface, and so a pressure difference in that direction cannot be balanced and maintained. And because the pressure is due to the weight of the overlying layers of the star, the density must also be the same along each equipotential surface in order to produce a constant pressure there.

The appearance of a binary star system depends on which equipotential surfaces are filled by the stars. Binary stars with radii much less than their separation are nearly spherical (as shown by the small circles in Fig. 17.3). This situation describes a **detached binary** in which the stars evolve nearly independently. Detached binary systems have already been described in Chapter 7 as a primary source of astronomical information about the basic properties of stars.

If one star expands enough to fill the "figure-eight" contour in Fig. 17.3, then its atmospheric gases can escape through the inner Lagrangian point $L_1$ to be drawn toward its companion. The teardrop-shaped regions of space bounded by this particular equipotential surface are called **Roche lobes**.[7] The transfer of mass from one star to the other can begin when one of the stars has expanded beyond its Roche lobe. Such a system is called a **semidetached binary**. The star that fills its Roche lobe and loses mass is usually called the **secondary star**, with mass $M_2$, and its companion the **primary star** has mass $M_1$. The primary star may be either more or less massive than the secondary star.

---

[5]We will not be concerned with the other equipotential contours on Fig. 17.3 that pass through the Lagrangian points $L_3$, $L_4$, and $L_5$. However, the *Trojan asteroids* that accumulate at two locations along Jupiter's orbit are collections of interplanetary rubble found at Lagrangian points $L_4$ and $L_5$. (If $M_1 > 24.96 M_2$, as for the Sun and Jupiter, then the Coriolis force is strong enough to cause $L_4$ and $L_5$ to be *stable* equilibrium points.) Each of the Lagrange points $L_4$ and $L_5$ forms an equilateral triangle with masses $M_1$ and $M_2$ in Fig. 17.3, so the Trojan asteroids are found 60° ahead of and behind Jupiter in its orbit.

[6]The mathematical statement of this is $\mathbf{F} = -m\nabla\Phi$. This is exactly analogous to an electric field vector being oriented perpendicular to an electrical equipotential surface, pointing from higher to lower voltage.

[7]The term *Roche lobe* is in honor of the nineteenth-century French mathematician Edouard Roche.

## 17.1 Gravity in a Close Binary Star System

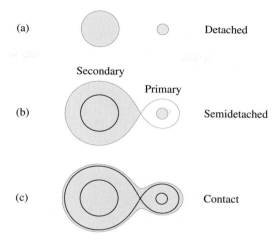

**Figure 17.4** The classification of binary star systems. (a) Shows a detached system, (b) shows a semidetached system in which the secondary star has expanded to fill its Roche lobe, and (c) shows a contact binary.

It may happen that *both stars* fill, or even expand beyond, their Roche lobes. In this case, the two stars share a common atmosphere bounded by a dumbbell-shaped equipotential surface, such as the one passing through the Lagrangian point $L_2$. Such a system is called a **contact binary**. Figure 17.4 illustrates the three classes of binary stars.

A crude estimate of the rate at which mass is transferred in a semidetached binary may be easily obtained for the case of two stars of equal mass. Let the radius of the star that has expanded beyond its Roche lobe be $R$. The equipotential surface at the radius of this star will be modeled by two spheres of radius $R$ that overlap slightly by a distance $d$, as shown in Fig. 17.5. We will assume that stellar gas will escape from the filled lobe through the circular opening of radius $x$. If the density of the stellar material at the opening is $\rho$ and its speed toward the opening of area $A = \pi x^2$ is $v$, then it is left as an exercise (Problem 17.3) to show that the rate at which mass leaves the filled lobe, the **mass transfer rate**, is

$$\dot{M} = \rho v A. \tag{17.10}$$

A bit of geometry shows that
$$x = \sqrt{Rd} \tag{17.11}$$
when $d \ll R$. Using Eq. (8.3) for the thermal velocity of the gas particles results in the estimate
$$\dot{M} \approx \rho v_{\text{rms}} \pi x^2$$

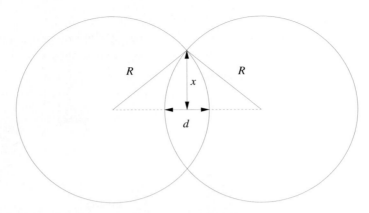

**Figure 17.5** Intersecting spheres used to estimate the mass transfer rate, $\dot{M}$.

or

$$\dot{M} \approx \pi R d\rho \sqrt{\frac{3kT}{m_H}}, \tag{17.12}$$

assuming a gas of hydrogen atoms. As the overfill distance $d$ becomes larger, the values of the density and temperature increase at the opening. In Problem 17.4, you are asked to show that the mass transfer rate increases rapidly with $d$; a more detailed calculation results in $\dot{M} \propto d^3$.

---

**Example 17.1** Suppose a star like the Sun is in a semidetached binary system with a companion of equal mass, and it slightly overfills its Roche lobe to a point just below its photosphere. Taking the third point in from the surface ($i = 3$) for the STATSTAR model in Table 9.3 on page 311, we have $d = 1.42 \times 10^8$ cm, $T = 6577$ K, and $\rho = 2.52 \times 10^{-10}$ g cm$^{-3}$. Using $R = 7.11 \times 10^{10}$ cm at the outermost point of the model, the rate at which this Sun-like star would lose its atmospheric gases would be roughly

$$\dot{M} \approx \pi R d\rho \sqrt{\frac{3kT}{m_H}} = 1.0 \times 10^{16} \text{ g s}^{-1} = 1.6 \times 10^{-10} \text{ M}_\odot \text{ yr}^{-1}.$$

This is typical of the mass transfer rates for semidetached binary systems. The values of $\dot{M}$ inferred from observations of various systems range from $10^{-11}$ to $10^{-7}$ M$_\odot$ yr$^{-1}$. For comparison, recall from Example 11.1 that the solar wind transports mass away from the Sun at a much smaller rate, approximately $3 \times 10^{-14}$ M$_\odot$ yr$^{-1}$.

---

## 17.1 Gravity in a Close Binary Star System

Before moving on to consider the consequences of the transfer of mass in semidetached binaries, it is worthwhile to consider the enormous energy that can be released when matter falls onto a star, especially onto a compact object such as a white dwarf or a neutron star.

---

**Example 17.2** Consider a mass $m = 1$ g that starts at rest infinitely far from a star of mass $M$ and radius $R$. The initial total mechanical energy of the mass $m$ is

$$E = K + U = 0.$$

Using conservation of energy, the kinetic energy of the mass when it arrives at the star's surface is

$$K = -U = G\frac{Mm}{R}.$$

This kinetic energy will be converted into heat and light upon impact with the star. If the star is a white dwarf with $M = 0.85$ M$_\odot$ and $R = 6.6 \times 10^8$ cm $= 0.0095$ R$_\odot$, then the energy released by one gram of infalling matter is

$$G\frac{Mm}{R} = 1.71 \times 10^{17} \text{ ergs}.$$

This is only 0.019% of the rest energy ($mc^2$) of one gram of material. For comparison, the amount of energy released by the thermonuclear fusion of one gram of hydrogen is

$$0.007mc^2 = 6.29 \times 10^{18} \text{ ergs}$$

(recall Example 10.4).

If the star is a neutron star with mass $M = 1.4$ M$_\odot$ and radius $R = 10$ km, then the energy released is much greater,

$$G\frac{Mm}{R} = 1.86 \times 10^{20} \text{ ergs}.$$

This is 21% of the rest energy of one gram, nearly 30 times greater than the energy that hydrogen fusion could provide! Obviously, infalling matter is capable of generating immense amounts of energy.

Observations of celestial x-ray sources have revealed objects with a steady x-ray luminosity of approximately $10^{37}$ ergs s$^{-1}$. If this radiation were produced by gases pulled from a companion star that then fell onto a neutron star's surface, the amount of mass per second transferred between the two stars that would be needed to account for the observed luminosity is

$$\dot{M} = \frac{10^{37} \text{ ergs s}^{-1}}{1.86 \times 10^{20} \text{ ergs g}^{-1}} = 5.38 \times 10^{16} \text{ g s}^{-1},$$

which is only about $10^{-9}$ M$_\odot$ yr$^{-1}$. This is similar to the mass transfer rate found in the previous example, a fortuitous agreement because $\dot{M}$ for semidetached systems can vary by several orders of magnitude.

## 17.2 Accretion Disks

The orbital motion of a semidetached binary can prevent the mass that escapes from the swollen secondary star from falling directly onto the primary star. The primary's movement is often enough to keep it out of the path of the gases that spill through the inner Lagrangian point. If the radius of the primary star is less than about 5% of the binary separation $a$, the mass stream will miss striking the primary's surface. Instead, the mass stream goes into orbit around the primary to form a thin **accretion disk** of hot gas in the orbital plane, as shown in Figs. 16.24 and 17.6.[8] **Viscosity**, an internal friction that converts the directed kinetic energy of bulk mass motion into random thermal motion, causes the orbiting gases to lose energy and slowly spiral inward toward the primary. The physical mechanism responsible for the viscosity in accretion disks is as yet poorly understood. The familiar molecular viscosity due to interparticle forces is far too weak to be effective. Other possibilities involve random motions of the gas, such as turbulence in the disk material caused by thermal convection or by a magnetohydrodynamic instability in the magnetic fields that interact with the differentially rotating disk (c.f., Section 11.2). Whatever the mechanism, the gas is heated throughout its descent to increasingly higher temperatures as the lost orbital energy is converted into thermal energy. Finally, the plunging gas ends its journey at the star's surface.

Just as a star may be treated as a blackbody as a rough first approximation, the assumption of an optically thick accretion disk radiating as a blackbody provides a simple, useful model. (The validity of this assumption will be discussed later.) At each radial distance, an optically thick disk emits blackbody radiation with a continuous spectrum corresponding to the local disk temperature at that distance. To estimate the temperature of a model accretion disk at a distance $r$ from the center of the primary star of mass $M_1$ and radius $R_1$, let's assume that the inward radial velocity of the disk gases is small compared with their orbital velocity. Then, to a good approximation, the gases follow circular Keplerian orbits, and the details of the viscous forces acting within the disk may be neglected. Furthermore, since the mass of the disk is very small

---

[8]Astronomers refer to the process of accumulating mass from an outside source as *accretion*.

## 17.2 Accretion Disks

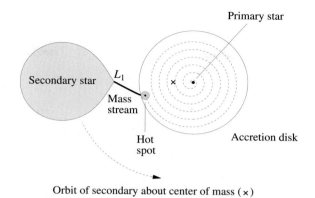

**Figure 17.6** A semidetached binary showing the accretion disk around the primary star and the hot spot where the mass streaming through the inner Lagrangian point impacts the disk. This system's parameters correspond to those of Z Chamaeleontis, described in Example 17.4.

compared with that of the primary, the orbiting material feels only the gravity of the central primary star. The *total energy* (kinetic plus potential) of a mass $m$ of orbiting gas is given by Eq. (2.33),

$$E = -G \frac{M_1 m}{2r}.$$

As the gas spirals inward, its total energy $E$ becomes more negative. The lost energy maintains the disk's temperature and is ultimately emitted in the form of blackbody radiation.

Now consider an annular ring of radius $r$ and width $dr$ within the disk, as shown in Fig. 17.7. If the rate at which mass is transferred from the secondary to the primary star is a constant $\dot{M}$, then in time $t$ the amount of mass that passes through the outer boundary of the circular ring shown in Fig. 17.7 is $\dot{M}t$. Assuming a *steady-state disk* that does not change with time, no mass is allowed to build up within the ring. Therefore during this time an amount of mass $\dot{M}t$ must also leave through the ring's inner boundary.

Conservation of energy requires that the energy $dE$ radiated by the ring in time $t$ be equal to the difference in the energy that passes through the ring's outer and inner boundaries:

$$dE = \frac{dE}{dr} dr = \frac{d}{dr}\left(-G\frac{M_1 m}{2r}\right) dr = G\frac{M_1 \dot{M} t}{2r^2} dr,$$

where $m = \dot{M}t$ has been used for the orbiting mass entering and leaving the ring. If the luminosity of the ring is $dL_{\text{ring}}$, then the energy radiated by the

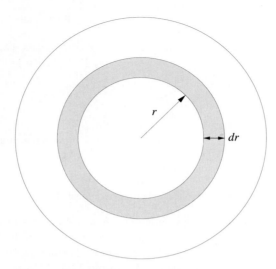

**Figure 17.7** One of the (imaginary) annular rings comprising the accretion disk.

ring in time $t$ is related to $dL_{\text{ring}}$ by

$$dL_{\text{ring}} t = dE = G\frac{M_1 \dot{M} t}{2r^2}\, dr.$$

Canceling the $t$'s and using the Stefan–Boltzmann law in the form of Eq. (3.16) with $A = 2(2\pi r\, dr)$ for the surface area of the ring (both sides) gives

$$dL_{\text{ring}} = 4\pi r \sigma T^4\, dr \tag{17.13}$$

$$= G\frac{M_1 \dot{M}}{2r^2}\, dr \tag{17.14}$$

for the luminosity of the ring. Solving for $T$, the disk temperature at radius $r$, results in

$$T = \left(\frac{GM\dot{M}}{8\pi\sigma R^3}\right)^{1/4} \left(\frac{R}{r}\right)^{3/4}. \tag{17.15}$$

The "1" subscript has been dropped, with the understanding that $M$ and $R$ are the mass and radius of the primary star, and that $\dot{M}$ is the mass transfer rate for the semidetached binary system.

A more thorough analysis would take into account the thin turbulent boundary layer that must be produced when the rapidly orbiting disk gases encounter the surface of the primary star. This results in a better estimate of

## 17.2 Accretion Disks

the disk temperature,

$$T = \left(\frac{3GM\dot{M}}{8\pi\sigma R^3}\right)^{1/4} \left(\frac{R}{r}\right)^{3/4} \left(1 - \sqrt{R/r}\right)^{1/4}$$

$$= T_{\text{disk}} \left(\frac{R}{r}\right)^{3/4} \left(1 - \sqrt{R/r}\right)^{1/4}, \tag{17.16}$$

where

$$T_{\text{disk}} \equiv \left(\frac{3GM\dot{M}}{8\pi\sigma R^3}\right)^{1/4} \tag{17.17}$$

is a characteristic temperature of the disk. Actually, $T_{\text{disk}}$ is roughly twice the maximum disk temperature (Problem 17.5),

$$T_{\max} = 0.488 \left(\frac{3GM\dot{M}}{8\pi\sigma R^3}\right)^{1/4} = 0.488 T_{\text{disk}}, \tag{17.18}$$

which occurs at $r = (49/36)R$; see Fig. 17.12.[9] When $r \gg R$, the last term on the right-hand side of Eq. (17.16) may be neglected, leaving

$$T = \left(\frac{3GM\dot{M}}{8\pi\sigma R^3}\right)^{1/4} \left(\frac{R}{r}\right)^{3/4} = T_{\text{disk}} \left(\frac{R}{r}\right)^{3/4} \quad (r \gg R). \tag{17.19}$$

This differs from our simple estimate, Eq. (17.15), by a factor of $3^{1/4} = 1.32$.

Integrating Eq. (17.14) for the luminosity of each ring from $r = R$ to $r = \infty$ results in an expression for the disk luminosity,

$$L_{\text{disk}} = G \frac{M\dot{M}}{2R}. \tag{17.20}$$

However, recall from Example 17.2 that without an accretion disk, the accretion luminosity (the rate at which falling matter delivers kinetic energy to the primary star) is twice as great,

$$L_{\text{acc}} = G \frac{M\dot{M}}{R}. \tag{17.21}$$

Thus, if half of the available accretion energy is radiated away as the gases spiral down through the disk, then the remaining half must be deposited at the surface of the star (or in the turbulent boundary layer between the rapidly rotating disk and the more slowly rotating primary star).[10]

---

[9]Including the boundary layer results in $T = 0$ where the disk meets the star's surface, an unrealistic artifact of the assumptions of the model.

[10]This result is just another consequence of the virial theorem.

**Example 17.3** The maximum disk temperature, $T_{\max}$, and the value of the disk luminosity for the white dwarf and neutron star used in Example 17.1 can now be evaluated. For a white dwarf with $M = 0.85$ M$_\odot$, $R = 0.0095$ R$_\odot$, and $\dot{M} = 10^{16}$ g s$^{-1}$ ($1.6 \times 10^{-10}$ M$_\odot$ yr$^{-1}$), Eq. (17.18) is

$$T_{\max} = 0.488 \left( \frac{3GM\dot{M}}{8\pi\sigma R^3} \right)^{1/4} = 2.62 \times 10^4 \text{ K}.$$

According to Wien's displacement law, Eq. (3.19), at this temperature the blackbody spectrum peaks at a wavelength of

$$\lambda_{\max} = \frac{(5000 \text{ Å})(5800 \text{ K})}{26,200 \text{ K}} = 1110 \text{ Å},$$

which is in the ultraviolet region of the electromagnetic spectrum (Table 3.1). From Eq. (17.20), the luminosity of the accretion disk is

$$L_{\text{disk}} = G \frac{M\dot{M}}{2R} = 8.55 \times 10^{32} \text{ ergs s}^{-1},$$

or about 0.22 L$_\odot$.

For a neutron star with $M = 1.4$ M$_\odot$, $R = 10$ km, and $\dot{M} = 10^{17}$ g s$^{-1}$ ($1.6 \times 10^{-9}$ M$_\odot$ yr$^{-1}$), the maximum disk temperature is

$$T_{\max} = 0.488 \left( \frac{3GM\dot{M}}{8\pi\sigma R^3} \right)^{1/4} = 6.86 \times 10^6 \text{ K}.$$

Its blackbody spectrum peaks at a wavelength of

$$\lambda_{\max} = \frac{(5000 \text{ Å})(5800 \text{ K})}{686,000 \text{ K}} = 4.23 \text{ Å},$$

which is in the x-ray region of the electromagnetic spectrum. The luminosity of the neutron star's accretion disk is

$$L_{\text{disk}} = G \frac{M\dot{M}}{2R} = 9.29 \times 10^{36} \text{ ergs s}^{-1},$$

over 2400 L$_\odot$. Thus the inner regions of accretion disks around white dwarfs should shine in the ultraviolet, while those around neutron stars will be strong x-ray sources.[11]

---

[11] Actually, as will be seen in Section 17.5, the accretion disk around a white dwarf or neutron star may be disrupted by the star's magnetic field, and so may not extend down to its surface. Such systems will be strong sources of x-rays.

## 17.2 Accretion Disks

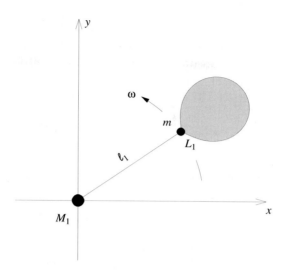

**Figure 17.8** A parcel of mass $m$ passing through the inner Lagrangian point $L_1$, seen from a frame of reference with the primary star at rest at the origin.

The radial extent of the accretion disk can be estimated by finding the value of $r = r_{\text{circ}}$ where a continuous stream of mass that passes through $L_1$ will settle into a circular orbit around the primary star. This may be done by considering the angular momentum of a parcel of mass $m$ about the primary star; see Fig. 17.8. Assuming that the motion of the mass at the inner Lagrangian point is due solely to the orbital motion of the binary system, the angular momentum, $L$, of the mass located there is

$$L = m\omega \ell_1^2 = m\ell_1^2 \sqrt{\frac{G(M_1 + M_2)}{a^3}},$$

where Eq. (17.6) has been used for the angular frequency of the orbit, and $\ell_1$ is given by Eq. (17.8).

The mass $m$ does not immediately enter into a circular orbit. Instead, the stream of mass to which $m$ belongs flows around the primary star and collides with itself after one orbit. The orbits of the mass parcels are made circular around the primary by the collisions as energy is lost while angular momentum is conserved. When the parcel of mass has settled into a circular orbit of radius $r_{\text{circ}}$ around $M_1$, its angular momentum is

$$L = m\sqrt{GM_1 r_{\text{circ}}},$$

where Eq. (2.32) was used for a circular orbit with $\mu = mM_1/(m+M_1) \simeq m$. Equating these two expressions for the angular momentum results in

$$r_{\text{circ}} = a \left(\frac{\ell_1}{a}\right)^4 \left(1 + \frac{M_2}{M_1}\right)$$

$$= a \left[0.500 - 0.227 \log_{10}\left(\frac{M_2}{M_1}\right)\right]^4 \left(1 + \frac{M_2}{M_1}\right). \quad (17.22)$$

Since the total angular momentum must be conserved when only internal and central forces act, the reader may wonder what happens to the angular momentum lost by the infalling material as it spirals through the accretion disk. As shown in Pringle (1981), orbiting material that is initially in the form of a narrow ring at $r = r_{\text{circ}}$ will spread, moving both inward and outward. The time for this migration of the disk material probably ranges from a few days to a few weeks. While most of the matter spirals inward, a small amount of the mass carries the "missing" angular momentum to the outer edge of the disk. From there, the angular momentum may be carried away from the system by wind-driven mass loss. If the accretion disk extends 80% to 90% of the way out to the inner Lagrangian point, angular momentum may also be returned to the orbital motion of the two stars by tides raised in the disk by the secondary star. Because of this outward migration of mass, we will adopt

$$R_{\text{disk}} \approx 2 r_{\text{circ}} \quad (17.23)$$

as a rough estimate of the outer radius of the accretion disk.

It is comforting to know that there is evidence, obtained from observing *eclipsing* semidetached binary systems, that the objects described above actually exist. Several studies have been made of a type of close binary called a **cataclysmic variable**, a semidetached system in which the primary is a white dwarf and the secondary is a late-type main-sequence star.[12] The light from the accretion disk around the white dwarf would dominate such a system. Cataclysmic variables are characterized by long quiescent intervals punctuated by outbursts in which the brightness of the system increases dramatically; the outbursts are believed to be due to a sudden increase in the rate at which mass flows down through the disk. As the eclipsed disk emerges from behind the secondary star, the radial variation in the disk's temperature can be determined. During an outburst, the disk does indeed appear to be optically thick, with $T \propto r^{-3/4}$, in agreement with Eq. (17.19). But during quiescence the observations are not consistent with the disk model described above, probably

---

[12]Cataclysmic variables will be discussed in more detail in Section 17.4.

## 17.2 Accretion Disks

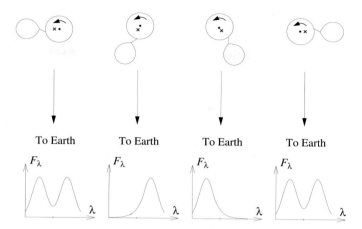

**Figure 17.9** A Doppler-shifted emission line at different stages of the eclipse of an accretion disk. The binary system orbits about its center of mass (the "×"), and is observed nearly edge-on. The disk rotates in the direction indicated by the arrow.

because the disk is not completely optically thick when it is cooler and contains less mass.

Additional evidence supporting this view comes from the strong, wide emission lines of hydrogen and helium that are seen in cataclysmic variables during quiescence. These lines are usually doubly peaked, as shown in Fig. 17.9. However, during an eclipse a single emission line is observed, either redshifted or blueshifted. This is what would be expected from a rotating disk of optically thin gas; the Doppler-shifted emission lines produced on the opposite sides of the disk disappear when one side or the other is hidden behind the secondary star.

The source of the emission lines that appear during a cataclysmic variable's quiescent phase is not yet clear. During an outburst, these lines appear in absorption, as would be expected from an optically thick disk that produces absorption lines in the same manner as an optically thick stellar atmosphere. But during quiescence, the rate at which mass flows down through the disk has presumably decreased, making the disk less dense and cooler. At larger radii the disk may then be optically thin and so produce emission lines. Alternatively, there may be a thin layer of hot gas above the disk that produces the emission lines.

Observations of light curves for other eclipsing semidetached binaries, such as shown in Fig. 17.10, indicate the presence of a hot spot where the mass transfer stream collides with the outer edge of the accretion disk. The light

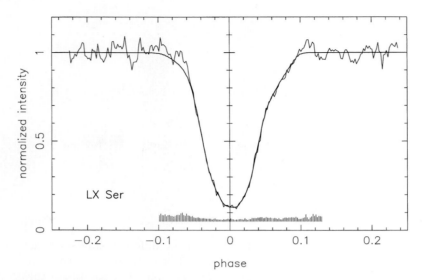

**Figure 17.10** The light curve of the eclipse of the accretion disk in the LX Serpentis binary system. The jagged line is the observed light curve and the smooth line shows the fit calculated from a reconstructed image of the accretion disk, shown in Fig. 17.11. (Figure from Rutten, van Paradijs, and Tinbergen, *Astron. Astrophys.*, *260*, 213, 1992.)

curve can be interpreted as the result of observing consecutive "slices" of the disk as they disappear and then reappear from behind the primary star. In fact, Fig. 17.10 can be used to re-create an image of the disk itself, shown in Fig. 17.11.[13] Because the hot spot is on the trailing side of the disk during the eclipse (see Fig. 17.9), more light is received from the disk near the beginning of the eclipse (when the hot spot is still visible) than near the end (when the hot spot is still hidden). This produces the light deficit on the right-hand side of the light curve in Fig. 17.10.

Perhaps the most important property of a semidetached binary is its orbital period. When used together with other quantities (any two of the separation $a$ and the masses $M_1$ and $M_2$), a picture of the system emerges.

**Example 17.4** Z Chamaeleontis is a type of cataclysmic variable called a *dwarf nova*. It consists of an $M_1 = 0.85$ $M_\odot$ white dwarf primary with a

---

[13] Using slices of the emerging disk to reconstruct an image of the accretion disk is somewhat analogous to using a CAT-scan (computerized axial tomography) in a hospital to mathematically reassemble x-ray slices of the human body. Because there is more than one model disk that will reproduce a given light curve, a technique called *maximum entropy* is used to choose the smoothest possible model for the final disk image.

## 17.2 Accretion Disks

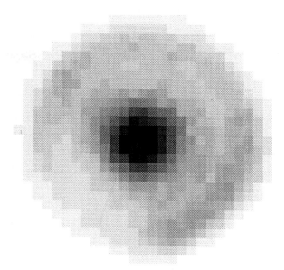

**Figure 17.11** A reconstructed negative image of the accretion disk in the LX Serpentis binary system. The hot spot is smeared out in the azimuthal direction and has the appearance of a partial ring. (Figure from Rutten, van Paradijs, and Tinbergen, *Astron. Astrophys.*, *260*, 213, 1992.)

radius of $R = 0.0095$ $R_\odot$ and a late M-type main-sequence secondary star of mass $M_2 = 0.17$ $M_\odot$. The orbital period of the system is $P = 0.0745$ day. What does this system look like?

From Kepler's third law, Eq. (2.35), the separation of the two stars is

$$a = \left[ \frac{P^2 G(M_1 + M_2)}{4\pi^2} \right]^{1/3} = 5.22 \times 10^{10} \text{ cm},$$

about 75% of the radius of the Sun. The distance between the white dwarf primary and the inner Lagrangian point $L_1$ is given by Eq. (17.8),

$$\ell_1 = a \left[ 0.500 - 0.227 \log_{10} \left( \frac{M_2}{M_1} \right) \right] = 3.44 \times 10^{10} \text{ cm}.$$

Because the secondary star fills its Roche lobe in a semidetached binary system, the distance between the secondary star and the inner Lagrangian point is a measure of the size of the secondary. For Z Cha,

$$R_2 \approx \ell_2 = a - \ell_1 = 1.78 \times 10^{10} \text{ cm},$$

which agrees quite well with the size of an M6 main-sequence star (see Appendix E).

The value of $r_{\text{circ}}$ for this system is, from Eq. (17.22),

$$r_{\text{circ}} = a \left(\frac{\ell_1}{a}\right)^4 \left(1 + \frac{M_2}{M_1}\right) = 1.18 \times 10^{10} \text{ cm},$$

and so a crude estimate of the outer radius of the disk is

$$R_{\text{disk}} \approx 2r_{\text{circ}} = 2.4 \times 10^{10} \text{ cm},$$

(Eq. 17.23), which is about two-thirds of the way to the inner Lagrangian point. This is in good agreement with observations that indicate that the Z Cha's disk emits very little light from beyond this radius.

The mass transfer rate inferred for Z Cha during an outburst is roughly

$$\dot{M} = 1.3 \times 10^{-9} \text{ M}_\odot \text{ yr}^{-1},$$

or $7.9 \times 10^{16}$ g s$^{-1}$ ($1.3 \times 10^{-9}$ M$_\odot$ yr$^{-1}$), which implies a maximum disk temperature of

$$T_{\text{max}} = 0.488 \left(\frac{3GM\dot{M}}{8\pi\sigma R^3}\right)^{1/4} = 4.4 \times 10^4 \text{ K},$$

using Eq. (17.18). Figure 17.12 shows the variation in the disk temperature with radius for Z Cha [calculated from Eq. (17.16)]. Moving from the inner to the outer regions of the disk, the temperature falls from 44,000 K to 8000 K. According to Wien's law, Eq. (3.15), this corresponds to an increase in the peak wavelength of the emitted radiation from 660 Å to 3630 Å (from the far to near portions of the ultraviolet spectrum).

The monochromatic luminosity, $L_\lambda$, for the entire disk can be calculated by integrating Eq. (3.20) for the Planck function, $B_\lambda$, over the disk area and over all directions (recall Section 3.5). The resulting graph of the energy emitted per second within wavelength intervals of 1 Å is shown in Fig. 17.13. According to Eq. (17.20), the total luminosity of the accretion disk (integrated over all wavelengths) is

$$L_{\text{disk}} = G \frac{M\dot{M}}{2R} = 6.8 \times 10^{33} \text{ ergs s}^{-1},$$

which exceeds the luminosity of the Sun by about 75%.

An artist's conception of the appearance of Z Cha is shown in Fig. 17.14; see also Fig. 17.6.

## 17.2 Accretion Disks

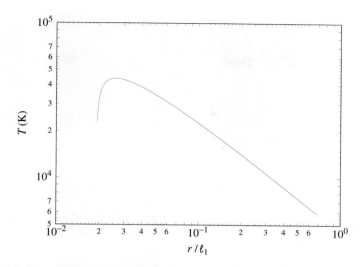

**Figure 17.12** The temperature of the accretion disk calculated for the dwarf nova Z Chamaeleontis. The radius $r$ is given in units of $\ell_1$, the distance from the white dwarf to the inner Lagrangian point. The sudden drop in temperature near the surface of the white dwarf primary is an unrealistic artifact of the assumptions.

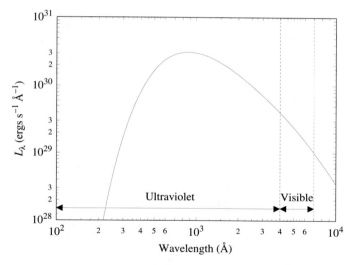

**Figure 17.13** The monochromatic luminosity calculated for the accretion disk of the dwarf nova Z Chamaeleontis.

**Figure 17.14** An artist's conception of Z Chamaeleontis. (Courtesy of Dale W. Bryner, Dept. of Visual Arts, Weber State University.)

## 17.3 A Survey of Close Binary Systems

The life history of a close binary system is quite complicated, with many possible variations depending on the initial masses and separation of the two stars involved. As mass passes from one star to the other, the mass ratio $M_2/M_1$ will change. The resulting redistribution of angular momentum affects the orbital period of the system as well as the separation of the two stars. The extent of the Roche lobes, given by Eqs. (17.8) and (17.9), depends on both the separation and the mass ratio of the stars, so it too will vary accordingly.

The effects of mass transfer can be illustrated by considering the total angular momentum of the system. The contribution of the stars' rotation to the total angular momentum is small and may be neglected. The orbital angular momentum is given by Eq. (2.32) with an eccentricity of $e = 0$ for a circular orbit,

$$L = \mu\sqrt{GMa}.$$

In this expression, $\mu$ is the reduced mass (Eq. 2.23),

$$\mu = \frac{M_1 M_2}{M_1 + M_2},$$

## 17.3 A Survey of Close Binary Systems

and $M = M_1 + M_2$ is the total mass of the two stars. Assuming (to a first approximation) that no mass or angular momentum is removed from the system via stellar winds or gravitational radiation, both the total mass and angular momentum of the system remain constant as mass is transferred between the two stars.[14] That is, $dM/dt = 0$ and $dL/dt = 0$. Some useful insights concerning the effect of the transfer of mass on the separation of the two stars can be gained by taking a time derivative of the expression for the angular momentum,

$$\frac{dL}{dt} = \frac{d}{dt}\left(\mu\sqrt{GMa}\right)$$

$$0 = \sqrt{GM}\left(\frac{d\mu}{dt}\sqrt{a} + \frac{\mu}{2\sqrt{a}}\frac{da}{dt}\right)$$

$$\frac{1}{a}\frac{da}{dt} = -\frac{2}{\mu}\frac{d\mu}{dt}. \tag{17.24}$$

Remembering that $M$ remains constant, we find that the time derivative of the reduced mass is

$$\frac{d\mu}{dt} = \frac{1}{M}\left(\frac{dM_1}{dt}M_2 + M_1\frac{dM_2}{dt}\right).$$

The mass lost by one star is gained by the other. Writing $\dot{M} \equiv dM/dt$, this means that $\dot{M}_1 = -\dot{M}_2$, and so

$$\frac{d\mu}{dt} = \frac{\dot{M}_1}{M}(M_2 - M_1).$$

Inserting this into Eq. (17.24) achieves our result,

$$\frac{1}{a}\frac{da}{dt} = 2\dot{M}_1\frac{M_1 - M_2}{M_1 M_2}. \tag{17.25}$$

Equation (17.25) describes the consequence of mass transfer on the separation of the binary system. The angular frequency of the orbit will also be affected,

---

[14]In fact, gravitational radiation, which will be discussed in Section 17.5, is primarily responsible for the loss of angular momentum in some short-period binary systems ($P < 14$ hours).

as shown by using Kepler's third law in the form of Eq. (17.6). Since $M_1 + M_2 =$ constant, Kepler's third law states that $\omega \propto a^{-3/2}$ so that

$$\frac{1}{\omega}\frac{d\omega}{dt} = -\frac{3}{2}\frac{1}{a}\frac{da}{dt}. \tag{17.26}$$

As the orbital separation decreases, the angular frequency increases.

The following description illustrates the probable evolution of a binary system that is destined to become a cataclysmic variable. The starting point is a widely separated binary system with main-sequence stars having an initial orbital period ranging from a few months to a few years. At the start, suppose that star 1 is more massive than star 2, so $M_1 - M_2 > 0$. Star 1 therefore evolves more rapidly, and becomes a red giant or supergiant before it begins to overflow its Roche lobe. This initiates the transfer of mass from star 1 to star 2 (so $\dot{M}_1 < 0$). According to Eqs. (17.25) and (17.26), in this situation $da/dt$ is negative and $d\omega/dt$ is positive; the stars spiral closer together with an increasingly shorter period. Furthermore, according to Eq. (17.8), as $a$ decreases and $M_2/M_1$ increases, the Roche lobe around star 1 shrinks (as measured by the distance of star 1 from the inner Lagrangian point). The mass transfer rate accelerates under the positive feedback of a shrinking Roche lobe, eventually producing an extended atmosphere around both stars as shown in Fig. 17.4(c). The system is now a contact binary, with the degenerate core of star 1 and the main-sequence star 2 sharing a common gaseous envelope. The two stars transfer angular momentum to this envelope as they slowly spiral inward to a much smaller separation and shorter period. If the cores of the two stars merge, the result is a single star.[15] On the other hand, the envelope surrounding the stars may be ejected.[16] (For the sake of the following discussion, we will assume that the latter occurs.)

After emerging from their gaseous cocoon, the system is a detached binary; star 2 (the secondary) lies inside its Roche lobe as star 1 (the primary) cools to become a white dwarf. Eventually, the less-massive secondary star evolves and fills its Roche lobe, and mass begins to flow in the opposite direction, with $\dot{M}_1 > 0$. In this case a negative feedback mitigates the mass transfer process, because, as Eq. (17.25) implies, the stars will now spiral farther apart (assuming that $M_1$ is still greater than $M_2$) as the Roche lobe around the secondary star expands according to Eq. (17.9). If the mass flow is to persist,

---

[15]This may explain the observations of stars called *blue stragglers* in stellar clusters, discussed in Section 13.4.

[16]Several systems have been observed in which a binary is found at the center of a planetary nebula. They may be the result of the ejection of a common envelope, although the situation is as yet uncertain.

## 17.3 A Survey of Close Binary Systems

either the secondary must expand faster than the Roche lobe grows or the stars must move closer together as angular momentum is removed from the system, either by torques due to stellar winds confined by magnetic fields or by gravitational radiation.[17] Whatever the mechanism, a steady rate of mass transfer from the secondary to the white dwarf is maintained, and the stage is set for the outbursts of a cataclysmic variable, as will be described in Section 17.4.

As the secondary star continues to evolve, another common envelope stage may occur. Figure 17.15 shows an example of the life history of a close binary system that begins with two intermediate-mass stars (between 5 and 9 $M_\odot$), and culminates with two carbon–oxygen white dwarfs in a very tight orbit, circling each other every 15 s to 3 s. The larger, less massive white dwarf (recall Eq. 15.13) overflows its Roche lobe and dissolves into a heavy disk that is accreted by the more massive dwarf. The accumulation of mass pushes the primary white dwarf toward the Chandrasekhar limit, and it explodes as a supernova.[18]

There are many types of close binary systems, too many to discuss in any detail. The following list[19] describes the main classes of interacting binaries together with some of the features that make these systems important for astronomers. Many of the classes are named after the prototype object for that class.

- **Algols**. These are two normal stars (main-sequence stars or subgiants) in a semidetached binary system. They provide checks on stellar properties and evolution, and they yield information on mass loss and mass exchange. Active Algols (the W Serpens stars) provide laboratories for studying rapid (short-lived) stages of stellar and binary star evolution. These systems are important for studying accretion processes and accretion disks. Mass loss from Algols may contribute to the chemical enrichment of the interstellar medium.

- **RS Canum Venaticorum** and **BY Draconis Stars**. These stars are chromospherically active binaries that are important systems for investigating dynamo-driven magnetic activity in cool stars (spectral type F and

---

[17] The reduction of the Sun's angular momentum by the solar wind was described in Section 11.2.

[18] In Section 17.4, we will find that nuclear reactions begin in the core of the white dwarf before the Chandrasekhar limit is reached.

[19] Quoted with permission from E. F. Guinan, *Evolutionary Processes in Interacting Binary Stars*, Kondo, Sisteró, and Polidan (eds.), Kluwer Academic Publishers, Dordrecht, 1992. Reprinted by permission of Kluwer Academic Publishers.

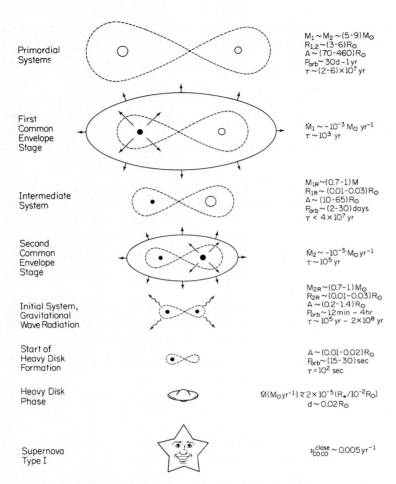

**Figure 17.15** One possibility for the evolution of a close binary system, ending in a supernova. The masses and radii of the stars, and their orbital separation ($A$), orbital period, and mass transfer rate are given for some stages, along with the duration ($\tau$). (Figure from Iben and Tutukov, *Ap. J. Suppl.*, **54**, 335, 1984.)

later). Manifestations of enhanced magnetic activity include starspots, chromospheres, coronae, and flares. These systems also contribute to our understanding of the magnetic activity of the Sun—the so-called *solar–stellar connection*.

- **W Ursae Majoris Contact Systems**. These short-period (0.2–0.8 day) contact binaries display very high levels of magnetic activity and are important stars for studying the stellar dynamo mechanism at ex-

## 17.3 A Survey of Close Binary Systems

treme levels. The drag of magnetic braking may cause these binaries to coalesce into single stars.

- **Cataclysmic Variables** and **Nova-like Binaries**. These systems have short periods and contain white dwarf components together with cool M-type secondaries that fill their Roche lobes. They provide valuable information on the final stages of stellar evolution. These binaries are also important for studying accretion phenomena and accretion disk properties.

- **X-ray Binaries with Neutron Star and Black Hole Components**. These systems are powerful ($L_x > 10^{35}$ ergs s$^{-1}$) x-ray sources that have neutron star or (more rarely) black hole components. The x-rays are due to the accretion of gas onto the degenerate component of the system from a nondegenerate companion. Observations of neutron star systems supplement the information on their structure and evolution that comes from pulsars (such as masses, radii, rotation, and magnetic fields). Systems such as A0620−00 and Cygnus X-1 provide evidence for the existence of black holes; see Section 17.5.

- **ζ Aurigae and VV Cephei Systems**. These long-period interacting binaries contain a late-type supergiant component and a hot (usually spectral type B) companion. ζ-Aur systems contain G or K supergiant stars, and VV Cep binaries contain M supergiants. Although not originally interacting binaries, they became so when the more massive star evolved to become a supergiant. When eclipses occur, the atmosphere of the cooler supergiant can be probed as the hotter star passes behind.

- **Symbiotic Binaries**. Symbiotic stars are long-period interacting binaries consisting of an M giant (sometimes a pulsating Mira-type variable) and an accreting component that can be a white dwarf, subdwarf, or low-mass main-sequence star. The common feature of these systems is the accretion of the cool component's wind onto its hot companion. Orbital periods of symbiotic stars typically range between 200 and 1500 days. Several of the symbiotic binaries have the cool component filling its Roche lobe, making them symbiotic Algol systems.

- **Barium and S-Star Binaries**. These stars[20] are thought to be long-period binaries in which the originally more massive component evolved and

---

[20]S-stars are peculiar red giants, similar to those of spectral type M but with unusual abundances of heavy metals such as zirconium and yttrium.

transferred some of its nuclear-processed gas to the present K or M giant companion. The giant stars are thought to have white dwarf companions that are often too cool to be seen in the ultraviolet. These objects are important for studying nucleosynthesis and mass loss in evolved stars.

- **Post-Common Envelope Binaries.** These binary systems usually contain hot white dwarf or subdwarf components and cooler secondary stars that have presumably passed through the common envelope phase of binary star evolution. The binary nuclei of planetary nebulae are examples of post-common envelope binaries. These systems are important for studying short-lived stages of stellar evolution.

## 17.4 White Dwarfs in Semidetached Binaries

When a white dwarf is the primary component of a semidetached binary system, the result may be a dwarf nova, a classical nova, or a supernova, in order of increasing brilliance. It is somewhat unfortunate that the term *nova* (Latin for "new") appears in each name because the three types of outbursts employ three very different mechanisms.

Dwarf novae and novae belong to the class of cataclysmic variables. They survive their release of energy (unlike supernovae), and the outburst process recurs. Cataclysmic variables are characterized by long quiescent intervals punctuated by outbursts in which the brightness of the system increases by a factor between 10 (for dwarf novae) and $10^6$ (for classical novae). Typically, the mass of the primary star is around 0.85 $M_\odot$, which is larger than the average of about 0.6 $M_\odot$ for isolated white dwarfs. The secondary star is usually a main-sequence star of spectral type G or later and is less massive than the primary. The two stars orbit each other with a period ranging from 76 min to 16 hours. Most have periods of less than 8 hours, although *none* of the 100 or so known systems have a period between 2.25 and 2.83 hours.[21]

The first observation of a **dwarf nova** (U Geminorum) was made in 1855. However, the basic nature of these objects remained elusive until 1974, when Brian Warner at the University of Cape Town showed that the outburst of an eclipsing dwarf nova, Z Chamaeleontis (recall Example 17.4), was due to a brightening of the accretion disk surrounding the white dwarf. Since most of the light from a dwarf nova comes from the accretion disk around the white dwarf, these systems provide astronomers with their best opportunity to study

---

[21] The reason for this period gap is not fully understood.

## 17.4 White Dwarfs in Semidetached Binaries

the dynamic structure of accretion disks.[22] Observations of the dwarf nova VW Hydri showed that the outburst at visible wavelengths preceded the ultraviolet brightening by about a day. This indicates that the outburst started in the cooler, outer part of the disk and then spread down to the hotter central regions. For these reasons, astronomers have concluded that the outbursts of dwarf novae are caused by a sudden increase in the rate at which mass flows down through the accretion disk.

To date, approximately 300 dwarf novae have been discovered. Characteristically, they brighten by between 2 and 6 magnitudes during outbursts that usually last from about 5 to 20 days. These eruptions are separated by quiet intervals of 30–300 days; see Fig. 17.16. Estimates of the rate of mass transfer through the disks of dwarf novae have been obtained by comparing theoretical models with observations of the amount of energy released at different wavelengths. Apparently, during the long quiescent intervals,

$$\dot{M} \approx 10^{15}\text{–}10^{16} \text{ g s}^{-1} \approx 10^{-11}\text{–}10^{-10} \text{ M}_\odot \text{ yr}^{-1},$$

which increases to

$$\dot{M} \approx 10^{17}\text{–}10^{18} \text{ g s}^{-1} \approx 10^{-9}\text{–}10^{-8} \text{ M}_\odot \text{ yr}^{-1}$$

during an outburst. Since the disk luminosity is proportional to $\dot{M}$ (Eq. 17.20), this increase in the mass transfer rate by a factor of 10–100 is consistent with the observed brightening of the system.[23] The mystery remaining to be solved by astronomers is the origin of the increased rate of mass transfer through the disk of a dwarf nova during an outburst. Possible explanations focus on either an instability in the mass transfer rate from the secondary to the primary star or an instability in the accretion disk itself that periodically dams up and releases the gases flowing through it.

A modulation of the mass transfer rate must depend on the details of the mass flow through the inner Lagrangian point, $L_1$. One possibility is an instability in the outer layers of the secondary star, causing it to periodically overflow its Roche lobe. Such an instability could be powered by the hydrogen partial ionization zone (at $T \approx 10,000$ K) damming up and releasing energy.[24]

---

[22]In some systems, the primary white dwarf has a magnetic field that is sufficiently strong (a few ten-million gauss) to prevent the formation of an accretion disk. Instead, the accretion takes place through a magnetically controlled column that funnels mass onto one (or both) of the white dwarf's magnetic poles. These *AM Herculis stars* (or *polars*) will be considered in Section 17.5.

[23]Recall that a difference of 5 magnitudes corresponds to a factor of 100 in brightness; see Eq. (3.4).

[24]This is somewhat reminiscent of the $\kappa$-mechanism that is involved in stellar pulsation; recall Section 14.2.

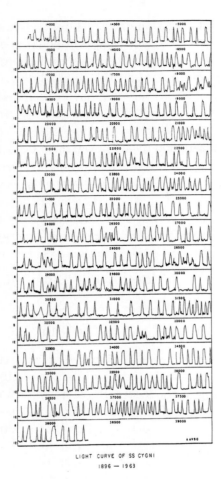

**Figure 17.16** Outbursts of the dwarf nova SS Cygni, about 95 pc away. This light curve, labeled by Julian day at 500-day intervals, covers the years 1896–1963, and was compiled by the American Association of Variable Star Observers. (Courtesy of Janet A. Mattei, AAVSO Director.)

When one gram of H II ions recombines with free electrons, as much as $1.3 \times 10^{13}$ ergs is released. If the ionization zone occurs close enough to the surface of the secondary, this could be sufficient to propel some of the overlying stellar material through the $L_1$ point and initiate a dwarf nova outburst. Recall, however, that the secondary star is usually a main-sequence star of spectral type G or later, so the ionization zone may well lie too deep to produce the instability.

## 17.4 White Dwarfs in Semidetached Binaries

The alternative explanation involving an instability in the outer part of the accretion disk also utilizes the hydrogen partial ionization zone. The viscosity of the disk material governs the rate at which mass spirals down through the disk. The lower the viscosity, the lower the resistance to the orbital motion of the disk gases; the inward drift of material decreases, and more matter accumulates in the disk. If the viscosity periodically switches from a low to a high value, the resulting wave of stored material plunging inward could produce the brightening of the disk observed for dwarf novae. Although the source of the viscosity in accretion disks is poorly understood, it has been suggested that the switch between low and high viscosity may be produced by an instability involving the periodic ionization and recombination of hydrogen in the outer part of the disk where $T \approx 10{,}000$ K. In such a scenario, the viscosity is roughly proportional to the disk temperature, which in turn depends on the opacity of the disk material. Below $10^4$ K, a plausible chain of reasoning then suggests

$$\text{neutral hydrogen} \rightarrow \text{low opacity}$$
$$\rightarrow \text{efficient cooling}$$
$$\rightarrow \text{low temperature}$$
$$\rightarrow \text{low viscosity}$$
$$\rightarrow \text{mass retained in the outer disk.}$$

On the other hand, above $10^4$ K,

$$\text{ionized hydrogen} \rightarrow \text{high opacity}$$
$$\rightarrow \text{inefficient cooling}$$
$$\rightarrow \text{high temperature}$$
$$\rightarrow \text{high viscosity}$$
$$\rightarrow \text{mass released to fall through the disk.}$$

The instability occurs because the accumulation of matter tends to slowly heat the outer disk, while its release results in a rapid cooling. This mechanism should operate only for low accretion rates ($< 10^{15}$ g s$^{-1}$ $\approx 10^{-11}$ M$_\odot$ yr$^{-1}$), so dwarf novae outbursts should not occur for systems with larger values of $\dot{M}$. This limit is in fact observed, and is one reason why most astronomers favor the disk instability explanation of dwarf novae outbursts.

Higher accretion rates are associated with **classical novae**. The earliest recorded appearance of a nova (CK Vulpeculae) occurred in 1670, and hundreds

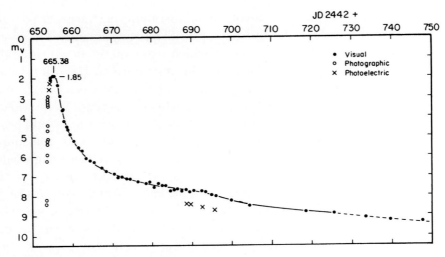

**Figure 17.17** The light curve of V1500 Cyg, a fast nova. (Figure adapted from Young, Corwin, Bryan, and De Vaucouleurs, *Ap. J.*, *209*, 882, 1976.)

of others have been observed since then. About 30 novae are detected in the Andromeda galaxy (M31) each year, but only two or three per year can be seen in those regions of our own Milky Way Galaxy that are unobscured by dust. Novae are characterized by a sudden increase in brightness of between 7 and 20 magnitudes, with an average brightening of about 10–12 magnitudes. The rise in luminosity is very rapid, taking only a few days, with a brief pause or *standstill* when the star is about two magnitudes from its maximum brilliance. At its peak, a nova may shine with about $10^5$ $L_\odot$, and release roughly $10^{45}$ ergs (integrated across all wavelengths) over $\sim 100$ days.

The subsequent decline occurs more slowly over several months, with the rate of decline defining the *speed class* of a nova. A *fast nova* takes a few weeks to dim by two magnitudes, while a *slow nova* may take nearly 100 days to decline by the same amount from maximum; see Figs. 17.17 and 17.18. The declines are sometimes punctuated by large fluctuations in brightness, which in extreme cases may take the form of the complete absence of visible light from the nova for a month or so before it reappears. Fast novae are typically three magnitudes brighter than slow novae, but in either case a nova falls to nearly its pre-eruption appearance after a few decades.

During the first few months, the decline in brightness occurs only at visual wavelengths. When observations at infrared and ultraviolet wavelengths are included, the bolometric luminosity of a nova is found to remain approximately constant for several months following its outburst; see Fig. 17.19. In

## 17.4 White Dwarfs in Semidetached Binaries

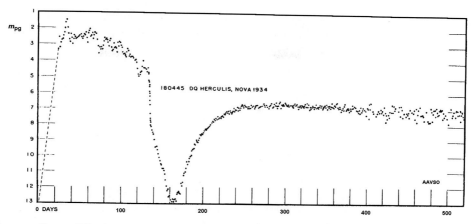

**Figure 17.18** The light curve of DQ Her, a slow nova. The *photographic magnitude*, $m_{pg}$, is measured from the nova's image on photographic plates. (Courtesy of Janet A. Mattei, AAVSO Director.)

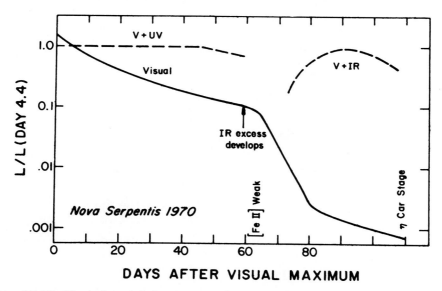

**Figure 17.19** The bolometric luminosity of nova FH Serpentis, in terms of its luminosity at day 4.4. Note that during the first 60 days, the decline in visible energy was almost exactly offset by an increase at ultraviolet wavelengths. Thereafter, the infrared rose as the visible light output was redistributed to infrared wavelengths. (Figure from Gallagher and Starrfield, *Annu. Rev. Astron. Astrophys.*, **16**, 171, 1978. Reproduced with permission from the *Annual Review of Astronomy and Astrophysics*, Volume 16, ©1978 by Annual Reviews Inc.)

addition, spectra of novae show that they are accompanied by the ejection of $10^{-5}$–$10^{-4}$ M$_\odot$ of hot gases at velocities between several hundred and several thousand km s$^{-1}$. The speed of the gases is roughly three times greater for fast novae, but the total mass ejected is about the same for both speed classes. We will see that the changing characteristics of this expanding shell of gas are responsible for the features seen in Fig. 17.19.

The average value of the absolute visual magnitude of a nova in its quiescent state is $M_V = 4.5$. Assuming that the light from such a system comes primarily from the accretion disk around the white dwarf, an estimate of the mass transfer rate for a typical nova can be obtained. (For the purposes of this estimation, visual magnitudes will be used instead of bolometric magnitudes. This means that the mass transfer rate will be slightly underestimated.) From Eq. (3.7), the luminosity of the system is

$$L = 100^{(M_{\text{Sun}} - M_V)/5} \, L_\odot = 1.3 \, L_\odot = 4.9 \times 10^{33} \text{ ergs s}^{-1}.$$

With this result, the luminosity of an accretion disk (Eq. 17.20) can be solved for the mass transfer rate, giving

$$\dot{M} = \frac{2RL}{GM} = 5.7 \times 10^{16} \text{ g s}^{-1},$$

or about $9.0 \times 10^{-10}$ M$_\odot$ yr$^{-1}$.

This is in good agreement with the accepted theoretical model of a nova, which incorporates a white dwarf in a semidetached binary system that accretes matter at a rate of about $10^{-8}$–$10^{-9}$ M$_\odot$ yr$^{-1}$. The hydrogen-rich gases accumulate on the surface of the white dwarf, where they are compressed and heated. At the base of this layer, turbulent mixing enriches the gases with the carbon, nitrogen, and oxygen of the white dwarf. (Without this mixing, the ensuing explosion would be too feeble to eject the mass observed for the expanding shell of hot gases.) Spectroscopic analysis of the shell shows an enrichment of carbon, nitrogen, and oxygen by a factor of 10 to 100 times the solar abundance of these elements.

At the base of this enriched layer of hydrogen, the material is supported by electron degeneracy pressure. When about $10^{-4}$–$10^{-5}$ M$_\odot$ of hydrogen has accumulated and the temperature at the base reaches a few million kelvins, a shell of hydrogen-burning develops, using the CNO cycle. For highly degenerate matter the pressure is independent of the temperature, so the shell source cannot dampen the reaction rate by expanding and cooling. The result is a runaway thermonuclear reaction, with temperatures reaching $10^8$ K before the electrons lose their degeneracy.[25] When the luminosity exceeds the Eddington

---

[25]This mechanism is similar to the helium core flash that was described in Section 13.2.

## 17.4 White Dwarfs in Semidetached Binaries

limit of about $10^{38}$ ergs s$^{-1}$ (recall Eq. 12.20), radiation pressure can lift the accreted material and expel it into space. The fast and slow speed classes of novae are probably due to variations in the mass of the white dwarf and in the degree of CNO enrichment of the hydrogen surface layer. The brief standstill that occurs before maximum luminosity is probably an effect of the changing opacity of the ejecta.

The energy that would be released in the complete fusion of a hydrogen layer of $m = 10^{-4}$ M$_\odot$ is $0.007mc^2 \approx 10^{48}$ ergs, roughly $10^3$ times larger than the energies actually observed. If all of the hydrogen were in fact consumed, the nova would shine for several hundred years. Most of the accumulated material must therefore be propelled into space by the explosion. However, the kinetic energy of the ejecta (far from the nova) is much smaller than the gravitational binding energy of the surface layer, indicating that the total energy given to the ejecta is just barely enough to allow it to escape from the system.

Only about 10% of the hydrogen layer is ejected by the nova explosion. Following this initial *hydrodynamic ejection phase*, which dominates for fast novae, hydrostatic equilibrium is established and the *hydrostatic burning phase* begins. During this prolonged stage of hydrogen burning, which is most important for slow novae, energy is produced at a constant rate approximately equal to the Eddington luminosity. The layer above the shell of CNO burning becomes fully convective and expands by a factor of 10 to 100, extending to some $10^{11}$ cm.[26] At the surface of the convective envelope the effective temperature is about $10^5$ K, much less than the $4 \times 10^7$ K in the active CNO shell source below.

Finally, the last of the accreted surface layer is ejected, a few months up to about a year after the hydrostatic burning phase began. Deprived of fuel, the hydrostatic burning phase ends and the white dwarf begins to cool. Eventually the binary system reverts to its quiescent configuration, and the accretion process begins anew. For accretion rates of $10^{-8}$–$10^{-9}$ M$_\odot$ yr$^{-1}$, it will take some $10^4$–$10^5$ years to build up another surface layer of $10^{-4}$ M$_\odot$.

The physical character of the ejected gases passes through three distinct phases following the nova explosion. Initially, during the *fireball expansion phase*, the material ejected in the hydrodynamic ejection phase forms an optically thick "fireball" that radiates as a hot blackbody of 6000–10,000 K. The observed light originates in the "photosphere" of the expanding fireball; at this point, the spectrum of the nova resembles that of an A or F supergiant.

---

[26]The white dwarf remnant may overflow its Roche lobe. The consequences of the resulting disruption of the close binary system are not yet clear.

The fireball expansion phase will be examined in Problem 17.14, where you will consider a simple model of a nova for which mass is ejected at a constant rate of $\dot{M}_{\text{eject}}$ at a constant speed $v$. The expanding model photosphere has a radius that initially increases linearly with time and then approaches a constant value of

$$R_\infty = \frac{3\overline{\kappa}\dot{M}_{\text{eject}}}{8\pi v}. \qquad (17.27)$$

If the luminosity, $L$, of the nova is also assumed to be constant, then from Eq. (3.17) the effective temperature of the model photosphere approaches

$$T_\infty = \left(\frac{L}{4\pi\sigma}\right)^{1/4} \left(\frac{8\pi v}{3\overline{\kappa}\dot{M}_{\text{eject}}}\right)^{1/2}. \qquad (17.28)$$

For an opacity of $\overline{\kappa} = 0.4$ cm$^2$ g$^{-1}$ [Eq. (9.21) for electron scattering, with pure hydrogen assumed for convenience], a mass ejection rate of $\dot{M}_{\text{eject}} \approx 10^{22}$ g s$^{-1}$ (about $10^{-4}$ M$_\odot$ yr$^{-1}$), and an ejection speed of $v \approx 1000$ km s$^{-1}$, the fireball's photosphere approaches a limiting radius of about $5 \times 10^{12}$ cm, or 1/3 AU. Taking $L$ to be the Eddington limit of about $10^{38}$ ergs s$^{-1}$, the effective temperature of the model photosphere approaches a value of nearly 9000 K.

The optically thick fireball phase ends in a few days, at the point of maximum visual brightness. Then, as the shell of gas thrown off by the nova continues to expand, it becomes less and less dense. The rate of mass ejection, $\dot{M}_{\text{eject}}$, has also declined in the hydrostatic burning phase. The result, according to Eqs. (17.27) and (17.28), is that the location of the photosphere moves inward and its temperature increases slightly. Although these general trends are correct, the opacity is in fact very sensitive to the temperature for $T < 10^4$ K (recall Fig. 9.10), and our model is too simplistic to describe the evolution of the nova. More advanced arguments show that as the visual brightness declines, more light is received from the nova at ultraviolet wavelengths. Finally, the shell becomes transparent and the *optically thin phase* begins. The central white dwarf, swollen by its hydrostatic burning phase, now has the appearance of a blue horizontal-branch object located just blueward of the RR Lyrae stars on the H–R diagram. The white dwarf envelope may burn irregularly, resulting in the substantial fluctuations in brightness observed for some novae.

After a few months, when the temperature of the expanding envelope of gases has fallen to about 1000 K, carbon in the ejecta can condense to form

## 17.4 White Dwarfs in Semidetached Binaries

**Figure 17.20** A 1949 photo of Nova Persei, which exploded in 1901. (Courtesy of Palomar/Caltech.)

dust consisting of graphite grains.[27] This initiates the *dust formation phase*. The resulting dust shell becomes optically thick in roughly 50% of all novae. The visible light from a nova is undiminished by an optically thin shell, but the formation of an optically thick cocoon of dust obscures or completely hides the central white dwarf. In the latter case, the output of visible light suddenly plunges, as seen in Fig. 17.19. The light from the white dwarf is absorbed and re-emitted by the graphite grains, so the optically thick dust shell radiates as a ~900 K blackbody at infrared wavelengths. In this way, the nova's bolometric luminosity remains constant as long as the white dwarf continues to produce energy at roughly the Eddington rate while in its hydrostatic burning phase. Figure 17.20 shows that the expanding shell may remain visible for years after the hydrostatic burning phases has ended, its gases and dust enriching the interstellar medium.

We have seen that there are many differences among the characteristics of individual novae. The peak luminosity, the rate of decline, the presence of rapid fluctuations, and/or the complete disappearance of the nova at visible wavelengths—all of these vary greatly from system to system. On the other

---

[27]The identification of the grain composition comes in part from an infrared emission "bump" at a wavelength of 5 $\mu$m; recall Section 12.1. Novae are natural laboratories for testing theories of grain formation.

hand, we will find that another type of cataclysmic variable, the **Type Ia supernova**, varies so little that it may be possible to use these exploding stars as standard candles, allowing astronomers to establish the distances to the systems in which they are found. A Type Ia supernova occurs when a white dwarf in a close binary system accretes enough mass from the secondary star to initiate a runaway nuclear reaction in the core of the white dwarf.

Before launching into a discussion of Type I supernova, it is important to recognize that the basic classification scheme for supernovae is based on the characteristics of their spectra near the time of maximum brilliance. Those displaying spectral lines of hydrogen are classified as *Type II supernovae*. As discussed in Section 13.3, this is the final act in the life of a single massive star. The collapse of its degenerate core results in an immense explosion, and the shape of its light curve leads to its identification as a Type II-L (linear) or a Type II-P (plateau).

Supernovae that do not show prominent hydrogen lines are classified as **Type I**. Those Type I spectra that show a strong Si II line at 6150 Å are called Type Ia. The others are designated **Type Ib** or **Type Ic**, depending on the presence (Ib) or absence (Ic) of strong helium lines.[28] The lack of hydrogen lines indicates that the stars involved have been stripped of their hydrogen envelopes. The differences in the spectral signatures between Type Ia and Types Ib or Ic indicate that different physical mechanisms are at work. This is reflected in the different environments observed for these outbursts. Type Ia supernovae are found in all types of galaxies, including ellipticals that show very little evidence of recent star formation. On the other hand, Types Ib and Ic have been seen only in spiral galaxies, near sites of recent star formation (H II regions). This implies that short-lived massive stars are probably involved with Types Ib and Ic, but not with Type Ia.

All Type I supernovae show similar rates of decline of their brightness after maximum, about 0.065 ($\pm 0.007$) magnitude per day at 20 days. After about 50 days, the rate of dimming slows and becomes constant, with Type Ia's declining 50% faster than the others (0.015 mag d$^{-1}$ vs. 0.010 mag d$^{-1}$). Figure 17.21 shows a composite light curve [at blue ($B$) wavelengths] for Type I's.

Initially, the supernova ejecta forms an optically thick shell. As the expansion progresses, the shell becomes stratified. Each layer of the shell coasts outward with a different constant speed, with the inner regions moving more slowly than the outer ones. The spectrum consists of lines showing P Cygni profiles (recall Fig. 12.12) formed in the shell's faster-moving optically thin outer layers, superimposed on a blackbody continuum from the underlying

---

[28]Like any classification scheme, the one for supernovae is constantly being refined.

## 17.4 White Dwarfs in Semidetached Binaries

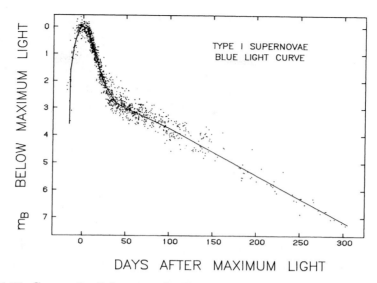

**Figure 17.21** Composite light curve for Type I supernovae at blue wavelengths. All magnitudes are relative to $m_B$ at maximum. (Figure from Doggett and Branch, *Astron. J.*, *90*, 2303, 1985.)

photosphere of the shell. One hundred days or so after the explosion, the shell gradually becomes optically thin. The supernova ejecta now becomes a nebula, with a spectrum dominated by emission lines.

The constant rate of decline of 0.010 mag d$^{-1}$ for Types Ib and Ic is similar to that observed for Type II supernovae and implies that the later light curves of both types of supernovae are sustained by the radioactive decay of $^{56}_{28}$Ni $\rightarrow$ $^{56}_{27}$Co $\rightarrow$ $^{56}_{26}$Fe; see Eqs. (13.17) and (13.18). The beta decay of cobalt into iron has a half-life of 77.7 days, which is consistent (via Eq. 13.22) with the observed 0.010 mag day$^{-1}$ decline of the light curve. The more rapid fading of Type Ia supernovae is probably due to the declining efficiency with which the energy of the gamma rays and positrons produced by the beta decay of $^{56}_{27}$Co is converted into visible light.

For Type Ia supernovae, the best agreement between the theoretical and observed spectra is obtained by modeling the explosion of an accreting carbon–oxygen white dwarf in a close binary system. This is consistent with the lack of hydrogen lines and the absence of an association with regions of star formation. When the mass of the white dwarf reaches about 1.30 M$_\odot$, carbon burning begins at the center of the star.[29] As in the case of the helium core

---

[29]Note that carbon burning begins before the Chandrasekhar limit of 1.4 M$_\odot$ is reached (see Section 15.4).

flash (see Section 13.2), the increase in temperature does not produce an expansion of the degenerate core that would slow the reaction rate. This means that a front of vigorous carbon burning moves toward the surface.[30] A model in which the front moves somewhat more slowly than the local sound speed (a **deflagration** rather than a supersonic **detonation**) is most successful at reproducing the characteristics of the observed light curve and spectrum of a Type Ia supernova. About one-half of the white dwarf's mass is processed into iron before the degeneracy is removed. The subsequent expansion produces a cooling that eventually dampens the nuclear burning, leaving a mixture of partially processed intermediate-mass elements surrounding a nickel–iron core. The energy released in this process is enough to completely disrupt the star, resulting in a Type Ia supernova. The inner part of the star is ejected at low speeds, while the outer layers are expelled at speeds exceeding $10^4$ km s$^{-1}$ ($0.1c$). This agrees with spectral evidence for the incomplete burning of carbon into iron in Type Ia supernovae. It is Type Ia supernovae, rather than Type II's, that are responsible for most of the iron-peak elements found in the enriched interstellar medium.

Many astronomers argue that the light curves of Type Ia supernovae are sufficiently similar that they could be used as standard candles to determine intergalactic distances.[31] At maximum light, the value of a supernova's absolute blue magnitude is $M_B = -19.6 \pm 0.2$, according to Branch and Tammann (1992). Those questioning this use of supernovae cite the fact that most supernovae are observed after peak luminosity, which introduces some uncertainty into the value of $M_B$. There are also hints that the characteristics of Type Ia supernovae may not be perfectly homogeneous; for example, some spectra show significantly higher ejection speeds than others.

To date, the most distant supernova discovered is SN 1992bi in the constellation Hercules. The supernova appeared as a faint 22nd-magnitude star in a galaxy that has a redshift of $z = 0.457$ (see Eq. 4.36). The light from this supernova began its journey some 5 billion years ago, about the time our own solar system was forming. The prospect of measuring distances across billions of light years to determine the global geometry of the universe is a powerful incentive for astronomers to further develop and justify the use of Type Ia's as standard candles.

---

[30] Fractal aficionados will be interested to learn that the burning front is so convoluted that it may have a fractal dimension of $\sim 2.5$. If so, the crinkled area of a front of radius $r$ may be much greater than $4\pi r^2$.

[31] This is similar to the use of Cepheid variables as standard candles; see Section 14.1. Supernovae are much more luminous, and so a reliable value of their maximum brilliance would be a powerful tool for surveying the universe.

The light curves of Types Ib and Ic supernovae are fainter than those for Type Ia's by 1.5 to 2 magnitudes in blue light, but are otherwise similar. As already mentioned, the significant difference is in their spectra. Although the carbon deflagration model can provide a reasonable reproduction of the observations of Types Ib and Ic, their close association with regions of active star formation suggest that short-lived, massive stars must be involved. The presently favored model holds that they are explosions of stars with an initial main-sequence mass of $\sim 20$ $M_\odot$. As the star evolves, its hydrogen envelope is stripped away, either through mass transfer in a close binary or by stellar winds. The result is an exposed core having a surface layer of about 4–6 $M_\odot$ of helium, with succeedingly heavier elements found toward the iron center. Just as with a Type II supernova, the explosion is triggered by the collapse of the degenerate iron core. The absence of a hydrogen envelope would explain the lack of hydrogen lines in the spectra of these supernovae.[32] Furthermore, it is calculated that the explosion of a more massive star would produce less $^{56}_{26}$Ni, resulting in a fainter light curve for Types Ib and Ic. The difference between the spectra of Types Ib (strong He I lines) and Ic (weak He I lines) may be explained by the explosion of a helium-poor progenitor. More work is needed to determine whether this model can satisfactorily reproduce the behavior of supernovae of Types Ib and Ic. It may be that alternative models (such as the carbon deflagration model or one involving helium burning on the surface of a carbon–oxygen white dwarf) will prove to be more robust at explaining the observations.

## 17.5 Neutron Stars and Black Holes in Binaries

If one of the stars in a close binary system explodes as a supernova, the result may be either a neutron star or a black hole orbiting the companion star. In a semidetached system, hot gas spills through the inner Lagrangian point from the distended atmosphere of the companion star. A variety of intriguing phenomena are powered by the energy released when the gas falls down the deep gravitational potential well onto the compact object. As will be seen shortly, many of these systems emit copious quantities of x-rays. In fact, these **binary x-ray systems** shine most strongly in the x-ray region of the electromagnetic spectrum. Other systems may consist of *two* compact objects,

---

[32]This view is supported by SN 1993J in the spiral galaxy M81 in Ursa Major. This supernova initially displayed strong hydrogen emission lines (i.e., Type II), but within a month the hydrogen lines were replaced by helium and its appearance changed to that of a Type Ib. Apparently a thin layer of hydrogen remained on the star's surface when it exploded.

such as the binary pulsars. At the end of this section, we will describe the use of a binary pulsar system as a powerful test of Einstein's general theory of relativity.

Whether a binary system survives the supernova explosion of one of its component stars depends on the amount of mass ejected from the system. Consider a system initially containing two stars of mass $M_1$ and $M_2$ separated by a distance $a$, that are in circular orbits about their common center of mass. Using Eq. (2.33), the total energy of the system is

$$E_i = \frac{1}{2}M_1 v_1^2 + \frac{1}{2}M_2 v_2^2 - G\frac{M_1 M_2}{a} = -G\frac{M_1 M_2}{2a}. \quad (17.29)$$

The speeds of the two stars are related by Eq. (7.4), $M_1 v_1 = M_2 v_2$. Now suppose that star 1 explodes as a supernova, leaving a remnant of mass $M_R$. For a spherically symmetric explosion, there is no change in the velocity of star 1. Before the spherical shell of ejecta reaches star 2, its mass acts gravitationally as though it were still on star 1 (recall Example 2.2). So far, the supernova has had no effect on the binary. However, as soon as the shell has swept beyond star 2, the gravitational influence of the ejecta is no longer detectable.

Thus the main consequence of the supernova on the orbital dynamics of the binary system arises from the ejection of mass, the removal of some of the gravitational glue that was binding the stars together.[33] Since the velocity of star 2 is unchanged and the separation of the two stars remains the same, the total energy of the system after the explosion is now

$$E_f = \frac{1}{2}M_R v_1^2 + \frac{1}{2}M_2 v_2^2 - G\frac{M_R M_2}{a}. \quad (17.30)$$

If the explosion results in an *unbound* system, then $E_f \geq 0$. It is left as an exercise to show that the mass of the remnant must satisfy

$$\frac{M_R}{M_1 + M_2} \leq \frac{1}{(2 + M_2/M_1)(1 + M_2/M_1)} < \frac{1}{2} \quad (17.31)$$

for an unbound system. That is, *at least* one-half of the total mass of the binary system must be ejected if the supernova explosion of star 1 is to disrupt the system. If one-half or more of the system's mass is retained, the result will be a neutron star or a black hole gravitationally bound to a companion star. For a massive companion star ($M_2 \gg M_1$), this is a likely result. But when low-mass companions are involved (as with Type Ib and Ic supernovae),

---

[33]The direct impact of the supernova blast on the companion star has been neglected, although this too will contribute to disrupting the system.

## 17.5 Neutron Stars and Black Holes in Binaries

other mechanisms must be considered for producing a neutron star in a binary system.

Isolated neutron stars formed by isolated supernovae may be gravitationally captured during a chance encounter with another star. Because the total energy of two unbound stars is initially greater than zero, some of the excess kinetic energy must be removed for a capture to occur. If the proximity of the two objects raises a tidal bulge on the nondegenerate star, energy may be dissipated by the damping mechanisms discussed in Section 14.2 for pulsating stars.[34] The outcome of such a **tidal capture** depends on the nearness of the passage and the type of star involved. Although a direct hit would destroy a main-sequence star, the penetration of a neutron star into a giant star would bring it close to the star's degenerate core. The result, a neutron star orbiting inside the giant star, is known as a **Thorne–Żytkow object**.[35] It is thought that the envelope of the giant star would be quickly expelled, producing a neutron star–white dwarf binary with an orbital period of about 10 minutes. More frequently, if the neutron star passes between about 1 to 3 times the radius of the other star, then a capture will occur. The resulting binary system would have a period ranging from several hours (with a main-sequence star) to several days (with a giant).

This tidal capture process is most effective in regions that are extremely densely populated with stars, such as the centers of globular clusters (Section 13.4). It is estimated that in a compact globular cluster, tidal capture could produce up to about ten close binary systems containing a neutron star over a period of some $10^{10}$ years. This is consistent with the number of x-ray sources observed in globular clusters. (The estimated lifetime of a binary x-ray system is on the order of $10^9$ years, so only the most compact globular clusters would be expected to harbor even one x-ray source at a given time; see Problem 17.22.)

Another mechanism capable of producing a neutron star is the **accretion-induced collapse** of a white dwarf in a close binary system. If the accreting white dwarf could surpass the Chandrasekhar limit *without* exploding as a Type Ia supernova, the resulting gravitational collapse could produce a neutron star. Observations of binary x-ray pulsars (described below) with low-mass companions indicate that this does in fact occur. Theory suggests, however, that the ignition of explosive carbon burning should always occur

---

[34] Alternatively, a capture may involve three (or more) stars. One of the stars would be gravitationally flung from the system, removing energy and so allowing the capture to take place.

[35] The properties of Thorne–Żytkow objects were first calculated in 1977 by Kip Thorne and Anna Żytkow of Caltech. To date, these objects remain hypothetical.

before the Chandrasekhar limit is reached, and it is not yet understood how the detonation of a Type Ia supernova can be avoided.

Close binary systems containing neutron stars were first identified by their energetic emission of x-rays. The first source of x-rays beyond the solar system was discovered in 1962 in the constellation Scorpius by a Geiger counter arcing above Earth's atmosphere in a sounding rocket. (X-rays cannot penetrate the atmosphere, so detectors and telescopes designed for x-ray wavelengths must make their observations from space.[36]) This object, called Sco X-1, is now known to be an x-ray pulsar. The periodic eclipse of another x-ray pulsar, Cen X-3 in the constellation Centaurus, revealed its binary nature.

X-ray pulsars are powered by the gravitational potential energy released by accreting matter. Recall from Example 17.2 that when mass falls from a great distance to the surface of a neutron star, about 20% of its rest energy is released, an amount that far exceeds the fraction of a percent that would be produced by fusion. The observed x-ray luminosities range up to $10^{38}$ ergs s$^{-1}$ [the Eddington limit; see Eq. (12.20)]. For a neutron star with a radius of 10 km, the Stefan–Boltzmann equation in the form of Eq. (3.17) shows that the temperature associated with this luminosity is about $2 \times 10^7$ K. According to Wien's law, Eq. (3.19), the spectrum of a blackbody with this temperature would peak at an x-ray wavelength of about 1.5 Å.

The reader should recall from Section 15.6 that neutron stars are often accompanied by powerful magnetic fields. In fact, these fields may be sufficiently strong to prevent the accreting matter from reaching the star's surface. The strength of the neutron star's magnetic dipole field is proportional to $1/r^3$, so the plunging gases encounter a rapidly increasing field. When the magnetic energy density $u_m = B^2/8\pi$ (Eq. 11.12) becomes comparable to the kinetic energy density $u_K = \frac{1}{2}\rho v^2$, the magnetic field will channel the infalling ionized gases toward the poles of the neutron star; see Fig. 17.22. This occurs at a distance from the star known as the **Alfvén radius**, $r_A$, where

$$\frac{1}{2}\rho v^2 = \frac{B^2}{8\pi}. \tag{17.32}$$

For the special case of spherically symmetric accretion, with the gases starting at rest at a great distance, the free-fall velocity is $v = \sqrt{2GM/r}$ for a star of mass $M$ (from energy conservation). Furthermore, the density and velocity are related to the mass accretion rate, $\dot{M}$, by Eq. (11.4),

$$\dot{M} = 4\pi r^2 \rho v, \tag{17.33}$$

---

[36]The first x-ray detector was designed to look for x-rays from the lunar surface, produced when solar wind particles cause the lunar soil to fluoresce. The presence of enormously stronger cosmic x-ray sources came as a surprise to astronomers at the time.

## 17.5 Neutron Stars and Black Holes in Binaries

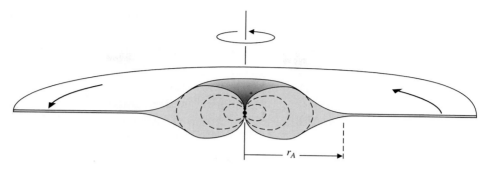

**Figure 17.22** Accreting gas channeled onto a neutron star's magnetic poles, where $r \approx r_A$.

and the radial dependence of the magnetic dipole field strength may be expressed as

$$B(r) = B_s \left(\frac{R}{r}\right)^3, \tag{17.34}$$

where $B_s$ is the surface value of the magnetic field. Inserting these expressions into Eq. (17.32) and solving for the Alfvén radius, we obtain

$$r_A = \left(\frac{B_s^4 R^{12}}{2GM\dot{M}^2}\right)^{1/7} \tag{17.35}$$

(the proof is left as an exercise). Of course, the accretion will not actually be spherically symmetric. However, the magnetic field increases so rapidly as the falling matter approaches the star that a more realistic calculation yields nearly the same result: the flow will be disrupted at a *disruption radius* $r_d$,

$$r_d = \alpha r_A, \tag{17.36}$$

with $\alpha \sim 0.5$.

Before considering the details of channeled accretion onto a neutron star, let's look at the case of accretion onto the white dwarf considered in Example 17.3, for which $M = 0.85$ $M_\odot$, $R = 0.0095$ $R_\odot = 6.6 \times 10^8$ cm, and $\dot{M} = 10^{16}$ g s$^{-1}$ ($1.6 \times 10^{-10}$ $M_\odot$ yr$^{-1}$). Assume that its magnetic field has a surface strength of $B_s = 10^7$ G, about 100 times stronger than the typical value for a white dwarf. Then the Alfvén radius is

$$r_A = \left(\frac{B_s^4 R^{12}}{2GM\dot{M}^2}\right)^{1/7} = 6.07 \times 10^{10} \text{ cm}.$$

This is comparable to the separation of the stars in a cataclysmic variable (see Example 17.4), so an accretion disk cannot form around a white dwarf with an extremely strong magnetic field. Instead, the mass spilling through the inner Lagrangian point is confined to a stream that narrows as it is magnetically directed toward one (or both) of the poles of the white dwarf. In the absence of an accretion disk, all of the accretion energy will be delivered to the pole(s) of the star, with an accretion luminosity of (Eq. 17.21)

$$L_{\text{acc}} = G\frac{M\dot{M}}{R} = 1.71 \times 10^{33} \text{ ergs s}^{-1}.$$

This is the case for the **AM Herculis stars** (also called **polars**), which are semidetached binaries containing a white dwarf with a magnetic field of about $2 \times 10^7$ G. The torque produced by the white dwarf's field interacting with the secondary star's envelope results in a nearly synchronous rotation; the two stars perpetually face each other, connected by a stream of hot gas.[37] As this gas approaches the white dwarf, it moves almost straight down toward the surface and forms an accretion column a few tens of kilometers across. A shock front occurs above the white dwarf's photosphere, where the gas is decelerated and heated to a temperature of several $10^8$ K. The hot gas emits hard x-ray photons; some escape, and some are absorbed by the photosphere and re-emitted at soft x-ray and ultraviolet wavelengths.

The visible light observed from these systems is in the form of *cyclotron radiation* emitted by the (nonrelativistic) electrons spiraling along the magnetic field lines of the accretion column. This is the nonrelativistic analog of the synchrotron radiation emitted by relativistic electrons; see Fig. 15.17. In contrast to the continuous spectrum of synchrotron radiation, most of the energy of cyclotron radiation is emitted at the *cyclotron frequency*,

$$\nu_c = \frac{eB}{2\pi m_e c} \qquad \text{(cgs units)}. \tag{17.37}$$

For $B_s = 10^7$ G, $\nu_c = 2.8 \times 10^{13}$ Hz, which is in the infrared. However, a small fraction of the energy is emitted at higher harmonics (multiples) of $\nu_c$ and may be detected at visible wavelengths by astronomers on Earth. The cyclotron radiation is circularly polarized when observed parallel to the direction of the magnetic field lines, and linearly polarized when viewed perpendicular to the

---

[37] If the white dwarf has a somewhat weaker field ($B_s < 10^7$ G), or if the stars are farther apart, an accretion disk may form, only to be disrupted near the star (as shown in Fig. 17.22). These systems, called **DQ Herculis stars**, or **intermediate polars**, do not exhibit synchronous rotation.

## 17.5 Neutron Stars and Black Holes in Binaries

field lines.[38] Thus, as the two stars orbit each other (typically every 1 to 2 hours), the measured polarization changes smoothly between being circularly and linearly polarized. In fact, it is this strong variable polarization (up to 30%) that gives polars their name.

Now consider the case of accretion onto the neutron star described in Example 17.3. For this star, $M = 1.4$ M$_\odot$, $R = 10$ km, and $\dot{M} = 10^{17}$ g s$^{-1}$ ($1.6 \times 10^{-9}$ M$_\odot$ yr$^{-1}$). Furthermore, take the value of the magnetic field at the neutron star's surface to be $B_s = 10^{12}$ G. The value of the Alfvén radius is then

$$r_A = \left( \frac{B_s^4 R^{12}}{2GM\dot{M}^2} \right)^{1/7} = 3.09 \times 10^8 \text{ cm}.$$

Although 300 times the radius of the neutron star itself, it is much less than the value of $r_{\text{circ}}$ (Eq. 17.22) that describes the extent of an accretion disk. Thus an accretion disk will form around the neutron star but will be disrupted near the neutron star's surface as shown in Fig. 17.22 (unless the magnetic field is quite weak, roughly $< 10^8$ G). As the accreting gas is funneled onto one of the magnetic poles of the neutron star, it forms an accretion column similar to the one described for white dwarfs. In this case, however, the accretion luminosity (Eq. 17.21) is four orders of magnitude greater,

$$L_{\text{acc}} = G \frac{M\dot{M}}{R} = 1.86 \times 10^{37} \text{ ergs s}^{-1},$$

close to the Eddington limit of $\sim 10^{38}$ ergs s$^{-1}$; see Eq. (12.20). As $L_{\text{acc}}$ approaches $L_{\text{Ed}}$, radiation pressure elevates the shock front to heights reaching $r \sim 2R$. As a result, x-rays are emitted over a large solid angle. If the neutron star's magnetic and rotation axes are not aligned, as in Fig. 15.25, then the x-ray-emitting region may be eclipsed periodically.

The result is a **binary x-ray pulsar**.[39] Figure 17.23 shows the signal received from Hercules X-1, which exhibits a pulse of x-rays every 1.245 s (the rotation period of the neutron star). Note that the broad pulse (due to the large solid angle of the emission) may occupy $\sim 50\%$ of the pulse period of a radio pulsar, compared to the sharper radio pulses shown in Fig. 15.19, which take up only 1% to 5% of the pulse period. To date, about 20 binary x-ray pulsars have been found, with periods ranging from 0.069 s to 835 s. As noted on pages 611–612, white dwarfs cannot rotate as rapidly as the lower end of this

---

[38]The electric field vector of linearly polarized light oscillates in a single plane, while for circularly polarized light this plane of polarization rotates about the direction of travel.

[39]The Crab pulsar also emits x-rays, but it is primarily a radio pulsar that radiates in every region of the electromagnetic spectrum.

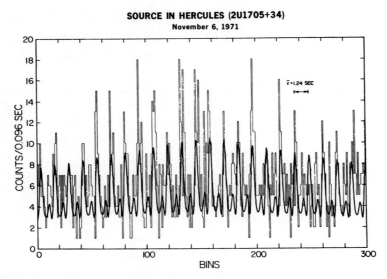

**Figure 17.23** X-ray pulses from Her X-1, with a period of 1.245 s. The peaks are the x-ray counts received from Her X-1 grouped in bins that are 0.096 s wide, and the heavier curve is a fit to the data using sine functions. (Figure from Tananbaum et al., *Ap. J. Lett.*, *174*, L143, 1972.)

period range without breaking up. This is one indication that x-ray pulsars are indeed accreting neutron stars.

Further confirmation that x-ray pulsars are accreting neutron stars comes from the observation that the periods of these objects are slowly decreasing. As time passes, they spin *faster*.[40] The time derivative of the star's rotation period, $\dot{P} \equiv dP/dt$, is related to the rate of change of its angular momentum, $L = I\omega$, by

$$\frac{dL}{dt} = I\frac{d\omega}{dt} = I\frac{d}{dt}\left(\frac{2\pi}{P}\right) = -2\pi I \frac{\dot{P}}{P^2},$$

where $I$ is the moment of inertia of the neutron star. Near the disruption radius, the angular momentum of the gas parcels orbiting in the accretion disk ($L = mvr$) is transferred to the neutron star via magnetic torques. The time derivative of the neutron star's angular momentum is just the rate at which angular momentum arrives at the disruption radius, so at $r = r_d$ we set

$$\frac{dL}{dt} = \dot{M}vr_d,$$

---

[40] Recall from Section 15.7 that the periods of radio pulsars *increase* with time as they lose energy due to magnetic dipole radiation.

## 17.5 Neutron Stars and Black Holes in Binaries

where the orbital velocity at $r = r_d$ is $v = \sqrt{GM/r_d}$ [Eq. (2.30) or (2.31) with $e = 0$ and $a = r_d$ for a circular orbit]. Equating these expressions for $dL/dt$ and using the definitions of the Alfvèn and disruption radii, Eqs. (17.35) and (17.36), results in

$$\frac{\dot{P}}{P} = -\frac{P\sqrt{\alpha}}{2\pi I} \left( \frac{B_s^2 R^6 G^3 M^3 \dot{M}^6}{\sqrt{2}} \right)^{1/7}. \tag{17.38}$$

---

**Example 17.5** The x-ray pulsar Centaurus X-3 has a period of 4.84 s and an x-ray luminosity of about $L_x = 5 \times 10^{37}$ ergs s$^{-1}$. Assuming that it is a 1.4 M$_\odot$ neutron star with a radius of 10 km, its moment of inertia (assuming for simplicity that it is a uniform sphere) is

$$I = \frac{2}{5} MR^2 = 1.11 \times 10^{45} \text{ g cm}^2.$$

Using Eq. (17.21) for the accretion luminosity, the mass transfer rate is found to be

$$\dot{M} = \frac{RL_x}{GM} = 2.69 \times 10^{17} \text{ g s}^{-1},$$

or $4.27 \times 10^{-9}$ M$_\odot$ yr$^{-1}$. Then, for an assumed magnetic field of $B_s = 10^{12}$ G and $\alpha = 0.5$, Eq. (17.38) gives the fractional change in the period per second and per year,

$$\frac{\dot{P}}{P} = -\frac{P\sqrt{\alpha}}{2\pi I} \left( \frac{B_s^2 R^6 G^3 M^3 \dot{M}^6}{\sqrt{2}} \right)^{1/7}$$

$$= -2.74 \times 10^{-11} \text{ s}^{-1}$$

$$= -8.64 \times 10^{-4} \text{ yr}^{-1}.$$

That is, the characteristic time for the period to change is $P/\dot{P} = 1160$ years.

The measured value for Cen X-3 is $\dot{P}/P = 2.8 \times 10^{-4}$ yr$^{-1}$, smaller than our estimate by a factor of 3 but in good agreement with this simple argument. The reader may verify that if a 0.85 M$_\odot$ white dwarf with a radius of $6.6 \times 10^8$ cm and $B_s = 10^7$ G is used for the accreting star, rather than a neutron star, then $\dot{P}/P = 1.03 \times 10^{-5}$ yr$^{-1}$. The measured value is larger by a factor of 27. A white dwarf is hundreds of times larger than a neutron star, so it has a much larger moment of inertia and is more difficult to spin up. The substantially better agreement between the neutron star model and the observations obtained for these systems is compelling evidence that neutron stars are the accreting objects in binary x-ray pulsars.

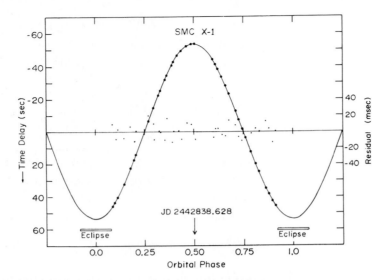

**Figure 17.24** Measured pulse arrival times (dots) for the binary x-ray pulsar SMC X-1 as a function of its orbital phase. The curve is for the best-fit circular orbit, and the dots about the straight line show the residuals from the best-fit orbit. (Figure from Primini et al., *Ap. J.*, *217*, 543, 1977.)

As an x-ray pulsar orbits its binary companion, the distance from the pulsar to Earth constantly changes. This results in a cyclic variation in the measured pulse period that is analogous to the Doppler shift of a spectral line observed for a spectroscopic binary (see Section 7.3). Figure 17.24 shows the shift in pulse arrival times as a function of the orbital phase for the x-ray pulsar SMC X-1 in the Small Magellanic Cloud, a small satellite galaxy of the Milky Way. The orbit for this system is almost perfectly circular, with a radius of 53.5 light-seconds = 0.107 AU, less than one-third the size of Mercury's orbit around the Sun.

A complete description of the binary system has been obtained for six *eclipsing* x-ray pulsars with visible companions. Such systems are analogous to double-line, eclipsing, spectroscopic binaries. For example, in the SMC X-1 system the mass of the secondary star is 17.0 $M_\odot$ (with an uncertainty of about 4 $M_\odot$), and its radius is 16.5 $R_\odot$ ($\pm 4\ R_\odot$). The masses of the neutron stars have also been determined for these six systems. The results are consistent with a neutron star mass of 1.4 $M_\odot$ ($\pm 0.2\ M_\odot$), in good agreement with the Chandrasekhar limit.

If the magnetic field of the neutron star is too weak ($\ll 10^{12}$ G) to completely disrupt the accretion disk and funnel the accreting matter onto its

## 17.5 Neutron Stars and Black Holes in Binaries

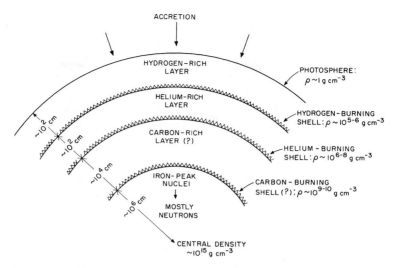

**Figure 17.25** Surface layers on an accreting neutron star. (Figure from Joss, *Comments Astrophys.*, *8*, 109, 1979.)

magnetic poles, these gases will settle over the surface of the star. Without an accretion column to produce a hot spot, x-ray pulses cannot be produced by the rotation of the neutron star. Instead, calculations indicate that when a layer of hydrogen a few meters thick accumulates on the surface, a shell of hydrogen slowly begins burning about a meter below the surface, with a shell of helium-burning ignited another meter below that; see Fig. 17.25.[41] This fusion of helium is explosive and releases a total of $\sim 10^{39}$ ergs in just a few seconds, with the surface reaching a temperature of about $3 \times 10^7$ K (twice the Sun's central temperature). The resulting blackbody spectrum peaks at x-ray wavelengths, and a flood of x-rays is liberated by this **x-ray burster**. Some of the x-rays may be absorbed by the accretion disk and re-emitted as visible light, so an optical flash is sometimes seen a few seconds after the x-ray burst. As the burst luminosity declines in a matter of seconds, the spectrum matches that of a cooling blackbody with a radius of $\sim 10$ km, consistent with the presence of a neutron star. After a time that can vary from a few hours to a day or more, another layer of hydrogen accumulates and another x-ray burst is triggered.[42] More than 40 x-ray bursters have been found so far. Most are

---

[41]This mechanism is reminiscent of the helium shell flashes discussed in Section 13.2.

[42]It is thought that the gases accreting on x-ray pulsars are constantly undergoing fusion. However, recall from Example 17.2 that the energy released in the accretion column will be about 30 times larger, so the energy from fusion will be lost in the glare of the accretion energy.

concentrated near the Galactic plane, toward the center of our Galaxy, with some 20% located in old globular clusters.

From these and other results, astronomers have identified two classes of binary x-ray systems. The more common type are those with low-mass secondary stars (late spectral-type stars with $M_2 \leq 2\ M_\odot$). These systems belong to the **low-mass x-ray binaries** (LMXBs). LMXBs produce x-ray bursts rather than pulses, indicating that the neutron star's magnetic field is relatively weak. Because low-mass stars are small, the two stars must orbit more closely if mass is to be transferred from one star to the other. For this reason, the LMXBs have short orbital periods, from 9.8 days down to 41 minutes.[43] About one-quarter of these systems are found within globular clusters, where the high number density of stars makes the gravitational capture of a neutron star more likely. The neutron stars in LMXBs may also have been formed by the accretion-induced collapse of a white dwarf.

Systems with higher-mass secondaries are called **massive x-ray binaries** (MXRBs). About half of the $\sim 40$ MXRBs are x-ray pulsars. With giant or supergiant O and B stars available to fill their Roche lobes, the separation of the stars can be larger, and the orbital periods correspondingly longer, from 2.1 days up to 581 days. Even if the secondary star's envelope does not overflow its Roche lobe, the vigorous stellar winds of these stars may still provide the mass transfer rate needed to sustain the production of x-rays. The MXRBs are found near the plane of our Galaxy, where there are young massive stars and ongoing star formation. This is consistent with the idea that an MXRB is the product of the normal evolution of a binary system containing a massive star that survived the supernova explosion of its companion.

So far, only neutron stars have been considered as the accreting object in binary x-ray systems. However, the gravitational potential well is even deeper for matter falling toward a black hole. In this case, up to about 30% of the rest energy of the falling disk material may be emitted as x-rays. In fact, as was discussed in Section 16.3, these systems provide the best evidence for the existence of stellar-mass black holes. The gas spilling through the inner Lagrangian point is heated to millions of kelvins as it spirals down through the black hole's accretion disk (see Fig. 16.24), and so emits x-rays. The identification of a black hole rests on determining that the mass of a compact, x-ray-emitting object exceeds the 3 $M_\odot$ upper limit for the mass of a neutron star. Thus the procedure for detecting a black hole in a binary x-ray system is similar to that used to measure the masses of neutron stars in these systems.

---

[43]The x-ray binary with the shortest period discovered to date is 4U1820−30 in the galaxy NGC 6624. This x-ray burster has an orbital period of only 11.4 minutes.

## 17.5 Neutron Stars and Black Holes in Binaries

At present, there are only a handful of x-ray binaries that allow such a dynamical determination of the masses involved. The best cases at the time of this writing are A0620−00, V404 Cygni, Cygnus X-1, and LMC X-3. Since none of these system exhibit eclipses, the resulting uncertainty about their orbital inclinations means that the masses calculated are lower limits (see Section 7.3). A0620−00 is an **x-ray nova**, powered by the sporadic accretion of material from its companion, a K5 main-sequence star. The relative faintness of the secondary star allows the measurement of the radial velocity of *both* the accretion disk and the companion star. The identification of A0620−00 as a $3.82 \pm 0.24$ $M_\odot$ black hole seems secure, as was described in Section 16.3 and Problem 16.21. V404 Cyg is also an x-ray nova, where recent measurements persuasively document the presence of a 6.3 $M_\odot$ black hole.

The arguments for the other two systems, although strong, are not as conclusive. Neither has a fully developed accretion disk, and so the velocities of both members cannot be determined. Cygnus X-1, perhaps the best-known black hole candidate, is a bright MXRB. Because almost all of the light comes from the secondary, Cyg X-1 is essentially a single-line spectroscopic binary. The identification of Cyg X-1 as a black hole therefore depends on the identification of the secondary star (HDE 226868) as a typical B supergiant. The most likely result, making reasonable assumptions about this binary system, is that the mass of the compact object in Cyg X-1 is between 11 $M_\odot$ and 21 $M_\odot$. A worst-case argument results in a secure lower limit of 3.4 $M_\odot$, providing the evidence that Cyg X-1 is a black hole.

The secondary star in the LMC X-3 system is a B3 main-sequence star that is orbiting an unseen, more massive companion. Although the lower limit on the mass of the compact companion is 3 $M_\odot$, a more probable mass range is 4–9 $M_\odot$—again, solid evidence for a black hole. Other x-ray binary systems may contain black holes, such as Nova Mus 1991 in the southern constellation Musca (the Fly), LMC X-3 in the Large Magellanic Cloud, and CAL 87 (in the direction of the Large Magellanic Cloud), but the evidence in these cases is not yet as strong.

One more x-ray binary and possible black hole candidate should be mentioned: SS 433, one of the most bizarre objects known to astronomers.[44] In 1978, it was discovered that this object displays *three* sets of emission lines. One set of spectral lines was greatly blueshifted, another set was greatly redshifted, and a third set lacked a significant Doppler shift. Here was an object with three components: Two were approaching and receding, respectively, at

---

[44] "SS" stands for the catalog of peculiar emission-line stars compiled by Bruce Stephenson and Nicholas Sanduleak.

one-quarter the speed of light while the third stayed nearly still! The wavelengths of the shifted lines vary with a period of 164 days, while the wavelengths of the nearly stationary lines show a smaller shift with a 13.1 day period. Furthermore, the position of SS 433 lies at the center of a diffuse, elongated shell of gas known as W50, which is probably a supernova remnant. Although the mystery of this object has not been fully unraveled, an overall picture of the system has emerged.[45] The 13.1-day period describes the orbit of a compact object (most probably a neutron star, but perhaps a black hole) around the primary. The primary is thought to be a 10–20 $M_\odot$ early-type star with a stellar wind that produces the broad stationary emission lines.[46] Surrounding the compact object is an accretion disk that contributes to the visible light from the system equally with the secondary. A tidal interaction between the disk and the two stars could be responsible for a precessional wobble of the disk that has a period of 164 days, analogous to Earth's 25,770-year precessional wobble discussed in Section 1.3. There is broad agreement that the varying Doppler-shifted emission lines, shown in Fig. 17.26, come from two *relativistic jets* that expel particles at $0.26c$ in opposite directions along the axis of the disk. The jets are probably powered by the accretion of matter at a rate exceeding the Eddington limit, generating x-rays at a prodigious rate. This could produce a radiation pressure sufficient to expel a portion of the accreting gases at relativistic speeds in the direction of least resistance—perpendicular to the disk. As the disk precesses, two oppositely directed jets sweep out a cone in space every 164 days, resulting in cyclic variations in both the radial velocity of the jets and the observed Doppler shift. The collimation of the jets could be the result of the ionized gases moving along magnetic field lines. The axis of the precessional cone makes an angle of 79° with the line of sight; the cone's axis is also closely aligned with the long axis of the probable supernova remnant, W50. In fact, there are two regions that have been observed to emit x-rays, presumably where the jets collide with the remnant's gases and heat them to about $10^7$ K. Figure 17.27 shows the general features of this incredible system.

What is the fate of a binary x-ray system? As it reaches the endpoint of its evolution, the secondary star will end up as a white dwarf, neutron star,

---

[45] The interested reader is referred to Clark (1985) for the fascinating story of SS 433.

[46] For example, one recent measurement of SS 433 favors a 0.8 $M_\odot$ neutron star orbiting a 3.2 $M_\odot$ companion. Some astronomers have suggested that the primary may be a Wolf–Rayet star (briefly described in Section 13.3) to account for the broad stationary emission lines. Although a substantial percentage of Wolf–Rayet stars are found in binaries, these stars' own energetic winds, rather than the transfer of mass in a close binary system, seem to be responsible for removing most of their hydrogen envelopes.

## 17.5 Neutron Stars and Black Holes in Binaries

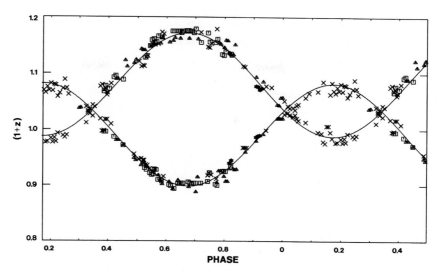

**Figure 17.26** Doppler shifts measured for the emission lines in SS 433. From Eq. (4.38), $z = 0.1$ and $z = 0.2$ correspond to speeds of 28,500 km s$^{-1}$ and 54,100 km s$^{-1}$, respectively. (Figure from Margon, Grandi, and Downes, *Ap. J.*, *241*, 306, 1980.)

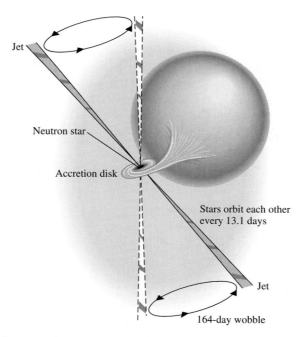

**Figure 17.27** SS 433. The axis of the cone swept out by the precessing jets makes a 79° angle with the line of sight.

or black hole. The effect on the system depends on the mass of the secondary star. In low-mass systems (LMXBs), the companion star will become a white dwarf without disturbing the circular orbit of the system. On the other hand, the higher-mass secondary in a MXRB may explode as a supernova. If more than half of the system's mass is retained (Eq. 17.31), a pair of neutron stars will circle each other in orbits that probably have been elongated by the blast. Otherwise, the supernova may disrupt the system and hurl the solitary neutron stars into space. This is consistent with observations that pulsars (like MXRBs) are concentrated near the plane of our Galaxy, and have high space velocities of 100–200 km s$^{-1}$.

The only way that a binary system containing two neutron stars can be detected is if at least one of them is a pulsar. Astronomers therefore search for cyclic variations in the measured periods of radio pulsars, analogous to the effect described here for the x-ray pulsars. Although half of all stars in the sky are actually multiple systems, however, *none* of the first one hundred pulsars discovered belonged to a binary. The first binary pulsar, PSR 1913+16, was discovered in 1974 by American astronomers Russell Hulse and Joseph Taylor using the Arecibo radio telescope. The search strategy for binary pulsars changed with the 1982 discovery by Donald Backer of UC Berkeley and his colleagues of the fastest known pulsar, PSR 1937+214. With a period of 1.558 milliseconds, this pulsar spins 642 times each second.[47] Although this astounding rotation rate seemed to indicate a young pulsar, the very small value of the period derivative ($\dot{P} = 1.051054 \times 10^{-19}$) implies a weak magnetic field ($\approx 8.6 \times 10^8$ G; see Example 15.6) and a very old pulsar. From Problem 15.22, the age of the pulsar may be estimated as $P/2\dot{P} = 235$ million years, an order of magnitude older than previously discovered pulsars.[48] Although PSR 1937+214 is an isolated pulsar, the paradox of the oldest pulsar also being the fastest quickly brought astronomers to a surprising conclusion: PSR 1937+214 must once have been a member of a low-mass x-ray binary system. (Recall that, like PSR 1937+214, LMXBs have weak magnetic fields.) Accretion from the secondary star could have spun up the neutron star to its present rapid rate. The neutron star's magnetic field may also have been rejuvenated by this process, although the details of how this might occur are not yet clear.

A likely evolutionary picture has emerged that brings the observations of binary x-ray sources and binary pulsars together. In this scenario, there are two

---

[47]Middle C on a piano has an audible frequency of 262 Hz. The pulsar's rotation frequency is more than an octave higher, between D$^{\#}$ and E!

[48]$P/2\dot{P}$ is an estimate of a pulsar's age only if the pulsar's spin has not been affected by accretion.

## 17.5 Neutron Stars and Black Holes in Binaries

classes of binary pulsars. Those with high-mass companions (neutron stars) have shorter periods and eccentric orbits, and are likely the result of the evolution of a massive x-ray binary system. (Recall that an MXRB that managed to retain more than half of its mass following the supernova of the companion star would produce such a pair of neutron stars with elongated orbits.) The other class of binary pulsars are characterized by low-mass companions (white dwarfs), longer orbital periods, and circular orbits. These are probably the descendants of low-mass x-ray binary systems.

Because the LMXBs are common in globular clusters, radio astronomers slued their telescopes toward these targets and discovered more binary and millisecond pulsars (those with periods less than approximately 10 ms). During the four years preceding 1991, 13 millisecond pulsars were discovered in 12 different globular clusters. In 1991, ten millisecond pulsars were discovered in a single globular cluster, 47 Tucanae, bringing the total number known to reside there to eleven.[49] Their periods range between 1.79 ms and 5.76 ms, six (possibly seven) being members of binary systems. The mounting statistics make it clear that most of the globular cluster pulsars are members of binaries, and most (but not all) are millisecond pulsars.[50] If these pulsars are the evolutionary product of LMXBs, then how can the absence of a white dwarf companion be explained for a significant minority of them?

An answer may be found from observations of PSR 1957+20. It is a rarity: a binary millisecond pulsar that eclipses its companion, a meager 0.025 $M_\odot$ white dwarf. However, the eclipses last for some 10% of the orbit, implying that the light is blocked by an object larger than the Sun. Significantly, the dispersion of the pulsar signal (see page 617) increases just before and after the eclipse, indicating that the white dwarf is surrounded by ionized gas. The pulsar seems to be *evaporating* its white dwarf companion with its energetic beam of photons and charged particles. Within a few million years, the white dwarf may disappear, devoured by this "black widow pulsar"; see Fig. 17.28.

A second example of the ablation of an eclipsing millisecond pulsar's companion has recently been found for PSR 1744−24A in the globular cluster Terzan 5, where the eclipses last for half of the orbital period. It is possible that some of the evaporated material may form a disk of gas and dust around the pulsar that could eventually (after a million years or so) condense and form planets around the pulsar. Or, if the evaporation of the companion star is incomplete, a planet-size remnant could be left orbiting the pulsar. Mechanisms

---

[49] 47 Tuc certainly contains many more pulsars than eleven. It is estimated that another globular cluster, Terzan 5, holds more than 100 pulsars.

[50] Conversely, most of the known millisecond pulsars have been found in globular clusters. As of January 1993, just seven millisecond pulsars have been found to reside elsewhere.

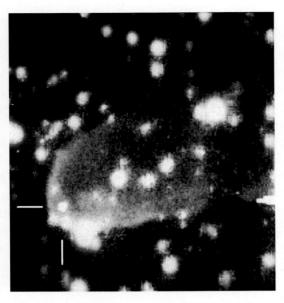

**Figure 17.28** Gas being removed by the "black widow pulsar," PSR 1957+20. The pulsar is at the intersection of the white lines. (Photo courtesy of S. Kulkarni and J. Hester, Caltech.)

such as these may be responsible for the three planets thought to be traveling in circular orbits around PSR 1257+12, some 500 pc away in the constellation Virgo. As determined from a careful analysis of pulse arrival times, the innermost planet has a mass of 0.015 $M_\oplus$ that is 0.19 AU from the pulsar, followed by a 3.4 $M_\oplus$ object that is at a distance of 0.36 AU. The outermost planet's mass is 2.8 $M_\oplus$, and it is at a distance of 0.47 AU.

As more millisecond pulsars are discovered, it should become clear whether the foregoing evolutionary picture is correct. Meanwhile, because the distances to the globular clusters are known, the radio beams of millisecond pulsars can be used to investigate the space lying between clusters containing pulsars and Earth. The dispersion of the pulsar signals have been used to map the distribution of ionized gas and magnetic fields in our Galaxy. Also, pulsars in clusters are accelerated by the cluster's mass, so measurements of the changes in the pulse periods can be used to investigate the distribution of mass in the cluster.

The best example of the use of pulsars as natural laboratories is the first binary pulsar discovered, the Hulse–Taylor pulsar, PSR 1913+16. It consists of two neutron stars whose close orbit is just a little larger than the Sun's diameter. A 20-year study of this system has led to a confirmation of one of

## 17.5 Neutron Stars and Black Holes in Binaries

the most dramatic predictions of Einstein's general theory of relativity: the existence of **gravitational waves**. According to general relativity, mass acts on spacetime, telling it how to curve. If the distribution of a system's mass varies, the resulting changes in the surrounding spacetime curvature may propagate outward as a gravitational wave, carrying energy and angular momentum away from the system.[51] When applied to a close binary system, general relativity shows that the emission of gravitational radiation will cause the stars to spiral together. If the orbital period is under about 14 hours, the loss of energy via gravitational waves governs the subsequent evolution of a system with solar-mass components. For example, as a white dwarf and a neutron star spiral closer together, the white dwarf may break up and donate some of its mass and angular momentum to its companion. The result could be an isolated millisecond pulsar.

Nearly everything is known about the Hulse–Taylor system with incredible precision. The pulse frequency on January 14, 1986, was

$$\omega = 16.940539184253 \text{ Hz},$$

when the frequency was changing at a rate of

$$\dot{\omega} \equiv d\omega/dt = -2.47583 \times 10^{-15} \text{ Hz s}^{-1}.$$

In both $\omega$ and $\dot{\omega}$, there is an uncertainty of 1 in the last digit. The system consists of two neutron stars with masses of $M_1 = 1.4410 \pm 0.0005$ M$_\odot$ (the pulsar) and $M_2 = 1.3874 \pm 0.0005$ M$_\odot$ (the companion) in an orbit with an eccentricity of $e = 0.6171308 \pm 0.0000004$. The period of the binary orbit (1990 value) is $P_{\text{orb}} = 27906.9808968 \pm 0.0000016$ s (or about 7.75 hours). This binary system provides an ideal natural laboratory for testing Einstein's theory of gravity. For example, recall from Section 16.1 that as Mercury passes through the curved spacetime near the Sun, the position of perihelion in its orbit is shifted by 43″ per century; see Fig. 16.1. For PSR 1913+16, general relativity predicts is a similar shift in the point of periastron, where the two neutron stars are nearest each other. The theoretical value is in good agreement with the measurement of 4.22626° per year (35,000 times Mercury's rate of shift). This effect on the orbit is cumulative; with every orbit, the pulsar arrives later and later at the point of periastron. Figure 17.29 shows the incredible agreement between theoretical and observed values of the accumulating time delay.

---

[51]The spherically symmetric collapse of a star will not produce gravitational waves; there must be a departure from spherical symmetry.

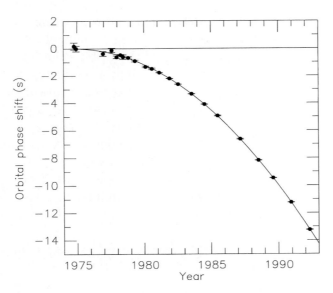

**Figure 17.29** Observations (dots) of the delay in the time of periastron for PSR 1913+16, compared with the prediction of the theory of general relativity (solid line). (Courtesy of J. Taylor, Princeton University.)

The most spectacular aspect of the studies of PSR 1913+16 is the confirmation of the existence of gravitational radiation. As the two neutron stars move in their orbits, gravitational waves carry energy away from the system and the orbital period decreases. According to general relativity, the rate at which the orbital period changes due to the emission of gravitational quadrupole radiation[52] is

$$\dot{P}_{\text{orb}} = \frac{dP_{\text{orb}}}{dt} = -\frac{96}{5} \frac{G^3 M^2 \mu}{c^5} \left(\frac{4\pi^2}{GM}\right)^{4/3} \frac{f(e)}{P_{\text{orb}}^{5/3}} , \qquad (17.39)$$

where

$$M = M_1 + M_2$$

$$\mu = \frac{M_1 M_2}{M_1 + M_2}$$

---

[52]The term *quadrupole* describes the geometry of the emitted gravitational radiation, just as electric dipole radiation describes the electromagnetic radiation emitted by two electric charges moving around each other.

and $f(e)$ describes the effect of the eccentricity of the orbit,

$$f(e) = \left(1 + \frac{73}{24}e^2 + \frac{37}{96}e^4\right)\left(1 - e^2\right)^{-7/2}.$$

(There are also higher-order correction terms that have been neglected here.) Inserting the preceding values for the masses and eccentricity, the theoretical rate of orbital period decay is calculated to be $\dot{P}_{\text{orb}} = -2.40259 \times 10^{-12}$, which agrees with the measured value to within 0.3%. In presenting the results of an earlier calculation of the orbital period decay in 1984, Joel Weisberg and Joseph Taylor wrote: "It now seems an inescapable conclusion that gravitational radiation exists as predicted by the general relativistic quadrupole formula." Astronomers are fortunate to have caught this superb natural laboratory before it disappears. As the separation of the neutron stars shrinks by about 3 mm per orbit, the system will coalesce some 300 million years in the future.

## Suggested Readings

### General

Backer, Donald C., and Kulkarni, Shrinivas R., "A New Class of Pulsars," *Physics Today*, March 1990.

Cannizzo, John K., and Kaitchuck, Ronald H., "Accretion Disks in Interacting Binary Stars," *Scientific American*, January 1992.

Clark, David H., *The Quest for SS 433*, Viking Penguin Inc., New York, 1985.

Kleppner, Daniel, "The Gem of General Relativity," *Physics Today*, April 1993.

Margon, Bruce, "Exploring the High-Energy Universe," *Sky and Telescope*, December 1991.

Piran, Tsvi, "Binary Neutron Stars," *Scientific American*, May 1995.

Starrfield, Sumner, and Shore, Steven N., "Nova Cygni 1992: Nova of the Century," *Sky and Telescope*, February 1994.

van den Heuvel, Edward P. J., and van Paradijs, Jan, "X-ray Binaries," *Scientific American*, November 1993.

Wheeler, J. Craig, and Harkness, Robert P., "Helium-rich Supernovas," *Scientific American*, November 1987.

TECHNICAL

Backer, D. C., et al., "A Millisecond Pulsar," *Nature, 300*, 615, 1982.

Branch, David, and Tammann, G. A., "Type Ia Supernovae as Standard Candles," *Annual Review of Astronomy and Astrophysics, 30*, 359, 1992.

Cowley, Anne P., "Evidence for Black Holes in Stellar Binary Systems," *Annual Review of Astronomy and Astrophysics, 30*, 287, 1992.

Damour, Thibault, and Taylor, J. H., "On the Orbital Period Change of the Binary Pulsar PSR 1913+16," *The Astrophysical Journal, 366*, 501, 1991.

Frank, Juhan, King, Andrew, and Raine, Derek, *Accretion Power in Astrophysics*, Second Edition, Cambridge University Press, Cambridge, 1992.

Gallagher, J. S., and Starrfield, S., "Theory and Observations of Classical Novae," *Annual Review of Astronomy and Astrophysics, 16*, 171, 1978.

Gehrz, Robert D., "The Infrared Temporal Development of Classical Novae," *Annual Review of Astronomy and Astrophysics, 26*, 377, 1988.

Horne, Keith, and Cook, M. C., "*UBV* Images of the Z Cha Accretion Disc in Outburst," *Monthly Notices of the Royal Astronomical Society, 214*, 307, 1985.

Iben, Icko, Jr., "The Life and Times of an Intermediate Mass Star—in Isolation/in a Close Binary," *Quarterly Journal of the Royal Astronomical Society, 26*, 1, 1985.

Iben, Icko, Jr., "Single and Binary Star Evolution," *The Astrophysical Journal Supplement, 76*, 55, 1991.

Joss, Paul C., and Rappaport, Saul A., "Neutron Stars in Interacting Binary Systems," *Annual Review of Astronomy and Astrophysics, 22*, 537, 1984.

Lyne, A. G., and Graham-Smith, F., *Pulsar Astronomy*, Cambridge University Press, Cambridge, 1990.

Margon, Bruce, "Observations of SS 433," *Annual Review of Astronomy and Astrophysics, 22*, 507, 1984.

Petschek, Albert G., *Supernovae*, Springer-Verlag, New York, 1990.

Pringle, J. E., "Accretion Discs in Astrophysics," *Annual Review of Astronomy and Astrophysics, 19*, 137, 1981.

Pringle, J. E., and Wade, R. A. (eds.), *Interacting Binary Stars*, Cambridge University Press, Cambridge, 1985.

Verbunt, Frank, "Origin and Evolution of X-ray Binaries and Binary Radio Pulsars," *Annual Review of Astronomy and Astrophysics, 31*, 93, 1993.

Weisberg, J. M., and Taylor, J. H., "Observations of Post-Newtonian Timing Effects in the Binary Pulsar PSR 1913+16," *Physical Review Letters*, *52*, 1348, 1984.

## Problems

17.1 Use the ideal gas law to argue that, in a close binary system, the temperature of a star's photosphere is approximately constant along an equipotential surface. What effect could the proximity of the other star have on your argument?

17.2 Each of the Lagrange points $L_4$ and $L_5$ forms an equilateral triangle with masses $M_1$ and $M_2$ in Fig. 17.3. Use this to confirm the value of the effective gravitational potential at $L_4$ and $L_5$ given in the figure caption.

17.3 (a) Consider a gas of density $\rho$ moving with velocity $v$ across an area $A$ perpendicular to the flow of the gas. Show that the rate at which mass crosses the area is given by Eq. (17.10).

(b) Derive Eq. (17.11) for the radius of the intersection of two identical overlapping spheres, when $d \ll R$.

17.4 **Computer Problem**

(a) Use the STATSTAR model data on page 311 and Eq. (17.12) to make a graph of $\log_{10} \dot{M}$ (vertical axis) vs. $\log_{10} d$ (horizontal axis). Use the slope of your graph to find how the mass transfer rate, $\dot{M}$, depends on $d$.

(b) Use Eqs. (H.1) and (H.2) to show that $\dot{M} \propto d^{4.75}$ near the surface, and so verify that the mass transfer rate increases rapidly with the overlap distance $d$ of two stars. Note that your answer to part (a) will be slightly different from this because of the density-dependence of TOG_BF (the ratio of the guillotine factor to the gaunt factor) calculated in the STATSTAR's equation-of-state subroutine EOS.

17.5 Use Eq. (17.16) to show that the maximum disk temperature is found at $r = (49/36)R$, and is equal to $T_{\max} = 0.488 T_{\text{disk}}$.

17.6 Integrate Eq. (17.13) for the ring luminosity from $r = R$ to $r = \infty$ [with Eq. (17.16) for the disk temperature]. Does your answer agree with Eq. (17.20) for the disk luminosity?

17.7 Consider an "average" dwarf nova that has a mass transfer rate of

$$\dot{M} = 10^{16.5} \text{ g s}^{-1} = 5 \times 10^{-10} \text{ M}_\odot \text{ yr}^{-1}$$

during an outburst that lasts for 10 days. Estimate the total energy released and the absolute magnitude of the dwarf nova during the outburst. Use values for Z Cha's white dwarf from Example 17.4.

17.8 Assume that the absolute bolometric magnitude of a dwarf nova during quiescence is 7.5 and that it brightens by three magnitudes during outburst. Using values for Z Cha, estimate the rate of mass transfer through the accretion disk.

17.9 When the accretion disk in a cataclysmic variable is eclipsed by the secondary star, the blueshifted emission line is the first to disappear at the beginning of the eclipse, and the redshifted emission line is the last to reappear when the eclipse ends. What does this have to say about the directions of rotation of the binary system and the accretion disk?

17.10 (a) In a close binary system where angular momentum is conserved, show that the change in orbital period produced by mass transfer is given by

$$\frac{1}{P}\frac{dP}{dt} = 3\dot{M}_1 \frac{M_1 - M_2}{M_1 M_2}.$$

(b) U Cephei (an Algol system) has an orbital period of 2.49 days that has increased by about 20 s in the past 100 years. The masses of the two stars are $M_1 = 2.9$ M$_\odot$ and $M_2 = 1.4$ M$_\odot$. Assuming that this change is due to the transfer of mass between the two stars in this Algol system, estimate the mass transfer rate. Which of these stars is gaining mass?

17.11 Algol (the "demon" star, in Arabic) is a semidetached binary. Every 2.87 days, its brilliance is reduced by more than half as it undergoes a deep eclipse, its apparent magnitude dimming from 2.1 to 3.4. The system consists of a B8 main-sequence star and a late-type (G or K) subgiant; the deep eclipses occur when the larger, cooler star (the subgiant) moves in front of its smaller, brighter companion. The "Algol paradox," which troubled astronomers in the first half of the twentieth century, is that according to the ideas of stellar evolution discussed in Section 10.6, the more massive B8 star should have been the first to evolve off the main sequence. What is your solution to this paradox? (The Algol system actually contains a third star that orbits the other two every 1.86 years, but this has nothing to do with the solution to the Algol paradox.)

Algol may be easily found in the constellation Perseus (the Hero, who rescued Andromeda in Greek mythology). *Sky and Telescope* provides a monthly listing of the minima of Algol, which last about 2 hours.

17.12 Consider a $10^{-4}$ $M_\odot$ layer of hydrogen on the surface of a white dwarf. If this layer were completely fused into helium, how long would the resulting nova last (assuming a luminosity equal to the Eddington luminosity)? What does this say about the amount of hydrogen that actually undergoes fusion during a nova outburst?

17.13 Consider a layer of $10^{-4}$ $M_\odot$ of hydrogen on the surface of a white dwarf. Compare the gravitational binding energy before the nova outburst to the kinetic energy of the ejected layer when it has traveled far from the white dwarf and has a speed of 1000 km s$^{-1}$.

17.14 In this problem, you will examine the fireball expansion phase of a nova shell. Suppose that mass is ejected by a nova at a constant rate of $\dot{M}_{\text{eject}}$ and at a constant speed $v$.

(a) Show that the density of the expanding shell at a distance $r$ is $\rho = \dot{M}_{\text{eject}}/4\pi r^2 v$.

(b) Let the mean opacity, $\overline{\kappa}$, of the expanding gases be a constant. Suppose that at some time $t = 0$, the outer radius of the shell was $R$, and the radius of the photosphere, where $\tau = 2/3$, was $R_o$. Show that
$$\frac{1}{R} = \frac{1}{R_o} - \frac{1}{R_\infty},$$
where
$$R_\infty \equiv \frac{3\overline{\kappa}\dot{M}_{\text{eject}}}{8\pi v}.$$

(The reason for the "$\infty$" subscript will become clear below.)

(c) At some later time $t$, the radius of the shell will be $R + vt$ and the radius of the photosphere will be $R(t)$. Show that

$$\frac{1}{R + vt} = \frac{1}{R(t)} - \frac{1}{R_\infty}.$$

(d) Combine the results from parts (b) and (c) to write

$$R(t) = R_\circ + \frac{vt(1 - R_\circ/R_\infty)^2}{1 + (vt/R_\infty)(1 - R_\circ/R_\infty)},$$

(e) Argue that terms containing $R_\circ/R_\infty$ are very small and can be ignored, and so obtain

$$R(t) \simeq \frac{vt}{1 + vt/R_\infty}.$$

(f) Show that initially the fireball's photosphere expands linearly with time, and then approaches the limiting value of $R_\infty$, in agreement with Eq. (17.27).

(g) Using the data given in the text following Eq. (17.28), make a graph of the $R(t)$ vs. $t$ for the five days after nova explodes. The "knee" in the graphs marks the end of the linear expansion period; estimate when this occurs. How does this compare with the duration of the optically thick fireball phase of the nova?

17.15 Use Eq. (17.28) to estimate the photospheric temperature of a nova fireball, adopting the Eddington luminosity for the luminosity of the fireball.

17.16 Assuming that the hydrostatic burning phase of a nova lasts for 100 days, find the (constant) rate at which mass is ejected, $\dot{M}_{\text{eject}}$, for a surface layer of $10^{-4}$ $M_\odot$.

17.17 If the linear decline of a supernova light curve is powered by the radioactive decay of the ejecta, find the rate of decline (in mag d$^{-1}$) produced by the decay of $^{56}_{27}\text{Co} \rightarrow ^{56}_{26}\text{Fe}$, with a half-life of 77.7 days.

17.18 For each gram of a carbon–oxygen composition (30% $^{12}_{6}\text{C}$) that is burned to produce iron, $7.3 \times 10^{17}$ ergs of energy is released. Assuming an initial 1.38 $M_\odot$ white dwarf with a radius of 1600 km, how much iron would have to be produced to cause the star to be gravitationally unbound? How much additional iron would have to be manufactured to produce a Type Ia supernova with an average ejecta speed of 5000 km s$^{-1}$? Take the gravitational potential energy to be $-5.1 \times 10^{50}$ ergs (for a realistic white dwarf model), and express your answers in units of M$_\odot$.

17.19 Use Eqs. (7.4), (17.29), and (17.30) to derive Eq. (17.31), the condition for a supernova to disrupt a binary system.

17.20 (a) Show that the Alfvèn radius is given by Eq. (17.35).

(b) Show that $\dot{P}/P$ for the spin-up of an x-ray pulsar is given by Eq. (17.38).

17.21 Find the value of the magnetic field for which the Alfvèn radius is equal to the radius of the white dwarf found in Example 17.3. Do the same thing for the neutron star used in that example.

17.22 Estimate the lifetime of a binary x-ray system using the information in Example 17.3. Take the lifetime to be the time required to transfer a mass of 1 M$_\odot$.

17.23 The x-ray pulsar 4U0115+63 has a period of 3.61 s and an x-ray luminosity of about $L_x = 3.8 \times 10^{36}$ ergs s$^{-1}$. Assuming that it is a 1.4 M$_\odot$ neutron star with a radius of 10 km and a surface magnetic field of $10^{12}$ G, find its mass transfer rate, $\dot{M}$, and the value of $\dot{P}/P$. Repeat these calculations assuming that this object is a 0.85 M$_\odot$ white dwarf with a radius of $6.6 \times 10^8$ cm and a surface magnetic field of $10^7$ G. For which of these models do you obtain the better agreement with the measured value of $\dot{P}/P = -3.2 \times 10^{-5}$ yr$^{-1}$?

17.24 (a) Use Eq. (17.21) to show that the spin-up rate can be written as

$$\log_{10}\left(-\frac{\dot{P}}{P}\right) = \log_{10}\left(PL_{\text{acc}}^{6/7}\right) + \log_{10}\left[\frac{\sqrt{\alpha}}{2\pi I}\left(\frac{B_s^2 R^{12}}{\sqrt{2}\,G^3 M^3}\right)^{1/7}\right].$$

The term on the left and the first term on the right consist of quantities that can be measured observationally. The second term on the right depends on the specific model (neutron star or white dwarf) of the x-ray pulsar.

(b) Make a graph of $\log_{10}(-\dot{P}/P)$ (vertical axis) vs. $\log_{10}\left(PL_{\text{acc}}^{6/7}\right)$ (horizontal axis). Use the values from Example 17.5 to plot two lines, one for a neutron star and one for a white dwarf. Let $\log_{10}\left(PL_{\text{acc}}^{6/7}\right)$ run from 31 to 35.

| System | $P$ (s) | $L_{\text{acc}}$ ($10^{37}$ ergs s$^{-1}$) | $-\dot{P}/P$ (yr$^{-1}$) |
|---|---|---|---|
| SMC X-1 | 0.714 | 50 | $7.1 \times 10^{-4}$ |
| Her X-1 | 1.24 | 1 | $2.9 \times 10^{-6}$ |
| Cen X-3 | 4.84 | 5 | $2.8 \times 10^{-4}$ |
| A0535+26 | 104 | 6 | $3.5 \times 10^{-2}$ |
| GX301−2 | 696 | 0.3 | $7.0 \times 10^{-3}$ |
| 4U0352+30 | 835 | 0.0004 | $1.8 \times 10^{-4}$ |

**Table 17.1** X-ray Pulsar Data, for Problem 17.24. (Data from Rappaport and Joss, *Nature*, *266*, 683, 1977, and Joss and Rappaport, *Annu. Rev. Astron. Astrophys.*, *22*, 537, 1984.)

(c) Use the data in Table 17.1 to plot the positions of six binary x-ray pulsars on your graph. (You will have to convert $-\dot{P}/P$ into units of s$^{-1}$.)

(d) Which model of a binary x-ray pulsar is in better agreement with the data? Comment on the position of Her X-1 on your graph.

17.25 (a) Consider an x-ray burster that releases $10^{39}$ ergs in 5 seconds. If the shape of its peak spectrum is that of a $2 \times 10^7$ K blackbody, estimate the radius of the underlying neutron star.

(b) In Problem 16.9 you showed that using the Stefan–Boltzmann formula to find the radius of a compact blackbody can lead to an overestimate of its radius. Use Eq. (16.28) to find a more accurate value for the radius of the neutron star.

17.26 Make a scale drawing of the SMC X-1 binary pulsar system, including the size of the secondary star. Assuming that the primary is a 1.4 $M_\odot$ neutron star, locate the system's center of mass and its inner Lagrangian point, $L_1$. (You can omit the accretion disk.)

17.27 The relativistic ($v/c = 0.26$) jets coming from the accretion disk in SS 433 sweep out cones in space as the disk precesses. The central axis of these cones makes an angle of 79° with the line of sight, and the half-angle of each cone is 20°. This means that at some point in the precession cycle, the jets are moving perpendicular to the line of sight. Yet, from Fig. 17.26, the radial velocities obtained from the Doppler-shifted spectral lines do *not* cross at zero radial velocity, but at $\sim 10{,}000$ km s$^{-1}$. Use Eq. (4.32) to explain this discrepancy in terms of a transverse Doppler shift. (You can ignore the speed of the SS 433 binary system itself, which is only about 70 km s$^{-1}$.)

17.28 The distance to SS 433 is about 5.5 kpc, and the angular separation of SS 433 and the x-ray emitting regions (where the jets interact with the gases of W50) extends as far as $44'$. Estimate a lower limit for the amount of time the jets have been active.

17.29 **Computer Problem** Use Eq. (17.16) and Wien's law to make two log-log graphs: (1) the disk temperature, $T(r)$, and (2) the peak wavelength $[\lambda_{\max}(r)]$ of the blackbody spectrum, for the accretion disk around the black hole A0620−00 as a function of the radial position $r$. For this system, the mass of the black hole is $3.82$ M$_\odot$, the mass of the secondary star $0.36$ M$_\odot$, and the period of the orbit is $0.3226$ day. Assume $\dot{M} = 10^{17}$ g s$^{-1}$ (about $10^{-9}$ M$_\odot$ yr$^{-1}$), and use the Schwarzschild radius, $R_S$ (Eq. 16.23), for the radius of the black hole. (On your graph, plot $r/R_S$ rather than $r$.) For a nonrotating black hole, the last stable orbit for a massive particle is at $3R_S$, so use this as the inner edge of the disk. Let the outer edge of the disk be determined by Kepler's third law along with Eqs. (17.22) and (17.23). On your log-log graph of $\lambda_{\max}$ vs. $r/R_S$, identify the regions of the disk that emit x-ray, ultraviolet, visible, and infrared radiation.

17.30 The smallest period derivative discovered to date is $\dot{P} = 3 \times 10^{-20}$ for the pulsar PSR 1953+29, which has a period of 6.133 ms. Use Problem 15.22 to estimate the age of this millisecond pulsar, assuming that no accretion has occurred to alter the pulsar's spin. Also, use Eq. (15.30) to estimate the value of this pulsar's magnetic field.

17.31 Integrate Eq. (17.38) for the spin-up of an x-ray pulsar to estimate the time for a millisecond pulsar to be spun up to a final period of 1 ms from an initial period of 100 s (longer than the longest pulsar period of 4.3 s, and within the range of x-ray pulsar periods). Assume a 1.4 M$_\odot$ neutron star with a radius of 10 km. Use $B_s = 10^8$ G for the magnetic field, and $\dot{M} = 10^{17}$ g s$^{-1}$ for the mass transfer rate. How much mass is transferred in that time (in grams and in solar masses)?

17.32 The three planets thought to be orbiting PSR 1257+12 have orbital periods of 25.34 d, 66.54 d, and 98.22 d. Verify that these objects obey Kepler's third law.

17.33 (a) Use Kepler's third law to find the semimajor axis of the orbit of the binary pulsar PSR 1913+16.

(b) What is the change in the semimajor axis after one orbital period of the pulsar?

# Appendix A

# ASTRONOMICAL AND PHYSICAL CONSTANTS

### Astronomical Constants

| | | |
|---|---|---|
| Solar mass | $1\ M_\odot$ | $= 1.989 \times 10^{33}$ g |
| Solar luminosity | $1\ L_\odot$ | $= 3.826 \times 10^{33}$ ergs s$^{-1}$ |
| Solar radius | $1\ R_\odot$ | $= 6.9599 \times 10^{10}$ cm ~ 700,000 km |
| Solar effective temperature | $T_\odot$ | $= 5770$ K |
| Earth mass | $1\ M_\oplus$ | $= 5.974 \times 10^{27}$ g ≈ $3 \times 10^{-6} M_\odot$ |
| Earth radius | $1\ R_\oplus$ | $= 6.378 \times 10^{8}$ cm ≈ 6400 km |
| Light year | $1$ ly | $= 9.4605 \times 10^{17}$ cm |
| Parsec | $1$ pc | $= 3.0857 \times 10^{18}$ cm |
| | | $= 3.2616$ ly |
| Astronomical unit | $1$ AU | $= 1.4960 \times 10^{13}$ cm |
| Sidereal day | | $= 23^\mathrm{h}\ 56^\mathrm{m}\ 04.09054^\mathrm{s}$ |
| Solar day | | $= 86400$ s ~ $10^5$ s |
| Sidereal year | | $= 3.155815 \times 10^{7}$ s |
| Tropical year | | $= 3.155693 \times 10^{7}$ s |

1 radian = $2.06 \times 10^5$ arcsecs.

## Physical Constants

| | | |
|---|---|---|
| Gravitational constant | $G$ | $= 6.67259 \times 10^{-8}$ dyne cm$^2$ g$^{-2}$ |
| Speed of light (exact) | $c$ | $= 2.99792458 \times 10^{10}$ cm s$^{-1}$ |
| Planck's constant | $h$ | $= 6.6260755 \times 10^{-27}$ erg s |
| | $\hbar$ | $\equiv h/2\pi$ |
| | | $= 1.05457266 \times 10^{-27}$ erg s |
| Boltzmann's constant | $k$ | $= 1.380658 \times 10^{-16}$ erg K$^{-1}$ |
| Stefan–Boltzmann constant | $\sigma$ | $= 5.67051 \times 10^{-5}$ erg cm$^{-2}$ s$^{-1}$ K$^{-4}$ |
| Radiation constant | $a$ | $= 4\sigma/c$ |
| | | $= 7.56591 \times 10^{-15}$ erg cm$^{-3}$ K$^{-4}$ |
| Proton mass | $m_p$ | $= 1.6726231 \times 10^{-24}$ g |
| Neutron mass | $m_n$ | $= 1.674929 \times 10^{-24}$ g |
| Electron mass | $m_e$ | $= 9.1093897 \times 10^{-28}$ g |
| Hydrogen mass | $m_H$ | $= 1.673534 \times 10^{-24}$ g |
| Atomic mass unit | 1 u | $= 1.6605402 \times 10^{-24}$ g |
| | | $= 931.49432$ MeV/$c^2$ |
| Coulomb law constant (cgs) | $k_C$ | $\equiv 1$ |
| (SI) | | $= 8.9875518 \times 10^9$ N m$^2$ C$^{-2}$ |
| Electric charge (cgs) | $e$ | $= 4.803206 \times 10^{-10}$ esu |
| (SI) | | $= 1.60217733 \times 10^{-19}$ C |
| Electron volt | 1 eV | $= 1.60217733 \times 10^{-12}$ erg |
| Avagadro's number | $N_A$ | $= 6.0221367 \times 10^{23}$ mole$^{-1}$ |
| Gas constant | $R$ | $= 8.314510 \times 10^7$ ergs mole$^{-1}$ K$^{-1}$ |
| Bohr radius | $a_o$ | $= \hbar^2/m_e e^2$ |
| | | $= 5.29177249 \times 10^{-9}$ cm |
| Rydberg constant | $R_H$ | $= \mu e^4/4\pi\hbar^3 c$ |
| | | $= 1.09677585 \times 10^5$ cm$^{-1}$ |

# Suggested Readings

TECHNICAL

Cohen, E. Richard, and Taylor, Barry N., "The 1986 Adjustment of the Fundamental Physical Constants," *Reviews of Modern Physics*, 59, 1121, 1987.

Lang, Kenneth R., *Astrophysical Data: Planets and Stars*, Springer-Verlag, New York, 1992.

# Appendix B

# SOLAR SYSTEM DATA

## Planetary Physical Data

| Planet | Mass ($M_\oplus$) | Equatorial Radius ($R_\oplus$) | Average Density (g cm$^{-3}$) | Sidereal Rotation Period (d) | Oblateness | Albedo |
|---|---|---|---|---|---|---|
| Mercury | 0.0553 | 0.382 | 5.43 | 58.65 | 0.0 | 0.06 |
| Venus | 0.8150 | 0.949 | 5.25 | 243.01 | 0.0 | 0.77 |
| Earth | 1.0000 | 1.000 | 5.52 | 0.997 | 0.0034 | 0.30 |
| Mars | 0.1074 | 0.533 | 3.93 | 1.026 | 0.0052 | 0.15 |
| Jupiter | 317.894 | 11.19 | 1.33 | 0.414 | 0.0648 | 0.51 |
| Saturn | 95.184 | 9.46 | 0.71 | 0.444 | 0.1076 | 0.50 |
| Uranus | 14.537 | 4.01 | 1.24 | 0.718 | 0.030 | 0.66 |
| Neptune | 17.132 | 3.81 | 1.67 | 0.671 | 0.022 | 0.62 |
| Pluto | 0.0022 | 0.182 | 2.1 | 6.387 | 0.0 | 0.6 |

## Planetary Orbital Data

| Planet | Semimajor Axis (AU) | Orbital Eccentricity | Sidereal Orbital Period (yr) | Orbital Inclination to Ecliptic (°) | Equatorial Inclination to Orbit (°) |
|---|---|---|---|---|---|
| Mercury | 0.3871 | 0.2056 | 0.2408 | 7.004 | 7.0 |
| Venus | 0.7233 | 0.0068 | 0.6152 | 3.394 | 177.4 |
| Earth | 1.0000 | 0.0167 | 1.0000 | 0.000 | 23.45 |
| Mars | 1.5237 | 0.0934 | 1.8809 | 1.850 | 23.98 |
| Jupiter | 5.2028 | 0.0483 | 11.8622 | 1.308 | 3.08 |
| Saturn | 9.5388 | 0.0560 | 29.4577 | 2.488 | 26.73 |
| Uranus | 19.1914 | 0.0461 | 84.0139 | 0.774 | 97.92 |
| Neptune | 30.0611 | 0.0097 | 164.793 | 1.774 | 28.8 |
| Pluto | 39.5294 | 0.2482 | 248.54 | 17.148 | 122.46 |

## Data of Selected Major Satellites

| Satellite | Parent Planet | Mass ($10^{25}$ g) | Radius ($10^3$ km) | Density (g cm$^{-3}$) | Orbital Period (d) | Orbital Distance ($10^3$ km) |
|---|---|---|---|---|---|---|
| Moon | Earth | 7.35 | 1.738 | 3.34 | 27.322 | 384.4 |
| Io | Jupiter | 8.92 | 1.815 | 3.55 | 1.769 | 421.6 |
| Europa | Jupiter | 4.87 | 1.569 | 3.04 | 3.551 | 670.9 |
| Ganymede | Jupiter | 14.9 | 2.631 | 1.93 | 7.155 | 1070 |
| Callisto | Jupiter | 10.8 | 2.400 | 1.83 | 16.689 | 1880 |
| Titan | Saturn | 13.5 | 2.575 | 1.88 | 15.945 | 1222 |
| Triton | Neptune | 2.14 | 1.355 | 2.05 | 5.877 | 354.8 |

# Appendix C

# THE CONSTELLATIONS

| Latin Name<br>Translation | Genitive | Abbrev. | R. A.<br>h | Dec.<br>° |
|---|---|---|---|---|
| Andromeda<br>  Princess of Ethiopia | Andromedae | And | 1 | +40 |
| Antlia<br>  Air Pump | Antliae | Ant | 10 | −35 |
| Apus<br>  Bird of Paradise | Apodis | Aps | 16 | −75 |
| Aquarius<br>  Water Bearer | Aquarii | Aqr | 23 | −15 |
| Aquila<br>  Eagle | Aquilae | Aql | 20 | + 5 |
| Ara<br>  Altar | Arae | Ara | 17 | −55 |
| Aries<br>  Ram | Arietis | Ari | 3 | +20 |
| Auriga<br>  Charioteer | Aurigae | Aur | 6 | +40 |
| Boötes<br>  Herdsman | Boötis | Boo | 15 | +30 |
| Caelum<br>  Chisel | Caeli | Cae | 5 | −40 |
| Camelopardalis<br>  Giraffe | Camelopardis | Cam | 6 | +70 |
| Cancer<br>  Crab | Cancri | Cnc | 9 | +20 |

| Latin Name<br>Translation | Genitive | Abbrev. | R. A.<br>h | Dec.<br>° |
|---|---|---|---|---|
| Canes Venatici<br>  Hunting Dogs | Canum Venaticorum | CVn | 13 | +40 |
| Canis Major<br>  Big Dog | Canis Majoris | CMa | 7 | −20 |
| Canis Minor<br>  Little Dog | Canis Minoris | CMi | 8 | + 5 |
| Capricornus<br>  Goat | Capricorni | Cap | 21 | −20 |
| Carina<br>  Ship's Keel | Carinae | Car | 9 | −60 |
| Cassiopeia<br>  Queen of Ethiopia | Cassiopeiae | Cas | 1 | +60 |
| Centaurus<br>  Centaur | Centauri | Cen | 13 | −50 |
| Cepheus<br>  King of Ethiopia | Cephei | Cep | 22 | +70 |
| Cetus<br>  Sea Monster (whale) | Ceti | Cet | 2 | −10 |
| Chamaeleon<br>  Chameleon | Chamaeleontis | Cha | 11 | −80 |
| Circinus<br>  Compass | Circini | Cir | 15 | −60 |
| Columba<br>  Dove | Columbae | Col | 6 | −35 |
| Coma Berenices<br>  Berenice's Hair | Comae Berenices | Com | 13 | +20 |
| Corona Australis<br>  Southern Crown | Coronae Australis | CrA | 19 | −40 |
| Corona Borealis<br>  Northern Crown | Coronae Borealis | CrB | 16 | +30 |
| Corvus<br>  Crow | Corvi | Crv | 12 | −20 |
| Crater<br>  Cup | Crateris | Crt | 11 | −15 |
| Crux<br>  Southern Cross | Crucis | Cru | 12 | −60 |

# Appendix C  The Constellations

| Latin Name<br>Translation | Genitive | Abbrev. | R. A.<br>h | Dec.<br>° |
|---|---|---|---|---|
| Cygnus<br>  Swan | Cygni | Cyg | 21 | +40 |
| Delphinus<br>  Dolphin, Porpoise | Delphini | Del | 21 | +10 |
| Dorado<br>  Swordfish | Doradus | Dor | 5 | −65 |
| Draco<br>  Dragon | Draconis | Dra | 17 | +65 |
| Equuleus<br>  Little Horse | Equulei | Equ | 21 | +10 |
| Eridanus<br>  River Eridanus | Eridani | Eri | 3 | −20 |
| Fornax<br>  Furnace | Fornacis | For | 3 | −30 |
| Gemini<br>  Twins | Geminorum | Gem | 7 | +20 |
| Grus<br>  Crane | Gruis | Gru | 22 | −45 |
| Hercules<br>  Son of Zeus | Herculis | Her | 17 | +30 |
| Horologium<br>  Clock | Horologii | Hor | 3 | −60 |
| Hydra<br>  Water Snake | Hydrae | Hya | 10 | −20 |
| Hydrus<br>  Sea Serpent | Hydri | Hyi | 2 | −75 |
| Indus<br>  Indian | Indi | Ind | 21 | −55 |
| Lacerta<br>  Lizard | Lacertae | Lac | 22 | +45 |
| Leo<br>  Lion | Leonis | Leo | 11 | +15 |
| Leo Minor<br>  Little Lion | Leonis Minoris | LMi | 10 | +35 |
| Lepus<br>  Hare | Leporis | Lep | 6 | −20 |

| Latin Name<br>Translation | Genitive | Abbrev. | R. A.<br>h | Dec.<br>° |
|---|---|---|---|---|
| Libra<br>  Balance, Scales | Librae | Lib | 15 | −15 |
| Lupus<br>  Wolf | Lupi | Lup | 15 | −45 |
| Lynx<br>  Lynx | Lyncis | Lyn | 8 | +45 |
| Lyra<br>  Lyre, Harp | Lyrae | Lyr | 19 | +40 |
| Mensa<br>  Table, Mountain | Mensae | Men | 5 | −80 |
| Microscopium<br>  Microscope | Microscopii | Mic | 21 | −35 |
| Monoceros<br>  Unicorn | Monocerotis | Mon | 7 | − 5 |
| Musca<br>  Fly | Muscae | Mus | 12 | −70 |
| Norma<br>  Square, Level | Normae | Nor | 16 | −50 |
| Octans<br>  Octant | Octantis | Oct | 22 | −85 |
| Ophiuchus<br>  Serpent-bearer | Ophiuchi | Oph | 17 | 0 |
| Orion<br>  Hunter | Orionis | Ori | 5 | + 5 |
| Pavo<br>  Peacock | Pavonis | Pav | 20 | −65 |
| Pegasus<br>  Winged Horse | Pegasi | Peg | 22 | +20 |
| Perseus<br>  Rescuer of Andromeda | Persei | Per | 3 | +45 |
| Phoenix<br>  Phoenix | Phoenicis | Phe | 1 | −50 |
| Pictor<br>  Painter, Easel | Pictoris | Pic | 6 | −55 |
| Pisces<br>  Fish | Piscium | Psc | 1 | +15 |

# Appendix C  The Constellations

| Latin Name / Translation | Genitive | Abbrev. | R. A. h | Dec. ° |
|---|---|---|---|---|
| Piscis Austrinus / Southern Fish | Piscis Austrini | PsA | 22 | −30 |
| Puppis / Ship's Stern | Puppis | Pup | 8 | −40 |
| Pyxis / Ship's Compass | Pyxidis | Pyx | 9 | −30 |
| Reticulum / Net | Reticuli | Ret | 4 | −60 |
| Sagitta / Arrow | Sagittae | Sge | 20 | +10 |
| Sagittarius / Archer | Sagittarii | Sgr | 19 | −25 |
| Scorpius / Scorpion | Scorpii | Sco | 17 | −40 |
| Sculptor / Sculptor | Sculptoris | Scl | 0 | −30 |
| Scutum / Shield | Scuti | Sct | 19 | −10 |
| Serpens / Serpent | Serpentis | Ser | 17 | 0 |
| Sextans / Sextant | Sextantis | Sex | 10 | 0 |
| Taurus / Bull | Tauri | Tau | 4 | +15 |
| Telescopium / Telescope | Telescopii | Tel | 19 | −50 |
| Triangulum / Triangle | Trianguli | Tri | 2 | +30 |
| Triangulum Australe / Southern Triangle | Trianguli Australis | TrA | 16 | −65 |
| Tucana / Toucan | Tucanae | Tuc | 0 | −65 |
| Ursa Major / Big Bear | Ursae Majoris | UMa | 11 | +50 |
| Ursa Minor / Little Bear | Ursae Minoris | UMi | 15 | +70 |

| Latin Name<br>  Translation | Genitive | Abbrev. | R. A.<br>h | Dec.<br>° |
|---|---|---|---|---|
| Vela<br>  Ship's Sail | Velorum | Vel | 9 | −50 |
| Virgo<br>  Maiden, Virgin | Virginis | Vir | 13 | 0 |
| Volans<br>  Flying Fish | Volantis | Vol | 8 | −70 |
| Vulpecula<br>  Little Fox | Vulpeculae | Vul | 20 | +25 |

# Appendix D

# THE BRIGHTEST STARS

| Name | Star | Spectral Class A | Spectral Class B | $V^a$ A | $V^a$ B | $M_V$ A | $M_V$ B |
|---|---|---|---|---|---|---|---|
| Sirius | α CMa | A1V | wd[b] | −1.46 | + 8.7 | +1.4 | +11.6 |
| Canopus | α Car | F0Ib–II | | −0.72 | | −3.1 | |
| Rigel Kentaurus | α Cen | G2V | K0V | −0.01 | + 1.3 | +4.4 | + 5.7 |
| Arcturus | α Boo | K2IIIp | | −0.06 | | −0.3 | |
| Vega | α Lyr | A0V | | +0.04 | | +0.5 | |
| Capella[c] | α Aur | GIII | M1V | +0.05 | +10.2 | −0.7 | + 9.5 |
| Rigel | β Ori | B8Ia | B9 | +0.14 | + 6.6 | −6.8 | − 0.4 |
| Procyon | α CMi | F5IV–V | wd[b] | +0.37 | +10.7 | +2.6 | +13.0 |
| Betelgeuse | α Ori | M2Iab | | +0.41v | | −5.5 | |
| Achernar | α Eri | B5V | | +0.51 | | −1.0 | |
| Hadar | β Cen | B1III | ? | +0.63 | + 4 | −4.1 | − 0.8 |
| Altair | α Aql | A7IV–V | | +0.77 | | +2.2 | |
| Acrux | α Cru | B1IV | B3 | +1.39 | + 1.9 | −4.0 | − 3.5 |
| Aldebaran | α Tau | K5III | M2V | +0.86 | +13 | −0.2 | +12 |
| Spica | α Vir | B1V | | +0.91v | | −3.6 | |
| Antares | α Sco | MIIb | B4eV | +0.92v | + 5.1 | −4.5 | − 0.3 |
| Pollux | β Gem | K0III | | +1.16 | | +0.8 | |
| Fomalhaut | α PsA | A3V | K4V | +1.19 | + 6.5 | +2.0 | + 7.3 |
| Deneb | α Cyg | A2Ia | | +1.26 | | −6.9 | |
| Mimosa | β Cru | B0.5IV | | +1.28v | | −4.6 | |

[a] Values labeled v designate variable stars.
[b] wd represents a white dwarf star.
[c] Capella has a third member of spectral class M5V, $V = +13.7$, and $M_V = +13$.

| Name | R.A.[a] h | m | Dec.[a] ° | ′ | Distance (pc) | Proper Motion ($''\,\text{yr}^{-1}$) | Radial Velocity ($\text{km s}^{-1}$) |
|---|---|---|---|---|---|---|---|
| Sirius | 6 | 42.9 | −16 | 39 | 2.6 | 1.33 | − 7.6 |
| Canopus | 6 | 22.8 | −52 | 40 | 30 | 0.02 | +20.5 |
| Rigel Kentaurus | 14 | 36.2 | −60 | 38 | 1.3 | 3.68 | −24.6 |
| Arcturus | 14 | 13.4 | +19 | 27 | 11 | 2.28 | − 5.2 |
| Vega | 18 | 35.2 | +38 | 44 | 8.0 | 0.34 | −13.9 |
| Capella | 5 | 13.0 | +45 | 57 | 14 | 0.44 | +30.2 |
| Rigel | 5 | 12.1 | −08 | 15 | 250 | 0.00 | +20.7 |
| Procyon | 7 | 36.7 | +05 | 21 | 3.5 | 1.25 | − 3.2 |
| Betelgeuse | 5 | 52.5 | +07 | 24 | 150 | 0.03 | +21.0 |
| Achernar | 1 | 35.9 | −57 | 29 | 20 | 0.10 | +19 |
| Hadar | 14 | 00.3 | −60 | 08 | 90 | 0.04 | −12 |
| Altair | 19 | 48.3 | +08 | 44 | 5.1 | 0.66 | −26.3 |
| Acrux | 12 | 23.8 | −62 | 49 | 120 | 0.04 | −11.2 |
| Aldebaran | 4 | 33.0 | +16 | 25 | 16 | 0.20 | +54.1 |
| Spica | 13 | 22.6 | −10 | 54 | 80 | 0.05 | + 1.0 |
| Antares | 16 | 26.3 | −26 | 19 | 120 | 0.03 | − 3.2 |
| Pollux | 7 | 42.3 | +28 | 09 | 12 | 0.62 | + 3.3 |
| Fomalhaut | 22 | 54.9 | −29 | 53 | 7.0 | 0.37 | + 6.5 |
| Deneb | 20 | 39.7 | +45 | 06 | 430 | 0.00 | − 4.6 |
| Mimosa | 12 | 44.8 | −59 | 24 | 150 | 0.05 | |

[a] Right ascension and declination are given in epoch 1950.0.

## Suggested Readings

TECHNICAL

Hoffleit, Dorrit, and Warren, Wayne H. Jr., *The Bright Star Catalogue*, Fifth Edition, Yale University Observatory, New Haven, 1991.

Lang, Kenneth R., *Astrophysical Data: Planets and Stars*, Springer-Verlag, New York, 1992.

# Appendix E

# Stellar Data

## Main-Sequence Stars (Luminosity Class V)

| Sp. Type | $T_e$ (K) | $L/L_\odot$ | $R/R_\odot$ | $M/M_\odot$ | $M_{\text{bol}}$ | BC | $M_V$ | $U-B$ | $B-V$ |
|---|---|---|---|---|---|---|---|---|---|
| O5 | 44500 | 790000 | 15 | 60 | −10.1 | −4.40 | −5.7 | −1.19 | −0.33 |
| O6 | 41000 | 420000 | 13 | 37 | −9.4 | −3.93 | −5.5 | −1.17 | −0.33 |
| O7 | 38000 | 260000 | — | — | −8.9 | −3.68 | −5.2 | −1.15 | −0.32 |
| O8 | 35800 | 170000 | 11 | 23 | −8.4 | −3.54 | −4.9 | −1.14 | −0.32 |
| O9 | 33000 | 97000 | — | — | −7.8 | −3.33 | −4.5 | −1.12 | −0.31 |
| B0 | 30000 | 52000 | 8.4 | 17.5 | −7.1 | −3.16 | −4.0 | −1.08 | −0.30 |
| B1 | 25400 | 16000 | — | — | −5.9 | −2.70 | −3.2 | −0.95 | −0.26 |
| B2 | 22000 | 5700 | — | — | −4.7 | −2.35 | −2.4 | −0.84 | −0.24 |
| B3 | 18700 | 1900 | 4.2 | 7.6 | −3.5 | −1.94 | −1.6 | −0.71 | −0.20 |
| B5 | 15400 | 830 | 4.1 | 5.9 | −2.7 | −1.46 | −1.2 | −0.58 | −0.17 |
| B6 | 14000 | 500 | — | — | −2.1 | −1.21 | −0.9 | −0.50 | −0.15 |
| B7 | 13000 | 320 | — | — | −1.6 | −1.02 | −0.6 | −0.43 | −0.13 |
| B8 | 11900 | 180 | 3.2 | 3.8 | −1.0 | −0.80 | −0.2 | −0.34 | −0.11 |
| B9 | 10500 | 95 | — | — | −0.3 | −0.51 | +0.2 | −0.20 | −0.07 |
| A0 | 9520 | 54 | 2.7 | 2.9 | +0.3 | −0.30 | +0.6 | −0.02 | −0.02 |
| A1 | 9230 | 35 | — | — | +0.8 | −0.23 | +1.0 | +0.02 | +0.01 |
| A2 | 8970 | 26 | — | — | +1.1 | −0.20 | +1.3 | +0.05 | +0.05 |
| A3 | 8720 | 21 | — | — | +1.3 | −0.17 | +1.5 | +0.08 | +0.08 |
| A5 | 8200 | 14 | 1.9 | 2.0 | +1.7 | −0.15 | +1.9 | +0.10 | +0.15 |
| A7 | 7850 | 10.5 | — | — | +2.1 | −0.12 | +2.2 | +0.10 | +0.20 |
| A8 | 7580 | 8.6 | — | — | +2.3 | −0.10 | +2.4 | +0.09 | +0.25 |
| F0 | 7200 | 6.5 | 1.6 | 1.6 | +2.6 | −0.09 | +2.7 | +0.03 | +0.30 |
| F2 | 6890 | 3.2 | — | — | +3.4 | −0.11 | +3.5 | 0.00 | +0.35 |
| F5 | 6440 | 2.9 | 1.4 | 1.4 | +3.5 | −0.14 | +3.6 | −0.02 | +0.44 |
| F8 | 6200 | 2.1 | — | — | +3.8 | −0.16 | +4.0 | +0.02 | +0.52 |

## Main-Sequence Stars (Luminosity Class V)

| Sp. Type | $T_e$ (K) | $L/L_\odot$ | $R/R_\odot$ | $M/M_\odot$ | $M_{bol}$ | BC | $M_V$ | $U-B$ | $B-V$ |
|---|---|---|---|---|---|---|---|---|---|
| G0      | 6030 | 1.5    | 1.1  | 1.05      | +4.2  | −0.18 | +4.4  | +0.06 | +0.58 |
| G2      | 5860 | 1.1    | —    | —         | +4.5  | −0.20 | +4.7  | +0.12 | +0.63 |
| Sun[a]  | 5780 | 1.00   | 1.00 | 1.00      | +4.64 | −0.19 | +4.83 | +0.17 | +0.68 |
| Sun[b]  | 5770 | 1.00   | 1.00 | 1.00      | +4.76 | −0.07 | +4.83 | +0.16 | +0.64 |
| G5      | 5770 | 0.79   | 0.89 | 0.92      | +4.9  | −0.21 | +5.1  | +0.20 | +0.68 |
| G8      | 5570 | 0.66   | —    | —         | +5.1  | −0.40 | +5.5  | +0.30 | +0.74 |
| K0      | 5250 | 0.42   | 0.79 | 0.79      | +5.6  | −0.31 | +5.9  | +0.45 | +0.81 |
| K1      | 5080 | 0.37   | —    | —         | +5.7  | −0.37 | +6.1  | +0.54 | +0.86 |
| K2      | 4900 | 0.29   | —    | —         | +6.0  | −0.42 | +6.4  | +0.64 | +0.91 |
| K3      | 4730 | 0.26   | —    | —         | +6.1  | −0.50 | +6.6  | +0.80 | +0.96 |
| K4      | 4590 | 0.19   | —    | —         | +6.4  | −0.55 | +7.0  | —     | +1.05 |
| K5      | 4350 | 0.15   | 0.68 | 0.67      | +6.7  | −0.72 | +7.4  | +0.98 | +1.15 |
| K7      | 4060 | 0.10   | —    | —         | +7.1  | −1.01 | +8.1  | +1.21 | +1.33 |
| M0      | 3850 | 0.077  | 0.63 | 0.51      | +7.4  | −1.38 | +8.8  | +1.22 | +1.40 |
| M1      | 3720 | 0.061  | —    | —         | +7.7  | −1.62 | +9.3  | +1.21 | +1.46 |
| M2      | 3580 | 0.045  | 0.55 | 0.40      | +8.0  | −1.89 | +9.9  | +1.18 | +1.49 |
| M3      | 3470 | 0.036  | —    | —         | +8.2  | −2.15 | +10.4 | +1.16 | +1.51 |
| M4      | 3370 | 0.019  | —    | —         | +8.9  | −2.38 | +11.3 | +1.15 | +1.54 |
| M5      | 3240 | 0.011  | 0.33 | 0.21      | +9.6  | −2.73 | +12.3 | +1.24 | +1.64 |
| M6      | 3050 | 0.0053 | —    | —         | +10.3 | −3.21 | +13.5 | +1.32 | +1.73 |
| M7      | 2940 | 0.0034 | —    | —         | +10.8 | −3.46 | +14.3 | +1.40 | +1.80 |
| M8      | 2640 | 0.0012 | 0.17 | 0.06[c]   | +11.9 | −4.1  | +16.0 | +1.53 | +1.93 |

[a] Values adopted by Schmidt-Kaler (1982).

[b] Values adopted in this book.

[c] Schmidt-Kaler (1982) uses a value of 0.06 $M_\odot$ for the M8 star, which is below the lower limit of 0.085 $M_\odot$ for stable hydrogen burning on the main sequence. However, stars in the range of 0.06–0.08 $M_\odot$ may undergo an extended period ($10^9$–$10^{10}$ years) of hydrogen fusion before failing to stabilize as main-sequence stars; see Liebert and Probst (1987).

## Giant Stars (Luminosity Class III)

| Sp. Type | $T_e$ (K) | $L/L_\odot$ | $R/R_\odot$ | $M/M_\odot$ | $M_{\text{bol}}$ | BC | $M_V$ | $U-B$ | $B-V$ |
|---|---|---|---|---|---|---|---|---|---|
| O5 | 42500 | 990000 | 18 | — | −10.3 | −4.05 | −6.3 | −1.18 | −0.32 |
| O6 | 39500 | 650000 | — | — | −9.9 | −3.80 | −6.1 | −1.17 | −0.32 |
| O7 | 37000 | 440000 | — | — | −9.5 | −3.58 | −5.9 | −1.14 | −0.32 |
| O8 | 34700 | 340000 | — | — | −9.2 | −3.39 | −5.8 | −1.13 | −0.31 |
| O9 | 32000 | 220000 | — | — | −8.7 | −3.13 | −5.6 | −1.12 | −0.31 |
| B0 | 29000 | 110000 | 13 | 20 | −8.0 | −2.88 | −5.1 | −1.08 | −0.29 |
| B1 | 24000 | 39000 | — | — | −6.8 | −2.43 | −4.4 | −0.97 | −0.26 |
| B2 | 20300 | 17000 | — | — | −5.9 | −2.02 | −3.9 | −0.91 | −0.24 |
| B3 | 17100 | 5000 | — | — | −4.6 | −1.60 | −3.0 | −0.74 | −0.20 |
| B5 | 15000 | 1800 | 6.3 | 7 | −3.5 | −1.30 | −2.2 | −0.58 | −0.17 |
| B6 | 14100 | 1100 | — | — | −2.9 | −1.13 | −1.8 | −0.51 | −0.15 |
| B7 | 13200 | 700 | — | — | −2.5 | −0.97 | −1.5 | −0.44 | −0.13 |
| B8 | 12400 | 460 | — | — | −2.0 | −0.82 | −1.2 | −0.37 | −0.11 |
| B9 | 11000 | 240 | — | — | −1.3 | −0.71 | −0.6 | −0.20 | −0.07 |
| A0 | 10100 | 106 | 3.4 | 4 | −0.4 | −0.42 | +0.0 | −0.07 | −0.03 |
| A1 | 9480 | 78 | — | — | −0.1 | −0.29 | +0.2 | +0.07 | +0.01 |
| A2 | 9000 | 65 | — | — | +0.1 | −0.20 | +0.3 | +0.06 | +0.05 |
| A3 | 8600 | 53 | — | — | +0.3 | −0.17 | +0.5 | +0.10 | +0.08 |
| A5 | 8100 | 43 | 3.3 | — | +0.6 | −0.14 | +0.7 | +0.11 | +0.15 |
| A7 | 7650 | 29 | — | — | +1.0 | −0.10 | +1.1 | +0.11 | +0.22 |
| A8 | 7450 | 26 | — | — | +1.1 | −0.10 | +1.2 | +0.10 | +0.25 |
| F0 | 7150 | 20 | 2.9 | — | +1.4 | −0.11 | +1.5 | +0.08 | +0.30 |
| F2 | 6870 | 17 | — | — | +1.6 | −0.11 | +1.7 | +0.08 | +0.35 |
| F5 | 6470 | 17 | 3.3 | — | +1.6 | −0.14 | +1.6 | +0.09 | +0.43 |
| F8 | 6150 | — | — | — | — | −0.16 | — | +0.10 | +0.54 |
| G0 | 5850 | 34 | 5.7 | 1.0 | +0.8 | −0.20 | +1.0 | +0.21 | +0.65 |
| G2 | 5450 | 40 | — | — | +0.6 | −0.27 | +0.9 | +0.39 | +0.77 |
| G5 | 5150 | 43 | 8.3 | 1.1 | +0.6 | −0.34 | +0.9 | +0.56 | +0.86 |
| G8 | 4900 | 51 | — | — | +0.4 | −0.42 | +0.8 | +0.70 | +0.94 |
| K0 | 4750 | 60 | 11 | 1.1 | +0.2 | −0.50 | +0.7 | +0.84 | +1.00 |
| K1 | 4600 | 69 | — | — | +0.1 | −0.55 | +0.6 | +1.01 | +1.07 |
| K2 | 4420 | 79 | — | — | −0.1 | −0.61 | +0.5 | +1.16 | +1.16 |
| K3 | 4200 | 110 | — | — | −0.5 | −0.76 | +0.3 | +1.39 | +1.27 |
| K4 | 4000 | 170 | — | — | −0.9 | −0.94 | 0.0 | — | +1.38 |
| K5 | 3950 | 220 | 32 | 1.2 | −1.2 | −1.02 | −0.2 | +1.81 | +1.50 |
| K7 | 3850 | 280 | — | — | −1.5 | −1.17 | −0.3 | +1.83 | +1.53 |

## Giant Stars (Luminosity Class III)

| Sp. Type | $T_e$ (K) | $L/L_\odot$ | $R/R_\odot$ | $M/M_\odot$ | $M_{\text{bol}}$ | BC | $M_V$ | $U-B$ | $B-V$ |
|---|---|---|---|---|---|---|---|---|---|
| M0 | 3800 | 330 | 42 | 1.2 | −1.6 | −1.25 | −0.4 | +1.87 | +1.56 |
| M1 | 3720 | 430 | — | — | −1.9 | −1.44 | −0.5 | +1.88 | +1.58 |
| M2 | 3620 | 550 | 60 | 1.3 | −2.2 | −1.62 | −0.6 | +1.89 | +1.60 |
| M3 | 3530 | 700 | — | — | −2.5 | −1.87 | −0.6 | +1.88 | +1.61 |
| M4 | 3430 | 880 | — | — | −2.7 | −2.22 | −0.5 | +1.73 | +1.62 |
| M5 | 3330 | 930 | 92 | — | −2.8 | −2.48 | −0.3 | +1.58 | +1.63 |
| M6 | 3240 | 1070 | — | — | −2.9 | −2.73 | −0.2 | +1.16 | +1.52 |

## Supergiant Stars (Luminosity Class Approximately Iab)

| Sp. Type | $T_e$ (K) | $L/L_\odot$ | $R/R_\odot$ | $M/M_\odot$ | $M_{\text{bol}}$ | BC | $M_V$ | $U-B$ | $B-V$ |
|---|---|---|---|---|---|---|---|---|---|
| O5 | 40300 | 1100000 | 22 | 70 | −10.5 | −3.87 | −6.6 | −1.17 | −0.31 |
| O6 | 39000 | 900000 | 21 | 40 | −10.2 | −3.74 | −6.5 | −1.16 | −0.31 |
| O7 | 35700 | 710000 | — | — | −10.0 | −3.48 | −6.5 | −1.14 | −0.31 |
| O8 | 34200 | 620000 | 22 | 28 | −9.8 | −3.35 | −6.5 | −1.13 | −0.29 |
| O9 | 32600 | 530000 | — | — | −9.7 | −3.18 | −6.5 | −1.13 | −0.27 |
| B0 | 26000 | 260000 | 25 | 25 | −8.9 | −2.49 | −6.4 | −1.06 | −0.23 |
| B1 | 20800 | 150000 | — | — | −8.3 | −1.87 | −6.4 | −1.00 | −0.19 |
| B2 | 18500 | 110000 | — | — | −8.0 | −1.58 | −6.4 | −0.94 | −0.17 |
| B3 | 16200 | 76000 | — | — | −7.6 | −1.26 | −6.3 | −0.83 | −0.13 |
| B5 | 13600 | 52000 | 41 | 20 | −7.2 | −0.95 | −6.2 | −0.72 | −0.10 |
| B6 | 13000 | 49000 | — | — | −7.1 | −0.88 | −6.2 | −0.69 | −0.08 |
| B7 | 12200 | 44000 | — | — | −7.0 | −0.78 | −6.2 | −0.64 | −0.05 |
| B8 | 11200 | 40000 | — | — | −6.9 | −0.66 | −6.2 | −0.56 | −0.03 |
| B9 | 10300 | 35000 | — | — | −6.7 | −0.52 | −6.2 | −0.50 | −0.02 |
| A0 | 9730 | 35000 | 66 | 16 | −6.7 | −0.41 | −6.3 | −0.38 | −0.01 |
| A1 | 9230 | 35000 | — | — | −6.7 | −0.32 | −6.4 | −0.29 | +0.02 |
| A2 | 9080 | 36000 | — | — | −6.7 | −0.28 | −6.5 | −0.25 | +0.03 |
| A3 | 8770 | 35000 | — | — | −6.7 | −0.21 | −6.5 | −0.14 | +0.06 |
| A5 | 8510 | 35000 | 86 | 13 | −6.7 | −0.13 | −6.6 | −0.07 | +0.09 |
| A7 | 8150 | 33000 | — | — | −6.7 | −0.06 | −6.6 | 0.00 | +0.12 |
| A8 | 7950 | 32000 | — | — | −6.6 | −0.03 | −6.6 | +0.11 | +0.14 |
| F0 | 7700 | 32000 | 100 | 12 | −6.6 | −0.01 | −6.6 | +0.15 | +0.17 |
| F2 | 7350 | 31000 | — | — | −6.6 | 0.00 | −6.6 | +0.18 | +0.23 |
| F5 | 6900 | 32000 | 130 | 10 | −6.6 | −0.03 | −6.6 | +0.27 | +0.32 |
| F8 | 6100 | 31000 | — | — | −6.6 | −0.09 | −6.5 | +0.41 | +0.56 |
| G0 | 5550 | 30000 | 190 | 10 | −6.6 | −0.15 | −6.4 | +0.52 | +0.76 |
| G2 | 5200 | 29000 | — | — | −6.5 | −0.21 | −6.3 | +0.63 | +0.87 |
| G5 | 4850 | 29000 | 240 | 12 | −6.5 | −0.33 | −6.2 | +0.83 | +1.02 |
| G8 | 4600 | 29000 | — | — | −6.5 | −0.42 | −6.1 | +1.07 | +1.15 |
| K0 | 4420 | 29000 | 290 | 13 | −6.5 | −0.50 | −6.0 | +1.17 | +1.24 |
| K1 | 4330 | 30000 | — | — | −6.6 | −0.56 | −6.0 | +1.28 | +1.30 |
| K2 | 4250 | 29000 | — | — | −6.5 | −0.61 | −5.9 | +1.32 | +1.35 |
| K3 | 4080 | 33000 | — | — | −6.6 | −0.75 | −5.9 | +1.60 | +1.46 |
| K4 | 3950 | 34000 | — | — | −6.7 | −0.90 | −5.8 | — | +1.53 |
| K5 | 3850 | 38000 | 440 | 13 | −6.8 | −1.01 | −5.8 | +1.80 | +1.60 |
| K7 | 3700 | 41000 | — | — | −6.9 | −1.20 | −5.7 | +1.84 | +1.63 |

## Supergiant Stars (Luminosity Class Approximately Iab)

| Sp. Type | $T_e$ (K) | $L/L_\odot$ | $R/R_\odot$ | $M/M_\odot$ | $M_{\text{bol}}$ | BC | $M_V$ | $U-B$ | $B-V$ |
|---|---|---|---|---|---|---|---|---|---|
| M0 | 3650 | 41000 | 510 | 13 | −6.9 | −1.29 | −5.6 | +1.90 | +1.67 |
| M1 | 3550 | 44000 | — | — | −7.0 | −1.38 | −5.6 | +1.90 | +1.69 |
| M2 | 3450 | 55000 | 660 | 19 | −7.2 | −1.62 | −5.6 | +1.95 | +1.71 |
| M3 | 3200 | 56000 | — | — | −7.7 | −2.13 | −5.6 | +1.95 | +1.69 |
| M4 | 2980 | 160000 | — | — | −8.3 | −2.75 | −5.6 | +2.00 | +1.76 |
| M5 | 2800 | 300000 | 2300 | 24 | −9.1 | −3.47 | −5.6 | +1.60 | +1.80 |
| M6 | 2600 | 450000 | — | — | −9.5 | −3.90 | −5.6 | — | — |

Except for the stellar radii, the data in the foregoing tables were taken from Schmidt-Kaler (1982). The values of the stellar radii were calculated using

$$\frac{R}{R_\odot} = \left(\frac{T_\odot}{T}\right)^2 \sqrt{\frac{L}{L_\odot}}.$$

# Suggested Readings

TECHNICAL

Liebert, James, and Probst, Ronald G., "Very Low Mass Stars," *Annual Review of Astronomy and Astrophysics*, *25*, 473, 1987.

Schmidt-Kaler, Th., "Physical Parameters of the Stars," *Landolt-Börnstein Numerical Data and Functional Relationships in Science and Technology*, New Series, Group VI, Volume 2b, Springer-Verlag, Berlin, 1982.

# Appendix F

# The Messier Catalog

| M | NGC | Name | Const. | $m_V{}^a$ | R. A.[b] h | R. A.[b] m | Dec.[b] ° | Dec.[b] ′ | Type[c] |
|---|---|---|---|---|---|---|---|---|---|
| 1 | 1952 | Crab | Tau | 8.4: | 5 | 34.5 | +22 | 01 | SNR |
| 2 | 7089 | | Aqr | 6.5 | 21 | 33.5 | −0 | 49 | GC |
| 3 | 5272 | | CVn | 6.4 | 13 | 42.2 | +28 | 23 | GC |
| 4 | 6121 | | Sco | 5.9 | 16 | 23.6 | −26 | 32 | GC |
| 5 | 5904 | | Ser | 5.8 | 15 | 18.6 | + 2 | 05 | GC |
| 6 | 6405 | | Sco | 4.2 | 17 | 40.1 | −32 | 13 | OC |
| 7 | 6475 | | Sco | 3.3 | 17 | 53.9 | −34 | 49 | OC |
| 8 | 6523 | Lagoon | Sgr | 5.8: | 18 | 03.8 | −24 | 23 | N |
| 9 | 6333 | | Oph | 7.9: | 17 | 19.2 | −18 | 31 | GC |
| 10 | 6254 | | Oph | 6.6 | 16 | 57.1 | − 4 | 06 | GC |
| 11 | 6705 | | Sct | 5.8 | 18 | 51.1 | − 6 | 16 | OC |
| 12 | 6218 | | Oph | 6.6 | 16 | 47.2 | − 1 | 57 | GC |
| 13 | 6205 | | Her | 5.9 | 16 | 41.7 | +36 | 28 | GC |
| 14 | 6402 | | Oph | 7.6 | 17 | 37.6 | − 3 | 15 | GC |
| 15 | 7078 | | Peg | 6.4 | 21 | 30.0 | +12 | 10 | GC |
| 16 | 6611 | | Ser | 6.0 | 18 | 18.8 | −13 | 47 | OC |
| 17 | 6618 | Swan[d] | Sgr | 7: | 18 | 20.8 | −16 | 11 | N |
| 18 | 6613 | | Sgr | 6.9 | 18 | 19.9 | −17 | 08 | OC |
| 19 | 6273 | | Oph | 7.2 | 17 | 02.6 | −26 | 16 | GC |
| 20 | 6514 | Trifid | Sgr | 8.5: | 18 | 02.6 | −23 | 02 | N |
| 21 | 6531 | | Sgr | 5.9 | 18 | 04.6 | −22 | 30 | OC |
| 22 | 6656 | | Sgr | 5.1 | 18 | 36.4 | −23 | 54 | GC |
| 23 | 6494 | | Sgr | 5.5 | 17 | 56.8 | −19 | 01 | OC |
| 24 | 6603 | | Sgr | 4.5: | 18 | 16.9 | −18 | 29 | OC |
| 25 | | | Sgr | 4.6 | 18 | 31.6 | −19 | 15 | OC |
| 26 | 6694 | | Sct | 8.0 | 18 | 45.2 | − 9 | 24 | OC |
| 27 | 6853 | Dumbbell | Vul | 8.1: | 19 | 59.6 | +22 | 43 | PN |

# Appendix F  The Messier Catalog

| M | NGC | Name | Const. | $m_V{}^a$ | R. A.$^b$ h | m | Dec.$^b$ ° | ′ | Type$^c$ |
|---|---|---|---|---|---|---|---|---|---|
| 28 | 6626 |  | Sgr | 6.9: | 18 | 24.5 | −24 | 52 | GC |
| 29 | 6913 |  | Cyg | 6.6 | 20 | 23.9 | +38 | 32 | OC |
| 30 | 7099 |  | Cap | 7.5 | 21 | 40.4 | −23 | 11 | GC |
| 31 | 224 | Andromeda | And | 3.4 | 0 | 42.7 | +41 | 16 | SbI–II |
| 32 | 221 |  | And | 8.2 | 0 | 42.7 | +40 | 52 | cE2 |
| 33 | 598 | Triangulum | Tri | 5.7 | 1 | 33.9 | +30 | 39 | Sc(s)II–III |
| 34 | 1039 |  | Per | 5.2 | 2 | 42.0 | +42 | 47 | OC |
| 35 | 2168 |  | Gem | 5.1 | 6 | 08.9 | +24 | 20 | OC |
| 36 | 1960 |  | Aur | 6.0 | 5 | 36.1 | +34 | 08 | OC |
| 37 | 2099 |  | Aur | 5.6 | 5 | 52.4 | +32 | 33 | OC |
| 38 | 1912 |  | Aur | 6.4 | 5 | 28.7 | +35 | 50 | OC |
| 39 | 7092 |  | Cyg | 4.6 | 21 | 32.2 | +48 | 26 | OC |
| 40 |  |  | UMa | 8: | 12 | 22.4 | +58 | 05 | DS |
| 41 | 2287 |  | CMa | 4.5 | 6 | 47.0 | −20 | 44 | OC |
| 42 | 1976 | Orion$^e$ | Ori | 4: | 5 | 35.3 | − 5 | 23 | N |
| 43 | 1982 |  | Ori | 9: | 5 | 35.6 | − 5 | 16 | N |
| 44 | 2632 | Praesepe | Cnc | 3.1 | 8 | 40.1 | +19 | 59 | OC |
| 45 |  | Pleiades | Tau | 1.2 | 3 | 47.0 | +24 | 07 | OC |
| 46 | 2437 |  | Pup | 6.1 | 7 | 41.8 | −14 | 49 | OC |
| 47 | 2422 |  | Pup | 4.4 | 7 | 36.6 | −14 | 30 | OC |
| 48 | 2548 |  | Hya | 5.8 | 8 | 13.8 | − 5 | 48 | OC |
| 49 | 4472 |  | Vir | 8.4 | 12 | 29.8 | + 8 | 00 | E2 |
| 50 | 2323 |  | Mon | 5.9 | 7 | 03.2 | − 8 | 20 | OC |
| 51 | 5194 | Whirlpool$^f$ | CVn | 8.1 | 13 | 29.9 | +47 | 12 | Sbc(s)I–II |
| 52 | 7654 |  | Cas | 6.9 | 23 | 24.2 | +61 | 35 | OC |
| 53 | 5024 |  | Com | 7.7 | 13 | 12.9 | +18 | 10 | GC |
| 54 | 6715 |  | Sgr | 7.7 | 18 | 55.1 | −30 | 29 | GC |
| 55 | 6809 |  | Sgr | 7.0 | 19 | 40.0 | −30 | 58 | GC |
| 56 | 6779 |  | Lyr | 8.2 | 19 | 16.6 | +30 | 11 | GC |
| 57 | 6720 | Ring | Lyr | 9.0: | 18 | 53.6 | +33 | 02 | PN |
| 58 | 4579 |  | Vir | 9.8 | 12 | 37.7 | +11 | 49 | Sab(s)II |
| 59 | 4621 |  | Vir | 9.8 | 12 | 42.0 | +11 | 39 | E5 |
| 60 | 4649 |  | Vir | 8.8 | 12 | 43.7 | +11 | 33 | E2 |
| 61 | 4303 |  | Vir | 9.7 | 12 | 21.9 | + 4 | 28 | Sc(s)I |
| 62 | 6266 |  | Oph | 6.6 | 17 | 01.2 | −30 | 07 | GC |
| 63 | 5055 | Sunflower | CVn | 8.6 | 13 | 15.8 | +42 | 02 | Sbc(s)II–III |
| 64 | 4826 | Evil Eye | Com | 8.5 | 12 | 56.7 | +21 | 41 | Sab(s)II |
| 65 | 3623 |  | Leo | 9.3 | 11 | 18.9 | +13 | 05 | Sa(s)I |
| 66 | 3627 |  | Leo | 9.0 | 11 | 20.2 | +12 | 59 | Sb(s)II |
| 67 | 2682 |  | Cnc | 6.9 | 8 | 50.4 | +11 | 49 | OC |
| 68 | 4590 |  | Hya | 8.2 | 12 | 39.5 | −26 | 45 | GC |
| 69 | 6637 |  | Sgr | 7.7 | 18 | 31.4 | −32 | 21 | GC |

## Appendix F  The Messier Catalog

| M | NGC | Name | Const. | $m_V{}^a$ | R.A.$^b$ h | m | Dec.$^b$ ° | ′ | Type$^c$ |
|---|---|---|---|---|---|---|---|---|---|
| 70 | 6681 | | Sgr | 8.1 | 18 | 43.2 | −32 | 18 | GC |
| 71 | 6838 | | Sge | 8.3 | 19 | 53.8 | +18 | 47 | GC |
| 72 | 6981 | | Aqr | 9.4 | 20 | 53.5 | −12 | 32 | GC |
| 73 | 6994 | | Aqr | 9.1 | 20 | 58.9 | −12 | 38 | OC |
| 74 | 628 | | Psc | 9.2 | 1 | 36.7 | +15 | 47 | Sc(s)I |
| 75 | 6864 | | Sgr | 8.6 | 20 | 06.1 | −21 | 55 | GC |
| 76 | 650/651 | | Per | 11.5: | 1 | 42.3 | +51 | 34 | PN |
| 77 | 1068 | | Cet | 8.8 | 2 | 42.7 | − 0 | 01 | Sb(rs)II |
| 78 | 2068 | | Ori | 8: | 5 | 46.7 | + 0 | 03 | N |
| 79 | 1904 | | Lep | 8.0 | 5 | 24.5 | −24 | 33 | GC |
| 80 | 6093 | | Sco | 7.2 | 16 | 17.0 | −22 | 59 | GC |
| 81 | 3031 | | UMa | 6.8 | 9 | 55.6 | +69 | 04 | Sb(r)I–II |
| 82 | 3034 | | UMa | 8.4 | 9 | 55.8 | +69 | 41 | Amorph |
| 83 | 5236 | | Hya | 7.6: | 13 | 37.0 | −29 | 52 | SBc(s)II |
| 84 | 4374 | | Vir | 9.3 | 12 | 25.1 | +12 | 53 | E1 |
| 85 | 4382 | | Com | 9.2 | 12 | 25.4 | +18 | 11 | S0 pec |
| 86 | 4406 | | Vir | 9.2 | 12 | 26.2 | +12 | 57 | S0/E3 |
| 87 | 4486 | Virgo A | Vir | 8.6 | 12 | 30.8 | +12 | 24 | E0 |
| 88 | 4501 | | Com | 9.5 | 12 | 32.0 | +14 | 25 | Sbc(s)II |
| 89 | 4552 | | Vir | 9.8 | 12 | 35.7 | +12 | 33 | S0 |
| 90 | 4569 | | Vir | 9.5 | 12 | 36.8 | +13 | 10 | Sab(s)I–II |
| 91 | 4548 | | Com | 10.2 | 12 | 35.4 | +14 | 30 | SBb(rs)I–II |
| 92 | 6341 | | Her | 6.5 | 17 | 17.1 | +43 | 08 | GC |
| 93 | 2447 | | Pup | 6.2: | 7 | 44.6 | −23 | 52 | OC |
| 94 | 4736 | | CVn | 8.1 | 12 | 50.9 | +41 | 07 | RSab(s) |
| 95 | 3351 | | Leo | 9.7 | 10 | 44.0 | +11 | 42 | SBb(r)II |
| 96 | 3368 | | Leo | 9.2 | 10 | 46.8 | +11 | 49 | Sab(s)II |
| 97 | 3587 | Owl | UMa | 11.2: | 11 | 14.8 | +55 | 01 | PN |
| 98 | 4192 | | Com | 10.1 | 12 | 13.8 | +14 | 54 | SbII |
| 99 | 4254 | | Com | 9.8 | 12 | 18.8 | +14 | 25 | Sc(s)I |
| 100 | 4321 | | Com | 9.4 | 12 | 22.9 | +15 | 49 | Sc(s)I |
| 101 | 5457 | Pinwheel | UMa | 7.7 | 14 | 03.2 | +54 | 21 | Sc(s)I |
| 102 | 5866 | | UMa | 10.5 | 15 | 06.5 | +55 | 46 | S0 |
| 103 | 581 | | Cas | 7.4: | 1 | 33.2 | +60 | 42 | OC |
| 104 | 4594 | Sombrero | Vir | 8.3 | 12 | 40.0 | −11 | 37 | Sa/Sb |
| 105 | 3379 | | Leo | 9.3 | 10 | 47.8 | +12 | 35 | E0 |
| 106 | 4258 | | CVn | 8.3 | 12 | 19.0 | +47 | 18 | Sb(s)II |
| 107 | 6171 | | Oph | 8.1 | 16 | 32.5 | −13 | 03 | GC |
| 108 | 3556 | | UMa | 10.0 | 11 | 11.5 | +55 | 40 | Sc(s)III |
| 109 | 3992 | | UMa | 9.8 | 11 | 57.6 | +53 | 23 | SBb(rs)I |
| 110 | 205 | | And | 8.0 | 0 | 40.4 | +41 | 41 | S0/E pec |

$^a$ : indicates approximate apparent visual magnitude.
$^b$ Right ascension and declination are given in epoch 2000.0.
$^c$ Type abbreviations correspond to: SNR = supernova remnant, GC = globular cluster, OC = open cluster, N = diffuse nebula, PN = planetary nebula, DS = double star.
Galaxies are indicated by their morphological Hubble types.
$^d$ M17, the Swan nebula, is also known as the Omega nebula.
$^e$ M42 also corresponds to the Trapezium H II region.
$^f$ M51 also includes NGC 5195, the satellite to the Whirlpool galaxy.

## Suggested Readings

TECHNICAL

Hirshfeld, Alan, Sinnott, Roger W., and Ochsenbein, Francois, *Sky Catalogue 2000.0*, Second Edition, Cambridge University Press and Sky Publishing Corporation, New York, 1991.

Sandage, Allan, and Bedke, John, *The Carnegie Atlas of Galaxies*, Carnegie Institution of Washington, Washington, D.C., 1994.

# Appendix G

# A Planetary Orbit Code

```
      Program ORBIT
      real*8 a, aau, e, time, tyears, dt, LoM, period, Pyears
      real*8 Mstrsun, Mstar, Msun, theta, dtheta, r, x, y
      real*8 G, AU, spyr
      integer*4 n, k, kmax, i
c define basic constants
      data G, Msun, AU, spyr
     1 / 6.67259d-08, 1.989d33, 1.4960d13, 3.155815d7 /
c open output file for orbital parameters
      open(unit=10,file='orbit.dat',form='formatted',status='unknown')
c enter physical parameters for the system
      write(*,*) ' Enter the mass of the parent star (in solar masses):'
      read(*,*) Mstrsun
      write(*,*) ' Enter the semimajor axis of the orbit (in AU):'
      read(*,*) aau
      write(*,*) ' Enter the eccentricity:'
      read(*,*) e
c calculate orbital period using Kepler's third law (Eq. 2.35)
      Pyears = sqrt(aau**3/Mstrsun)
      write(*,*) ' '
      write(*,100) Pyears
c enter the number of time steps and the time interval to be printed
      write(*,*) ' '
      write(*,*) ' '
```

```fortran
      write(*,*) ' You may now enter the number of time steps to ',
     1 'be calculated and the'
      write(*,*) ' frequency with which you want time steps printed.'
      write(*,*) ' Note that taking too large a time step during the',
     1 ' calculation will'
      write(*,*) ' produce inaccurate results.'
      write(*,*) ' '
      write(*,*) ' Enter the number of time steps desired for the ',
     1 'calculation:  '
      read(*,*) n
      write(*,*) ' '
      write(*,*) ' How often do you want time steps printed?'
      write(*,*) '     1 = every time step'
      write(*,*) '     2 = every second time step'
      write(*,*) '     3 = every third time step'
      write(*,*) '              etc.'
      read(*,*) kmax
c convert to cgs units
      period = Pyears*spyr
      dt = period/float(n-1)
      a = aau*AU
      Mstar = Mstrsun*Msun
c initialize print counter, angle, and elapsed time
      k = 0
      theta = 0.0d0
      time = 0.0d0
c print output header
      write(10,200) Mstrsun, aau, Pyears, e
c start main time step loop
      do 10 i = 1,n
c increment print counter
         k = k+1
c use Eq. (2.3) to find r
         r = a*(1.0d0-e**2)/(1.0d0+e*cos(theta))
c convert to cartesian coordinates and print time and position
c also print last point to close ellipse
         if (k.eq.1 .or. i.eq.n) then
            x = r*cos(theta)/AU
            y = r*sin(theta)/AU
            tyears = time/spyr
```

# Appendix G  A Planetary Orbit Code

```
              write(10,300) tyears, x, y
         end if
c calculate the angular momentum per unit mass L/m (Eq. 2.32)
         LoM = sqrt(G*Mstar*a*(1.0d0-e**2))
c calculate the next value of theta by combining Eqs. (2.27)
c and (2.29) (Kepler's second law)
         dtheta = LoM/r**2*dt
         theta = theta + dtheta
c update the elapsed time
         time = time + dt
c check print counter then return to the top of the loop
         if (k.eq.kmax) k = 0
   10 continue
      write(*,*) ' '
      write(*,*) ' The calculation is finished and listed in ',
     1 'orbit.dat'
      stop
c formats
  100 format(1x,' The period of this orbit is ',f10.3,' years')
  200 format(1x,25x,'Elliptical Orbit',//,
     1 26x,'Mstar = ',f10.3,' Mo',/,
     2 26x,'a = ',f10.3,' AU',/,
     3 26x,'P = ',f10.3,' yr',/,
     4 26x,'e = ',f10.3,///,
     5 11x,'t (yr)',14x,'x (AU)',12x,'y (AU)')
  300 format(7x,f10.3,10x,f10.3,8x,f10.3)
      end
```

# Appendix H

# STATSTAR, A STELLAR STRUCTURE CODE

The FORTRAN computer program listed here (STATSTAR) is based on the equations of stellar structure and the constitutive relations developed in Chapters 9 and 10. An example of the output generated by STATSTAR is given in Appendix I.

STATSTAR is designed to illustrate as clearly as possible many of the most important aspects of numerical stellar astrophysics. To accomplish this goal, STATSTAR models are restricted to a fixed composition throughout (i.e., homogeneous main-sequence models).

The four basic stellar structure equations are computed in the functions dPdr, dMdr, dLdr, and dTdr. These functions calculate the derivatives $dP/dr$ (Eq. 10.7), $dM_r/dr$ (Eq. 10.8), $dL_r/dr$ (Eq. 10.45), and $dT/dr$ [Eqs. (10.61) and (10.81)].

The density $[\rho(r) = \text{rho}]$ is calculated directly from the ideal gas law and the radiation pressure equation (Eq. 10.26), given local values for the pressure $[P(r) = \text{P(i)}]$, temperature $[T(r) = \text{T(i)}]$, and the mean molecular weight ($\mu = \text{mu}$, assumed here to be for a completely ionized gas only); note that i is the *number* of the mass shell currently being calculated ($\text{i} \equiv 1$ at the surface, see Fig. 10.11). Once the density is determined, both the opacity $[\overline{\kappa}(r) = \text{kappa}]$ and the nuclear energy generation rate $[\epsilon(r) = \text{epslon}]$ may be calculated. The opacity is determined using the bound–bound and bound–free opacity formulae [Eqs. (9.19) and (9.20), respectively], together with electron scattering (Eq. 9.21). The energy generation rate is computed from the equation for the total pp chain (Eq. 10.49) and for the CNO cycle (Eq. 10.53). Each of these calculations is carried out in the equation-of-state subroutine, EOS.[1]

The program begins by asking the user to supply the desired stellar mass (Msolar, in solar units), the trial luminosity (Lsolar, also in solar units), the trial effective temperature (Te, in kelvin), and the mass fractions of hydrogen (X) and metals (Z). Using the stellar structure equations, the program proceeds to integrate from the surface of the star toward the center, stopping when a problem is detected or when a satisfactory solution is obtained. If the inward integration is not successful, a new trial luminosity and/or effective temperature must be chosen. Recall that the Vogt–Russell theorem states that a unique stellar structure exists for a given mass and composition. Satisfying the central boundary conditions therefore requires specific surface boundary conditions. It is for this reason that a well-defined main sequence exists.

Since it is nearly impossible to satisfy the central boundary conditions exactly by the crude *shooting method* employed by STATSTAR, the calculation is terminated when the core is approached. The stopping criteria used here are that the interior mass $M_r < 0.01 M_s$ and the interior luminosity $L_r < 0.1 L_s$, when the radius $r < 0.02 R_s$, where $M_s$, $L_s$, and $R_s$ are the surface mass, luminosity, and radius, respectively. [Within STATSTAR, $M_r = $ M_r(i), $M_s = $ Ms, $L_r = $ L_r(i), $L_s = $ Ls, $r = $ r(i), and $R_s = $ Rs.] Once the criteria for halting the integration are detected, the conditions at the center of the star are estimated by an extrapolation procedure.

Since the pressure, temperature, and density are all assumed to be zero at the surface of the star, it is necessary to begin the calculation with approximations to the basic stellar structure equations. This can be seen by noting that the mass, pressure, luminosity, and temperature gradients are all proportional to the density and are therefore exactly zero at the surface. It would appear that applying these gradients in their usual form implies that the fundamental physical parameters cannot change from their initial values since the density would remain zero at each step!

One way to overcome this problem is to assume that the interior mass and luminosity are both constant through a number of surface zones. In the case of the luminosity, this is clearly a valid assumption since temperatures are not sufficient to produce nuclear reactions near the surfaces of main-sequence stars. For the interior mass, the assumption is not quite as obvious. However, we will see that in realistic stellar models, the density is so low near the surface that

---

[1] State-of-the-art research codes use much more sophisticated prescriptions for the equations of state.

# Appendix H  STATSTAR, A Stellar Structure Code

the approximation is indeed valid. Of course, it is important to verify that the assumption is not being violated to within some specified limit.

Assuming that the surface zone is radiative, and given the surface values $M_r = M_s$ and $L_r = L_s$, dividing Eq. (10.7) by Eq. (10.61) leads to

$$\frac{dP}{dT} = \frac{16\pi ac}{3} \frac{GM_s}{L_s} \frac{T^3}{\bar{\kappa}}.$$

Since relatively few free electrons exist in the atmospheres of stars, electron scattering can be neglected and $\bar{\kappa}$ may be replaced by the bound–free and free–free Kramers' opacity laws [Eqs. (9.19) and (9.20)] expressed in the forms $\bar{\kappa}_{bf} = A_{bf}\rho/T^{3.5}$ and $\bar{\kappa}_{ff} = A_{ff}\rho/T^{3.5}$, respectively. Defining $A \equiv A_{bf} + A_{ff}$ and using Eq. (10.14) to express the density in terms of the pressure and temperature through the ideal gas law (assuming that radiation pressure may be neglected),

$$\frac{dP}{dT} = \frac{16\pi}{3} \frac{GM_s}{L_s} \frac{ack}{A\mu m_H} \frac{T^{7.5}}{P}.$$

Integrating with respect to temperature and solving for the pressure, we find that

$$P = \left( \frac{1}{4.25} \frac{16\pi}{3} \frac{GM_s}{L_s} \frac{ack}{A\mu m_H} \right)^{1/2} T^{4.25}. \qquad \text{(H.1)}$$

It is now possible to write $T$ in terms of the independent variable $r$ through Eq. (10.61), again using the ideal gas law and Kramers' law, along with Eq. (H.1) to eliminate the dependence on pressure. Integrating,

$$T = GM_s \left( \frac{\mu m_H}{4.25k} \right) \left( \frac{1}{r} - \frac{1}{R_s} \right). \qquad \text{(H.2)}$$

Equation (H.2) is first used to obtain a value for $T(r)$, then Eq. (H.1) gives $P(r)$. At this point it is possible to calculate $\rho$, $\bar{\kappa}$, and $\epsilon$ from the usual equation-of-state routine, EOS.

A very similar procedure is used in the case that the surface is convective. In this situation Eq. (10.81) may be integrated directly if $\gamma$ is constant. This gives

$$T = GM_s \left( \frac{\gamma - 1}{\gamma} \right) \left( \frac{\mu m_H}{k} \right) \left( \frac{1}{r} - \frac{1}{R_s} \right). \qquad \text{(H.3)}$$

Now, since convection is assumed to be adiabatic in the interior of our simple model, the pressure may be found from Eq. (10.75). Subroutine STARTMDL computes Eqs. (H.1), (H.2), and (H.3).

The conditions at the center of the star are estimated by extrapolating from the last zone that was calculated by direct numerical integration. Beginning with Eq. (10.7), and identifying $M_r = 4\pi\rho_0 r^3/3$, where $\rho_0$ is taken to be the average density of the central ball (the region inside the last zone calculated by the usual procedure),[2]

$$\frac{dP}{dr} = -G\frac{M_r \rho_0}{r^2} = -\frac{4\pi}{3}G\rho_0^2 r.$$

Integrating,

$$\int_{P_0}^{P} dP = -\frac{4\pi}{3}G\rho_0^2 \int_0^r r\, dr$$

and solving for the central pressure results in

$$P_0 = P + \frac{2\pi}{3}G\rho_0^2 r^2. \tag{H.4}$$

Other central quantities can now be found more directly. Specifically, the central density is estimated to be $\rho_0 = M_r/(4\pi r^3/3)$, where $M_r$ and $r$ are the values of the last zone calculated. Neglecting radiation pressure, $T_0$ may be determined from the ideal gas law. Finally, the central value for the nuclear energy generation rate is computed using $\epsilon_0 = L_r/M_r$.

The numerical integration technique employed here is a fourth-order Runge–Kutta algorithm, which is accurate through fourth order in the step size $\Delta r =$ `deltar`. This means that if $\Delta r/r = 0.01$, the solutions for $P$, $M_r$, $L_r$, and $T$ are accurate to approximately a few parts in $0.01^4 = 10^{-8}$, assuming that the results of the previous zone were exact. To accomplish this accuracy, the Runge–Kutta algorithm evaluates derivatives at several intermediate points between mass shell boundaries. Details of the Runge–Kutta method are given in many numerical analysis texts and will not be discussed further here.

STATSTAR execution times vary depending on the machine being used. For instance, on PCs with 486 33MHz chips or higher, a model can be completed in a few seconds; on faster machines only a fraction of a second is required. It should be pointed out that if STATSTAR is compiled on a VAX computer running VMS, the \G_FLOAT option should be invoked. This option provides the large exponent range required of most astrophysical calculations.

---

[2] You might notice that $dP/dr$ goes to zero as the center is approached. This behavior is indicative of the smooth nature of the solution. Close inspection of the graphs in Section 11.1 showing the detailed interior structure of the Sun illustrates that the first derivatives of many physical quantities go to zero at the center.

# Appendix H STATSTAR, A Stellar Structure Code

```fortran
      Program STATSTAR
c
c This program will calculate a static stellar model using the
c equations developed in the text.  The user is expected to supply the
c star's mass, luminosity, effective temperature, and composition
c (X and Z). If the choices for these quantities are not consistent
c with the central boundary conditions, an error message will be
c generated and the user will then need to supply a different set of
c initial values.
c
      real*8 r(999), P(999), M_r(999), L_r(999), T(999), deltar, Te
      real*8 rho(999), kappa(999), epslon(999), dlPdlT(999)
      real*8 X, Y, Z, XCNO, mu
      real*8 Ms, Ls, Rs, T0, P0
      real*8 Pcore, Tcore, rhocor, epscor, rhomax
      real*8 Rsolar, Msolar, Lsolar, Qm, Rcrat, Mcrat, Lcrat
      real*8 deltam, dlPlim
      real*8 Rsun, Msun, Lsun
      real*8 f_im1(4), f_i(4), dfdr(4)
      real*8 dMdr, dPdr, dLdr, dTdr
      real*8 sigma, c, a, G, k_B, m_H, pi, gamma, gamrat, kPad, tog_bf,
     1 g_ff
      character clim, rcf
      common /cnstnt/ sigma, c, a, G, k_B, m_H, pi, gamma, gamrat, kPad,
     1 g_ff
c
c  deltar = radius integration step
c  idrflg = set size flag
c         = 0 (initial surface step size of Rs/1000.)
c         = 1 (standard step size of Rs/100.)
c         = 2 (core step size of Rs/5000.)
c
c  Nstart = number of steps for which starting equations are used
c           (the outermost zone is assumed to be radiative)
c  Nstop = maximum number of allowed zones in the star
c  Igoof = final model condition flag
c         = -1 (number of zones exceeded; also the initial value)
c         = 0 (good model)
c         = 1 (core density was extreme)
c         = 2 (core luminosity was extreme)
```

```
c          = 3 (extrapolated core temperature is too low)
c          = 4 (mass became negative before center was reached)
c          = 5 (luminosity became negative before center was reached)
c    X, Y, Z = mass fractions of hydrogen, helium, and metals
c    T0, P0 = surface temperature and pressure (T0 = P0 = 0 is assumed)
c    Ms, Ls, Rs = mass, luminosity, and radius of the star (cgs units)
c
      data Nstart, Nstop, Igoof, ierr / 10, 999, -1, 0 /
      data P0, T0, dlPlim / 0.0d0, 0.0d0, 99.9d0 /
      data Rsun, Msun, Lsun / 6.9599d+10, 1.989d+33, 3.826d+33 /
      data sigma, c, a, G, k_B, m_H, pi, gamma, tog_bf, g_ff
     1 / 5.67051d-5, 2.99792458d+10, 7.56591d-15, 6.67259d-8,
     2 1.380658d-16, 1.673534d-24, 3.141592654d0, 1.6666667d0, 0.01d0,
     3 1.0d0 /
c
c   Assign values to constants (cgs units)
c      Rsun = radius of the Sun
c      Msun = mass of the Sun
c      Lsun = luminosity of the Sun
c      sigma = Stefan-Boltzmann constant
c      c = speed of light in vacuum
c      a = 4*sigma/c (radiation pressure constant)
c      G = universal gravitational constant
c      k_B = Boltzmann constant
c      m_H = mass of hydrogen atom
c      pi = 3.141592654
c      gamma = 5/3 (adiabatic gamma for a monatomic gas)
c      gamrat = gamma/(gamma-1)
c      kPad = P/T**(gamma/(gamma-1)) (adiabatic gas law constant)
c      tog_bf = bound-free opacity constant
c            (ratio of guillotine to gaunt factors)
c      g_ff = free-free opacity gaunt factor (assumed to be unity)
c
      open(unit=20,file='starmodl.dat',form='formatted',
     1 status='unknown')
c
c   Enter desired stellar parameters
c
      write(*,*) ' Enter the mass of the star (in solar units):'
      read(*,*) Msolar
```

## Appendix H  STATSTAR, A Stellar Structure Code

```fortran
            write(*,*) ' Enter the luminosity of the star (in solar units):'
            read(*,*) Lsolar
            write(*,*) ' Enter the effective temperature of the star (in K):'
            read(*,*) Te
   10   continue
            write(*,*) ' Enter the mass fraction of hydrogen (X):'
            read(*,*) X
            write(*,*) ' Enter the mass fraction of metals (Z):'
            read(*,*) Z
            Y = 1.d0 - X - Z
            if (Y.lt.0.0d0) then
                write(*,100)
                go to 10
            end if
c
c   Select the mass fraction CNO to be 50% of Z.
c
            XCNO = Z/2.0d0
c
c   Calculate the mass, luminosity, and radius of the star.
c   The radius is calculated from Eq. (3.17).
c
            Ms = Msolar*Msun
            Ls = Lsolar*Lsun
            Rs = sqrt(Ls/(4.d0*pi*sigma))/Te**2
            Rsolar = Rs/Rsun
c
c   Begin with a very small step size since surface conditions vary
c   rapidly.
c
            deltar = -Rs/1000.0d0
            idrflg = 0
c
c   Calculate mean molecular weight mu assuming complete ionization
c   (see Eq. 10.21).
c
            mu = 1.0d0/(2.0d0*X + 0.75d0*Y + 0.5d0*Z)
c
```

```fortran
c     Calculate the delimiter between adiabatic convection and radiation
c     (see Eq. 10.87).
c
      gamrat = gamma/(gamma - 1.0d0)
c
c     Initialize values of r, P, M_r, L_r, T, rho, kappa, and epslon at
c     the surface.  The outermost zone is assumed to be zone 1.  The zone
c     number increases toward the center.
c
      r(1)    = Rs
      M_r(1)  = Ms
      L_r(1)  = Ls
      T(1)    = T0
      P(1)    = P0
      if (P0.le.0.0d0 .or. T0.le.0.0d0) then
          rho(1)    = 0.0d0
          kappa(1)  = 0.0d0
          epslon(1) = 0.0d0
      else
          call EOS(X, Z, XCNO, mu, P(1), T(1), rho(1), kappa(1),
     1         epslon(1), tog_bf, 1 ,ierr)
          if (ierr.ne.0) stop
      end if
c
c     Apply approximate surface solutions to begin the integration,
c     assuming radiation transport in the outermost zone (do 20 loop).
c     irc = 0 for radiation, irc = 1 for convection.
c     Assume arbitrary initial values for kPad, and dlPdlT.
c     dlPdlT = dlnP/dlnT (see Eq. 10.87)
c
      kPad = 0.3d0
      irc = 0
      dlPdlT(1) = 4.25d0
      do 20 i = 1, Nstart
          ip1 = i + 1
          call STARTMDL(deltar, X, Z, mu, Rs, r(i), M_r(i), L_r(i),
     1         r(ip1), P(ip1), M_r(ip1), L_r(ip1), T(ip1), tog_bf, irc)
          call EOS(X, Z, XCNO, mu, P(ip1), T(ip1), rho(ip1), kappa(ip1),
     1         epslon(ip1), tog_bf, ip1, ierr)
```

## Appendix H  STATSTAR, A Stellar Structure Code  A-35

```
              if (ierr.ne.0) then
                  write(*,400) r(i)/Rs, rho(i), M_r(i)/Ms, kappa(i), T(i),
     1                epslon(i), P(i), L_r(i)/Ls
                  stop
              end if
c
c     Determine whether convection will be operating in the next zone by
c     calculating dlnP/dlnT numerically between zones i and i+1 (ip1).
c     Update the adiabatic gas constant if necessary.
c
              if (i.gt.1) then
                  dlPdlT(ip1) = log(P(ip1)/P(i))/log(T(ip1)/T(i))
              else
                  dlPdlT(ip1) = dlPdlT(i)
              end if
              if (dlPdlT(ip1).lt.gamrat) then
                  irc = 1
              else
                  irc = 0
                  kPad = P(ip1)/T(ip1)**gamrat
              end if
c
c     Test to see whether the surface assumption of constant mass is still
c     valid.
c
              deltaM = deltar*dMdr(r(ip1), rho(ip1))
              M_r(ip1) = M_r(i) + deltaM
              if (abs(deltaM).gt.0.001d0*Ms) then
                  write(*,200)
                  if (ip1.gt.2) ip1 = ip1 - 1
                  go to 30
              end if
   20     continue
c
c     This is the main integration loop.  The assumptions of constant
c     interior mass and luminosity are no longer applied.
c
   30 Nsrtp1 = ip1 + 1
      do 40 i = Nsrtp1, Nstop
          im1 = i - 1
```

```
c     Initialize the Runge-Kutta routine with zone i-1 quantities
c     and their derivatives.  Note that the pressure, mass, luminosity,
c     and temperature are stored in the memory locations f_im1(1),
c     f_im1(2), f_im1(3), and f_im1(4), respectively.  The derivatives of
c     those quantities with respect to radius are stored in dfdr(1),
c     dfdr(2), dfdr(3), and dfdr(4).  Finally, the resulting values for
c     P, M_r, L_r, and T are returned from the Runge-Kutta routine in
c     f_i(1), f_i(2), f_i(3), and f_i(4), respectively.
c
c     The stellar structure equations dPdr (Eq. 10.7), dMdr (Eq. 10.8),
c     dLdr (Eq. 10.45), and dTdr (Eq. 10.61 or Eq. 10.81) are calculated
c     in function calls, defined later in the code.
c
          f_im1(1) = P(im1)
          f_im1(2) = M_r(im1)
          f_im1(3) = L_r(im1)
          f_im1(4) = T(im1)
          dfdr(1)  = dPdr(r(im1), M_r(im1), rho(im1))
          dfdr(2)  = dMdr(r(im1), rho(im1))
          dfdr(3)  = dLdr(r(im1), rho(im1), epslon(im1))
          dfdr(4)  = dTdr(r(im1), M_r(im1), L_r(im1), T(im1), rho(im1),
     1         kappa(im1), mu, irc)
          call RUNGE(f_im1, dfdr, f_i, r(im1), deltar, irc, X, Z, XCNO,
     1         mu, i, ierr)
          if (ierr.ne.0) then
             write(*,300)
             write(*,400) r(im1)/Rs, rho(im1), M_r(im1)/Ms, kappa(im1),
     1            T(im1), epslon(im1), P(im1), L_r(im1)/Ls
             stop
          end if
c
c     Update stellar parameters for the next zone, including adding
c     dr to the old radius (note that dr < 0 since the integration is
c     inward).
c
          r(i)    = r(im1) + deltar
          P(i)    = f_i(1)
          M_r(i)  = f_i(2)
          L_r(i)  = f_i(3)
          T(i)    = f_i(4)
```

## Appendix H STATSTAR, A Stellar Structure Code

```
c     Calculate the density, opacity, and energy generation rate for
c     this zone.
c
          call EOS(X, Z, XCNO, mu, P(i), T(i), rho(i), kappa(i),
     1         epslon(i), tog_bf, i, ierr)
          if (ierr.ne.0) then
              write(*,400) r(im1)/Rs, rho(im1), M_r(im1)/Ms, kappa(im1),
     1             T(im1), epslon(im1), P(im1), L_r(im1)/Ls
              stop
          end if
c
c     Determine whether convection will be operating in the next zone by
c     calculating dlnP/dlnT and comparing it to gamma/(gamma-1)
c     (see Eq. 10.87).  Set the convection flag appropriately.
c
          dlPdlT(i) = log(P(i)/P(im1))/log(T(i)/T(im1))
          if (dlPdlT(i).lt.gamrat) then
              irc = 1
          else
              irc = 0
          end if
c
c     Check to see whether the center has been reached.  If so, set Igoof and
c     estimate the central conditions rhocor, epscor, Pcore, and Tcore.
c     The central density is estimated to be the average density of the
c     remaining central ball, the central pressure is determined by
c     applying Eq. (H.4), and the central value for the energy
c     generation rate is calculated to be the remaining interior
c     luminosity divided by the mass of the central ball.  Finally, the
c     central temperature is computed by applying the ideal gas law
c     (where radiation pressure is neglected).
c
          if (r(i).le.abs(deltar) .and.
     1       (L_r(i).ge.0.1d0*Ls .or. M_r(i).ge.0.01d0*Ms)) then
              Igoof = 6
          else if (L_r(i).le.0.0d0) then
              Igoof = 5
              rhocor = M_r(i)/(4.0d0/3.0d0*pi*r(i)**3)
              if (M_r(i).ne.0.0d0) then
                  epscor = L_r(i)/M_r(i)
```

```fortran
              else
                  epscor = 0.0d0
              end if
              Pcore = P(i) + 2.0d0/3.0d0*pi*G*rhocor**2*r(i)**2
              Tcore = Pcore*mu*m_H/(rhocor*k_B)
          else if (M_r(i).le.0.0d0) then
              Igoof  = 4
              Rhocor = 0.0d0
              epscor = 0.0d0
              Pcore  = 0.0d0
              Tcore  = 0.0d0
          else if (r(i).lt.0.02d0*Rs .and. M_r(i).lt.0.01d0*Ms .and.
     1         L_r(i).lt.0.1d0*Ls) then
              rhocor = m_r(i)/(4.0d0/3.0d0*pi*r(i)**3)
              rhomax = 10.0d0*(rho(i)/rho(im1))*rho(i)
              epscor = L_r(i)/M_r(i)
              Pcore  = P(i) + 2.0d0/3.0d0*pi*G*rhocor**2*r(i)**2
              Tcore  = Pcore*mu*m_H/(rhocor*k_B)
              if (rhocor.lt.rho(i) .or. rhocor.gt.rhomax) then
                  Igoof = 1
              else if (epscor.lt.epslon(i)) then
                  Igoof = 2
              else if (Tcore.lt.T(i)) then
                  Igoof = 3
              else
                  Igoof = 0
              end if
          end if
          if (Igoof.ne.-1) then
              istop = i
              go to 50
          end if
c
c     Is it time to change the step size?
c
          if (idrflg.eq.0 .and. M_r(i).lt.0.99d0*Ms) then
              deltar = -Rs/100.0d0
              idrflg = 1
          end if
```

# Appendix H  STATSTAR, A Stellar Structure Code

```fortran
              if (idrflg.eq.1 .and. deltar.ge.0.5*r(i)) then
                  deltar = -Rs/5000.0d0
                  idrflg = 2
              end if
              istop = i
   40     continue
c
c     Generate warning messages for the central conditions.
c
          rhocor = M_r(istop)/(4.0d0/3.0d0*pi*r(istop)**3)
          epscor = L_r(istop)/M_r(istop)
          Pcore  = P(istop) + 2.0d0/3.0d0*pi*G*rhocor**2*r(istop)**2
          Tcore  = Pcore*mu*m_H/(rhocor*k_B)
   50     continue
          if (Igoof.ne.0) then
              if (Igoof.eq.-1) then
                  write(*,5000)
                  write(*,5100)
              else if (Igoof.eq.1) then
                  write(*,6000)
                  write(*,5200) rho(istop)
                  if (rhocor.gt.1.0d10) write(*,5300)
              else if (Igoof.eq.2) then
                  write(*,6000)
                  write(*,5400) epslon(istop)
              else if (Igoof.eq.3) then
                  write(*,6000)
                  write(*,5500) T(istop)
              else if (Igoof.eq.4) then
                  write(*,5000)
                  write(*,5600)
              else if (Igoof.eq.5) then
                  write(*,5000)
                  write(*,5700)
              else if (Igoof.eq.6) then
                  write(*,5000)
                  write(*,5800)
              end if
```

```
          else
              write(*,7000)
          end if
c
c   Print the central conditions.  If necessary, set limits for the
c   central radius, mass, and luminosity to avoid format field overflows.
c
          Rcrat = r(istop)/Rs
          if (Rcrat.lt.-9.999d0) Rcrat = -9.999d0
          Mcrat = M_r(istop)/Ms
          if (Mcrat.lt.-9.999d0) Mcrat = -9.999d0
          Lcrat = L_r(istop)/Ls
          if (Lcrat.lt.-9.999d0) Lcrat = -9.999d0
          write( *,2000) Msolar, Mcrat, Rsolar, Rcrat, Lsolar, Lcrat, Te,
         1 rhocor, X, Tcore, Y, Pcore, Z, epscor, dlPdlT(istop)
          write(20,1000)
          write(20,2000) Msolar, Mcrat, Rsolar, Rcrat, Lsolar, Lcrat, Te,
         1 rhocor, X, Tcore, Y, Pcore, Z, epscor, dlPdlT(istop)
          write(20,2500) Ms
c
c   Print data from the center of the star outward, labeling convective
c   or radiative zones by c or r, respectively.  If abs(dlnP/dlnT)
c   exceeds 99.9, set a print warning flag (*) and set the output limit
c   to +99.9 or -99.9 as appropriate to avoid format field overflows.
c
          write(20,3000)
          do 60 ic = 1, istop
              i = istop - ic + 1
              Qm = 1.0d0 - M_r(i)/Ms
              if (dlPdlT(i).lt.gamrat) then
                  rcf = 'c'
              else
                  rcf = 'r'
              end if
              if (abs(dlPdlT(i)).gt.dlPlim) then
                  dlPdlT(i) = sign(dlPlim,dlPdlT(i))
                  clim = '*'
              else
                  clim = ' '
              end if
```

## Appendix H  STATSTAR, A Stellar Structure Code

```fortran
             write(20,4000) r(i), Qm, L_r(i), T(i), P(i), rho(i), kappa(i),
     1            epslon(i), clim, rcf, dlPdlT(i)
   60 continue
      write(*,9000)
c
c  Format statements
c
  100 format(' ',/,' You must have X + Z <= 1',/,
     1 ' please reenter composition',/)
  200 format(' ',/,
     1 ' The variation in mass has become larger than 0.001*Ms',/,
     2 ' leaving the approximation loop before Nstart was reached',/)
  300 format(' ',/,' The problem occurred in the Runge-Kutta routine',/)
  400 format(' Values from the previous zone are:',/,
     1 10x,'r/Rs    = ',1pe12.5,' ',
     2 12x,'rho     = ',1pe12.5,' g/cm**3',/,
     3 10x,'M_r/Ms = ',1pe12.5,' ',
     4 12x,'kappa   = ',1pe12.5,' cm**2/g',/,
     5 10x,'T       = ',1pe12.5,' K',
     6 12x,'epsilon = ',1pe12.5,' ergs/g/s',/,
     7 10x,'P       = ',1pe12.5,' dynes/cm**2',/,
     8 10x,'L_r/Ls = ',1pe12.5)
 1000 format(' ',15x,'A Homogeneous Main-Sequence Model',/)
 2000 format(' ',/,
     1 ' The surface conditions are:',10x,'The central conditions are:',
     2 //,
     3 ' Mtot = ',0pf13.6,' Msun',12x,'Mc/Mtot     = ',1pe12.5,/,
     4 ' Rtot = ',0pf13.6,' Rsun',12x,'Rc/Rtot     = ',1pe12.5,/,
     5 ' Ltot = ',0pf13.6,' Lsun',12x,'Lc/Ltot     = ',1pe12.5,/,
     6 ' Teff = ',0pf13.6,' K ',12x,'Density     = ',1pe12.5,
     7 ' g/cm**3',/,
     8 ' X    = ',0pf13.6,'      ',12x,'Temperature = ',1pe12.5,' K',/,
     9 ' Y    = ',0pf13.6,'      ',12x,'Pressure    = ',1pe12.5,
     1 ' dynes/cm**2',/,
     2 ' Z    = ',0pf13.6,'      ',12x,'epsilon     = ',1pe12.5,
     3 ' ergs/s/g',/,
     4 '           ',   13x ,'      ',12x,'dlnP/dlnT   = ',1pe12.5,//)
 2500 format(' ',/,' Notes: ',/,
     1 ' (1) Mass is listed as Qm = 1.0 - M_r/Mtot, where Mtot = ',
     2 1pe13.6,' g',/,
```

```
     3 ' (2) Convective zones are indicated by c, radiative zones by r',
     4 /,
     5 ' (3) dlnP/dlnT may be limited to +99.9 or -99.9; if so it is',
     6 ' labeled by *',//)
3000 format(' ',5x,'r',7x,'Qm',7x,'L_r',7x,'T',8x,'P',7x,
    1 'rho',6x,'kap',6x,'eps',3x,'dlPdlT')
4000 format(' ',1p8e9.2,2a1,0pf5.1)
5000 format(' ',/,15x,'Sorry to be the bearer of bad news, but...',/,
    1 15x,'     Your model has some problems',/)
5100 format(' ',  8x,'The number of allowed shells has been exceeded'
    1 ,/)
5200 format(' ', 14x,'The core density seems a bit off,'/,
    1 5x,' density should increase smoothly toward the center.',/,
    2 5x,' The density of the last zone calculated was rho = ',
    3 1pe10.3,' gm/cm**3',/)
5300 format(' ',  1x,'It looks like you will need a degenerate',
    1 ' neutron gas and general relativity',/,
    2 ' to solve this core.  Who do you think I am, Einstein?',/)
5400 format(' ', 14x,'The core epsilon seems a bit off,',/,
    1 9x,' epsilon should vary smoothly near the center.',/,
    2 9x,' The value calculated for the last zone was eps =',
    3 1pe10.3,' ergs/g/s',/)
5500 format(' ',8x,' Your extrapolated central temperature is too low'
    1 ,/,8x,' a little more fine tuning ought to do it.',/,
    2 8x,' The value calculated for the last zone was T = ',
    3 1pe10.3,' K',/)
5600 format(' ', 10x,'You created a star with a hole in the center!',
    1 /)
5700 format(' ', 10x,'This star has a negative central luminosity!',/)
5800 format(' ', 5x,'You hit the center before the mass and/or ',
    1 'luminosity were depleted!',/)
6000 format(' ',///,15x,'It looks like you are getting close,',/,
    1 12x,'however, there are still a few minor errors',/)
7000 format(' ',///,15x,'CONGRATULATIONS, I THINK YOU FOUND IT!',/,
    1 9x,'However, be sure to look at your model carefully.',//)
9000 format(' ',10x,'***** The integration has been completed *****',
    1 /,10x,'  The model has been stored in starmodl.dat',
    2 /)
     stop
     end
```

## Appendix H  STATSTAR, A Stellar Structure Code

```fortran
c     Subroutine STARTMDL computes values of M_r, L_r, P, and T, near the
c     surface of the star using expansions of the stellar structure
c     equations (M_r and L_r are assumed to be constant).
c
      Subroutine STARTMDL(deltar, X, Z, mu, Rs, r_i, M_ri, L_ri, r,
     1    P_ip1, M_rip1, L_rip1, T_ip1, tog_bf, irc)
      real*8 deltar, X, Z, mu, Rs, r_i, M_ri, L_ri, r, P_ip1, M_rip1,
     1    L_rip1, T_ip1, tog_bf
      real*8 A_bf, A_ff, Afac
      real*8 sigma, c, a, G , k_B, m_H, pi, gamma, gamrat, kPad, g_ff
      common /cnstnt/ sigma, c, a, G, k_B, m_H, pi, gamma, gamrat, kPad,
     1    g_ff
c
      r = r_i + deltar
      M_rip1 = M_ri
      L_rip1 = L_ri
c
c     This is the radiative approximation (neglect radiation pressure
c     and electron scattering opacity); see Eqs. (H.1), (H.2), (9.19),
c     and (9.20).
c
      if (irc.eq.0) then
          T_ip1 = G*M_rip1*mu*m_H/(4.25d0*k_B)*(1.0d0/r - 1.0d0/Rs)
          A_bf = 4.34d25*Z*(1.0d0 + X)/tog_bf
          A_ff = 3.68d22*g_ff*(1.0d0 - Z)*(1.0d0 + X)
          Afac = A_bf + A_ff
          P_ip1 = sqrt((1.0d0/4.25d0)*(16.0d0/3.0d0*pi*a*c)*
     1        (G*M_rip1/L_rip1)*(k_B/(afac*mu*m_H)))*T_ ip1**4.25d0
c
c     This is the convective approximation; see Eqs. (H.3) and (10.75).
c
      else
          T_ip1 = G*M_rip1*mu*m_H/k_B*(1.0d0/r - 1.0d0/Rs)/gamrat
          P_ip1 = kPad*T_ip1**gamrat
      end if
      return
      end
c
```

```fortran
c     Subroutine EOS calculates the values of density, opacity, the
c     guillotine-to-gaunt factor ratio, and the energy generation rate at
c     the radius r.
c
      Subroutine EOS(X, Z, XCNO, mu, P, T, rho, kappa, epslon, tog_bf,
     1     izone, ierr)
      real*8 X, Z, mu, P, T, rho, kappa, epslon, tog_bf, Prad, Pgas
      real*8 k_bf, k_ff, k_e
      real*8 T6, fx, fpp, epspp, epsCNO, XCNO, oneo3, twoo3, psipp, Cpp
      real*8 sigma, c, a, G, k_B, m_H, pi, gamma, gamrat, kPad, g_ff
      common /cnstnt/ sigma, c, a, G, k_B, m_H, pi, gamma, gamrat, kPad,
     1     g_ff
      data oneo3, twoo3 / 0.333333333d0, 0.666666667d0 /
c
c     Solve for density from the ideal gas law (remove radiation
c     pressure); see Eq. (10.26).
c
      if (T.le.0.0d0 .or. P.le.0.0d0) then
          ierr = 1
          write(*,100) izone, T, P
          return
      end if
      Prad = a*T**4/3.0d0
      Pgas = P - Prad
      rho=(mu*m_H/k_B)*(Pgas/T)
      if (rho.lt.0.0d0) then
          ierr = 1
          write(*,200) izone, T, P, Prad, Pgas, rho
          return
      end if
c
c     Calculate opacity, including the guillotine-to-gaunt factor ratio;
c     see Novotny (1973), p. 469. k_bf, k_ff, and k_e are the bound-free,
c     free-free, and electron scattering opacities, given by Eqs. (9.19),
c     (9.20), and (9.21), respectively.
c
      tog_bf = 2.82d0*(rho*(1.0d0 + X))**0.2d0
      k_bf = 4.34d25/tog_bf*Z*(1.0d0 + X)*rho/T**3.5d0
      k_ff = 3.68d22*g_ff*(1.0d0 - Z)*(1.0d0 + X)*rho/T**3.5d0
      k_e = 0.2d0*(1.0d0 + X)
```

## Appendix H  STATSTAR, A Stellar Structure Code

```
            kappa = k_bf + k_ff + k_e
c
c   Compute energy generation by the pp chain and the CNO cycle.  These
c   are calculated using Eqs. (10.49) and (10.53), which come from
c   Fowler, Caughlan, and Zimmerman (1975).  The screening factor for
c   the pp chain is calculated as fpp; see Clayton (1968), p. 359ff.
c
            T6 = T*1.0d-06
            fx = 0.133d0*X*sqrt((3.0d0 + X)*rho)/T6**1.5d0
            fpp = 1.0d0 + fx*X
            psipp = 1.0d0 + 1.412d8*(1.0d0/X - 1.0d0)*exp(-49.98*T6**(- oneo3))
            Cpp = 1.0d0 + 0.0123d0*T6**oneo3 + 0.0109d0*T6**twoo3
         1       + 0.000938d0*T6
            epspp = 2.38d6*rho*X*X*fpp*psipp*Cpp*T6**(-twoo3)
         1       *exp(-33.80d0*T6**(-oneo3))
            CCNO = 1.0d0 + 0.0027d0*T6**oneo3 - 0.00778d0*T6**twoo3
         1       - 0.000149d0*T6
            epsCNO = 8.67d27*rho*X*XCNO*CCNO*T6**(-twoo3)
         1       *exp(-152.28d0*T6**(-oneo3))
            epslon = epspp + epsCNO
c
c   Formats
c
      100 format(' ',/,' Something is a little wrong here.',
     1  /,' You are asking me to deal with either a negative temperature'
     2  /,' or a negative pressure.  I am sorry but that is not in my'
     3    ' contract!',/,' You will have to try again with different',
     4    ' initial conditions.',/,
     5    ' In case it helps, I detected the problem in zone ',i3,
     6    ' with the following',/,' conditions:',/,
     7  10x,'T = ',1pe10.3,' K',/,
     8  10x,'P = ',1pe10.3,' dynes/cm**2')
      200 format(' ',/,' I am sorry, but a negative density was detected.',
     1  /,' my equation-of-state routine is a bit baffled by this new',
     2  /,' physical system you have created.  The radiation pressure',
     3  /,' is probably too great, implying that the star is unstable.'
     4  /,' Please try something a little less radical next time.',/,
     5    ' In case it helps, I detected the problem in zone ',i3,
     6    ' with the following',/,' conditions:',/,
     7  10x,'T     = ',1pe10.3,' K',/,
```

```
      8 10x,'P_total = ',1pe10.3,' dynes/cm**2',/,
      9 10x,'P_rad   = ',1pe10.3,' dynes/cm**2',/,
      1 10x,'P_gas   = ',1pe10.3,' dynes/cm**2',/,
      2 10x,'rho     = ',1pe10.3,' g/cm**3')
        return
        end
c
c   The following four function subprograms calculate the gradients of
c   pressure, mass, luminosity, and temperature at r.
c
        real*8 Function dPdr(r, M_r, rho)
        real*8 r, M_r, rho
        real*8 sigma, c, a, G, k_B, m_H, pi, gamma, gamrat, kPad, g_ff
        common /cnstnt/ sigma, c, a, G, k_B, m_H, pi, gamma, gamrat, kPad,
     1     g_ff
c   Eq. (10.7)
        dPdr = -G*rho*M_r/r**2
        return
        end
c
        real*8 Function dMdr(r, rho)
        real*8 r, rho
        real*8 sigma, c, a, G, k_B, m_H, pi, gamma, gamrat, kPad, g_ff
        common /cnstnt/ sigma, c, a, G, k_B, m_H, pi, gamma, gamrat, kPad,
     1     g_ff
c   Eq. (10.8)
        dMdr = 4.0d0*pi*rho*r**2
        return
        end
c
        real*8 Function dLdr(r, rho, epslon)
        real*8 r, rho, epslon
        real*8 sigma, c, a, G, k_B, m_H, pi, gamma, gamrat, kPad, g_ff
        common /cnstnt/ sigma, c, a, G, k_B, m_H, pi, gamma, gamrat, kPad,
     1     g_ff
c   Eq. (10.45)
        dLdr = 4.0d0*pi*rho*epslon*r**2
        return
        end
c
```

## Appendix H  STATSTAR, A Stellar Structure Code

```fortran
      real*8 Function dTdr(r, M_r, L_r, T, rho, kappa, mu, irc)
      real*8 r, M_r, L_r, T, rho, kappa, mu
      real*8 sigma, c, a, G, k_B, m_H, pi, gamma, gamrat, kPad, g_ff
      common /cnstnt/ sigma, c, a, G, k_B, m_H, pi, gamma, gamrat, kPad,
     1    g_ff
c  This is the radiative temperature gradient (Eq. 10.61).
      if (irc.eq.0) then
          dTdr = - (3.0d0/(16.0d0*pi*a*c))*kappa*rho/T**3*L_r/r**2
c  This is the adiabatic convective temperature gradient (Eq. 10.81).
      else
          dTdr = -1.0d0/gamrat*G*M_r/r**2*mu*m_H/k_B
      end if
      return
      end
c
c  This is a fourth-order Runge-Kutta integration routine.
c
      Subroutine RUNGE(f_im1, dfdr, f_i, r_im1, deltar, irc, X, Z, XCNO,
     1    mu, izone, ierr)
      real*8 f_im1(4), dfdr(4), f_i(4), df1(4), df2(4), df3(4),
     1    f_temp(4)
      real*8 r_im1, r_i, deltar, dr12, dr16, r12, X, Z, XCNO, mu
c
      dr12 = deltar/2.0d0
      dr16 = deltar/6.0d0
      r12  = r_im1 + dr12
      r_i  = r_im1 + deltar
c
c  Calculate intermediate derivatives from the fundamental stellar
c  structure equations found in Subroutine FUNDEQ.
c
      do 10 i = 1, 4
          f_temp(i) = f_im1(i) + dr12*dfdr(i)
   10 continue
      call FUNDEQ(r12, f_temp, df1, irc, X, Z, XCNO, mu, izone, ierr)
      if (ierr.ne.0) return
c
      do 20 i = 1, 4
          f_temp(i) = f_im1(i) + dr12*df1(i)
   20 continue
```

```
            call FUNDEQ(r12, f_temp, df2, irc, X, Z, XCNO, mu, izone, ierr)
            if (ierr.ne.0) return
c
            do 30 i = 1, 4
                f_temp(i) = f_im1(i) + deltar*df2(i)
   30       continue
            call FUNDEQ(r_i, f_temp, df3, irc, X, Z, XCNO, mu, izone, ierr)
            if (ierr.ne.0) return
c
c     Calculate the variables at the next shell (i + 1).
c
            do 40 i = 1, 4
                f_i(i) = f_im1(i) + dr16*(dfdr(i) + 2.0d0*df1(i)
     1               + 2.0d0*df2(i) + df3(i))
   40       continue
            return
            end
c
c     This subroutine returns the required derivatives for RUNGE, the
c     Runge-Kutta integration routine.
c
            Subroutine FUNDEQ(r, f, dfdr, irc, X, Z, XCNO, mu, izone, ierr)
            real*8 f(4), dfdr(4), X, Z, XCNO, r, M_r, P, T, L_r, rho, kappa,
     1          epslon, tog_bf, mu
            real*8 dPdr, dMdr, dLdr, dTdr
c
            P   = f(1)
            M_r = f(2)
            L_r = f(3)
            T   = f(4)
            call EOS(X, Z, XCNO, mu, P, T, rho, kappa, epslon, tog_bf, izone,
     1          ierr)
            dfdr(1) = dPdr(r, M_r, rho)
            dfdr(2) = dMdr(r, rho)
            dfdr(3) = dLdr(r, rho, epslon)
            dfdr(4) = dTdr(r, M_r, L_r, T, rho, kappa, mu, irc)
            return
            end
```

# Suggested Readings

### Technical

Clayton, Donald D., *Principles of Stellar Evolution and Nucleosynthesis*, McGraw-Hill, New York, 1968.

DeVries, Paul L., *A First Course in Computational Physics*, John Wiley and Sons, New York, 1994.

Fowler, William A., Caughlan, Georgeanne R., and Zimmerman, Barbara A., "Thermonuclear Reaction Rates, II," *Annual Review of Astronomy and Astrophysics*, *13*, 69, 1975.

Hansen, C. J., and Kawaler, S. D., *Stellar Interiors: Physical Principles, Structure, and Evolution*, Springer-Verlag, New York, 1994.

Kippenhahn, Rudolf, and Weigert, Alfred, *Stellar Structure and Evolution*, Springer-Verlag, Berlin, 1990.

Koonin, Steven E., and Meredith, Dawn C., *Computational Physics: FORTRAN Version*, Addison-Wesley, Reading, MA, 1990.

Novotny, Eva, *Introduction to Stellar Atmospheres and Interiors*, Oxford University Press, New York, 1973.

Press, William H., Flannery, Brian P., Teukolsky, Saul A., and Vetterling, William T., *Numerical Recipes: The Art of Scientific Computing (FORTRAN Version)*, Second Edition, Cambridge University Press, Cambridge, 1992.

# Appendix I

# STATSTAR STELLAR MODELS

STATSTAR (see Appendix H) can be used to generate a theoretical main sequence once the composition is chosen. Although the results differ somewhat from those of more sophisticated codes, many of the important features of stellar interiors are present.[1] Problems 10.18, 10.19, and 10.20 are designed to investigate the changes in stellar structure on the main sequence caused by variations in mass and/or composition.

The STATSTAR main sequence for the composition $X = 0.7$, $Y = 0.292$, and $Z = 0.008$ is shown in Fig. I.1. Furthermore, Table I.1 shows the dependence of effective temperature on mass for the same composition.

A sample STATSTAR session and output file are also presented. The user is asked to input the mass of the star, an estimate of its luminosity, the desired effective temperature, and the mass fractions of hydrogen and metals. STATSTAR will then compute a model stellar interior, integrating from the surface toward the center, as described in Appendix H.

---

[1]The careful reader may notice, for instance, that the 1 $M_\odot$ model presented here has a convective core and a radiative envelope, which is actually the reverse of a real main-sequence star like the Sun (see the extended discussion of the solar interior in Section 11.1). The reasons for these discrepancies reside largely in our approximate treatments of the opacity (particularly at lower temperatures) and thermodynamics (our assumption of adiabatic convection). Both of these approximations are poor ones near the surfaces of cooler stars.

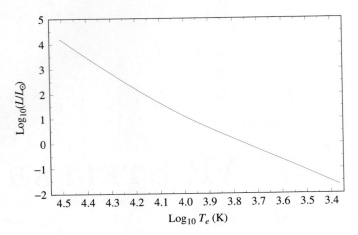

**Figure I.1** The STATSTAR main sequence for $X = 0.7$, $Y = 0.292$, and $Z = 0.008$.

| $M/M_\odot$ | $T_e(\text{K})$ | $M/M_\odot$ | $T_e(\text{K})$ | $M/M_\odot$ | $T_e(\text{K})$ | $M/M_\odot$ | $T_e(\text{K})$ |
|---|---|---|---|---|---|---|---|
| 0.50 | 2321.4  | 2.25 | 12260.0 | 4.75 | 19730.0 | 9.00  | 27061.2 |
| 0.60 | 2910.8  | 2.50 | 13240.0 | 5.00 | 20302.0 | 9.50  | 27712.0 |
| 0.70 | 3523.0  | 2.75 | 14170.8 | 5.50 | 21354.0 | 10.00 | 28263.6 |
| 0.80 | 4163.3  | 3.00 | 15007.3 | 6.00 | 22310.0 | 10.50 | 28845.2 |
| 0.90 | 4832.8  | 3.25 | 15790.8 | 6.50 | 23217.0 | 11.00 | 29414.6 |
| 1.00 | 5500.2  | 3.50 | 16525.0 | 7.00 | 24074.0 | 11.50 | 29964.8 |
| 1.25 | 7203.6  | 3.75 | 17252.0 | 7.50 | 24880.0 | 12.00 | 30496.5 |
| 1.50 | 8726.4  | 4.00 | 17904.0 | 8.00 | 25613.6 | 12.50 | 31009.0 |
| 1.75 | 10090.0 | 4.25 | 18546.8 | 8.50 | 26332.0 | 13.00 | 31493.0 |
| 2.00 | 11218.4 | 4.50 | 19153.6 |      |         |       |         |

**Table I.1** The Variation of Effective Temperature with Mass Along the STATSTAR Main Sequence, Assuming $X = 0.7$, $Y = 0.292$, and $Z = 0.008$.

# Appendix I  STATSTAR Stellar Models

Here is an example of the required input, together with STATSTAR's response:

```
 Enter the mass of the star (in solar units):
1.0
 Enter the luminosity of the star (in solar units):
0.86071
 Enter the effective temperature of the star (in K):
5500.2
 Enter the mass fraction of hydrogen (X):
0.70
 Enter the mass fraction of metals (Y):
0.008
```

```
              CONGRATULATIONS, I THINK YOU FOUND IT!
         However, be sure to look at your model carefully.
```

```
  The surface conditions are:           The central conditions are:

Mtot =       1.000000 Msun           Mc/Mtot    =    4.00418E-04
Rtot =       1.020998 Rsun           Rc/Rtot    =    1.90000E-02
Ltot =        .860710 Lsun           Lc/Ltot    =    7.67225E-02
Teff =    5500.200000 K              Density    =    7.72529E+01 g/cm**3
X    =        .700000                Temperature =   1.41421E+07 K
Y    =        .292000                Pressure   =    1.46284E+17 dynes/cm**2
Z    =        .008000                epsilon    =    3.17232E+02 ergs/s/g
                                     dlnP/dlnT  =    2.49808E+00
```

```
         ***** The integration has been completed *****
             The model has been stored in starmodl.dat
```

# Appendix I  STATSTAR Stellar Models

The following is a portion of the output file, `STARMODL.DAT`, that is produced by STATSTAR for the input values used in the previous example session.

A Homogeneous Main-Sequence Model

The surface conditions are:          The central conditions are:

| | | | | | | |
|---|---|---|---|---|---|---|
| Mtot | = | 1.000000 Msun | | Mc/Mtot | = | 4.00418E-04 |
| Rtot | = | 1.020998 Rsun | | Rc/Rtot | = | 1.90000E-02 |
| Ltot | = | .860710 Lsun | | Lc/Ltot | = | 7.67225E-02 |
| Teff | = | 5500.200000 K | | Density | = | 7.72529E+01 g/cm**3 |
| X | = | .700000 | | Temperature | = | 1.41421E+07 K |
| Y | = | .292000 | | Pressure | = | 1.46284E+17 dynes/cm**2 |
| Z | = | .008000 | | epsilon | = | 3.17232E+02 ergs/s/g |
| | | | | dlnP/dlnT | = | 2.49808E+00 |

Notes:
(1) Mass is listed as Qm = 1.0 - M_r/Mtot, where Mtot = 1.989000E+33 g
(2) Convective zones are indicated by c, radiative zones by r
(3) dlnP/dlnT may be limited to +99.9 or -99.9; if so it is labeled by *

| r | Qm | L_r | T | P | rho | kap | eps | | dlPdlT |
|---|---|---|---|---|---|---|---|---|---|
| 1.35E+09 | 1.00E+00 | 2.53E+32 | 1.40E+07 | 1.45E+17 | 7.72E+01 | 1.40E+00 | 1.51E+01 | c | 2.5 |
| 2.06E+09 | 9.99E-01 | 2.83E+32 | 1.39E+07 | 1.43E+17 | 7.66E+01 | 1.42E+00 | 1.46E+01 | c | 2.5 |
| 2.77E+09 | 9.97E-01 | 3.40E+32 | 1.38E+07 | 1.40E+17 | 7.57E+01 | 1.44E+00 | 1.39E+01 | c | 2.5 |
| 3.48E+09 | 9.93E-01 | 4.28E+32 | 1.37E+07 | 1.36E+17 | 7.45E+01 | 1.46E+00 | 1.31E+01 | c | 2.5 |
| 4.19E+09 | 9.88E-01 | 5.51E+32 | 1.35E+07 | 1.32E+17 | 7.31E+01 | 1.49E+00 | 1.21E+01 | c | 2.5 |
| 4.90E+09 | 9.82E-01 | 7.06E+32 | 1.33E+07 | 1.27E+17 | 7.15E+01 | 1.53E+00 | 1.11E+01 | c | 2.5 |
| 5.61E+09 | 9.73E-01 | 8.90E+32 | 1.31E+07 | 1.22E+17 | 6.97E+01 | 1.58E+00 | 1.00E+01 | c | 2.4 |
| 6.32E+09 | 9.62E-01 | 1.10E+33 | 1.28E+07 | 1.16E+17 | 6.77E+01 | 1.63E+00 | 8.87E+00 | r | 2.5 |
| 7.03E+09 | 9.49E-01 | 1.32E+33 | 1.25E+07 | 1.10E+17 | 6.55E+01 | 1.70E+00 | 7.79E+00 | r | 2.7 |
| 7.75E+09 | 9.33E-01 | 1.54E+33 | 1.22E+07 | 1.03E+17 | 6.30E+01 | 1.76E+00 | 6.76E+00 | r | 2.7 |
| 8.46E+09 | 9.15E-01 | 1.77E+33 | 1.19E+07 | 9.65E+16 | 6.04E+01 | 1.83E+00 | 5.81E+00 | r | 2.8 |
| 9.17E+09 | 8.94E-01 | 1.99E+33 | 1.16E+07 | 8.96E+16 | 5.76E+01 | 1.91E+00 | 4.93E+00 | r | 2.9 |
| 9.88E+09 | 8.71E-01 | 2.19E+33 | 1.13E+07 | 8.28E+16 | 5.46E+01 | 1.99E+00 | 4.14E+00 | r | 3.0 |
| 1.06E+10 | 8.46E-01 | 2.38E+33 | 1.10E+07 | 7.60E+16 | 5.16E+01 | 2.07E+00 | 3.45E+00 | r | 3.1 |
| 1.13E+10 | 8.19E-01 | 2.55E+33 | 1.07E+07 | 6.95E+16 | 4.86E+01 | 2.16E+00 | 2.83E+00 | r | 3.1 |
| 1.20E+10 | 7.91E-01 | 2.70E+33 | 1.04E+07 | 6.31E+16 | 4.55E+01 | 2.25E+00 | 2.31E+00 | r | 3.2 |
| 1.27E+10 | 7.60E-01 | 2.82E+33 | 1.00E+07 | 5.70E+16 | 4.24E+01 | 2.35E+00 | 1.86E+00 | r | 3.3 |
| 1.34E+10 | 7.29E-01 | 2.92E+33 | 9.71E+06 | 5.13E+16 | 3.94E+01 | 2.45E+00 | 1.48E+00 | r | 3.3 |

## Appendix I  STATSTAR Stellar Models

The structure near the surface is:

| r | Qm | L_r | T | P | rho | kap | eps | dlPdlT |   |
|---|---|---|---|---|---|---|---|---|---|
| 6.85E+10 | 1.69E-07 | 3.29E+33 | 1.15E+05 | 1.57E+08 | 1.01E-05 | 3.78E+01 | 3.61E-29 | r | 4.6 |
| 6.86E+10 | 1.48E-07 | 3.29E+33 | 1.12E+05 | 1.37E+08 | 9.11E-06 | 3.85E+01 | 1.71E-29 | r | 4.6 |
| 6.86E+10 | 1.30E-07 | 3.29E+33 | 1.09E+05 | 1.20E+08 | 8.20E-06 | 3.91E+01 | 7.87E-30 | r | 4.6 |
| 6.87E+10 | 1.14E-07 | 3.29E+33 | 1.06E+05 | 1.04E+08 | 7.35E-06 | 3.99E+01 | 3.52E-30 | r | 4.6 |
| 6.88E+10 | 9.89E-08 | 3.29E+33 | 1.02E+05 | 9.05E+07 | 6.58E-06 | 4.06E+01 | 1.53E-30 | r | 4.6 |
| 6.89E+10 | 8.57E-08 | 3.29E+33 | 9.93E+04 | 7.82E+07 | 5.86E-06 | 4.14E+01 | 6.42E-31 | r | 4.6 |
| 6.89E+10 | 7.39E-08 | 3.29E+33 | 9.60E+04 | 6.72E+07 | 5.21E-06 | 4.22E+01 | 2.60E-31 | r | 4.6 |
| 6.90E+10 | 6.34E-08 | 3.29E+33 | 9.28E+04 | 5.74E+07 | 4.61E-06 | 4.31E+01 | 1.01E-31 | r | 4.6 |
| 6.91E+10 | 5.42E-08 | 3.29E+33 | 8.96E+04 | 4.89E+07 | 4.06E-06 | 4.40E+01 | 3.80E-32 | r | 4.6 |
| 6.91E+10 | 4.60E-08 | 3.29E+33 | 8.64E+04 | 4.14E+07 | 3.56E-06 | 4.49E+01 | 1.36E-32 | r | 4.6 |
| 6.92E+10 | 3.88E-08 | 3.29E+33 | 8.32E+04 | 3.48E+07 | 3.11E-06 | 4.60E+01 | 4.65E-33 | r | 4.6 |
| 6.93E+10 | 3.26E-08 | 3.29E+33 | 8.01E+04 | 2.91E+07 | 2.70E-06 | 4.70E+01 | 1.50E-33 | r | 4.6 |
| 6.94E+10 | 2.72E-08 | 3.29E+33 | 7.69E+04 | 2.41E+07 | 2.34E-06 | 4.82E+01 | 4.60E-34 | r | 4.6 |
| 6.94E+10 | 2.25E-08 | 3.29E+33 | 7.37E+04 | 1.99E+07 | 2.01E-06 | 4.94E+01 | 1.32E-34 | r | 4.6 |
| 6.95E+10 | 1.84E-08 | 3.29E+33 | 7.05E+04 | 1.63E+07 | 1.71E-06 | 5.07E+01 | 3.55E-35 | r | 4.6 |
| 6.96E+10 | 1.50E-08 | 3.29E+33 | 6.74E+04 | 1.32E+07 | 1.45E-06 | 5.21E+01 | 8.83E-36 | r | 4.6 |
| 6.96E+10 | 1.21E-08 | 3.29E+33 | 6.42E+04 | 1.06E+07 | 1.22E-06 | 5.37E+01 | 2.02E-36 | r | 4.6 |
| 6.97E+10 | 9.65E-09 | 3.29E+33 | 6.11E+04 | 8.40E+06 | 1.02E-06 | 5.53E+01 | 4.20E-37 | r | 4.6 |
| 6.98E+10 | 7.62E-09 | 3.29E+33 | 5.80E+04 | 6.59E+06 | 8.45E-07 | 5.71E+01 | 7.87E-38 | r | 4.6 |
| 6.99E+10 | 5.94E-09 | 3.29E+33 | 5.48E+04 | 5.11E+06 | 6.92E-07 | 5.90E+01 | 1.31E-38 | r | 4.6 |
| 6.99E+10 | 4.57E-09 | 3.29E+33 | 5.17E+04 | 3.90E+06 | 5.61E-07 | 6.11E+01 | 1.91E-39 | r | 4.6 |
| 7.00E+10 | 3.46E-09 | 3.29E+33 | 4.86E+04 | 2.93E+06 | 4.48E-07 | 6.34E+01 | 2.39E-40 | r | 4.6 |
| 7.01E+10 | 2.58E-09 | 3.29E+33 | 4.55E+04 | 2.16E+06 | 3.53E-07 | 6.60E+01 | 2.50E-41 | r | 4.6 |
| 7.01E+10 | 1.90E-09 | 3.29E+33 | 4.24E+04 | 1.56E+06 | 2.74E-07 | 6.89E+01 | 2.14E-42 | r | 4.6 |
| 7.02E+10 | 1.36E-09 | 3.29E+33 | 3.93E+04 | 1.10E+06 | 2.09E-07 | 7.21E+01 | 1.44E-43 | r | 4.6 |
| 7.03E+10 | 9.63E-10 | 3.29E+33 | 3.62E+04 | 7.58E+05 | 1.55E-07 | 7.57E+01 | 7.27E-45 | r | 4.6 |
| 7.03E+10 | 6.68E-10 | 3.29E+33 | 3.32E+04 | 5.04E+05 | 1.13E-07 | 7.99E+01 | 2.61E-46 | r | 4.7 |
| 7.04E+10 | 4.17E-10 | 3.29E+33 | 2.98E+04 | 3.07E+05 | 7.64E-08 | 8.48E+01 | 4.25E-48 | r | 4.7 |
| 7.05E+10 | 2.47E-10 | 3.29E+33 | 2.65E+04 | 1.76E+05 | 4.94E-08 | 9.06E+01 | 3.64E-50 | r | 4.7 |
| 7.06E+10 | 1.37E-10 | 3.29E+33 | 2.31E+04 | 9.41E+04 | 3.02E-08 | 9.75E+01 | 1.34E-52 | r | 4.7 |
| 7.06E+10 | 6.92E-11 | 3.29E+33 | 1.98E+04 | 4.55E+04 | 1.70E-08 | 1.06E+02 | 1.54E-55 | r | 4.7 |
| 7.07E+10 | 3.11E-11 | 3.29E+33 | 1.65E+04 | 1.92E+04 | 8.61E-09 | 1.17E+02 | 3.36E-59 | r | 4.7 |
| 7.08E+10 | 1.18E-11 | 3.29E+33 | 1.32E+04 | 6.64E+03 | 3.72E-09 | 1.30E+02 | 5.65E-64 | r | 4.8 |
| 7.08E+10 | 3.40E-12 | 3.29E+33 | 9.88E+03 | 1.66E+03 | 1.24E-09 | 1.48E+02 | 1.20E-70 | r | 4.9 |
| 7.09E+10 | 6.08E-13 | 3.29E+33 | 6.58E+03 | 2.26E+02 | 2.52E-10 | 1.72E+02 | 3.79E-81 | r | 4.8 |
| 7.10E+10 | 3.94E-14 | 3.29E+33 | 3.29E+03 | 7.97E+00 | 1.75E-11 | 2.30E+02 | 1.67-102 | r | 4.3 |
| 7.11E+10 | 0.00E+00 | 3.29E+33 | 0.00E+00 | 0.00E+00 | 0.00E+00 | 0.00E+00 | 0.00E+00 | r | 4.3 |

# INDEX

Aberration, 161, 169, 170
   astigmatism, 170
   chromatic, 169
   coma, 170
   curvature of field, 170
   distortion of field, 170
   spherical, 169
Absorption, 265
Absorption coefficient (opacity), 265
   and convection, 360, 398
   and stellar pulsation, 553, 554
   bound–bound, 269
   bound–free, 270, 274
   electron scattering, 270, 274
   free–free, 270, 274
   Kramers law, 274
   sources, 269–273
Absorption lines, 126, 141, 269, 278, 399
Abundances, 350
   solar, 305, 306, 526
Accretion disks, 471–475, 672, 692–702, 710–713
   angular momentum, 698
   eclipses of, 698–700
   luminosity, 695
   magnetic fields in, 726–728
   radius, 698
   spectrum, 703
   temperature, 695
   viscosity, 692, 713
Accretion luminosity, 695
Acoustic frequency, 563
Adaptive optics, 179
Adiabatic process, 354, 355
Airy disk, 164
Albedo, 383
Alfvén radius, 726, 727
Alfvén speed, 414
Alfvén wave, 414
Algol paradox, 746
Algols, 707
AM Herculis stars (polars), 728
Angstrom, 71
Angular momentum, 47–51, 135, 150, 415, 416, 456, 475, 698
   spin, 152
Antenna pattern, 185
Antimatter, 153, 342
Ap stars, 157
Aperture, 163, 173
Aphelion, 30
Astronomical unit, 27, 63
Atmospheric transparency, 187
Atomic mass unit, 332
Aurora, 408

Balmer jump, 272
Balmer lines (series), 135, 140, 142, 236
Barium binaries, 709

Barrier penetration, 148, 335
Beta decay, 517
Big Bang,
    nucleosynthesis, 525
Binary stars, 201–218
    angle of inclination, 206
    astrometric, 202
    eclipsing, 202, 208–218
    line of nodes, 207
    masses, 207, 210, 211
    optical double, 202
    radii, 212
    spectroscopic, 205, 208–218
    spectrum binaries, 203
    temperatures, 215
    visual, 202
Binary stars, close, 683–743
    and gravitational waves, 705, 741–743
    classes of, 707–710
    contact binaries, 689
    detached binaries, 688
    effect of supernova on, 724, 725
    mass transfer, 704–707
    mass transfer rate, 689, 690
    primary star, 688
    secondary star, 688
    semidetached binaries, 688
    tidal capture, 725
    see also binary x-ray pulsars; binary x-ray systems; cataclysmic variables; dwarf novae; novae; polars; SS 433; supernovae, Type I; Thorne–Żytkow objects; x-ray bursters; x-ray novae
Binary x-ray pulsars, 726, 729–732
    spin-up, 730, 731
Binary x-ray systems, 709, 723–738
    high-mass (MXRB), 734, 738
    low-mass (LMXB), 734, 738
Binding energy, 332, 348, 349, 510
Black holes, 514, 661–674
    evaporation of, 673, 674
    event horizon, 663
    in close binaries, 672, 673, 734, 735
    "no hair" theorem, 668
    observations of, 672, 673, 734, 735
    photon sphere, 681
    primordial, 667, 673, 674
    rotating, 668, 669
        ergosphere, 669
        frame dragging, 668, 669
        static limit, 669
    Schwarzschild radius, 662
    singularity, 663, 664, 670
        Law of Cosmic Censorship, 664
    tidal forces, 666
Black widow pulsar, 739
Blackbody, 76
Blackbody radiation, 76
    energy density, 257, 258
    radiation pressure, 261
    source function, 282
    Wien's displacement law, 76, 78
    see also Planck function
Blue stragglers, 533
Blueshift, 109
Bohr atomic model, 135–141
Bohr radius, 139
Bok globules, 446, 447
Bolometric correction, 83
Boltzmann equation, 231
Boltzmann factor, 230

## Index

Bose–Einstein distribution, 327
Bosons, 153, 327
Bound–bound transition, 269, 273
Bound–free absorption, 270, 274
Boundary conditions, 367,
    A-28–A-30
Bremsstrahlung, 270
Broadening (of spectral lines),
    Doppler, 295, 296
    natural, 157
    pressure and collisional, 297,
        298
Brunt–Väisälä frequency, 566
Buoyancy frequency, 566
BY Draconis stars, 707

Carbon burning reactions, 348
Carbon stars, 504
Cataclysmic variables, 706, 709,
    710
    accretion disk, 698–700,
        710–713
Celestial equator, 12
Celestial poles, 4
Celestial sphere, 4
Centaurus A, 182
Center of mass, 43–45
Center-of-mass frame, 43–45
    orbital angular momentum,
        47, 49–51
    total energy, 47, 51
Chandrasekhar limit, 505, 590
Charge-coupled device (CCD), 181
Chromosphere, 400, 401, 416
Clusters (of stars), 529–534
    color–magnitude diagram,
        531–534
CNO cycle, 345, 346, 361, 373
    nuclear energy generation rate,
        346

Color index, 83, 272
Color temperature, 262
Color–color diagram, 85
Color–magnitude diagram, 531
Column density, 299, 439
Comets,
    tail, 408
Compton scattering, 133, 271
Conduction, 350, 592
Conic sections, 30
Conservation laws, 342, 343
Constitutive relations, 366
Continuous spectrum, 76, 126, 141,
    269
Continuum, 269
Convection, 350–352, 356–365, 565
    and ionization, 360, 364
    and nuclear energy generation
        rate, 360
    condition for, 359, 360, 566
    in protostars, 459, 462
    in pulsating stars, 365, 555
    in Sun, 387–390, 396, 420, 568
    mixing length theory, 361–365
    mixing length, 361, 363
Cooling time scale,
    white dwarfs, 594, 595
Coordinate speed, 656
    of light, 662, 663
Coordinate systems,
    altitude–azimuth (horizon),
        10, 11
    equatorial, 14, 15, 19
Corona, 403–408, 416
    coronal hole, 406
Cosmic rays, 421, 423
Coulomb's law, 136, 137
Crab Nebula, 517, 518, 612–616
Crab pulsar, 612–616, 622, 729
Cross section, 336–339

collision, 264
Curvature radiation, 614
Curve of growth, 300–305
Cyclotron frequency, 728
Cyclotron radiation, 728

Dark matter, 394
de Broglie wavelength, 144
Declination, 14
Deferent, 5
Degeneracy, 150, 230, 489, 499, 584
    condition for, 586
Difference equations, 369
Diffraction grating, 129
    resolving power, 130
Diffusion,
    of elements, 382, 383
    of photons, 280, 381
Distance modulus, 68
Diurnal motion, 10
Doppler shift, 126
    nonrelativistic, 110
    relativistic, 108, 109
    sound, 107
    transverse, 109
DQ Herculis stars (intermediate polars), 728
Dwarf novae, 700, 710–713
    mass transfer rate, 711
Dwarf stars, 241, 243
Dynamical instability, 561, 590

Ecliptic, 11
Eddington approximation, 287, 288
Eddington limit, 463, 716, 717
Eddington–Barbier relation, 309
Effective temperature, 77
Einstein–Rosen bridge, 670
Electromagnetic spectrum, 73
Electron capture, 511, 599, 600
Electron degeneracy pressure, 489, 584–588, 590
Electron pressure, 234, 237
Electron scattering, 270, 274
Electron screening, 340, 341
Electroweak theory, 394
Ellipse, 28, 29
    eccentricity, 28
    focal points, 28
        principal focus, 29
    semimajor axis, 28, 51
    semiminor axis, 28
Elongation, 7
Emission, 276
Emission coefficient, 280
Emission lines, 126, 141, 400
Endothermic reaction, 348
Energy density, 258
    specific, 257
Entropy, specific, 366
Epicycle, 5
Equant, 5
Equation of state, 320, 366
Equinox, 13
Equivalent width, 293
Ergosphere, 669
Escape velocity, 42, 43
$\eta$ Carinae, 528
Ether (luminiferous), 93, 96
Eulerian codes, 369
Event horizon, 663
Excitation temperature, 262
Exothermic reaction, 348
Extinction coefficient, 440

Fermi energy, 584
Fermi–Dirac distribution, 327
Fermions, 153, 327
Fission, 332
Flare stars, 428

# Index

Flux,
    monochromatic, 82
    radiative or radiant, 66, 258, 259
Focal length, 160, 173, 174
Focal plane, 161, 174
Focal point, 160
Focal ratio, 173
Forbidden transition, 404
Frame dragging, 668, 669
Fraunhofer lines, 125, 127, 128, 399, 400
Free–free absorption, 270, 274
Free-fall time scale, 451, 483
Fusion, 332–350
    in protostars, 459–462
    temperature required for, 335

Galactic (open) clusters, 530
Galilean transformations, 95
Gamow peak, 339, 340
Gaunt factor, 274
Geocentric system, 4
Geodesic, 655
Giant molecular clouds, 446
Giant stars, 241
Globular clusters, 530
Grand unified theories (GUTs), 394
Granulation, 396
Gravitation, law of, 36
Gravitational potential,
    effective, 685
        equipotential surfaces, 687
Gravitational potential energy, 41, 330
Gravitational quadrupole radiation, 742
Gravitational redshift, 645, 646, 658
Gravitational time dilation, 647, 658
Gravitational waves, 705, 741–743
Gray atmosphere, 285
Guillotine factor, 274

H I clouds, diffuse, 445
H II regions, 464–467
    Strömgren sphere and radius, 467
Hawking radiation, 673, 674
Hayashi track, 459, 501
Headlight effect, 112
Heisenberg's uncertainty principle, 146, 294, 335, 587
Heliocentric system, 6
Heliopause, 410
Helioseismology, 567–571
    160-minute oscillations, 568
    five-minute oscillations, 567
    g-modes, 568, 570
    p-modes, 567–570
Helium core flash, 499, 591
Helium shell flashes, 501, 509
Herbig–Haro objects, 471
Hertzsprung gap, 533
Hertzsprung–Russell diagram, 241–250
    asymptotic giant branch (AGB), 501
    DA white dwarfs, 581
    evolutionary tracks,
        0.6 $M_\odot$ star, 509
        18 $M_\odot$ star, 522
        5 $M_\odot$ star, 495
        post-main-sequence, 485
        pre-main-sequence, 460
        protostars, 457
    horizontal branch (HB), 500
    instability strip, 546, 547

boundaries, 555
red giant branch (RGB), 498
subgiant branch, 488
variable white dwarfs, 583
zero-age main sequence, 464
Homologous collapse, 451, 513
Hour angle, 14
Hour circle, 14
Hulse–Taylor pulsar, 738, 740–743
Hydrodynamic equations, 412
Hydrogen,
Bohr model, 135–141
$H^-$ ion, 273, 398, 399, 459
in the interstellar medium, 443–447
quantum states, 149, 231
tracers of $H_2$, 446
Hydrostatic equilibrium, 318
Hyperbola, 30

Ideal gas law, 320, 323
Illumination, 172, 173
Inertial reference frame, 34, 45, 94
local, 641
Inferior planets, 7
Initial mass function (IMF), 464, 465
Instability strip, 546, 547
boundaries, 555
Intensity, 170, 282
mean, 256
specific, 255, 258
Interference of light
circular aperture, 164–166
double-slit, 70, 71
single-slit, 164, 165
Internal gravity waves, 564
Interstellar dust, 441–443
polarization by, 443
Interstellar extinction, 438–442

coefficient, 440
Interstellar medium (ISM), 437–447
Interstellar reddening, 272, 440
Inverse square law, 67
Ionization, 138, 233
and convection, 360, 364
Ionization temperature, 262
Isochrone, 531
Isothermal core, 487–490
Isotope, 331

Jeans criterion, 448, 449
Jeans length, 449
Jeans mass, 449
Jets, 471
in SS 433, 736
Julian day, 542

Kelvin–Helmholtz time scale, 330, 331, 458, 462, 483
Kepler's laws, 26–28
first, 27, 43, 45
second, 27, 49
third, 27, 52, 206
Kinetic temperature, 262
Kirchhoff's laws, 126, 141, 278
Kramers opacity law, 274

Lagrangian codes, 369
Lagrangian points, 686
distances to, 687
Law of Cosmic Censorship, 664
Length contraction, 105
Lensmaker's formula, 161
Leptons, 342
Light cone, 649
Light cylinder, 621, 622
Light-year, 64
Limb darkening, 215, 279, 289–293
Line blanketing, 262, 272

Line of nodes, 207
Line profile, 294
    damping profile, 297
    Doppler broadening, 295, 296
    equivalent width, 293
    natural broadening, 294, 295
    P Cygni profile, 470
    pressure and collisional broadening, 297, 298
    Voigt profile, 298, 300
Lorentz factor, 100
Lorentz force, 407
Lorentz transformations, 98–100
    inverse, 101
    velocity, 111
Luminosities, stellar, 244, 372
Luminosity, 66
    interior, 342
    monochromatic, 81, 82
Luminosity class, 248, 297
Luminosity equation, 342
Lyman lines (series), 135

Magellanic Clouds, 504, 542
Magnetic dipole radiation, 621
Magnetic dynamo, 425–428
Magnetic energy density, 413
Magnetic flux, 606
Magnetic pressure, 414
Magnetohydrodynamics (MHD), 413
Magnification, angular, 174
Magnitude,
    *UBV*, 82, 83
    absolute, 67, 68
    apparent, 66, 68
    bolometric, 67, 82
    photographic, 715
Main sequence, 243, 371–373
    lifetime, 373
    stellar evolution on, 483–494
    turn-off point, 533
    zero-age, 464
Main-sequence fitting, 530
Maser emission, 506
Mass conservation equation, 319
Mass fraction, 324
Mass function, 211
Mass loss, 505, 506
Mass number, 331
Mass shells, 369
Mass–luminosity relation, 211, 212
Masses, stellar, 245, 246, 372, 462, 463
Maxwell–Boltzmann distribution, 225, 229, 230, 296, 323, 336, 339
Mean free path, 264
Mean molecular weight, 323–326, 346
    ionized gas, 326
    neutral gas, 325
Mercury,
    perihelion shift, 633, 637
Meridian, 11
Meridional currents, 605
Messier catalog, A-19–A-22
Metals, 248, 324
Metastable energy level, 404
Metric,
    for flat spacetime, 653
    Schwarzschild, 657
Mie scattering, 440, 441
Mikheyev–Smirnov–Wolfenstein (MSW) effect, 394
Milky Way Galaxy,
    disk, 437
    age of, 597
Mixing length theory, 361–365
    mixing length, 361, 363

Molecular clouds, 445
    Bok globules, 446, 447
    collapse of, 447–458
        adiabatic collapse, 453
        and angular momentum, 456, 475
        fragmentation, 453–455
        homologous collapse, 451
        isothermal collapse, 453
    giant, 446
    translucent, 446

Neutrinos, 342, 343, 390–394, 429, 500, 513, 514, 524, 607
Neutron degeneracy pressure, 598
Neutron drip, 601
Neutron stars, 514, 598–607
    composition, 599–603
    cooling of, 607
    in close binaries, 723–743
    magnetic field, 606
    mass–volume relation, 604
    maximum mass, 604
    model of, 603, 604
    radii, 598
    rotation, 604, 605
Neutronization, 601
Novae, classical, 709, 710, 713–719
    fast and slow, 714
    photosphere, 717, 718
Nuclear energy generation rate, 341
    and convection, 360
    CNO cycle, 346
    pp chain, 344, 345
    triple alpha process, 347
Nuclear reaction rates, 335, 336, 341
Nuclear reactions, *see* carbon burning reactions; CNO cycle; nucleosynthesis; oxygen burning reactions; proton–proton chain; silicon burning reactions; triple alpha process
Nuclear time scale, 333, 483, 510
Nuclei (notation), 331, 343
Nucleon, 331
Nucleosynthesis,
    Big Bang, 525
    stellar, 349, 515, 525–527, 722

OB associations, 468
Observatories,
    optical,
        Cerro Tololo, 167
        European Southern, 179
        Keck, 173, 179
        Kitt Peak, 167
        Mauna Kea, 167
        Mount Palomar, 177
        Special Astrophysical, 178
        Yerkes, 175
    radio,
        Arecibo, 183
        Very Large Array, 186
        Very Long Baseline Array, 187
    satellite,
        Advanced Satellite for Cosmology and Astrophysics, 190
        Advanced X-ray Astrophysics Facility, 190
        Compton Gamma Ray Observatory, 190, 191
        Cosmic Background Explorer, 188
        Einstein Observatory, 190

Extreme Ultraviolet
    Explorer, 189
Hubble Space Telescope,
    179, 180
Infrared Astronomy
    Satellite, 188
Infrared Space Observatory,
    188
International Ultraviolet
    Explorer, 189
Roentgen Satellite, 190
Space Infrared Telescope
    Facility, 188
OH/IR sources, 506
Opacity (absorption coefficient),
    265
  and convection, 360, 398
  and stellar pulsation, 553, 554
  bound–bound, 269
  bound–free, 270, 274
  electron scattering, 270, 274
  free–free, 270, 274
  Kramers law, 274
  sources, 269–273
Open (galactic) clusters, 530
Opposition, 7
Optical axis, 160
Optical depth, 126, 266, 267, 278,
    289, 439
  vertical, 284
Optically thick and thin, 268
Oscillator strength ($f$-value), 300
Oxygen burning reactions, 348

P Cygni profile, 470
Pair production, 623
Parabola, 30
Paradigm, 10, 79, 96
Parallax,
  secular, 544
  spectroscopic, 250
  trigonometric, 63
Parallax angle, 64
Parsec, 64
Partial ionization zone, 354,
    553–556
  He II, 554, 556
  hydrogen, 235, 554, 556, 711
Particle–antiparticle annihilation,
    153
Partition function, 233
Paschen lines (series), 135
Pauli exclusion principle, 152, 586
Perihelion, 30
Period–luminosity relation, 544,
    545
Period–mean density relation, 549
Photodisintegration, 511
Photodissociation, 273
Photoelectric effect, 131, 132
Photoionization, 270
Photon sphere, 681
Photons, 131
Photosphere, 278
  solar, 237, 263, 395–400
Pions, 602
Plages, 422
Planck function, 76, 79, 81, 256
Plane-parallel atmosphere, 284,
    285
Planetary nebula, 507, 508
Plasma, 410
Plate scale, 162
Polarization, 72, 728
Polars (AM Herculis stars), 728
  intermediate (DQ Herculis
      stars), 728
Polycyclic aromatic hydrocarbons
    (PAHs), 443
Polytropes, 369

polytropic index, 369
Population I and II stars, 529
Position angle, 20
Post-common envelope binaries, 710
Potential energy, 40
    gravitational, 41, 330
    nuclear, 333
Poynting vector, 73
Precession, 15, 16
Pressure integral, 322
Pressure scale height, 351, 352
Prime focus, 176
Proper distance, 652
Proper length, 105, 652
Proper motion, 18, 19, 63, 128
Proper time, 103, 651, 653
Proplyds, 472
Proton–proton (pp) chain, 343–345, 373
    nuclear energy generation rate, 344, 345
    PP I, 343
    PP II, 344
    PP III, 344
Protostars, 447–458
    collapse of,
        fragmentation, 464
    convection in, 459, 462
    fusion in, 459–462
Pulsars, 518, 608–624
    basic model, 617–624
    binary, 738–743
    binary x-ray, 726, 729–732
        spin-up, 730, 731
    black widow pulsar, 739
    characteristics, 608, 609
    Crab pulsar, 612–616, 622, 729
    emission mechanism, 622–624
    glitches, 612, 624
    Hulse–Taylor pulsar, 738, 740–743
    lifetime, 609
    magnetic field, 621
    magnetosphere, 623
    millisecond, 612, 618, 622, 738–740
    period derivatives, 609
    periods, 608, 610
    pulses, 616, 617
        dispersion, 617, 618
        integrated pulse profile, 617
    spatial distribution, 617, 619
    subpulses, 617
        drifting, 617, 620
        nulling, 617
Pulsating stars, 69, 366, 371, 500, 541–567
    adiabatic calculation of, 559
    and convection, 365, 555
    as heat engines, 551–553
    $\beta$ Cephei stars, 548, 556
    Cepheids, classical, 542, 545, 546, 550
        as standard candles, 544
    DAV stars (ZZ Ceti stars), 548, 556, 582
    DBV stars, 582
    $\delta$ Scuti stars, 547
    DOV stars, 583
    $\epsilon$-mechanism, 372, 553
    $\gamma$-mechanism, 554
    instability strip, 546, 547
        boundaries, 555
    $\kappa$-mechanism, 554, 556
    linearized calculation of, 557–560
    long period variables (LPVs), 503, 541, 548, 550, 556
    luminous blue variables

(LBVs), 528
  Mira variables, 503
  nonadiabatic calculation of, 561
  nonlinear calculation of, 557
  one-zone model of, 558–560
  partial ionization zones, 553–556
  period–luminosity relation, 544, 545
  period–mean density relation, 549
  phase lag, 546, 556
  PNNV stars, 583
  RR Lyrae stars, 509, 547, 550
  types of, 548
  valve mechanism, 553
  W Virginis stars, 547, 550
  ZZ Ceti stars (DAV stars), 548, 556, 582
  *see also* helioseismology
Pulsation modes,
  nonradial, 549, 561–567
    f-mode, 563
    g-modes, 564, 566, 567, 582
    p-modes, 563
    rotational splitting, 564, 570
  radial, 549–551

Quantum, 80
Quantum mechanics, 79–81, 130–153
Quantum numbers,
  angular momentum, 150
  principal, 139
  spin, 152

r-process (rapid), 527
Radial velocity, 18, 126

Radiation pressure, 74, 259–261, 279, 280, 327, 328, 463, 716, 717
Radiative transfer, equation of, 282, 285
Radii, stellar, 244, 245
Radio interferometry, 184–187
Radioactive decay, 515–517
Random walk, 276, 277
Rayleigh criterion, 166
Rayleigh scattering, 271
Redshift, 109
  gravitational, 645, 646, 658
Redshift parameter, 109
Reduced mass, 46, 47
Reflection nebula, 440
Relativistic energy
  kinetic, 114
  total, 114
Relativistic invariance, 113
Relativistic momentum, 113, 116–118
Relativity,
  general, 633–661
    coordinates, 656
    principle of equivalence, 641
    three fundamentals, 655
    *see also* gravitational quadrupole radiation; gravitational waves
  spacetime interval, 122, 651, 652
  special, 96–118
    headlight effect, 112
    inverse Lorentz transformations, 101
    length contraction, 105
    Lorentz transformations, 98–100
    postulates, 97

relativistic Doppler shift, 108, 109
simultaneity, 102
time dilation, 103
velocity transformations, 111
Resolution, 163, 166
Rest energy, 114
Rest frame, 103
Retrograde motion, 4, 7, 8
Right ascension, 14
Roche lobes, 688
Rosseland mean (opacity), 273, 275
RS Canum Venaticorum stars, 707
Rydberg constant, 135, 140

s-process (slow), 527
S-star binaries, 709
Saha equation, 234
Scale heights,
    pressure, 351, 352
    temperature, 263
Schönberg–Chandrasekhar limit, 488–494
Schuster–Schwarzschild model, 299, 301
Schwarzschild metric, 657
Schwarzschild radius, 662
Schwarzschild throat, 670
Sensitivity function, 84, 86, 182
Shock front, 413
Shock wave, 413
    in Type II supernovae, 513, 514
Sidereal period, 9
Sidereal time, 12
Silicon burning reactions, 510
Simultaneity, 102
Singularity, 663, 664, 670
Sirius B, 577–579
SN 1987A, 519–524
    neutrinos from, 524
Snell's law, 159, 160
Solar constant, 67
Solar cycle, 416–429
Solar flares, 422, 423
Solar lithium problem, 526
Solar neutrino problem, 390–394, 429
Solar neutrino unit (SNU), 391
Solar prominences, 423, 424
Solar time, 12
Solar wind, 406–411
Solid angle, 79, 170
Solstice, 13
Sound speed, 355, 412
Source function, 281, 282
Spacetime diagram, 648
Spacetime interval, 122, 651
    lightlike or null, 651
    spacelike, 652
    timelike, 651
Specific heats, 353
    ratio of, 354, 554
Spectral flux density, 182
Spectral lines, 125, 153
    equivalent width, 293
    optically thin, 294
    strengths, 240, 247
    *see also* line profile
Spectral types, 223, 226–228, 241
Spectrograph, 129
Spicules, 401
Spin,
    electron, 152, 443
SS 433, 735, 736
Starspots, 428
Static limit, 669
Statistical weight, 230
Stefan–Boltzmann equation, 77
Stellar clusters, 529–534

color–magnitude diagram, 531–534
Stellar evolution,
   on main sequence, 483–494
      high-mass stars, 490
      low-mass stars, 484–490
   post-main-sequence, 494–509
      carbon–oxygen core flash, 505
      first dredge-up, 498
      helium core flash, 499
      helium shell flashes, 501, 508
      mass loss, 505, 506
      on asymptotic giant branch (AGB), 501
      on horizontal branch (HB), 500
      on red giant branch (RGB), 498
      second dredge-up, 500
      third dredge-up, 504
Stellar models, 365–370
   polytropes, 369
Stellar pulsation, *see* pulsating stars
Stellar structure, 315–373
   1 $M_\odot$ star, 487, 488
   15 $M_\odot$ star, 512
   5 $M_\odot$ star, 496
   7 $M_\odot$ star, 503
   main sequence, 373
Stellar structure equations, 365
   boundary conditions, 367, A-28–A-30
   hydrostatic equilibrium, 318
   luminosity, 342
   mass conservation, 319
   numerical solution, 369, 370, A-27–A-30
   temperature gradients,
      adiabatic, 356, 357
      radiation, 351
Steradian, 170
Strömgren sphere and radius, 467
Strong force, 334, 513
Sun, 381–430
   absolute magnitude, 68
   age, 382
   atmosphere, 394–416
   central composition, 382
   central conditions, 384
   chromosphere, 400, 401
   convection zone, 387–390, 396, 420, 568
   corona, 403–408
   coronal hole, 405
   flash spectrum, 400
   Fraunhofer lines, 125, 127, 128, 399, 400
   interior, 381–394
   magnetic field, 406–416, 418–420, 425–428
   mass loss, 409, 410
   photosphere, 237, 263, 395–400
   pulsation of, 567–571
      160-minute oscillations, 568
      five-minute oscillations, 567
      g-modes, 568, 570
      p-modes, 567–570
   rotation, 400, 415, 416
   solar wind, 406–411
   spectral type, 381
   standard solar model, 382–390
   surface composition, 381, 570
   transition region, 401
   *see also* solar neutrino problem
Sunspots, 416–422
   butterfly diagram, 417
   Maunder minimum, 421
   umbra and penumbra, 418

Superconducting fluid, 601
Superfluid, 601
Supergiant stars, 244
Supergranulation, 400
Superior planets, 7
Supernova remnants, 517–519
Supernovae,
    deflagration and detonation, 722
    Type I, 515, 710, 720–723
        as standard candles, 722
        nucleosynthesis, 722
        radioactive decay, 721
        Type Ia, 720–722
        Types Ib and Ic, 720, 723
    Type I$\frac{1}{2}$, 505
    Type II, 511–515, 720
        delayed hydrodynamic explosion, 514
        nucleosynthesis, 515
        prompt hydrodynamic explosion, 514
        radioactive decay, 515–517
        shock wave, 513, 514
        Types II-L and II-P, 515
Superwind, 506
Symbiotic binaries, 709
Synchrotron radiation, 112, 614
Synodic period, 9

T Tauri stars, 468–471
Telescopes,
    Cassegrain, 176
    coudé, 176
    mounts, 178
    Newtonian, 176
    radio, 181–187
    reflecting, 159, 175–179
    refracting, 159, 173–175
    Schmidt, 177
    *see also* observatories
Temperature gradients,
    adiabatic, 356, 357
    radiation, 351
    superadiabatic, 357, 361, 364
Temperature of plane-parallel gray atmosphere in LTE, 289
Temperature scale height, 263
Thermodynamic equilibrium, 263, 282
    local (LTE), 263, 266, 282, 283, 289
Thermodynamics, 352–355
    first law, 352
Thomson scattering, 270
    cross section, 271
Thorne–Żytkow objects, 725
Tidal capture, 725
Tidal forces, 666
Time dilation, 103
    gravitational, 647, 658
Time scales,
    free-fall, 451, 483
    Kelvin–Helmholtz, 330, 331, 458, 462, 483
    nuclear, 333, 483, 510
    white dwarf cooling, 594, 595
Transfer equation, 282, 285
Transverse (tangential) velocity, 18, 128
Triple alpha process, 347, 361
    nuclear energy generation rate, 347
Tunneling, 148, 335
21-cm line, 443–445

URCA process, 607

Van Allen radiation belts, 408

Virial theorem, 51, 53, 56, 138, 365, 447, 448, 695
Vogt–Russell theorem, 368, 371
Volume, specific, 354
VV Cephei systems, 709

W Ursae Majoris contact systems, 708
Wave–particle duality, 144
Weak force, 344
White dwarfs, 243, 244, 506, 579–583, 592–597
  composition, 581, 591
  cooling of, 592–597
    time scale, 594, 595
  crystallization, 595
  in close binaries, 698, 700, 710–723
    accretion-induced collapse, 725
  mass–radius relation, 591
  mass–volume relation, 589
  masses, 581
  radii, 589
  Sirius B, 577–579
  spectral types, 580
  variable, 582, 583
Wien's displacement law, 76, 78
WIMPs, 393
Wolf–Rayet stars, 527
Work function, 132
Worldline, 649
Wormholes, 671

X-ray binaries, *see* binary x-ray systems
X-ray bursters, 733, 734
X-ray novae, 735
X-ray pulsars, binary, 726, 729–732
  spin-up, 730, 731

Zeeman effect, 150–152, 418
Zenith, 10
Zero-age main sequence, 464
$\zeta$ Aurigae systems, 709
Zodiacal light, 404